KB275400

軍政廳法令集

在朝鮮美國陸軍司令部

英文版

（上）

韓國學資料院

在朝鮮美國陸軍司令部

軍政廳法令集

英文版

刊 行 辞

우리民族의 피는 数十世紀에 걸쳐 保族을 하기 爲하여 経済와 같이 再生産되어져 連綿하게

오늘에 이어졌고 또한 未来에도 傳하여져 永久히 그 피의 保族을 하기爲한 經濟生活의 營爲

는 來日도 繼續되어 火食의 斷絶을 除去할 것이다.

이에는 必然的으로 어떠한 某一種의 規範이 定하여져 온 것은 否認할 수 없는 事實로 우리

國家制度에 外來의 文物制度가 強制로 加工되어 아무리 많은 變遷을 가져왔다고는 하지만 固

有의 各種制度上 줄기는 連綿하게 名脈을 維持하여 傳하여져 왔고 또 그 줄기는 이어주고 있

어 우리는 이를 다시 찾아 보아 不足 하였던 点을 補充하고 長点은 再善用하여 未來 法生活의

發展을 期하고자 旧代의 法令을 再問世시켜 斯道에 이바지 하여 보겠다는 信念에서 이 大全을

出刊하는 바이오니 江湖 諸賢諸位의 叱正과 鞭撻을 바라마지 않는바이다.

解　題

在朝鮮美國陸軍司令部

軍政廳法令集

이 法令集은 元來 太平洋美陸軍總司令部가 布告한 布告文과 在朝鮮美陸軍司令部軍政廳이 그
官報에 揭載 公布하였던 法令을 編輯한 것이다.

이 布告文과 法令은 一九四五年 八月 十日 밤 八時에 日本이 聯合國에게 無條件 降伏을 宣言한
지 四週가 되는 九月 七日에 太平洋美陸軍總司令部가 「朝鮮住民에게 布告함」이라는 布告 第一
號外 三件을 飛行機로 撤布한 布告文을 비롯하여 滿二年 十一個月 六日間인 一九四八年 八月
二日까지 즉 今年 八月 十五日 大韓民國政府가 樹立되기 前에 在朝鮮美陸軍總司令部 軍政廳이
布告六件、法令二一九件、行政命令二四件、部令 및 指令(處·局令 및 規則包含)二一五件、在朝
鮮過渡政府法律一四件、全立法議院立法決議案四件、其他一一件、計三九七件을 法令으로서 公布
하였던 것이다.(當時의 官報는 紙類의 求得難으로 後期에는 馬糞紙로까지 發行 되었음.)

그리고 이러한 法令은 우리 民族이 異民族의 굴레에서 벗어난 時代에 豫想하지 못했던 遺物
이라고 하겠으나 한 나라의 經綸에 있어 間斷이 있을 수 없기 때문에 近三年間이나 이 땅위에
서 施行되었던 制度로서 法史學研究에 重要한 資料가 된다.

그런데 이러한 法令은 美軍政法令集이란 名稱으로나 軍政廳官報綴로서 法制處、內務部、國會、
韓國銀行 및 韓國産業銀行 등의 各 圖書舘과 圖書室에 또는 資料로서 故 鄭光鉉 博士의 書

齋에 散在해 있었다。 이와같이 散藏되어 있던 美軍政 當時의 法令을 拾遺集録하여 在朝鮮

美國陸軍司令部 軍政廳法令集이라고 題名한 것이다.

이번에 平素 民族文化遺産의 普及에 至大한 關心을 보여왔던 民族文化의 金容和 代表가

法史學과 더불어 斯道에 이바지하고자 하는 뜻에서 이 法令集을 影印 出刊하게 된것이라고

보아져 諸賢諸位의 鞭撻과 以諒을 乞하는 바이다.

東亞大學校 法政大學 教授 (法制史)

尹 載 秀 識

TABLE OF CONTENTS

CHAPTER I PROCLAMATION

(G.H.Q.U.S. ARMY FORCES, PACIFIC)

CHAPTER 2 KILA PUBLIC ACTS.

CHAPTER 4 EXECUTIVE ORDERS
(USAMGIK)

CHAPTER 5 USAMGIK DEPARTMENT ORDERS AND DIRECTIVES
(INCLUDING THOSE ISSUED BY BUREAUS, BOARDS, OFFICES
OR SIMILAR AUTHORITIES)
I. DEPARTMENT OF AGRICULTURE ORDERS

9. NATIONAL FOOD ADMINISTRATION REGULATION

- END -

CHAPTER 1   PROCLAMATION

G.H.Q. U.S. ARMY FORCES, PACIFIC
OFFICE OF THE COMMANDING GENERAL
YOKOHAMA, JAPAN, 7 SEPTEMBER 1945

TO THE PEOPLE OF KOREA:

As Commander-in-chief, United States Army Forces, Pacific, I do hereby proclaim as follows:

By the terms of the Instrument of Surrender, signed by command and in behalf of the Emperor of Japan and the Japanese Government and by command and in behalf of the Japanese Imperial General Headquarters, the victorious military forces of my command will today occupy the territory of Korea south of 38 degrees north latitude.

Having in mind the long enslavement of the people of Korea and the determination that in due course Korea shall become free and independent, the Korean people are assured that the purpose of the occupation is to enforce the Instrument of Surrender and to protect them in their personal and religious rights. In giving effect to these purposes, your active aid and compliance are required.

By virtue of the authority vested in me as Commander-in-Chief, United States Army Forces, Pacific, I hereby establish military control over Korea south of 38 degrees north latitude and the inhabitants thereof, and announce the following conditions of the occupation:

ARTICLE I

All powers of Government over the territory of KOREA south of 38 degrees north latitude and the people thereof will be for the present exercised under my authority.

ARTICLE II

Until further orders, all governmental, public and honorary functionaries and employees, as well as all officials and employees, paid or voluntary, of all public utilities and services, including public welfare and public health, and all other persons engaged in essential services, shall continue to perform their usual functions and duties, and shall preserve and safeguard all records and property.

ARTICLE III

All persons will obey promptly all my orders and orders issued under my authority. Acts of resistance to the occupying forces or any acts which may disturb public peace and safety will be punished severely.

ARTICLE IV

Your property rights will be respected. You will pursue your normal occupations, except as I shall otherwise order.

ARTICLE V

For all purposes during the military control, English will be the official language. In event of any ambiguity or diversity of interpretation or definition between any English and Korean or Japanese text, the English text shall prevail.

ARTICLE VI

Further proclamations, ordinances, regulations, notices, directives and enactments will be issued by me or under my authority, and will specify what is required of you.

Given under my hand at YOKOHAMA
THIS SEVENTH DAY OF SEPTEMBER 1945

DOUGLAS MacARTHUR

General of the Army of the United States
Commander-in-Chief, United States Army Forces, Pacific

G.H.Q U.S. ARMY · FORCES. PACIFIC
OFFICE OF THE COMMANDING GENERAL
YOKOHAMA, JAPAN, 7 SEPTEMBER 1945

TO THE PEOPLE OF KOREA:

In order to make provision for the security of the armed forces under my command and for the maintenance of public peace, order and safety in the occupied area, as Commander-in-Chief, United States Army Forces, Pacific, I do hereby proclaim as follows:

ANY PERSON WHO:

Violates the provisions of the Instrument of Surrender, or any proclamation, order, or directive given under the authority of the Commander-in-Chief, United States Army Forces, Pacific, or does any act to the prejudice of good order or the life, safety, or security of the persons or property of the United States or its Allies, or does any act calculated to disturb public peace and order, or prevent the administration of justice, or willfully does any act hostile to the Allied Forces, shall, upon conviction by a Military Occupation Court, suffer death or such other punishment as the Court may determine.

Given under my hand at YOKOHAMA
THIS SEVENTH DAY OF SEPTEMBER 1945

DOUGLAS MacARTHUR

General of the Army of the United States
Commander-in-Chief, United States Army Forces, Pacific

G.H.Q. U.S. ARMY FORCES, PACIFIC
OFFICE OF THE COMMANDING GENERAL
YOKOHAMA, JAPAN, 7 SEPTEMBER 1945

TO THE PEOPLE OF KOREA:

As Commander-in-Chief, United States Army Forces, Pacific, I hereby proclaim as follows:

ARTICLE I

LEGAL TENDER

1. Supplemental military yen currency, marked "A", issued by the Military Occupation Forces is legal tender in Korea, south of 38° north latitude, for the payment of all yen debts, public or private.

2. Supplemental military yen currency, marked "A", issued by the Military Occupation Forces, and regular yen currencies now legal tender in Korea south of 38° north latitude, except Bank of Japan and Bank of Taiwan notes, are interchangeable at face value without distinction.

3. No other currencies shall be legal tender in Korea south of 38° north latitude.

ARTICLE II

JAPANESE MILITARY YEN

4. All military and all occupational currency which has been issued by the Imperial Japanese Government, Army or Navy, is void and valueless and the giving or accepting of such currency in any transaction is prohibited.

ARTICLE III

EXPORT AND IMPORT OF CURRENCY PROHIBITED

5. All foreign financial transactions, including the export and import of currency, coin and securities are prohibited except as authorized by me.

6. All financial transactions shall be deemed to be foreign except those taking place solely within the area of Korea south of 38° north latitude.

ARTICLE IV

REGULATION OF OTHER CURRENCY

7. The delivery or acceptance of any currency other than the supplemental military and regular yen currency now legal tender in Korea south of 38° north latitude in any transaction is prohibited, except as authorized by me.

ARTICLE V

PENALTIES

8. Any person violating the provisions of this proclamation shall, upon conviction by a Military Occupation Court, suffer such punishment as the Court shall determine.

Given under my hand at YOKOHAMA
THIS SEVENTH DAY OF SEPTEMBER 1945

DOUGLAS MacARTHUR

General of the Army of the United States
Commander-in-Chief, United States Army Forces, Pacific

G.H.Q. U.S. ARMY FORCES. PACIFIC
Office of the Commanding General
Tokyo. Japan, 1 July 1946

TO THE PEOPLE OF KOREA

As Commander in Chief. United States Army Forces, Pacific. I hereby proclaim as follows:

ARTICLE I
LEGAL TENDER

1. Effective 10 July 1946. supplemental military yen currency marked "A" is no longer legal tender in Korea south of 38° North Latitude. for the payment of yen debts. public or private.

2. Possession of supplemental military yen currency marked "A", after 10 July 1946, is prohibited. Supplemental yen currency marked "A" will be redeemed as provided for by Ordinance. All provisions of Proclamation Number 3 dated 7 September 1945. inconsistent herewith. are hereby rescinded.

ARTICLE II
PENALTIES

Any person violating the provisions of this Proclamation shall. upon conviction by a Military Occupation Court. suffer such punishment as the court shall determine.

Given under my hand this 1st day of July 1946.

DOUGLAS MacARTHUR

General of the Army of the United States
Commander-in-Chief. United States Army Forces, Pacific

CHAPTER 2 KLLA PUBLIC ACTS

HEADQUARTERS
UNITED STATES ARMY MILITARY
GOVERNMENT IN KOREA
Office of the Military Governor
Seoul, Korea

PUBLIC ACT

NUMBER 1   6 May 1947

AMENDING SECTION VII OF ORDINANCE NO 102
(ESTABLISHMENT OF SEOUL NATIONAL UNIVERSITY)

SECTION I. Section VII of Ordinance No 102, dated 22 August 1946, is amended in the following respects:

a. Section VII *a (1)* thereof is renumbered Section VII *a* and amended to read as follows:

"*a.* The Board of Regents of Seoul National University is hereby established. It shall consist of the President of Seoul National University (*ex officio*) and one member appointed by the Military Governor in Korea (but, after the period of military occupation, by the chief executive official of the government) for each of the several colleges, schools and learned faculties constituting the University. Such appointments shall be made upon recommendation of the majority of the directors of the several national departments and offices, subject to confirmation by the Korean Interim Legislative Assembly. Each of the appointive members of the Board shall be a Korean selected as preeminent in the active practice of the particular profession or calling he is deemed qualified to represent on the Board; *provided, however,* that no member of the faculty or staff employed by the University and no governmental employee shall at any time be eligible to serve on the Board as an appointive member thereof. The term of office of each appointive member shall be six years; *provided, however,* that one-third of the initial appointments to the Board shall be for a period of two years, one-third for a period of four

years, and the remaining one-third for a period of six years; and *further
provided*, that appointments to fill vacancies on the Board shall be made only
for the unexpired portion of the term of the vacancy filled; and *further provided*, that upon creation of new seats on the Board by the establishment of
additional colleges and learned faculties, each appointment thereto shall be
made in such manner as to carry out the intent of this provision, which is
to insure, so far as possible; the rotation of one-third of the total membership
of the Board each and every two years. The Board shall elect a chairman
to serve for a period of two years. A quorum of the Board shall consist of
a majority of the members thereof, and such quorum shall be required for
the transaction of all business of the Board. No member of the Board shall
at any time during his or her membership engage in any political activity
other than the exercise of the right to vote. Members of the Board may
be removed for cause at any time by the Military Governor in Korea (but,
after the period of military occupation, by the chief executive official of the
government)."

b. Section VII a (3) thereof is deleted.

c. Section VII *b (3)* thereof is amended to read as follows:

"(3) To tender appointment to any qualified Korean as President
of Seoul National University who, upon his acceptance in
writing, shall be formally appointed by the Military Governor
in Korea (but, after the period of military occupation, by
chief executive official of the government)."

d. Section VII *c* thereof is amended to read as follows:

"*c. Remuneration.* The remuneration of appointive members of the
Board of Regents of Seoul National University shall be determined by the
Military Governor in Korea (but, after the period of military occupation, by
the chief executive official of the government). *Ex officio* members of the
Board shall receive no remuneration for such service."

— 2 —

SECTION II. *Effective Date.* This Public Act shall be effective on the tenth day after the date appearing hereon.

| | Duly enacted by the Korean Interim |
| APPROVED : | Legislative Assembly on 20 March 1947 : |

ARCHER L LERCH
Major General United States Army
Military Governor in Korea

KIM KIU SIC
Chairman, Korean Interim
Legislative Assembly

HEADQUARTERS
UNITED STATES ARMY MILITARY
GOVERNMENT IN KOREA
Office of the Military Governor
Seoul, Korea

PUBLIC ACT
NUMBER 2

8 May 1947

COLLECTION OF SUMMER GRAINS

SECTION 1 The summer grains collection program, submitted by the National Food Administration, is approved, subject to the modifications set out in Section II These modifications shall be made part of the program which shall be published simultaneously herewith.

SECTION II. *Modifications.*

a. Since summer grains are customarily used for food by the farmers, the collection shall be limited to one-fifth of the estimated total production. Individual quotas must be just and fair and based on a careful estimate of production.

b. In order to insure just quotas, each eup and myun head shall organize an investigation committee in each dong or ri in order to check farmers' production estimates.

c. Compulsory methods shall be prohibited in making the collection. Only civilian administrative officals shall be used; they shall encourage the farmers to turn in their quotas as a matter of conscience and duty to their country, *provided, however,* that penalties set out in National Food Regulation No 5, "Collection of Summer Grains" shall be enforced by the duly constituted authorities in accordance with law.

d. Prices for summer grains shall be as fair as possible and shall take into consideration costs of production and the latest price index.

e. Farmers who have turned in their quota shall be accorded preferential rights in the rationing of consumers' goods.

— 1 —

f. Summer grains collected shall be adequately protected and provision shall be made for proper transportation, warehousing and administration.

SECTION III. This Public Act shall be effective on the date appearing hereon.

APPROVED:

Duly enacted by the Korean Interim Legislative Assembly on 19 April 1947:

ARCHER L LERCH

Major General United States Army

Military Governor In Korea

KIM KIU SIC

Chairman, Korean Interim

Legislative Assembly

S E A L

USAMGIK

SOUTH KOREAN INTERIM GOVERNMENT

Seoul, Korea

PUBLIC ACT
NUMBER 3   27 May 1947

ELEVATION OF TOWN OF IRI TO CITY STATUS

SECTION I. *City of Iri Created.* The town of Iri, in North Cholla Province, is hereby elevated to the status of a city, and shall have the right to act as such under all applicable ordinances and other provisions of law.

SECTION II. *Effective Date.* This Public Act shall be effective as of the 23rd day of February, 1947.

APPROVED:

ARCHER L LERCH
Major General United States Army
Military Governor in Korea

Duly enacted by the Korean
Interim Legislative Assembly
on 20 February 1947

KIM KIU SIC
Chairman, Korean Interim
Legislative Assembly

S E A L
USAMGIK

SOUTH KOREAN INTERIM GOVERNMENT
Seoul, Korea

PUBLIC ACT
NUMBER 4

16 May 1947

CHILD LABOR LAW

SECTION I. *Purpose.* The purpose of this Act is to regulate child labor and to prohibit, limit, improve and mediate conditions of child labor so as to protect children from harmful or dangerous occupation or heavy labor, guarantee healthy development of life and body as well as just compensation, and thus secure the welfare of society.

Rules and regulations necessary to implement this Act shall be issued and enforced in accordance with the terms of this Act by the Director of the Department of Labor, hereinafter referred to as "the appropriate Director".

SECTION II. *Definition of "Children".* For purposes of this Act the term "children" means children, both male and female, under the age of eighteen (18) years. In the computation of age, a child shall be considered one year old one full calendar year after date of birth.

SECTION III. *Prohibition or Limitation of Employment or Labor. a.* Employment of children under the age of twelve (12) years in any occupation or under any labor conditions is prohibited.

b. Children under the age of fourteen (14) years shall not be engaged or employed in any business or work listed below; *provided, however,* that the provisions of paragraph *a* of Section III of this Act shall apply to children under fourteen (14) attending public school.

c. Children under the age of sixteen (16) years shall not be engaged or employed in any business, work or occupation dangerous to life or limb or injurious to health or morals.

d. The provisions of paragraph *c* of Section III of this Act shall also apply to children under the age of eighteen (18) years.

e. The scope of occupations prohibited or limited by the several paragraphs of this Section is as follows:

(1) *Occupations Prohibited for Children under 14 Years*. Occupations in any of the following workshops or facilities (except assistance in or about the child's own home): (*a*) mill, (*b*) factory, (*c*) workshop, (*d*) mercantile or mechanical establishment, (*e*) store, (*f*) office, (*g*) office building, (*h*) restaurant, (*i*) bakery, (*j*) barber shop, (*k*) hotel, (*l*) apartment house. (*m*) bootblack or shoeshine stand or establishment, (*n*) public stable, (*o*) garage. (*p*) laundry, (*q*) place of amusement, (*r*) club, (*s*) occupation as a driver. (*t*) occupation in any brick or lumber yard, (*u*) occupation in the construction or repair of buildings, (*v*) occupation in the transmission of messages.

(2) *Occupations Prohibited for Children under 16 Years*. (*a*) Adjusting any belt to any moving machinery, (*b*) sewing or lacing machine belts in any workshop or factory, (*c*) oiling, wiping or cleaning machinery in motion or assisting therein, (*d*) operating or assisting in operating any of the following machines: (*1*) circular or band saws, (*2*) wood shapers, (*3*) wood jointers, (*4*) planers, (*5*) sandpaper or wood-polishing machines, (*6*) wood-turning or boring machinery, (*7*) picker machines or machines used in picking wool. cotton, hair or any other material, (*8*) carding machines, (*9*) paperlace machines, (*10*) leather-burnishing machines, (*11*) job or cylinder printing presses operated by power other than foot power, (*12*) boring or drill presses, (*13*) stamping machines used in sheet metal and tinware or in paper and leather manufacturing, or in washer and nut factories, (*14*) metal- or paper-cutting machines, (*15*) corner-staying machines in paper box factories, (*16*) corrugating rolls, such as are used in corrugated paper, roofing or washboard factories, (*17*) steam boilers, (*18*) dough brakes or cracker machinery of any description, (*19*) wire or iron straightening or drawing machinery, (*20*) rolling mill machinery, (*21*) power punches or shears, (*22*) washing, grinding or mixing machinery, (*23*) calendar rolls in paper manufacturing, (*24*) laundering machinery, (*e*) occupation in proximity to any hazardous or unguarded belts, machinery or gearing, (*f*) occupation upon any railroad, whether steam electric or hydraulic, (*g*) occupation upon any vessel or boat engaged in navigation or commerce, (*h*) occupation in, about or in connection with any processes in which dangerous or poisonous acids are used, (*i*) the manufacture or use of dangerous or poisonous dyes. (*j*) the manufacture or packing of paints. white or red lead. (*k*) occupation causing dust in injurious quantities, (*l*) the manufacture or use of compositions of lye in which the quantity thereof is injurious

— 2 —

to health, (*m*) occupation on scaffolding, (*n*) occupation in any distillery, brewery or any other establishment where malt or alcoholic liquors are manufactured, packed, wrapped or bottled, (*o*)occupation on or in connection with any dam, smeltery, reverberatory furnace, refinery, dressing and reduction plant, X-ray machinery; or any activity involving use of or exposure to any radioactive element or substance, (*p*) heavy work in the building trades, (*q*) occupation in any tunnel or excavation, (*r*) occupation in, about or in connection with any mine, coal breaker, coke oven or quarry, (*s*) assorting, manufacturing or packing tobacco, (*t*) operating any automobile, motor car or truck, (*u*) occupation in any bowling alley, pool or billiard room, (*v*) occupation upon the stage of any theatre or concert hall or in connection with any theatrical performance or other exhibition or show, (*w*) any other occupation dangerous to the life or limb, or injurious to the health or morals of such child.

(*3*) *Heavy Industry or Work Prohibited for Children under 18 Years.* (*a*) Occupation in, about or in connection with blast furnaces, docks or wharves, (*b*) occupation in the outside erection and repair of electric wires, (*c*) occupation in the running or management of elevators or lifts (unless equipped with suitable safety devices), (*d*) occupation in, about or in connection with hoisting machinery, or dynamos, (*e*) occupation in oiling or cleaning machinery in motion, (*f*) occupation in the operation of emory wheels or any abrasive, polishing or buffing wheel where articles of the baser metals or iridium are manufactured, (*g*) switch tending, (*h*) gate tending, (*i*) track repairing, (*j*) occupation as brakeman, fireman, engineer, motorman or conductor upon a railroad, (*k*) occupation as railroad telegraph operator, (*l*) occupation as pilot, fireman or engineer upon any boat or vessel, (*m*) occupation in or about any establishment wherein nitroglycerin, dynamite, dualin, guncotton, gunpowder or other high or dangerous explosives are manufactured, compounded or stored, (*n*) occupation in the manufacture of white or yellow phosphorus or phosphorus matches, (*o*) occupation in any cement mill, (*p*) occupation in any hotel, theatre, concert hall, place of amusement, or any establishment where intoxicating liquors are sold, (*q*) occupation in, about or in connection with any race track or any gaming, betting or gambling establishment or enterprise. No female child under 18 shall be employed in any mine, quarry or coal breaker (except in the office thereof).

— 3 —

SECTION IV.. *Prohibition of Burdensome Work.* No individual, group of individuals, or juridical person, hereinafter called "employer", employing children shall impose burdensome work upon such children, irrespective of the consent of the parties or the conditions of work. Employers shall give due regard to the physical development of such employees and shall permit proper rest. More especially, no night work or overtime work shall be imposed upon children; provided, however, that hours for night work may be determined for seasonal occupations by the appropriate Director. For purposes of this Act, with reference to burdensome work of children under twelve (12) years, the parents, guardian or custodian of such children, hereinafter referred to as the "person in parental authority", shall be deemed to be employers.

SECTION V. *Limitation of Working Hours.* Working hours for children under sixteen (16) years, including children under fourteen (14) years, shall not exceed six (6) days in any one week or seven (7) hours in any one day. Working hours for children under eighteen (18) years shall not (exceed six (6) days in any one week or nine (9) hours in any one day) *provided*, that within the working hours prescribed by this Act the employer shall permit proper rest hours and meal hours in accordance with the character of the work or labor

SECTION VI. *Employment Contract or Agreement. a.* All employment contracts of children (including collective agreements) shall be accompanied by a permit issued in accordance with this Act (including regulations issued by the appropriate Director pursuant to the authority granted in Section I of this Act). Such contract shall be legally effective from the date the permit is issued.

b. Any employment where the consideration for the work is an advance made to the employee or any other person prior to the work, or whereby the employee is required to engage in personal relations, is hereby prohibited notwithstanding any contract.

c. The term of an employment contract shall not exceed one year. When a contract is renewed a new permit shall be required.

SECTION VII.. *Application for Permit.* Any employee who applies for a permit to enter into an employment contract in accordance with Section VI hereof shall present the application for such permit to the factory inspector who has jurisdiction over that district or the government official in charge of

the enforcement of this Act (hereinafter referred to as "supervising authorities".
in accordance with the following procedure:

 a. The applicant, and the person in parental authority shall be present
in person.

 b. The application shall contain the following items and be accompanied
by the following documents:-

  (1) *Bonjuk* (permanent domicile), address, name, sex and date of
  birth of employee.

  (2) *Bonjuk*, present address, name, occupation and relation to the
  employee, of the person in parental authority.

  (3) Address, name (trade name or juridical name), occupation (or
  profession) of the employer.

  (4) Duration of the employment contract, wage (in case of piecework, the method and amount of compensation), nature and
  description of the work and work place, any special conditions
  connected with the contract, and whether the contract is a
  renewal contract and if so the period of renewal.

  (5) Date of application.

  (6) Written consent of the person in parental authority.

  (7) Proof of age (an abstract of *Hojuk*, birth certificate, or similar
  certificate issued by the appropriate government official).

  (8) School record or certificate, or *curriculum vitae.*

  (9) Medical certificate.

  (10) Labor contract, or preparatory agreement or letter of consent.

 c. The application shall contain a statement, signed, sealed and sworn
to by the parties and the person in parental authority, that they will not
violate the provisions of this Act.

 d. The application for renewal of an employment contract may omit
all or part of the accompanying documents provided for in section *b,* items
6-10, of this Section VII.

 e. When the name of the employer is changed, the application to

— 5 —

change the name of the employer in the permit may also omit all or part of the accompanying documents.

SECTION VIII. *Employment Permit. a.* Upon approval of the application the appropriate supervising authority shall, within thirty (30) days from the date of receiving the application, issue the permit herein described.

b. The permit may be revoked for illegality or fraud.

c. The permit shall be retained by the employer for the duration of the employment.

d. If the employer fails to retain the permit or to show it to the supervising authority or any other person, the employment pursuant to such permit shall be illegal.

e. The permit shall contain the following items:

(1) Address, name, sex, date of birth, and physical characteristics of the employee.

(2) Address, name, occupation, and relation to the employee, of the person in parental authority.

(3) Address, name (trade name or juridical person), occupation (or profession) of the employer.

(4) Duration of employment contract, wage (in case of piecework, method and amount of compensation), kind and nature of the work and work place, any special condition connected with the contract, and whether it is a renewal contract and if so the duration of the renewal.

(5) Date of receiving the application.

(6) Date of issuing the permit.

f. The permits shall be numbered in a single annual series, and the supervising authority shall state under seal that he examined the series and found it in accordance with this Act.

g. The permit shall be made in duplicate. One copy shall be given to the applicant and the other shall be kept by the supervising authority. When a new permit is needed because of the loss or defacement of a permit, the party may apply for the reissuance of the permit.

h. When a party applies for reissuance, he shall state the permit number, date of issuance. and the reason for application. He shall apply in writing to the supervising authority, attaching a copy of the original, where available.

When the supervising authority examines the case and finds good reason to reissue, he shall compare the copy attached to the application with the original copy kept by the supervising authority and, if there is no discrepancy, he shall endorse under seal a verification on the copy. This shall constitute the reissuance.

In the case of application for renewal of permit and for change of name in the permit, the examination of the contents of the application and procedure for issuance may be simplified.

k. In the application of this Act to children employed by the head of a government office or public juridical person, the permit of employment of that organization shall be regarded as the permit prescribed in this Section.

SECTION IX. *Termination, Cancellation or Dissolution of the Employment Contract.* *a.* The employment contract shall be terminated or cancelled for any of the following reasons:

(1) The expiration of the term of the contract.
(2) The cancellation of the permit.
(3) The death, bankrupty, or dissolution of the employer.
(4) The disappearance of the enterprise establishment as result of a disaster.
(5) The confirmation of dissolution of the contract by consent of the parties.

b. A party may request the dissolution of the contract for any of the following reasons:

(1) When the other party does not fulfill the contract.
(2) When the employee is unable to work because of sickness or other physical hindrance.
(3) When the employee misbehaves and there is no hope of repentance.
(4) When the employee is absent without notice for more than twenty (20) days.
(5) When the employer requests burdensome labor or treats the

employee so cruelly that the employee cannot endure the employment. or when the employer does not fulfill his duties prescribed in this Act.

c. A request for dissolution of the contract must be answered in writing within twenty (20) days.

d. The employer shall. within sixty (60) days prior to the expiration of an employment contract. notify the employee whether he will agree to a renewal.

e. When the parties fail to reach agreement as to the dissolution of a contract, or when the reason for the request for dissolution by one party is considered to be improper by the other party. the matter may be submitted to arbitration by the supervising authority in accordance with regulations prescribed by the appropriate Director.

f. When a contract has expired or has been cancelled. the employer, stating that fact, shall return the permit to the supervising authority within seven (7) days after such expiration or cancellation.

SECTION X. *Employer's Duty to Employees. a.* Every employer who employs two or more children shall post and keep posted in a conspicuous place in every room where. any child is employed, a list of the names, addresses, sex. age, nature of occupation and the date of commencement of employment of each child. He shall keep a duplicate copy as part of his records.

b. Every employer who employs five (5) or more children shall post and keep posted, in the same manner. the working regulations showing the weekly and daily working hours of each child, the hours of commencing and stopping work, the hours allowed for meals, and other important matters. The employer shall keep a copy as part of his records.

c. An employer who employs ten (10) or more children (hereinafter referred to as an "authorized factor") shall keep, as part of his records, the detailed plan for the education and protection of his employees.

d. An employer who provides a dormitory establishment shall post, in a conspicuous place in the dormitory, the employees' general schedule during non-working hours.

— 5 —

e. The employer shall not request the child to stay in the dormitory, factory or in any designated place on a non-working day, holiday, or during non-working hours; *provided, however,* that proper supervision or restriction of behavior, for the sake of all the employees, may be imposed.

SECTION XI. *Physical Examination and Medical Treatment. a.* Irrespective of the nature of the work, the employer shall provide physical examinations at the beginning of the employment and at intervals of three (3) months thereafter by a physician designated by the supervising authority.

b. When physical examination discloses that any sickness or ill health is caused by such employment, the employer shall provide treatment and relief therefor, and shall report the results to the supervising authority.

c. All expenses necessitated by the provisions of this Section shall be borne in full by such employer.

d. The employer shall file in his records the reports of physical examinations.

SECTION XII. *Sick Leave, Pregnancy, Child-bearing and Nursing. a.* Employees shall be entitled to not less than two and one-half (2 1/2) days of sick leave per month at full pay (regular wage, allowance and other pay), which sick leave may be accumulated during the employment period. Employees may, in appropriate cases, be required to notify the employer in writing one day prior to the date of absence where the absence is expected to be for two days or more.

b. Pregnant employees shall be granted sixty (60) days leave without pay for purposes of rest, *provided, however,* that an employer shall be required to grant such pregnancy leave only in case the employee has been in his employ for at least six (6) months.

c. In case of child-bearing, a female employee shall be entitled to sixty (60) days leave with regular pay immediately before and after the child-bearing.

d. As to nursing required for the infant, appropriate facilities and opportunities shall be afforded to such women employees.

Employees may be required to furnish documentary proof in case of any claim of absence due to sickness, pregnancy, or childbearing.

SECTION XIII. *Education, Sports and Recreation for Employees.* a. Authorized factories shall provide education, sports and recreation for employees under eighteen (18).

b. At least four (4) hours a week shall be devoted to general educational instruction not connected with the employment. At least two (2) hours a week shall be devoted to sports and recreation.

c. All the expenses and facilities necessitated by the provisions of this Section shall be borne or provided by the employer.

d. The detailed regulations for the education and recreation necessitated by this Section shall be prescribed by the appropriate Director in cooperation with the Director of the Department of Education and the Director of the Department of Public Health and Welfare

SECTION XIV. *Issuance of Detailed Regulations by the Appropriate Director.* In addition to the special provisions for the detailed regulations in this Act, the appropriate Director shall enact enforcement regulations concerning the following matters:

    a. Form of application for employment permit for children (Section VII).

    (b) Form of employment permit. (Section VIII).

    (c) Form of list to be posted (Section X).

    (d) Form of working regulations (Section X).

    (e) Form of report of the result of physical examinations (Section XI).

    (f) Details of education and recreation in the authorized factories (Section XIII).

    (g) Forms of report concerning arbitration (Section IX).

    (h) Forms of report concerning violations (Sections XVII and XVIII).

    (i) Any other matter requiring detailed regulations.

SECTION XV. *Appointment, Powers, Duties and Reports of the Supervising Authority.* a. The supervising authority shall be appointed by the appropriate Director. He shall be supervised primarily by the local administrative official, and secondarily by the appropriate Director. The organisation of the supervising authority shall be determined separately by the appropriate Director.

b. The supervising authority shall have the power to issue orders, conduct investigations, and make dispositions necessitated by the provisions of this Act. In addition, he shall have the following duties:

(1) The investigation of local customs concerning the employment of children.

(2) The investigation of the maximum working hours of specific industries and enterprises, or generally.

(3) The investigation of working conditions and maximum and minimum wages.

(4) The inspection of industrial and commercial businesses.

(5) The inspection of facilities for, and practice concerning, education and recreation in authorized factories.

(6) Concerning matters of improvement of working conditions.

(7) Concerning other matters necessary for the enforcement of this Act.

c. The supervising authority shall, before the tenth of each month, report to the appropriate Director on the matters set forth in paragraph e of this Section. The supervising authority may, at any time, present his opinion concerning matters pertinent to his duties.

d. The foregoing report and opinions shall be presented through the local administrative officials.

e. Reports shall contain the following information:
(1) Name-list of employees.
(2) Disposition and results of violations.
(3) Results of inspection of businesses, industries, and authorized factories.
(4) Results of physical examinations.
(5) Other comments pertinent to duties of the supervising authority.

f. The supervising authority shall keep whatever records and files are necessary for the performance of his duties, and shall make such records and files available to the local administrative officials and the appropriate Director, upon their request.

— 71 —

SECTION XVI. *Rights of Recovery.* a. Any person injured by the violation of this Act may recover from the offender such damages as he may sustain thereby. In any suit by or on behalf of any child illegally employed hereunder, contributory negligence, assumption of risk, or negligence by fellow employees shall not be a defense and may not be pleaded or proved either by way of defense or mitigation, nor shall misrepresentation of age by the person securing the employment or by the person employed be a defense or matter in mitigation.

This shall not apply in case of fraud or tort to gain the benefit of recovery.

b. Any person illegally employed hereunder shall be entitled to recover a sum equivalent to ten times the agreed amount of wages earned or ten times the legal minimum wage, whichever is greater.

c. In case the employer is other than a juridical person, all members of such employer-body shall be liable as employers under the provisions of this Section.

d. None of the provisions of this Section may be waived, and any release of any rights hereunder shall be ineffective to bar any recovery hereunder.

e. No suit may be maintained under this Section for any period of employment which occurred two years prior to the filing of the suit.

SECTION XVII. *Juvenile Delinquency.* a. When a child violates any provisions of this Act, or refuses to obey the order or report of, or permit investigation by the supervising authority, or makes any false statements, the supervising authority may take the following steps depending on the nature of the offense:

(1) Admonish and exhort the child to repent.
(2) Request that the schoolmaster and/or person in parental authority admonish the child and furnish a letter of guarantee.
(3) Request a written oath of repentance.
(4) Cancel the employment permit or reject the application.

The above steps may be taken either separately or jointly.

b. In case the steps mentioned above do not correct the behavior of the child, such child shall be conducted forthwith before the juvenile court which has jurisdiction over that district; or, if there is no such court in that place, before any court or judicial officer having jurisdiction over offenses committed by children.

c. In the treatment of offenses committed by children, special care and caution shall be taken and a constant effort made to guide the child.

SECTION XVIII. *General Penalties.* *a.* Any person who permits, procures or suffers the violation of the provisions of Sections III and IV of this Act shall be punished by penal servitude not exceeding six (6) months or a fine of not less than one thousand won (¥ 1,000).

b. Any employer who violates the provisions of Sections V or VI-*b*, or who refuses to permit investigation by, or obey the order or report of the appropriate Director or the supervising authority, or makes any false statements, shall be punished by confinement of less than thirty (30) days or a fine of not less than five hundred won (¥ 500).

c. Any employer who wilfully or continuously violates any of the provisions of this Act shall be punished with penal servitude of not less than one year.

SECTON XIX. *Interim Provisions:* *a.* Any employer employing children in permitted occupations at the effective date of this Act shall, within sixty (60) days, apply for an employment permit, and may continue such permitted employment while the application is pending.

b. Where children over fourteen (14) years are employed in work prohibited or limited in Section III of this Act, on the effective date thereof, the employer or the supervising authority shall within sixty (60) days allocate them to permitted work consistent with the purpose of this Act.

c. If such allocation is impossible because of the nature of the work and the difficulty in providing other work, an application for a permit shall be filed in accordance with Section VII of this Act. The application may be granted on a temporary basis by special action of the appropriate Director, regardless of the provisions of Section III, *provided, however*, that such temporary employment shall not exceed one year.

軍政廳　官報　法律　第四號　　一九四七年五月一六日

(1) Persons who were advisors, members or vice-chairmen of the Central Advisory Council under the Japanese regime.

(2) Persons who were members of an advisory or deciding council of PU or DO (Province) under the Japanese regime.

(3) Persons who held positions of the third class or higher of the "Kotokan" or who received a medal (Kun) of the seventh class or higher. However, educators and technical officials are not included in this category.

(4) Persons who under the Japanese regime held rank of "Hanninkan" or higher in the civilian police force or who served as "Kempei" or "Kempei-Ho" in the Japanese military police force, or persons who held positions in the police in charge of "thought control" or those who acted as spies for the latter.

Section 3. Government officials shall not, while in office, hold membership in the Assembly in addition to such position, except Jungmukwan.

Section 4. The members of local self-governing bodies shall not, while in office, hold membership in the Assembly in addition to such position.

Section 5. The officials participating in election procedures and the members of election committees shall not be eligible to be elected members in the Assembly within the districts concerned.

Section 6. Expenses incurred in the election of Assembly members shall be defrayed by the national treasury.

CHAPTER II. ELECTORAL DISTRICTS AND REPRESENTATION

Section 7. Each administrative district (i. e., PU. GUN. DO. island and KU of the city of Seoul, hereinafter referred to as district) shall constitute one electoral district for the purpose of electing members of the Assembly.

Section 8. Each PU, EUP, MYUN and KU (administrative districts) shall constitute one voting district, respectively. The election committee of each electoral district may divide the voting district into voting sub-districts and establish polling places for the sake of convenience in carrying out the voting procedure.

The establishment of voting sub-districts should be publicized before the beginning of voters' registration.

Section 9. Each electoral district shall be represented in the Assembly by one member. In case the population of an electoral district is over 100,000, members shall be elected in the ratio of one member for each 100,000, and one additional member shall be elected for the fraction of 50.000 or more thereof.

Numbers of members to be elected from each electoral district shall be determined and publicized by the Central Election Committee in accordance with population statistics nearest to the election date.

CHAPTER III. POLL REGISTERS

Section 10. The voters shall register by signing and sealing (or by thumbprint for a seal) on registration paper at the registration place designated by the chairman of the electoral district election committee, some time during a period of ten days beginning sixty days before the date of election.

Each voter shall register at only one registration place.

Section 11. The election committee of the voting district (or in case of a voting sub-district, the election committee of that sub-district) shall prepare poll registers indicating the name, address, date of birth, sex, date of registration, and other pertinent data with respect to all voters who have been in residence in that district since the date sixty days prior to the date of the election concerned. The registers shall be open to the public at the registration places for a period of fourteen days beginning forty days prior to the date of the election. The poll registers shall become final on the tenth day before the date of election.

Section 12. When there are incorrect recordings or omissions, or when a person who has no right to vote is registered in the poll register, each voter may object to the election committee concerned during the period of public notification.

Examination and decision in regard to these objections shall be made within three days by the election committee concerned.

In case objections to the foregoing decision be raised, the voters may request, within five days, the electoral district election committee to review the decision.

— 3 —

The election committee of the electoral district shall on a request for review in accordance with the preceding paragraph review and decide each case within three days and shall notify in writing the person who asked for review and other persons concerned.

CHAPTER IV. ELECTION COMMITTEES

Section 13. The Chief Executive shall organize a central election committee eighty days before the date of election and shall order the organization of provincial election committees, election committees of the electoral districts, and election committees of the voting districts and voting subdistricts.

Section 14. The central election committee shall consist of fifteen committee members, and its chairman shall be elected by and from among the members of the committee.

The provincial and Seoul City election committees shall consist of nine committee members each, and their chairman shall be elected by and from among the members of the individual committees.

The election committees of the electoral districts shall consist of nine committee members each, and their chairmen shall be elected by and from among the members of the individual committees. The election committees of the voting districts shall consist of seven committee members each, and their chairmen shall be elected by and from among the members of the individual committees.

The election committees of the voting subdistricts shall consist of seven members each, and their chairmen shall be elected by and from among the members of the individual committees.

Section 15. The heads of the administrative districts concerned will organize the election committees for their respective districts, appointing the members from among those persons eligible to vote. One-third or more of the members of any committee may not be chosen from the same party or organization.

A candidate for election is not eligible for appointment to the election committee in the electoral district in which he is a candidate.

Each type of election committee may employ several secretaries and clerks.

Section 16. Each election committee shall act in accordance. with laws and regulations and in accordance with the orders and instructions of higher election committees and shall render various reports on its activities to the higher election committes and shall supervise the activities of the lower election committees.

Upon request, each election committee shall show its various records and documents to a court.

Section 17. The quorum of each election committee shall be a majority of its members, and all the actions of the committee shall be determined by majority vote of the members present. In case of a tie in voting, the chairman of the committee shall decide. The chairman may take necessary measures when the members do not respond to calling, or in case of emergency.

CHAPTER V. CANDIDATES AND ELECTION CAMPAIGNS

Section 18. A person who wishes to stand for election to the Assembly shall register with the election committee of the electoral district some time between the date of the beginning of voters' registration and a date twenty-five days before election day, and submit, at the same time, a recommendation signed by one hundred or more registered voters.

In case of recommendation of one person by another to be a candidate, the person making the recommendation must secure the signature of one hundred or more registered voters and the written consent of the person being recommended and register with the election committee of the electoral district.

The registration of a candidate who is registered in two or more electoral districts shall be invalid.

Section 19. In case the number of registered candidates falls short of the number of Assembly members specified for the electoral district concerned, or in case of the death or withdrawal of any candidate, the registration of other candidates may take place until a date ten days before election day.

Section 20. Election committees of electoral districts shall give public notice of the name, address, age, occupation, and party or organization affiliation of every candidate within three days after each candidate's registration.

Election committees of electoral districts shall also give public notice immediately of the withdrawal or death of any candidate.

— 5 —

Section 21. Registered candidates may freely conduct campaigns for election. but they may not give. or promise to give, money, goods or other property benefits.

Members of each election committee. public officials connected with election matters and any other public official may not take part in campaigns within the districts concerned.

CHAPTER VI. ELECTION PROCEDURE AND SUCCESSFUL CANDIDATES

Section 22. The election of Assembly members shall take place on the same day throughout the area in which this law is applicable.

The Chief Executive shall announce the date eighty days before the election day.

The election committees of the electoral districts shall make official announcements of the date. time, polling places and number of Assembly members to be elected sixty days before election day.

Polling places shall be the same places in which the voters register; however, when any polling place must be changed on account of a calamity or other unavoidable emergency, the electoral district election committee concerned shall immediately give public notice.

Section 23. The voting shall be accomplished by (the casting of) secret (unsigned) ballots. each of which shall designate one candidate only.

Every voter shall cast one ballot.

Section 24. Voters shall present themselves in person at the polls. and each voter shall cast his own ballot.

In case of doubt as to the identity of any voter. the election committee of the voting distrinct (in case voting subdistricts are established, the election committee of the voting subdistrict) shall decide the question.

In such case, the head of the Dong or Pan in which the voter resides may be called upon as a witness.

Section 25. Balloting shall being at 8:00 A. M. and close at 6:00 P. M. on election day.

Section 26. When the voting time begins, the election committee of the voting district (in case voting subdistricts have been established, the election

— 6 —

committees of voting subdistricts) shall give each voter one ballot after he has sealed (or thumbprinted) on the poll register.

Section 27. Persons whose names are not registered on the poll register cannot vote; provided, however, that a person may vote who brings with him the notification of a decision rendered by the election committee of the electoral district or voting district certifying that the person shall be registered on the poll register.

Section 28. Each voter shall himself write the name of the candidate on the ballot at a place established at the polling place and shall himself cast his ballot into the ballot box.

Section 29. Besides voters, no one except police or election officials whom the appropriate election committee has recognized, may enter the polling place. The chairman of the election committee concerned may restrain speech, discussion, clamor, negotiations concerning the election, electioneering or any other disturbance at the polling place or in its neighborhood. He may expel from the polling place any person who disobeys such restraining orders.

Voters who have been expelled from the polling place in accordance with the preceding paragraph may vote at the last. However, when the chairman of the election committee concerned finds that there is no longer any danger of disturbance of the polling place, he may let such person cast his vote.

Section 30. The polling places shall be closed when the polling hours are ended and the ballot boxes shall be locked after the last voter in the polling place has cast his vote. No votes shall be cast after the ballot boxes are locked.

Section 31. The election committes of voting districts shall deliver the ballot boxes and records to their respective election committees of electoral districts within six hours after the voting has been closed.

In the case of a voting subdistrict, the election records and poll registers shall be delivered to the appropriate election committee of the voting district without delay after the voting has been closed.

The election committees of the voting districts shall deliver the poll registers and all other documents pertaining to the election to the appropriate heads of PU, heads of EUP, heads of MYUN, or heads of KU after the voting has been closed. The officials concerned shall have those records in their safekeeping during the terms of the elected Assembly members.

Section 32; In case there is difficulty in meeting the time requirements specified in the foregoing section, on account of the distance between islands or the lack of means for rapid communication in some areas, the election committees of the voting districts shall deliver them according .to a schedule fixed by the election committee of the electoral district.

Section 33. In case the voting cannot be held on account of some calamity or unavoidable accident or in case of loss or destruction of ballot boxes, which makes a partial re-election necessary, the election committee of the electoral district concerned shall secure permission from the appropriate provincial or Seoul City election committee and hold a second balloting. The date of this second balloting shall be announced at least five days before it is to take place.

Section 34. When the election committee of the electoral district has received all the ballot boxes in its district, each box shall be opened and the number of ballots counted to see whether the number of ballots corresponds with the number of registered voters who received ballots, according to the poll records. The ballots shall then be opened and the number of votes for each candidate counted.

In case ballot boxes are lost or destroyed the counting of ballots cannot be accomplished properly for other reasons, the election committee of the electoral district concerned by permission of the appropriate provincial Seoul City election committee may open only those boxes which it received in proper conditions.

Section 35. The following types of ballots shall be null and void:

(1) Ballot other than the regular, official ballot.

(2) Blank ballot, or ballot which records the name of a candidate other than one of those officially registered in the electoral district concerned.

(3) Ballot which records names of two or more candidates.

(4) Ballot by which it cannot be told which name is written.

(5) Ballot which records matter other than the name of a candidate; provided, however, that designation of occupation, rank, the address or honorific title of the candidate concerned may be included.

Section 36. . The candidate who receives the greatest number of valid votes shall be elected.

In an electoral district to which more than one Assembly seat is allocated, the winners shall correspond in number to the available seats, and shall be those who receive the highest number of votes.

In case of a tie, the chairman of the electoral district election committee concerned shall decide the winner by lot.

Section 37. If the number of registered candidates in an electoral district does not exceed the number of Assembly seats specified for that district, the candidates will be elected automatically, without an actual vote.

Section 38. After the winners have been determined, the election committees of the electoral districts shall, without delay, inform the successful candidates of their election, and give public notice of the results of the election.

Section 39. In the following cases, the candidates running next in the number of votes shall be elected :

(1) In the event the elected person refuses to assume his post.

(2) In the event the elected person dies.

(3) In the event the elected person loses the right to be elected after the date of election.

(4) In the event the election of the winner becomes void.

Section 40. In the following cases the provincial or Seoul City election committee shall decide the date for a second election and announce it publicly at least twenty days before such date, and shall order the electoral district concerned to hold a second election :

(1) In case the number of elected candidates is less than the number of Assembly seats specified for the electoral district concerned.

(2) In case all or part of the election becomes void.

Section 41. After completion of the ballot tabulation, the election committees of the electoral districts shall deliver their election records without delay to the appropriate provincial or Seoul City election committee.

The election committees of the electoral districts shall separate and identify valid and invalid ballots after the completion of the election, and shall deliver them, together with all pertinent documents, to the heads of PU, GUN, Islands or, KU, who shall keep them during the term of the elected Assembly members.

Section 42. As soon as the provincial and Seoul City election committees receive all the election records from their respective electoral districts, they shall submit the reports on the election without delay to the central election committee.

CHAPTER VII.   SPECIAL ELECTORAL DISTRICT

Section 43. A special electoral district covering all of South Korea shall be established for the benefit of voters who have their family register (BONJUK) in North Korea and residence in South Korea and who wish a BONJUK district election.

Voting districts, subdistricts, and registration places of the special electoral district shall be the same as those of the general electoral districts.

Section 44. Persons who wish to vote in the special electoral district must register at their appropriate registration places.

Poll registers for the special electoral district shall be prepared separately.

Section 45. The election committee of the special electoral district shall consist of nine members and its chairman shall be eletced by and from among its members. Provincial, Seoul City, and all lower election committees shall handle election matters for the special electoral district as well as for general electoral districts.

Section 46. The special electoral district voters shall vote only in the special electoral district election, and shall not vote in any general electoral district election.

Polling places for the special electoral district shall be attached to those established for general electoral districts, but the ballot boxes shall be separate.

Voting records and ballot tabulations of the special electoral district shall be prepared separately from those for general districts.

Provincial and Seoul City election committees shall receive the ballot count for the special electoral district from each lower election committee and

軍政廳   官報   法令   過渡院   一九四七年九月三日

shall submit the combined figures to the election committee of the special electoral district.

Section 47. Besides the special regulations stated in this chapter, all provisions of this law for the election in the general electoral district are applicable *mutatis mutandis* to the election in the special electoral district.

CHAPTER VIII. TERM OF OFFICE AND BY-ELECTIONS

Section 48. The term of office for Assembly members shall continue until such time as the Provisional Government of united Korea is established.

Section 49. When a vacancy occurs, a by-election shall be held to fill it.

The Chief Executive shall hold the by-election within seventy days after receipt of notification from the chairman of the Assembly that the vacancy exists.

The date of the by-election shall be announced at least fifty-days before it is to be held.

If the by-election is held within six months after general election day, the poll registers which were used for the general election shall be used for the by-election also.

Section 50. Besides the regulations stated in this chapter, all provisions of this law are applicable to by-elections.

CHAPTER IX. LITIGATION CONCERNING ELECTIONS

Section 51. Contests of the validity of elections against the chairman of the electoral district election committee as defendant may be filed with the Supreme Court by candidates or voters concerned within twenty days after the election day.

Section 52. Any defeated candidate may bring actions concerning the validity of the election of the winner in the Supreme Court against the chairman of the electoral district election committee and the successful candidate concerned as defendants within 10 days after the successful candidate has been officially designated.

Section 53. The Supreme Court shall declare all or part of the election void if such election is found by the Court to involve violations of the provisions of law and such violations may have caused a change in the result of the election.

In litigations in accordance with the preceding section, the Supreme Court shall declare all or part of the election void if such election involves violations of the provisions of law and such violations may have caused a change in the result of the election.

Section 54. Prosecutors shall be present at the Supreme Court when it holds trials concerning the election.

Section 55. The Supreme Court shall give priority to election cases.

Section 56. In addition to the regulations contained in this chapter, pertinent provisions of ordinary civil procedure shall be applicable to election cases.

Section 57. The Chief Justice of the Supreme Court shall inform the Chief Executive, the central election committee, and the local election committees concerned of the election cases filed with it. The Chief Justice of the Supreme Court shall send copies of the judgments, when it renders judgments on cases concerning the election, to the Chief Executive, the central election committee, the election committees concerned and the chairman of the Legislative Assembly.

CHAPTER X. PENAL REGULATIONS

Section 58. Any of the following violators of the law shall be punished by penal servitude for not more than five years, or a fine, of not more than 100,000 won, provided that both penal servitude and fine may be imposed in case that special circumstances require it:

(1) Any person who either registers on the poll register or votes by fraudulent means.

(2) Any person who gives or receives or promises to give or receive money, goods, entertainment, or any other property gain, or gives or promises to give positions of honor on favorable terms, for the promise of votes or the abandonment of right to vote as a condition.

(3) Any person who makes false statements concerning candidates in speeches or publications in order to be elected, or to promote or prevent the election of another candidate.

(4) Any person who tries to prevent anyone from voting or forces anyone not to vote by the use of violence, threats, arrest or confinement, or any other method.

(5) Any person who, for the purpose of hindering the election, uses violence or threats against the election committee members or public officials, or captures or destroys the ballot boxes or the election records.

(6) Any person who interferes with the free exercise of the right to vote or with the election process in general, by mass distrubances or demonstrations at or near the polling place.

(7) Any person who interferes with the voting of a voter in the polling place without proper reasons.

(8) Any person who forces entrance into the polling place carrying firearms, a sword, cudgel, or any other weapon.

(9) Any member of an election committee or public official who violates any laws or reguletions pertaining to the election.

Section 59. Any person who has been punished for the crimes described in this chapter shall be deprived of the rights to vote and to be elected for a period of three years from the date on which he completes the serving of his sentence. In case he is the person elected, his election shall become void.

Section 60. The statute of limitations for public prosecution with respect to the offenses described in this chapter shall be one year.

Supplementary Rules.

Section 61. The Chief Executive may enact such detailed regulations as may be necessary to carry out this law.

Section 62. This law shall be effective on and after the date of its promulgation.

APPROVED 3 September 1947.

Duly enacted by the Korean Interim Legislative Assembly 12 August 1947.

C. G. HELMICK
Brigadier General United States Army
Acting Military Governor

YUN KI SOP
Acting Chairman
Korean Interim Legislative Assembly

SOUTH KOREAN INTERIM GOVERNMENT
Seoul, Korea

PUBLIC ACT
NUMBER 6

27 September 1947

COLLECTION OF RICE

SECTION I. The purpose of this Act is to collect rice in order to ration it to nonsuppliers, thereby guaranteeing food to the nation.

SECTION II. Persons in the following categories are required to deliver rice to the government collection point:

1. Landlords; however, the tenants shall deliver the rental rice for the landlords to the government.

2. Owner-farmers who till more than three tanbo of rice land.

3. Tenant-farmers who till more than five tanbo of rice land.

As for those who are owner-farmers and, at the same time, tenant-farmers, their tenant-acreage shall be converted and decided by the ratio of paragraph 2 and this paragraph (i. e., three-fifths).

4. Those farmers who till less than the fixed limitation of tanbo under Section Number II, paragraphs 2 and 3, above, and have surplus food after deducting family requirements and seed rice.

5. New Korea Company. Those farmers who are owner-farmers in addition to tenant-farmers for New Korea Company, and those who are solely New Korea Company tenant-farmers shall deliver all their rental rice regardless of planted acreage.

All landlords shall report to the eup chang, myun chang (shi chang or pu yun) in whose area they reside, the crop estimates and amount of rental rice of their farms. In case the amount of the contracted rental rice is less than one-third of the rice crop production, then the tenant-farmer shall deliver directly to the government the difference between the contracted amount and the legal amount (one-third of the crop).

— 1 —

The New Korea Company shall be treated in the same manner as any other landlord. The amount of the individual rice quotas assigned to New Korea Company tenants shall be comparable to the quotas assigned to farmers who till similar farms adjacent to the New Korea Company farms.

SECTION III. Individual quotas for the farmers shall be decided by the eup chang or myun chang (shi chang or pu yun) with the advice of the rice collection committee established in each eup, myun (shi, pu) and each ri (jung dong).

SECTION IV. The rice collection committee shall consist of landlords, owner-farmers, tenant-farmers, representatives of big farms, farming experts and those persons who are well acquainted with the local situation. The number of members in each rice collection committee in a eup and myun (shi, pu) shall be not more than 15 persons. A rice collection committee in a ri (jung dong) shall be not more than 10 persons. The rice collection committee shall advise and assist the eup chang, myun chang (shi chang, pu yun) or ku chang (jung, dong hoi chang) by performing the following duties:

1. Investigate the crop estimates.
2. Investigate the food needs of the people and workers on each farm.
3. Estimate the individual farm quotas.
4. Perform any other important matters pertaining to the rice collection program.

When a rice collection committee believes that a eup chang, myun chang (shi chang, pu yun), or ku chang (jung, dong hoi chang) has made an unreasonable or unfair decision in regard to the rice collection, it is hereby authorized to object in the following manner:

Each ri (jung, dong) rice collection committee shall present its objections to the appropriate eup chang, myun chang (shi chang, pu yun). Each eup or myun rice collection committee shall present its objections to the appropriate gun soo or do sa (island-head). All objections shall state in detail the reasons for the objections. The pu yun, gun soo, do sa or the eup and myun changs (shi chang, pu yun) shall decide on and answer any such objections raised by a rice collection committee within 10 days.

— 2 —

SECTION V. When the rice collection program has been completed or when the government has ceased to collect rice, any farmer who has delivered his entire quota of rice to the government shall be authorized to sell the remainder of his rice in the open market and/ or transport it wherever he pleases.

SECTION VI. In determining the purchase price for the collected rice, the government shall take into consideration the general price level and the production costs. Those farmers who have delivered their entire quota of rice to the collection point shall have the first priority to buy a goodly amount of the available incentive goods.

SECTION VII. Those persons who conceal rice in order to avoid delivering it to the government, and those persons who decline to submit or wrongfully refuse or avoid delivering rice to the government shall be punished by imprisonment for a period not exceeding 6 months or by a fine not exceeding 50,000 won. In all these cases the rice in question shall be confiscated by the government.

SECTION VIII. Those persons who illegally buy, sell or transport rice during the collection period; those persons who either before, during or after the collection period buy rice for hoarding or refuse to sell rice to the government in the hope of profiteering thereby, shall be punished by imprisonment for a period not exceeding 1 year, or by a fine not exceeding 100,000 won. In all these cases the rice in question shall be confiscated by the government.

SECTION IX. Those persons who hamper or prevent the rice collection committee or officials from investigating the amount of the rice production in order to obstruct the rice collection program, or those persons who spread false propaganda or threaten the farmers in order to lessen the farmers' willingness to deliver rice to the government shall be punished by imprisonment for a period not less than 3 months or by a fine not exceeding 50,000 won.

SECTION X. Any official who intentionally or through neglect levies on a farmer an unjust rice quota or violates any of the provisions of this Act shall be punished by imprisonment for a period not exceeding 6 months.

— 3 —

SECTION XI. In order to carry out this Act any necessary regulations may be issued as administrative instructions.

SECTION XII. This Act shall be effective on the date appearing hereon.

APPROVED: 27 September 1947.

Duly enacted by the Korean Interim Legislative Assembly.

C G HELMICK
Brigadier General United States Army
Acting Military Governor

KIMM KIUSIC
Chairman
Korean Interim Legislative Assembly

SOUTH KOREAN INTERIM GOVERNMENT
Seoul, Korea

PUBLIC ACT
NUMBER 7

14 November 1947

ABOLISHMENT OF THE PUBLIC PROSTITUTION LAW

SECTION I. In accordance with a democratic point of view this law is created for the purpose of abolishing and prohibiting public prostitution—a malicious custom which was established under the Japanese regime and tended to infect Korean morality.

SECTION II. Ordinance No. 4 of the Director of the Department of Police Affairs, dated March 1916, is hereby abolished.

Prostitution and brothel licenses and the authorization for the establishment of the Prostitute's Business Association, which were issued under the foregoing ordinance, are hereby declared null and void.

SECTION III. Any person found guilty of the following shall be subject to imprisonment for a term of not more than two years, or suffer a fine of not more than W 50,000, or both:

a. Any person who continues to carry on or engages in the business of prostitution.

b. Any person who is a prostitute or who is an intermediary or who provides places for prostitution.

c. Any person who traffics in women for the purpose of prostitution.

d. Any person who has tainted another with a venereal disease.

Supplementary Rules: This law shall go into effect three months after the date appearing hereon.

警政廳 ─官 報─ 法 律 第七號   一九四七年十一月十四日

Granting of licenses for prostitution or brothel business shall cease on the date appearing hereon.

APPROVED:
C G HELMICK
Brigadier General United States Army
Acting Military Governor

Duly enacted by the Korean
Interim Legislative Assembly
KIMM KIUSIC
Chairman, Korean Interim
Legislative Assembly

SOUTH KOREAN INTERIM GOVERNMENT
Seoul, Korea

PUBLIC ACT
NUMBER 8

14 January 1948

ACT TO PROHIBIT AIDING AMERICAN PERSONNEL TO VIOLATE MILITARY DIRECTIVES

SECTION I. *Purpose.* The purpose of this act is to prohibit aiding in the violation of military directives by military and civilian personnel under the jurisdiction of the United States Army Forces in Korea.

SECTION II. *"Off Limits" Established:* No person who owns or operates an establishment marked "Off Limits" by any authority under the United States Army Forces in Korea shall permit or suffer any military or civilian personnel under the jurisdiction of the United States Army Forces in Korea to enter or remain in such establishment.

SECTION III. *Prostitution.* No person shall prostitute themselves, procure or aid in procuring a prostitute for any military or civilian personnel under the jurisdiction of the United States Army Forces in Korea.

SECTION IV. *Penalties.* Any person who violates the provisions of this act shall be punished by penal servitude not exceeding one (1) year or a fine of not more than fifty thousand won (¥50,000).

SECTION V. *Effective Date.* This act shall be effective on the date appearing hereon.

APPROVED:

WILLIAM F DEAN
Major General, United States Army
Military Governor in Korea

Duly enacted by the Korean
Interim Legislative Assembly
KIMM KIUSIC
Chairman, Korean Interim
Legislative Assembly

軍政廳　官報　法律　第八號　　一九四八年一月十四日

SOUTH KOREAN INTERIM GOVERNMENT
Seoul, Korea

PUBLIC ACT
NUMBER 9   12 February 1948

PUBLIC ACT NUMBER 7 AMENDED

SECTION I. Section III d of Public Act Number 7, dated 14 November 1947 (*Abolishment of the Public Prostitution Law*), is hereby repealed.

SECTION II. This act shall be effective on the date appearing hereon.

APPROVED: Duly enacted by the
                                   Korean Interim Legislative Assembly

WILLIAM F DEAN   **KIMM KIUSIC**
Major General, United States Army   Chairman, Korean Interim
Military Governor in Korea   Legislative Assembly

SOUTH KOREAN INTERIM GOVERNMENT
Seoul, Korea

PUBLIC ACT
NUMBER 10

31 March 1948

ARBOR DAY AS NATIONAL HOLIDAY

The fifth day of April each year is established as a national holiday to be designated as "Arbor Day."

APPROVED:

Duly enacted by the

Korean Interim Legislative Assembly

W. F. DEAN
Major General. United States Army
Military Governor in Korea

SHIN IK HI
Chairman
Korean Interim Legislative Assembly

軍政廳　官報　法律　第一〇號　　一九四八年三月三一日

SOUTH KOREAN INTERIM GOVERNMENT
Seoul, Korea

**PUBLIC ACT
NUMBER 11**   11 May 1948

TEMPORARY PROVISIONS CONCERNING THE LAW OF NATIONALITY

SECTION I. The purpose of this Act is to define those who are Korean nationals, in order to clarify legal relationships.

SECTION II. Persons who come under the following categories shall have Korean nationality:

1. Any person whose father is Korean.

2. Any person whose mother is Korean and whose father is unknown or has no nationality.

3. Any person born in Korea whose father and mother are unknown or who have no nationality.

4. Any alien woman who married a Korean; but if there is a dissolution of the marriage, her former nationality shall be restored if the law of her former nationality so provides.

5. Any alien who became naturalized in Korea. The requirements as to naturalization and rights of naturalized persons shall be provided by another law.

SECTION III. The rights of those persons who come under the categories "2" "3" and "4" of Section II hereof, shall be the same as those receiving naturalization.

SECTION IV. Any person who comes under the following categories shall lose Korean nationality:

1. Any person who becomes a naturalized citizen of another country.

2. A woman who marries an alien or any person adopted by an alien.

SECTION V. Any person who obtained foreign nationality or was

— 1 —

軍政廳　官報　法律　第一一號　　　一九四八年五月一一日

entered in a Japanese family register and has waived such nationality or cancelled such Japanese family registry on the effective date, or shall waive or cancel hereafter, shall be deemed to be restored to Korean nationality as of, and prior to, 9 August 1945.

SECTION VI. *Effective date.* This Act shall be effective on the date appearing hereon.

APPROVED:

Duly enacted by the Korean Interim Legislative Assembly

W. F. DEAN
Major General, United States Army
Military Governor in Korea

SHIN IK HI
Chairman
Korean Interim Legislative Assembly

SOUTH KOREAN INTERIM GOVERNMENT
Seoul, Korea

PUBLIC ACT
NUMBER 12   19 May 1948

DISSOLUTION OF THE
KOREAN INTERIM LEGISLATIVE ASSEMBLY

The 20th day of May 1948 at 1400 hours is hereby fixed for the date and
hour of dissolution of the Korean Interim Legislative Assembly.

APPROVED: Duly enacted by the
                                   Korean Interim Legislative Assembly

W. F. DEAN   SHIN IK HI
Major General, United States Army   Chairman
Military Governor in Korea   Korean Interim Legislative Assembly

UNITED STATES ARMY FORCES IN KOREA
Office of the Military Governor of Korea
Seoul, Korea

ORDINANCE
NUMBER 1   24 September 1945

ESTABLISHMENT OF THE BUREAU OF PUBLIC HEALTH

1. The Public Health Section of the Bureau of Police is hereby abolished.

2. The Bureau of Public Health is hereby established with the same duties and functions heretofore performed by the Public Health Section of the Bureau of Police.

3. All funds, records, and property of the Public Health Section of the Bureau of Police are hereby transferred to the Bureau of Public Health.

BY DIRECTION OF THE COMMANDING GENERAL
UNITED STATES ARMY FORCES IN KOREA

V. ARNOLD
Major General, United States Army
Military Governor of Korea.

CHAPTER 3   ORDINANCES

(USAMGIK)

UNITED STATES ARMY FORCES IN KOREA
Office of the Military Governor of Korea
Seoul, Korea

ORDINANCE
NUMBER 2

25 September 1945

CONCERNING PROPERTY TRANSFERS

Section 1. The purchase, sale, acquisition, transfer, payment, withdrawal, disposition, importation, exportation, or any dealing in or the exercise of any right, power, or privilege with respect to any gold, silver, platinum, currency, securities, accounts in financial institutions, credits, valuable papers, and any other property owned or controlled directly or indirectly, in whole or part on or since 9 August 1945, by any of the Governments of Japan, Germany, Italy, Bulgaria, Rumania, Hungary and Thailand, or any agencies thereof, or by any of their nationals, corporations, societies, associations, or any other organization of such governments or incorporated or regulated by them, are hereby prohibited, except in accordance with this ordinance.

Section 2. All custodians, curators, officials, banks or trust companies, or other persons having possession, custody or control of property belonging to governments referred to in Section 1, any Bureau, section or agency thereof, including control societies, associations, corporations or other organization controlled, directed or supervised by such governments, in which such governments or the Government of Korea, any agency, society, association, or corporate subsidiary thereof has any direct or indirect, vested or contingent financial interest, the Political Association of Greater Japan, the Imperial Rule Assistance Association, the Imperial Rule Assistance Society, their affiliates and agencies or any successor organizations, the League of Great Japan Women's Association, and any Japanese ultra-nationalistic, terroristic and secret societies, and their affiliates are required:

a. (1) To hold the same, subject to the directions of the Military Government, and pending such direction not to transfer or otherwise dispose of the same.

(2) To preserve, maintain and safeguard and to prevent any action which will impair the value or utility of such property.

(3) To maintain accurate records and accounts.

b. When and as directed by the Military Government:

(1) File reports furnishing such data as may be required with respect to such property and all receipts and expenditures in connection therewith on and after 9 August 1945.

(2) Deliver custody and control of such property and all books, records, and accounts; and

(3) Account for the property and all income and proceeds.

Section 3. Transactions affecting any property not coming within th e purview of Section 2 are permitted provided:

a. They are made to or for the benefit of nationals of Korea or of any of the United Nations, agencies of such governments, corporations, societies, associations or any other organizations of such governments, or incorporated or regulated by them.

b. Adequate consideration is paid as subsequently determined by duly constituted agencies of the Government of Korea.

c. Prior to the consummation of any such transaction the owner selling such property shall report in writing to the Government of Korea the proposed financial terms of such transactions, the location and nature of such property, including full details of any corporation, society, association or other organization owning such property.

d. Unless prohibited by an order of the Government of Korea, or by any duly authorized division thereof, issued within sixty (60) days of the date of the report referred to in Par c, above, the sale may be consummated, and

e. The proceeds of such sale shall be delivered simultaneously with the consummation of the transaction to the Bank of Chosen or the nearest branch or agency thereof for the account of the Government of Korea. All such property and funds received by the Bank of Chosen will be impounded by the Government of Korea for proper accounting to former owners.

f. Withdrawals from such impounded funds will be permitted for reasonable living expenses of persons having property interest in such funds.

Section 4. Any transactions of the character described in this ordinance consummated after 9 August 1945 are hereby declared void as of this date provided, however, any such transactions may be validated upon application to the Government of Korea for such validation and the submission of appropriate data regarding the financial details of the transaction, the consideration paid, the persons dealt with and involved in the transaction, the property involved in the transaction and all other pertinent facts.

Section 5. This ordinance is effective immediately.

Section 6. Any person violating the provisions of this ordinance, shall, upon conviction by a Military Occupation Court, suffer such punishment as the court shall determine.

BY DIRECTION OF THE COMMANDING GENERAL
UNITED STATES ARMY FORCES IN KOREA

A. V. ARNOLD
Major General, United States Army
Military Governor of Korea.

HEADQUARTERS
UNITED STATES ARMY FORCES IN KOREA
Office of the Military Governor
Seoul, Korea

ORDINANCE
NUMBER 3   23 September 1945

SECTION

DISARMING OF CIVILIANS I
ARMS CONSIDERED FAMILY HEIRLOOMS OR RELICS......... II
PENALTY.. III

Section I. All civilians in Korea of all races will surrrender, at the times and places fixed by orders issued by the Government of Korea, or any authorized subdivision thereof, all swords and hara kiri knives.

Section II. All swords and hara kiri knives which are valued by their present owners as family heirlooms or historic relics will be tagged by the owners with names and addresses of the owners. Such owners will file at the times and places fixed by orders issued by the Government of Korea, or by any authorized subdivision thereof, a statement of facts which caused them to consider such weapons family heirlooms or historic relics.

Section III. Any person violating the provisions of this Ordinance shall upon conviction by a Military Occupation Court suffer such punishment as the court shall determine.

BY DIRECTION OF THE COMMANDING GENERAL
UNITED STATES ARMY FORCES IN KOREA

A. V. ARNOLD
Major General, United States Army
Military Governor of Korea.

55

HEADQUARTERS
UNITED STATES ARMY FORCES IN KOREA
Office of the Military Governor
Seoul, Korea

ORDINANCE
NUMBER 4   28 September 1945

CONCERNING JAPANESE MILITARY AND NAVAL PROPERTY

Section I. The purchase, sale, acquisition, and transfer of Japanese Military and Naval property of all kinds are hereby prohibited.

Section II. All such property in Korea having become the property of the United States, the possession thereof except under the authority of the United States or the Allied Powers is unlawful. All persons having the possession of such property are directed to:

a. File a report, at the times and places fixed by orders issued by the Government of Korea, or by any authorized subdivision thereof, furnishing a complete description of such property, the persons dealt with in the securing of such property, and all other pertinent facts.

b. Preserve, maintain, safeguard and prevent any action which will impair the value or utility of such property.

c. Deliver possession of such property at the times and places fixed by orders issued by the Government of Korea, or by any authorized subdivision thereof.

Section III. This ordinance is effective at midnight 28 September 1945.

Section IV. Any person violating the provisions of this ordinance shall, upon conviction of a Military Occupation Court, suffer such punishment as the court shall determine.

BY DIRECTION OF THE COMMANDING GENERAL
UNITED STATES ARMY FORCES IN KOREA

A. V. ARNOLD
Major General, United States Army
Military Governor of Korea.

HEADQUARTERS
UNITED STATES ARMY FORCES IN KOREA
Office of the Military Governor
Seoul, Korea

ORDINANCE

NUMBER 5   29 September 1945

General Order Number 3 is superseded and amended to read:

SECTION

DISARMING OF CIVILIANS.. I

UNLAWFUL POSSESSION OF ARMS, AMMUNITION

    OR EXPLOSIVES.. II

SECTION I DISARMING OF CIVILIANS

All civilians in Korea, of all races, will surrender to the Police Station nearest their residence, all firearms, ammunition and explosives, on or before 1200 noon Sunday, 23 September 1945.

SECTION II UNLAWFUL POSSESSION OF ARMS, AMMUNITION

OR EXPLOSIVES

The unlawful possession or control of any firearms, ammunition, or explosives after 1200 noon Sunday, 23 September 1945, will be punishable by such penalty as the court may impose.

BY DIRECTION OF THE COMMANDING GENERAL
UNITED STATES ARMY FORCES IN KOREA

A. V. ARNOLD
Major General, United States Army
Military Governor of Korea.

57

HEADQUARTERS
UNITED STATES ARMY FORCES IN KOREA
Office of the Military Governor
Seoul, Korea

ORDINANCE
NUMBER 6

29 September 1945

General Order Number 4 is superseded and amended to read:

SECTION

SECTION I REOPENING OF PUBLIC SCHOOLS

The public primary schools of Korea will reopen on Monday, 24 September 1945. All Korean children between the ages of six and twelve will register. The reopening of other public schools will be ordered in later directives.

SECTION II PRIVATE SCHOOLS

Private schools may reopen when they have obtained permission from the Bureau of Education.

SECTION III RACE AND RELIGION

There will be no racial or religious discrimination in the schools of Korea.

SECTION IV LANGUAGE OF INSTRUCTION

The language of instruction in the schools of Korea will be the Korean language. Until adequate materials are available in the Korean language, foreign languages may be used.

SECTION V CURRICULUM

No subject will be taught or practice observed that is inimical to the interests of Korea.

<u>SECTION VI TEACHERS</u>

All Korean primary school teachers will register immediately at the nearest educational office and hold themselves ready to begin work on Monday, 24 September 1945. All other Korean teachers will register at the nearest educational office between 24 September and 29 September 1945.

<u>SECTION VII SCHOOL BUILDINGS</u>

All school buildings now used for other than educational purposes. except those occupied by United States Army Forces, will be vacated at once, cleaned, and left in good condition for use as schools.

BY DIRECTION OF THE COMMANDING GENERAL
UNITED STATES ARMY FORCES IN KOREA

A. V. ARNOLD
Major General United States Army
Military Governor of Korea

HEADQUARTERS
UNITED STATES ARMY FORCES IN KOREA
Office of the Military Governor.
Seoul, Korea

ORDINANCE
NUMBER 7

1 October 1945

1. The Local Affairs Section of the Secretariat of the Government General of Korea is hereby abolished.

2. All duties, functions, records and property of the Local Affairs Section of the Secretariat are hereby transferred to the Planning Section of the Secretariat of the Government of Korea.

3. This order is effective at midnight 1 October 1945.

BY DIRECTION OF THE COMMANDING GENERAL
UNITED STATES ARMY FORCES IN KOREA

A. V. ARNOLD
Major General, United States Army
Military Governor of Korea

ORDINANCE
NUMBER 8   1 October 1945

1. The Foreign Affairs Section of the Secretariat of the Government of Korea is hereby established.

2. ·All duties, functions, records and property of the Foreign Affairs Department of the General Affairs Section of the Secretariat are hereby transferred to the Foreign Affairs Section of the Secretariat of the Government of Korea.

3. The Property Custody Section of the Secretariat of the Government of Korea is hereby established.

4. All duties, functions, records and property of the Property Custody Section of the Financial Bureau are hereby transferred to the Property Custody Section of the Secretariat of the Government of Korea.

5. This order is effective at midnight 1 October 1945.

BY DIRECTION OF THE COMMANDING GENERAL
'UNITED STATES ARMY FORCES IN KOREA

A. V. ARNOLD
Major General, United States Army
Military Governor of Korea

HEADQUARTERS
UNITED STATES ARMY FORCES IN KOREA
Office of the Military Governor
Seoul, Korea

ORDINANCE
NUMBER 9

5 October 1945

MAXIMUM TENANT FARMERS RENTS FIXED

SECTION I DECLARATION OF NATIONAL EMERGENCY

A national emergency in Korea is hereby declared to exist by reason of oppressive rents and interest rates payable under existing contracts by tenants of farm lands and the resulting semi-enslavement and a standard of living below the standard which is the object of the Military Government. The standard which Military Government will enforce is that standard which will provide prosperity and security for the peoples of Korea.

SECTION II FIXING MAXIMUM FARM RENTS

Upon the effectiveness of this order all existing contracts, agreements and understandings of any kind and however entered into or evidenced or recorded between any tenant on farm lands and any landowner is by force of law hereby amended so that the maximum rent in kind, money or in any form payable by any type of tenant to any person or persons, including natural and legal persons, for the occupancy or use of farm lands and property, including all existing structures, works, appurtenances, stock, equipment and implements now appertaining to such property and the occupancy and the use thereof will hereafter not exceed one-third (1/3) of the total of the natural crops, produce and fruits of such land cultivated and tended by any such tenant and thereafter harvested by anyone from such land.

<u>SECTION III FARM RENT CONTRACTS FOR LESS THAN FIXED</u>

<u>MAXIMUM NOT DISTURBED</u>

All existing rental agreements or arrangements, if any, which provide for rental payments lower than the maximum of one-third (1/3) hereby fixed will continue in full force and effect without any change until their normal termination pursuant to their existing provisions and arrangements.

<u>SECTION IV UNILATERAL OPTIONS TO TERMINATE</u>

<u>FARM LEASES VOIDED</u>

No outstanding option or power to terminate any tenancy by any landowner during any fixed duration of the existing tenancy may lawfully be exercised except for just cause as determined by the decision after due process of trail by the duly constituted District Court or other tribunal having proper jurisdiction in disputes affecting such lands or tenancy agreements.

<u>SECTION V NEW CONTRACTS FOR FARM RENT EXCEEDING</u>

<u>FIXED MAXIMUM RENTS</u>

No new, extended or renewed agreement or arrangement for the occupancy or use of any farm lands by any person for any form of agriculture purpose, including sericulture, and the culture, breeding and growth of any specie of the plant or animal kingdoms will be lawful or enforcible in any court or other tribunal which provides by any device for a rental payment in excess of the maximum one-third (1/3) hereby fixed. Any attempt to circumvent this order will be considered a violation of this order. Any agreement or arrangement which violates this order will continue, for the benefit of the tenant, as providing against the property owner.

<u>SECTION VI FILING OF EXISTING FARM LEASE CONTRACTS</u>

Within sixty (60) days after the effective date of this order all persons who by reason of any property interest in farm lands has or claims any legal right under any agreement or arrangement to collect any rent or return for any use by another of such property will file with the office of land registry where such land is registered, a copy certified under oath by such property holder as correct of all agreements with any person affecting any occupancy or use of such property, if such agreement is in writing. If any such agreement or arrangement in whole or in part was never written or signed by the parties, the property holders shall file in such office within such period a written statement of all the provisions of such arrangement signed and certified under oath by him as correct. The penalty for failure by any person whitin Korea to make such filing whitin such period will be that until such filing has been duly made under this order the existing occupants and users of such property under any oral or written agreement or arrangement existing on the effective date of this order will continue in such occupancy and use for the balance of the duration of such agreement, or arrangement without any obligation to pay any rent or return to any such property holder, but in such event shall pay the rent or return due under the existing agreement or arrangement to the Government of Korea at the nearest duly constituted agency of the Government for the collection of taxes on such land and will receive a duly signed receipt by such agency for each such payment.

SECTION VII CONCERNING FARM RENTS IN KIND

Payments of rents or returns by any tenant required to be made under any existing agreement or arrangement as modified by this order shall be paid at the times and in the manner provided by such agreement or arrangement whether made to any property holder or to the Government of Korea under Section VI of this order. If under any existing agreement or arrangement any provision requires or permits any rental payment in kind to be commuted to a payment in money such payment will be made in legal tender currency in a commuted amount not exceeding the then existing lowest bid price of the Government of Korea for the produce commuted to a money rental, if the Government of Korea has then outstanding a bid price for such produce, and otherwise not exceeding the existing reasonable market price of such produce at the place where such rental payment in kind would be payable if it were not commuted in money. In the event of any disagreement between any tenant and any property holder as to the proper amount of any rental payment in money the obligation to make such payment will be satisfied by the tender in good faith to the nearest warehouse of The Korean Commodity Company for the account of the land owner in an amount of such produce not exceeding the provisions of the agreement or arrangement and not exceeding the maximum amount of one third (1/3) fixed by this order.

SECTION VIII EFFECTIVE DATE OF ORDER

This order is effective at midnight 5 October 1945.

SECTION IX PENALTY

Any violation of this order, including any attempt to evade its intent and any false statement or report in connection with any provision of this ordinance, shall be punished in such manner as the Military Occupation Court shall determine.

BY DIRECTION OF THE COMMANDING GENERAL
UNITED STATES ARMY FORCES IN KOREA

A. V. ARNOLD
Major General United States Army
Military Governor of Korea

HEADQUARTERS
UNITED STATES ARMY FORCES IN KOREA
Office of the Military Governor
Seoul, Korea

ORDINANCE
NUMBER 10   8 October 1945

REGISTRATION OF JAPANESE NATIONALS

Section I. All persons of Japanese nationality are prohibited from travelling a distance in excess of ten (10) kilos from their Block Association office without written permission from the Military Government through the Civil Police. Such persons are prohibited from travelling any distance whatsoever except for a temporary purpose and such persons must return to their abode within their Block Association area bofore curfew on the evening of the same day.

Section II. When announced by the Military Government all persons of Japanese nationality will register in the office of their Block Association. The Block Association leader will accumulate such registrations daily and file them by 1200 noon on each succeeding day, at the city, town or village office. The city, town or village office head will file with the Secretariat of the Provincial Government the accumulated registrations by 1200 noon on each Monday for the preceding week. The Provincial Government immediately will file all such registrations with the Foreign Affairs Section of the Secretariat of the Government of Korea at Seoul.

Section III. This order will be placed in effect by announcement of Provincial Governors.

Section IV. Any person violating the provisions of this ordinance, shall, upon conviction thereof suffer such punishment as the court shall determine.

BY DIRECTION OF THE COMMANDING GENERAL
UNITED STATES ARMY FORCES IN KOREA

A. V. ARNOLD
Major General United States Army
Military Governor of Korea

HEADQUARTERS
UNITED STATES ARMY FORCES IN KOREA
Office of the Military Governor
Seoul, Korea

ORDINANCE
NUMBER 11

9 October 1945

General Order Number 5 is superseded and amended to read.

SECTION I SPECIFIC LAWS REPEALED

1. In order to eliminate from the laws applicable to the people of Korea and the administration thereof, within the territory occupied south of North Latitude 38 degrees, the policies and doctrines which discriminate against and are oppressive to the Korean people, and to restore to the Korean people the rule of Justice and Equality before the law, the following laws, decrees or orders having the force of law are hereby repealed:

a. Act of Punishing Political Convicts, Vol 6, Sec 14, p 1020 of the General Code of Korea enacted 15 April 1919.

b. The Preliminary Imprisonment Act, Vol 2. Sec 8, p 26 of the General Code of Korea enacted 15 May 1941.

c. Act of Preserving Public Order, Vol 2, Sec 8, p 16 of the General Code of Korea enacted 8 May 1925.

d. Act of Publication, Vol 2, Sec 8, p 355 of the General Code of Korea enacted February 1910.

c. The Decree for the Protection of Political Convicts, Vol 2, Sec 8, p 23 of the General Code of Korea enacted 12 December 1936.

f. The Act of Shrine, Vol 2, Sec 6, p 1-88 of the General Code of Korea enacted 18 1919.

g. The Judicial Power of Police Chiefs, Vol 6, Sec 3, p 939-940 General Code of Korea.

66

SECTION II GENERAL REPEALING CLAUSE

1.   All other laws, decrees or orders having the force of law are hereby repealed. the Judicial or Administrative enforcement of which would cause discriminations on grounds of race, nationality, creed or political opinions.

SECTION III LIMITATIONS ON PUNISHMENTS

1.   No charge shall be preferred or sentence imposed or punishment inflicted against any person, for an act, unless such act is expressly made punishable by law in force at the time of its commission.

2.   The detention of any person not charged with a specific crime or offense and the punishment of any person without lawful trial and conviction are prohibited.

SECTION IV PENALTY

1.   Any person violating the provisions of this order, shall, upon conviction by a Military Occupation Court, suffer such punishment as the court shall determine.

BY DIRECTION OF THE COMMANDING GENERAL
UNITED STATES ARMY FORCES IN KOREA

A. V. ARNOLD
Major General United States Army
Military Governor in Korea

ORDINANCE
NUMBER 12

9 October 1945

1. All duties, functions, records and property of the Forestry Section of the Bureau of Mining and Industry, except for the duties, functions, records and property of the Lumbering sub-section thereof, are hereby transferred to the Bureau of Agriculture and commerce.

2. All duties, functions, records and property of the Manufacturing of Agricultural Implements Section of the Doparnment of General Agriculture Production of the Bureau of Agriculture and Commerce are hereby transferred to the Bureau of Mining and Industry.

3. All duties, functions, records and property of the Textile Industry Section (Manufacturing) of the Department of Miscellaneous Products and Household Goods of the Bureau of Agriculture and Commerce are hereby transferred to the Bureau of Mining and Industry.

4. All duties, functions, records and property of the Food Processing Section (including fish) of the Department of Grain Processing and Distribution of the Bureau of Agriculture and Commerce are hereby transferred to the Bureau of Mining and Industry.

5. This order is effective at midnight 9 October 1945.

BY DIRECTION OF THE COMMANDING GENERAL
UNITED STATES ARMY FORCES IN KOREA

A. V. ARNOLD
Major General United States Army
Military Governor of Korea

HEADQUARTERS
UNITED STATES ARMY FORCES IN KOREA
Office of the Military Governor
Seoul, Korea

ORDINANCE
NUMBER 13   10 October 1945

FREE POSTAGE

SECTION I. Effective this date, all mail delivered to the Korean Postal System by members of the Armed Forces of the United States will be accepted without the payment of postage. All mail so delivered will be marked "Mailed by USAFIK", followed by the name and official title of the sender. Such mail will be promptly processed and forwarded in the same manner as ordinary mail.

SECTION 2. Any laws or regulations in conflict herewith are hereby repealed.

BY DIRECTION OF THE COMMANDING GENERAL
UNITED STATES ARMY FORCES IN KOREA

A. V. ARNOLD
Major General, United States Army
Military Governor of Korea

ORDINANCE
NUMBER 14

10 October 1945

WAGES FOR CIVILIAN LABOR

Section I. Civilian labor employed by corporations, businesses, trusts, societies, or other organizations owned or controlled in whole or in part by the Government of Korea, civil or military, except Civil Service labor, will be compensated as follows:

Type of Laborer	Basic Daily Wage	Differential	Total Wage Per Day
Apprentice	1– 2 yen	7 yen	8– 9 yen
Unskilled	3– 5 yen	7 yen	10-12 yen
Semi-skilled	5– 7 yen	7 yen	12–14 yen
Skilled	7–10 yen	7 yen	14–17 yen

Section II. Further classification of skill within the four major skill classes will be made by the using agency in order to establish the proper basic wage. This spread in basic wages is to permit the customary oriental economic practice of differentiating between wage payments to male and female workers, young and old and supervisory workers within the same skill group.

Section III. All wages will be paid in legal tender currency.

Section IV. This ordinance is effective at midnight, 10 October 1945.

Section V. Any person violating the provisions of this ordinance shall, upon conviction of a Military Occupation Court, suffer such punishment as the court shall determine.

BY DIRECTION OF THE COMMANDING GENERAL
UNITED STATES ARMY FORCES IN KOREA

A. V. ARNOLD
Major General United States Army
Military Governor of Korea

HEADQUARTERS
UNITED STATES ARMY FORCES IN KOREA
Office of the Military Governor
Seoul, Korea

ORDINANCE
NUMBER 15   16 October 1945

1. The name of Keijo Imperial University is hereby changed to Seoul University.

2. The name of the Confucian temple, Kyung Hak Won, is hereby changed to Sung Kyoon Kwan.

3. This order is effective at midnight 16 October 1945

BY DIRECTION OF THE COMMANDING GENERAL
UNITED STATES ARMY FORCES IN KOREA

A. V. ARNOLD
Major General, United States Army
Military Governor of Korea.

ORDINANCE
NUMBER 16

24 October 1945

SECTION

DISPLAY OF NAMES.. I
TEMPORARY APPOINTMENT OF OFFICIALS............ II

SECTION I. DISPLAY OF NAMES. When and as directed by the Provincial Governor of each Province of Korea south of North Latitude 38°, every office or place of business owned, operated or maintained by any company, corporation, partnership, political pary, group of persons, person in business, association, or organization of any nature whatsoever, shall have the name thereof in English and Korean prominently displayed at a place readily visible to the public and legible from a distance of thirty (30) feet.

SECTION II. TEMPORARY APPOINTMENT OF OFFICIALS. Appointments of Officials of the Government of Korea or any agency or subdivision thereof by the Military Governor or by any official under the direction of the Military Governor are temporary appointments and are subject to revocation at any time by the Military Governor or by his direction.

BY DIRECTION OF THE COMMANDING GENERAL
UNITED STATES ARMY FORCES IN KOREA

A. V. ARNOLD
Major General, United States Army
Military Governor of Korea.

ORDINANCE
NUMBER 17   24 October 1945

1. The Economic Police Department of the Bureau of Police of the Government General of Korea is hereby dissolved and all duties and functions thereof are hereby eliminated from the Government.

2. This order is effective at midnight 24 October 1945.

BY DIRECTION OF THE COMMANDING GENERAL
UNITED STATES ARMY FORCES IN KOREA

A. V. ARNOLD
Major General, United States Army
Military Governor of Korea.

HEADQUARTERS
UNITED STATES ARMY FORCES IN KOREA
Office of the Military Governor
Seoul, Korea

ORDINANCE
NUMBER 18

27 October 1945

1. The name of the Bureau of Public Health is hereby changed to the Bureau of Public Health and Welfare.

2. The following duties and functions are hereby assigned to the Bureau of Public Health and Welfare in addition to those previously assigned to the Bureau of Public Health:

    a. Emergency and disaster relief.

    b. Public assistance to permanently needy individuals.

    c. Child welfare and other institutional care.

    d. Employee welfare and pension systems.

    e. Housing.

    f. Care and relocation of repatriated and displaced Koreans, in coordination with Foreign Affairs Section of the Secretariat.

    g. Such other public welfare programs and agencies now in existence as may be required to aid in achieving the aims of the occupational forces in Korea.

3. All duties, functions, records, property and civilian personnel of the following organizations are hereby transferred to the Bureau of Public Health and Welfare:

    a. Social Welfare Department of the Bureau of Education.

    b. War Refugee Division of the Defense Department of the Bureau of Police.

    c. Protection Division of the Foreign Affairs Section of the Secretariat.

    d. Relief Society of Chosen for Laborers and War Sufferers of the Bureau of Mining and Industry.

4. This order is effective at midnight 27 October 1945.

BY DIRECTION OF THE COMMANDING GENERAL
UNITED STATES ARMY FORCES IN KOREA

A. V. ARNOLD
Major General, United States Army
Military Governor of Korea.

HEADQUARTERS
UNITED STATES ARMY FORCES IN KOREA
Office of the Military Governor
Seoul, Korea

ORDINANCE
NUMBER 19   30 October 1945

Reprinted Copy

SECTION

SECTION I DECLARATION OF NATIONAL EMERGENCY

After four long years of war, from which they emerged victorious, American Forces landed upon your shores the friends and protectors of the Korean people. They came for the avowed purpose of requiring the complete and permanent eviction of all Japanese military forces from Korea and eliminating all Japanese militaristic and nationalistic ideology. In addition to that objective, the Military Government was instructed to take steps to effect complete political and administrative separation of Korea from Japanese social, economic, and financial control; to facilitate development of a sound Korean economy and to work towards the restoration of a free, independent and responsible Korea. The program of Military Government included taking over all Japanese property as rapidly as possible for the benefit of the Korean people, relieving labor from the condition of absolute servitude under which it had existed for the last forty years, returning to the farmers the lands which had been wrested from them by Japanese guile and treachery, and giving to the farmer a fair and just proportion of the fruits of his sweat and labor, restoring the principles of a free market, giving to every man, woman and child within the country equal opportunity to enjoy his just and fair share of the great wealth with which this beautiful nation has been endowed.

Upon the arrival of the American Forces the Americans found that to support the war Japan drained food and other living commodities from Korea until she was weak from starvation. Production of consumer goods had all but ceased. There was a wholesale embezzlement of Government funds. The currency had been deliberately inflated.

Law and order was immediately established. All seized property that could possibly be of benefit to the Korean people was preserved for their use. Grain was moved to areas where starvation existed.

Because of the long continued exploitation of Korea's wealth and the oppression of her people, and the fact that the people had been prohibited from carrying out during the years of Japanese oppression those pursuits of life and happiness which are normal to a rich and aggressive race, the people have not been able to produce goods fast enough to prevent widespread malnutrition, disease, and suffering during the coming winter. In addition, certain groups, with the sole idea of acquiring the wealth of the Korean people for themselves, have prevented labor from returning to employment, children from returning to school, and farmers from selling the produce of their lands. Such conditions have created within Korea an emergency which for the sole protection of the people can only be met by extraordinary measures which will inure to their benefit. As all of the people know the American Nation is a powerful nation. Its people, however, are gentle with the true gentleness that comes only through an appreciation of their own good fortune and in their desire to protect others against adversity. At this time the representatives of the American people in Korea foresee that their friends, the Korean people, are threatened in their security and right to be warm and well fed during the coming winter. Facilities, raw materials, and labor are available within the country which, if utilized properly, will prevent suffering. Rigid emergency controls are therefore hereby established, in order to prevent such conditions from existing which will harm the people. Such temporary measures will be withdrawn as quickly as the emergency passes.

SECTION II PROTECTION OF LABOR

The right of any individual or group of individuals to accept employment and to work unmolested shall be respected and protected. Any interference with this right is unlawful.

In the interest of preventing suspension or reduction of output in industries which are declared by the Military Government of Korea to be essential to the people's livelihood disputes arising over rights and conditions

of labor will be settled by a mediation board set up by the Military
Government of Korea whose decision is final and binding on all parties.
Until the matter is presented to the Labor Mediation Board and a decision
is reached, production will continue.

SECTION III PROTECTION AGAINST PROFITEERING

To aid in assuring an equitable distribution of commodities necessary
to the livelihood of the people, at a price within the financial resources of
the people, the hoarding of essential commodities and/or their sale at an
excessive price, which would result in profiteering at the expense of the
people, is hereby declared unlawful.

SECTION IV PROTECTION OF THE PUBLIC AGAINST ACTS

PREJUDICIAL TO THE PEOPLE'S WELFARE

The following acts are hereby declared unlawful:

a. Knowingly making any false statement orally or in writing to
any member of or person acting under the authority of USAFIK or the
Military Government of Korea in a matter of official concern or in any manner
attempting to defraud the Military Government of Korea.

b. Since the program of the Military Government of Korea is for
the people's welfare, the obstructing, attempting to obstruct, or contravening any
orders or announced program of the Military Government of Korea whether
by acts, conspiracies, intimidation, corruption, or bribery of any person.

c. Prejudicing the people's welfare by directing or participating
in acts of discipline, threats, coercion, or any other form of intimidation or
victimization (including boycotting) against any person cooperating in any
form directly or indirectly with USAFIK or the Military Government of
Korea.

SECTION V REGISTRATION OF NEWSPAPERS AND OTHER

PUBLICATIONS

In order that freedom of speech and freedom of the press may be
preserved and safeguarded without being perverted to unlawful and subversive
purposes, the registration of every organization engaged in printing of
books, pamphlets, papers or other reading materials in Korea south of 38°
North Latitude sponsored, owned, directed, controlled, or managed by
any natural or juridical person is hereby ordered. Such registration will be
accomplished within ten (10) days after the effective date of this order or in
the case of new publications not less than ten (10) days before publication, by

filing in duplicate by registered mail addressed to the Military Government of Korea, Capitol, Seoul, the name of the organization; the places where printed and published; the names of the persons employed in the printing and publishing; the name of each person sponsoring, owning, directing, controlling, managing or otherwise influencing the organization, its activities or operations or the material printed and published; an accurate description of every form, printing and publishing, used or proposed to be used by the organization, including exact facsimiles of the format, character and size of each type of printed matter; the financial backing, capital, funds, property, and assets of the organization including the amount of critical essential supplies on hand and intended future sources in such detail as to reflect the financial ability of the organization to operate and maintain the news and editorial services proposed to be rendered, the name and address of each person financially interested in the organization or enterprise.

SECTION VI PENALTY

Any person violating the provisions of this ordinance shall, upon conviction by a Military Occupation Court, suffer such punishment as the court shall determine.

SECTION VII EFFECTIVE DATE OF ORDINANCE

This ordinance is effective immediately.

BY DIRECTION OF THE COMMANDING GENERAL
UNITED STATES ARMY FORCES IN KOREA

A. V ARNOLD
Major General United States Army
Military Governor of Korea

Date of publication: 28 December 1945.

HEADQUARTERS
UNITED STATES ARMY FORCES IN KOREA
Office of the Military Governor
Seoul, Korea

ORDINANCE
NUMBER 20   30 October 1945

SECTION

SECTION I. The Department of Criminal Investigation is hereby established in the Police Bureau of the Government of Korea to perform the following functions:

a. Conducting investigations of criminal matters which are referred to it by Military Government.

b. Assisting and cooperating with the Military Police, Counter Intelligence Corps and the Police of Korea, when requested, in the investigation and apprehension of criminals.

c. Establishing and maintaining a National Finger Print Record and Criminal Investigation System.

SECTION II. All duties, functions, records, property, and civilian personnel of the Finger Print Section of the Criminal Department of the Bureau of Justice are hereby transferred to the Bureau of Police.

SECTION III. This order is effective 30 October 1945.

BY DIRECTION OF THE COMMANDING GENERAL
UNITED_STATES ARMY FORCES IN KOREA

A. V. ARNOLD
Major General, United States Army
Military Governor of Korea.

ORDINANCE
NUMBER 21

2 November 1945

SECTION

SECTION I RETENTION OF LAWS

Until further ordered, and except as previously repealed or abolished, all laws which were in force, regulations, orders, notices or other documents issued by any former government of Korea having the force of law on 9 August 1945 will continue in full force and effect until repealed by express order of the Military Government of Korea. All local laws and customs will continue in force until repealed by competent authority. The provisions of law with respect to the organization of the Government General of Korea, the Provincial Governments, the cities, towns, and villages, including the directors of bureaus, cities, towns, and villages, including the directors of bureaus, the secretaries of sections of the secretariat, the mayor, county head, police station chief, tax office chief, head of town and village, and their subordinates are continued in force until abolished by competent authority, except as they have been modified or repealed by the orders of the Military Governor. Subject to directives of higher authority, any and all authority heretofore exercised by the Governor-General of Korea may be exercised by the Military Governor.

SECTION II ENFORCEMENT OF PROCLAMATIONS,

ORDINANCES AND ORDERS

All courts of Korea south of 38 degrees north latitude shall take cognizance of and enforce all laws of Korea, the provisions of the proclamations of Commander-in-Chief, United States Army Forces, Pacific, and all orders and ordinances of the Military Governor of Korea. For such purpose all such

courts are hereby constituted Military Occupation Courts. Nothing herein
contained shall confer upon such courts jurisdiction over any military personnel
or official of the United States or any of the United Nations, or deprive
Military Commissions, Provost Courts, or other military courts established
by the United States Army Forces in Korea of the jurisdiction conferred upon
them

SECTION III EFFECTIVE DATE OF ORDINANCE

This ordinance is effective at midnight 2 November 1945.

BY DIRECTION OF THE COMMANDING GENERAL
UNITED STATES ARMY FORCES IN KOREA

A. V. ARNOLD
Major General United States Army
Military Governor of Korea

ORDINANCE
NUMBER 22

3 November 1945

1. For purposes of governmental administration and control, the following temporary transfers of jurisdiction over districts, villages, towns, counties, and cities located south of and contiguous to North Latitude 38° are hereby effected:

    a. In Kyung Kei Do (京畿道):

      (1) District of Yongnam Village (嶺南面), Kaepung County (開豐郡) to Jinso Village (津西面), Changtan County (長湍郡).

      (2) District of Buk Village (北面) to Tosong Village (土城面), Kaepung County (開豐郡).

      (3) District of Changdo Village (長道面), Changdan County (長湍郡) to Jinso Village (津西面), Changdan County (長湍郡).

      (4) District of Yonchon County (漣川郡) to Paju County (坡州郡).

      (5) District of Chongok Village (全谷面) to Choksong Village (積城面), Paju County (坡州郡).

      (6) District of Paekhak Village (百鶴面) to Choksong Village (積城面), Paju County (坡州郡).

      (7) District of Idong Village (二東面) to Ildong Village (一東面).

      (8) Districts of Changsu Village (蒼水面) and Chongsan Village (靑山面) joined forming a new Village with the Village Office at Chosengri (硝城里).

      (9) District of Yongjung Village (永中面) to Shinbuk Village (新北面).

    b. In Whang Hae Do (黃海道):

      (1) Districts of Whang Hae Do (黃海道) to Kyung Kei Do (京畿道).

      (2) Villages of Chuhwa (秋花面), Naesong (來城面), Ilshin (日新面), and Chongyong (靑龍面), all of Pyoksong County (碧城郡), to Yonpaek County (延白郡).

      (3) Villages of Dongkang (東江面), Haenam (海南面), and Songrim (松林面), all of Pyoksong County (碧城郡), to Ongjin County (甕津郡).

(4) District of Hwasong Village (花城面), Yonpaek County (延白郡) to Soksan Village (石山面).

(5) District of Unsan Village (雲山面), Yonpaek County (延白郡) to Unchon Village (銀川面).

(6) District of Yongchon Village (泳泉面), Pyoksong County (碧城郡) to Naesong Village (來城面).

2. This order is effective at midnight 3 November 1945.

BY DIRECTION OF THE COMMANDING GENERAL
UNITED STATES ARMY FORCES IN KOREA

A. V. ARNOLD
Major General, United States Army
Military Governor of Korea.

HEADQUARTERS
UNITED STATES ARMY FORCES IN KOREA
Office of the Military Governor
Seoul, Korea

ORDINANCE
NUMBER 23

2 November 1945

PROHIBITING SALES OF ALCOHOLIC BEVERAGES

SECTION I. Effective immediately, all sales, gifts, or transfers by any means of any alcoholic beverage of any strength or proportion of alcohol, except beer which does not contain any proportion of any distilled alcoholic substance, to members of the Armed Forces or civilians of any of the United Nations under the jurisdiction of the United States Army Forces in Korea, are hereby prohibited.

SECTION II. Any person violating the provisions of this ordinance shall, upon conviction of a Military Occupation Court, suffer such punishment as the court shall determine.

BY DIRECTION OF THE COMMANDING GENERAL
UNITED STATES ARMY FORCES IN KOREA

A. V. ARNOLD
Major General, United States Army
Military Governor of Korea.

ORDINANCE
NUMBER 24   5 November 1945

MATERIALS CONTROL CORPORATION

SECTION I.   The Korean Important Materials Control Corporation created by the law of 28 December 1943, is continued as a body- corporate and an instrumentality of the Government of Korea, but is hereby renamed and known as the Materials Control Corporation.

SECTION II.   The Corporation is hereby constituted an independent agency of the Government of Korea for the purpose of aiding in and facilitating the procurement, receiving, handling, storing, safe-guarding, custody, accounting, sale and other distribution of all civilian relief and rehabilitation supplies, surplus war materials, and all surrendered or abandoned Japanese movable and tangible personal property which is the subject of market operations.

a.   Existing provisions having legal or regulatory effect on the corporation are hereby conformed to this order.

b.   The Corporation is hereby directed to use all its assets and funds under its control in the exercise of its functions.

SECTION III.   The main office of the Corporation will be in Seoul, Korea.   The Corporation will establish branch offices in each of the provincial capitals and at such other places as it deems necessary.

SECTION IV.   The affairs of the Corporation will be under the direction and management of a Director appointed by the Military Governor.

a.   The Director will appoint and fix the compensation of officers and employees of the Corporation.

b.   The Director will receive from the Military Governor of Korea procurement, priority, and allocation instructions.

c.   The Director will prescribe necessary regulations for carrying out the purposes of this Ordinance.

SECTION V.   This order is effective at midnight 5 November. 1945.

BY DIRECTION OF THE COMMANDING GENERAL
UNITED STATES ARMY FORCES IN KOREA

A. V. ARNOLD
Major General, United States Army
Military Governor of Korea.

ORDINANCE
NUMBER 25

7 November 1945

1. The Department of Public Health and Welfare is hereby established within each Provincial Government of Korea for the purpose of performing the following functions:

a. Study, protection, and improvement of health of the provincial population.

b. Health Education.

c. Epidemiology, preventative medicine, and control of communicable diseases.

d. Maternal and child health.

e. Hospitals and instituions.

f. Nursing affairs.

g. Veterinary affairs.

h. Dental affairs.

i. Sanitation.

j. Laboratories.

k. Medical supply.

l. Vital Statistics.

m. Emergency and disaster relief.

n. Public assistance to permanently needy individuals.

o. Child Welfare and other institutional care.

p. Employee Welfare and Pension systems.

q. Housing.

r. Care and relocation of repatriated and displaced Koreans.

s. Such other public Welfare programs as may be required to aid in achieving the aims of the occupational forces in Korea.

2. All duties, functions, records, property, civilian personnel of the following organization are hereby transferred to the Public Health and Welfare Department of the Provincial Governments within Korea:

a. The Public Health Section of the Police Departments of the Provincial Governments.

b. The Sanitary Sections of the Police Stations of the Police Departments of the Provincial Governments.

c. The District Public Doctors of the Police Departments of the Provincial Governments.

d. The Soldier's Welfare Sections of the Police Departments of the Provincial Governments.

e. The Social Welfare Sections of the Departments of the Interior of the Provincial Governments.

f. The Social Welfare Sections of the Labor Departments of the Provincial Governments.

3. A National Veterinary Service Department within the Bureau of Public Health and Welfare is hereby established for the purpose of regulating and directing a National Veterinary Service within Korea.

4. All duties functions, records, property, and civilian personnel of the Bureau of Agriculture and Commerce that are devoted exclusively to the conduction of veterinary services in Korea including the Fusan Cattle Sanitation Laboratory in Fusan, Kyung Sang Nam Do and the Branch there of in Anyang, Kyung Kei Do are hereby transferred to the Bureau of Public Health and Welfare.

5. The order is effective at midnight 7 November 1945.

BY DIRECTION OF THE COMMANDING GENERAL
UNITED STATES ARMY FORCES IN KOREA

A. V. ARNOLD
Major General, United States Army
Military Governor of Korea

HEADQUARTERS
UNITED STATES ARMY FORCES IN KOREA
Office of the Military Governor
Seoul, Korea

ORDINANCE
NUMBER 26

8 November 1945

1. Section 4 of General Notice Number 2 dated 20 October 1945 is hereby repealed.

2. The name of the Prince Lee Household is hereby changed to Koo Wang Koong.

3. This order is effective upon its publication in the Official Gazette.

BY DIRECTION OF THE COMMANDING GENERAL
UNITED STATES ARMY FORCES IN KOREA

A. V. ARNOLD
Major General United States Army
Military Governor of Korea

ORDINANCE
NUMBER 27

9 November 1945

FISHING

1. All fishing rights within Korean waters granted by the Japanese Imperial Government, the Governor General of Korea or any other Japanese or Korean governmental subdivision or agency prior to 9 August 1945 to persons. natural or juridical, are hereby declared null and void.

2. After the effective date of this ordinance all fishing rights within Korean waters will be exercised in accordance with regulations issued by the Bureau of Agriculture and Commerce of the Government of Korea.

3. Any person violating the provisions of this ordinance, shall, upon conviction of a Military Occupation Court, suffer such punishment as the court shall determine.

4. This ordinance is effective upon its publication in the Official Gazette.

BY DIRECTION OF THE COMMANDING GENERAL
UNITED STATES ARMY FORCES IN KOREA

A. V. ARNOLD
Major General, United States Army
Military Governor of Korea.

ORDINANCE
NUMBER 23

13 November 1945

SECTION I. In order to prepare for the eventual independence of Korea, to progressively provide the necessary forces to defend, protect and safeguard its sovereign rights and prerogatives among the Nations of the world, to assist civilian police agencies in the maintenance of civil peace and security and the defense of the rights of the people against civil disorder and to maintain freedom of religion, freedom of speech and property rights, to inaugerate the recruiting, organization, training and equipping of the requisite armed forces on land and sea, and to guarantee the evolution of administration of nationhood, the Office of the Director of National Defense of the Military Government of Korea is hereby established.

SECTION II. The Bureau of Armed Forces of the Government of Korea is hereby created as a Bureau of the Government. Within the Bureau of Armed Forces the Army Department and Navy Department are hereby established. The existing Bureau of Police and the Bureau of Armed Forces are hereby placed under the supervision and direction of the Office of the Director of National Defense.

SECTION III. No persons or group of persons will engage in any form of recruiting, training, organizing or equipping of any persons for any form of police, military or naval activities, or engage in any activities which are within the jurisdiction of police or armed forces, except under the written authorization of the Director of National Defense, or any agency delegated by him to grant such authority.

SECTION IV. Any person violating the provisions of this order, shall, upon conviction by a Military Occupation Court, suffer such punishment as the court shall determine.

SECTION V. This ordinance is effective at midnight 13 November 1945.

BY DIRECTION OF THE COMMANDING GENERAL
UNITED STATES ARMY FORCES IN KOREA

A. V. ARNOLD
Major General United States Army
Military Governor of Korea

HEADQUARTERS
UNITED STATES ARMY FORCES IN KOREA
Office of the Military Governor
Seoul, Korea

ORDINANCE
NUMBER 29

16 November 1945

CREATION OF KOREAN PERSONNEL REVIEW BOARD

1. The Korean Personnel Review Board is hereby created. The Board will consist of seven members, appointed by the Civil Administrator. They will serve as such members for such periods of time as the Civil Administrator directs.

2. The Board will function as an investigative agency of the Military Government of Korea and will have the following duties, powers, and jurisdiction :

    a. To inquire into pro-Japanese or pro-enemy activity of Koreans who hold any office under the Military Government, or who may apply for any such office.

    b. To make such inquiry at the direction of the Military Governor. on the request of the Director of any Bureau or Secretary of any Section of the Secretariat of the Military Government, any Provincial Governor, or any Military Government Team Commander, or upon its own initiative.

    c. To convene meetings of the Board on the call of the senior officer member, to hear testimony, to subpoena witnesses, documents or other evidence, and to make findings and recommendations to the Military Governor. All hearings will be conducted in a fair and impartial manner to the end that a full and complete investigation will be made. Testimony will be reduced to writing and signed and sealed by the witnesses. Hearings will be conducted in closed session, and the testimony, evidence, findings and recommendations will be made available only to the Military Governor or to such person or persons as the Military Governor may, in each instance. specifically direct.

3. This Ordinance will become effective upon publication in the Official Gazette.

BY DIRECTION OF THE COMMANDING GENERAL
UNITED STATES ARMY FORCES IN KOREA

A. V. ARNOLD
Major General United States Army
Military Governor of Korea

92

ORDINANCE
NUMBER 30

30 November 1945

Ordinance Number 14, WAGES FOR CIVILIAN LABOR, dated 10 October 1945, is hereby repealed, effective 1 December 1945.

BY DIRECTION OF THE COMMANDING GENERAL
UNITED STATES ARMY FORCES IN KOREA

A. V. ARNOLD
Major General United States Army
Military Governor of Korea

ORDINANCE
NUMBER 32

2 November 1945

1. The name of the Intelligence and Information Section of the Secretariat of the Government of Korea is hereby changed to the Public Information Section of the Secretariat of the Government of Korea, effective as of 23 November 1945.

2. All duties, functions, records, property, and civilian personnel of the Central Liaison Office are hereby transferred to the Foreign Affairs Section of the Secretariat of the Government of Korea, effective as of 15 November 1945.

3. All duties, functions, records, property, and civilian personnel of the Colonization Section of the Bureau of Agriculture and Commerce of the Government of Korea are hereby transferred to the Foreign Affairs Section of the Secretariat of the Government of Korea, effective as of 29 November 1945.

BY DIRECTION OF THE COMMANDING GENERAL
UNITED STATES ARMY FORCES IN KOREA

A. V. ARNOLD
Major General, United States Army
Military Governor of Korea.

ORDINANCE
NUMBER 33

6 December 1945

VESTING TITLE TO JAPANESE PROPERTY WITHIN KOREA

SECTION I. Ordinance Number 31 having never been published in the Official Gazette is hereby declared null and void as though it were never issued.

SECTION II. The title to all gold, silver, platinum, currency, securities, accounts in financial institutions, credits, valuable papers, and any other property located within the jurisdiction of this Command, of any type and description, and the proceeds thereof, owned or controlled, directly or indirectly, in whole or part, on or since 9 August 1945, by the Government of Japan, or any agency thereof, or by any of its nationals, corporations, societies, associations, or any other organization of such government or incorporated or regulated by it is hereby vested in the Military Government of Korea as of 25 September 1945, and all such property is owned by the Military Government of Korea. It is illegal for any person, without the authority of the Military Government of Korea, to enter upon or take possession of any such property, remove any part of such property, or injure or impair the value or utility of any such property.

SECTION III. All custodians, curators, officials, banks, trust companies and all other individuals, organizations or associations having possession, custody, or control of property vested in the Military Government of Korea by Section II of this ordinance are required:

    a. (1) To hold the same, subject to the directions of the Military Governor, and pending such direction not to transfer or otherwise dispose of the same.

        (2) To preserve, maintain and safeguard and to prevent any action which will impair the value or utility of such property.

        (3) To maintain accurate records and accounts.

    b. When and as directed by the Military Governor:

(1) File reports furnishing such data as may be required with respect to such property and all receipts and expenditures in connection therewith on and after 9 August 1945.

(2) Deliver custody and control of such property and all books, records and accounts; and

(3) Account for the property and all income and proceeds.

SECTION IV. Any person violating the provisions of this ordinance or of any license or order issued thereunder, shall, upon conviction by a Military Occupation Court, suffer such punishment as the court shall determine.

SECTION V. This ordinance is effective upon publication in the Official Gazette.

BY DIRECTION OF THE COMMANDING GENERAL
UNITED STATES ARMY FORCES IN KOREA

A. V. ARNOLD
Major General United States Army
Military Governor of Korea

ORDINANCE
NUMBER 34

8 December 1945

ESTABLISHING LABOR MEDIATION BOARDS

SECTION I. In order to effectuate the policy of protection of labor and mediation of labor disputes in the existing emergency declared by Sections I and II of Ordinance Number 19, there are hereby established:

a. A National Labor Mediation Board, consisting of five voting members and a permanent non-voting Secretary. An officer of the National Labor Department of Korea will be the Advisor. A staff of seven Field Representatives will be appointed and all of the foregoing will hold office until replaced by the Military Governor. The Board will meet in Seoul.

b. A Provincial Labor Mediation Board in each province, consisting of three or five voting members and a permanent non-voting Secretary. A local Military Government Labor Officer will be the Advisor, and all of the foregoing will hold office until replaced by the Provincial Governor. The Governor is authorized to establish such additional Boards as he deems necessary in his province and to appoint, compensate, suspend and remove additional officials and employees of the Boards. The Boards will meet in places designated by the Governor.

SECTION II. The Boards will have the following powers:

a. To adopt, amend and require the observance of rules and regulations of the Board.

b. To require the attendance of witnesses and the production of documents or other evidence, to punish for contempt, and to issue process for enforcing orders or decisions of the Board.

SECTION III. The Boards will have the following jurisdiction:

a. National Board: Mediation of labor disputes extending over two or more provinces and of all other labor disputes referred to it by the National Labor Department.

b. Provincial Boards: Mediation of labor disputes in their respective provinces, in places designated by the Governor.

SECTION IV. The Boards will observe procedures prescribed by National Labor Department, Bureau of Mining and Industry of the Military Government of Korea.

SECTION V. This Ordinance is effective upon its publication in the Official Gazette.

BY DIRECTION OF THE COMMANDING GENERAL
UNITED STATES ARMY FORCES IN KOREA

A. V. ARNOLD
Major General, United States Army
Military Governor of Korea.

Date of Publication : 24 December, 1945

ORDINANCE
NUMBER 35   17 December 1945

1. The supervision and administration of hospitals operated by medical schools in Korea are hereby transferred from the Bureau of Public Health and Welfare of the Government of Korea to the Bureau of Education of the Government of Korea.

2. Hospitals operated by medical schools in Korea are subject to the same regulations and directives of the Bureau of Public Health and Welfare of the Government of Korea as all civilian hospitals.

3. This order is effective upon publication in the Official Gazette.

BY DIRECTION OF THE COMMANDING GENERAL
UNITED STATES ARMY FORCES IN KOREA

A. V. ARNOLD
Major General United States Army
Military Governor of Korea

Date of Publication: 26 December 1945

HEADQUARTERS
UNITED STATES ARMY FORCES IN KOREA
Office of the Military Governor
Seoul, Korea

ORDINANCE

NUMBER 36   20 December 1945

1. The Provincial-Affairs Section of the Secretariat of the Government of Korea is hereby established.

2. All duties, functions, records, property and civilian personnel of the Local Administration sub-section of the Planning Section of the Secretariat are hereby transferred to the Provincial Affairs Section of the Secretariat.

3. This order is effective upon publication in the Official Gazette.

BY DIRECTION OF THE COMMANDING GENERAL
UNITED STATES ARMY FORCES IN KOREA

A. V. ARNOLD
Major General United States Army
Military Governor of Korea

Date of publication: 28 December 1945.

ORDINANCE
NUMBER 37   19 December 1945

Dissolution of the Finance Control Corporation of Chosen

SECTION I. The Finance Control Corporation of Chosen, a member control association of the National Financial Control Association, is hereby dissolved.

SECTION II. The Chosen Savings Bank is appointed liquidator of the Finance Control Corporation of Chosen. It will distribute the funds of the corporation to its members in the same ratio as the amounts assessed against them and will take all other steps necessary to settle the affairs of the corporation. Upon the final settlement of the affairs of the corporation the Chosen Savings Bank will file a report with the Director of the Financial Bureau of the Government of Korea.

SECTION III. This ordinance is effective upon its publication in the Official Gazette.

BY DIRECTION OF THE COMMANDING GENERAL
UNITED STATES ARMY FORCES IN KOREA

A. V. ARNOLD
Major General, United States Army
Military Governor of Korea

Date of publication: 12 January 1946

HEADQUARTERS

UNITED STATES ARMY FORCES IN KOREA

Office of the Military Governor

Seoul, Korea

ORDINANCE
NUMBER 38   27 December 1945

NAME OF PERSONNEL SECTION CHANGED TO KOREAN

CIVIL SERVICE SECTION..".... I

NAME OF MARUNE TRUCK COMPANY CHANGED TO

KOREAN TRUCK COMPANY.................................... II

<u>SECTION I.</u> The name of the Personnel Section of the Secretariat of the Government of Korea is, hereby changed to the Korean Civil Service Section of the Secretariat of the Government of Korea, effective as of 20 December 1945.

<u>SECTION II.</u> The name of the juridical person formerly known as the Marune Truck Company and registered as the Pan Korean Truck Transportation Union Limited is hereby changed to the Korean Truck Company, effective 3 December 1945.

BY DIRECTION OF THE COMMANDING GENERAL
UNITED STATES ARMY FORCES IN KOREA

A. V. ARNOLD
Major General, United States Army
Military Governor of Korea

ORDINANCE
NUMBER 39   3 January 1946

REGULATIONS OF FOREIGN TRADE

SECTION I.   The transportation of all property including goods or valuables of any kind to or from Korea by air, land or by water is prohibited except by persons duly authorized to perform such function by the Military Governor or by such agency or agencies as may be designated by him as the authorized agent of the Military Government of Korea.

SECTION II.   The transportation of all property including goods or valuables of any kind to and from Korea by air, land or by water is prohibited where transported by any airplane, boat, cart, vehicle or mechanical conveyance of any kind except in those cases where such airplane, boat, cart, vehicle or mechanical conveyance is duly licensed for such purpose by the Military Governor or by such agency or agencies as may be designated by him as the authorized agent of the Military Government of Korea.

SECTION III.   The transportation of all property including goods or valuables of any kind to and from Korea is prohibited where transported on the person of the traveler or upon any animal or animals controlled by him except in those cases where such person is duly authorized to transport such goods to or from Korea by the Military Governor or by such agent or agencies as may be designated by him as the authorized agent of the Military Government of Korea.

SECTION IV.   The transportation of all property including goods or valuables of any kind to and from Korea is prohibited except by way of such ports or airports of entry as may from time to time be designated as open ports by the Military Governor or by such agent or agencies as may be designated by him for this purpose.

SECTION V.   All property including goods or valuables of any kind found in the course of unauthorized transportation to or from a foreign country shall be considered as contraband and shall be turned over by the apprehending authority to the custody of the Military Government of Korea, to be disposed of in accordance with existing provisions relative to the sale or other disposition of government property.

SECTION VI. Any private vessel, airplane, cart, animal, motor vehicle or any other conveyance or instrumentality not owned by any government which is utilized in violation of this ordinance shall upon the conviction of the violator or violators be turned over to the custody of the Military Government to be disposed of in accordance with the existing provisions relative to the sale or other disposition of government property. Such instrumentalities may be impounded by the apprehending agency pending the outcome of the trial.

SECTION VII. This ordinance does not apply to the transportation of the personal effects of any officer or employee of the Military Government of Korea or of the United States. It shall not be construed to apply to the transportation of the personal effects, goods or valuables of any persons subject to the military or naval laws of the United States or any of the United Nations nor shall it be construed as a prohibition of the transportation to or from Korea of the personal effects, goods or valuables of persons duly authorized to enter or depart from Korea under repatriation programs approved by the Military Government of Korea or of any persons in the course of performance of official business or such Government.

SECTION VIII. In this ordinance "Korea" means all of the area of Korea subject to this command.

SECTION IX. Any person violating the provisions of this ordinance shall, upon conviction by a Military Occupation Court, suffer such punishment as the court shall determine.

SECTION X. This ordinance is effective 2400, 12 January 1946.

BY DIRECTION OF THE COMMANDING GENERAL
UNITED STATES ARMY FORCES IN KOREA

A. V. ARNOLD
Major General, United States Army
Military Governor of Korea.

HEADQUARTERS
UNITED STATES ARMY FORCES IN KOREA
Office of the Military Governor
Seoul, Korea

ORDINANCE
NUMBER 40

10 January 1946

SALE OF PROPERTY FOR TAX DELINQUENCY

1. Laws relating to the sale of property for tax delinquency are continued in full force and effect, except as to the classes of property provided for in paragraphs 3 and 4 hereof. The proceeds of such sales will be deposited as provided by existing law. The portion of such proceeds which constitutes national taxes, as determined by the tax official, will be deposited for transmission to the Bank of Chosen, Seoul, for the account of the Military Government of Korea.

2. In addition to reports of tax sales now required by existing law, the tax official, conducting such sales, will forward by registered mail a special report within ten (10) days after the date of the sale addressed to the Tax Section of the Financial Bureau of the Military Government of Korea, Seoul. Where the consent of the Property Custodian is required, a copy of the same report will be sent by registered mail addressed to the Property Custodian of the Military Government of Korea, Seoul. The following information will be included in the special report:

   a. The Court and Order Number directing the sale.

   b. Name and address of former owner of property.

   c. Description of Property.

   d. Amount of taxes, interest, penalties, cost and fees, and amount of money received in excess thereof.

   e. Dates, amounts, location and name of financial institution in which each deposit was made and other disposition of the proceeds.

3. All property formerly owned by Japanese nationals (natural and juridical persons), became vested in the Military Government of Korea as of 25 September 1945 by Ordinance No. 33. Sales of such property for taxes is prohibited.

4. The following classes of property will not be sold for taxes except with the written consent of the Property Custodian:

a. Property the transfer of which is prohibited by Ordinance No. 2.
b. Property the ownership of which is unknown.
c. Property requisitioned by the Military Government of Korea.
d. Abandoned property.

5. Any person violating the provisions of this Ordinance will, upon conviction by a Military Occupation Court, suffer such punishment as the court shall determine.

5. This Ordinance becomes effective at 2400, 10 January 1946.

BY DIRECTION OF THE COMMANDING GENERAL
UNITED STATES ARMY FORCES IN KOREA

Archer L. Lerch
Major General, United States Army
Military Governor of Korea.

ORDINANCE
NUMBER 41   10 January 1946

SPECIAL JUDICIAL OFFICERS

1. A system of special judicial officers is established.

2. Each special judicial officer shall have original jurisdiction of cases in which the punishment imposed shall not exceed confinement in a penal institution for a period of thirty (30) days or a fine of three hundred yen (¥300), or both.

3. Upon recommendation of a chief judge of a district court, the Director of the Bureau of Justice shall appoint special judicial officers who shall exercise the powers of a judge with reference to the class of cases mentioned in paragraph two.

4. The number of special judicial officers in each province whose appointments are authorized by this ordinance shall not exceed the total number of district and branch court judges in such provinces.

5. Special judicial officers shall be paid the sum of six hundred one yen and ninety-nine sen (¥601. 99) for each month of service.

6. Proceedings before special judicial officers shall be summary in nature. The special judicial officer shall hear evidence on behalf of the prosecution and the accused and render his decision thereon.

7. An accused who is not satisfied with the decision of a special judicial officer may demand trial in a district or branch court.

8. The decision shall contain the name, age and address of the accused, a brief statement of the facts, the sentence imposed, the name and title of the person making the decision and the date thereof. A permanent record of the decision shall be preserved.

9. An accused who demands trial by district or branch court shall present a written or oral application therefor to the special judicial officer within five (5) days after the rendition of the decision. If the application is oral the special judicial officer shall make a written notation of the same.

10. A special judicial officer who receives the application mentioned in the preceding paragraph shall send the record of the case and all pertinent records promptly to the public prosecutor of the district or branch court.

11. In a case where a sentence of penal servitude or imprisonment is pronounced, the special judicial officer may issue the writ of custody for a period of five (5) days unless appeal is perfected earlier but the period of custody shall not exceed the sentence imposed.

12. In a case where a fine or lower fine is pronounced the accused shall deposit the amount immediately. If he does not do so he shall be held in custody for a term calculated by converting ten yen (¥ 10) into one day and for this purpose any amount less than ten yen (¥ 10) shall be taken as one day.

13. When an accused has appealed from the decision of a special judicial officer and the appeal has been perfected and a writ of summons has been issued he shall be released immediately pending the appeal.

14. The number of days of custody mentioned in paragraph 12 shall accordingly be considered as days served. The number of days of custody mentioned in paragraph 13 shall accordingly be considered as yen paid on the basis of ten yen (¥ 10) for one day, toward the discharge of a fine or lower fine.

15. At any time prior to the rendition of his decision, the special judicial officer may determine that the nature of the case requires trial before a judge of the district court or branch court. In such case, he shall stay all proceedings before him and forward the record of the case and all pertinent papers, together with the reason for his opinion, to the public prosecutor of the branch or district court.

16. It shall be the duty of the chief judge of each district court to assign special judicial officers within the province to such places and for such periods of time as he deems necessary to accomplish the trial of cases the jurisdiction of which are vested in special judicial officers by the terms of this ordinance.

17. This ordinance is effective at midnight, 2400, 10 January 1946.

BY DIRECTION OF THE COMMANDING GENERAL
UNITED STATES ARMY FORCES IN KOREA

Archer L. Lerch
Major General, United States Army
Military Governor of Korea.

ORDINANCE
NUMBER 42

14 January 1946

SECTION I. All duties, functions, records, property and personnel of the Coast Guard of the Marine Section of the Bureau of Transportation of the Government of Korea except such as relate to Aids to Navigation (lighthouse service) are hereby transferred to the Office of the Director of National Defense of the Government of Korea.

SECTION II All duties, functions, records, property and civilian personnel of the Historical Research Division and the Modern Korean Code Drafting Division of the Bureau of Justice of the Government of Korea are hereby transferred to the General Affairs Section of the Secretariat of the Government of Korea.

SECTION III. This order is effective at 2400, 18 January 1946.

BY DIRECTION OF THE COMMANDING GENERAL
UNITED STATES ARMY FORCES IN KOREA

ARCHER L. LERCH
Major General, United States Army
Military Governor of Korea

ORDINANCE
NUMBER 43

16 January 1946

DISSOLUTION OF THE CHOSUN STOCK EXCHANGE

SECTION I.   The Chosun Stock Exchange is hereby dissolved

SECTION II.   The Chosun Trust Company is appointed liquidator of the Chosun Stock Exchange. Upon the final settlement of the affairs of the Chosun Stock Exchange the Chosun Trust Company will file a report with the Director of the Financial Bureau of the Government of Korea.

SECTION III.   This ordinance is effective 2400, 16 Jaunary 1946.

BY DIRECTION OF THE COMMANDING GENERAL
UNITED STATES ARMY FORCES IN KOREA

ARCHER L. LERCH
Major General United States Army
Military Governor of Korea

HEADQUARTERS
UNITED STATES ARMY MILITARY
GOVERNMENT IN KOREA
Office of the Military Governor
Seoul, Korea

ORDINANCE
NUMBER 44

22 January 1946

SECTION

SECTION I

The Patent Office is hereby established in the Bureau of Mining and Industry of the Government of Korea to perform the following functions:

a. Administer the laws relating to:

(1) Patents

(2) Utility Models

(3) Trade Marks

(4) Designs

(5) Copyrights

b. Administering applicable portions of the laws relating to patent attorneys.

c. Keeping and preserving all records, books, models, drawings, specifications and other papers and things pertaining to the subjects mentioned in paragraphs *a* and *b*.

SECTION II

All records and property of the Principal Office in Korea of the Japanese Inventors' Association are hereby transferred to the Patent Office.

SECTION III

This ordinance is effective 2400, 28 January 1946.

ARCHER L. LERCH
Major General United States Army
Military Governor in Korea

111

ORDINANCE
NUMBER 45

25 January 1946

NATIONAL RICE COLLECTION

SECTION I. Purpose. To insure against wide scale starvation, malnutrition, disease and civil unrest, the Military Government of Korea will collect and pay a reasonable price for all rice in Korea south of 38° north latitude except that each family may retain polished or semi-polished rice in an amount equal to forty five hundredths of one suk (sixty-seven and one half kilograms) multiplied by the number of persons regularly residing as a member of each family household. In lieu of polished or semi-polished rice, unhulled rice may be retained in an amount equivalent to all or part.

SECTION II. Surrender of Rice. Every person, other than the heads of governmental, religious, educational, medical, and welfare institutions as exempted by the Military Governor of the province in which such institution is located, holding more than such an amount shall surrender to the head of his village, town, or city, or his agents, such excess immediately upon the demand of the head of the village, town, or city in which such rice is located and accept therefor payment in legal currency or a written order for payment in legal currency signed by the head of the village, town, or city or an authorized agent of such head, at the official price payable for such rice as fixed by public notice.

SECTION III. Duties of Heads of Blocks, Villages, Towns, Cities, and Counties.

a. Each head of each village, town, and city shall ascertain and report in writing to the Military Governor of his province, as ordered by him (or by the head of his county):

(1) The amount of rice and other grains (separately listed) harvested on each farm in his village, town, or city at the last preceding harvest.

2) The number of persons without families, the number of families and the number of persons in each such family regularly residing in his village, town, or city.

— 1 —

(3) The quantity of unhulled, semi-polished, and polished rice, and other grain, listed separately, and the location thereof in the possession of each person or each family residing in his village, town, or city.

(4) The amount of rice and other grain required for the planting of the following season's rice crop in his village, town, or city.

b. When ordered by the Military Governor of his province (or by the head of his county), each head of each village, town, and city will collect and transport to the nearest official rice-collection point such excess rice except that he shall retain, store, and safeguard rice suitable for use as seed in an amount not exceeding the amount required for planting the next rice crop in his village, town, or city. In collecting and transporting rice he shall place in separate bags plainly marked as such, rice suitable for use as seed. Each head of each county shall determine the points in his county to which rice will be transported by village and town heads. He shall execute the orders of the Military Governor of his province, transmit necessary orders to heads of villages and towns and require performance by them. Each head of each city will establish rice collection points.

c. Each head of each block in each town and city shall:

(1) Obtain and report to the head of his town or city such information regarding his block as the head of the town or city shall require in connection with this program.

(2) Collect and deliver the excess rice which is in his block as ordered by the head of his town or city.

(3) Perform other duties as ordered by the head of the town or city.

SECTION IV. Duties of Other Officials and Agencies.

In connection with this program every official and employee of the Military Government of Korea, its instrumentalities and agencies, and of the Military Government of the provinces, cities, towns, and villages will execute the orders of the Military Governor of Korea, and the orders of the Military Governor of the province in which such officials or employees are employed.

SECTION V. Violations.

In addition to any other violation of law, the following acts are violations of this ordinance:

(1) To fail to make any report required in connection with this program.

(2) To fail or refuse to give information required by any head
of any block, village, town, county, or city, in performing his duties under
this program or to make a report which is false.

(3) To fail to surrender excess rice when and as ordered. The
rice involved shall be forfeited in addition to other penalties. No dispute as
to proper payment for rice will be a defense for failure to surrender rice.

(4) To hide or attempt to hide, to destroy or attempt to
destroy rice, or to transport rice except as ordered from any village, town, or
city to any point beyond such village, town, or city. The rice involved shall
be forfeited in addition to other penalties.

(5) To fail or refuse to execute any order issued by any
official in connection with this program.

(6) Any act or threat which encourages the disobedience of
this ordinance.

SECTION VI. Removal and Appointment Power.

The Military Governor of each province is authorized to remove and
appoint the successor to the head of any city, county, town, village, or block,
or any official or employee of any governmental division or agency, for failure
efficiently to execute orders issued in connection with this program. The
services of any person may be requisitioned by the Military Governor of any
province or his authorized agents to carry out this program.

SECTION VII. Penalties.

Any person violating the provisions of this ordinance shall upon
conviction by a Military Occupation Court, suffer such punishment as the
court shall determine.

SECTION VIII. Effective Date.

This ordinance is effective at midnight on 31 January 1946.

ARCHER L. LERCH
Major General, United States Army
Military Governor in Korea

ORDINANCE
NUMBER 46

7 February 1946

SECTION I. The Special Criminal Investigating Committee of Korea and the Office of Secretary of the Special Criminal Investigating Committee of the Bureau of Justice of the Government of Korea are hereby abolished.

SECTION II. Section 9 of Appointment Number 12, dated 11 October 1945, and Section 4 of Appointment Number 36, dated 19 November 1945, are hereby rescinded.

SECTION III. This ordinance shall be effective on the tenth day after the date appearing hereon.

ARCHER L. LERCH
Major General United States Army
Military Governor in Korea

HEADQUARTERS
UNITED STATES ARMY MILITARY
GOVERNMENT IN KOREA
Office of the Military Governor
Seoul, Korea

ORDINANCE
NUMBER 47

13 February 1946

SECTION I. The Public Information Section of the Secretariat of the Government of Korea is hereby established as a Bureau of the Government of Korea and is hereby renamed the Bureau of Public Information of the Government of Korea.

SECTION II. This ordinance shall be effective on the tenth day after the date appearing herein.

ARCHER L. LERCH
Major General, United States Army
Military Governor in Korea.

ORDINANCE
NUMBER 48

13 February 1946

SECTION I. The name of the Bureau of Mining and Industry of the Government of Korea is hereby changed to the Bureau of Commerce of the Government of Korea.

SECTION II. The name of the Bureau of Agriculture and Commerce in the Government of Korea is hereby changed to the Bureau of Agriculture of the Government of Korea.

SECTION III. All duties, functions, records and property of the Domestic Commerce Section and the Foreign Commerce Section of the Bureau of Agriculture are hereby transferred to the Bureau of Commerce.

SECTION IV. All duties, functions, records and property of the Finance Sub-section of the Agricultural Economy Section of the Bureau of Agriculture are hereby transferred to the Bureau of Finance.

SECTION V. All duties, functions, records and property pertaining to lumbering and the processing of fish and marine products in the Lumbering and Processed Foods Sub-section of the Bureau of Commerce are hereby transferred to the Bureau of Agriculture.

SECTION VI. This ordinance is effective at 2400, 13 February 1946.

ARCHER L. LERCH
Major General, United States Army
Military Governor in Korea

ORDINANCE
NUMBER 49

19 February 1946

CONTROLLING AND RECORDING MOVEMENTS OF
PERSONS ENTERING OR LEAVING KOREA

SECTION I. All persons, except those enumerated in Section 5 hereof, who desire to leave Korea for any reason are required to obtain a Letter of Identity to authorize travel outside Korea.

SECTION II. Letters of Identity will be issued by the Foreign Affairs Section of the Secretariat of the Military Government of Korea. The Foreign Affairs Section is hereby authorized to implement this Ordinance by regulations to insure:

a. That all requirements for entry of the nation or nations which such person is to visit are met.

b. That such persons receive the necessary health inoculations, exchange of funds and currency, transportation, and such information and advice as may be necessary.

c. That accurate records are kept of such persons abroad, in case of emergency.

SECTION III. Applications for Letters of Identity may be made to any Military Government Team Commander. who will forward all such applications to the Foreign Affairs Section. Such applications will include:

a. Name and address of applicant.

b. Two photographs taken within six months prior to the date of application.

c. Reason for making application to leave. (If travelling on orders of the Military Government, a copy of such orders will be attached).

d. Birth certificate, or hochok certificate.

e. Proposed itinerary.

f. Designation of address or agent in all foreign countries to be visited. for the receipt of all notices.

g. Period of time for which Letter of Identity is to be valid.

SECTION IV. All persons entering Korea will give the following information to the official representing the Foreign Affairs Section at the point of entry:

    a. Passport or other credentials.

    b. Nature and purpose of visit.

    c. Proposed itinerary and length of visit.

    d. Address in Korea, or designation of agent for receipt of all notices.

SECTION V. The provisions of this Ordinance do not apply to:

    a. Persons included in mass repatriation movements.

    b. Persons who are members of or attached to the armed forces of any of the United Nations and who are travelling on official orders.

SECTION VI. Any person violating the provisions of this Ordinance will, upon conviction by a competent court, suffer such penalty as this court may direct.

SECTION VII. This Ordinance shall be effective on the tenth day after the date appearing hereon.

ARCHER L. LERCH
Major General United States Army
Military Governor in Korea

ORDINANCE
NUMBER 50

16 February 1946

SECTION I. Paragraph "2" of Section III of Ordinance Number 11, dated 9 October 1945, is hereby amended to read as follows:

"2. The punishment of any person without lawful trial and conviction is prohibited."

SECTION II. This ordinance shall be effective on the tenth day after the date appearing hereon.

ARCHER L. LERCH
Major General United States Army
Military Governor in Korea

ORDINANCE
NUMBER 51

20 February 1946

INCREASING FINES

SECTION I. All existing provisions of law and regulatory measures which fix or impose a monetary fine for the violation thereof, are hereby amended so that each monetary fine is increased five (5) times; provided, that this enactment shall not be applicable to offenses consummated prior to the effective date hereof.

SECTION II. Paragraph 2 of Ordinance Number 41, dated 10 January 1946, is amended to read as follows:

"Each Special Judicial Officer shall have original jurisdiction of cases in which the punishment imposed shall not exceed confinement in a penal institution for a period of thirty (30) days or a fine of seven hundred and fifty yen (¥750), or both."

SECTION III. Paragraph 12 of Ordinance Number 41, dated 10 January 1946, is amended to read as follows:

"In a case where a fine or net fine is pronounced, the accused shall deposit the amount immediately. If he does not do so, he shall be held in custody for a term calculated by converting twenty five yen (¥25) into one day and for this purpose any amount less than twenty five yen (¥25) shall be taken as one day."

SECTION IV. Paragraph 14 of Ordinance Number 41, dated 10 January 1946, is amended to read as follows:

"The number of days of custody mentioned in paragraph 11 shall accordingly be considered as days served. The number of days of custody mentioned in paragraph 12 shall accordingly be considered as yen paid on the basis of twenty five yen (¥25) for one day, toward the discharge of a fine or net fine."

SECTION V. This ordinance shall be effective on the tenth day after the date appearing hereon.

ARCHER L. LERCH
Major General, United States Army
Military Governor in Korea

HEADQUARTERS
UNITED STATES ARMY MILITARY
GOVERNMENT IN KOREA
Office of the Military Governor
Seoul, Korea

ORDINANCE
NUMBER 52

21 February 1946

CREATION OF THE NEW KOREA COMPANY LIMITED

SECTION I. The New Korea Company, Limited, is hereby created as an independent agency of the Government of Korea. The corporation will be directed by ten directors who will hold office until replaced by the Military Governor or his nominee. Such Board of Directors will operate the corporation under by-laws and regulations prescribed by it with the approval of the Military Governor or his nominee. The United States Army Officer acting as President of the corporation will have full power to decide questions of policy affecting United States interests, subject to the approval of the Military Governor. The principal office of the corporation will be in Seoul, Korea, and will have such branch offices as the Board of Directors shall determine with the approval of the Military Governor or his nominee.

SECTION II. The directors will appoint an Advisory Board consisting of the Chairman of the Board of Directors of the Corporation, who will serve as Chairman, and one or more members from each province nominated by the Military Governor or his nominee. The Advisory Board will meet at the call of the Chairman. The Advisory Board will make recommendations to the Board of Directors. The Board of Directors will consult the Advisory Board on major questions of policy.

SECTION III. The corporation will have a capital stock of one hundred million yen (¥ 100,000,000) divided into shares of fifty yen (¥ 50) each. The total amount of the capital stock will be subscribed, exclusively owned, and paid in full by the Military Government of Korea by the physical transfer of property of the Oriental Development Company, Limited. Its liability will be limited to its capital stock. It may be dissolved only by the Military Government of Korea. To qualify the directors, a certificate for one share of stock will be transferred on the books of the Company to the name of each Director,

SECTION IV. All property owned on or since 9 August 1945 by The

— 1 —

Oriental Development Company, Limited, and all Japanese property in juridical persons in Korea, in which juridical persons Oriental Development Company, Limited, had a financial interest on or since 9 August 1945. is hereby vested in the New Korea Company, Limited.

SECTION V. The corporation will have the following powers :

a. To acquire and own its own shares of stock at a price not less than the par value thereof.

b. The capital stock may be reduced or increased by the Board of Directors by the issuance of a public notice signed by the Military Governor.

c. All of the powers of such a juridical person under the Commercial Code now or hereafter in force in Korea.

d. To issue debentures under the provisions of the Commercial Code except that debentures not in excess of ten times the amount of the paid-up capital may be issued, when the issuance, and terms of the debentures have been approved by the Military Governor or his nominee.

e. To appoint and compensate such officers and employees as the Board of Directors may determine.

f. To pay dividends and establish reserves under the provisions of the Commercial Code.

g. To transact all forms of industrial activity. including the extraction of metallic and non-metallic natural resources, the processing and fabrication of materials and chemicals. the manufacture of any articles for industrial or commercial purposes, machinery, motive and transportation equipment, ships and aircraft parts and equipment, electrical equipment and appliances, textiles, and food processing.

h. To transact all forms of commercial business at all levels of distribution, warehousing and storage.

i. To engage in all forms of agriculture, including ownership or operation of leased lands, agricultural experimental stations or laboratories, and in all forms of forestry, including lands used for the propagation of nursery stock and lands under reforestation.

j. To engage in all forms of fishing and acquisition of aquatic products, including the ownership or operation under lease or charter of fishing boats, docks, and other fishing gear and facilities for hauling, storing and preserving fish and aquatic products.

k. Under the directives of the Military Governor or his nominee, to regulate and supervise, including the furnishing of technical assistance to. any business engaged in activities within the scope of the powers of the New Korea Company, Limited.

— 2 —

125

l. To own property and dispose of the same.

m. All other powers incidental or necessary to the accomplishment of the objects of the corporation.

SECTION VI. The corporation, its directors, officers and employees will be subject to such duties and will exercise such powers as are provided by the Commercial Code insofar as they are pertinent or necessary to the business of the corporation.

SECTION VII. Any person who, for the purpose of inducing the corporation to transact any business with such person, makes any false statement orally or in writing, knowing it to be false, or who makes or circulates any statement or rumor orally or in writing which is untrue or which directly or by inference is derogatory to the financial condition or standing of the corporation, or who induces others to do so, shall upon conviction by a Military Occupation Court suffer such punishment as the Court shall determine.

SECTION VIII. The corporation will make its public notices by publication in a newspaper of general circulation.

SECTION IX. This ordinance shall be effective on the tenth day after the date appearing hereon.

ARCHER L. LERCH
Major General United States Army
Military Governor in Korea

ORDINANCE
NUMBER 53

20 February 1946

KOREAN ECONOMIC ADVISORY BOARD

SECTION I. Purpose. The purpose of this ordinance is to provide a representative Korean agency to advise concerning the stabilization of the cost of living and the fair distribution of essential commodities, the curbing of rising prices, the prevention of scarcities of consumer goods, the development of a coordinated program of production and distribution, and the elimination of economic conditions adverse to the welfare of the people of Korea.

SECTION II. Creation of Korean Economic Advisory Board. The Korean Economic Advisory Board is hereby established as an agency of the Government of Korea. It shall consist of seven members, one of whom shall be the Recording Secretary. Members will be appointed by the Military Governor to hold office during his pleasure. Members shall serve for such part of their time as the Chairman may require. They shall be compensated for service rendered, except that paid officials of the government shall receive no additional compensation.

SECTION III. Duties of the Board. The Korean Economic Advisory Board shall make studies and recommend to the Military Governor programs for establishing, administering and enforcing price control and rationing, and for allocations for production and methods of distribution.

SECTION IV. Assistance to the Board. The Director of the Bureau of Commerce shall furnish such facilities and assistance as the Board may require in carrying out its work.

SECTION V. Effective Date. This ordinance shall be effective on the tenth day after the date appearing hereon.

ARCHER L. LERCH
Major General United States Army
Military Governor in Korea

HEADQUARTERS
UNITED STATES ARMY MILITARY
GOVERNMENT IN KOREA
Office of the Military Governor
Seoul, Korea

ORDINANCE
NUMBER 54   25 February 1946

SECTION I. For purposes of tax collection, administration and control, the following transfers of jurisdiction over Counties (Gun) within Kyong Sang Nam Do are hereby effected:

a. Yangsan County (Gun) to Pusan District Tax Collection Office.

b. Tongnae County (Gun) to Pusan District Tax Collection Office.

SECTION II. This order is effective at 2400, 31 March 1946.

ARCHER L. LERCH
Major General United States Army
Military Governor in Korea

ORDINANCE
NUMBER 55

23 February 1946

REGULATION OF POLITICAL PARTIES

SECTION I. Registration of Political Parties

a. Who shall register.

In addition to any previous reports or registrations, each group of three or more persons who constitute any organization or association for the purpose of conducting and who engage in any form of political activities shall register such organization as a political party. Activities conducted in the name of and by any group or organization which involve public discussions or propaganda in any written or oral form, or public actions, are political when they tend to influence the laws, the form, the organization, the selection, nomination, election, appointment or removal of personnel or office holders, the administration, the procedures, or the policies of the government including the enactment, execution and enforcement of laws, foreign relations, and the rights, powers, duties, freedoms and privileges of the people in relation to government. For any group or organization secretly to conduct any activities which tend to exert any of such political influences is hereby prohibited.

b. The place of registration.

If such political party extends any of its activities beyond the limits of one province or is affiliated in any way with any other political party or any part of any political party which operates beyond the limits of that province, the place of registration shall be The Bureau of Public Information, Capitol Building, Seoul. If such political party does not extend by its own actions or by affiliation with any other political party or part of any political party beyond the limits of one province the place of registration shall be the Office of the Governor of that Province.

c. Information to be registered.

Each political party shall file by registered mail the following

— 1 —

information, which shall be certified as correct by the signature and seal of the leader or leaders of the party:

(1) The name or designation of the political party, and any symbol thereof.

(2) A copy of its charter or a declaration of its purposes, including a copy of every paper previously issued which discloses the purposes of the party and any or all persons openly or secretly affiliated with the party.

(3) The name and designation of position in the party of each person holding any office or exercising any function or influence in the party other than as a member. If the party act from more than one place, the name of each person, and his position, having any responsibility for actions taken at each different place, listed according to the different places where the party acts through branches or sections, or through any subordinate group affiliated with it, in any way controlled by it or furnished money or materials or other assistance by it.

(4) The exact address and description of each place used by the party or any part of it for offices, for meetings of any sort, for the preparation of material or for the transaction of any business, including every such place actually used in the past sixty days for any such purpose. The place or places at which mail is received by the party and which place is to be registered as the party headquarters shall be stated. The headquarters of each branch or subordinate chapter or section shall also be stated.

(5) The date of organization of the party, and each date on which political activity began.

(6) The name, the address, and a clear description of the organization of each political party with which it is affiliated, or with which it has cooperated in any way in the past sixty days.

(7) The number of members of the party. If subordinate branches exist, the number of members of each party branch. If the exact number is not ascertainable, an estimate of the number of actual members. The number of persons who have contributed any money to the political party or otherwise financially assisted its activities shall be separately reported.

d. The time to register.

Registration under this section shall be completed by placing in the Post Office as registered mail a letter or letters accompanying the

— 2 —

128

required information properly addressed to the proper place of registration on or before 28 February 1946.

SECTION II. Regulations Governing Political Parties.

a. Party Office.

Each political party shall have a publicly known name or designation and shall maintain a headquarters office at the place stated in its registration papers until it has filed by registered mail addressed to the proper place of registration the exact address and description of a new party headquarters. If the party acts through any branch or affiliated organization, it shall maintain a branch office at each place reported in its registration as its branch offices. Failure to maintain such party headquarters offices or any change of any such office without a prior registration of the new address shall be cause for an order dissolving the party to be issued by the Director of the Bureau of Public Information. After the issuance of such a dissolution order any action by any official or member of such party, except in dissolution or applying for renewal of registration of the party under regulations issued by the Bureau of Public Information shall be a violation of this ordinance. Each person who participates in any way with any group engaged in political activities without compliance with this section, publicly or secretly, is violating this ordinance.

b. Accounting for Party Funds.

Each political party shall maintain in account books an accurate accounting for all party funds and property. Such account books shall be securely kept at the Party Office and shall be available during business hours for governmental employees for examination by authorized representatives of the Bureau of Public Information who display signed and sealed authorization of such examination. An accurate report of the property and receipts and disbursements by the party shall be filed, by registered mail addressed to the proper place for registration on or before the last day of each quarter of the calendar year. Such reports shall show the name and address of each person (or other source of contributions) who has contributed any funds or things of value to the party and the amount of each contribution. It shall specify the actual source of all receipts and the actual and, if known, the ultimate recipient of all disbursements.

c. Additional Regulations.

The Bureau of Public Information may issue additional regulations governing political parties.

— 3 —

SECTION III. Party Membership.

a. Persons Disqualified.

No person shall be permitted to be a member of any political party who is disqualified by law from holding public office. No person of alien nationality shall be permitted membership. Secret membership affiliation is unlawful. No contribution or direct or indirect financial assistance may lawfully be accepted from any source other than from members of the party. Any person conniving to defeat this prohibition has violated this ordinance.

b. Date of Filing.

Each political party shall file by registered mail addressed to the Governor of each Province in which it has any members an accurate list of the members residing in the province with their correct addresses and each list shall be signed and sealed by each party member. The filing shall be completed by placing such registered mail in the Post Office on or before 31 March 1946.

SECTION IV. Civil and Criminal Liabilities.

The persons holding any official position or exercising any function or influence in the party other than as a member shall be jointly and severally liable in civil or criminal actions for any act in violation of this ordinance. Knowingly to make any false statement in any registration or report by the party is a violation of this ordinance.

SECTION V. Any person violating the provisions of this ordinance will, upon conviction by a competent court, suffer such penalty as the court may direct.

SECTION VI. This ordinance shall be effective at 2400 of the date hereof.

ARCHER L. LERCH
Major General United States Army
Military Governor in Korea

HEADQUARTERS
UNITED STATES ARMY MILITARY
GOVERNMENT IN KOREA
Office of the Military Governor
Seoul, Korea

ORDINANCE
NUMBER 56

_ March 194_

Ordinance Number 20 is amended to read as follows :

SECTION

CRIMINAL INVESTIGATION ESTABLISHED I
TRANSFER OF FINGER PRINT SECTION OF CRIMINAL
    DEPARTMENT OF BUREAU OF JUSTICE TO POLICE
    BUREAU II
EFFECTIVE DATE III

SECTION I. Bureau of Police is hereby authorized to establish criminal investigation including the following functions :

a. Conducting investigations of criminal matters which are referred to it by Military Government.

b. Assisting and cooperating with the Military Police, Counter Intelligence Corps and the Police of Korea, when requested, in the investigation and apprehension of criminals.

c. Establishing and maintaining a National Finger Print Record and Criminal Investigation System.

SECTION II. All duties, functions, records, property, and civilian personnel of the Finger Print Section of the Criminal Department of the Bureau of Justice are hereby transferred to the Bureau of Police.

SECTION III. This order is effective on the date appearing hereon.

ARCHER L. LERCH
Major General, United States Army
Military Governor in Korea

131

HEADQUARTERS
UNITED STATES ARMY MILITARY
GOVERNMENT IN KOREA
Office of the Military Governor
Seoul, Korea

ORDINANCE
NUMBER 57

21 February 1946

DEPOSIT OF NOTES OF BANK OF JAPAN
AND BANK OF TAIWAN

<u>SECTION I.</u> Deposit Required

All natural and juridical persons within Korea South of 38° North Latitude are hereby:

a. Ordered to deposit within the period 2 March to 7 March 1946 inclusive in one of the financial institutions designated in Section 2 all notes of the Bank of Japan of a denomination of one yen or more and all notes of the Bank of Taiwan owned or possessed by them.

b. Prohibited after 7 March 1946 from importing, exporting, receiving, paying out, knowingly owning or possessing, giving or otherwise transfering or dealing in or engaging in any transaction with respect to any such currency.

<u>SECTION II.</u> Designated Financial Institutions.

The following financial institutions are authorized and directed to receive Bank of Japan and Bank of Taiwan notes for deposit in Japanese yen accounts pursuant to this Ordinance in the name of the person making the deposit:

> Bank of Chosen
> Chosen Industrial Bank
> Choheung Bank
> Chosen Commercial Bank
> Chosen Trust Company
> Chosen Savings Bank
> Federation of Financial Association

Each financial institution shall:

a. Hold in safekeeping separate from any other currency the notes deposited under this ordinance.

132

b. Report to the Bureau of Finance, Military Government of Korea, Seoul, immediately after the deposit period the amount of each currency deposited.

SECTION III. Limitations on Deposit Accounts

Accounts resulting from the deposit of notes of the Bank of Japan and Bank of Taiwan pursuant to this Ordinance shall not have withdrawals therefrom, shall bear no interest, shall not be transferable or negotiable or used as collateral for any present or future loan or indebtedness.

SECTION IV. Penalty Clause

Any person violating the provisions of this Ordinance will, upon conviction by a Military Occupation Court, suffer such penalty as the court shall determine.

SECTION V. Effective Date.

This Ordinance shall be effective at midnight, (2400) 21 February 1946.

ARCHER L. LERCH
Major General United States Army
Military Governor in Korea

ORDINANCE
NUMBER 58

5 March 1946

SECTION I. All duties, functions, records, property and civilian personnel of the Inspection Sub-section of the Planning Section of the Secretariat are hereby transferred to the Office of the Civil Administrator of the Government of Korea.

SECTION II. This ordinance shall be effective on the tenth day after the date appearing hereon.

ARCHER L. LERCH
Major General United States Army
Military Governor in Korea

HEADQUARTERS
UNITED STATES ARMY MILITARY
GOVERNMENT IN KOREA
Office of the Military Governor
Seoul, Korea

ORDINANCE
NUMBER 59   7 March 1946

AMENDING ORDINANCE NUMBER 57 BY EXTENDING PERIOD FOR
DEPOSIT OF NOTES OF BANK OF JAPAN AND BANK OF TAIWAN

SECTION I. Extension of Period for Deposit.-Paragraphs "a" and "b" of Section I of Ordinance Number 57, dated 21 February 1946, are hereby amended by substituting the date "10 March 1946" for the date "7 March 1946" therein.

SECTION II. Effective Date.-This ordinance shall be effective at 2400, 7 March 1946.

ARCHER L. LERCH
Major General, United States Army
Military Governor in Korea.

UNITED STATES ARMY MILITARY
GOVERNMENT IN KOREA
Office of the Military Governor
Seoul, Korea

ORDINANCE
NUMBER 60

14 March 1946

PROVINCIAL COUNCILS

SECTION I. All Provincial Councils and councils of lower levels of government, including city, town, and village councils and school councils of counties and islands, which were in existence in Korea South of 38° North Latitude on or before 15 August 1945, are hereby dissolved.

SECTION II. Provincial Military Governors are directed to cause all funds, records and property of such councils within their respective jurisdictions to be taken into custody and control by designated Military Government or civilian personnel, and preserved, maintained and safeguarded as property of the Military Government, subject to further directions of the Military Governor in Korea.

SECTION III. This ordinance shall be effective on the tenth day after the date appearing hereon.

ARCHER L. LERCH
Major General United States Army
Military Governor in Korea

HEADQUARTERS
UNITED STATES ARMY MILITARY
GOVERNMENT IN KOREA
Seoul, Korea

ORDINANCE
NUMBER 61   29 March 1946

DISSOLUTION OF KOREAN ASSOCIATION FOR PROTECTION AND RELIEF OF CONSCRIPTED LABORERS AND WAR SUFFERERS; PAYMENT OF BENEFITS; ACCOUNTING

SECTION I. Dissolution of Association. The Korean Association for Protection and Relief of Conscripted Laborers and War Sufferers is hereby dissolved and its corporate existence terminated.

SECTION II. Payment of Benefits; Reimbursement. The Association shall forthwith proceed to pay out all moneys which are now or hereafter subject to its control and which represent payments actually made to or for the benefit of specific conscripted Korean laborers, their families or beneficiaries. The Association shall reimburse out of its funds under Section III hereof, governmental organizations which in good faith have made payments to conscripted Korean laborers, their families or beneficiaries, of the type described in this Section.

SECTION III. Disposition of Unidentifiable Balances. All remaining moneys of the Association not identifiable by it as due and owing to specific conscripted Korean laborers, their families or beneficiaries, shall be devoted to welfare purposes for needy Korean conscripted laborers, their families or beneficiaries. An inventory of all real and personal property other than moneys, of the Association, shall be made forthwith and reported to the Property Custodian of the Government of Korea. The proceeds of the sale of such property shall be similarly used.

SECTION IV. Collection of Indebtednesses. All continuing responsibilities and duties of the Association shall be fulfilled, including its duty to take all necessary steps for collection and payment of amounts due and owing to conscripted Korean laborers, their families or beneficiaries, from employers. All moneys so collected shall be distributed in accordance with the provisions of Section II or III hereof.

<u>SECTION V.</u> <u>Responsible Agency.</u> The Bureau of Public Health and
Welfare of the Government of Korea shall be responsible for carrying out the
provisions of this Ordinance.

<u>SECTION VI.</u> <u>Effective Date.</u> This Ordinance shall be effective on
the tenth day after the date appearing hereon.

                                        ARCHER L. LERCH
                                Major General, United States Army
                                    Military Governor in Korea

HEADQUARTERS
UNITED STATES ARMY MILITARY
GOVERNMENT IN KOREA
Office of the Military Governor
Seoul, Korea

ORDINANCE
NUMBER 62

29 March 19..

REGULATION OF DRUGS, MEDICINES,
PHARMACEUTICALS AND RELATED ARTICLES

SECTION I. Purpose and Effect. In accordance with paragraph 3 of General Notice Number 2, dated 20 October 1945, drugs, medicines, pharmaceuticals and other items of commerce used in the practice of medicine or public health activities, and derivatives, compounds, or substances containing any of such materials, are declared to be in critical demand and needed for the benefit of the people of Korea. This scarcity requires some measure of control by the Military Government. This ordinance amends the provisions of paragraph 2 of General Notice Number 2, dated 20 October 1945.

SECTION II. Regulation and Control. The Director of the Bureau of Public Health and Welfare of the Government of Korea is hereby designated as the sole controlling agency of the Military Government of Korea for drugs, medicines, pharmaceuticals and other items of commerce used in the practice of medicine or public health activities, and for derivatives, compounds, or substances containing any of such materials. He will establish all necessary controls, including but not limited to price, rationing, allocation, licensing, purchasing, selling, production, distribution, transportation, storage, and export and import controls for the aforementioned articles.

SECTION III. Penalties. Any person violating the provisions of this ordinance or any regulation or order issued by the Director of the Bureau of Public Health and Welfare pursuant thereto shall, upon conviction by a Military Occupation Court, suffer such punishment as the court shall determine.

SECTION IV. Effective Date. This ordinance shall be effective on the tenth day after the date appearing hereon.

ARCHER L. LERCH
Major General United States Army
Military Governor in Korea

ORDINANCE
NUMBER 63

29 March 1946

BUREAU OF POLICE REMOVED FROM SUPERVISION AND DIRECTION OF OFFICE OF DIRECTOR OF NATIONAL DEFENSE*

SECTION I. Bureau of Police Removed from Supervision and Direction of Office of Director of National Defense. The Bureau of Police of the Government of Korea is hereby removed from the supervision and direction of the Office of the Director of National Defense.

SECTION II Effective Date. This ordinance shall be effective on the tenth day after the date appearing hereon.

ARCHER L. LERCH
Major General, United States Army
Military Governor in Korea

---

* See Ordinance No 28, 13 November 1945, Sec II.

ORDINANCE
NUMBER 64   29 March 1946

NOMENCLATURE IN THE
GOVERNMENT OF KOREA

SECTION I. Scope. The principal elements of the Government of Korea at the national level, will hereafter be known as Departments or Offices. This Ordinance does not change existing powers, functions or duties. The classification and nomenclature fixed by this ordinance shall not be changed or supplemented except by ordinance.

SECTION II. Bureaus Renamed Departments. The following elements of the Government of Korea will hereafter be known as Departments:

 a. Department of Agriculture (formerly Bureau of Agriculture)
 b. Department of Commerce (formerly Bureau of Commerce)
 c. Department of Communications (formerly Bureau of Communications)
 d. Department of Education (formerly Bureau of Education)
 e. Department of Finance (formerly Bureau of Finance)
 f. Department of Justice (formerly Bureau of Justice)
 g. Department of National Defense (formerly Office of the Director of National Defense)
 h. Department of Police (formerly Bureau of Police)
 i. Department of Public Health and Welfare (formerly Bureau of Public Health and Welfare)
 j. Department of Public Information (formerly Bureau of Public Information)
 k. Department of Transportation (formerly Bureau of Transportation)

SECTION III. Sections Renamed Offices. The following elements of the Government of Korea will hereafter be known as Offices:

 a. Office of Accounts (formerly Accounts Section)
 b. Office of Foreign Affairs (formerly Foreign Affairs Section)

c. Office of General Affairs (formerly General Affairs Section)

d. Office of Korean Civil Service (formerly Korean Civil Service Section)

e. Office of Planning (formerly Planning Section)

f. Office of Property Custody (formerly Property Custody Section)

g. Office of Provincial Affairs (formerly Provincial Affairs Section)

SECTION IV. _Departments._ Departments will be divided as follows:

a. Departments into Bureaus.

b. Bureaus into Sections.

c. Sections into Subsections.

d. Subsections into Branches.

SECTION V. _Offices._ Offices will be divided as follows:

a. Offices into Divisions.

b. Divisions into Sections.

c. Sections into Subsections.

d. Subsections into Branches.

SECTION VI. _Titles of Officials in Charge._ Titles of individuals in charge will be as follows:

a. Director—for Departments and Offices.

b. Chief—for Bureaus, Divisions, Sections, Subsections and Branches.

SECTION VII. _Titles of Chief Assistants._ When the position of a chief assistant to an individual in charge has been authorized, the title therefor will be as follows:

a. Deputy Director.

b. Assistant Chief.

SECTION VIII. _Use of Authorized Titles._ Only authorized titles will be applied to both military and civilian individuals appointed to and occupying positions within the Government of Korea.

SECTION IX. _Effective Date._ This ordinance shall be effective on the date appearing hereon.

ARCHER L LERCH
Major General United States Army
Military Governor in Korea

ORDINANCE
NUMBER 65   29 March 1946

REGULATION OF VEHICULAR AND PEDESTRIAN TRAFFIC

SECTION I. <u>Vehicles to Keep to Right.</u> All vehicular traffic and animals proceeding along or upon any street, road, highway or bridge shall keep to the right side of the center thereof. Section 1a of Governor General Order No 231, November 1938, is hereby amended in accordance herewith.

SECTION II. <u>Vehicles Meeting on Single-Lane Roads.</u> When vehicles proceeding in opposite directions meet on a single-lane road or other place where simultaneous passage is hazardous, each shall bear as far as possible to the right side of the road, and pass with caution to the left of each other. Section 6 of Governor General Order No 231, November 1938, is hereby amended in accordance herewith.

SECTION III. <u>Passing Vehicles; Tramcars.</u> When any vehicle intends to pass another vehicle proceeding in the same direction, the rear vehicle shall sound a warning device, and proceed with caution to pass the leading vehicle to the left thereof. Tramcars proceeding along tracks in the center of any street may be passed to the right thereof. Section 7 of Governor General Order No 231, November 1938, is hereby amended in accordance herewith.

SECTION IV. <u>Parking.</u> All vehicles, unless otherwise restricted, shall be parked on the right side of any street. Section 13 of Governor General Order No 231, November 1938, and Section 92 of Governor General Order No 131, December 1934, are hereby amended in accordance herewith.

SECTION V. <u>Right of Way at Junctions and Intersections.</u> A vehicle passing along any primary or heavily-travelled street, road, or highway, shall have the right of way over any vehicle entering upon or crossing same from a secondary or lesser travelled steet, road or highway. When vehicles meet at any intersection or junction of streets, roads or highways equally travelled, the vehicle on the right shall have the right of way. Section 78 of Governor General Order No 131, December 1934, is hereby amended in accordance herewith.

SECTION VI. Right of Way of Fire Department Vehicles. When a fire engine or other fire department vehicle is proceeding along a street, road, highway, or bridge to a fire, a warning device, bell or siren shall be continuously sounded, and all vehicles travelling along the same street, road, highway, or bridge shall move as far to the right as possible and come to a stop. Vehicles on any cross-street, road, or highway shall not enter the street, road, or highway on which the fire department vehicle is proceeding, until same shall have passed. Section 79 of Governor General Order No 131, December 1934, is hereby amended in accordance herewith.

SECTION VII. Right of Way on Hills. When vehicles approach each other on a hill where there is single-lane traffic or where simultaneous passage is hazardous, the ascending vehicle shall move as far to the right of the road as possible and give the right of way to the descending vehicle. Section 80 of Governor General Order No 131, December 1934, is hereby amended in accordance herewith.

SECTION VIII. Penalties. Any person violating the provisions of this Ordinance shall, upon conviction by a Military Occupation Court or Special Judicial Officer, suffer such penalty as the court or Special Judicial Officer shall determine.

SECTION IX. Effective Date. This Ordinance shall be effective at 2400, 31 March 1946.

ARCHER L. LERCH
Major General, United States Army
Military Governor in Korea

HEADQUARTERS
UNITED STATES ARMY MILITARY
GOVERNMENT IN KOREA
Office of the Military Governor
Seoul, Korea

ORDINANCE
NUMBER 66   10 April 1946

CREATION OF FIRE DEPARTMENTS AND BOARDS

SECTION I.   City, Town and Village Fire Departments.

a. Creation.

·When and as directed by the Provincial Governors of each Province south of 38° North Latitude, and not later than 30 April 1946:

(1) Operation and control of Fire Departments by the Department of Police of the Government of Korea will cease;

(2) Independent Fire Departments will be established under the direct supervision, administration and control of the authorities in cities, towns and villages; and

(3) All duties, functions, records, property and civilian personnel of the Department of Police concerned with the operation and control of Fire Departments will be allocated and transferred to city, town and village Fire Departments.

b. Inventory.

To assist in accomplishing the foregoing allocation and transfer, the Director of the National Department of Police will furnish by 20 April 1946 to the Provincial Governors, a complete inventory of all property, equipment, funds, records, and personnel of the National Department of Police devoted primarily to the operation or control of Fire Departments in their respective provinces.

SECTION II.   National Board of Fire Commissioners.

a. Creation.

The National Board of Fire Commissioners is hereby created within the Public Works Bureau of the Department of Commerce. The Board shall consist of seven (7) members, one (1) of whom shall be the Recording Secretary. Members shall be appointed by Appointment Order signed by the Military Governor, as far as possible from among persons who possess expert

— 1 —

145

knowledge of fire protection and prevention and shall hold office during his pleasure. Members shall serve for such part of their time as the Chairman of the Board may require. They shall be compensated at civil service rates for services rendered, except that paid officials of the Government shall receive no additional compensation.

b. Duties of the Board.

(1) The Board of National Fire Commissioners shall prepare, in coordination with Provincial Affairs, the National Fire Budget and recommend allocation of funds to cities, towns and villages for the operation of their fire departments.

(2) The Board shall make a studies and prepare standards and regulations on the following matters and such other matters as are deemed essential to fire protection and fire prevention:

    (a) Standardization of specifications for fire-fighting equipment and appliances.

    (b) Organizational plans for city, town and village Fire Departments

    (c) Water supply and distribution for fire-fighting.

    (d) Fire alarms and communications systems.

When approved by the Military Governor of Korea such regulations shall be published in the Official Gazette.

c. Technical Assistants and Staff.

The Director of the Department of Commerce is authorized to appoint, compensate, suspend and remove technical assistants and personnel for the staff of the Board.

SECTION III. Provincial Boards of Fire Commissioners.

a. Creation.

A Provincial Board of Fire Commissioners is hereby established as an agency of each Province in Korea south of 38° North Latitude. It shall consist of five members appointed by Appointment Order signed by the Provincial Governor as far as possible from among persons who possess expert knowledge of fire protection and prevention, one of whom shall be the Recording Secretary. Members shall serve for such part of their time as the Provincial Governor may require and shall be compensated at civil service rates for services rendered, except that paid officials of the Government shall receive no additional compensation.

b. Duties of the Board.

The Provincial Boards of Fire Commissioners shall make studies of fire losses and fire hazards, and shall assist cities, towns and villages in developing adequate plans for fire protection and fire prevention. They shall assist the National Board of Fire Commissioners in matters of policy, planning reports and budget.

c. Technical Assistants and Staff.

Each Provincial Governor is authorized to appoint, compensate, suspend and remove technical assistants and personnel for the staff of the Board.

SECTION IV. Effective Date. This ordinance shall be effective on the tenth day after the date appearing hereon.

ARCHER L. LERCH
Major General, United States Army
Military Governor in Korea

HEADQUARTERS
UNITED STATES ARMY MILITARY
GOVERNMENT IN KOREA
Office of the Military Governor
Seoul, Korea

ORDINANCE
NUMBER 67

2 April 1946

SECTION I.   Legal Division of Office of General Affairs Transferred to Department of Justice.

All duties, functions, records, property and civilian personnel of the Legal Division of the Office of General Affairs are hereby transferred to the Department of Justice.

SECTION II.   Functions of Department of Justice.

The Director of the Department of Justice is the legal advisor of the Military Governor in Korea. He shall advise the Military Governor on policies affecting the national law establishment, and regarding the best qualified persons for appointment by the Military Governor as Justices of the Supreme Court and Judges of Courts of Review, and concerning the legality of governmental policies and proposed laws, ordinances, and regulatory measures. He shall supervise the administration of law and legal to the nation. He shall pass upon the language, form, and legal authority for the issuance and legal effectiveness of all official documents of the government that are sealed and documented, except orders and decisions of the courts. He shall represent the government in all cases before the courts.

He shall appoint, with the approval of the Military Governor and from nominations submitted by the Central Council of the National Bar Association, judges of courts other than the Supreme Court and Courts of Review, and all prosecutors. With the approval of the Military Governor he shall appoint and supervise the chief wardens and administrative officers of the penal and juvenile institutions, members of parole and probation boards, the Librarian of the National Law Library, and the Editor in Chief of Court Reports. He shall appoint and supervise a staff of lawyers to furnish legal advice to Departments and Offices of the Military Government, and to Provincial Governors, to conduct legal research, render legal opinions, and draft all legal documents of the government and its agencies. Within policies established by the Military Governor in Korea, he shall supervise the training and disciplining of lawyers, and the training of judicial and court clerk personnel. He shall certify the admissions of applicants to practice law who have met the requirements for admission to the legal profession.

SECTION III.   Apportionment of Books and Budgetary Funds to the National Law Library.

In order to enable the National Law Library to function adequately for the benefit of the legal establishment of the nation, all books of a legal character now deposited in the Capitol Library and in the National Library are hereby transferred to the National Law Library. Deliveries thereof will be made as requested by the Director of the Department of Justice. That proportion of the budget of the Capitol Library and of the National Library used on the average for the purchase during the past five years of legal material shall be transferred and added to the budget of the National Law Library.

SECTION IV.   Administrative Division of Office of General Affairs Transferred to Office of Accounts; Office of General Affairs Abolished.

All duties, functions, records, property and civilian personnel of the Administrative Division of the Office of General Affairs are hereby transferred to the Office of Accounts. All duties and functions of the Office of General Affairs having been transferred, the Office of General Affairs is hereby abolished.

<u>SECTION V.</u> <u>Name of Office of Accounts Changed to Office of Administration.</u>

The Office of Accounts shall hereafter be known as the Office of Administration.

<u>SECTION VI.</u> <u>Division of Accounts of Office of Administration Transferred to Finance Department; Name Changed to Bureau of Accounts.</u>

All duties, functions, records, property and civilian personnel of the Division of Accounts of the Office of Administration are hereby transferred to the Finance Department, and said Division shall hereafter be a Section of the Bureau of Accounts.

<u>SECTION VII.</u> <u>Effective Date.</u>

This Ordinance shall be effective on the date appearing hereon

ARCHER L. LERCH
Major General United States Army
Military Governor in Korea

ORDINANCE
NUMBER 68

12 April 1946

CONTROL OF MOTION PICTURES

SECTION I. Repeal of Former Law. The provisions of Paragraph 6, Article 18 of Government General Instruction No 18, dated 17 April 1945, which charged the Inspection Section of the former Bureau of Police with responsibility for inspection and control of motion picture films are hereby repealed.

SECTION II. Transfer of Responsibility. All duties, functions, records and property of the Department of Police of the Government of Korea concerned with supervision and control of the production, distribution and exhibition of motion pictures in Korea under existing laws are hereby transferred to the Department of Public Information of the Government of Korea.

SECTION III. Effective Date. This ordinance shall be effective on the date appearing hereon.

ARCHER L. LERCH
Major General United States Army
Military Governor in Korea

ORDINANCE
NUMBER 69

26 April 1946

DEFINING THE FUNCTIONS OF
THE OFFICE OF KOREAN CIVIL SERVICE

SECTION I. Repeal of Former Law. Paragraph 4 of Article 1 of Government General Instruction Number 18, dated 17 April 1945, defining the duties and functions of the former Personnel Section of the Secretariat of the Government of Korea, now known as the Office of Korean Civil Service, is hereby repealed and the following provisions are substituted therefor.

SECTION II. Duties and Functions of Office of Korean Civil Service. The Office of Korean Civil Service is hereby assigned the following duties and functions:

a. To provide for, direct, and administer a classified civil service system for the Government of Korea, including:

(1) Classification of each position in the government service according to kind and difficulty of work performed;

(2) Definition of grades, titles, compensation, hours, duties, responsibilities, and qualifications for each position.

b. To provide for, direct, and administer the recruitment, placement, and training of persons in the government service, including:

(1) Investigation of the character, loyalty, record, qualifications and fitness of applicants;

(2) Preparation and administration of examinations to determine the qualifications and fitness of applicants;

(3) Certification of lists of persons qualified on the basis of merit and fitness, as determined by investigations and examination, to appointing officers authorized by law or by directive of the Military Governor to appoint such personnel to office from such lists;

152

b. Training programs for government employees.

c. To design, install, and administer a program of retirement and disability benefits for government employees.

d. To determine the forms of ceremony to be observed by government officials and employees on important public occasions and to supervise their observance.

SECTION III. Establishment of Civil Service Rules and Regulations. The Director of the Office of Korean Civil Service is hereby directed to prepare civil service rules and regulations for publication in the Official Gazette upon approval of the Military Governor; and will establish staff and branch offices necessary to carry out the provisions of this Ordinance.

SECTION IV. Excepted Positions. All persons employed by the government shall be required to pass civil service examinations designed to establish their loyalty and fitness to hold positions to which they are appointed, except elected officials. lawyers in good standing in the National Bar Association (when appointed to legal positions in the Law Office of the Government, the Department of Justice, or as justices, judges, or prosecutors). expert advisors of foreign nationality, especially retained for temporary purposes, and other positions specifically excepted by law.

SECTION V. Appointing Authorities. Appointing authorities are persons of the Government who have been given the power to appoint by law.

SECTION VI. Effective Date. This Ordinance shall be effective on the tenth day after the date appearing hereon.

ARCHER L LERCH
Major General United States Army
Military Governor in Korea

153

HEADQUARTERS
UNITED STATES ARMY MILITARY
GOVERNMENT IN KOREA
Office of the Military Governor
Seoul, Korea

ORDINANCE
NUMBER 70

17 May 194

<u>Reprinted Copy</u>

SALE OR CONTRACTS FOR SALE OF FEMALE PERSONS PROHIBITED

SECTION I. *Sale or Contracts for Sale of Female Persons Prohibited.* All sales and contracts for the sale of any female person for any purpose are hereby prohibited. All such sales, contracts or agreements of any kind now existing or entered into heretofore or hereafter are hereby declared to be against public policy and null and void, and without any legal effect whatsoever.

SECTION II. *Collection of Debts Incurred in Connection with Sale of Female Persons.* No action at law, or any other proceeding of any type whatsoever shall be maintained for the collection of any debt incurred in connection with the sale or contract for sale of any female person, and all such debts are hereby declared to be against public policy and null and void, unenforceable, and uncollectible by any means whatsoever. Any attempt to collect such debts, or any payment or acceptance of money or other consideration by any person on account of such debts, is hereby declared to be a violation of this ordinance.

SECTION III. *All Parties to Sale Equally Guilty.* All persons who are agents, accessories or parties to any sale or contract for sale of any female person or any contract or agreement of any kind concerning same, or to the payment or collection of any debt incurred thereby, or who shall violate any of the provisions of this ordinance, shall be deemed to be equally guilty and shall be liable to punishment as principals.

SECTION IV. *Penalties.* Any person violating the provisions of this ordinance shall, upon conviction by a Military Occupation Court, suffer such penalty as the court shall determine.

SECTION V. *Effective Date.* This ordinance shall be effective on the tenth day after the date appearing hereon.

ARCHER L LERCH
Major General United States Army
Military Governor in Korea

154

ORDINANCE
NUMBER 71

18 April 1946

REPRINTED COPY

PUBLIC INFORMATION SECTON IN PROVINCIAL GOVERNMENTS

SECTION I. Establishment of Public Information Section in Provincial Governments. A Public Information Section is hereby established in the Home Affairs Bureau of each Provincial Government in Korea south of 38 North Latitude.

SECTION II. Functions and Duties. Each Provincial Public Information Section shall, within its Province, collect and evaluate information of a public nature on organizations, publications, individuals and activities; and shall disseminate information to the public by utilization of the press, radio, leaflets posters and other means for news release service. Each such Section shall be responsible for keeping its Provincial Military Governor informed regarding public opinion in its Province.

SECTION III. Transfer of Personnel, etc. All personnel records and property in the respective Provincial Governments which pertain to or are being used in public information activities, are hereby transferred to the jurisdiction and control of the respective Provincial Public Information Sections, except branch offices and substations of the Radio Corporation of Korea, and any other organization or agency which may be specifically exempted from the provisions hereof by the Military Governor.

SECTION IV. Effective Date. This ordinance shall be effective on the tenth day after the date appearing hereon.

ARCHER L LERCH
Major General United States Army
Military Governor in Korea

ORDINANCE
NUMBER 72

4 May 1946

OFFENSES AGAINST THE MILITARY GOVERNMENT. *

SECTION I OFFENSES ENUMERATED.

Without limiting the provisions of Proclamation Number 2, GHQ, USAFPAC, 7 September 1945, or the provisions of any ordinance heretofore published, the following are declared to be offenses against the Military Government:

1. Killing, assaulting or wounding any member of the occupying forces or person acting under their authority.

2. Armed or physical resistance to the occupying forces or person acting under their authority.

3. Making hostile or threatening acts or gestures toward any member of the occupying forces or person acting under their authority.

4. Kidnapping, abducting, wrongfully imprisoning, holding for ransom or other wrongful demand, or extorting or wrongfully demanding money or property from any member of the occupying forces or person acting under their authority.

5. Inviting or conducting any member of the occupying forces into a place or area designated "Off Limits" or "Out of Bounds", or supplying goods or services to such member in any such place or area.

6. Procuring, assisting, or advising any member of the occupying forces to be absent without leave, desert, mutiny, or serve as a spy.

7. Bribing or promising, offering or arranging for a bribe or gift of money, property or benefits of any kind to a member of the occupying forces or person acting under their authority; engaging in any similar corruption or attempted corruption; soliciting, receiving or offering to receive, a bribe or gift of money, property or benefits of any kind, for nonperformance of

---

* See Proclamation No 2, GHQ USAFPAC, 7 September 1945.

duty or for the performance of any act, authorized or unauthorized, by or on behalf of the Military Government.

8. Influencing or attempting to influenc the official action or conduct, or contemplated official action or conduc, of the occupying forces or any member thereof or person acting under their authority, by the use of force, duress, threat, promise or withholding or threat of withholding of economic or other benefits, boycott, or similar action.

9. Procuring, or participating, aiding or abetting in any act constituting a breach or violation of duty or orders by any member of the occupying forces or person acting under their authority.

10. Wilfully interfering with or misleading any member of the occupying forces or person acting under their authority, in the performance of his duties.

11. Plunder, pillage or looting.

12. Espionage.

13. Communicating information which may be harmful to the security or property of the occupying forces, or unauthorized possession of such information without promptly reporting it.

14. Communicating, without authority, by code, cipher or any other device or method concealing or disguising the meaning of such communication.

15. Communicating with any person outside of the occupied territory subject to the jurisdiction of USAMGIK, by any means other than those authorized by Military Government.

16. Possession or use of private telegraph apparatus or facilities.

17. Possession or use of unauthorized radio station, apparatus or facilities.

18. Possession or use of unauthorized private telephone line, apparatus or facilities.

19. Sending or receiving unauthorized telegraph, radio or telephone message.

20. Transmission by telegraph or radio operator of uncensored or unauthorized message or broadcast.

21. Sending, carrying or transporting any letter into or outside of the occupied territory subject to the jurisdiction of USAMGIK in any manner

other than through the official mails; failure to promptly report receipt of any letter in any manner other than through the official mails.

19. Acts or conduct in support of, or participating in the formation of, any organization or movement dissolved or declared illegal by, or contrary to the interests of, the occupying forces, including publication or circulation of matter printed or written in aid of any thereof or the possession thereof with intent to publish or circulate same, or the provocative display of flags, uniforms or insignia of any such organization or movement.

20. Assisting any prisoner of war or member of the enemy forces to escape or avoid capture, or harboring or failing to report or accurately disclose the whereabouts of any such person.

21. Escaping from detention, arrest or confinement imposed by the occupying forces or under their authority.

25. Furthering the escape of any person detained, arrested or confined by or under the authority of the occupying forces, or assisting or concealing any such person after escape, or misleading the authorities regarding his whereabouts.

26. Aiding or failing to report any person known to be wanted by the occupying forces or by persons acting under their authority.

27. Transporting, or seeking or obtaining transportation of, persons or property into or outside of the occupied territory subject to the jurisdiction of USAMGIK without authority of the Military Government.

28. Leaving shore in any vessel or otherwise except as authorized by Military Government.

29. Moving any ship or vessel other than on inland waterways exclusively, or any aircraft, except as authorized by Military Government.

30. Removing, obliterating, defacing or altering written, printed or typed matter posted by or under authority of Military Government.

31. Publishing, importing or circulating printed, typed or written matter which is detrimental or disrespectful to the occupying forces or to the United Nations or any member thereof, or tending to promote dissatisfaction or bad feeling with them or which has not been submitted as required; publication or circulation of scandalous matter or malicious libels or slanders against the occupying forces or any of the United Nations, any member thereof, or any person acting under their authority.

32. Disseminating any rumor calculated to alarm or excite the people

or to undermine the morale of the occupying forces or persons acting under their authority.

33. Inciting to or participating in rioting or public disorder.

34. Organizing, promoting, publicizing, aiding or attending any public gathering, parade or demonstration for which no permit has been granted, unless held for religious purposes or in the exercise of functions authorized by the Military Government.

35. Unauthorized possession, control or use of any firearm or other deadly weapon, ammunition, explosive, or similar material or apparatus, or of unauthorized apparatus or means for transmitting messages.

36. Manufacturing, selling, buying, possessing or using arms, ammunition or explosives without permit.

37. Sabotaging, destroying, or damaging any property of the occupying forces or any installation or property occupied by, or necessary or useful to the occupying forces or to any person acting under their authority.

38. Interfering with communication by mail, telegraph, telephone or radio; destruction, larceny or embezzlement of mail; destroying or damaging public telegraph, telephone or radio apparatus or facilities; cutting or otherwise destroying, damaging or interfering with communication wires or cables.

39. Damaging or impairing the utility of railroads, roads, canals, rivers, streams or other media of public transportation, or interfering with their operation or with travel or commerce upon them; damaging or interfering with the operation or utility of waterworks or supply, electric light or power installations or tranmisission lines, gasworks, or the like.

40. Damaging or interfering with the operation of any public service or utility.

41. Destroying, concealing, damaging, disposing of, buying, receiving, unauthorized possession of, or interference with any ship, installation, plant, equipment, property or other economic asset, or plans or records with respect thereto, of the Military Government.

42. Stealing, embezzling, misappropriating, concealing, selling, buying, receiving, pawning or receiving in pawn, obtaining by fraud or false pretenses, destroying, damaging, or wrongful possession or use of property of the occupying forces or of any member thereof or intended for their or his use.

— 4 —

43. Stealing, embezzling, misappropriating, wrongfully applying, negligently losing, or wrongfully receiving or possessing public funds, property, records or papers.

44. Stealing, concealing, transporting, destroying, damaging, transferring, buying, receiving, pawning or receiving in pawn, impairing the value or utility of, or wrongfully possessing property or funds of enemy nations or records pertaining thereto.

45. Wrongfully selling, purchasing, receiving, transporting, possessing or exercising control over, or ortherwise wrongfully engaging in any transaction with respect to any type of property or interest therein formerly owned or controlled by the Government of Japan or any agency, instrumentality or national thereof, title to which has vested in the Military Government.

46. Forcing a safeguard over property of or in the custody of the occupying forces or of persons acting under their authority.

47. Wilfully interfering with, destroying, damaging, removing, concealing, obliterating, falsifying, altering or concealing records or archives of any nature, public or private.

48. Wilfully destroying, damaging, stealing, obtaining by fraud, altering or concealing any work of art, monument or other cultural property created by another.

49. Knowingly making any false or misleading statement, orally or in writing, to any member of, or person acting under authority of, the occupying forces, in a matter of official concern: or in any manner defrauding, misleading or refusing to give information required by the Military Government.

50. Submitting a false, fraudulent, forged or exaggerated claim against or document to the occupying forces or to any government they represent: forging or altering any receipt, voucher, check or other like paper issued by or addressed to the occupying forces, or with which they have any official concern.

51. Making, issuing or knowingly having possession of any false permit, identity card or other document of official concern to the occupying forces; delivering any such matter, whether false or valid, to any unauthorized person or for any unauthorized purpose.

52. Forging or altering an identity card, permit, pass or other similar paper issued by or addressed to the occupying forces or with which they

have any official concern; or making improper use thereof.

53. Falsifying, altering without proper authority, or concealing any official book, document, deed, record or entry.

54. Falsifying, altering, or wrongful possession of a falsified or altered contract, deed or other record of property or evidence of ownership or interest therein or control thereof.

55. Counterfeiting or altering any currency, coin or stamps or having possession of or uttering, transferring or transporting any thereof having reason to believe it to be false or altered; or having possession of, disposing of or transporting any property for any such purpose.

56. Knowingly possessing any illegal or unauthorized currency, coin or stamps.

57. Making a false statement under oath, affirmation or certification, before a Military Occupation Court concerning a material matter.

58. Falsely pretending to be a member of the occupying forces; unlawfully wearing, possessing, disposing of or transporting any uniform or any part of any uniform, whether genuine or false, of the occupying forces or of any of the United Nations.

59. Assumption of governmental authority falsely or without permission of the occupying forces.

60. Impersonating or falsely pretending to be a member of the occupying forces, or falsely pretending to be an official, employee, agent or representative acting under their authority; wrongful use of a false or fictitious name or address.

61. Hoarding, illegally storing or concealing, wrongfully failing to deliver up, or otherwise illegally or improperly dealing in commodities rationed or declared by competent authority to be in critical need.

62. Selling, offering for sale, buying or offering to buy, merchandise or any other property at prices higher than the maximum prices fixed by competent authority.

63. Discriminating against the occupying forces or any member thereof in prices charged or asked for merchandise or services.

64. Charging, asking or collecting prices in excess of those marked.

65. Failing to comply with a demand for a correct bill for merchandise purchased.

66 Interfering in any manner with the free and unimpeded exercise of the franchise or right to vote in accordance with law; bribing voters, or other corrupt practices tending to interfere with or impair the effectiveness of legal, proper and orderly voting.

67. False or fraudulent voting, including multiple voting and assumption of false identity or residence for the purpose of false or fraudulent voting.

68. Failing to comply with orders or requirements of competent authority with respect to sanitation and public health.

69. Committing public nuisances or otherwise acting or forbearing to act so as to endanger injuriously affect public safety, health or welfare.

70. Engaging in or soliciting sexual intercourse with any member of the occupying forces, by a woman suffering from a venereal disease in an infectious stage.

71. Wrongfully growing, manufacturing, selling, buying, or making or accepting a gift of any habit-forming drug, including marijuana.

72. Wrongfully manufacturing, distilling, preparing, transporting, selling, offering or displaying for sale, making a gift of, serving or offering to serve, or otherwise wrongfully dealing in or possessing any beverage or foodstuff containing methyl alcohol or other poisonous, harmful or deleterious substance or ingredient.

73. Failing or refusing to collect taxes or other official misconduct or failure or wilful neglect of duty.

74. Circulating without a permit during curfew.

75. Initiating, carrying out, furthering or participating in any criminal prosecution, disciplinary measure or any other form of punishment, sanction, victimization, or collective action, including boycotting, against any person for cooperating with the occupying forces or any person acting under their authority.

76. Obstructing or contravening any announced program, policy, plan or orders of the Military Government with respect to United Nations nationals in Korea, or assaulting, despoiling or without justification confining any United Nations national or otherwise infringing the rights of such national.

77. Engaging in conduct hostile or disrespectful to the occupying forces, the members thereof or persons acting under their authority, or to any of the United Nations.

ORDINANCE No 72   — 7 —

78. Killing, assaulting, wounding, wrongfully imprisoning, kidnapping, abducting, holding for ransom or other wrongful demand, extorting or wrongfully demanding money or property from, or wrongfully threatening any public minister or accredited diplomatic, consular or military representative, or member or employee of his household, staff or retinue.

79. Acting in contravention of terms imposed by the Allies upon their enemies on their defeat or surrender, or of any orders supplementing such terms.

80. Violating or disobeying any proclamation, law, ordinance, notice, directive or order of the Military Government or of any person acting under the authority thereof or pursuant thereto, where a penalty is not expressly imposed.

81. Engaging in acts to the prejudice of good order or to the interests of the occupying forces or any member thereof.

82. Violating in any way any of the laws of war or acting in aid of the enemy or endangering the security, safety or operations of the occupying forces, any member therof, or any person acting under their authority in the performance of his duties.

SECTION II. PENALTIES.

Any person who commits any of the offences set forth in Section I hereof shall, upon conviction by a Military Occupation Court, suffer such punishment as the court shall determine.

SECTION III. ATTEMPTS AND CONSPIRACIES.

Any person who attempts to commit, or who conspires or agrees with another to commit, any offense enumerated in Section I hereof, or who advises, assists in, arranges or procures the commission of any such offense, or who, having knowlegge of such an alleged offense, fails to report it or assists an alleged offender to avoid arrest or escape detection, conviction or punishment, shall be punishable as a principal.

SECTION IV. RESPONSIBILITY FOR ACTS OF JURIDICAL PERSONS.

Every director, officer or employee of any juridical person, who in any such capacity, either alone or jointly with others, causes, directs, procures, advises, urges, approves, votes in favor of or participates in an act or omission which constitutes an offense for which such juridical person would be triable by a Military Occupation Court, shall be liable therefor as though such act or omission had been done or made in his individual capacity.

SECTION V. DEFENSES.

It shall not be a defense to any charge hereunder that the offense charged was committed under instructions or orders of any civil or military superior of any former government of Korea or of any person purporting to act as an official or member of any dissolved or illegal organization or movement, or that the offense was committed under duress, by reason of threats or through fear.

SECTION VI. DEFINITIONS.

1. The expression "occupying forces" as used herein includes persons subject to military, naval or air force law of any of the United Nations, and any military formation, organization or commission, or civilian agency composed in whole or in part of such persons.

2. The expression "enemy" includes the governments of Japan, Germany, Bulgaria, Hungary and Rumania, and all nationals thereof, whether entitled to belligerent status or not. It does not include Korea or persons resident in Korea who became subject to the sovereignty of Japan by virtue of the annexation of Korea by Japan.

SECTION VII. EFFECTIVE DATE.

This ordinance shall be effective on the tenth day after the date appearing hereon.

ARCHER L LERCH
Major General, United States Army
Military Governor in Korea

ORDINANCE
NUMBER 73   23 April 1946

PROVINCIAL PROPERTY CUSTODY ORGANIZATION

SECTION I. Establishment of Provincial Property Custody Offices. A Provincial Property Custody Office is hereby established in each Provincial Government of Korea south of 38° North Latitude.

SECTION II. Functions and Duties. The functions and duties of each Provincial Property Custody Office are as follows, within its Province:

a. To seek out and take possession of all property vested in the Military Government of Korea by Ordinance Number 33.

b. To operate, manage, utilize, and conserve such vested property, unless the property is under the direct control and jurisdiction of a National Office, Department, or agency.

c. To transfer the control and management of this vested property to a Provincial Bureau or any type of private agency authorized by the Office of Property Custody.

d. To maintain records and accounts of all vested property.

e. To sell, or arrange for the sale of, all perishable vested property whose value cannot otherwise be preserved and maintained.

f. To authorize temporary short-term arrangements for lease, occupancy, or use of vested property, in accordance with instructions of the Office of Property Custody.

SECTION III. Transfer of Duties. etc. All duties, functions, records, property, and civilian personnel of any other Provincial Bureau, Section, or agency Pertaining to, or forming a part of the operations of the Provincial Property Custody Office are hereby transferred to the Provincial Property Custody Office.

SECTION IV. Effective Date. This ordinance shall be effective on the tenth day after the date appearing hereon.

ARCHER L LERCH
Major General United States Army
Military Governor of Korea

ORDINANCE
NUMBER 74

27 April 1946

ABOLITION OF THE OFFICE OF PROVINCIAL AFFAIRS.*

SECTION I. Abolition of Office. The Office of Provincial Affairs of the Government of Korea is hereby abolished.

SECTION II. Transfer of Duties and Functions. The duties and functions of the Office of Provincial Affairs are hereby transferred as follows:

a. Supervision of school grants, subsidies, funds, budgets, and expenditures is transferred to the Department of Education of the Government of Korea.

b. Supervision of taxes, budgets, funds, and expenditures of Provincial and lower levels of government is transferred to the Department of Finance of the Government of Korea.

c. Supervision of national grants and subsidies pertaining to administration of Provincial and lower levels of government is transferred to the Department of Finance of the Government of Korea.

d. Supervision over liquidation of property of Shinto shrines is transferred to the Office of the Property Custodian of the Government of Korea.

e. Supervision of the Emperor's Endowment Fund, hereby renamed the Provincial Foundation Fund, is transferred to the Department of Education of the Government of Korea.

f. All other governmental functions and duties formerly performed by the Office of Provincial Affairs under existing laws will be assigned as the need arises to appropriate Departments or Offices. Such assignments will be by further ordinance or by directives of the Civil Administrator. If any such assignment involves action on more than specific matters, the directives

---

* See Ordinance No. 36, 20 December 1945.

making such assignment for continuing responsibility will be published in the Official Gazette.

SECTION III. <u>Transfer of Records, Property and Civilian Personnel.</u> The records, property, and civilian personnel of the Office of Provincial Affairs are hereby transferred as follows:

a. All copies and translations of Korean and other laws and regulations are transferred to the Department of Justice of the Government of Korea.

b. All records and property relating to schools are transferred to the Department of Education of the Government of Korea.

c. All records and property relating to National Government grants and subsidies to and taxes, budgets, funds and expenditures of Provincial and lower levels of government, except those concerning schools, are transferred to the Department of Finance of the Government of Korea.

d. All records and property relating to the Provincial Foundation Fund are transferred to the Department of Education of the Government of Korea.

e. All records and property relating to Shinto shrines are transferred to the Office of the Property Custodian of the Government of Korea.

f. All records and property relating to Korean civilian personnel are transferred to the Office of Korea Civil Service of the Government of Korea.

g. All other records and property are transferred to the Office of Administration of the Government of Korea.

h. All Korean civilian personnel are transferred as directed by the Office of Korea Civil Service of the Government of Korea for reassignment within the Government of Korea.

SECTION IV. <u>Effective Date.</u> This ordinance shall be effective on the tenth day after the date appearing hereon.

ARCHER L LERCH
Major General United States Army
Military Governor

167

HEADQUARTERS
UNITED STATES ARMY MILITARY
GOVERNMENT IN KOREA
Office of the Military Governor
Seoul, Korea

ORDINANCE
NUMBER 75

7 May 1946

UNIFICATION OF KOREAN RAILWAYS

SECTION I. Purpose. The purpose of this Ordinance is to develop an integrated national system of railway transportation in Korea by unification through government acquisition of certain private railroads for public use.

SECTION II. Condemnation. To effectuate the foregoing purpose, it is hereby declared that all property of the Chosun (Korean) Railroad Company, Limited, the Kyung Nam Railroad Company, Limited, and the Kyung Chun Railroad Company, Limited shall be and it is hereby condemned and taken for the public use of Korea under the supervision of the Department of Transportation of the Government of Korea. Title thereto is hereby vested in the Government of Korea, but fair compensation shall be paid, as hereinafter provided, to all persons entitled thereto.

SECTION III. Right of Compensation. Any person entitled to compensation by reason of the taking of such property of private railroad companies hereunder may file a written claim therefor, submitting evidence concerning the description, location and value of such property, the nature and extent of his right, title or interest thereto or therein, and the exact amount of compensation in yen claimed to be due to him for the taking of the same, with the Director of the Department of Transportation, within sixty (60) days after the effective date of this ordinance, or shall be deemed to have waived and abandoned all rights to such compensation and to the filing of any claim with respect thereto.

SECTION IV. Valuation Board. The Director of the Department of Transportation shall appoint three capable and disinterested persons as a Valuation Board to study the claims made, conduct necessary inspections and investigations, analyze the books, accounts and records of the companies and all other available pertinent evidence, and impartially and to the best of their judgment ascertain, appraise and recommend to the Director of the

168

Bureau of Transportation a fair and just valuation for such property as of the date of taking hereunder, and determine and direct who is entitled to receive compensation therefor, according to each claimant's respective ownership or other interest in such property. The Board shall report its findings as to the value of the Property taken, the persons entitled to compensation therefor, and the amount recommended as awards to each, to the Director of the Department of Transportation as soon as the facts can be ascertained. The Director shall have power to modify the valuations fixed by the Board in its report, to order the Board to make further inquiries or findings of facts, and to issue other necessary orders to carry into effect the purposes of this ordinance.

SECTION V. Certificates of Compensation. Upon final determination of the amount of compensation to be awarded to any person for the taking of property hereunder, the Director of the Department of Transportation shall deliver to each such person a non-interest bearing certificate expressed in yen, representing the amount awarded, payable by the Government of Korea within five (5) years, or payable in whole or in part at any earlier date at the option of the Government of Korea. The certificates shall contain a provision that payments shall be made in legal tender and that the amount of each payment may be adjusted by reason of any change in the officially-recognized purchasing power of such legal tender since the date of issuance of the certificate.

SECTION VI. Operation. The Director of the Department of Transportation shall allocate and assign all properties title to which is vested in the Government of Korea hereunder, for management and operation under the control and supervision of appropriate Bureaus of the Department.

SECTION VII. Employees. Until further orders, all employees of the private railroad companies affected hereby, and all employees of subsidiaries of such companies, are deemed persons engaged in essential services within the meaning of Article II of Proclamation No. 1, GHQ, USAFPAC, 7 September 1945, As such, they will continue to perform their usual functions and Duties, will preserve and safeguard all records and property of such railroads, and will obey promptly all orders of compenent authority.

SECTION VIII. Effective Date. This Ordinance shall be effective on the tenth day after the date appearing hereon.

ARCHER L LERCH
Major General United States Army
Military Governor in Korea

169

ORDINANCE
NUMBER 76

27 April 1946

CUSTOMS ADMINISTRATION

SECTION I. Purpose. The purpose of this ordinance is to provide a single customs administration in Korea south of 38° North Latitude.

SECTION II. Responsibility. To effectuate the foregoing purpose, all responsibility for customs control and operation and for supervision of all personnel connected therewith is hereby assigned to the Department of Finance of the Government of Korea.

SECTION III. Transfer of Customs Section. All duties, functions, records, property and civilian personnel of the Customs Section of the Marine Bureau of the Department of Transportation of the Government of Korea are hereby transferred to the Department of Finance of the Government of Korea.

SECTION IV. Effective Date. This ordinance shall be effective on the tenth day after the date appearing hereon.

ARCHER L LERCH
Major General United States Army
Military Governor in Korea

ORDINANCE
NUMBER 77   24 April 1946

SPECIAL RICE PROVISIONS FOR CITY OF SEOUL

SECTION I. Purpose and Scope. -- The purpose of this ordinance is to alleviate the existing critical rice shortage in the City of Seoul by providing a method for the legal transportion of rice into Seoul during such emergancy.

SECTION II. Procedure for Obtaining Permit to Transport Rice into City of Seoul.

a. Any person having a fixed and regular residence within the City of Seoul may obtain an application from his Chung head for a permit to bring a limited amount of rice into the City of Seoul. The amount specified on the application form, together with the quantity of rice in his possession in the City of Seoul, on the date of such application, shall not exceed two (2) hop per day from the date of such application until 15 October 1946 multiplied by the number of persons regularly and actually residing with such applicant as members of his household. The Chung head will certify to the accuracy of the statements contained in the application form by endorsing his written certificate thereon.

b. The applicant will then submit such certified application form to the appropriate Ban Chang for approval.

c. The applicant will then submit such certified and approved application form to the approriate Chung Office for approval and registration. If the application is approved by such Chung Office, the names of the applicant and such members of his household as are included and involved will be removed from the rice ration system for the period covered by the amount of rice, calculated at two (2) hop per individual per day as hereinabove provided, shown on the approved application form.

d. The applicant will present such approved application to the appropriate City District Office, which will issue a permit to allow the applicant to transport rice into Seoul in accordance with the terms of the approved

— 1 —

171

and registered application.

e. The Chung Office shall notify the Seoul Distributing Center as to the names and addresses of all applicants who have been issued permits, or on whose behalf permits have been issued, to transport rice into the City of Seoul, so that such Distributing Center will not make sales of rice to or for such persons during the periods covered by such permits.

f. The Chung head, the Chung Office, and the City District Office shall maintain accurate records of all permits to transport rice granted hereunder, so that the amounts specified therein and obtained by each applicant will be charged against the allotment available to such applicant (and the members of his household, as hereinabove described) upon any subsequent application.

g. Immediately upon delivery of any rice hereunder, the holder of such permit shall indorse thereon the exact quantity received. Failure to so indorse the permit will subject such permit, together with all rice in the possession of the holder, to confiscation and the annulment of all rights of such person to any further permit or purchase of rice at the City Distributing Center.

h. Two or more holders of permits to transport rice hereunder may transport such rice jointly. Employers, companies, organizations and associations may transport rice for their employees or members who hold permits hereunder.

SECTION III. <u>Duration and Surrender of Permits.</u> -- No permit to transport rice into the City of Seoul, issued hereunder, shall be issued or valid for a period in excess of fourteen days after the date of issuance thereof, unless otherwise provided in writing on the approved application form by the District Office for good cause shown at the time of the original approval thereof by such District Office. No permit shall be used or accepted after such fourteen-day period, unless a longer period has been specified by the District Office as herein provided. All permits shall be surrendered by the holders thereof to their Chung heads not later than the expiration of the term specified therein, together with a certificate of the holder as to the amount of rice actually obtained, as hereinabove provided. In every case where such holder fails to surrender such permit as herein provided, together with such certificate in proper form, it shall be conclusively presumed that the holder of such permit has actually obtained the full amount of rice specified in such permit.

SECTION IV. <u>Inspections: Confiscation: Penalties</u> -- All persons transporting rice shall submit to inspection of permits and rice at inspection stations established on highways and roads, at inspection centers at railroad

— 2 —

stations, and elsewhere as required by the police. Where the amount of rice transported exceeds the amount authorized by the permit, or where the permit is falsely or fraudulently indorsed or altered, all rice in the possession of the person transporting such excess or possessing such false or altered permit shall be confiscated by the police, upon receipt given to such person, and all such rice shall forthwith be delivered by the police to the nearest city or government warehouse for distribution and sale under the ration system, Receiving warehouses shall keep records of all such rice received from the police. Both police and receiving warehouses shall forthwith report the facts of all seizures to the appropriate Chung heads and the City Distributing Center. Persons transporting such excess rice, and Persons having in their possession a falsified or altered permit or a permit which they are not entitled to possess, shall, upon conviction by a Military Occupation Court, suffer such punishment as the Court shall determine.

SECTION V. Effect Upon Ordinance No 45. Dated 25 January 1946. — Legal operations under this ordinance shall not be deemed to be a violation of Ordinance No 45, dated 25 January 1946.

SECTION VI. Effective Date. — This ordinance shall be effective on the date appearing hereon.

ARCHER L LERCH
Major General United States Army
Military Governor in Korea

ORDINANCE
NUMBER 78   30 April 1946

TOLLING OF STATUTE OF LIMITATIONS

SECTION I. Suspension of Statute of Limitations. The running of the Statute of Limitations on or after 9 August 1945 against any claim or right to institute suit against any person (natural or juridical) is hereby suspended:

a. During the time of absence of such person from Korea south of 38° North Latitude, or

b. During the time such person cannot be located in Korea south of 38° North Latitude by the exercise of due diligence, or

c. During the time such person conceals his whereabouts.

ARCHER L. LERCH
Major General United States Army
Military Governor in Korea

HEADQUARTERS
UNITED STATES ARMY MILITARY
GOVERNMENT IN KOREA
Office of the Military Governor
Seoul, Korea

ORDINANCE
NUMBER 79

30 April 1946

AMENDING ORDINANCE NUMBER 55

SECTION I. Ordinance Number 55 Amended. -- Ordinance Number 55, dated 23 February 1946, is hereby amended by eliminating Section III b therefrom.

SECTION II Effective Date. -- This ordinance shall be effective on the date appearing hereon.

ARCHER L LERCH
Major General United States Army
Military Governor in Korea

HEADQUARTERS
UNITED STATES ARMY MILITARY
GOVERNMENT IN KOREA
Office of the Military Governor
Seoul, Korea

ORDINANCE
NUMBER 80

7 May 1946

Ordinance No 52 is amended to read as follows:

CREATION OF THE NEW KOREA COMPANY LIMITED

SECTION I. The New Korea Company, Limited, is hereby created as a corporate instrumentality of the Military Government of Korea. It is created for the sole purpose of promoting the economy of the people of Korea. The corporation will be directed by ten directors who will hold office until replaced by the Military Governor. Such Board of Directors will operate the corporation under by-laws and regulations prescribed by it with the approval of the Military Governor. The principal office of the corporation will be in Seoul, Korea, and it will have such branch offices as the Board of Directors shall determine with the approval of the Military Governor. Upon the termination of the Military Government of Korea the New Korea Company, Limited, will become the responsibility of the Korean government.

SECTION II. The directors will appoint an Advisory Board consisting of the Chairman of the Board of Directors of the Corporation, who will serve as Chairman, and one or more members from each province nominated by the Military Governor. The Advisory Board will meet at the call of the Chairman. The Advisory Board will make recommendations to the Board of Directors. The Board of Directors will consult the Advisory Board on major questions of policy.

SECTION III. The corporation will have a capital stock of one hundred million yen (¥ 100,000,000) divided into shares of fifty yen (¥ 50) each. The total amount of the capital stock will be subscribed, exclusively owned, and paid in full by the Military Government of Korea by the physical transfer of property of the Oriental Development Company, Limited. Its liability will be limited to its capital stock. It may be dissolved only by the Military Government of Korea or the future Korean government. To qualify the directors, a certificate for one share of stock will be transferred on the books

— 1 —

175

of the Company in the name of each Director.

SECTION IV. All property owned on or since 9 August 1945 by The Oriental Development Company, Limited, is here vested in the New Korea Company, Limited.

SECTION V. The corporation will have the following powers:

a. To acquire and own its own shares of stock at a price not less than the par value thereof.

b. The capital stock may be reduced or increased by the Board of Directors by the issuance of a public notice signed by the Military Governor.

c. All of the powers of a Limited Company under the Commercial Code now or hereafter in force in Korea.

d. To issue debentures under the provisions of the Commercial Code, except that debentures not in excess of ten times the amount of the paid-up capital may be issued, when the issuance, and terms of the debentures have been approved by the Military Governor.

e. To appoint and compensate such officers and employees as the Board of Directors may determine.

f. To engage in all forms of industrial activity, including the extraction of metallic and non-metallic natural resources, the processing and fabrication of materials and chemicals, the manufacture of any articles for industrial or commercial purposes, machinery, motive and transportation equipment, ships and aircraft parts and equipment, electrical equipment and appliances, textiles, and food processing.

g. To transact all forms of commercial business at all levels of distribution, warehousing and storage

h. To engage in all forms of agriculture, including ownership or operation of leased lands, agricultural experimental stations or laboratories, and in all forms of forestry, including lands used for the propagation of nursery stock and lands under reforestation.

i. To engage in all forms of fishing and acquisition of aquatic products, including the ownership or operation under lease or charter of fishing boats, docks, and other fishing gear and facilities for hauling, storing and preserving fish and aquatic products.

j. Under the directives of the Military Governor to regulate and supervise, including the furnishing of technical assistance to, any business engaged in activities within the scope of the powers of the New Korea Company, Limited.

k. To own property and dispose of the same.

l. All other powers incidental or necessary to the accomplishment of the objects of the corporation.

— 2 —

<u>SECTION VI.</u> The corporation, its directors, officers and employees will be subject to such duties and will exercise such powers as are provided by the Commercial Code insofar as they are pertinent or necessary to the business of the corporation.

<u>SECTION VII.</u> The corporation will issue its public notices by publication in a newspaper of general circulation.

<u>SECTION VIII.</u> This ordinance shall be effective on the date appearing hereon.

ARCHER L LERCH
Major General United States Army
Military Governor in Korea

ORDINANCE
NUMBER 81   17 May 1946

ABOLITION OF THE OFFICE OF PLANNING

SECTION I.   Abolition of Office.   The Office of Planning of the Government of Korea is hereby abolished.

SECTION II.   Transfer of Analysis Division.   All duties, functions, records and property of the Analysis Division of the Office of Planning are hereby transferred to the Office of the Civil Administrator of the Government of Korea. Korean personnel of the Analysis Division are hereby transferred to the National Food Administration.

SECTION III.   Transfer of Statistics, Research and Administration Divisions.   All duties, functions, records, property and Korean personnel of the Statistics, Research and Administration Divisions of the Office of Planning are hereby transferred to the Office of Administration of the Government of Korea.

SECTION IV.   Transfer of Civilian Supply Division.   All duties, functions, records and property of the Civilian Supply Division of the Office of Planning will be transferred by directives of the Civil Administrator to be published in the Official Gazette upon approval by the Military Governor.

SECTION V. Effective Date.   This ordinance shall be effective on the tenth day after the date appearing hereon.

ARCHER L. LERCH
Major General United States Army
Military Governor in Korea

ORDINANCE
NUMBER 82

17 May 1947

REGULATION OF FOREIGN TRADE

SECTION I. **Purpose.** The purpose of this ordinance is to implement Military Government control and licensing of foreign trade in accordance with Ordinance No. 39, dated 3 January 1946.

SECTION II. **Designation.** To effectuate this purpose the Bureau of Commerce of the Department of Commerce of the Government of Korea is hereby designated as the authorized agent of the Military Government of Korea for the control and licensing of foreign trade, in accordance with Sections I to IV of Ordinance No. 39.

SECTION III. **Duties.** The Bureau of Commerce will:

a. Maintain liaison with the Office of Foreign Affairs and the National Economic Board with respect to foreign trade agreements and export and import programs and policies.

b. Formulate and administer necessary regulations to control and license individuals and firms engaging in foreign commerce, within the scope of approved foreign trade agreements and export and import programs of the Military Government of Korea.

c. Use funds allotted to it to implement the processing of foreign trade transactions which cannot be handled by private capital, subject to the approval of the Military Governor in Korea.

SECTION IV. **Regulations.** All regulatory measures and directives of the Bureau of Commerce in connection with the control and licensing of foreign trade hereunder will be published in the Official Gazette upon approval by the Military Governor in Korea.

SECTION V. **Effective Date.** This ordinance will be effective on the tenth day after the date appearing hereon.

ARCHER L LERCH
Major General United States Army
Military Governor in Korea

HEADQUARTERS
UNITED STATES ARMY MILITARY
GOVERNMENT IN KOREA
Office of the Military Governor
Seoul, - Korea

ORDINANCE
NUMBER 83   13 May 1946

<u>Reprinted Copy</u>

LICENSING OF PUBLIC BATHS AND ESTABLISHMENTS
SERVING FOOD OR DRINK

SECTION I. *Transfer of Functions.* All duties, functions, records and property of all Divisions, Department of Police, pertaining to sanitary inspections, the licensing of public baths and the licensing of business establishments serving food or non-alcoholic beverages are hereby transferred to the Provincial Departments of Public Health and Welfare.

SECTION II. *Applications for Public Bath Licenses.* Applications for a public bath license will be accompanied by a certificate from the Department of Police that public morals are being properly safeguarded and protected in the establishment proposed to be licensed.

SECTION III. *Additional Licenses.* Establishments serving food or non-alcoholic beverages, engaged in activities in addition to the serving of food or non-alcoholic beverages, shall obtain additional and separate licenses for such additional activities, from the Department of Police or other appropriate authorities, and shall otherwise comply with all requirements of law.

SECTION IV. *Effective Date.* This ordinance shall be effective on the tenth day after the date appearing hereon.

ARTHUR S CHAMPENY
Colonel   Infantry
Acting Military Governor in Korea

HEADQUARTERS
UNITED STATES ARMY MILITARY
GOVERNMENT IN KOREA
Office of the Military Governor
Seoul, Korea

ORDINANCE
NUMBER 84

13 May 1946

Reprinted Copy

CHANGES IN LOCAL GOVERNMENT

SECTION I. *Elevation of Certain Towns to Cities.* The following towns are hereby elevated to the status of cities:

a. The town of Chongju in North Chungchong Province.

b. The town of Chunchon in Kangwon Province.

SECTION II. *Change of Certain County Names.* The names of the following counties are hereby changed as indicated:

a. The county of Chonju in North Chungchong Province to Chong-won County.

b. The county of Chunchon in Kangwon Province to Chunsong County.

SECTION III. *Special Status of Saju Village in North Chungchong Province.* The following special provisions shall apply to Saju Village in Chong-won County in North Chungchong Province:

a. The inhabitants of Saju Village are authorized to use the educational facilities of the city of Chongju.

b. School taxes shall be levied on inhabitants of Saju Village by the city of Chongju instead of by Chong-won County.

SECTION IV. *Effective Date.* This ordinance shall be effective on 1 June 1946.

ARTHUR S CHAMPENY
Colonel   Infantry
Acting Military Governor in Korea

ORDINANCE
NUMBER 85

AMENDING ORDINANCE NUMBER 67

Ordinance Number 67 is amended to read as follows:

SECTION I. *Legal Division of Office of General Affairs Transferred to Department of Justice.*

All duties, functions, records, property and civilian personnel of the Legal Division of the Office of General Affairs are hereby transferred to the Department of Justice.

SECTION II. *Functions of Department of Justice.*

The Director of the Department of Justice is the legal adviser of the Military Governor in Korea. He shall advise the Military Governor on policies affecting the national law establishment, and regarding the best qualified persons for appointment by the Military Governor as Justices of the Supreme Court and Judges of Courts of Review, and concerning the legality of governmental policies and proposed laws, ordinances and regulatory measures. He shall supervise the administration of law and legal institutions. He shall pass upon the language, form and legal authority for the issuance and legal effectiveness of all official documents of the government that are sealed and documented, except orders and decisions of the courts.

He shall represent the government in all cases before the courts.

He shall appoint, with the approval of the Military Governor and from nominations submitted by the Central Council of the National Bar Association, judges of courts other than the Supreme Court and Courts of Review, and all prosecutors. With the approval of the Military Governor and subject to Civil Service Rules and Regulations he shall appoint and supervise the chief wardens and administrative officers of the penal and juvenile institutions, members of parole and probation boards, the Librarian of the National Law Library, and the Editor in Chief of Court Reports. He shall appoint and supervise a staff of lawyers to furnish legal advice to Departments and Offices of the Military Government, and to Provincial Governors, to conduct legal research, render legal opinions and draft all legal documents of the government and its agencies. Within policies established by the Military Governor in Korea, he shall supervise the training and disciplining of lawyers, and the training of judicial and court clerk personnel. He shall certify the admission of applicants to practice law who have met the requirements for admission to the legal profession.

SECTION III. *Administrative Division of Office of General Affairs Transferred to Office of Accounts; Office of General Affairs Abolished.*

All duties, functions, records, property and civilian personnel of the Administrative Division of the Office of General Affairs are hereby transferred to the Office of Accounts. All duties and functions of the Office of General Affairs having been transferred, the Office of General Affairs is hereby abolished.

SECTION IV. *Name of Office of Accounts Changed to Office of Administration.*

The Office of Accounts shall hereafter be known as the Office of Administration.

SECTION V. *Division of Accounts of Office of Administration Transferred to Finance Department.*

All duties, functions, records, property and civilian personnel of the Division of Accounts of the Office of Administration are hereby transferred to the Department of Finance.

SECTION VI. *Effective Date.*

This ordinance shall be effective on the date appearing hereon.

ARCHER L. LERCH
Major General United States Army
Military Governor in Korea

ORDINANCE
NUMBER 86   15 June 1946
KOREAN CONSTABULARY
AND KOREAN COAST GUARD

SECTION I. *Department of National Defense Renamed**; *Bureau of Armed Forces Disestablished***. The Department of National Defense of the Government of Korea* is hereby renamed the Department of Internal Security. The Bureau of Armed Forces** of the Department of Internal Security of the Government of Korea is hereby abolished. Section II of Ordinance Number 28, dated 13 November 1945, is hereby repealed.

SECTION II. *Korean Constabulary*. The Korean Constabulary is hereby established and is activated, effective 14 January 1946, as a National Agency for the purpose of providing a reserve police force for the Government of Korea for maintaining internal security. The Korean Constabulary shall be subject to the control of the Korean Constabulary Bureau of the Department of Internal Security.

SECTION III. *Korean Coast Guard*. The Korean Coast Guard is hereby established and is activated, effective 14 January 1946, as a marine unit for the Government of Korea for maintaining inshore and interisland patrol of Korean coastal waters. The Korean Coast Guard shall be subject to the control of the Korean Coast Guard Bureau of the Department of Internal Security.

SECTION IV. *Articles for Government of Korean Constabulary*. The Articles for the Government of the Korean Constabulary are hereby established, and shall at all times and in all places govern the Korean Constabulary.

SECTION V. *Articles for Government of Korean Coast Guard*. The Articles for the Government of the Korean Coast Guard are hereby established, and shall at all times and in all places govern the Korean Coast Guard.

SECTION VI. *Effective Date*. This ordinance shall be effective on the date appearing hereon.

ARCHER L LERCH
Major General United States Army
Military Governor in Korea

* Ordinance No 64, 29 March 1946, Section IIg.
** Ordinance No 28, 13 November 1945, Section II.

HEADQUARTERS
UNITED STATES ARMY MILITARY GOVERNMENT IN KOREA
Office of the Military Governor
Seoul, Korea

ORDINANCE
NUMBER 87

20 May 1946

SPECIAL RICE PROVISIONS FOR CITY OF PUSAN

SECTION I. *Purpose and Scope.* The purpose of this ordinance is to alleviate the existing critical rice shortage in the City of Pusan by providing a method for the legal transportation of rice into Pusan during such emergency.

SECTION II. *Procedure for Obtaining Permit to Transport Rice into City of Pusan.*

a. Any person having a fixed and regular residence within the City of Pusan may obtain an application from his Block Association head for a permit to bring a limited amount of rice into the City of Pusan. The amount specified on the application form, together with the quantity of rice in his possession in the City of Pusan, on the date of such application, shall not exceed two (2) hop per day from the date of such application until 15 October 1946 multiplied by the number of persons regularly and actually residing with such applicant as members of his household. The Block Association head will certify to the accuracy of the statements contained in the application form by indorsing his written certificate thereon.

b. The applicant will then submit such certified application form to the Pusan Ration Board at the City Government Building, Pusan, which will issue a permit to allow the applicant to transport rice into Pusan in accordance with the terms of the approved and certified application. If the permit is issued by the Pusan Ration Board, the applicant will be required to surrender the ration cards of all persons on whose behalf the application has been made and the permit granted, to such Ration Board prior to delivery of such permit to the applicant. If such permit is issued, such Ration Board will remove the names of the applicant and all persons on whose behalf the application has been made and the permit granted, from the rice ration system for the period covered by the amount of rice, calculated at two (2) hop per individual per day as hereinabove provided, shown on the approved application form and permit.

c. The Block-Association head and the Pusan Ration Board shall maintain accurate records of all permits to transport rice granted hereunder, so that the amounts specified therein and obtained by each applicant will be charged against the allotment available to such applicant (and the members of his household, as hereinabove described) upon any subsequent application.

d. Immediately upon delivery of any rice hereunder, the holder of such permit shall indorse thereon the exact quantity received. Failure to so indorse the permit will subject such permit, together with all rice in the possession of the holder, to confiscation, revocation of ration card, and annulment of all rights of such person to any further permit, or ration card for the period covered by such ration card.

e. In the event the full amount of rice specified in any such permit is not actually obtained, the holder thereof may apply for the return of his ration card and those of all persons on whose behalf the application was made and the permit granted. In such case, each such card will be appropriately marked, prior to the return thereof by the Ration Board, so as to indicate the receipt of an amount of rice equal to the total amount actually received under the permit divided by the total number of ration cards surrendered by the applicant at the time of the issuance of the permit by the Ration Board. In the event the holder of any such ration card is not entitled to such amount by reason of rations already credited to him, an amount equal to the remainder to which he is entitled will be charged against him, the card will be retained or cancelled by the Ration Board, and the remaining amount of rice will be charged proportionately to the remaining ration card holders.

f. Two or more holders of permits to transport rice hereunder may transport such rice jointly. Employers, companies, organizations and associations may transport rice for their employees or members who hold permits hereunder.

SECTION III. *Duration and Surrender of Permits.* No permit to transport rice into the City of Pusan, issued hereunder, shall be issued or valid for a period in excess of fourteen days after the date of issuance thereof, and no permit shall be used or accepted after the expiration of the period specified therein. All permits shall be surrendered by the holders thereof to the Pusan Ration Board not later than the expiration of

— 2 —

the term specified therein, together with a certificate of the holder as to
the amount of rice actually obtained, as hereinabove provided. In every
case where such holder fails to surrender such permit as herein provided,
together with such certificate in proper form, it shall be conclusively presumed that the holder of such permit has actually obtained the full amount
of rice specified in such permit.

SECTION IV. *Inspections; Confiscation; Penalties.* All persons transporting rice shall submit to inspection of permits and rice at inspection
stations established on highways and roads, at inspection centers at railway
stations, and elsewhere as required by the police. Where the amount of rice
transported exceeds the amount authorized by the permit, or where the
permit is falsely or fraudulently indorsed or altered, all rice in the possession of the person transporting such excess or possessing such false or altered
permit shall be confiscated by the police, upon receipt given to such person,
and all such rice shall forthwith be delivered by the police to the nearest
city or government warehouse for distribution and sale under the ration system.
Receiving warehouses shall keep records of all such rice received from the
police. Both police and receiving warehouses shall forthwith report the facts
of all seizures to the appropriate Block Association heads and the Pusan
Ration Board. Persons transporting such excess rice, and persons having in
their possession a falsified or altered permit or a permit which they are not
entitled to possess, shall, upon conviction by a Military Occupation Court,
suffer such punishment as the Court shall determine.

SECTION V. *Effect Upon Ordinance No 45, Dated 25 January 1946.*
Legal operations under this ordinance shall not be deemed to be a violation
of Ordinance No 45, dated 25 January 1946.

SECTION VI. *Effective Date.* This ordinance shall be effective on the
date appearing hereon.

ARCHER L LERCH —
Major General United States Army
Military Governor in Korea

HEADQUARTERS
UNITED STATES ARMY MILITARY
GOVERNMENT IN KOREA
Office of the Military Governor
Seoul, Korea.

ORDINANCE
NUMBER 88   29 May 1946

LICENSING OF NEWSPAPERS
AND OTHER PERIODICALS

SECTION I. Publication of Newspapers or Other Periodicals Without License Unlawful. -- No person, natural or juridical, shall, personally or through any agent, print, publish, issue, circulate, distribute, sell or offer to sell, or post, exhibit or display any newspaper or other periodical which is not licensed as hereinafter provided.

SECTION II. Applications for License. -- a. Information Required in Applications. -- Every person, natural or juridical, desiring a license to publish a newspaper or other periodical, shall file a written application setting forth the following information:

    (1) Name and address of applicants;

    (2) Title, frequency of issue, designated location of principal office, and purposes of the publication;

    (3) Names and addresses of all members of the editorial and reportorial staff, business staff, owners, and, if the publication is owned by a juridical person, names and addresses of all stockholders; and

    (4) Names and addresses of all bondholders, mortgagees, and other security holders.

b. Place of Filing Applications; Licensing Authority. -- The application shall be filed with the Department of Commerce of the Government of Korea, Seoul. The licensing authority shall be the Director of the Department of Commerce.

c. Licensing of Existing Newspapers and Periodicals. -- All newspapers and other periodicals now being published or circulated shall be required to display a license as herein provided, at all times on and after 30 June 1946.

— i —

d. _Reports of Change._ Changes in any of the information furnished in the application will be filed in writing within 10 days with the licensing authority.

SECTION III. _Notice Requirements._ a. _Display of Licenses._ The license issued to any newspaper or other periodical shall at all times be conspicuously displayed in its office designated in the application.

b. _Publication of Statement of Compliance._ A statement shall be printed in a prominent place in each issue of every newspaper or other periodical, in the following form:

> " (NAME OF NEWSPAPER OR OTHER PERIODICAL,
> WITH DISTINCTIVE SYMBOL OR EMBLEM)
> " Published (daily, or other applicable period) by (name and address of publisher) at (name and address of printer).
> " Responsible Editor: (Name)
> " Editorial Staff: (Titles and names).
> " License Number: (Number)

c. _Delivery of Copy to Licensing Authority._ One copy of each issue of every newspaper or other periodical shall be delivered or mailed to the licensing authority on the day of publication thereof. No charge shall be made for such copies.

SECTION IV. _Revocation or Suspension of Licenses._ The license of any newspaper or other periodical may be revoked or suspended by the licensing authority for any of the following reasons:

a. The making of any false or misleading statement or omission in the application for a license ;

b. Failure to report any change in the information furnished in the application, as hereinabove required ;

c. Violation of law.

Upon receipt of notice of revocation or suspension of license, the newspaper or other periodical shall forthwith surrender and deliver up its license to the licensing authority, and shall immediately discontinue publication and circulation. Publication and circulation shall not be resumed until such license is returned, reinstated or reissued by the licensing authority.

SECTION V. _Foreign Newspapers and Periodicals._ Newspapers or other periodicals published outside of Korea south of 38° North Latitude may, under the direction of the Director of the Department of Commerce of the Government of Korea, on application of the publishers or their agent, be circulated as if licensed hereunder. Nothing herein contained shall be so

— 2 —

construed as to allow the circulation of any publication in violation of any valid and subsisting copyright.

SECTION VI. "Newspaper or Other Periodical" Defined. "Newspaper or other periodical" as used herein means any publications (other than a re-issued book or pamphlet) issued more than once in any year and devoted to the dissemination of information or opinion of a public character or of public interest. It includes any publication or publications issued by the same person or organization for the same or similar purposes, or under the same control. It does not include directories, catalogues, announcements or other regular publications designed exclusively for advertising purposes, or publications of the Government of Korea or any of its political subdivisions or instrumentalities.

SECTION VII. Penalties. Any person violating the provisions of this ordinance shall, upon conviction by a Military Occupation Court, suffer such punishment as the Court shall determine.

SECTION VIII. Effective Date. This ordinance shall be effective on the date appearing hereon.

ARCHER L LERCH
Major General United States Army
Military Governor in Korea

HEADQUARTERS
UNITED STATES ARMY MILITARY
GOVERNMENT IN KOREA
Office of the Military Governor
Seoul, Korea

ORDINANCE
NUMBER 89

20 April 1946

ISSUE OF KOREAN EMANCIPATION
COMMEMORATIVE POSTAGE STAMPS;
DISCONTINUANCE OF FORMER ISSUES

SECTION I. Korean Emancipation Commemorative Postage Stamp Issue
Authorized: The issuance and use of Korean Emancipation Commemorative
postage stamps are hereby authorized. Such postage stamps shall be placed
on sale in all Post Offices on 1 May 1946 and shall be in the following
denominations and distinctive colors:

Denomination	Color
2 cheun	Yellow
5 cheun	Green
10 cheun	[illegible]
20 cheun	[illegible]
50 cheun	Purple
1 weun	Brown

SECTION II. Imperial Japanese and Korean Overprinted Postage
Stamps Discontinued: The sale of Imperial japanese and Korean overprinted
postage stamps for postal use shall be discontinued at 2400, 30 April 1946.
Such postage stamps shall not be used for postal purposes after 2400, 30 June
1946.

SECTION III. Effective Date: This ordinance shall be effective on
the date appearing hereon.

ARCHER L LEECH
Major General United States Army
Military Governor in Korea.

ORDINANCE
NUMBER 90

28 May 1946

ECONOMIC CONTROLS

SECTION I. *Purpose.* The purpose of this ordinance is to clarify spheres of responsibility and to define functions and relationships of elements of the Government of Korea engaged in activities involving economic controls.

SECTION II. *National Economic Board.*

a. *Establishment and Organization.* The National Economic Board is hereby established. It shall consist of six members, including a Chairman and an Executive Secretary, and the Directors of the Departments of Agriculture, Commerce, Finance and Transportation. The Chairman and Executive Secretary shall be appointed by the Military Governor, to hold office at his pleasure. The Chairman is authorized to appoint, compensate, suspend and remove assistants and personnel for the staff of the Board. Paid officials of the government serving on or for the Board shall receive no additional compensation for such service.

b. *Duties and Functions.* The Board shall:

(1) Assist the Military Governor in harmonizing policies, plans and actions of all elements of the Government pertaining to economic controls;

(2) Conduct national economic planning, determining the potentials and goals of production, the character and volume of imports and exports, allocation of producers' materials, plant, machinery, equipment, power, fuel and labor, the distribution and utilization of imported and locally produced commodities, the stabilization of prices and the rationing of consumers' goods, and the collection or control of goods in short supply;

(3) Develop and supervise at the policy level any economic controls affecting prices, production, distribution, labor, and the coordination of such controls throughout Military Government, as may be authorized by the Military Government;

(4) Collect, analyze, correlate and make available to the Military Governor data necessary to the planning it undertakes;

(5) Issue, subject to the approval of the Military Governor, such regulations, directives and orders as are necessary to establish the policies governing economic action by all agencies of the Military Government and to designate means of accomplishing such action.

c. *Publication of Regulations.* Regulations of the National Economic Board shall be published in the Official Gazette upon approval by the Military Governor and shall have the force of law.

SECTION III. *Korean Economic Advisory Board*

a. *Section III of Ordinance Number 53 Repealed.* Section III of Ordinance Number 53, dated 20 February 1946, is hereby repealed.

b. *Duties and Functions.* The Korean Economic Advisory Board shall meet at the call of the Chairman, shall make studies and recommendations to the Military Governor on economic matters, and shall furnish the Military Governor and the National Economic Board with advice and assistance as required.

SECTION IV. *National Price Administration*

a. *Establishment and Organization.* The National Price Administration is hereby established. It shall consist of a National Price Administrator, appointed by the Military Governor, to hold office at his pleasure. The Administrator is authorized to appoint, compensate, suspend and remove assistants and personnel for the staff of the Administration. Paid officials of the government shall receive no additional compensation for such service.

b. *Duties and Functions.* The Administration shall:

(1) Conform its policies, plans and programs to the policies, plans and programs of the National Economic Board;

(2) Make recommendations for the maintenance of relative price stability and the protection of the consumer interest;

(3) Maintain liaison and coordinate with all political subdivisions, agencies and instrumentalities of the Government of Korea concerned with price policies and programs;

(4) Fix prices within pricing principles and formulae established by the National Economic Board, formulate other measures in connection with price policies and programs, and direct the execution of such measures.

c. *Publication of Regulations.* Regulations of the National Price

— 2 —

Administration shall be published in the Official Gazette upon approval by the Military Governor and shall have the force of law.

SECTION V. *National Food Administration.*

a. Establishment and Organization The National Food Administration is hereby established. It shall consist of a National Food Administrator, appointed by the Military Governor, to hold office at his pleasure. The Administrator is authorized to appoint, compensate, suspend and remove assistants and personnel for the staff of the Administration. Paid officials of the government shall receive no additional compensation for such service.

b. Duties and Functions. The Administration shall:

(1) Conform its policies, plans and programs to the policies, plans and programs of the National Economic Board;

(2) Make recommendations for the procurement, supply, allocation, distribution and rationing of all articles of human and animal food and drink, including collection, movement and stockpiling procedures for grains and other foods;

(3) Maintain liaison and coordinate with all political subdivisions, agencies and instrumentalities of the Government of Korea concerned with food policies and programs;

(4) Formulate measures in connection with food policies and programs and direct the execution of such measures.

c. Publication of Regulations. Regulations of the National Food Administration shall be published in the Official Gazette upon approval by the Military Governor and shall have the force of law.

SECTION VI. *Repeal of Laws; Saving Provisions.*

a. General Notices Numbers 1-6 Repealed. General Notices Numbers 1-6, inclusive, are hereby repealed; *provided, however*:

(1) That this provision is not intended and shall not be construed to revive any corporate entity dissolved, or to reenact any laws, regulations. provisions or measures abolished or repealed by any of such General Notices;

(2) That Chosen Coal Distributing Company (hereby renamed Korean Coal Distributing Company) is continued in existence as a Military Government agency. In accordance with the policies, plans and programs of the National Economic Board and the National Price Administration, and subject to other requirements of law, the Korean Coal Distributing Company shall:

Procure, store, allocate, ration, sell and distribute bituminous coal, anthracite coal and pitch (for fuel purposes), license distributors thereof, and engage in such related activities as are authorized by the Military Government.

(3) That Korean Fisheries Corporation is continued in existence as a Military Government agency. In accordance with the policies, plans and programs of the National Economic Board, the National Price Administration and the National Food Administration, and subject to other requirements of law, the Korean Fisheries Corporation shall:

Procure, store, allocate, ration, sell and distribute marine products (including all forms of marine life, animal and vegetable), and fishing equipment, appliances, accessories, goods and utensils used or applied in the catching or taking, preservation and processing of fish and all other forms of marine life; and engage in such related activities as are authorized by the Military Government.

(4) That Petroleum Distributing Agency is continued in existence as a Military Government agency. In accordance with the policies, plans and programs of the National Economic Board and the National Price Administration, and subject to other requirements of law, the Petroleum Distributing Agency shall:

Procure, store, allocate, ration, sell and distribute petroleum and petroleum products, license distributors thereof, and engage in such related activities as are authorized by the Military Government.

b. *Ordinance Number 62 Repealed; Maximum Prices of Drugs, &c, Retained in Effect.* Ordinance Number 62, dated 29 March 1946, is hereby repealed; *provided, however:*

(1) That Department Order Number 1, dated 2 May 1946, of the Department of Public Health and Welfare, shall not be affected hereby and that the provisions of such order and the prices set forth on the price list attached thereto shall remain in full force and effect until modified by regulation of the National Price Administration.

(2) That the Director of the Department of Public Health and Welfare, in accordance with the policies, plans and programs of the National Economic Board and the National Price Administration, and subject to other requirements of law, shall:

Establish all necessary controls relating to production, procure-

— 4 —

HEADQUARTERS
UNITED STATES ARMY MILITARY
GOVERNMENT IN KOREA
Office of the Military Governor
Seoul, Korea

ORDINANCE
NUMBER 91   5 October 1946

PATENT LAW

SECTION I. *Patent Law, 1946, Enacted.*   To promote the useful arts and sciences and to encourage the manufacture of new inventions and devices and their early utilization in Korea, *The Patent Law, 1946*, is hereby promulgated :

SECTION II. *Repeal of Inconsistent Provisions.*

a. All laws, decrees, ordinances, notices and regulations inconsistent with *The Patent Law, 1946*, are hereby repealed.

b. The following laws, ordinances and notices are specifically repeale

(*1*) The Japanese Patent Law, Law No 96 of 30 April 1921, as amended.

(*2*) The Japanese Utility Model Law, Law No 97 of 30 April 1921, as amended.

(*3*) The Japanese Design Law, Law No 98 of 30 April 1921 as amended.

(*4*) The Japanese Law Concerning Patent Attorneys, Law No 100 of 30 April 1921.

(*5*) Pars. 11, 12 and 12a of the Japanese Law No 27 of 1896, as amended.

(*6*) Imperial Ordinance No 335 of 29 August 1910.

(*7*) Imperial Ordinance No 336 of 29 August 1910.

(*8*) Imperial Ordinance No 460 of 16 December 1921, as amended.

(*9*) Imperial Ordinance No 461 of 16 December 1921, as amended.

(*10*) Imperial Ordinance No 462 of 16 December 1921.

(*11*) Imperial Ordinance No 463 of 16 December 1921.

(*12*) Imperial Ordinance No 465 of 16 December 1921, as amended.

(*13*) Imperial Ordinance No 466 of 16 December 1921, as amended.

(*14*) Imperial Ordinance No 9 of 9 January 1922.

(*15*) Imperial Ordinance No 52 of 29 January 1938.

(16) Ministry of Agriculture & Commerce Ordinance No 33 of December 1921, as amended.

(17) Ministry of Agriculture & Commerce Ordinance No 34 of December 1921, as amended.

(18) Ministry of Agriculture & Commerce Ordinance No 35 of December 1921, as amended.

(19) Ministry of Agriculture & Commerce Ordinance No 37 of December 1921, as amended.

(20) Ministry of Agriculture & Commerce Ordinance No 38 of December 1921, as amended.

(21) Ministry of Agriculture & Commerce Ordinance No 39 of December 1921, as amended.

(22) Ministry of Agriculture & Commerce Ordinance No 40 of December 1921.

(23) Ministry of Agriculture & Commerce Ordinance No 41 of December 1921.

(24) Ministry of Agriculture & Commerce Notice No 310 of 1921, as amended.

(25) Ministry of Agriculture & Commerce Extraordinary Ordinance No 6 of 1923, as amended.

(26) Ministry of Agriculture & Commerce Extraordinary Ordinance No 7 of 1923, as amended.

(27) Ministry of Agriculture & Commerce Extraordinary Ordinance No 8 of 1923, as amended.

(28) Ministry of Commerce and Industry Ordinance No 27 of 1938.

SECTION III. *Effective Date.* This ordinance and *The Patent Law, 1946,* shall be effective on the tenth day after the date appearing hereon.

ARCHER L LERCH
Major General United States A...
Military Governor in Kore...

ORDINANCE
NUMBER 92

6 June 1946

ELIMINATION OF INSPECTION FUNCTIONS
OF OFFICE OF CIVIL ADMINISTRATOR*

SECTION I. *Elimination of Inspection Functions.* All duties and functions of the former Inspection Subsection of the Planning Section of the Secretariat, transferred to the Office of the Civil Administrator of the Government of Korea by Ordinance Number 58, dated 5 March 1946, are hereby eliminated from the Government of Korea.

SECTION II. *Effective Date.* This ordinance shall be effective on the tenth day after the date appearing hereon.

ARCHER L LERCH
Major General United States Army
Military Governor in Korea

---

*See Ordinance Number 58, dated 5 March 1946.

ORDINANCE
NUMBER 93

4 July 1946

FOREIGN EXCHANGE CONTROL

SECTION I. *Unlicensed Transactions Prohibited.* The following transactions are prohibited, except as specifically authorized by the Military Government in Korea by license or otherwise:

a. Any transaction in which any person or political entity outside of Korea has any interest, direct or indirect.

b. Any transaction which involves property of any of the following types, owned, held or controlled by any person or political entity in Korea:

(1) Property (or evidence thereof) located outside of Korea;

(2) Gold, silver, platinum or other precious metals, including coin or bullion containing such metals;

(3) Money not legal tender in Korea;

(4) Securities, negotiable instruments and other evidences of ownership or indebtedness, wherever located, expressed or payable in money not legal tender in Korea, or issued or created by, or constituting or asserted to constitute an obligation of a person or political entity outside of Korea;

(5) Claims or evidence thereof, wherever located, against any person or political entity outside of Korea, or expressed in money not legal tender in Korea;

(6) Property, wherever located, in which a person or political entity outside of Korea has any interest of any nature whatsoever, direct or indirect.

c. Any transaction for the purpose or which has the effect of evading or avoiding any of the foregoing prohibitions.

SECTION II. *Declaration of Property and Obligations.* Within sixty days after the effective date of this ordinance all persons (*a*) owning, holding or controlling, directly or indirectly, in whole or in part, any property described in Section 1 *b* hereof, other than property in Japan, or any claim or evidence thereof against any person or property in Japan or against the Japanese Government, or (*b*) owing any obligation of payment or performance, whether matured or not, to any person or political entity outside of Korea, other than to a person in Japan or to the Japanese Government, shall file with the nearest branch or designated agent of the Bank of Chosen a written declaration of such asset or obligation on a form obtained from any such branch or agent.

SECTION III. *Maintenance and Inspection of Records.* The Director of the Department of Finance may require any person to keep a complete record of, and to furnish under oath or affirmation, in the form of reports or otherwise, from time to time, complete information relative to any property or transaction referred to in Section I hereof. Such information may include entries in books, contracts, letters or other papers in the custody of such person, and the Director of the Department of Finance may require the production thereof. Every person engaging in any transaction referred to in Section I hereof shall keep a complete record thereof, regardless of whether such transaction is effected pursuant to license or otherwise, and such record shall be preserved and be available for examination for at least two years after the date of such transaction.

SECTION IV. *Applications for Licenses: Licensing Authority.* Applications for licenses or for permission to engage in transactions prohibited by this ordinance shall be submitted to the Director of the Department of Finance who shall be the licensing authority; *provided,* that whenever a license or permit authorizing any foreign trade transaction has been issued by the Bureau of Commerce pursuant to Foreign Commerce Regulation No 1 no separate license hereunder shall be required, but such license or permit issued by the Bureau of Commerce shall be approved on the face thereof by the Department of Finance, prior to engaging in such foreign trade.

SECTION V. *Prohibited Transactions and Transfers Void.* Prohibited transfers or transactions under this ordinance and all arrangements made for the purpose or having the effect of evading or avoiding this ordinance, shall be null and void from their inception.

SECTION VI. *Conflicting Provisions Conformed* All laws and regulatory measures and parts thereof which are inconsistent herewith or in conflict with the provisions hereof are hereby repealed.

SECTION VII. *Definitions.* As used herein:

The term "person" includes any natural or juridical person;

b. The term "transaction" includes acquiring, importing, borrowing, withdrawing, earmarking, or receiving with or without consideration, remitting, selling, leasing, transferring, removing, exporting, hypothecating, lending, guaranteeing or otherwise dealing in any property mentioned in this ordinance:

c. The term "property" includes all movable and immovable property and all legal, equitable, or economic rights or interests in, or claims to, such property, whether matured or not, including, but not limited to, land and buildings; money, bank balances, checks, drafts, bills of exchange and other instruments of payment; stocks, shares, ownership certificates; claims, bonds, debentures and other evidences of indebtedness;

d. Property shall be deemed to be "owned held or controlled" by any person if such property is held in his name or for his account or benefit, or owed to him or to his nominee or agent or if such person has a right or obligation to purchase, receive or acquire such property.

e. "Korea" shall, for purposes of this ordinance, be deemed to mean that portion of Korea which is south of 38 degrees North Latitude.

f. "Political entity" includes any government or subdivision, instrumentality or agency thereof.

SECTION VIII. *Penalties.* Any person violating the provisions of this ordinance shall, upon conviction by a Military Occupation Court, suffer such punishment as the court shall determine.

SECTION IX. *Effective Date.* This ordinance shall be effective on the date appearing hereon.

ARCHER L LERCH
Major General United States Army
Military Governor in Korea

ORDINANCE
NUMBER 94   2 July 1946

ESTABLISHMENT OF CHEJU PROVINCE

SECTION I. *Cheju Island Removed From Jurisdiction of South Cholla Province.* Cheju Island, including such islands adjacent to it as are presently included within the jurisdiction of any town or village listed in Sections IV or V hereof and including the Chuja archipelago, is hereby removed from the territorial jurisdiction of South Cholla Province.

SECTION II. *Establishment of Cheju Province.* Cheju Island, together with the islands referred to in Section I hereof, is hereby constituted a province, with all the powers, duties, functions and rights thereof. The name of the province shall be Cheju Province.

SECTION III. *Counties within Cheju Province.* Cheju Province shall of two counties. to be known as North Cheju County and South Cheju County, respectively.

SECTION IV. *Jurisdiction of North Cheju County.* North Cheju County shall consist of the following town and villages:

			Location of District Office
a.	Cheju	(town)	Cheju town
b.	Kujwa	(village)	Pengte Ri
c.	Chuja	(village)	Yong Hung Ri
d.	Hallim	(village)	Hallim Ri
e.	Chochon	(village)	Chochon Ri
f.	Aewol	(village)	Aewol Ri

SECTION V. *Jurisdiction of South Cheju County.* South Cheju County shall consist of the following villages:

		Location of District Office
a.	Songsan	Songsan Ri
b.	Namwon	Namwon Ri
c.	Chung Mun	Chung Mun Ri

d. Taejong	Mosulpo
e. Pyoson	Pyoson Ri
f. Sogwi	Sogwi Ri
g. Andok	Masun Ri

SECTION VI. *Effective Date.* This ordinance shall be effective at 2400, 30 July 1946.

ARCHER L LERCH
Major General United States Army
Military Governor in Korea

ORDINANCE
NUMBER 95   1 July 1946

DEPOSIT OF SUPPLEMENTAL MILITARY YEN CURRENCY MARKED " A "

SECTION I. *Deposit Required.* All natural and juridical persons within Korea south of 38° North Latitude are hereby:

a. Ordered to deposit within the period 1 July to 10 July 1946, inclusive, in one of the financial institutions designated in Section II, all supplemental military yen currency, marked " A ", issued by the Military Occupation Forces, owned or possessed by them. The same procedures will be followed as in Ordinance No. 57.

b. Prohibited after 10 July 1946 from importing, exporting, receiving, paying out, knowingly owning or possessing, giving or otherwise transfering or dealing in or engaging in any transaction with respect to any such currency.

SECTION II. *Designated Financial Institutions.* The following financial institutions are authorized and directed to receive supplemental military yen currency, marked " A ", for deposit pursuant to this ordinance:

Bank of Chosen
Chosen Industrial Bank
Choheung Bank
Chosen Commercial Bank
Chosen Trust Company
Chosen Savings Bank
Federation of Financial Associations.

Each financial institution shall:

a. Forward to Head Office, Bank of Chosen, Seoul Korea, the notes deposited under this ordinance.

b. Report to the Department of Finance, Military Government of Korea, Seoul, immediately after the deposit period the amount of such currency deposited.

SECTION III. *Reimbursement*. Reimbursement for all supplemental
military yen currency marked " A " legally held and deposited within the
period 1 July to 10 July 1946, inclusive, will be made in Bank of Chosen yen
at par.

SECTION IV. *Penalty Clause*. Any person violating the provisions of this
ordinance shall, upon conviction by a Military Occupation Court, suffer such
penalty as the court shall determine.

SECTION V. *Scope*. The provisions of this ordinance shall not apply to members of the Armed Forces of the Allied Powers, accredited civilians
on duty with the forces of the Allied Powers, or their dependents, and other
personnel authorized by G. H. Q., U. S. Army Forces, Pacific, or higher
authority, to use supplemental military yen currency marked " A ".

SECTION VI. *Effective Date*. This ordinance shall be effective on the
date appearing hereon.

ARCHER L. LERCH

Major General United States Army

Military Governor in Korea.

ORDINANCE
NUMBER 96   25 July 1946

AMENDING DRUG AND DRUG BUSINESS REGULATIONS

SECTION I. *Purpose.* The purpose of this ordinance is to improve standards in drug vending and manufacturing by revision of existing regulations, and to consolidate licensing requirements within a single Department of the Government of Korea.

SECTION II. *Governor General Ordinance Number 22 Amended.* Governor General Ordinance Number 22, *Regulations of Drugs and Drug Business,* dated 28 March 1912, is hereby amended as follows:

a. The words *Korean General Governor* are hereby deleted wherever they occur, and the words *Director, Department of Public Health and Welfare* substituted therefor.

b. The words *Provincial Governor* are hereby deleted wherever they occur, and the words *Chief of the Provincial Bureau of Public Health and Welfare* substituted therefor.

c. The words *but will be required to re-register as directed by the Director, Department of Public Health and Welfare* are hereby added to the sentence constituting Paragraph 2 of Additional Rules.

SECTION III. *Government General Order Number 55 Amended.* Government General Order Number 55, dated March 1912, is hereby amended as follows:

a. The words *Korean General Governor* are hereby deleted wherever they occur, and the words *Director, Department of Public Health and Welfare* substituted therefor.

b. The words *Provincial Government Police Department* and *police station* are hereby deleted wherever they occur, and the words *Provincial Bureau of Public Health and Welfare* substituted therefor.

c. The words *Department of Public Health and Welfare of the Government of Korea* are hereby substituted for the Home Department in the second paragraph of Section I thereof.

d. Subsection 4 of Section I thereof is hereby deleted in its entirety, and the following substituted therefor:

Simultaneously with payment of the fee herein prescribed, the applicant shall submit to the Director of the Department of Public Health and Welfare the following data:

(1) *Name, sex, date of birth, permanent domicile and address;*

(2) *Record of personal history;*

(3) *Two copies of a recent photograph (name-card size);*

(4) *A certificate or copy of diploma of graduation from a recognized pharmacy school, or a certificate showing that the applicant has passed a Korean or foreign pharmacists' examination, or a copy of the applicant's foreign pharmacists' license;*

(5) *A fifty-yen registration fee in cash or postal money order payable to the Director of the Department of Public Health and Welfare.*

e. Section II thereof is hereby deleted in its entirety.

f. The last two paragraphs of Section V thereof are hereby deleted in their entirety, and the following substituted therefor:

Patent medicine merchants and medicine manufacturers shall submit the following information concerning medicine manufactured or imported to the Provincial Bureau of Public Health and Welfare in which their principal office is located : name, quantity of each and every ingredient, method of manufacture, and directions for use or dosage.

All licenses for medicine manufacturers and patent medicine merchants shall first be approved by the Department of Public Health and Welfare before any Provincial Bureau of Public Health and Welfare may issue such licenses.

g. Section VI thereof is hereby amended to read:

If the patent medicine merchant or medicine manufacturer desires to change any of the data required by subsection 4. of Section V hereof, he shall file an application containing the original data and the proposed change or changes with the Provincial Bureau of Public Health and Welfare in which their principal office is located, requesting permission to make such change or changes.

h. The words *but when the manufacturer has been approved by the Chief Supervisor of the Police Department, it is not necessary to write the quantity of the principle ingredients* are hereby deleted from Section VIII thereof.

i. Section XVII thereof is hereby deleted in its entirety and the following substituted therefor:

> *No permits shall be granted for patent medicine peddlers. Patent medicine peddling is hereby declared to be unlawful.*

All reference to or use of the words *patent medicine peddler* in any other section of such order is hereby deleted.

j. Section XX thereof is hereby amended to read:

> *All communications to the Department of Public Health and Welfare of the Government of Korea shall be submitted through the local Public Health Officer.*

k. The additional rule thereof is hereby deleted in its entirety and the following substituted therefor:

> *It shall be unlawful for any person licensed to compound, sell or manufacture a medicine or a patent medicine who fails to submit a copy of his present license and the data required by either Subsection 4 of Section I or Section V or Section VII of this order (as hereby amended), and who fails to obtain a new license as hereinabove provided within two months after the effective date of this ordinance, to conduct or participate in such business.*

SECTION IV. *Penalties.* Any person violating the provisions of this ordinance shall, upon conviction by a Military Occupation Court, suffer such penalty as the court shall determine.

SECTION V. *Effective Date.* This ordinance shall be effective on the date appearing hereon.

ARCHER L LERCH
Major General United States Army
Military Governor in Korea

ORDINANCE
NUMBER 97

23 July 1946

PUBLIC POLICY IN LABOR MATTERS DECLARED; DEPARTMENT OF LABOR ESTABLISHED

SECTION I. *Declaration of Policy.* The public policy of the United States Army Military Government in Korea in labor matters is hereby declared as follows:

a. The development of democratic labor organizations shall be encouraged;

b. Workers shall have the right through self-organization to form and join labor unions, to give aid to and receive assistance from other labor organizations, and to designate representatives of their own choosing for the purpose of negotiating terms and conditions of their employment contract, without interference from employers, or their agents;

c. Peaceful negotiations between employers and labor organizations resulting in written employment contracts specifying agreed wages, hours of work, and other conditions of employment shall be encouraged.

SECTION II. *Department of Labor Established.* To effectuate the foregoing labor policy and to assist the efficient functioning of the economy of Korea, the Department of Labor is hereby established in the Government of Korea. The Bureau of Labor of the Department of Commerce of the Government of Korea is hereby disestablished. All duties and functions of the Bureau of Labor are hereby transferred to the Department of Labor so as effectuate the carrying out of the purposes herein described. All records, funds, property and Korean personnel of the Bureau of Labor shall be transferred to the Department of Labor, in accordance with orders to be issued by the Director of the Department of Commerce in coordination with the Office of Korea Civil Service and the Budget Bureau of the Department of Finance of the Government of Korea. The purpose of the Department of Labor shall be to aid the efficient functioning of the economy of Korea, to foster, promote, and develop the welfare of the wage earners of Korea, to improve their working conditions, and to advance their opportunities for profitable employment.

SECTION III. *Functions and Duties of the Department of Labor*. The Department of Labor shall have the following functions and duties:

a. To formulate labor policies and programs and to exercise general supervision over conditions of employment, labor relations with employers and with the government, and any other factors affecting the welfare of labor, under the direction of the Military Governor and within the limitations of Ordinance Number 90, dated 28 May 1946, the declared public policy of Military Government in labor matters, and other requirements of law;

b. To supervise and control Labor Mediation Boards and all that pertains to the same so as to effectuate the mediation of labor disputes in accordance with Ordinance Number 34, dated 8 December 1945, the declared public policy of Military Government in labor matters, and other requirements of law;

c. To supervise or administer a national employment service and a labor information service;

d. To recommend to the Military Governor labor laws to effectuate the declared policy of the Military Government in labor matters, to assist the efficient functioning of the economy of Korea, and to promote the welfare of labor;

e. To prepare for issuance in the Official Gazette upon approval by the Military Governor all necessary regulations for the administration of this ordinance and other labor laws which it may be charged with administering, or for carrying into effect the purposes of the Department;

f. To perform all other functions and duties necessary to effectuate the purposes of the Department and the declared public policy of the Military Government in labor matters, under the supervision and direction of the Military Governor and within the scope of existing laws.

SECTION IV. *Penalties*. Any person violating the provisions of any regulation of the Department of Labor issued pursuant to this ordinance, and having the force and effect of law upon publication in the Official Gazette with approval of the Military Governor, shall, upon conviction by a Military Occupation Court, suffer such punishment as the court shall determine.

SECTION V. *Effective Date*. This ordinance shall be effective on the date appearing hereon.

ARCHER L. LERCH
Major General United States Army
Military Governor in Korea.

HEADQUARTERS
UNITED STATES ARMY MILITARY
GOVERNMENT IN KOREA
Office of the Military Governor
Seoul, Korea

ORDINANCE
NUMBER 98   11 July 1946

BUREAUS OF FOREIGN COMMERCE AND DOMESTIC COMMERCE CREATED IN DEPARTMENT OF COMMERCE: ORDINANCE NO 82 AMENDED

SECTION I. *Bureaus of Foreign Commerce and Domestic Commerce Established.* The Bureau of Foreign Commerce and the Bureau of Domestic Commerce are hereby created within the Department of Commerce of the Government of Korea. The Bureau of Commerce of the Department of Commerce of the Government of Korea is hereby disestablished. All duties and functions of the Bureau of Commerce are hereby transferred to the Bureaus created herein so as to effectuate the carrying out of the functions and duties hereinafter prescribed. All Korean personnel of the Bureau of Commerce shall be transferred to the Bureaus created herein, in accordance with orders to be issued by the Director of the Department of Commerce in coordination with the Office of Korea Civil Service

SECTION II. *Functions and Duties of Bureau of Foreign Commerce. Amendment of Ordinance No 82.* The Bureau of Foreign Commerce shall have the following functions and duties:

    a. To formulate foreign trade programs and to exercise general superintendence over foreign trade, within policies of the Military Governor and within the limitations of Ordinance Number 90, dated 28 May 1946, and other requirements of law;

    b. To act as the authorized agent of the Military Government of Korea for the control and licensing of foreign trade, in accordance with Sections I to IV of Ordinance Number 39, dated 3 January 1946. Sections II and IV of Ordinance Number 82, dated 17 May 1946, are hereby amended by substituting the words *Bureau of Foreign Commerce* for the words *Bureau of Commerce* therein.

SECTION III. *Functions and Duties of Bureau of Domestic Commerce.*
The Bureau of Domestic Commerce shall have the function and duty of
formulating domestic trade programs and exercising general superintendence
and control over domestic trade, including control of standards of accuracy for
weighing and measuring devices, within policies of the Military Government
and within the limitations of Ordinance Number 90, dated 28 May 1946, and
other requirements of law.

SECTION IV. *Effective Date.* This ordinance shall be effective on the
tenth day after the date appearing hereon.

ARCHER L LERCH
Major General United States Army
Military Governor in Korea

213

HEADQUARTERS
UNITED STATES ARMY MILITARY
GOVERNMENT IN KOREA
Office of the Military Governor
Seoul, Korea

ORDINANCE
NUMBER 96

14 July 1946

POSTAL, TELEGRAPH AND TELEPHONE RATES

SECTION I. *Modification of Existing Rates by Regulation.* All laws, ordinances, orders, regulations, directives and instructions relating to postal, telegraph and telephone rates, shall remain in full force and effect until superseded, modified or repealed by regulations, issued by the Director of the Department of Communications pursuant hereto.

SECTION II. *Preparation and Publication of Regulations.* The Director of the Department of Communications of the Government of Korea shall prepare regulations for postal, telegraph and telephone rates. Such regulations, upon approval by the Military Governor, shall be published in the Official Gazette and shall have the force of law.

SECTION III. *Effective Date.* This ordinance shall be effective on the date appearing hereon.

ARCHER L LERCH
Major General United States Army
Military Governor in Korea

HEADQUARTERS
UNITED STATES ARMY MILITARY
GOVERNMENT IN KOREA
Office of the Military Governor
Seoul, Korea

ORDINANCE
NUMBER 100

25 July 1946

REPEAL OF COMMERCIAL INDUSTRIAL
CONFERENCE ASSOCIATION ORDINANCE

SECTION I. *Purpose.* The purpose of this ordinance is to eliminate existing provisions relating to compulsory membership in designated commercial and industrial associations, and to permit the formation thereof and membership therein on a voluntary basis.

SECTION II. *Acts Repealed: Saving Provision.*

a. *Acts Repealed.*

(1) Government General Ordinance Number 30, dated 18 August 1944 (*Chosun Commercial-Industrial Conference Association Ordinance*) is hereby repealed.

(2) Government General Order 300, dated 18 August 1944 (*Government General Order Concerning the Enforcement of the Chosun Commercial-Industrial Conference Association Ordinance*) is hereby repealed.

b. *Saving Provision.* This ordinance is not intended and shall not be construed to reenact Government General Ordinance Number 4, dated 10 May 1931 (*Korean Commerce and Industry Chamber Ordinance*) or any of the provisions thereof.

SECTION III. *Effective Date.* This ordinance shall be effective on the tenth day after the date appearing hereon.

ARCHER L LERCH
Major General United States Army
Military Governor in Korea

HEADQUARTERS
UNITED STATES ARMY MILITARY
GOVERNMENT IN KOREA
Office of the Military Governor
Seoul, Korea

ORDINANCE
NUMBER 101

31 August 1946

AMENDMENT OF SALES TAX LAWS

SECTION I. *Revision of Tax Laws and Rates.*

  a. *Commodity Tax Revised.*

(*1*) The tax rates provided in Tax Ordinance Number 18, dated 31 March 1940. as revised, are hereby further revised and shall be as follows:

(*a*) 1st class:	Sub-class (a)	30%
	Sub-class (b)	20%
	Sub-class (c)	10%
	Sub-class (d)	5%
(*b*) 2d class:	Sub-class (a)	30%
	Sub-class (b)	20%
	Sub-class (c)	10%
	Sub-class (d)	5%

(*2*) The schedule of taxable items listed in Tax Ordinance Number 18, dated 31 March 1940, as revised, is hereby amended and shall be as follows:

  (*a*) 1st class: Sub-class (b): Items 8, 17, 20 and 21 are hereby eliminated in their entirety.

          Sub-class (c): Items 23, 24. 25 and 27 are hereby eliminated in their entirety. *Headwear* is hereby deleted from Item 22.

          Sub-class (d): Items 28 and 29 are hereby eliminated in their entirety.

b) 2d class: Sub-class (a): Item 15 is hereby amended toread *toilet articles except soap, toothpowder, toothpaste and toothbrushes.*

Sub-class (b): Items 18, 24, 30, 34 and 35 are hereby eliminated in their entirety. *Baskets* is hereby deleted from Item 28. *Children's* is hereby deleted from Item 29. Item 37 is hereby amended to read *confectionery and candy, except that containing rice.* Item 31 is hereby amended to read *ornamental lacquer, ceramic and glass wares.*

Sub-class (c): Items 41, 53, 54, 55, 56, 58 and 59 are hereby eliminated in their entirety.

Sub-class (d): Items 64, 66, 68, 70 and 71 are hereby eliminated in their entirety.

(c) 3d class: Item 1 is hereby eliminated in its entirety. Item 2 is hereby amended to read *wheat gluten, grape sugar and maltose, except confectionery and candy containing rice.* Item 5 is hereby added, to read *confectionery and candy containing rice shall be taxed at the rate of 100% of the retail price thereof.*

b. *Sugar Consumption Tax Repealed.* The tax rates provided in Tax Ordinance Number 13, dated 30 March 1902, as revised, are hereby repealed.

c. *Amusement, Drinking and Eating Tax Revised.* The tax rates provided in Tax Ordinance Number 19, dated 31 March 1940, as revised, are hereby further revised and shall be as follows:

(1) Class A: Restaurant with *kisang* license: charge of over ¥ 20 per person, 30% of entire bill.

(2) Class B: Restaurant without *kisang* license: charge of over ¥ 20 per person, 20% of entire bill. Restaurant without *kisang* license: patron providing own *kisang* entertainment, 20% of entire bill.

(3) Class C: Hotel charge over ¥ 30 for one day per person, 10% of entire bill.

d. *Admission Tax Revised.* The tax rates provided in Tax Ordinance Number 22, dated 31 March 1940, as revised, are hereby further revised and shall be as follows:

> *(1)* 1st class: Admission charge of over ¥ 2.50. 30% of entire charge.
>
> *(2)* 2d class: Admission charge to billiards game, 30% of entire charge. Admission to mahjong club, dance halls, golf clubs, 50% of entire charge. Taxes on admission charges to skating rinks, ping pong games, bow and arrow and shooting practice are hereby repealed.
>
> *(3)* 3d class: Special admission tax on events given to the public without profit (non-professional) is hereby repealed.

e. Special Action Tax Revised. The tax rates provided in Tax Ordinance Number 8, dated 31 March 1945, as revised, are hereby further revised and shall be as follows:

> *(1)* 1st class: Charge of over ¥ 20 for photographing, developing, printing and reproduction, 20% of entire bill.
>
> *(2)* 2d class: Barber and beauty shops: charge of over ¥ 15, 20% of entire bill.
>
> *(3)* 3d class: Dyeing and embroidery of cloth: charge of over ¥ 70. 10% of entire bill.
>
> *(4)* 4th class: Tailoring: charge of over ¥ 70, 10% of entire bill.
>
> *(5)* 5th class: Picture framing and mounting: 20% of entire bill.
>
> *(6)* 6th class: Printing and binding rates are hereby repealed.
>
> *(7)* 7th class: Repairing of musical instruments, cameras, phonographs, etc: 10% of entire bill.
>
> *(8)* 8th class: Safe deposit vaults: 20% of entire bill.

f. Advertising Tax Repealed. The tax rates provided in Tax Ordinance Number 20, dated 24 March 1942, as revised, are hereby repealed.

g. Card and Dice Tax Revised. The tax rates provided in Tax Ordinance Number 1, dated 15 April 1931, as revised, are hereby further revised and shall be as follows:

> *(1)* Manufacture of playing cards: ¥ 5 per deck.
>
> *(2)* Manufacture of kolpai (dice): ¥ 15 per set.
>
> *(3)* Manufacture of mahjong: ¥ 100 per set.

h. ***Construction Tax Repealed.*** The tax rates provided in Tax Ordinance Number 12, dated 31 March 1940, as revised, are hereby repealed.

SECTION II. *Effective Date.* This ordinance shall be effective on the date appearing hereon.

ARCHER L LERCH
Major General United States Army
Military Governor in Korea

HEADQUARTERS
UNITED STATES ARMY MILITARY
GOVERNMENT IN KOREA
Office of the Military Governor
Seoul, Korea

ORDINANCE
NUMBER 102

22 August 1946

ESTABLISHMENT OF SEOUL NATIONAL UNIVERSITY

SECTION I. *Purpose.* The purpose of this ordinance is to provide for and make available to the people of Korea improved facilities for higher education, so that the youth of Korea may take advantage of the benefits and opportunities accruing therefrom, for the betterment of themselves as individuals and the Korean people as a nation in modern society.

SECTION II. *Method of Accomplishment of Objective.* The objective of this ordinance shall be achieved through the merger of certain existing educational facilities into a reorganized *Seoul National University*, the progressive amalgamation of additional educational facilities therein as necessary or desirable from time to time, the establishment therein of required additional colleges, schools and learned faculties, the recruitment of the best available teaching staff, the elimination of inefficiencies in present operations, the institution of economies in administration, and the evolution of a system of controls designed (by competitive examinations and otherwise) to raise the academic standards of the University to the point where they are equivalent to those of leading institutions of higher learning throughout the world.

SECTION III. *Entrance Requirements.* Entrance requirements to the University shall be prescribed from time to time by the Board of Regents thereof; *provided*, that at no time and in no case shall any consideration of race, nationality, religion, sex, caste, or economic position or condition be made or deemed a requirement for admission to or attendance at the University or for the granting or withholding of any degree, diploma, certificate, scholarship, grant, emolument, honor or award by or in the University; *provided, however,* that the Regents may at any time, when the public interest of Korea requires, by written announcement establish quotas by percentage or number, of non-residents of Korea who may be admitted to matriculation or attendance at the University.

SECTION IV. *Existing Provisions of Law Repealed; Saving Provision.* The following existing provisions of law are hereby repealed, insofar as they or any of them, or any part of any thereof, pertain to Seoul National University and/or to any of the colleges, schools, institutions of learning or learned faculties referred to in this ordinance:

a. Imperial Ordinance Number 388, 6 December 1918 (*University Ordinance*), as revised by Imperial Ordinance Number 7, January 1928 and Imperial Ordinance Number 40, January 1943. and as further revised;

b. Imperial Ordinance Number 12, 7 February 1919 (*Regulations for Imperial Universities*), as revised;

c. Government General Ordinance Number 30, 1 April 1926 (*Regulations for Universities*). as revised by Government General Ordinance Number 34, March 1934, Government General Ordinance Number 79, April 1940. and Government General Ordinance Number 137, May 1943, and as further revised;

d. Imperial Ordinance Number 47, 1 April 1926 (*Faculty of Keijo Imperial University*), as revised by Imperial Ordinance Number 154, June 1927, Imperial Ordinance Number 58. April 1928, Imperial Ordinance Number 81. April 1929, Imperial Ordinance Number 66; April 1931. Imperial Ordinance Number 219. April 1940, Imperial Ordinance Number 934. December 1940, Imperial Ordinance Number 444, April 1941. and Imperial Ordinance Number 400, April 1942, and as further revised;

e. Government General Ordinance Number 138, 7 May 1943 (*Regulations for Preparatory School of Keijo Imperial University*), as revised;

f. Government College Order Number 2. January 1932 (*Conferring Degrees upon Graduates of Keijo Imperial University*), as revised;

g. Imperial Ordinance Number 61, 27 March 1903 (*Ordinance Relating to Colleges*), as revised by Imperial Ordinance Number 8, January 1923. and as further revised;

h. Government General Ordinance Number 21, 7 March 1922 (*Regulations for Public and Private Colleges*), as revised by Government General Ordinance Number 81, April 1940, Government General Ordinance Number 139, May 1943, and enforcing regulations. and as further revised;

Government General Ordinance Number 49, 1 April 1922 (*Regulations for Keijo Law College*), and enforcing regulations, as revised;

j. Government General Ordinance Number 50, 1 April 1922 (*Regulations for the Government Medical College*), as revised by Government General Ordinance Number 262. 30 June 1944. and enforcing regulations, and as further revised;

k. Government General Ordinance Number 52, 1 April 1922 (*Regulations for Government Agricultural Colleges*). as revised by Government General Ordinance Number 261, 30 June 1944, and enforcing regulations, and as further revised;

l. Government General Ordinance Number 53, 1 April 1922 (*Regulations for the Government Commercial College*), as revised by Government General Ordinance Number 263. 30 June 1944, and enforcing regulations, and as further revised;

m. Government General Ordinance Number 51, 1 April 1922 (*Regulations for the Government Technical College*). as revised by Government General Ordinance Number 260, 30 June 1944, and enforcing regulations, and as further revised;

n. Imperial Ordinance Number 109, 8 March 1943 (*Ordinance Relating to Normal Schools*), and enforcing regulations, as revised;

o. Imperial Ordinance Number 104, 2 May 1924 (*Faculties in Keijo Imperial University*), as revised;

p. Imperial Ordinance Number 105, 2 May 1924 (*Ordinance Concerning Keijo Imperial University*), as revised;

q. Government General Ordinance Number 31, 1 April 1926 *Establishment of Law and Literature Departments in Keijo Imperial University*), as revised by Government General Ordinance Number 15, March 1927;

r. Government General Ordinance Number 65, April 1939 (*Regulations for Seoul Mining College*), as revised by Government General Ordinance Number 84, April 1940, and enforcing regulations, and as further revised;

s. Imperial Ordinance Number 151, 31 March 1922 (*Regulation for Seoul Normal School and Seoul Girls' Normal School*), as revised;

t. All other provisions of law, and parts thereof, which are inconsistent with this ordinance or in conflict with any of the provisions hereof, to the extent they are inconsistent herewith or in conflict with any of the provisions hereof;

provided, however, that this ordinance is not intended and shall not be construed to revive any corporate entity extinguished, or to reenact any laws, regulations, provisions or measures abolished or repealed by any of the provisions of law or measures herein enumerated or described.

— 3 —

SECTION V. *Disestablishment of Certain Sch...ls by Merger Thereof into Seoul National University.* The following colleges...d institutions of learning, and installations affiliated with each thereof, are hereby disestablished, and merged into Seoul National University, which is hereby created as a body corporate and continued in existence in perpetuity as a public instrumentality of the Government of Korea free from all forms of public and private taxation, assessment and levy, and devoted to the purposes set forth in this ordinance:

a. Seoul Commercial College;

b. Seoul Dental College;

c. Seoul Law College;

d. Seoul Medical College;

e. Seoul Mining College;

f. Seoul Normal College;

g. Seoul Technical College;

h. Seoul University;

i. Seoul Women's Normal College;

j. Suwon Agricultural and Forestry College.

The corporate existence of each of such colleges and institutions of learning is hereby terminated, and the property, equipment, records, funds and personnel of each thereof and of their affiliated installations are hereby transferred to the control of Seoul National University. Specific transfers, allocations of property, equipment, records, funds and personnel, and other necessary changes, shall be made by the Board of Regents of Seoul National University acting through the President thereof, in accordance with policy determinations and principles established by the Director of the Department of Education of the Government of Korea.

SECTION VI. *Establishment of Colleges and Faculties within Seoul National University.* The following colleges, schools and learned faculties are hereby established within Seoul National University:

a. College of Agriculture and Forestry ;

b. College of Commerce ;

c. College of Dentistry ;

d. College of Education ;

e. College of Engineering ;

f. College of Fine Arts ;

g. College of Law ;

h. College of Liberal Arts and Sciences;

i. College of Medicine, including School of Nursing and Nursing

Education :

j.   Graduate School.

SECTION VII.   *Board of Regents of Seoul National University: Organization, Functions, Duties and Remuneration.*

a.   *Organization.*

1   The Board of Regents of Seoul National University is hereby established.   It shall consist of the Director of the Department of Education *(ex officio)*, the President of Seoul National University *(ex officio)*, and one member appointed upon recommendation of the Director of the Department of Education by the Military Governor of Korea for each of the several colleges, schools and learned faculties constituting the University.   Each of the appointive members of the Board shall be a Korean selected as preeminent in the active practice of the particular profession or calling he is deemed qualified to represent on the Board ; *provided, however,* that no member of the faculty or staff or employed by the University and no governmental employee shall at any time be eligible to serve on the Board as an appointive member thereof.   The term of office of each appointive member shall be six years; *provided, however,* that one-third of the initial appointments to the Board shall be for a period of two years, one-third for a period of four years, and the remaining one-third for a period of six years ; and *further provided,* that appointments to fill vacancies on the Board shall be made only for the unexpired portion of the term of the vacancy filled ; and *further provided,* that upon creation of new seats on the Board by the establishment of additional colleges and learned faculties, each appointment thereto shall be made in such manner as to carry out the intent of this provision, which is to insure, so far as possible, the rotation of one-third of the total membership of the Board each and every two years.   The Board shall elect a Chairman to serve for a period of six years.   A quorum of the Board shall consist of a majority of the members thereof, and such quorum shall be required for the transaction of all business of the Board.   No member of the Board shall at any time during

his or her membership engage in any political activity other than the exercise of the right to vote. Members of the Board may be removed for cause at any time by the Military Governor of Korea.

(2) Notwithstanding any of the provisions of the foregoing paragraph (1). during the temporary period of military control, the Board of Regents of Seoul National University shall not function (unless otherwise ordered by the Military Governor), and such Board of Regents shall be replaced by a Temporary Board of Regents appointed by the Military Governor to hold office during his pleasure. Such Temporary Board of Regents shall consist of the Director of the Department of Education, the Korean Director of the Department of Education, the Deputy Director of the Department of Education, the Korean Deputy Director of the Department of Education, the Chief of the Bureau of Higher Schools of the Department of Education, and the Korean Chief of the Bureau of Higher Schools of the Department of Education. Such members of the Temporary Board of Regents shall receive no additional pay for such service, and the Temporary Board of Regents shall have the same powers and duties as the Board of Regents under this ordinance.

b. *Functions and Duties.* The Board of Regents of Seoul National University shall have the following functions and duties:

(1) To establish overall policy for the government of Seoul National University;

(2) To initiate studies forthwith, and to prepare and submit detailed plans and recommendations on the basis thereof, to the Director of the Department of Education, for a permanent location, buildings, campus. dormitories and other facilities, for the combined colleges, schools and learned faculties of Seoul National University, within or in the immediate vicinity of the city limits of the City of Seoul. Necessary land shall thereafter be acquired and taken by condemnation or other suitable proceedings initiated by the Director of the Department of Education in the name of the Government of Korea in the manner required by law;

(3) To tender appointment to any qualified Korean as President of

Seoul National University, which. when accepted in writing shall constitute appointment to office ; *provided, however,* that during the temporary period of military con rol, the President shall be appointed by the Military Governor of Korea ;

(4) To approve the merger of other schools and institutions of learning (public and private), and installations affiliated therewith, into Seoul National University and to establish additional colleges, schools, institutions and learned faculties within the University as required, upon consent of the Director of the Department of Education ;

(5) To recommend to the Director of the Department of Education classification scales and rates of pay for the President, members of the faculty, and executive and administrative staffs and employees of the University ;

(6) To establish scholarships, grants and endowments for needy students in the University. and to establish scholastic and athletic awards ;

(7) To establish entrance, matriculation. attendance, graduation and other requirements. including requirements for degrees diplomas and certificates ;

(8) To establish academic standards and scholastic requirements to be administered under the supervision of the President of the University ;

(9) To establish detailed regulations for the government of the colleges. schools and learned faculties constituting Seoul National University. and the members thereof ;

(10) To prescribe curricula and courses of study required to be pursued by students in attendance at the University ;

(11) To accept gifts. endowments. testamentary and other bequests and trusts, grants and donations on behalf of the University ; to direct the disposition and investment of the principal and interest thereof in a manner consistent with the optimum realization of the purposes of the University :

(12) To make available funds of the University to the respective

colleges, schools and learned faculties of the University for disposition by the respective Deans thereof in accordance with requirements established and supervised by the President, for the furtherance of research, scholarship and learning; including, but not limited to, the granting of fellowships and the publication of professional or learned journals, books, treatises and monographs;

(13) To make recommendations to the Director of the Department of Education regarding proposed leases by or for the University; and to similarly recommend the sale and acquisition of real estate and other property of or for the University; *provided, however*, that at no time shall any property of the University be mortgaged or otherwise encumbered;

(14) To create scholarships and endowments for students and faculty members admitted to the University from outside of Korea;

(15) To establish tuition rates and other proper charges for those students admitted to the University who are able to defray such expenses;

(16) To establish, operate, maintain and supervise such schools as are necessary for the professional education of teachers-in-training in the University;

(17) To establish rules for the conduct and guidance of the Board.

c. Remuneration. The remuneration of appointive members of the Board of Regents of Seoul National University shall be determined by the Director of the Office of Korea Civil Service, upon recommendation of the Director of the Department of Education. *Ex officio* members of the Board shall receive no remuneration for such service.

SECTION VIII. *Functions and Duties of President of Seoul National University.* The President of Seoul National University shall have the following functions and duties:

a. To execute the policies of the Board of Regents of the University;

b. To supervise the administration and operation of the University and the maintenance of its properties;

c. To preside at meetings of the faculty of the University;

d. To appoint the various Deans of the respective colleges, schools and learned faculties of the University, by and with the consent of the Board of Regents;

e. To nominate and pass upon prospective members of the faculty, on his own initiative or upon recommendations received from the various Deans of the several colleges, schools and learned faculties of the University, and, by and with the consent of the Board of Regents and subject to requirements of law, to make appointments to the faculty of the University;

f. To appoint and discharge executive and administrative assistants and personnel and clerical and other employees in the University, within budgetary and other limitations established by the Board of Regents, and subject to requirements of law;

g. To suspend, prefer charges against and recommend to the Board of Regents the removal of any member of the faculty; *provided, however,* that such member of the faculty shall be served notice in writing and shall have opportunity to appear before the Board of Regents and show cause why he should not be removed. The decision of the Board of Regents shall be final in all cases;

h. To preside at commencement and graduation exercises, and to grant degrees and sign diplomas and certificates in accordance with regulations and policies prescribed by the Board of Regents;

i. To prescribe requirements for scholastic and athletic honors and to award such honors, awards, medals and certificates of proficiency or excellence;

j. To suspend or dismiss students from the University for failure in scholarship or for improper conduct;

k. To prescribe detailed regulations, within policies of the Board of Regents, for the operation of the University and for the conduct of faculty, students and employees;

l. To prepare and submit periodic and special reports and budgets to the Board of Regents relative to the activities and operation of the University; to prepare and submit to the Board of Trustees the annual report of the University for each year, at the end of the school year;

m. To make recommendations to the Board of Regents, at any time, on his own initiative, relative to any aspect of the operation of the University, or for the furtherance of the objectives of the University;

n. To appoint faculty committees and boards to make continuing or special studies relative to proposed revisions of curricula, courses of study or other matters affecting the University, for submission by him to the Board of Regents with his recommendations.

SECTION IX. *Appropriation of Funds.* The Department of Finance of the Government of Korea is authorized and directed to appropriate, from time to time, such sums as are necessary to carry out the provisions and objectives of this ordinance.

SECTION X. *Effective Date.* This ordinance shall be effective on the date appearing hereon.

ARCHER L LERCH
Major General United States Army
Military Governor in Korea

HEADQUARTERS
UNITED STATES ARMY MILITARY
GOVERNMENT IN KOREA
Office of the Military Governor

Seoul, Korea

ORDINANCE
NUMBER 103

31 August 1946

ESTABLISHMENT OF
PROPERTY CLAIMS COMMISSION·

SECTION I. *Establishment of Commission.* A Property Claims Commission is hereby established in the Military Government of Korea.

SECTION II. *Jurisdiction and Powers of Commission.* The Property Claims Commission shall have the following jurisdiction, functions and powers :

a. To ascertain and determine the ownership of property affected or claimed to be affected by the provisions of Ordinance Number 2, dated 25 September 1945, and Ordinance Number 33, dated 6 December 1945 :

b. To conduct investigations, hold hearings, and make interlocutory and final determinations with respect to such property ;

c. To make, issue and publish decisions, judgments and orders with respect to such property, in the name of the Military Governor in Korea, which shall be final and binding as the determinations of a court of last resort and have the full force and effect of law ;

d. To stay and remove cases and proceedings with respect to such property which are now or which may hereafter be pending in any court in Korea ; to direct matters to be referred to or brought before the Commission by any Property Custodian or other official, political subdivision, agency or instrumentality of the Government of Korea, or any individual. for adjudication or other disposition ; and to direct any court in Korea, any Register. and any Property Custodian or other official, political subdivision. agency or instrumentality of the Government of Korea and any individual to proceed in such matters in a manner not inconsistent with the decision. judgment or order of the Commission.

— : —

e. To entertain applications, in the discretion of the Commission, for declaratory judgments, and to act as an arbitration or conciliation board in matters involving such property;

f. To issue process to compel the attendance of parties and witnesses and the production of evidence, and to enforce its process and its determinations through contempt proceedings and otherwise;

g. To take testimony and receive evidence, by deposition or otherwise, to issue interrogatories or commissions for the taking of depositions, and to be a tribunal of record in proceedings before it;

h. To make, issue and promulgate rules for its procedure, which shall be followed in all proceedings before it and which shall have the full force and effect of law;

i. To constitute and designate the members of branch commissions as and when necessary or desirable, with such jurisdiction and under such rules, as the Commission shall determine;

j. To employ and compensate necessary personnel for carrying out its functions.

SECTION III. *Determinations of the Commission.* The Commission shall not be bound by existing rules of evidence and procedure, which may, however, be employed by the Commission to the extent deemed practicable or desirable in the work of the Commission. The Commission may apply the existing substantive law of Korea, unless it is considered by the Commission to be contrary to public policy; *provided, however,* that the Commission shall have the power in all cases and proceedings to apply general considerations of justice, equity and fairness in all of its deliberations and decisions, notwithstanding the existence of any specific law or laws on the subject.

SECTION IV. *Composition of Commission.* The Commission shall consist of seven members, four of whom shall be appointed from time to time by the Director of the Department of Justice from among qualified legal personnel of the Legal Opinions, Legal Drafting, Legal Research and Claims Bureaus of the Department of Justice, and the remainder of whom shall be qualified Korean lawyers appointed by the Military Governor on recommendation of the Director of the Department of Justice. Members of the Commission shall serve during the pleasure of the Military Governor. Paid officals of the government serving on or for the Commission shall receive no additional compensation for such service.

SECTION V. *Appropriation of Funds.* The Department of Finance of the Government of Korea is directed to appropriate such sums as may be required by the Commission from time to time to effectuate the provisions of this ordinance, for disposition in accordance with determinations of the Commission, including, but not limited to, allowances for counsel appointed by the Commission for indigent parties. All political subdivisions, agencies, instrumentalities and personnel of the government are directed to furnish assistance and facilities (including, but not limited to, transportation) to the Commission as it may require.

SECTION VI. *Effective Date.* This ordinance shall be effective on the date appearing hereon.

ARCHER L LERCH
Major General United States Army
Military Governor in Korea

HEADQUARTERS
UNITED STATES ARMY MILITARY
GOVERNMENT IN KOREA
Office of the Military Governor
Seoul, Korea

ORDINANCE
NUMBER 104   7 August 1946

ESTABLISHMENT OF THE DEPARTMENT OF PUBLIC WORKS

SECTION I. *Department of Public Works Established; Transfer of Functions, etc.; Purpose.* The Department of Public Works is hereby established in the Government of Korea. The Bureau of Public Works of the Department of Commerce of the Government of Korea is hereby disestablished. All duties and functions of the Bureau of Public Works and of the National Board of Fire Commissioners* are hereby transferred to the Department of Public Works so as to effectuate the carrying out of the purposes hereinafter described. All records, funds, property and Korean personnel of the Bureau of Public Works and of the National Board of Fire Commissioners* shall be transferred to the Department of Public Works, in accordance with orders to be issued by the Director of the Department of Commerce in coordination with the Office of Korea Civil Service and the Budget Bureau of the Department of Finance of the Government of Korea. The purpose of the Department of Public Works shall be to maintain and develop public works necessary and appropriate to the proper functioning of the economy of Korea and the health and welfare of the people of Korea.

SECTION II. *Functions and Duties of the Department of Public Works.* The Department of Public Works shall have the following functions and duties:

a. Under the direction of the Military Governor, to prepare a comprehensive program for public works, which shall include, among other things, the following:

(1) Construction, repair and improvement of public highways and roads, bridges, canals, locks, utilities, and other publicly

---

*See Section II of Ordinance Number 66, dated 10 April 1946.

owned instrumentalities and facilities;

(2) Conservation and development of natural resources, including control, utilization and purification of waters, and flood control;

b. To carry out the foregoing and such other public works programs as the Military Governor may direct;

c. To institute engineering surveys and mapping service;

d. To conduct city planning, including general street improvement, zoning, waterworks, sewage and waste disposal, and fire prevention and protection ;

e. To initiate and recommend policies, laws and regulations concerning public works for enactment and publication in the manner prescribed by law;

f. To perform all other functions and duties necessary to effectuate the provisions of this ordinance, under the supervision and direction of the Military Governor and in accordance with the requirements of law.

SECTION III. *Effective Date.* This ordinance shall be effective on the date appearing hereon.

S E A L
USAMGIK

ARCHER L LERCH
Major General United States Army
Military Governor in Korea

HEADQUARTERS
UNITED STATES ARMY MILITARY
GOVERNMENT IN KOREA
Office of the Military Governor
Seoul, Korea

ORDINANCE
NUMBER 105

31 August 1946

REPEAL OF RICE COLLECTION ORDINANCE;
REVOCATION OF RICE TRANSPORTATION LICENSES

SECTION I. *Purpose.* The purpose of his ordinance is to abrogate certain existing provisions of law, and to annul licenses issued under authority thereof, in view of the issuance of National Food Regulation Number 2, dated 12 August 1946 (*Collection of Rice*) pursuant to the provisions of Section *Vb* of Ordinance Number 90, dated 23 May 1946 (*Economic Controls*).

SECTION II. *Ordinances Repealed.* The following ordinances are hereby repealed:

a. Ordinance Number 45, dated 25 January 1946 (*National Rice Collection*);

b. Ordinance Number 77, dated 24 April 1946 (*Special Rice Provisions for City of Seoul*);

c. Ordinance Number 87, dated 20 May 1946 (*Special Rice Provisions for City of Pusan*).

SECTION III. *Abrogation of Orders, Regulations, &c; Revocation of Licenses to Transport Rice.* All orders, regulations, instructions, memoranda and requirements issued pursuant to any of the ordinances enumerated in Section II hereof are hereby rescinded and abrogated, and all permits issued thereunder are hereby revoked and declared null and void as of the effective date of this ordinance.

SECTION IV. *Surrender of Permits; Restoration of Rice Ration Benefits.* All permits issued pursuant to Ordinance Number 77 or Ordinance Number 78 shall be surrendered by the holders thereof to the issuing authority not later than ten days after the effective date of this ordinance, together with a certificate of the holder as to the amount of rice actually obtained thereunder. In every case where such holder fails to surrender such permit as herein provided, together with such certificate in proper form, it shall be conclusively presumed that the holder of such permit has actually obtained the full amount of rice specified in such permit.

In the event the issuing authority is satisfied that the full amount of rice specified in any surrendered permit was not actually obtained, such authority shall cause appropriate rice ration benefits to be restored to all persons on whose behalf such permit was granted.

SECTION V. *Effect on Prosecutions and Offenses.* This ordinance is not intended and shall not be construed to extinguish offenses under any of the ordinances or measures herein repealed. rescinded or abrogated, committed prior to the effective date hereof, nor to abate or bar existing or future prosecutions based on such offenses.

SECTION VI. *Effective Date.* This ordinance shall be effective on the date appearing hereon.

S E A L
USAMGIK

ARCHER L LERCH
Major General United States Army
Military Governor in Korea

HEADQUARTERS
UNITED STATES ARMY MILITARY
GOVERNMENT IN KOREA
Office of the Military Governor
Seoul, Korea

ORDINANCE
NUMBER 106

18 September 1946

ESTABLISHMENT OF THE
INDEPENDENT CITY OF SEOUL

SECTION I. *Removal of Seoul from Jurisdiction of Kyonggi.* The City of Seoul is hereby removed from the jurisdiction of Kyonggi Province.

SECTION II. *Seoul Established as Independent City.* The City of Seoul is hereby established as an independent city, the capital of Korea, on the governmental level of a province, with all the powers, duties, functions and rights of a province.

SECTION III. *Transfers to Effectuate Purpose of Ordinance.* All duties, functions and records of Kyonggi Province relating primarily to the City of Seoul are hereby transferred to the City of Seoul. All Korean personnel of Kyonggi Province concerned primarily with the City of Seoul shall be transferred to the City of Seoul in the manner provided by law, upon recommendation of the Office of Korea Civil Service. The Office of Korea Civil Service shall consult with the Governor of Kyonggi Province and the Mayor of Seoul prior to making such recommendation. All funds of and appropriations for Kyonggi Province, relating primarily to the City of Seoul, shall be transferred to the City of Seoul in the manner provided by law, upon recommendation of the Chief of the Budget Bureau of the Department of Finance. The Chief of the Budget Bureau shall consult with the Governor of Kyonggi Province and the Mayor of Seoul prior to making such recommendation.

SECTION IV. *Districts within Seoul.* The City of Seoul shall consist of the eight districts now included within the city limits, to wit:

Chorgno District   Sodaemun (Westgate) District

Chung (Central) District   Dongdaemun (Eastgate) District

Mapo District

Yongsan District

Songtong District

Yongdungpo District

SECTION V. *Effective Date*. This ordinance shall be effective on the tenth day after the date appearing hereon.

ARCHER L LERCH

Major General United States Army

Military Governor in Korea

HEADQUARTERS
UNITED STATES ARMY MILITARY
GOVERNMENT IN KOREA
Office of the Military Governor
Seoul, Korea

ORDINANCE
NUMBER 107

14 September 1946

ESTABLISHMENT OF WOMEN'S BUREAU

SECTION I. *Women's Bureau Established.* The Women's Bureau is hereby established within the Department of Public Health and Welfare of the Government of Korea. The Chief of the Women's Bureau shall be a woman, appointed by the Military Governor, to hold office during his pleasure. The appointment of a necessary staff and personnel, and the procurement of suitable quarters and supplies for the Bureau are authorized.

SECTION II. *Functions and Duties.* The functions and duties of the Women's Bureau shall be:

a. To advise the Military Governor on policies affecting the social, economic, political and cultural amelioration of Korean women;

b. To compile and analyze data and to make and publish continuing studies concerning matters affecting the position and general welfare of Korean women;

c. To formulate standards and policies for recommendation to the appropriate political subdivisions, officials and instrumentalities of the Government of Korea, pertaining to the promotion of the welfare and interests of Korean women, including but not limited to:

(1) Improvement of the working conditions of women;

(2) Advancing the opportunities of women for profitable employment;

(3) The welfare of women in industry, agriculture, education, the arts and professions, and the home;

(4) The activities of women in government service;

(5) Health, prenatal care and maternity confinement;

(6) Woman suffrage;

(7) Control and elimination of prostitution;

239

(8) Female delinquency and institutional care;

(9) Travelers' aid for women and children.

SECTION III. *Appropriation of Funds*. The Department of Finance of the Government of Korea is directed to appropriate funds necessary to the carrying out of the provisions of this ordinance.

SECTION IV. *Effective Date*. This ordinance shall be effective on the tenth day after the date appearing hereon.

S E A L
USAMGIK

ARCHER L LERCH
Major General United States Army
Military Governor in Korea

HEADQUARTERS
UNITED STATES ARMY MILITARY
GOVERNMENT IN KOREA
Office of the Military Governor
Seoul, Korea

ORDINANCE
NUMBER 108

18 September 1946

REPEAL OF ENTERPRISE LAW

SECTION I. *Purpose.* The purpose of this ordinance is to encourage
and stimulate the development of industry and commerce through the abrogation of the Japanese Enterprise Law*, which restricted the development of
Korean business enterprise and activity, in the interest of the Japanese war
program.

SECTION II. *Repeal of Laws.* The following laws are hereby repealed:

a. Imperial Ordinance Number 1084, 11 December 1941 (*Enterprise
Law*:

b. Government General Order Number 338, 26 December 1941 (*Regulations Enforcing the Enterprise Law*), as revised;

c. All other provisions of law, and parts thereof, issued pursuent to,
or designed to enforce or effectuate the *Enterprise Law.**

SECTION III. *Effective Date.* This ordinance shall be effective on the
the date appearing hereon.

ARCHER L LERCH
Major General United States Army
Military Governor in Korea

---

* Imperial Ordinance Number 1084, dated 11 December 1941.

HEADQUARTERS
UNITED STATES ARMY MILITARY
GOVERNMENT IN KOREA
Office of the Military Governor
Seoul, Korea

ORDINANCE
NUMBER 109

15 October 1946

REPRINTED COPY*

REVISION OF PROVINCIAL AND LOCAL TAXES

SECTION I. *Purpose.* The purpose of this ordinance is to increase provincial and local revenues by revision of tax rates and simplification of the tax structure, so as to promote the more efficient functioning of provincial and local governments and the economy of Korea.

SECTION II. *Household Income Tax. a. Rates.* Commencing with the calendar year 1945, provincial and local tax rates on household income shall be:

Taxing Authority	Tax Rate
Province	¥ 11 per unit
City	¥ 11 per unit
Town	¥ 11 per unit
Village	¥ 11 per unit
School	¥ 30 per unit

b. Units. Units for the various income grades shall be those indicated by the household income tax schedules now established by law and in effect in the provinces; *provided,* however, that the unit for Grade 133 is hereby amended to read:

Grade	Amount of Income	Unit
133	¥ 2,000,000 and over	25,700; plus 10,000 for every additional ¥ 500,000 of income

c. Exemptions. The following earned income exemptions shall be the only permissible deductions from taxable income under the provincial and local household income tax:

---

*Pages 1-2 only reprinted. All former copies of these pages should be removed and destroyed.

(1) ¥ 1,000 for the head of the household.

(2) ¥ 200 for each member of the household under 18 or over 60 years of age, but in no case to exceed a total of ¥ 2,000.

d. Reduction. In any case where the aggregate of provincial and local household income taxes payable by the taxpayer added to the total of the national income taxes paid by him and the members of his household would exceed 90% of the total net income of such household, the provincial and local household income taxes payable by such taxpayer shall be reduced proportionately by an amount sufficient to make the aggregate of the national, provincial and local income taxes paid by him and by the members of his household equal to 90% of the total net income of such household.

SECTION III. *Building Tax. a. Rates.* Commencing with the calendar year 1946, the provincial tax and local surtax rates on buildings shall be:

Taxing Authority	Tax Rate
Province	¥ 1.5 per unit
City	240% surtax on the provincial tax
Town	200% surtax on the provincial tax
Village	160% surtax on the provincial tax

b. Units. Units for the various classes and grades of buildings shall be those indicated by the building tax schedules now established by law and in effect in the provinces.

SECTION IV. *Forestry Tax.* Commencing with the calendar year 1946, the provincial tax rate on forest areas shall be ¥ 1.5 per unit of 3000 pyong or fraction thereof.

SECTION V. *Slaughter Tax.* Commencing with the calendar year 1946, the provincial tax rate on the slaughter of cows and pigs shall be 50 yen per cow and 5 yen per pig.

SECTION VI. *Fisheries and Fishing Taxes.* Commencing with the fiscal year 1945* or the calendar year 1946 (whichever is applicable under existing laws, in the case of any particular tax), provincial tax rates on fisheries shall be:

---

*i.e.: 1 April 1945 to 31 March 1946.

Fish Culture.

Type Of Area	Tax Rate Per 1,000 Square Meters
(1) Area enclosed by stationary bamboo poles	(1) ¥ 5 per year
(2) Area enclosed by single bamboo net	(2) ¥ 5 per year
(3) Oyster culture area	(3) ¥ 5 per year
(4) Culture area for other forms of marine life	(4) ¥ 5 per year

Trap Net Fishing.

Three percent of the value of the total yearly catch.

Exclusive Fishing Right.

Three percent of the value of the total yearly catch of fish, and two percent of the value of the total yearly catch of other forms of marine life.

Whale Fishing.

Species	Tax Rate
(1) Ballaena	(1) ¥ 600 per whale
(2) Humpback, sperm and fin	(2) ¥ 400 per whale
(3) California Grey (devil fish)	(3) ¥ 100 per whale

Bottom Drag Net Fishing (by motor driven boat

Value Of Yearly Catch	Tax Rate Per Boat
(1) Under ¥ 5,000	(1) ¥ 200 per year
(2) ¥ 5,000 to 10,0 0	(2) ¥ 300 per year
(3) ¥ 10,001 to 15,000	(3) ¥ 400 per year
(4) ¥ 15,001 to 25,000	(4) ¥ 700 per year
(5) ¥ 25,000 and over	(5) ¥ 700 per year; plus ¥ 200 for each additional ¥ 5,000 value of catch

f. Purse Seine Fishing.

Value of Yearly Catch	Tax Rate Per Boat Equipped With Purse Seine
(1) Under ¥ 10,000	(1) ¥ 200 per year.
(2) ¥ 10,000 to ¥ 15,000	(2) ¥ 300 per year
(3) ¥ 15,001 to ¥ 20,000	(3) ¥ 400 per year
(4) ¥ 20,001 to ¥ 25,000	(4) ¥ 500 per year
(5) ¥ 25,001 to ¥ 35,000	(5) ¥ 600 per year
(6) Over ¥ 35,000	(6) ¥ 600 per year ; plus ¥ 200 for each additional ¥ 5,000 value of catch

g. Submarine Armor Fishing.

¥ 1,000 per year for each submarine armor diving suit.

h. Shore Dragnet Fishing (including shore seine and running tuck net).

Length of Floating Net	Tax Rate Per Net
(1) Under 300 meters	(1) ¥ 10 per year
(2) 301 to 500 meters	(2) ¥ 15 per year
(3) 501 to 800 meters	(3) ¥ 25 per year
(4) Over 800 meters	(4) ¥ 50 per year

i. Drag Net Fishing (by boat without motor).

Length of Boat	Tax Rate Per Net
(1) Under 15 meters	(1) ¥ 40 per year
(2) 15 meters or over	(2) ¥ 100 per year

j. Scoop Net Fishing.

Method	Tax Rate Per Net
(1) One boat and light	(1) ¥ 10 per year
(2) More than 4 boats	(2) ¥ 25 per year
(3) Other	(3) ¥ 5 per year

k. Roe Net Fishing.

Method	Tax Rate Per Net
(1) With boat	(1) ¥ 40 per year
(2) Without boat	(2) ¥ 10 per year

l. Swig Net Fishing (angler type net).

Beam Of Boat	Tax Rate Per Boat
(1) Under 6 meters	(1) ¥ 20 per year
(2) 6 to 8 meters	(2) ¥ 30 per year
(3) Over 8 meters	(3) ¥ 40 per year

m. Set Net Fishing.

Length of Main Net Floater	Tax Rate Per Net
(1) Under 90 meters	(1) ¥ 10 per year
(2) 90 to 180 meters	(2) ¥ 20 per year
(3) Over 180 meters	(3) ¥ 30 per year

n. Dam Net Fishing.

Length of Net	Tax Rate Per Net
(1) Under 200 meters	(1) ¥ 10 per year
(2) 200 to 400 meters	(2) ¥ 15 per year
(3) Over 400 meters	(3) ¥ 20 per year

o. Haul Net Fishing (reeled by hand or hauled by rowing).

Length Of Boat	Tax Rate Per Boat
(1) 8 meters or under	(1) ¥ 10 per year
(2) Over 8 meters	(2) ¥ 20 per year

— 5 —

p. Dredge Net Fishing.

Length Of Boat	Tax Rate Per Boat
(1) 9 meters or under	(1) ¥ 10 per year
(2) Over 9 meters	(2) ¥ 20 per year

q. Ring Net Fishing.

Species	Tax Rate Per Net
(1) Sardine, horse mackerel, mackerel, grey mullet, herring	(1) ¥ 100 per year
(2) Others	(2) ¥ 25 per year

r. Defensive Net (including bind net and scoop net into which fish are driven by stones).

¥ 30 per year per net

s. Drift Net (with sailboat).

Species	Tax Rate Per Boat
(1) Sardine, Alaska pollock, crab, mackerel, Spanish mackerel	(1) ¥ 20 per year
(2) Other	(2) ¥ 25 per year

t. Drift Net (with motorboat).

Species	Type of Motor	Tax Rate Per Boat
(1) Mackerel or Spanish mackerel	(a) Under 20 hp.	(a) ¥ 60 per year
	(b) 20 to 40 hp.	(b) ¥ 100 per year
	(c) Over 40 hp.	(c) ¥ 200 per year
(2) Others	(a) Under 20 hp.	(a) ¥ 50 per year
	(b) 20 to 40 hp.	(b) ¥ 80 per year
	(c) Over 40 hp.	(c) ¥ 150 per year

u. Trawling Line (artificial bait).

Length Of Boat	Tax Rate Per Boat
(1) 9 meters or under	*(1)* ￥ 10 per year
(2) Over 9 meters	*(2)* ￥ 40 per year

v. Diving (without submarine armor).

￥ 10 per year per diver

w. Harpooning (by harpoon gun).

￥ 60 per year per gun

x. Angling (including spearing or trawling from boat with natural bait).

Method	*Length of Boat*	*Type Of Motor*	*Tax Rate*
(1) Angling with single rod and line	Any	Any	￥ 10 per year per rod
(2) Trawling with natural bait or spearing	*(a)* 9 meters or under	*(a)* None	*(a)* ￥ 20 per year per boat
	(b) Over 9 meters	*(b)* None	*(b)* ￥ 40 per year per boat
(3) Trawling with natural bait or spearing	*(a)* Any	*(a)* Under 10 hp	*(a)* ￥ 20 per year per boat
	(b) Any	*(b)* 10-20 hp	*(b)* ￥ 40 per year per boat
	(c) Any	*(c)* 21-30 hp	*(c)* ￥ 60 per year per boat
	(d) Any	*(d)* Over 30 hp	*(d)* ￥ 100 per year per boat

y. Other Boat Fishing (not specified above).

Length Of Boat	Tax Rate Per Boat
(1) 6 meters or under	*(1)* ￥ 10 per year
(2) Over 6 meters	*(2)* ￥ 20 per year

SECTION VII. *Vehicle Tax.* Commencing with the calendar year 1946, the provincial tax and local surtax rates on vehicles shall be:-

Description			Provincial Tax Rate Per Vehicle		
Type Of Vehicle	Passenger Capacity	Capacity Pay Load In Kilograms	Common Carrier	Business Use	Private Use
(1) Trolley -Car	(a) 50 or under		(a) ¥6,000		
	(b) 51 to 80		(b) 7,200		
	(c) 81 or over		(c) 8,000		
(2) Automobile	(a) Bantam type			(a) ¥ 400	(a) ¥600
	(b) 3 or under			(b) 800	(b) 1,200
	(c) 4 to 8			(c) 1,200	(c) 1,800
(3) Passenger Bus	(a) 15 or under			(a) 1,600	(a) 2,400
	(b) 16 to 23			(b) 2,000	(b) 3,000
	(c) 24 or over			(c) 2,600	(c) 4,000
(4) Truck		(a) under 1,000		(a) 800	(a) 1,200
		(b) 1,000 to 2,000		(b) 1,200	(b) 1,800
		(c) 2,001 to 3,000		(c) 1,600	(c) 2,400
		(d) 3,001 to 4,000		(d) 2,000	(d) 3,000
		(e) Over 4,000		(e) 2,600	(e) 4,000
(5) Motorcycle			200	200	1,000
(6) Bicycle			30	30	30
(7) Trailer			16	16	16

Type Of Vehicle	Passenger Capacity	Capacity Pay Load In Kilograms	Common Carrier	Business Use	Private Use
(8) Carriage	(*a*) 1 horse (*b*) 2 or more horses		(*a*) 300 (*b*) 600	(*a*) 300 (*b*) 600	(*a*) 900 (*b*) 1,200
(9) Rickshaw			120	120	300
(10) Horse Cart	(*a*) 2 wheeled (drawn by horse or cow) (*b*) 4 wheeled (drawn by horse or cow)		(*a*) 120 (*b*) 200	(*a*) 120 (*b*) 200	(*a*) 120 (*b*) 200
(11) Cart (hand drawn)			60	60	60

City, town and village surtaxes shall be 170% of the provincial tax on automobiles and motorcycles, and 250% of the provincial tax on other vehicles.

SECTION VIII. *Immovable Property Purchase Tax.* Commencing with the calendar year 1946, the provincial tax and local surtax rates on purchases of immovable property shall be:

Taxing Authority	Tax Rate
Province	4% of purchase price
City, town or village tax	90% surtax on provincial tax

SECTION IX. *Restaurant Attendance Tax.* Commencing with the calendar year 1946 the rate of provincial tax on attendance in restaurants shall be 5￦ per person, of which two-sevenths shall be paid by the province and the city, town or village in which the tax is collected.

SECTION X *Residence Tax.* Commencing with the calendar year 1946 the city, town or village residence tax rates shall be

Taxing Authority	Tax Rate
City	￥ 20 per head of the household
Town	￥ 20 per head of the household
Village	￥ 10 per head of the household
City, town or village	Three tenths of one percent
(where main office is	of the paid up capital of each
located)	juridical person

SECTION XI. *Changes in Miscellaneous Local Taxes. a. Dog Taxes and Local Boat Taxes.* Commencing with the calendar year 1946, the city, town and village dog tax rates shall be ￥ 30 per year per dog, and the city, town and village boat tax rates shall be as follows:

Capacity of Boat	Tax Rate Per Boat
(1) 1 to 5 tons	(1) ￥ 50 per year
(2) 5 to 10 tons	(2) ￥ 100 per year
(3) Over 10 tons	(3) ￥ 100 per year ; plus ￥ 10 per additional ton

b. Taxes Repealed. The following miscellaneous city, town and village taxes are hereby repealed and shall not be levied commencing with the calendar year 1946 :

Actors or actresses	Female entertainers
Billiard tables	Pianos
Domestic servants	Safes
Electric fans	Waiters or waitresses
Electric poles	

SECTION XII. *Repeal of Inconsistent Laws.* All laws, ordinances orders, regulations, directives and instructions, or parts thereof, which are inconsistent herewith or in conflict with the provisions hereof are hereby repealed.

SECTION XIII. *Collection Procedures.* Existing collection procedure for provincial and local taxes shall remain in full force and effect until superseded, modified, or repealed in the manner prescribed by law.

— 10 —

SECTION XIV. *Effective Date.* This ordinance shall be effective on the date appearing hereon.

ARCHER L LERCH
Major General United States Army
Military Governor in Korea

HEADQUARTERS
UNITED STATES ARMY MILITARY
GOVERNMENT IN KOREA
Office of the Military Governor
Seoul, Korea

ORDINANCE
NUMBER 110

14 September 1946

ISSUE OF KOREAN REVENUE STAMPS;
DISCONTINUANCE OF FORMER ISSUES

SECTION I. *Korean Revenue Stamp Issue Authorized.* The issuance and use of Korean revenue stamps are hereby authorized. Such stamps shall be used in all cases where the use of revenue stamps is required. They shall be in the following denominations and distinctive colors:

Denomination	Color
1 weun	Orange
5 weun	Green
10 weun	Dark red
50 weun	Light red
100 weun	Blue

SECTION II. *Imperial Japanese (Korean Overprinted) Revenue Stamps Discontinued.* The sale by post offices of Imperial Japanese (Korean overprinted) revenue stamps shall be discontinued on the effective date hereof, and on and after such date only Korean revenue stamps shall be sold or used for revenue stamp purposes.

SECTION III. *Exchange of Imperial Japanese (Korean Overprinted) Revenue Stamps.* Within thirty days after the effective date hereof, unused Imperial Japanese (Korean overprinted) revenue stamps may be exchanged at any post office for Korean revenue stamps equal in face value. Within sixty days after the effective date hereof, unused Imperial Japanese (Korean overprinted) revenue stamps may be exchanged by any post office at the Department of Finance of the Government of Korea for Korean revenue stamps equal in face value.

SECTION IV. *Administrative Procedures.* The Director of the Department of Communications and the Director of the Department of Finance of the Government of Korea shall consult together and adopt administrative procedures and safeguards with respect to the exchange of Imperial Japanese (Korean overprinted) revenue stamps for Korean revenue stamps.

SECTION V. *Penalties.* Any person who fraudulently exchanges or attempts to exchange any counterfeit Imperial Japanese (Korean overprinted) revenue stamps for Korean revenue stamps, or who violates any provision hereof, shall, upon conviction by a Military Occupation Court, suffer such punishment as the court shall determine.

SECTION VI. *Effective Date.* This ordinance shall be effective on the tenth day after the date appearing hereon.

SEAL
USAMGIK

ARCHER L LERCH
Major General United States Army
Military Governor in Korea

HEADQUARTERS
UNITED STATES ARMY MILITARY
GOVERNMENT IN KOREA
Office of the Military Governor
Seoul, Korea

ORDINANCE
NUMBER 111   18 September 1946

GRAIN AND STRAW BAG INSPECTION

SECTION I. *Modification of Law by Regulation.* All laws, ordinances, orders, regulations, directives and instructions relating to the inspection, grading and packing of grains and straw bags shall remain in full force and effect until superseded, modified or repealed by regulations issued by the Director of the Department of Agriculture pursuant hereto.

SECTION II. *Preparation and Publication of Regulations.* The Director of the Department of Agriculture of the Government of Korea shall prepare regulations for the inspection, grading and packing of grains and straw bags. Such regulations, upon approval by the Military Governor, shall be published in the *Official Gazette* and shall have the force of law.

SECTION III. *Grain and Straw Bag Inspection.* The Director of the Department of Agriculture shall use existing Grain Inspection Offices, facilities and personnel, and may employ additional personnel necessary in the administration of published regulations issued pursuant to this ordinance.

SECTION IV. *Penalties.* Any person violating the provisions of any regulation issued pursuant to this ordinance shall, upon conviction by a Military Occupation Court, suffer such punishment as the court shall determine.

SECTION V. *Effective Date.* This ordinance shall be effective on the date appearing hereon.

SEAL
USAMGIK

ARCHER L LERCH
Major General United States Army
Military Governor in Korea

HEADQUARTERS
UNITED STATES ARMY MILITARY
GOVERNMENT IN KOREA
Office of the Military Governor
Seoul, Korea

ORDINANCE
NUMBER 112

18 September 1946

REGULATION OF CHILD LABOR

SECTION I. PURPOSE. THE PURPOSE OF THIS ORDINANCE IS TO REGULATE CHILD LABOR IN ACCORDANCE WITH HUMANITARIAN, ENLIGHTENED PRINCIPLES ACCEPTED BY CIVILIZED NATIONS THROUGHOUT THE WORLD, SO AS TO ENABLE THE CHILDREN OF KOREA TO ATTAIN MATURITY PREPARED FOR THE RESPONSIBILITIES OF CITIZENSHIP IN MODERN SOCIETY.

SECTION II. CHILDREN UNDER FOURTEEN. NO CHILD UNDER 14 FULL YEARS OF AGE SHALL BE EMPLOYED, PERMITTED OR SUFFERED TO WORK IN, ABOUT OR IN CONNECTION WITH ANY PUBLIC OR PRIVATE INDUSTRIAL OR MERCANTILE UNDERTAKING OR ENTERPRISE.

As used herein, *public or private industrial or mercantile undertaking or enterprise* is defined to include any: (*1*) mill, (*2*) factory, (*3*) workshop, (*4*) mercantile or mechanical establishment, (*5*) house or home manufactory or workshop, (*6*) store, (*7*) office, (*8*) office building, (*9*) restaurant, (*10*) bakery, (*11*) barber shop, (*12*) hotel, (*13*) apartment house, (*14*) bootblack or shoeshine stand or establishment, (*15*) public stable, (*16*) garage, (*17*) laundry, (*18*) place of amusement, (*19*) club, (*20*) occupation as a driver, (*21*) occupation in any brick or lumber yard, (*22*) occupation in the construction or repair of buildings, (*23*) occupation in the distribution, transmission or sale of merchandise, (*24*) occupation in the transmission of messages.

This provision shall not be effective with respect to presently existing contracts or agreements of hire or employment of children between the ages of 12 and 14 full years, until the 15th day of February 1947; the hiring or

— 1 —

employment of children under the age of 14 full years under new contracts or agreements for hire or employment is, however, prohibited on and after the date appearing hereon; and all existing contracts or agreements of hire or employment of children under the age of 12 full years are hereby abrogated, and the continued employment of any such children on or after the 30th day after the date appearing hereon is hereby prohibited.

SECTION III. *EMPLOYMENT OF CHILDREN UNDER FOURTEEN DURING SCHOOL HOURS.* IT SHALL BE UNLAWFUL FOR ANY PERSON, FIRM OR CORPORATION TO EMPLOY, PERMIT OF SUFFER TO WORK ANY CHILD UNDER 14 FULL YEARS OF AGE IN ANY BUSINESS OR SERVICE WHATEVER DURING ANY OF THE HOURS WHEN THE PUBLIC SCHOOLS OF THE DISTRICT IN WHICH THE CHILD RESIDES ARE IN SESSION, IF SUCH CHILD IS REQUIRED TO ATTEND AT ANY SUCH SCHOOL.

SECTION IV. *CHILDREN UNDER SIXTEEN.* NO CHILD UNDER THE AGE OF 16 FULL YEARS SHALL BE EMPLOYED, PERMITTED OR SUFFERED TO WORK IN, ABOUT OR IN CONNECTION WITH ANY PUBLIC OR PRIVATE HEAVY INDUSTRIAL OR HARMFUL UNDERTAKING OR ENTERPRISE.

As used herein, *public or private heavy industrial or harmful undertaking or enterprise* is defined to include any of the following: (*1*) adjusting any belt to any moving machinery, (*2*) sewing or lacing machine belts in any workshop or factory, (*3*) oiling, wiping or cleaning machinery in motion or assisting therein, (*4*) operating or assisting in operating any of the following machines (*a*) circular or band saws, (*b*) wood shapers, (*c*) wood jointers, (*d*) planers, (*e*) sandpaper or wood-polishing machinery, (*f*) wood-turning or boring machinery; (*g*) picker machines or machines used in picking wool, cotton, hair or any other material, (*h*) carding machines, (*i*) paperlace machines, (*j*) leather-burnishing machines, (*k*) job or cylinder printing presses operated by power other than foot power, (*l*) boring or drill presses, (*m*) stamping machines used in sheet-metal and tinware or in paper and leather manufacturing, or in washer and nut factories, (*n*) metal or paper cutting machines, (*o*) corner-staying machines in paper box factories, (*p*) corrugating rolls, such as are used in corrugated paper, roofing or washboard factories, (*q*) steam boilers, (*r*) dough brakes or cracker machinery of any description, (*s*) wire or iron straightening or drawing machinery, (*t*) rolling mill

— 2 —

machinery, (*u*) power punches or shears. (*v*) washing, grinding or mixing machinery. (*w*) calendar rolls in paper manufacturing, (*x*) laundering machinery; (5) occupation in proximity to any hazardous or unguarded belts, machinery or gearing, (6) occupation upon any railroad, whether steam, electric or hydraulic, (7) occupation upon any vessel or boat engaged in navigation or commerce, (8) occupation in, about or in connection with any processes in which dangerous or poisonous acids are used, (9) the manufacture or use of dangerous or poisonous dyes, (10) the manufacture or packing of paints, white or red lead, (11) occupation causing dust in injurious quantities, (12) the manufacture or use of compositions with dangerous or poisonous gases, (13) the manufacture or use of compositions of lye in which the quantity thereof is injurious to health, (14) occupation on scaffolding. (15) occupation in any distillery, brewery or any other establishment where malt or alcoholic liquors are manufactured, packed, wrapped or bottled, (16) occupation on or in connection with any dam, smeltery, reverberatory furnace. refinery, dressing and reduction plant, x-ray machinery, or any activity involving use of or exposure to any radioactive element or substance, (17) heavy work in the building trades, (18) occupation in any tunnel or excavation. (19) occupation in, about or in connection with any mine, coal breaker, coke oven or quarry, (20) assorting, manufacturing or packing tobacco, (21) operating any automobile, motor car or truck, (22) occupation in any bowling alley, pool or billiard room, (23) occupation upon the stage of any theatre or concert hall or in connection with any theatrical performance or other exhibition or show, (24) any other occupation dangerous to the life or limb. or injurious to the health or morals. of such child

SECTION V. CHILDREN UNDER EIGHTEEN. NO CHILD UNDER THE AGE OF 18 FULL YEARS SHALL BE EMPLOYED, PERMITTED OR SUFFERED TO WORK IN, ABOUT OR IN CONNECTION WITH ANY PUBLIC OR PRIVATE UNDERTAKING OR ENTERPRISE WHICH IS DANGEROUS TO LIFE, OR INJURIOUS TO HEALTH OR MORALS.

As used herein, *public or private undertaking or enterprise which is dangerous to life or limb, or injurious to health or morals* is defined to include any of the following: (1) occupation in, about or in connection with blast furnaces, docks or wharves, (2) occupation in the outside erection and repair of electric wires, (3) occupation in the running or management of elevators or lifts (unless equipped with suitable safety devices), (4) occupation in, about or in connection with hoisting machinery, or dynamos, (5) occupation in oiling or cleaning machinery in motion, (6) occupation in the operation of emory wheels or any abrasive, polishing or buffing wheel where articles of the baser metals or iridium are manufactured, (7) switch tending, (8) gate tending. (9) track repairing, (10) occupation as brakeman, fireman, engineer, motorman or conductor upon a railroad, (11) occupation as railroad telegraph operator. (12) occupation as pilot, fireman or engineer upon any boat or vessel, (13) occupation in or about any establishment wherein nitroglycerine, dynamite. dualin, guncotton, gunpowder or other high or dangerous explosives are manufactured, compounded or stored, (14) occupation in the manufacture of white or yellow phosphorus or phosphorus matches. (15) occupation in any cement mill, (16) occupation in any hotel, theater, concert hall, place of amusement or any establishment where intoxicating liquors are sold, (17) occupation in, about or in connection with any racetrack or any gaming. betting or gambling establishment or enterprise.

SECTION VI.   FEMALE PERSONS UNDER TWENTY-ONE.   NO FEMALE PERSON UNDER 21 FULL YEARS OF AGE SHALL BE EMPLOYED, PERMITTED OR SUFFERED TO WORK IN OR ABOUT ANY: (1 MINE, (2) QUARRY, (3) COAL BREAKER. EXCEPT IN THE OFFICE THEREOF, (4 OR IN OILING OR CLEANING MACHINERY WHILE IN MOTION.

SECTION VII.   MAXIMUM WORK HOURS: CHILDREN UNDER 16. NO CHILD UNDER THE AGE OF 16 FULL YEARS SHALL BE EMPLOYED. PERMITTED OR SUFFERED TO WORK IN, ABOUT, OR IN CONNECTION WITH ANY ESTABLISHMENT OR OCCUPATION: (1) FOR MORE THAN SIX DAYS IN ANY ONE WEEK, (2) NOR MORE THAN FORTY-EIGHT HOURS IN ANY ONE WEEK, (3) NOR MORE THAN EIGHT HOURS IN ANY ONE DAY, EXCLUSIVE OF LUNCH HOUR, (4) OR BEFORE THE HOUR OF 7 O'CLOCK IN THE MORNING OR AFTER THE HOUR OF 7 O'CLOCK IN THE EVENING. THE PRESENCE OF SUCH CHILD IN ANY ESTABLISHMENT DURING WORKING HOURS SHALL BE *PRIMA FACIE* EVIDENCE OF HIS EMPLOYMENT THEREIN.

THIS PROVISION SHALL NOT BE EFFECTIVE UNTIL THE 18th DAY OF FEBRUARY 1947.

SECTION VIII. MAXIMUM WORK HOURS: PERSONS UNDER 18. NO CHILD UNDER THE AGE OF 18 FULL YEARS SHALL BE EMPLOYED. PERMITTED OR SUFFERED TO WORK IN. ABOUT OR IN CONNECTION

— 4 —

WITH ANY ESTABLISHMENT OR OCCUPATION: (1) FOR MORE THAN SIX DAYS IN ANY ONE WEEK, (2) NOR MORE THAN FIFTY-FOUR HOURS IN ANY ONE WEEK, (3) NOR MORE THAN TEN HOURS IN ANY ONE DAY, (4) OR BEFORE THE HOUR OF 6 O'CLOCK IN THE MORNING OR AFTER THE HOUR OF 10 O'CLOCK IN THE EVENING.

THIS PROVISION SHALL NOT BE EFFECTIVE UNTIL THE 15th *DAY OF FEBRUARY 1947.*

SECTION IX. WAGES FOR PIECE WORK. NO PERSON UNDER THE AGE OF 18 FULL YEARS SHALL BE EMPLOYED, PERMITTED OR SUFFERED TO WORK IN, ABOUT OR IN CONNECTION WITH ANY ESTABLISHMENT OR OCCUPATION IN WHICH THE AMOUNT OF HIS WAGES IS DETERMINED BY THE NUMBER OF ARTICLES, FINISHED OR SEMI-FINISHED, WHICH HE PRODUCES, UNLESS SUCH METHOD AND AMOUNT OF COMPENSATION HAVE BEEN APPROVED IN ADVANCE IN WRITING BY THE DEPARTMENT OF LABOR OF THE GOVERNMENT OF KOREA.

SECTION X. LIMITATION ON DURATION OF CERTAIN CONTRACTS OF EMPLOYMENT. NO CONTRACT OR AGREEMENT FOR THE EMPLOYMENT OF PERSONS UNDER THE AGE OF 18 FULL YEARS SHALL BE MADE OR BE ENFORCEABLE FOR A PERIOD EXCEEDING SIX MONTHS FROM THE DATE OF THE CONTRACT OR AGREEMENT.

SECTION XI. SICK LEAVE. PREGNANCY AND NURSING. SICKNESS NOT EXCEEDING TWO WEEKS SHALL NOT BE DEEMED CAUSE FOR THE TERMINATION OF ANY CONTRACT OR AGREEMENT OF ANY PERSON UNDER 21 FULL YEARS OF AGE. ALL EMPLOYEES UNDER THE AGE OF 21 FULL YEARS SHALL BE ENTITLED TO NOT LESS THAN TWO AND ONE-HALF DAYS OF SICK LEAVE PER MONTH AT FULL PAY, WHICH MAY BE ACCUMULATED TO THIRTY DAYS PER YEAR. EMPLOYEES MAY BE REQUIRED TO FURNISH DOCUMENTARY PROOF IN CASE OF ANY CLAIM OF ABSENCE EXCEEDING TWO DAYS DUE TO SICKNESS. PREGNANT EMPLOYEES UNDER THE AGE OF 21 FULL YEARS SHALL BE GRANTED NINETY DAYS LEAVE WITHOUT PAY (EXCEPT THAT PAY SHALL BE GIVEN FOR ACCRUED REGULAR SICK LEAVE) FOR PURPOSES OF NORMAL CONFINEMENT, *PROVIDED, HOWEVER,* THAT SUCH PREGNANCY LEAVE SHALL BE REQUIRED TO BE GRANTED ONLY IN CASE SUCH EMPLOYEE HAS BEEN IN THE

— 5 —

EMPLOY OF THE EMPLOYER FOR AT LEAST ONE YEAR. WHERE INFANTS ARE BROUGHT TO THE PLACE OF EMPLOYMENT BY WOMEN EMPLOYEES UNDER THE AGE OF 21 FULL YEARS, APPROPRIATE OPPORTUNITIES SHALL BE AFFORDED FOR REASONABLE NURSING REQUIREMENTS FOR SUCH INFANTS.

SECTION XII. EDUCATION AND RECREATION FOR CHILDREN UNDER 16. AT LEAST 6 HOURS AND 2 HOURS, RESPECTIVELY, OF THE AUTHORIZED WORKING TIME DURING EACH WEEK, OF ANY CHILD UNDER THE AGE OF 16 FULL YEARS, EMPLOYED IN ANY AUTHORIZED ESTABLISHMENT OR OCCUPATION, SHALL BE DEVOTED TO EDUCATIONAL INSTRUCTION NOT CONNECTED WITH SUCH EMPLOYMENT, AND TO RECREATION AND PHYSICAL EXERCISE, AT FULL PAY. *PROVIDED, HOWEVER,* THAT THIS PROVISION SHALL BE APPLICABLE ONLY IN ENTERPRISES IN WHICH NOT LESS THAN FIFTEEN CHILDREN UNDER THE AGE OF 16 FULL YEARS ARE EMPLOYED.

SECTION XIII. PROHIBITIONS ON RESTRAINT OF MOVEMENT OF EMPLOYEES. NO EMPLOYEE SHALL BE REQUIRED TO REMAIN WITHIN ANY DORMITORY, FACTORY, FACTORY GROUNDS OR OTHER FIXED PLACE OR AREA DURING ANY NON-WORK DAY OR HOLIDAY; NOR DURING ANY NON-WORK HOURS, *PROVIDED* THAT EMPLOYERS MAY, SUBJECT TO THE SUPERVISION OF THE DEPARTMENT OF LABOR, ANNOUNCE REASONABLE CURFEWS FOR SUCH EMPLOYEES UNDER THE AGE OF 18 FULL YEARS WHO LIVE IN DORMITORIES MAINTAINED BY SUCH EMPLOYERS.

SECTION XIV. PHYSICAL EXAMINATIONS. ALL PERSONS UNDER THE AGE OF 18 FULL YEARS, EMPLOYED IN ANY AUTHORIZED ESTABLISHMENT OR OCCUPATION, SHALL BE REQUIRED TO UNDERGO A PHYSICAL (INCLUDING DENTAL) EXAMINATION, AT THE BEGINNING OF THEIR EMPLOYMENT IN ANY SUCH ESTABLISHMENT OR OCCUPATION, AND AT INTERVALS OF SIX MONTHS THEREAFTER, BY PHYSICIANS AND DENTISTS PROVIDED BY THEIR EMPLOYERS AND APPROVED BY THE DEPARTMENT OF LABOR. THE EXPENSES OF SUCH EXAMINATION, AS WELL AS OF ANY TREATMENT, OPERATION OR APPLIANCES NECESSITATED BY THE NATURE OF THE EMPLOY-MENT, SHALL BE BORNE IN FULL BY SUCH EMPLOYERS.

— 6 —

SECTION XV. EMPLOYMENT CERTIFICATES. No person under the the age of 21 full years shall be employed, permitted or suffered to work in, about or in connection with any establishment or occupation described in this ordinance (or in supplementary regulations issued by the Department of Labor pursuant to Section XIX hereof) unless the person, firm or corporation employing such individual procures and keeps on file, and accessible to any inspector of factories or other authorized official charged with the enforcement of this ordinance, the employment certificate as hereinafter provided, issued to such child; and keeps two complete lists of the names together with the ages of all boys under 16 full years of age and all girls under 18 full years of age employed in or for such establishment or in such occupation, one on file and one conspicuously posted near the principal entrance of the place or establishment in which such children are employed. Such inspectors and officials may require that the employment certificates and lists provided for herein be produced for their inspection.

On termination of the employment of a child whose employment certificate is on file, such certificate shall be returned by the employer within two days to the official who issued same, with a statement of the reasons for the termination of said employment.

Employment certificates shall be in the form prescribed by the Department of Labor and shall be issued only by officials authorized by the Director of the Department of Labor in writing. Where possible, such authorized officials shall be the local school authorities in the city, town or village in which the child is to be employed. Such employment certificates shall be issued only upon the application in person of the parent or guardian or custodian of the child desiring such employment; *provided*, that no person authorized to issue employment certificates shall have authority to issue such certificate for any child then in or about to enter such person's own employment or the employment of a firm or corporation of which he is a member, officer or employee.

The person authorized to issue an employment certificate shall not issue such certificate until he has received, approved and filed the following papers, duly executed: (1) The written pledge or promise of the person, firm or corporation to employ the child legally and also the written agreement to return the employment certificate within two days after the termination of such employment, as hereinabove provided; (2) The school record of such child, properly filled out and signed by the appropriate school officials; (3) A

— 7 —

certificate signed by a physician appointed by the official authorized to issue such employment certificate. stating that such child has been examined by him and, in his opinion, has reached the normal development of a child of its age. and is in sufficiently sound health and physically able to be employed in any of the occupations or processes in which a child between 14 and 16 full years of age legally, may be employed, (4) Evidence that the child is 14 full years old or upwards, which shall consist of one of the following proofs of age and shall be required in the order herein designated as follows: (a) a duly attested transcript of the birth certificate filed according to law with a registrar of vital statistics, or other officer charged with the duty of recording births, which certificate shall be *prima facie* evidence of the age of such child; (b) a duly attested transcript from the family register (*kong-chuk*) showing the place and date of birth of the child; (c) a passport or a duly attested transcript of a certificate of baptism showing the date of birth and place of baptism of such child; (d) in case none of the above proofs of age can be produced, other documentary evidence of age which shall appear to be satisfactory to the official issuing the certificate (aside from the school record of such child or the affidavit of parent, guardian or custodian), may be accepted in lieu thereof; (e) in case no documentary proof of age of any kind can be produced. the official issuing the certificate may receive and file an application signed by the parent, guardian or custodian of the child for physicians' certificates. Such application shall contain the name, alleged age, place and date of birth, and present residence of the child, together with such further facts as may be of assistance in determining the age of such child, and shall contain a statement certifying that the parent, guardian or custodian signing such application is unable to produce any of the documentary proofs of age hereinabove specified. Such application shall be filed for not less than sixty days for an examination to be made of the statements contained therein, and in case no facts appear within such period or by such examination tending to discredit or contradict any material statement of such application, the official issuing the certificate may direct such child to appear thereafter for physical examination before two physicians officially designated by him, and in case such physicians shall certify in writing that they have separately examined such child and that, in their opinion such child is at least 14 full years of age, such official shall accept such certificates as sufficient proof of the age of such child for the purposes of this provision. In case the opinions of such physicians do not concur, the child shall be examined by a third physician and the concurring opinions shall be sufficient for the purposes of

— 3 —

this provision as to the age of such child. The official issuing the certificate shall require the evidence of age specified in subdivision (*a*) in preference to that specified in any subsequent subdivision and shall not accept the evidence of age permitted by any subsequent subdivision unless he shall receive and file in addition thereto a certificate of the parent, guardian or custodian showing that no evidence of age specified in any preceding subdivision or subdivisions or this provision can be produced. Such certificate shall contain the age, date and place of birth, and present residence of such shild, and shall be signed, sealed and affirmed in the presence of the official authorized to issue the employment certificate.

No employment certificate shall be issued until the child in question has personally appeared before and been examined by the official issuing the certificate, nor until such official, after making such examination, has signed and filed in his office a statement that the child can read intelligently and write legibly simple sentences in the Korean language; and has complied with all existing educational requirements for children of his age, and that, in his opinion, the child is 14 full years of age or upwards.

Every such employment certificate shall state the name, sex, date and place of birth and place of residence of the child, and describe the color of the hair and eyes, the height and weight and any distinguishing facial marks of such child, and shall contain a statement of the proof of age accepted and shall certify that the papers required by the foregoing provisions have been duly examined, approved and filed, and that the child named in such certificate has appeared before the official issuing the certificate and has been examined.

Every such certificate shall be signed, in the presence of the official issuing the same, by the child in whose name it is issued. It shall show the date of its issue. A record giving all the facts contained on every certificate issued shall be kept on file in the office issuing the same, and also a record of the names and addresses of the children to whom certificates have been refused, together with the names of the schools which such children should attend and the reasons for refusal.

No employment certificate shall be issued to any child required to attend school unless, during the twelve months previous to applying for such certificate, he has attended such school for at least 130 days.

The blank certificate and other papers required in the issuing of employment certificates shall be formulated by or under the immediate supervision of the Director of the Department of Labor and furnished by him to officials authorized to issue such certificates. The Director of the Department of Labor may by regulation, as provided in Section XIX hereof, designate authorized officials for such purpose.

Each official authorized to issue employment certificates shall transmit between the first and tenth days of each month, to the Director of the Department of Labor, Seoul, Korea, upon blanks furnished by him, a list of the names of the children to whom such certificates have been issued. Each such certificate shall give the name and address of the prospective employer and the nature of the occupation the child intends to engage in.

SECTION XVI. Miscellaneous Provisions.

a. Additional Proof of Age by Employers. An inspector of factories or other official charged with the enforcement of this ordinance may make demand on any employer in, or about whose place or establishment a child apparently under the age of 16 full years is employed or permitted or suffered to work, and whose employment certificate is not filed as hereinabove required, that such employer shall either furnish him, within ten days, satisfactory evidence that such child is in fact 16 full years of age, or shall cease to employ or permit or suffer such child to work in such place or establishment. The inspector of factories or other official charged with the enforcement of this ordinance shall require from such employer the same evidence of age of such child as is required upon the issuance of an employment certificate, and the employer furnishing such evidence shall not be required to furnish any further evidence of the age of the child.

In case any employer shall fail to produce and deliver to a factory inspector or other official charged with the enforcement of this provision, the evidence of age herein required, and shall thereafter continue to employ such child or permit or suffer such child to work in such place or establishment, proof of the making of such demand and of such failure to produce and file such evidence shall be *prima facie* evidence of the illegal employment of such child in any prosecution brought therefor.

b. Additional Powers and Duties of Factory Inspectors and Enforcement Officials. Inspectors of factories and other officials charged with the enforcement of this ordinance may, within their respective districts or jurisdictions, visit and inspect at any time any place of employment mentioned in this ordinance (or in supplementary regulations issued by the Department of Labor pursuant to the provisions of Section XIX hereof), and shall ascertain whether any persons affected by this ordinance are employed therein contrary to the provisions hereof; and they shall report weekly to the school authorities and to the Director of the Department of Labor all cases of children under 16 full years of age discharged from illegal employment. It shall be the duty of such factory inspectors and other officials charged with the enforcement of this ordinance to make complaints against any person violating any of the provisions of this ordinance and to effect the prosecution thereof. This provision shall not be construed as a limitation upon the right of other persons or organizations or unions to make and prosecute such complaints.

c. *Effect of Failure to Provide Certificate or Lists.* A failure by an employer to produce to a factory inspector or other official charged with the enforcement of this ordinance, any employment certificate or list required in this ordinance shall be *prima facie* evidence of the illegal employment of any child whose employment certificate is not produced or whose name is not listed.

d. *Effect on Industrial Education.* Nothing in this ordinance shall prevent children of any age from receiving industrial education furnished by the Government of Korea or any political subdivision or school thereof, provided such education is first duly approved by the Department of Labor in coordination with the Department of Education of the Government of Korea.

e. *JUVENILE DELINQUENCY.* ANY CHILD WORKING IN OR IN CONNECTION WITH ANY OF THE ESTABLISHMENTS OR PLACES OR IN ANY OF THE OCCUPATIONS MENTIONED IN THIS ORDINANCE (OR IN ANY SUPPLEMENTARY REGULATIONS OF THE DEPARTMENT OF LABOR ISSUED PURSUANT TO THE PROVISIONS OF SECTION XIX HEREOF) WHO REFUSES TO GIVE TO THE FACTORY INSPECTOR OR OTHER OFFICIAL CHARGED WITH THE ENFORCEMENT OF THIS ORDINANCE, HIS OR HER NAME, AGE AND PLACE OF RESIDENCE, SHALL BE CONDUCTED FORTHWITH BY SUCH INSPECTOR OR OFFICIAL BEFORE THE JUVENILE COURT IN THE PLACE WHERE SUCH CHILD IS EMPLOYED OR RESIDES, OR IF THERE IS NO SUCH COURT IN SUCH PLACE, BEFORE ANY COURT OR JUDICIAL OFFICER HAVING JURISDICTION OVER OFFENSES COMMITTED BY CHILDREN, FOR EXAMINATION AND TO BE DEALT WITH ACCORDING TO LAW.

THIS PROVISION SHALL BE EFFECTIVE ON THE DATE APPEARING HEREON.

f. PROHIBITED SALES. ANY PERSON WHO. EITHER FOR HIMSELF OR HERSELF OR AS AGENT OF ANY OTHER PERSON OR OF ANY FIRM OR CORPORATION, FURNISHES OR SELLS TO ANY MINOR ANY ARTICLE OF ANY DESCRIPTION WITH THE KNOWLEDGE THAT SAID MINOR INTENDS TO SELL SAID ARTICLE IN VIOLATION OF THE PROVISIONS OF THIS ORDINANCE OR WHO SHALL CONTINUE TO FURNISH OR SELL ARTICLES OF ANY DESCRIPTION TO A MINOR AFTER HAVING RECEIVED WRITTEN NOTICE FROM ANY OFFICER CHARGED WITH THE ENFORCEMENT OF THIS ORDINANCE, OR WHO KNOWINGLY AND WILFULLY SELLS, FURNISHES OR TRANSPORTS ANY ARTICLE PRODUCED BY A MINOR IN VIOLATION OF THIS ORDINANCE. SHALL BE GUILTY OF A VIOLATION OF THIS ORDINANCE.

g. RIGHTS OF RECOVERY SAVED. ANY PERSON INJURED BY THE VIOLATION OF THIS ORDINANCE MAY RECOVER FROM THE OFFENDER SUCH DAMAGES AS HE MAY SUSTAIN THEREBY. IN ANY SUIT BY OR ON BEHALF OF ANY MINOR ILLEGALLY EMPLOYED HEREUNDER. CONTRIBUTORY NEGLIGENCE, ASSUMPTION OF RISK, OR NEGLIGENCE BY FELLOW-EMPLOYEES SHALL NOT BE A DEFENSE AND MAY NOT BE PLEADED OR PROVED EITHER BY WAY OF DEFENSE OF MITIGATION; NOR SHALL MISREPRESENTATION OF AGE BY THE PERSON SECURING THE EMPLOYMENT OR BY THE PERSON EMPLOYED BE A DEFENSE OR MATTER IN MITIGATION.

IN ANY SUIT BY OR ON BEHALF OF ANY PERSON ILLEGALLY EMPLOYED HEREUNDER, FOR THE RECOVERY OF WAGES DUE. SUCH PERSON SHALL (UNLESS THE EMPLOYMENT IS THE RESULT OF HIS OWN FRAUD OR WILFUL ILLEGAL ACT TO GAIN THE BENEFITS OF RECOVERY HEREINAFTER PROVIDED) BE ENTITLED TO RECOVER A SUM EQUIVALENT TO TEN TIMES THE AGREED AMOUNT OF WAGES DUE OR TEN TIMES THE LEGAL MINIMUM WAGE DUE, WHICHEVER IS GREATER. IN ALL SUCH SUITS, ALL OFFICERS, DIRECTORS AND STOCKHOLDERS OF CORPORATE EMPLOYERS SHALL BE PERSONALLY LIABLE TO ANY PLAINTIFF. NONE OF THE PROVISIONS OF THIS PARAGRAPH MAY BE WAIVED. AND ANY RELEASE OF ANY RIGHTS HEREUNDER SHALL BE INEFFECTIVE TO BAR ANY RECOVERY

HEREUNDER. NO EMPLOYER OR OTHER PERSON SHALL, AS A CONDITION OF EMPLOYMENT, ASK FOR OR REQUIRE ANY EMPLOYEE OR PROSPECTIVE EMPLOYEE TO WAIVE OR RELEASE ANY RIGHTS UNDER THIS ORDINANCE, OR ANY REQUIREMENT OF THIS ORDINANCE.

THIS PROVISION SHALL BE EFFECTIVE ON THE DATE APPEARING HEREON.

DEFINITION OF "FULL YEARS," "FULL AGE," &c. THE EXPRESSIONS *FULL YEARS, FULL AGE, &c.* AS USED HEREIN, MEAN AGE IN YEARS CALCULATED FROM THE ACTUAL DATE OF BIRTH, IN SUCH MANNER THAT A CHILD IS SAID TO BE ONE YEAR OLD ONE FULL CALENDAR YEAR AFTER THE DATE OF BIRTH.

SECTION XVII. Conspicuous Posting of Notices. Every employer shall post and keep posted in a conspicuous place in every room where any boy under the full age of 18, or any girl under the age of 21 full years is employed, permitted or suffered to work, a printed notice stating the maximum number of hours such person may be required or permitted to work on each day of the week, the hours of commencing and stopping work, the day or days off and the educational and recreational hours during each day, and the hours allowed for dinner or for other meals; and shall also post other notices as required by the Director of the Department of Labor. The employment of any person affected thereby for a longer time in any day than stated, or at any time other than as stated in said printed notice, shall be deemed a violation of the provisions of this ordinance.

SECTION XVIII. APPLICATION TO GOVERNMENT EMPLOYMENT AND TO GOVERNMENT AGENCIES AND CORPORATIONS. ALL PROVISIONS OF THIS ORDINANCE SHALL APPLY AS WELL TO THE GOVERNMENT OF KOREA, ALL OF ITS POLITICAL SUBDIVISIONS, AGENCIES, INSTRUMENTALITIES, OFFICES, AND CORPORATIONS AND ENTERPRISES WHICH IT OWNS OR IN WHICH IT HAS AN INTEREST OR WHICH ARE UNDER ITS MANAGEMENT OR CONTROL. NONE OF THE FOREGOING SHALL AT ANY TIME AWARD ANY CONTRACT TO OR ENTER INTO ANY AGREEMENT WITH ANY EMPLOYER, WHICH DOES NOT PROVIDE THAT SUCH EMPLOYER SHALL OBEY ALL REQUIREMENTS OF THIS ORDINANCE (AND OF ALL SUPPLEMENTARY REGULATIONS ISSUED PURSUANT HERETO BY THE DEPARTMENT OF

— 1 3 —

LABOR) AND SHALL SAVE THE GOVERNMENT (OR APPROPRIATE CONTRACTING AGENCY THEREOF) HARMLESS IN CASE OF ANY VIOLATION THEREOF.

SECTION XIX. Implementation of Ordinance by Regulations of Department of Labor. The Department of Labor of the Government of Korea is authorized to add to the categories of prohibited employments enumerated in Sections II, III, IV, V, VI, and VII hereof, from time to time, in regulations which, when approved by the Military Governor and published in the *Official Gazette* in the manner prescribed by law, shall have the force and effect of law. The Department of Labor is further authorized to issue such regulations, in the manner prescribed by law, as are necessary to implement the purpose or provisions of this ordinance, including but not limited to the following matters:

a. Maximum work hours, generally or in specified industries and enterprises;

b. Working conditions;

c. Health, sanitation, welfare and recreation requirements;

d. Food and housing requirements;

e. Factory and mercantile inspections;

f. Reports by employers relative to working conditions;

g. Minimum wages; *provided, however,* that collective agreements and determinations of Labor Mediation Boards shall take precedence over such minimum wage rates to the extent they are not lower than such minimum wage rates;

h. Any other matters dealing with the amelioration of labor conditions.

SECTION XX. Repeal of Inconsistent Laws. All ordinances, laws, edicts, rescripts, regulations, orders and measures which are inconsistent herewith or in conflict with the provisions hereof are, to the extent they are inconsistent or in conflict herewith, hereby repealed.

SECTION XXI. PENALTIES. ANY PERSON VIOLATING ANY OF THE PROVISIONS OF THIS ORDINANCE OR OF ANY REGULATION ISSUED BY THE DEPARTMENT OF LABOR PURSUANT HERETO, SHALL, UPON CONVICTION BY A MILITARY OCCUPATION COURT, SUFFER SUCH PUNISHMENT AS THE COURT SHALL DETERMINE.

SECTION XXII. EFFECTIVE DATE. THIS ORDINANCE SHALL BE EFFECTIVE ON THE 30th DAY AFTER THE DATE APPEARING HEREON EXCEPT AS OTHERWISE HEREIN SPECIFICALLY PROVIDED.

ARCHER L LERCH
Major General United States Army
Military Governor in Korea

HEADQUARTERS
UNITED STATES ARMY MILITARY
GOVERNMENT IN KOREA
Office of the Military Governor
Seoul, Korea

ORDINANCE
NUMBER 113   8 October 1946

LICENSING OF GRAIN MILLS

SECTION I. *Purpose.* The purpose of this ordinance is to provide a method for effectuating the enforcement of National Food Regulations.

SECTION II. *Repeal of Former Law.* Governor General Ordinance Number 207, dated 11 December 1939 (*Rice Cleaning Regulation*), is hereby repealed.

SECTION III. *Licensing Grain Mills.*

a. Each provincial governor shall immediately establish regulations, which shall be submitted to the Military Governor of Korea as required before becoming effective, for the licensing of all the grain mills within his province. Such regulations shall include a provision giving applicants for licenses the right to appeal adverse decisions of agencies of the provincial governor to the provincial governor, a system for periodic inspections of such mills and their records, and the periodic submission of mill records to an authorized official for inspection.

b. On and after the fifteenth day following the effective date of each such provincial regulation no person, natural or juridical, shall, personally or through any agent, operate a grain mill which is not licensed or for which an application for a license is not pending.

SECTION IV. *Records.* On and after the effective date of each such provincial regulation, each mill shall keep a complete set of records of each receipt of grain for milling or polishing. Each such receipt shall set forth the following information:

a. Person in whose name the grain was delivered for milling or polishing;

— 1 —

b. Address of person by whom so delivered;

c. Date of such delivery;

d. Amount delivered, expressed in bags of 54 kg net weight of unhulled rice or rye, 42 kg net weight of barley, and 60 kg net weight of naked barley or wheat;

e. Amount retained by the mill as toil, expressed in bags of the aforementioned weights.

In addition, a complete record shall be [kept by each mill of all deliveries by the mill to the Myun Collection Point or to private owners. Such record shall set forth the date of delivery, the amount of grain delivered and the price received. Such mill records shall be open for inspection at all times by the Military Government of Korea and its authorized agents.

SECTION V. *Compensation for Milling and Polishing.* Charges for milling and polishing shall be established by regulations issued by the National Price Administration pursuant to Ordinance Number 90, dated 28 May 1946.

SECTION VI. *Milling and Polishing.*

a. The provincial governor or his authorized agents shall issue to the owners of grain a certificate, stating the amount of grain permitted to be milled or polished. In no case shall the total amount stated on such certificates of grain exceed the lawful amount of grain permitted to be in the posession of any applicant. No amount in excess of that stated on such certificate, or grain without such certificate, shall be milled or polished. If such is delivered to the mill, a receipt shall be given to the owner and such excess or unauthorized quantity shall be turned over to the authorized agent of the provincial governor for confiscation at the discretion of the provincial governor. Such certificates shall be retained as part of the records of each mill.

b. Rice shall be cleaned so that the percentage by weight of cleaned rice obtained from brown rice is not less that 96%.

c. Wheat shall be milled so that the percentage by weight of flour obtained from rough wheat is not less than 80%.

d. Rye shall be milled so that the percentage by weight of flour obtained from rough rye is not less than 65%.

e. Barley shall be milled so that the percentage by weight of cleaned barley obtained from rough barley is not less than 74%.

— 2 —

f. Naked barley shall be milled so that the percentage by weight of cleaned naked barley obtained from rough naked barley is not less than 87%.

g. Deviations from these fixed percentages may be made only with prior approval of the provincial governor or his authorized agents and only if the grain is allocated by the National Food Administration for medicinal purposes, alcoholic beverages, or commercial bakeries or confection manufacturing.

h. All grain milled or polished in violation of this section shall be confiscated by the provincial governor or his authorized agent and delivered to the Myun Collection Point.

SECTION VII. *Delivery of Toll Grain.* Each mill shall deliver once weekly all grain retained as toll to the designated Myun Collection Point and accept the price there posted for such grain.

SECTION VIII. *Revocation of Licenses.* For any violation of this ordinance or provincial regulations issued pursuant hereto, the provincial governor or his authorized agent may revoke the license of the violator. Grain on hand at the time of any such revocation shall be disposed of in accordance with instructions issued by the provincial governor or his authorized agents.

SECTION IX. *Insurance.* Wherever and whenever available, each mill shall have adequate insurance to compensate the owners of the grain for any loss due to fire, theft, flood or the elements.

SECTION X. *"Grain" Defined. Grain,* as used herein, means rice, wheat, rye, barley and naked barley.

SECTION XI. *Penalties.* Any person violating the provisions of this ordinance or of any provincial regulation issued pursuant hereto shall, upon conviction by a Military Occupation Court, suffer such punishment as the court shall determine.

SECTION XII. *Effective Date.* This ordinance shall be effective on the tenth day after the date appearing hereon.

ARCHER L LERCH
Major General United States Army
Military Governor in Korea

HEADQUARTERS
UNITED STATES ARMY MILITARY
GOVERNMENT IN KOREA
Office of the Military Governor
Seoul, Korea

ORDINANCE
NUMBER 114

23 October 1946

REORGANIZATION OF PROVINCIAL GOVERNMENTS

SECTION I. *Purpose.* The purpose of this ordinance is to define the organization and duties and functions of provincial governments in Korea south of 38° north latitude and to clarify their relationship to the national government.

SECTION II. *Nomenclature in Provincial Governments.* The principal elements of each provincial government shall hereafter be known as bureaus, and shall be divided as follows:

a. Bureaus into sections;

b. Sections into subsections.

In addition, branches of political subdivisions, agencies, instrumentalities, and boards or commissions of the Government of Korea may, from time to time, be established by law in the provincial governments.*

SECTION III. *Titles of Personnel.* The title of any individual in charge of a provincial bureau, section, subsection or office shall be *Chief.* The title of his authorized principal assistant shall be *Assistant Chief.* Only authorized titles shall be applied to both military and civilian personnel appointed to and occupying positions within any provincial government.

SECTION IV. *Abolition of Provincial Secretariats.* The secretariat in each provincial government is hereby abolished. All records, property and Korean personnel of each such secretariat shall be transferred forthwith by the

---

*See Ordinance Number 73, dated 23 April 1946, establishing provincial Property Custody Offices; Civil Service Circular Number 1-1, dated 15 May 1946, establishing provincial Civil Service Offices; Ordinance Number 34, dated 8 December 1945, establishing provincial Labor Mediation Boards; and Ordinance Number 66, dated 10 April 1946, establishing provincial Boards of Fire Commissioners.

— 1 —

respective provincial governors to appropriate provincial bureaus established by this ordinance, or to the provincial Property Custod or Civil Service Offices. Duties and functions of such secretariats not specifically assigned to any other provincial bureau established herein or to the provincial Property Custody or Civil Service Offices are hereby transferred to the respective provincial Bureaus of Home Affairs.

SECTION V. *Abolition of Provincial Departments of Police.* Provincial Departments of Police are hereby abolished. All duties, functions, records, property and Korean personnel of the provincial Department of Police in each provincial government are hereby transferred to the lettered Division of the Department of Police of the Government of Korea which has jurisdiction coextensive with the province concerned. The Chief of each such Division of Police shall be responsible for the maintenance of law and order in the province, and shall coordinate and cooperate with the provincial governor in carrying out such responsibilty so as to insure the safety and protection of the people in the province.

SECTION VI. *Organization of Provincial Governments; Transfer of Property Records and Personnel.* Each provincial government shall consist of a provincial governor, other officials, boards and commissions established by law, and the following bureaus:

- *a.* Bureau of Home Affairs;
- *b.* Bureau of Agriculture;
- *c.* Bureau of Banking and Taxation;
- *d.* Bureau of Commerce;
- *e.* Bureau of Education;
- *f.* Bureau of Labor;
- *g.* Bureau of Public Health and Welfare;
- *h.* Bureau of Public Works.

Records, property and Korean personnel of existing departments, bureaus, sections, or other subdivisions or elements of each provincial government shall be transferred forthwith by the respective provincial governors to one or more of such provincial bureaus, or to the provincial Property Custody or Civil Service Offices.

SECTION VII. *The Provincial Governor.* The provincial governor is the chief executive in his province. He is reponsible for carrying out policies and directives of the Government of Korea, within his province.

Each provincial governor is hereby assigned the following duties and functions within his province:

a. Supervision and direction of the operation and administration of the provincial government;

b. Recommending to the Military Governor in Korea, appointment, transfer and dismissal of bureau and office chiefs, county, island and city heads, and other government officials and employees as provided by law;

c. Appointment of government officials and employees to, and removal from, positions within his province excepted from civil service classification, other than those mentioned in paragraph VII *b* hereof;

d. Disbursement of and accountability for provincial funds;

e. Convocation and adjournment of advisory councils or other provincial public bodies, as provided by law;

f. Coordination of policies and activities of the provincial government with those of the police, courts and other instrumentalities or agencies of the national government operating within his province, and supervision of such agencies to the extent and in the manner required by law;

g. Issuance of provincial ordinances, regulations, orders and directives on administrative matters concerned exclusively with his own province and not affecting any other province and not in conflict with existing ordinances, regulations, orders, directives or other regulatory measures of the Military Government in Korea; *provided, however,* that except in case of emergency, each such proposed provincial ordinance, regulation, order or directive shall be forwarded in triplicate prior to issuance, to the Military Government in Korea (Attention: Department of Justice); and *further provided,* that (except in case of emergency) no such provincial ordinance, regulation, order or directive may be issued prior to approval thereof by or under the direction of the Military Governor in Korea; and *further provided,* that each and any such provincial ordinance, regulation, order or directive shall be rescinded

— 3 —

forthwith by the provincial governor upon receipt of an order to that effect from or by direction of the Military Governor in Korea;

h. Assurance of coordination and unity of policy between the provincial government and the national government as provided in Section XVII *a* hereof;

i. Supervision of registration, election and voting procedures in accordance with the requirements of law.

SECTION VIII. *Bureau of Home Affairs.*

a. Organization. Each provincial Bureau of Home Affairs shall consist of the following sections:

(1) Accounts Section;
(2) Administration Section;
(3) Legal Section;
(4) Local Affairs Section;
(5) Public Information Section*;
(6) Transportation Section.

Duties and Functions. Each provincial Bureau of Home Affairs is hereby assigned the following duties and functions within its province:

(1) Disbursing and accounting for funds under the control of the provincial government;

(2) Supervision of provincial property and supply matters, including maintenance, repair, and safeguarding of such property and keeping of appropriate records pertaining thereto;

(3) Inspection of accounts of cities, counties, islands, towns and villages;

(4) Coordination with the provincial Civil Service Office in advising the provincial governor on appointments and removals of officials in excepted positions;

(5) Supervision and control of handling of official communications, including the publishing, issuing, recording, filing and mailing thereof;

(6) Receipt, distribution and tabulation of reports, records and governmental statistics;

*See Ordinance Number 71, dated 13 April 1946.

(7) Supervision and control of translation of official documents;

(8) Maintenance of an official provincial library consisting of laws, ordinances, directives, reports, statistics and necessary records;

(9) Furnishing counsel for the provincial government and handling the investigation and processing of claims;

(10) Drafting of proposed ordinances, proclamations, orders and directives for the provincial government;

(11) Supervision of administration of cities, counties, islands, towns and villages;

(12) Supervision and control of provincial finances and budgets and of finances and budgets of cities, counties, islands, towns and villages;

(13) Collection and evaluation of public information on organizations, publications, individuals and activities;

(14) Dissemination of information to the public by utilization of the press, radio, leaflets, posters and other suitable means;

(15) Supervision of transportation owned, operated or controlled by the province and by cities, counties, islands, towns and villages therein;

(16) Assumption of responsibility for other governmental matters not specifically assigned herein to any other provincial bureau or political subdivision, agency or instrumentality of the Government of Korea, and performance of all necessary duties and functions in connection therewith.

SECTION IX. *Bureau of Agriculture.*

a Organization. Each provincial Bureau of Agriculture shall consist of the following sections:

(1) Agricultural Economics Section;

(2) Agricultural Production Section;

(3) Fisheries Section (except North Chungchong Province);

(4) Food Section;

(5) Forestry Section;

(6) Reclamation and Irrigation Section.

— 5 —

b. *Duties and Functions.* Each provincial Bureau of Agriculture is hereby assigned the following duties and functions within its province:

(1) Conducting agricultural economic planning and statistical activities, and making studies and recommendations concerning agricultural landlord and tenant relationships;

(2) Supervision of inspection and grading of grains;

(3) Furnishing advice and assistance in matters regarding production of crops and livestock, and sericulture;

(4) Furnishing advice and assistance in matters regarding production, processing, inspection and marketing of fishery products;

(5) Supervision, control and guidance of fishery associations and guilds;

(6) Allocation of foodstuffs for the manufacture of sake and other alcoholic beverages;

(7) Supervision and control of the collection, distribution and rationing of foods in accordance with policies and programs of the National Food Administration;

(8) Supervision, guidance and execution of measures regarding reforestation, protection of forests and production of lumber and charcoal;

(9) Supervision, guidance and execution of measures to prevent soil erosion;

(10) Supervision and guidance of forestry associations;

(11) Supervision and guidance of guilds and associations charged with or engaged in the operation and maintenance of irrigation projects; planning, supervision and construction of irrigation projects under 300 chongbu in area, and execution of extensions, improvements or flood damage repair work on such projects.

SECTION X *Bureau of Banking and Taxation.*

a. *Organization.* Each provincial Bureau of Banking and Taxation shall consist of the following sections:

    (1) Direct Tax Section;

    (2) Finance Section;

    (3) Indirect Tax Section.

b. Duties and Functions. Each provincial Bureau of Banking and Taxation is hereby assigned the following duties and functions within its province:

    (1) Supervision of matters pertaining to national and local taxes;

    (2) Investigation, classification and assessment of taxable properties;

    (3) Maintenance of necessary tax and financial records;

    (4) Supervision and direction of procedure for the collection of national and local taxes and other revenues;

    (5) Supervision and administration of tax districts;

    (6) Inspection of books and records of factories producing alcoholic beverages, to insure proper payment and collection of taxes;

    (7) Furnishing advice and assistance to financial associations and local banking institutions;

    (8) Auditing of all expenditures and disbursements of the province and of cities, counties, islands, towns, and villages;

    (9) Furnishing advice and assistance to local insurance agencies.

SECTION XI. *Bureau of Commerce.*

a. Organization. Each provincial Bureau of Commerce shall consist of the following sections:

    (1) Commerce Section;

    (2) Mining and Industry Section.

b. Duties and Functions. Each provincial Bureau of Commerce is hereby assigned the following duties and functions within its province:

    (1) Furnishing advice and assistance in matters pertaining to the supply and distribution of consumer goods, the supervision and development of domestic and foreign trade, and control of weights and measures;

— 7 —

(2) Supervision and guidance of economic and commercial associations and guilds;

(3) Furnishing assistance in procuring local markets for industrial and commercial, goods and mineral resources;

(4) Stimulation of mining and industrial production;

(5) Furnishing mining and industrial, producers with accurate statistics;

(6) Supervision, guidance and direction of policies for the future development of local mining and industry;

(7) Enforcement and recommendation of standards pertaining to products of mines and of industry;

(8) Supervision and guidance of mining and industrial organizations and associations;

(9) Supervision and control of manufacture and distribution of alcoholic beverages.

SECTION XII. *Bureau of Education.*

a. *Organization.* Each provincial Bureau of Education shall consist of the following sections:

(1) Administration Section;

2) Research and Planning Section ;

(3) Schools Section;

(4) Social Education Section.

b. *Duties and Functions.* Each provincial Bureau of Education is hereby assigned the following duties and functions within its province:

(1) Administration of public secondary schools and normal schools below the college level;

(2) Granting of teachers' certificates to elementary school teachers;

(3) Recommending to the provincial governor appointment and dismissal of elementary and secondary school teachers and of elementary school principals;

(4) Recommending to the Department of Education of the Government of Korea appointment and dismissal of secondary school principals;

— 8 —

(5) Supervision of elementary public and private schools and secondary private schools;

(6) Financial administration of secondary schools and financial supervision of elementary schools;

(7) Establishment of new schools below the college level;

(8) Administration of adult, physical, social, vocational and technical education programs, including agricultural schools;

(9) Maintenance of provincial libraries (other than the official provincial library), museums, monuments and art treasures; and maintenance of such national libraries, museums, monuments and art treasures as are designated from time to time by the Department of Education of the Government of Korea;

(10) Supervision of normal colleges and other schools at the college level which are specifically placed under provincial jurisdiction and control by the Department of Education of the Government of Korea.

SECTION XIII. *Bureau of Labor.*

a. *Organization.* Each provincial Bureau of Labor shall consist of the following sections:

(1) Administration Section;
(2) Employment and Information Section;
(3) Labor Relations Section.

b. *Duties and Functions.* Each provincial Bureau of Labor is hereby assigned the following duties and functions within its province:

(1) Supervision and administration of employment and labor information services in accordance with the policies and instructions of the Department of Labor of the Government of Korea, including initiation and supervision of the collection and dissemination of information relative to wages, hours, conditions of employment, labor relations, labor unions, and labor matters not directly pertaining to civil service;

(2) Supervision and control of the relations of labor unions with the government;

(3) Furnishing advice on labor policies, programs and activities, on other matters affecting the welfare of labor, and exercis-

ing general supervision over conditions of employment, relations of employees with employers and with the government;

(4) Supervision and control of the provincial Labor Mediation Board and the activities thereof.

SECTION XIV. *Bureau of Public Health and Welfare.**

a. *Organization.** Each provincial Bureau of Public Health and Welfare shall consist of the following sections:

(1) Medical Services Section;

(2) Pharmaceutical Affairs Section;

(3) Preventive Medicine and Vital Statistics Section;

(4) Sanitation Section;

(5) Veterinary Affairs Section;

(6) Welfare Section.

b. *Duties and Functions**: Each provincial Bureau of Public Health and Welfare is hereby assigned the following duties and functions within its province:

(1) Inspectional supervision of city hospitals, private hospitals, clinics, dispensaries and health centers;

(2) Supervision of registration of physicians, limited physicians, herb physicians and dentists;

(3) Coordination with the provincial Bureau of Education regarding school health services and plans for health education;

(4) Supervision and control of the activities of public health nurses and institutional nurses, and registration of nurses and midwives in accordance with directives of the Department of Public Health and Welfare of the Government of Korea;

(5) Supervision and direction of the operation of laboratories under the control of the provincial Bureau of Public Health and Welfare;

(6) Furnishing advice and assistance in maternal and child care;

(7) Supervision and control of administration of provincial hospitals, except those administered by provincial medical colleges;

---

*Supersedes Ordinance Number 25, dated 7 November 1945.

(8) Supervision, inspection and control of production, procurement, storage, rationing, allocation and distribution of drugs, medicines, pharmaceuticals and other items of commerce used in the practise of medicine or public health activities (including derivatives, compounds or substances containing any of such materials); licensing of producers, manufacturers, vendors, distributors and dispensers thereof as required by law, and rendering reports in connection therewith to the Department of Public Health and Welfare of the Government of Korea when and as required;

(9) Supervision of control of communicable diseases;

(10) Supervision of registration of births, deaths, stillbirths and marriages, and preparation and transmission of transcripts thereof to the Census Division of the Office of Administration of the Government of Korea;

(11) Collection, tabulation, analysis and interpretation of vital statistics;

(12) Sanitary inspection of all products of animal origin, including meat, poultry, fish, dairy products, fat, glue, leather, fur, and fertilizer; sanitary inspection of meat and fish markets and all establishments producing or processing the above, including slaughter houses, dairy farms, pasteurizing plants, tanneries, and canning and other factories;

(13) Supervision and direction of sewage and garbage disposal, water supply, insect and rodent control and malaria control; inspection of barber shops and barbers; control of industrial hygiene; sanitary control of breeding and sale of shell fish; inspection and supervision of establishments for the manufacture, processing, distribution and sale of foods and beverages; *provided*, that establishments engaged in activities in addition to the serving of food or non-alcoholic beverages, shall obtain additional licenses in accordance with Section III of Ordinance Number 83*;

(14) Control of animal diseases and institution of measures for care of sick and injured animals;

---

*See Ordinance Number 83, dated 25 May 1946.

(15) Supervision and direction of registration of veterinarians and farriers;

(16) Preparation of reports required for planning and administration of public welfare activities, and conducting of surveys and research in connection therewith;

(17) Supervision and direction of assistance to Korean refugees and displaced persons, and refugee relocation and resettlement;

(18) Supervision and direction of assistance and relief to the unemployed or destitute, to victims of disaster and to orphans;

(19) Supervision of private welfare associations and agencies engaged in general relief activities;

(20) Supervision and direction of public welfare institutions;

(21) Supervision of private welfare institutions;

(22) Supervision and direction of requisition and distribution of public welfare supplies.

SECTION XV. *Bureau of Public Works.*

a. Organization. Each provincial Bureau of Public Works shall consist of the following sections:

(1) Administration Section;

(2) City Planning Section;

(3) Highway Section;

(4) Water Resources Section.

b. Duties and Functions. Each provincial Bureau of Public Works is hereby assigned the following duties and functions within its province:

(1) City planning, including general street improvement, zoning, waterworks, sewage and waste disposal, and fire prevention and protection, in coordination with the provincial Board of Fire Commissioners*;

(2) Institution of engineering surveys;

---

*See Section III of Ordinance Number 66, dated 10 April 1946.

    (3) Construction, repair and improvements of public highways and roads, bridges, canals, locks, utilities, and other publicly owned instrumentalities and facilities;

    (4) Conservation and development of natural resources, including control, utilization and purification of waters, and flood control;

    (5) Execution of the foregoing and other public works programs in accordance with instructions of the Department of Public Works of the Government of Korea.

SECTION XVI. *Appropriation of Funds.* The Director of the Department of Finance of the Government of Korea shall immediately make available to the provinces out of moneys not otherwise appropriated, funds necessary to accomplish the reorganization provided herein.

SECTION XVII *Relationship of Provincial Governments to National Government.*

    a. Coordination of Provincial Governments with National Government.
It shall be the responsibility of the respective provincial governors to assure coordination and unity of policy between their respective governments and the plans, policies, programs and activities of the departments, offices and agencies of the Government of Korea at the national level.

    b. Powers of National Government over Provincial Government. The Military Governor in Korea shall exercise his general powers of superintendence, direction and control of the provincial governments by ordinances, by directives issued by the Office of Civil Administrator, by orders issued, as authorized from time to time, by directors of departments or offices of the Government of Korea,* or in such other manner as may be appropriate.

SECTION XVIII. *Changes in Organization, Nomenclature, Titles or Functions Established Herein.* No change shall be made in the organization, nomenclature, titles or functions established herein without amendment of this ordinance, except that provincial governors are authorized to combine sections or create new sections within any provincial bureau, *provided*, that the functions of the bureau are not changed thereby, and that such changes can be financed out of funds allocated for such bureau, and *further provided*, that advance approval in writing has been obtained from the Office of Civil Administrator of the Government of Korea for each such proposed change.

---

* See Ordinance Number 74, dated 27 April 1946.

SECTION XIX. *Repeal of Inconsistent Laws.* All laws, ordinances, orders, regulations, directives and instructions, or parts thereof, which are inconsistent herewith or in conflict with the provisions hereof are, to the extent they are inconsistent or in conflict herewith, hereby repealed.

SECTION XX. *Effective Date.* This ordinance shall be effective on the tenth day after the date appearing hereon.

ARCHER L LERCH
Major General United States Army
Military Governor in Korea

HEADQUARTERS
UNITED STATES ARMY MILITARY GOVERNMENT IN KOREA
Office of the Military Governor
Seoul, Korea

ORDINANCE
NUMBER 115

8 October 1946

LICENSING OF FILMS

SECTION I. *Purpose.* The purpose of this ordinance is to eliminate the former Japanese governmental control of 'motion picture production and exhibition, which restricted the scope of artistic entertainment in order to serve the ends of Japanese nationalistic propaganda; and to achieve an orderly administration of the motion picture industry of Korea, under minimum controls, to assure the basic propriety of the content of films exhibited.

SECTION II. *Laws Repealed.* The following are hereby repealed:

a. Governor General Ordinance Number 1, dated 4 January 1940, which applied to Korea the provisions of Law 66, dated April 1939 (*Motion Picture Law*);

b. Government General Order Number 180, dated 25 July 1940 (*Order Enforcing Chosen Motion Picture Ordinance*);

c. Government General Order Number 181, dated 25 July 1940 (*Regulations Governing Enforcement of Chosen Motion Picture Ordinance*);

d. All other laws, orders, ordinances, regulations, directives and instructions, or parts thereof, relating to the production, distribution or exhibition of motion pictures, or censorship or licensing thereof; *provided, however,* that the provisions of Ordinance Number 68, dated 12 April 1946 (*Control of Motion Pictures*), shall not be affected hereby; and *further provided,* that this ordinance is not intended and shall not be construed to reenact any laws, orders, ordinances, regulations, directives, instructions or measures repealed by any of the provisions of law or measures herein repealed.

— 1 —

SECTION III. *Licensing of Motion Picture Films.*

a. Licensing Authority. The Department of Public Information of the Government of Korea is authorized and directed to pass upon the propriety of all motion picture films prior to their public exhibition, and to approve and license all films which meet standards to be determined by it, before such films are exhibited to the public.

b. Prohibition of Unlicensed Films. No person, natural or juridical, shall, personally or through any agent, distribute for public exhibition or exhibit to the public any motion picture film unless the film is first duly approved and licensed as hereinabove provided. Films unlawfully distributed or exhibited shall be seized and confiscated. This ordinance shall not apply to any film exhibited by the United States Armed Forces or any agency thereof.

a. Public Exhibition Defined. Public exhibition as used herein means any exhibition before more than fifteen persons, whether or not an admission fee is charged.

d Licensing Procedure. Films submitted to the Department of Public Information for approval and licensing shall be accompanied by the following information in writing:

(1) The name and address of the person, firm or corporation making the application for the license;

(2) The title of the film;

(3) The number of reels in the film;

(4) The name and address of the producer of the film;

(5) A short summary, in Korean and English, of the contents of the film;

(6) An exact copy of all titles other than English superimposed on the film, and of all sound dialogue and commentary in languages other than English on the film, together with an exact translation into English of such titles, dialogue and commentary.

e. Approval in Whole or in Part. The Department of Public Information may approve or disapprove any film in its entirety, or, in appropriate cases, may approve any film subject to deletion or change of specified part or parts thereof.

f. Leaders and Certificates for Approved and Licensed Films. The Director of Public Information shall issue a certificate of approval of each film licensed by him. He shall also provide a film leader, which will be shown in its entirety on the screen immediately before each exhibition of the film.

SECTION IV.   *Specific Offenses Enumerated.*   Without limiting the provisions of this ordinance or of any other existing provision of law, the following are specifically declared to be violations hereof:

a. The misstatement of any fact in an application for the licensing of any film;

b. The exhibition of any licensed film with any additions to it in title, dialogue or commentary (written or oral) or with the addition of any film footage not contained in the film at the time of the granting of the license;

c. The public exhibition of any film not licensed as herein provided.

SECTION V.   *License Fees.*   The Director of the Department of Public Information shall establish a uniform schedule of fees for licensing of films. Educational films used exclusively and without charge in the school system shall be exempt from the payment of such fees. All fees collected for film licenses shall periodically be turned over to the Department of Finance.

SECTION VI.   *Penalties.*   Any person violating the provisions of this ordinance shall, upon conviction by a Military Occupation Court, suffer such penalty as the court shall determine.

SECTION VII.   *Effective Date.*   This ordinance shall be effective on the tenth day after the date appearing hereon.

ARCHER L LERCH
Major General United States Army
Military Governor in Korea

HEADQUARTERS
UNITED STATES ARMY MILITARY
GOVERNMENT IN KOREA
Office of the Military Governor
Seoul, Korea

ORDINANCE
NUMBER 116   8 October 1946

REVISION OF CUSTOMS LAWS

SECTION 1. *Purpose.* The purpose of this ordinance is to promote foreign trade with Korea by reducing existing import tariff rates, and to suppress smuggling by providing rewards to informants and to seizing officials.

SECTION II. *Ceiling on Customs Duties.* Customs duties shall be levied upon articles imported from foreign countries in accordance with rates fixed by existing import tariff laws; *provided, however,* that in no case shall the customs duty collected upon any imported article exceed ten per centum *ad valorem.*

SECTION III. *Exemptions.* In addition to the exemptions provided under existing laws, the following articles shall be exempt from import duties:

a. Articles imported by the United States Armed Forces in Korea, or by the Government of Korea or any political subdivision, agency or instrumentality thereof, for official use;

b. Articles imported for official or personal use by members of the Armed Forces of the Allied Powers in Korea, by accredited civilians on duty with such forces, or by dependents of any of the foregoing.

SECTION IV. *Rewards to Informants and Seizing Officials.* A reward not exceeding twenty-five per centum of the net amount recovered (or fine or forfeiture incurred) and not exceeding ¥50,000 in any case, may be awarded by the Director of the Department of Finance of the Government of Korea to:

a. Any person giving original information concerning any fraud upon the customs revenue, or concerning any violation of the customs laws or regulations, perpetrated or contemplated, which leads to a fine, penalty or forfeiture;

b. Any official or employee paid from appropriated funds of the Government of Korea who, in the course of his official duties, seizes any property being transported or stored in violation of the customs laws or regulations, which leads to a fine, penalty or forfeiture.

Determinations of the Director of the Department of Finance of the Government of Korea under this Section shall in all cases be conclusive and final for all purposes. Rewards shall be paid out of the fine, penalty or proceeds of sale of forfeited property.

SECTION V.   *Repeal of Inconsistent Laws.*   All laws and regulatory measures and parts thereof, which are inconsistent herewith or in conflict with the provisions hereof, are, to the extent they are inconsistent or in conflict herewith, hereby repealed.

SECTION VI.   *Effective Date.*   This ordinance shall be effective on the tenth day after the date appearing hereon.

S E A L

USAMGIK

ARCHER L LERCH
Major General United States Army
Military Governor in Korea

HEADQUARTERS
UNITED STATES ARMY MILITARY
GOVERNMENT IN KOREA
Office of the Military Governor
Seoul, Korea

ORDINANCE
NUMBER 117

18 September 1946

AMENDING ORDINANCE NUMBER 93 BY EXTENDING PERIOD FOR FILING DECLARATIONS OF PROPERTY OR OBLIGATIONS

SECTION I. *Extension of Period for Filing.* The period of time for filing declarations of property or obligations under Section II of Ordinance Number 93, dated 4 July 1946 (Foreign Exchange Control), is hereby extended sixty days after the effective date of this ordinance.

SECTION II. *Effective Date.* This ordinance shall be effective at 2400, 2 September 1946.

ARCHER L LERCH
Major General United States Army
Military Governor in Korea

SEAL
USAMGIK

HEADQUARTERS
UNITED STATES ARMY MILITARY GOVERNMENT IN KOREA
Office of the Military Governor
Seoul, Korea

**ORDINANCE
NUMBER 118** **REPRINTED COPY** 24 August 1946

ESTABLISHMENT OF KOREAN INTERIM LEGISLATIVE ASSEMBLY

SECTION I. *Purpose.* The purpose of this ordinance is to increase the participation of democratic elements in the government by the establishment of an interim legislative body and thus to foster the development of the country on democratic principles pending the early establishment of a unified Korean State with a provisional democratic government of all Korea as provided in the Moscow Agreement.

SECTION II. *Purpose of Legislative Body.* It is declared to be the purpose of Military Government to establish through broad electoral processes a legislative body which shall have the duty of formulating and presenting to the Military Governor draft laws to be used as the basis for political, economic and social reforms, pending the establishment of a Provisional Korean Democratic Government. Initially, the assembly will be composed of members partly selected and partly elected. Eventually all will be elected.

SECTION III. *Establishment.* A Korean legislative body to be known as the Korean Interim Legislative Assembly is hereby established as an organ of government. It shall consist of 90 members, 45 of whom shall be elected. Elective members shall be chosen in each province and in the independent City of Seoul in the manner hereinafter prescribed on the basis of one member for each 550,000 population or major fraction thereof ; *provided*, that each province shall be represented by at least one elective member. In addition there shall be elected from each province and the independent City of Seoul a representative at large. The remaining members shall be selected so as to represent equally and fairly major democratic elements of Korean life, economic, political and intellectual.

SECTION IV. *Pay.* Members of the ssembly shall receive the same compensation as judges of the Supreme Court.   Payment of compensation of necessary employees shall be subject to Civil Service rules and regulations.

SECTION V. *Duties and Powers of the Korean Interim Legislative Assembly.* It shall be the duty of the Assembly to enact ordinances on matters affecting general welfare and interest and on such matters as may be referred to it by the Military Governor. It shall also have the power to review all past appointments to the Military Government above the Civil Service Status of Class 4 and to confirm and consent to all such future appointments. Ordinances enacted by the Assembly shall have the force and effect of law when concurred in by the Military Governor, duly signed by him and sealed and published in the *Official Gazette.* In case the Military Governor does not concur in any ordinance, he will return it to the legislature with a written statement of reasons for nonconcurrence.

SECTION VI. *Quorum; Immunity of Members for Utterances.* All action of the Korean Interim Legislative Assembly shall be by a majority vote of the quorum, and three-fourths of the members shall constitute a quorum unless otherwise determined by the Assembly. Debates in the Assembly, on the floor thereof, and in committees which may be established, shall be free, and the members shall not be questioned for their utterances during the course of such debate in any other place; but the Assembly shall have power to adopt rules of conduct and measures for punishment of its members for disorderly behavior.

SECTION VII. *Qualification of Members.* No person may be elected as a member of the Korean Interim Legislative Assembly unless he or she is a Korean who shall have passed his or her twenty-fifth birthday on the date of assuming office, and unless he or she has been a *bona fide* resident of the province or area represented for at least one year immediately preceding his or her election.

The date of assuming office shall, for the purpose of this section, be deemed to be the first day of the session of the Assembly following such election.

The following shall not be eligible for membership in the Assembly: persons who occupied positions as central, provincial or municipal councilors, or persons of the rank of *chikimkwan* or above, under the Japanese regime; and all those who collaborated with the Japanese, for gain, to the detriment of the Korean people.

— 2 —

SECTION VIII   *Method of Election of Members of the Assembly.* Elective members of the Korean Interim Legislative Assembly shall be elected by and for the people of each province of Korea south of the 38° parallel and the independent City of Seoul, by universal suffrage regardless of sex, according to rules to be established by an ordinance to be enacted by the Assembly and approved by the Military Governor. Until the enactment of such ordinance elective members shall be chosen by elective procedures established by the Military Governor. Such elective procedures will provide, in substance, that each hamlet, village and district shall elect two representatives, who shall in turn elect two representatives from their respective *myun*; that these *myun* representatives shall elect two representatives for their respective *gun*; that the *gun* representatives shall elect provincial representatives to the Assembly in accordance with provincial population ratios; and that all balloting shall be secret. Until the establishment of an official census, population figures shall be those determined from time to time by the National Economic Board. For the purpose of this ordinance, the part of Hwanghae Province which lies south of the 38° parallel shall be considered a part of Kyonggi Province.

SECTION IX. *Duration of Term.* Original members of the Assembly shall, unless reelected or reappointed for subsequent periods, hold office only until the seating of members elected in the first general election held pursuant to ordinances hereafter enacted.

SECTION X. *Initial Session of the Assembly; Adoption of Rules; Election of Chairman.* The initial session of the Assembly shall be held on the day fixed by the Military Governor at a place in Seoul designated by him. The Assembly shall elect from its number a presiding chairman. Election of the presiding chairman shall be by written secret ballot of a majority of the quorum. He shall hold office until his successor is elected. The Assembly shall adopt rules of procedure which shall include provisions for adjournments, committees and the conduct of members.

SECTION XI. *Authority of Military Government Unimpaired.* All duties and powers of the Korean Interim Legislative Assembly under this ordinance shall be exercised under the authority of the Military Government in Korea, until the establishment of a Provisional Korean Democratic Government. The Military Governor retains power to dissolve the Assembly, approve new members and require new elections.

SECTION XII. *Effective Date.* This ordinance shall be effective on the date appearing hereon.

ARCHER L LERCH
Major General United States Army
Military Governor in Korea

HEADQUARTERS
UNITED STATES ARMY MILITARY
GOVERNMENT IN KOREA
Office of the Military Governor
Seoul, Korea

ORDINANCE
NUMBER 119

11 November 194·

NARCOTICS CONTROL

SECTION I. *Purpose.* The purpose of this ordinance is to control all sources of and activities connected with narcotic drugs in Korea south of 38° north latitude, and to limit the supply and uses thereof to legitimate medical and scientific requirements.

SECTION II. *Transfer of Narcotics Control Section; Duties and Functions.* The Narcotics Control Section of the Monopoly Bureau of the Department of Finance of the Government of Korea, together with all records, property and Korean personnel thereof, is hereby transferred to the Bureau of Pharmaceutical Affairs of the Department of Public Health and Welfare of the Government of Korea. All duties and functions thereof, not inconsistent with provisions of this ordinance, are likewise transferred. The Narcotics Control Section shall have the duty and function of administering the provisions of this ordinance and of all rules and regulations issued pursuant hereto by the Director of the Department of Public Health and Welfare.

SECTION III. *Definition of Narcotic Drugs:* The term "narcotic drugs", as used herein, includes opium, morphine, heroin, codeine, cocaine, marihuana, and any component, derivative or preparation of any thereof.

SECTION IV. *Opium Poppy and Marihuana Control.* Planting, cultivation, growth, harvesting, and any other activity which facilitates the growth of marihuana (*Cannabis sativa*) or of the opium poppy (*Papaver somniferum*) or any other plant which is the source of opium or cocaine is hereby prohibited.

a. It shall be unlawful for any person to produce or attempt to produce the opium poppy or marihuana, or to permit the production thereof in or upon any place owned, occupied, used or controlled by him.

b. It shall be unlawful for any person to sell, import, export, transfer, convey any interest in, give away, purchase or otherwise obtain opium poppy or marihuana seed for the purpose of opium poppy   marihuana production.

c. It shall be unlawful for any person   sell, import, export, transfer, convey any interest in, give away, purchase or otherwise obtain opium poppy or marihuana (or parts or seeds thereof); or to manufacture, compound or extract opium products from the opium poppy or products from marihuana (or parts or seeds thereof); except to surrender and deliver up the same to the police or to the Director of the Department of Public Health and Welfare, or their duly authorized agents.

d. Any opium poppies or marihuana (or parts or seeds thereof) which have been produced or otherwise obtained heretofore, and which may be produced or otherwise obtained hereafter in violation of the provisions hereof, shall be seized by and forfeited to the Government of Korea.

e. The failure, upon demand by the police or the Director of the Department of Public Health and Welfare or their duly authorized agents, of the person in occupancy or control of land or premises upon which opium poppies or marihuana (or parts or seeds thereof) are being produced or stored, to surrender or deliver up such poppies or marihuana (or parts or seeds thereof) shall be sufficient basis for the seizure and forfeiture thereof.

f. The police or the Director of the Department of Public Health and Welfare, or their duly authorized agents, shall have authority to enter upon any land (but not a dwelling, unless pursuant to a search warrant issued according to law) where opium poppies or marihuana (or parts or seeds thereof) are being produced or stored, for the purpose of enforcing the provisions of this section.

g. Any opium poppies or marihuana (or parts or seeds thereof), the owner or owners of which are unknown, seized by or coming into the possession of the Government of Korea in the enforcement of this section, shall be forfeited to said Government.

h. The police, the Director of the Department of Public Health and Welfare and their duly authorized agents, are hereby directed to destroy any opium poppies or marihuana (or parts or seeds thereof) seized by and forfeited to the Government of Korea under this section, or to deliver for medical or scientific purposes such opium poppies or marihuana (or parts or seeds thereof) to any political subdivision, agency or instrumentality of the Government, upon proper application therefor under such regulations as may be prescribed by the Director of the Department of Public Health and Welfare.

SECTION V. *Control of Heroin, Crude Opium, Opium Marihuana, and of Implements for Smoking Opium.* Possession, control, production, manufacture, purchase, procurement, importation, receipt, use, transportation, distribution, sale, gift, transfer, exportation of or any other activity connected with heroin (*diacetylmorphine hydrochloride*) or marihuana, or any salt, compound, preparation, derivative, or combination thereof, crude, semiprocessed, or processed opium, smoking opium or opium prepared for smoking, or any implement for smoking opium or marihuana (except the surrendering and delivering up of such drugs or implements to the police, the Director of the Department of Public Health and Welfare or their duly authorized agents), are hereby prohibited.

a. It shall be unlawful for any person to smoke opium, opium tobacco or marihuana, or to take or administer to himself or another, heroin or marihuana (or any compound, salt, derivative or preparation thereof).

b. It shall be unlawful for any person to attempt, cause or facilitate any of the activities prohibited in this section, or to permit any such activities in or upon any place owned, occupied, used or controlled by him.

c. Any heroin or marihuana (or salt, compound, derivative or preparation thereof), crude, processed or semi-processed opium, smoking opium, or opium prepared for smoking, or any implements for smoking opium or marihuana which have been produced or otherwise obtained heretofore, and which may be produced or otherwise obtained hereafter in violation of the provisions hereof, shall be seized by and forfeited to the Government of Korea, and the provisions of paragraphs *e* through *h* of Section IV hereof shall apply, *mutatis mutandis*, to the seizure, forfeiture and disposition thereof.

SECTION VI. *Unlicensed Transactions in Narcotic Drugs Prohibited: Exceptions.* In addition to the foregoing prohibitions, it shall be unlawful for any person without an appropriate license or other proper authority from the Director of the Department of Public Health and Welfare to possess, produce, manufacture, compound, purchase or in any manner obtain, or to sell, transfer, send, ship, carry, transport or deliver, convey any interest in, or give away any narcotic drug, or to attempt, offer to do, cause or facilitate any of such acts or any other act prohibited under this ordinance, or to permit any of said acts in or upon any place owned, occupied, used, maintained or controlled by him, with the following exceptions:

a. Any patient who possesses narcotic drugs dispensed to him for medical purposes by a duly licensed physician, dentist, veterinarian or other authorised practitioner in the course of professional practice, or by a duly licensed pharmacist pursuant to a duly issued prescription of a licensed practitioner, or any patient to whom such drugs are administered or prescribed by such practitioner in the course of professional practice;

— 3 —

b. Any public official or employee who possesses narcotic drugs incidental to or within the scope of his official duties;

c. Any common carrier or warehouseman who transports or stores narcotic drugs pursuant to agreement with a person duly licensed under the provisions of this ordinance, or any employee of such common carrier or warehouseman, acting within the scope of his employment;

d. Any employee of a person duly licensed under the provisions of this ordinance, while acting within the scope of his employment.

SECTION VII. *Applications for Licenses; Qualifications of Applicants.* Any person who desires to procure a license to manufacture, import, sell, deal in, dispense, administer, use or possess any narcotic drugs, shall apply therefor in such manner and form and pay such fees as the Director of the Department of Public Health and Welfare shall prescribe in regulations published as hereinafter provided.

Licenses shall be issued only to persons who, in the opinion of the Director of the Department of Public Health and Welfare:

a. Possess good moral character; and

b. Possess (*1*) such experience in importing, manufacturing, administering, distributing, marketing or handling narcotic or other medical drugs as a wholesale or retail dealer, practitioner, pharmacist, or laboratory worker duly licensed and lawfully entitled to engage in such activities, and (*2*) such means and facilities for manufacturing, handling, and safeguarding narcotic drugs, as to render reasonably probable the orderly and lawful distribution of narcotic drugs of suitable quality to supply medical and scientific needs, without diversion to illicit channels; and

c. Have complied with such additional requirements as the Director of the Department of Public Health and Welfare shall prescribe as reasonably necessary for the controlled importation, production, manufacture, and distribution of narcotic drugs. Such licenses shall be non-transferable, shall be effective for a period of one year from the date of issue, and may be renewed at the discretion of the Director of the Department of Public Health and Welfare, for a like period.

SECTION VIII. *Issuance of Regulations.* The Director of the Department of Public Health and Welfare shall prepare necessary regulations for carrying out the provisions of this ordinance. Such regulations shall, upon approval of the Military Governor, be published in the *Official Gazette*, and shall have the force and effect of law.

SECTION IX. *Limitation of Licenses.* All licenses issued under authority of this ordinance shall be limited in number, locality and area, as the Director of the Department of Public Health and Welfare shall determine to be appropriate to supply medical and scientific needs for narcotic drugs, with due regard to provision for reasonable reserves; *provided, however*, that this ordinance shall not be construed to require the Director of the Department of Public Health and Welfare to issue or renew any license or licenses.

SECTION X. *Revocation of Licenses.* The Director of the Department of Public Health and Welfare may at any tine suspend, revoke or refuse to renew any license issued under this ordinance if, after due notice and opportunity for hearing, he finds such action to be in the public interest, or that the licensee has failed to comply with regulations or requirements of law or has failed to maintain requisite qualifications.

SECTION XI. *Seizure and Forfeiture of Narcotic Drugs.* Any narcotic drugs which have been produced or otherwise obtained heretofore, and which may be produced or otherwise obtained hereafter in violation of any of the provisions of this ordinance, shall be seized by and forfeited to the Government of Korea, and the provisions of paragraphs *e* through *h* of Section IV hereof shall apply, *mutatis mutandis*, to the seizure, forfeiture and disposition thereof

SECTION XII. *Miscellaneous Provisions.*

a. *Emergency Requirements.* It shall be the duty of the Director of the Department of Public Health and Welfare, whenever in his opinion medical and scientific needs will not be met by licensed importation or production of narcotic drugs, to provide for the acquisition, production, manufacture, use, sale, giving away or other proper distribution of narcotic drugs by the Government of Korea either directly or through and with the approval of the head of any agency of the government, including any government-owned or controlled company.

b. *Assistance by Government Agencies.* It shall be the duty of all political subdivisions, agencies and instrumentalities of the Government of Korea, whan requested by the Director of the Department of Public Health and Welfare, to furnish such assistance (including technical advice) as will aid in carrying out the purposes of this ordinance.

c. *Authorized Activities of Narcotics Control Officials.* None of the prohibitions in this ordinance shall apply to any officer or employee of the Narcotics Control Section who, in the performance of his official duties and within the scope of his authority, engages in any business or activity herein described, nor to any other officer or employee of the Government of Korea who, in the performance of his official duties and within the scope of his authority and with the approval of the Director of the Department of Public Health and Welfare, engages in any business or activity herein described.

SECTION XIII. *Repeal of Inconsistent Laws, &c.* All laws, ordinances, orders, regulations, directives and instructions, and parts thereof which are inconsistent or in conflict with the provisions of this ordinance are hereby repealed.

SECTION XIV. *Penalties.* Any person violating the provisions of this ordinance or of any regulation issued pursuant hereto by the Director of the Department of Public Health and Welfare, and having the effect of law, shall, upon conviction by a Military Occupation Court, suffer such punishment as the Court shall determine.

SECTION XV. *Effective Date.* This ordinance shall be effective on the tenth day after the date appearing hereon.

S E A L
USAMGIK

ARCHER L LERCH
Major General United States Army
Military Governor in Korea

HEADQUARTERS
UNITED STATES ARMY MILITARY GOVERNMENT IN KOREA
Office of the Military Governor
Seoul, Korea

ORDINANCE
NUMBER 120

24 October 1946

INCREASING FINES AND JURISDICTION OF SPECIAL JUDICIAL OFFICERS; PUNISHMENTS FOR VIOLATIONS OF PRICE AND FOOD MEASURES

SECTION I. *Repeal of Ordinance Number 51.* Ordinance Number 51, dated 21 February 1946 *(Increasing Fines)*, is hereby repealed; *provided, however,* that notwithstanding such repeal, the provisions of Ordinance Number 51 shall continue to apply to all offenses consummated prior to the effective date of this ordinance.

SECTION II. *Increase of Criminal Fines.** All existing provisions of law and regulatory measures which fix or impose a monetary fine for the violation thereof, are hereby amended so that each monetary fine is increased fifty (50) times.

SECTION III. *Jurisdiction of Special Judicial Officers Increased.* Paragraph 2 of Ordinance Number 41, dated 10 January 1946 *(Special Judicial Officers)*, is hereby amended to read as follows:

"Each Special Judicial Officer shall have original jurisdiction of cases in which the punishment which may be imposed by law does not exceed detention, imprisonment or penal servitude for a period of thirty (30) days or a fine of ten thousand yen (¥ 10,000), or both; provided, however, that in cases involving a violation of any price or food law or regulation, each Special Judicial Officer shall have jurisdiction over first offenses only, with jurisdiction enlarged in such cases to include those in which the maximum punishment which may be imposed by law does not exceed detention, imprisonment or penal servitude for a period of three months or a fine of twenty thousand yen (¥ 20,000), or both."

SECTION IV. *Length of Custody Where Fine Not Paid.* Paragraph 12 of

---

* This section replaces the provisions of *Section 1* of Ordinance No 51.

Ordinance Number 41, dated 10 January 1946 (*Special Judicial Officers*), is hereby amended to read as follows:

"In a case where a fine or net fine is pronounced, the accused shall deposit the amount immediately. If he does not do so, he shall be held in custody for a term calculated by converting one hundred and twenty-five yen (¥ 125) into one day, and for this purpose any amount less than one hundred and twenty-five yen (¥ 125) shall be taken as one day."

SECTION V. *Effect of Custody Where Fine Not Paid.* Paragraph 14 of Ordinance Number 41, dated 10 January 1946 (*Special Judicial Officers*), is hereby amended to read as follows:

"The number of days of custody mentioned in paragraph 11 shall accordingly be considered as days served. The number of days of custody mentioned in paragraph 12 shall accordingly be considered as yen paid, on the basis of one hundred and twenty five yen (¥ 125) for one day, toward the discharge of a fine or net fine."

SECTION VI. *Minimum and Maximum Punishments in Price Violation Cases.* Any person convicted of violating any price or food law or regulation shall be punished by fine, by detention, imprisonment or penal servitude, or by both. In the following cases, upon conviction of violating any price or food law or regulation, there shall be imposed not less than the minimum nor more than the maximum punishments hereinafter set forth:

a. *Upon Conviction of a First Offense:*

Amount Involved in Transaction	Minimum Fine	Maximum Fine	Maximum Detention, Imprisonment or Penal Servitude
¥ 500 or less	¥ 1,000	¥ 5,000	1 month
Over ¥ 500 and not more than ¥ 1,000	¥ 2,000	¥ 10,000	2 months
Over ¥ 1,000 and not more than ¥ 5,000	¥ 4,000	¥ 20,000	3 months
Over ¥ 5,000 and not more than ¥ 10,000	¥ 20,000	¥ 40,000	6 months
More than ¥ 10,000	¥ 40,000	No limit	3 years

Upon conviction of a first offense where no monetary amount is involved, the minimum fine shall be ¥ 1,000, the maximum fine shall be ¥ 5,000, and the maximum imprisonment or penal servitude shall be one month.

b. *Second Offenses:* Upon conviction of a second offense, the minimum and maximum punishments listed above shall be increased three times; *provided, however,* that the punishment imposed shall in no case be less than three times the fine imposed upon the first conviction plus not less than the maximum imprisonment or penal servitude listed above for the first offense (whether or not detention, imprisonment or penal servitude was imposed for the first offense).

    c.   Third and Subsequent Offenses: Upon conviction of any third or further offense, the minimum and maximum punishments listed above for the first offense shall be increased ten times; *provided, however,* that the punishment imposed shall in no case be less than five times the fine imposed upon the first conviction plus three times the maximum imprisonment or penal servitude listed above for the first offense (whether or not imprisonment or penal servitude was imposed for the first or any subsequent offense).

    SECTION VII. *Prospective Effect Only.* The provisions of the various sections contained herein shall not be applicable to offenses consummated prior to the effective date hereof.

    SECTION VIII. *Effective Date.* This ordinance shall be effective on the tenth day after the date appearing hereon.

S E A L
USAMGIK

ARCHER L LERCH
Major General United States Army
Military Governor in Korea

HEADQUARTERS
UNITED STATES ARMY MILITARY
GOVERNMENT IN KOREA
Office of the Military Governor
Seoul, Korea

ORDINANCE
NUMBER 121.

7 November 1946

MAXIMUM WORKING HOURS

SECTION I. *Purpose.* The purpose of this ordinance is to regulate and limit maximum working hours in industry, commerce and governmental work, so as to maintain and protect the health, efficiency and general well-being of workers.

SECTION II. *Maximum Hours.* No employer shall, except as otherwise provided in this section, employ any of his employees engaged in industry, commerce or governmental work for more than forty-eight (48) working hours in any workweek.

An employee may be employed for not more than sixty (60) working hours in any workweek, *provided*, that in all such cases where employment is in excess of forty-eight (48) hours in any workweek, the employee shall receive compensation for all hours in excess of forty-eight (48) at a rate not less than one and a half times his regular rate of compensation, and *further provided*, that:

a. Employment for such time in excess of 48 hours is pursuant to a collective bargaining agreement entered into between the employer and a labor union or organization duly authorized to represent the employee, *or*, in the absence of a collective agreement,

b. Employment for such time in excess of 48 hours is pursuant to specific agreement between the employer and the employee.

Each legal holiday, except Sunday or other legal day of rest, shall count as eight hours of the permitted total hours of employment. Every employee subject to the provisions of this ordinance shall be entitled to not less than twenty-four consecutive hours of rest in each workweek.

— 1 —

SECTION III. *Exemptions.* There shall be exempt from the provisions hereof, to the extent set forth:

a. Employees engaged in industries which are seasonal in nature. Such exemption shall be in whole or in part. The Director of the Department of Labor shall determine which industries are so exempt, and the extent of such exemption;

b. Employees engaged in agriculture, as fishermen or sailors, as domestic servants or in piece work in their own homes;

c. Employees employed in a *bona fide* executive, administrative or professional capacity, as hereinafter defined;

d. Employees employed on the national railway system; but the Director of the Department of Labor shall, after consultation with the Director of the Department of Transportation, formulate and issue published regulations in accordance with the spirit of this ordinance, fixing maximum hours of employment for employees of the national railway system, and providing that all employment in excess of forty-eight (48) hours per workweek shall be compensated at the rate of one and a half times the regular rate of payment;

e. Employment in excess of sixty (60) hours per workweek, where such work is necessary in cases of extreme emergency to protect life or property; *provided*, that the employer shall, before requiring any employment in excess of sixty (60) hours, in each case first secure the consent of the Director of the Department of Labor or his authorized representative. The Director of the Department of Labor shall have the power, by written order, in the event of any emergency caused by fire, famine or flood, or by other extraordinary event or condition creating danger to life or property, to suspend either generally or in stated areas the provisions of this ordinance limiting the maximum number of working hours, for the duration of such emergency. All employment in excess of sixty (60) hours per workweek, where permissible pursuant to this subsection, shall be compensated at the rate of one and a half times the regular rate of payment.

SECTION IV. *"Executive"*, *"Administrative"* and *"Professional"* Defined.

a. The term "employee employed in a *bona fide* executive *** capacity" means any employee:

(1) Whose primary duty consists of the management of the establishment in which he is employed or a customarily recognized department or

subdivision thereof, *and*

(*2*) Who customarily and regularly directs the work of other employees therein, *and*

(*3*) Who has the authority to hire or discharge other employees or whose suggestions and recommendations as to the hiring or discharge and as to the advancement and promotion or other change of status of other employees will be given particular weight, *and*

(*4*) Who customarily and regularly exercises discretionary powers.

b. The term "employee employed in a *bona fide***** administrative*** capacity" means any employee :

(*1*) Who regularly and directly assists an employee employed in a *bona fide* executive or administrative capacity (as such terms are defined in this ordinance), where such assistance is non-manual in nature and requires the exercise of discretion and independent judgment; *or*

(*2*) Who performs under only general supervision, responsible non-manual office or field work, directly related to management policies or general business operations, along specialized or technical lines requiring special training, experience or knowledge, and which requires the exercise of discretion and independent judgment; *or*

(*3*) Whose work involves the execution under only general supervision of special non-manual assignments and tasks directly related to management policies or general business operations involving the exercise of discretion and independent judgment.

c. The term "employee employed in a *bona fide**** professional capacity" means any employee who is engaged in work :

(*1*) (*a*) Predominantly intellectual and varied in character as opposed to routine mental, manual, mechanical or physical work. *and*

(*b*) Requiring the consistent exercise of discretion and judgment in its performance, *and*

(*c*) Of such a character that the output produced or the result accomplished cannot be standardized in relation to a given period of time, *and*

(*2*) (*a*) Requiring knowledge of an advanced type in a field of science or learning customarily acquired by a prolonged course of specialized intellectual instruction and study, as distinguished from a general academic education and from an apprenticeship and from training in the performance of routine mental, manual or physical processes, *or*

(*b*) Predominantly original and creative in character in a recognized field of artistic endeavor as opposed to work which can be produced by a person endowed with general manual or intellectual ability and training, and the result of which depends primarily on the invention, imagination or talent of the employee.

SECTION V. *Enforcement, Investigations, Inspection and Records.*

a. Enforcement of the provisions of this ordinance shall be under the supervision of the Department of Labor. The Director of the Department of Labor or his designated representatives may investigate and gather data regarding the hours and other conditions and practices of any employment subject to this ordinance, and may enter and inspect such places and such records (and make transcripts thereof and excerpts therefrom), question such employees, and investigate such facts, conditions, practices, or matters as he may deem necessary or appropriate to determine whether any person has violated any provision of this ordinance, or which may aid in the enforcement of the provisions of this ordinance.

b. Every employer subject to any provision of this ordinance or of any regulation or order issued pursuant to this ordinance shall make, keep and preserve such records of the persons employed by him and of the wages, hours and other conditions and practices of employment maintained by him, and shall preserve such records for such periods of time, and make such reports to the Director of the Department of Labor as he shall prescribe by regulations published in the *Official Gazette* upon approval of the Military Governor, or by order directed to any specific employer.

SECTION VI. *Regulations.*

a. In addition to the other powers and duties of the Director of the Department of Labor under this ordinance, he shall:

(*1*) Make such regulations or orders as are necessary or appropriate to carry out any of the provisions hereof;

(*2*) Regulate the maximum number of hours any employee may work in any workday;

(*3*) Define such terms not defined herein as may require definition, and amplify or amend any terms, definitions or amended definitions as may be necessary;

b. Such regulations shall, when ⸱lished in the *Official Gazette* upon approval of the Military Governor, have   force and effect of law.

SECTION VII. *Definitions.* As used in this ordinance:

a. *Person* means an individual, partnership, association, corporation, legal representative, governmental department, bureau or agency, or any organized group of persons.

b. *Commerce* includes trade, transportation, transmission or communication.

c. *Industry* includes a trade, handicraft, business, industry or branch thereof, or group of industries, in which individuals are gainfully employed.

d. *Agriculture* includes farming in all of its branches and, among other things, includes the cultivation and tillage of the soil, dairying, the production, cultivation, growing and harvesting of any agricultural or horticultural commodities, the raising of livestock, bees, or poultry, and any practices (including any forestry or lumbering operations) performed by a farmer or on a farm as an incident to such farming operations, including preparation for market, delivery to storage or to market or to carriers for transportation to market.

e. *Employer* includes any person acting directly or indirectly in the interests of an employer in relation to an employee, including governmental departments, bureaus and agencies, at the national, provincial and local levels.

f. *Employee* includes any individual suffered or permitted to work by an employer.

g. *Employ* includes to suffer or permit to work.

h. *Wage* paid to any employee includes the reasonable cost to the employer of furnishing such employee with board, lodging or other facilities, if such board, lodging or other facilities are customarily furnished by such employer to any of his employees.

i. *Workday* means any period of twenty-four consecutive hours.

j. *Workweek* means any period of seven consecutive days.

k. *Working hours* includes all hours during which the employee is present at his place of employment at the direction of his employer, whether or not work is actually performed (except meal and rest periods), and all hours during which, after reporting to his usual place of employment, the

— 5 —

employee is in transit to a place other than his usual place of employment, in order to perform labor or other required duty, and all hours during which the employee is required by an employer to travel to report for work at a place other than his usual place of employment.

SECTION VIII. *Relation to Other Laws.* No provision of this ordinance shall excuse non-compliance with any other law establishing a maximum workweek lower than the maximum workweek established under this ordinance or established by collective agreement.

SECTION IX. *Criminal Penalties.* Any person who wilfully violates any of the provisions of this ordinance or of any regulation issued pursuant hereto and having the effect of law shall, upon conviction of a first offense, be subject to a fine of not less than ¥20,000 and not more than ¥100,000, and shall, upon conviction of a second offense, be subject to a fine of not less than ¥20,000 and not more than ¥200,000 or to imprisonment for not more than six months, or both. Upon conviction of a third or subsequent offense, he shall be subject to such punishment as a Military Occupation Court shall determine, but in no event less than three times the fine or imprisonment, or both, imposed upon his last previous conviction under this ordinance or any regulation issued pursuant hereto and having the effect of law.

SECTION X. *Effective Date.*— This ordinance shall be effective on the tenth day after the date appearing hereon.

S E A L
USAMGIK

ARCHER L LERCH
Major General United States Army
Military Governor in Korea

HEADQUARTERS
UNITED STATES ARMY MILITARY
GOVERNMENT IN KOREA
Office of the Military Governor
Seoul, Korea

ORDINANCE
NUMBER 122

23 October 1946

RESTORATION OF KOREAN NAMES

SECTION I. *Purpose.* The purpose of this ordinance is to provide a simple procedure for the restoration of Korean names to those individuals whose names were changed to Japanese-style names during the Japanese domination of Korea.

SECTION II. *Japanese-Style Names Invalidated.* All changes of Korean-style names of individuals to Japanese-style names, made in accordance with laws in force during the Japanese domination of Korea, are hereby declared null and void *ab initio,* and all acts done under such changed names shall be considered, in legal effect, to have been done under the Korean names of such individuals prior to such change; *provided, however,* that no alterations shall be made in entries in family registers to restore such Korean names until sixty days after the effective date of this ordinance; and *provided, further,* that any individual desiring to retain a Japanese-style name may file an application for such retention with the registrar keeping his family register, within such sixty days, and upon due receipt of application, entries in the family register of such individual shall be left unaltered, and his Japanese-style name shall remain in legal force and effect.

Except where such applications for retention of Japanese-style names are duly received, registrars keeping family registers shall forthwith proceed, after such sixty days, to make alterations in the entries necessary to restore the former Korean names in accordance with procedures established by law.

SECTION III. *Change of Japanese-Style First Names.* Individuals never having had Korean-style first names, whose Japanese-style first names were recorded in family registers in Korea in accordance with laws in force during the Japanese domination of Korea, may file an application for change

313

of such Japanese-style first names to Korean-style first names, with the registrar keeping their family registers, within six months after the effective date of this ordinance, and upon due receipt of such applications, entries in the family registers of applicants shall be altered to accomplish such change.

After the expiration of such six-month period, such individuals desiring to change Japanese-style first names may apply to the appropriate court for such change, in accordance with procedures established by law.

SECTION IV. *Repeal of Inconsistent Laws.* All laws, ordinances, regulations, directives and instructions, or parts thereof, which are inconsistent herewith or in conflict with the provisions hereof are hereby declared null and void *ab initio.*

SECTION V. *Effective Date.* This ordinance shall be effective on the date appearing hereon.

ARCHER L LERCH
Major General United States Army
Military Governor in Korea

HEADQUARTERS
UNITED STATES ARMY MILITARY
GOVERNMENT IN KOREA
Office of the Military Governor
Seoul, Korea

ORDINANCE
NUMBER 123   November 1946

ESTABLISHMENT OF KOREAN NATIONAL HOUSING ADMINISTRATION

SECTION I. *Purpose.* The purpose of this ordinance is to create a national agency to encourage improvement in housing standards and conditions, and to provide a coordinated national program of low-cost housing for Koreans of low income.

SECTION II. *Korean National Housing Administration Established.* The Korean National Housing Administration is hereby established in the Government of Korea. It shall consist of a National Housing Administrator, appointed by the Military Governor to hold office during his pleasure, and such staff and assistants as may be necessary. The Administrator is authorized, subject to civil service rules and regulations and other requirements of law, to appoint, compensate, remove and suspend the necessary staff and assistants for the Administration.

SECTION III. *Functions and Duties.* The Korean National Housing Administration shall:

a. Survey and study the housing situation in Korea;

b. Prepare plans, designs and projects related to housing and housing conditions throughout Korea, in both rural and urban areas, and recommend to the Military Governor such plans, designs and projects as are essential to the improvement of housing conditions in Korea;

c. Within existing appropriations and budgetary limitations, take title to land, buildings and personal property by purchase or otherwise as provided by law, make construction and repairs thereon, and remodel or otherwise improve or remove such buildings and personal property, to the extent necessary or desirable to improve housing standards;

d. Mortgage, lease, sell or otherwise dispose of or convey title to any land, building, structure or project to which title has been legally obtained by the Korean National Housing Administration;

e. Enter into any and all transactions including contracts, financial transactions and all other procedures necessary to carry out the provisions of this ordinance, within the scope of policies approved by the Military Governor;

f. Advise and assist political subdivisions, agencies and instrumentalities of the government, quasi-public and private firms and organizations and persons regarding improvement of housing conditions throughout Korea;

g. Determine and define by regulation the groups and categories of persons eligible to share in the benefits of national housing programs of the Administration;

h. Establish and maintain model housing developments and projects in suitable locations throughout Korea;

i. Disseminate, in its discretion, data of general interest on housing conditions and problems in Korea;

j. Coordinate, as necessary, with other agencies and instrumentalities of the Government and with political subdivisions in the achievement of the purposes of this ordinance.

Korea, as used herein, means Korea south of 38° north latitude.

SECTION IV. *Publication of Regulations.* The Administration is authorized to prepare national housing regulations which, when approved by the Military Governor and published in the *Official Gazette,* shall have the force and effect of law.

SECTION V. *Appropriation of Funds.* The Department of Finance of the Government of Korea is directed to appropriate funds necessary for carrying out the provisions of this ordinance, and shall maintain regular audits of the funds and financial transactions of the Administration.

SECTION VI. *Effective Date.* This ordinance shall be effective on the date appearing hereon.

S E A L

USAMGIK

ARCHER L LERCH
Major General United States Army
Military Governor in Korea

HEADQUARTERS
UNITED STATES ARMY MILITARY GOVERNMENT IN KOREA
Office of the Military Governor
Seoul, Korea

ORDINANCE
NUMBER 124

13 November 1946

REPRINTED COPY

AMENDING DRUG AND DRUG BUSINESS REGULATIONS*

SECTION I. *Purpose.* The purpose of this ordinance is to amend and supplement Ordinance Number 96, dated 25 July 1946 (*Amending Drug and Drug Business Regulations*).

SECTION II. *Governor General Ordinance Number 22 Further Amended.* Governor General Ordinance Number 22, dated 28 March 1912 (*Regulations of Drugs and Drug Business*), heretofore amended by Ordinance Number 96, dated 25 July 1946, is hereby further amended as follows:

a. The words *or the pharmacist license issued by internal minister* in *Section II* are hereby deleted wherever they occur.

b. The words *police office* in *Section XVII* are hereby deleted and the words *health office* substituted therefor.

c. The words *police office* in *Section III* and in *Section XVIII* are hereby deleted and the words *Director of the Provincial Bureau of Public Health and Welfare* substituted therefor.

SECTION III. *Ordinance Number 96 Amended. Section III k* of Ordinance Number 96, dated 25 July 1946, is hereby amended by substituting the words *within five months* for the words *within two months*; the intent of this provision being to extend the date specified in said *Section III k* of Ordinance Number 96 to five months after the effective date of *Ordinance Number 96.*

SECTION IV. *Effective Date.* This ordinance shall be effective on the date appearing hereon.

ARCHER L LERCH

Major General United States Army

Military Governor in Korea

---

*See *Ordinance Number 96*, dated 25 July 1946.

軍政廳　官報　法令　第一二四號　一九四六年十一月十三日

HEADQUARTERS
UNITED STATES ARMY MILITARY GOVERNMENT IN KOREA
Office of the Military Governor
Seoul, Korea

ORDINANCE
NUMBER 125

1 November 1946

REPRINTED COPY
AMENDING ORDINANCE NUMBER 103

SECTION I. *Section IV of Ordinance No 103 Amended.* Section IV of Ordinance No 103, dated 31 August 1946, is amended to read as follows:

SECTION IV. Composition of Commission. The Commission shall consist of five qualified lawyers who shall be appointed by the Military Governor and will serve as members of the Commission at his pleasure. One of the members shall be designated President of the Commission. The Commission shall have a Recorder, an Interpreter, and personnel required to perform the clerical work of the Commission, who shall be appointed by the President of the Commission with the approval of the Military Governor. Paid officials of the Government of the United States or of the United States Army Military Government in Korea serving on or for the Commission shall receive no additional compensation for such service.

SECTION II. *Effective Date.* This ordinance shall be effective on the date appearing hereon.

ARCHER L LERCH
Major General United States Army
Military Governor in Korea

軍政廳　官報　法令　第一二五號　　　一九四六年十一月一日

HEADQUARTERS
UNITED STATES ARMY MILITARY GOVERNMENT IN KOREA
Office of the Military Governor
Seoul, Korea

ORDINANCE
NUMBER 126

15 November 1946

PROVINCIAL AND LOCAL ELECTIVE OFFICIALS AND COUNCILS

SECTION I. *Purpose.* The purpose of this ordinance is to provide for the election of important provincial and local officials and provincial and local councils through the freely expressed choice of a majority of Koreans, thereby fostering the development of the country on principles of democratic local self-government.

SECTION II. *Designation of Elective Officials.* The following are hereby designated as elective officials:

a. Provincial governor

b. Mayor

c. County headman

d. Town headman

e. Village headman

f. Island headman

g. Provincial councillor

h. City councillor

i. Town councillor

j. Village councillor

Other provincial and local officials so designated by law from time to time shall also be elective officials.

SECTION III. *Method of Election.* Elective officials designated in Section II hereof shall be elected by and for the people of their respective provinces, cities, counties, towns, villages, islands or election districts, by universal suffrage regardless of sex, according to rules to be established by an ordinance to

— 1 —

be enacted by the Korean Interim Legislative Assembly and approved by the Military Governor.   Until the enactment of such ordinance elective officials shall be chosen by elective procedures established by the Military Governor. Such elective procedures will provide, in substance, that each village, town, island, and county will elect a headman, and each village and town a council; that each city will elect a mayor and a council; and that each province south of 33° north latitude shall elect a governor and a council.   Village and town councils shall be elected on the basis of one member for each 5,000 population or major fraction thereof; provided that each such council shall have at least five members.   City councils to be elected shall consist of seven members plus an additional member for each 30,000 population or major fraction thereof.   Provincial councils shall be elected on the basis of one member for each 55,000 population or major fraction thereof; provided that each provincial council shall have at least ten members.   All balloting shall be secret. Until the establishment of an official census, population figures shall be those determined from time to time by the National Economic Board.   For the purpose of this ordinance, the part of Hwanghae Province which lies south of 38° north latitude shall be considered a part of Kyonggi Province.   Nothing contained herein shall be construed to apply to elective offices in the independent City of Seoul.

SECTION IV. *Duration of Term.* Persons appointed or elected to offices designated in Section II hereof shall, prior to the enactment of ordinances for a general election, hold office until officials elected pursuant to such general election ordinances assume office, unless sooner removed or unless their terms of office sooner terminate.

SECTION V. *Pay.* Elective officials designated in this ordinance shall receive the compensation provided by law unless it is provided that such services shall be without pay.

SECTION VI. *Duties and Powers.* Elective officials designated in this ordinance shall have the duties and powers provided by law.

The councillors hereinabove enumerated, acting through their respective councils, shall have the following powers:

a. Enactment and amendment of provincial and city ordinances and provincial and local regulations, subject to the provisions of Section VII *g* of Ordinance Number 114, dated 23 October 1946, and other requirements of law;

— 2 —

    b. Approval of provincial or local budgets and reports of final and special accounts;

    c. Determination of matters relating to provincial or local property, contracts, funds. taxes, rents, fees, charges, loans, gifts, subsidies and other property and contractual interests, revenues and expenditures, including continuing expenditures;

    d. Approval or disapproval of past and future provincial or local appointments within Civil Service Classes 7, 8 and 9, and recommendation of transfer or dismissal of any provincial or local official;

    e. Recommendations to appropriate authorities in connection with any provincial or local matter affecting the public interest;

    f. Advice in connection with any provincial or local matter referred to them by any public official in accordance with law, practice or custom;

    g. Other powers granted to councils by law, or existing by reason of practice or custom, not inconsistent with the provisions of this ordinance.

    Exercise of the foregoing powers shall be in accordance with requirements of law and within the scope of policies established by ordinances, regulations, orders and directives of higher authority.

    SECTION VII. *Qualifications for Office.* No person may be elected to any office designated in Section II hereof unless he or she is a Korean who shall have passed his or her twenty-fifth birthday on the date of assuming office; unless he or she has been a *bona fide* resident of the province or area concerned for at least one year preceding his or her election; and unless he or she meets all other conditions or qualifications provided by law.

    The following shall not be eligible for election to any office designated in Section II hereof: persons who occupied positions as central, provincial or municipal councillors, or persons of the rank of *chikimkwan* or above, under the Japanese regime; and all those who collaborated with the Japanese, for gain, to the detriment of the Korean people.

    SECTION VIII. *Authority of Military Government Unimpaired.* All duties and powers of officials and councils under this ordinance shall be exercised under the authority of the Military Government in Korea, until the establishment of a Provisional Korean Democratic Government. Military Government retains power to remove any elective official designated in this ordinance or

— 3 —

any person appointed by such official, to dissolve any elected council or other provincial or local public body, to require new elections or appointments, and to nullify any enactment, resolution, action or decision of any elective official or council.

SECTION IX. *Funds, Records and Property of Provincial and Local Councils.* All funds, records and property of the former provincial and local councils dissolved by Ordinance Number 60, dated 14 March 1946, and all funds, records and property of provincial and local advisory councils established by Military Government to replace them, shall be turned over to the custody of provincial and local councils elected in accordance with the provisions of this ordinance, immediately upon the convening of such elected councils.

SECTION X. *Repeal of Inconsistent Law.* All laws, ordinances, orders, regulations, directives and instructions, or parts thereof, which are inconsistent herewith or in conflict with the provisions hereof are, to the extent they are inconsistent or in conflict herewith, hereby repealed.

SECTION XI. *Effective Date.* This ordinance shall be effective on the date appearing hereon.

ARCHER L LERCH
Major General United States Army
Military Governor in Korea

軍政廳　官報　法令　第一二六號　　一九四六年十一月十五日

HEADQUARTERS
UNITED STATES ARMY MILITARY
GOVERNMENT IN KOREA
Office of the Military Governor
Seoul, Korea

ORDINANCE
NUMBER 127

15 November 1946

PENALTIES FOR RICE SMUGGLING AND CONNECTED ACTS

SECTION I. *Purpose.* The purpose of this ordinance is to provide severe penalties for the smuggling of rice out of Korea and criminal acts incidental thereto, because of the urgent necessity that this vital food be retained within Korea to meet the needs of the Korean people.

SECTION II. *Penalties.* The smuggling of rice out of Korea being prohibited by law*, the following penalties are prescribed:

a. Any person smuggling rice shall, upon a first conviction therefor be fined not less than ¥ 100,000 and be sentenced to not less than ten years penal servitude; upon a second conviction, not less than ¥ 200,000 and not less than twenty years penal servitude; and upon conviction of a third or further offense, not less than ¥ 500,000 and penal servitude for life.

b. In any case of conviction of smuggling rice, where the amount of rice involved exceeds seventy (70) *suk*, the minimum penalties above set forth (except penal servitude for life) shall be doubled.

c. Accessories and accomplices of persons engaged in smuggling rice shall, upon conviction, be subject to the same minimum penalties as principals.

---

* Ordinance Number 29, *Regulations of Foreign Trade*, dated 3 January 1946; National Food Regulation Number 2, *Collection of Rice*, dated 12 August 1946; Foreign Commerce Regulation Number 1, *Import and Export Licenses and Permits*, dated 12 July 1946; *etc.*

軍政廳　官報　法令　第一二七號　　一九四六年十一月十五日

d. Any person who is financially interested in the smuggling of rice shall, upon conviction, be subject to the same minimum penalties as a principal.

e. Any person who, with knowledge or intent that any ship or other means of transportation will be used in rice smuggling, secures, prepares, outfits, sells, leases or otherwise permits the use thereof; or who secures, collects, buys, sells or conceals rice intending to smuggle such rice or knowing that such rice is intended to be smuggled, shall, upon conviction, be fined not less than ¥ 50,000 and sentenced to not less than three years penal servitude. In addition, such property shall, if seized, be forfeited as provided by law, to the Government of Korea.

f. Any rice which is the subject of smuggling and any private ship, boat, motor vehicle, cart, animal, airplane or other conveyance or instrumentality used in rice smuggling shall be forfeited as provided by law, to the Government of Korea.*

SECTION III. *Rewards to Informants.*

a. Any person giving original information concerning rice smuggling, perpetrated, attempted or prepared for, which leads to a fine, penalty or forfeiture, shall be rewarded with twenty-five percent of the net amount recovered as fine, penalty or forfeiture, without monetary limit as to the amount that may be paid. If the information given results in the conviction of any person or persons for rice smuggling, the informant shall be rewarded with an additional ten percent of the net amount of any fine, penalty or forfeiture recovered by reason of the information given.

b. The determination of the Director of the Department of Finance of the Government of Korea as to persons entitled to payment under this Section shall be conclusive and final for all purposes. Rewards shall be paid by the Department of Finance of the Government of Korea out of amounts actually received as fines, penalties, or proceeds of sales of forfeited property.

c. Rewards pursuant to this Section shall not be payable to any public official, as herein defined.

---

*Ordinance Number 39, *Regulations of Foreign Trade*, dated 3 January 1946; *etc.*

軍政廳　官報　法令　第一二七號　　一九四六年十一月十五日

d. The identity of informants shall not be given publicity and every effort shall be made to conceal their identity.

e. The provisions of Section IV of Ordinance Number 116 (*Revision of Customs Laws*) dated 8 October 1946, shall remain in full force and effect, except as modified herein with respect to information given concerning rice smugglers and rice smuggling; *provided, however,* that rewards shall not be paid both under this ordinance and under Ordinance Number 116 to the same person or persons, for the same act or acts.

SECTION IV. *Punishment of Public Officials Concealing Violations of Laws against Smuggling Rice.* Any public official who knows, or has good reason to believe, that smuggling of rice has been, is being or is about to be committed and who omits to disclose such information, or wilfully fails to prevent any such act or apprehend any person engaged therein or to enforce the penalties therefor if within his jurisdiction or power, shall, upon conviction, be dismissed from the public service, be forever ineligible to hold any public office or trust, and be fined not less than ¥50,000 and sentenced to not less than five years penal servitude.

SECTION V. *Punishment of Public Officials for Gross Negligence in Enforcing Laws against Smuggling Rice.* Any public official who fails, through gross negligence, to prevent rice smuggling or apprehend or prosecute any person engaged therein, shall, upon conviction, be dismissed from the public service, be fined not less than ¥5,000 and be sentenced to not less than one year imprisonment.

SECTION VI. *Offering Bribes to Public Officials to Permit Rice Smuggling.* Any person who gives or offers, or procures or causes to be given or offered, a bribe directly or indirectly, to any public official to influence him to violate his duty to prevent rice smuggling or to apprehend, prosecute or convict rice smugglers, shall, upon conviction, be fined not less than ¥50,000 and be sentenced to not less than five years penal servitude. In addition, any bribe offered shall, if seized, be forfeited, as provided by law, to the Government of Korea.

SECTION VII. *Bribery of Public Officials to Permit Rice Smuggling.* Any public official who receives or agrees to receive a bribe directly or indirectly, upon any agreement or understanding, express or implied, that he will violate or deviate from his duty to prevent rice smuggling or to apprehend, prosecute or convict persons smuggling rice shall, upon conviction, be dismissed from the public service, be forever ineligible to hold any public office or trust, be fined not less than ¥ 50,000 and sentenced to not less than ten years penal servitude. A public official who solicits a bribe or who, upon his own request or solicitation, receives or agrees to receive a bribe, directly or indirectly, on his statement or promise, express or implied, that he will or may violate or deviate from his duty to prevent rice smuggling or to apprehend, prosecute or convict rice smugglers shall, upon conviction, be dismissed from the public service, be forever ineligible to hold any public office or trust, be fined not less than ¥ 75,000 and sentenced to not less than fifteen years penal servitude. In addition, in the cases above described, any bribe offered shall, if seized, be forfeited, as provided by law, to the Government of Korea.

SECTION VIII. *Obstructing Public Officials in Enforcement of Laws against Rice Smuggling.* A person who wilfully obstructs or hinders, or attempts to obstruct or hinder, a public official in the enforcement of the laws against rice smuggling, or in the apprehension, investigation, prosecution, conviction, imprisonment of or collection of fines or penalties from rice smugglers, shall, upon conviction, be fined not less than ¥ 10,000 and be sentenced to not less than two years penal servitude. Where obstruction or interference, or attempted obstruction or interforence, is by violence or threats of violence, the offender, upon conviction, shall be fined not less than ¥ 50,000 and sentenced to not less than five years penal servitude.

SECTION IX. *Mitigation of Punishment Not Permitted.* There shall be no suspension or mitigation by any trial or appellate judge or court of any fines, punishment, forfeiture or penalties provided in this ordinance.

SECTION X. *Maximum Penalties.* The specific penalties provided in this ordinance are *minimum* penalties. Any person violating any of the provisions of this ordinance shall, upon conviction by a Military Occupation Court, be subject to such maximum punishment as the court shall determine.

SECTION XI. *Attempts and Conspiracies.* Any person who attempts to commit any offense enumerated herein, or who advises, assists in, arranges or procures the commission of any such offense, or who with knowledge of such an alleged offense, harbors an alleged offender or assists him to avoid

軍政廳　官報　法令　第一二七號　一九四六年十一月十五日

arrest or escape detection, conviction or punishment, shall be punishable as a principal.

SECTION XII. *Responsibility for Acts of Juridical Persons.* Every director, officer or employee of any juridical person, who in any such capacity, either alone or jointly with others, causes, directs, procures, advises, urges, approves, votes in favor of or participates in an act or omission which constitutes an offense hereunder for which such juridical person would be triable by a Military Occupation Court, shall be liable therefor as though such act or omission had been done or made in his individual capacity.

SECTION XIII. *Defenses.* It shall not be a defense to any charge hereunder that the offense was committed under duress, by reason of threats or through fear.

SECTION XIV. *Repeal of Inconsistent Laws.* All laws, ordinances, orders, regulations, directives and instructions, or parts thereof, which are inconsistent herewith or in conflict with the provisions hereof are, to the extent they are inconsistent or in conflict herewith, hereby repealed.

SECTION XV. *Definitions.* As used herein:

a. *Korea* means that portion of Korea which is south of 38 degrees north latitude.

b. *Smuggling* means shipment or transport out of Korea, except where duly authorized pursuant to law. *Smuggling* includes an attempt to smuggle.

c. *Smuggler* means a person shipping or transporting out of Korea, except where duly authorised pursuant to law. *Smuggler* includes a person attempting to smuggle.

d. *Smuggle* means to ship or transport out of Korea, except where duly authorized pursuant to law. *Smuggle* includes attempt to smuggle.

e. *Public official* means any government official, employee or agent authorized, directed or empowered to halt rice smuggling or arrest, investigate, prosecute or convict persons charged with rice smuggling. *Public official* includes, but is not limited to, police, coastguardsmen, customs officials, port authorities, revenue agents, members of the constabulary, public prosecutors and judges.

— 5 —

f. Gross negligence means wilful inattention to or heedless disregard of duty.

g. Bribe includes any promise or forbearance, or any money, property or value of any kind, or any promise or agreement therefor.

h. Rice includes polished rice, semi-polished rice, unhulled rice, hulled rice, semi-hulled rice, glutinous rice and upland rice.

i. Obstruct includes to prevent, interfere with or attempt to obstruct, prevent or interfere with.

j. Receive includes to receive through a third person or receipt by a third person through agreement or understanding, express or implied.

k. Offer includes to offer to or through third persons.

SECTION XVI. *Effective Date.* This ordinance shall be effective on the thirtieth day after the date appearing hereon.

ARCHER L LERCH
Major General United States Army
Military Governor in Korea

HEADQUARTERS
UNITED STATES ARMY MILITARY
GOVERNMENT IN KOREA
Office of the Military Governor
Seoul, Korea

ORDINANCE
NUMBER 128   31 December 1946

REVISION OF LAND TAXES AND SERVICE FEES

SECTION I. *Purpose.* The purpose of this ordinance is to increase land taxes and service fees, in order to provide the government with more adequate revenue and to combat inflation in Korea.

SECTION II. *Land Taxes.* Land taxes shall be levied at the present rates established by *Imperial Ordinance Number 6,* dated 31 March 1943 (*Korea Land Tax Law*), but at ten times the present assessed valuation determined in accordance with *Imperial Ordinance Number 37,* dated 4 December 1940 (*Concerning Investigation of Rental Values of Land in Korea*).

SECTION III. *Service Fees.* The following service fees are hereby established in connection with land surveys and land registers:

a. *Land Surveys*

(1) *Rural Lands*

Type of Survey	*Fee*
Division of registered lot	¥100 per lot
Registration of new lot	¥150 per lot
Readjustment of lot boundaries in accordance with *Imperial Ordinance Number 16,* dated December 1927 (*Korean Land Improvement Law*)	¥150 per lot

(2) *Urban Lands*

Type of Survey	*Fee*
Readjustment of lot boundaries in accordance with *Imperial Ordinance Number 18,* dated 20 June 1934 (*Korean Urban Land Planning Law*)	¥250 per lot

— 1 —

    (3) *Forest Lands*

Type of Survey	*Fee*
Division of registered lot	¥200 per lot of less than 5 *chongbo*, plus ¥100 for each additional 5 *chongbo* or fraction thereof
Registration of new lot	¥300 per lot of less than 5 *chongbo*, plus ¥150 for each additional 5 *chongbo* or fraction thereof

    b. Land Registers

Type of Service	*Fee*
Inspection of land register	¥5 per book per inspection
Extract copy of land register	¥10 per lot per extract
Inspection of land registry maps	¥5 per sheet per inspection
Extract copy of land registry maps	¥10 per lot per extract, plus an additional ¥5 in each case where more than two lots are copied from the same map

    Application for the foregoing services shall be made at the appropriate district tax office and fees shall be paid in revenue stamps at such office. No service fees shall be charged for government applications.

    SECTION IV. *Repeal of Inconsistent Laws.* All laws, ordinances, orders, regulations, directives and instructions, or parts thereof, which are inconsistent herewith or in conflict with the provisions hereof are, to the extent they are inconsistent or in conflict herewith, hereby repealed.

    SECTION V. *Effective Date.* This ordinance shall be effective on the tenth day after the date appearing hereon.

C G HELMICK
Brigadier General United States Army
Acting Military Governor

HEADQUARTERS

UNITED STATES ARMY MILITARY
GOVERNMENT IN KOREA
Office of the Military Governor
Seoul, Korea

ORDINANCE
NUMBER 129   11 December 1946

AMENDING ORDINANCE NUMBER 118

SECTION I. *Ordinance Number 118 Amended.*   The first sentence of Section VI of Ordinance Number 118, dated 24 August 1946, is hereby amended to read as follows:

> *All action of the Korean Interim Legislative Assembly shall be by a majority vote of the quorum, and a majority of the members shall constitute a quorum unless otherwise determined by the Assembly.*

SECTION II.   *Effective Date.*   This ordinance shall be effective on the date appearing hereon.

C G HELMICK
Brigadier General United States Army
Acting Military Governor

軍政廳　官報　法令　第一二九號　一九四六年十二月十一日

HEADQUARTERS
UNITED STATES ARMY MILITARY
GOVERNMENT IN KOREA
Office of the Military Governor
Seoul, Korea

ORDINANCE
NUMBER 130

11 December 1946

REVISION OF STAMP TAX AND REGISTRATION TAX LAWS

SECTION I. *Purpose.* The purpose of this ordinance is to increase the revenues of the Government of Korea and to adjust tax rates.

SECTION II. *Certain Stamp Tax Rates Revised.* Law Number 54 of 1899 (*Stamp Tax Law*), as amended, is hereby further amended as follows:

    a. *Article IV:*

    (1) The amounts stated as being subject to tax in Items Number 1 through 6 inclusive are increased ten (10) times and the tax rate thereon is increased fifty (50) times.

    (2) The tax on Item Number 7 is increased to ¥ 2.

    (3) The tax on Items Number 8 through 34 inclusive is increased to ¥ 5.

    b. *Article V:*

    (1) The exemption stated in Item Number 7 is increased to ¥ 500
    (2) The exemption stated in Item Number 9 is increased to ¥ 100.
    (3) The exemption stated in Item Number 10 is increased to ¥ 100.
    (4) The exemption stated in Item Number 12 is increased to ¥ 100.
    (5) The exemption stated in Item Number 14 is increased to ¥ 500.
    (6) The exemption stated in Item Number 25 is increased to ¥ 100.

    c. *Article XI:*

    The minimum fine for failure to pay the *Stamp Tax* is increased to ¥ 250.

— 1 —

SECTION III. *Certain Registration Tax Rates Revised.* Ordinance Number 16, dated March 1912 (*Korean Registration Tax Ordinance*), as amended, is hereby further amended as follows:

a. *Article I, Section 1:*

(1) The tax on Items Number 16 and 17 is increased to ¥ 5.

(2) The tax on Items Number 18 and 19 is increased to ¥ 3, with a maximum of ¥ 30 in each case.

b. *Article I, Section 2:*

(1) The tax on Item Number 12 is increased to ¥ 10.

(2) The tax on Items Number 13 through 16, inclusive, is increased to ¥ 5 as to each item.

c. *Article I, Section 4:*

(1) In Item Number 1, the tax rate is increased to ¥ 25.

(2) In Items Number 2 through 4, inclusive, the tax rate is increased to ¥ 5.

d. *Article I, Section 5:*

(1) All tax rates in Item Number 1 are increased five (5) times.

(2) The charge for each change of registration in Item Number 2 is increased to ¥ 5.

e. *Article I, Section 6:*

All tax rates listed in Items Number 6 through 9, inclusive, are increased five (5) times.

f. *Article I, Section 7:*

The tax rate listed in Item Number 3 is increased five (5) times.

g. *Article III, Section 1:*

(1) In Items Number 1, 3, 5-2 and 6, the minimum tax is increased to ¥ 1,000.

(2) The tax on Item Number 9 is increased to ¥ 500.

(3) The tax on Items Number 10 through 21, inclusive, is increased to ¥ 100, in each case.

(4) The tax on Items Number 22 through 26, inclusive, is increased to ¥ 20, in each case.

(5) The charge for sub-office registration as to matters contained in Items Number 1 through 26, inclusive, is increased to ¥ 20 in each case.

h. Article III, Section 2:

The tax on Items Number 1 and 2 is increased to ¥ 20, in each case.

i. Article III, Section 3:

(1) The tax on Item Number 2 is increased to ¥ 75.

(2) The charge for sub-office registration as to matters contained in Items Number 1 and 2 is increased to ¥ 20, in each case.

j. Article IV, Section 1.

(1) The tax on Items Number 1 through 3, inclusive, is increased to ¥ 100, in each case.

(2) The tax on Items Number 4 through 7, inclusive, is increased five (5) times, in each case.

(3) The charge for sub-office registration as to matters contained in Items Number 1 through 7, inclusive, is increased to ¥ 10, in each case.

k. Article IV, Section 3:

The tax on Items Number 1 through 3-5, inclusive, on Items Number 5 through 7, inclusive, and on Items Number 9 through 12, inclusive, is increased five (5) times, in each case.

l. Article IV, Section 4:

(1) The tax on Items Number 1 through 3, inclusive, is increased ten (10) times, in each case.

(2) The specific taxes in Items Number 5 and 5-2 are increased ten (10) times, with the percentage rate in Item Number 5 remaining unchanged.

(3) The tax on Items Number 10 through 13, inclusive, is increased ten (10) times.

SECTION IV. Fractional Parts of Yen. Under Law Number 54 of 1899 (*Stamp Tax Law*), as heretofore and herein amended, and in Ordinance Number 16, dated March 1912, (*Korean Registration Tax Ordinance*) as heretofore and

軍政廳　官報　法令　第一三〇號　一九四六年十二月十一日

herein amended, whenever the tax to be paid is or contains a fractional part of one yen, such fractional part shall be comput   and paid as one yen.

SECTION V. Continuation of Taxes. All existing taxes, tax rates and exemptions not herein amended or changed shall continue in full force and effect.

SECTION VI. Effective Date. This ordinance shall be effective on 1 February 1947.

C G HELMICK
Brigadier General United States Army
Acting Military Governor

軍政廳　官報　法令　第一三〇號　　一九四六年十二月十一日

335

HEADQUARTERS
UNITED STATES ARMY MILITARY
GOVERNMENT IN KOREA
Office of the Military Governor
Seoul, Korea

ORDINANCE
NUMBER 131

6 March 1947

AMENDING ORDINANCE NUMBER 118

SECTION I. *Section VI of Ordinance Number 118 Amended.* Section VI of Ordinance Number 118, dated 24 August 1946 (*Establishment of Korean Interim Legislative Assembly*) is hereby amended by adding thereto the following:

While the Assembly is in session, no member thereof shall be arrested without its consent, unless he has committed treason or is apprehended in the act of committing a crime.

SECTION II. *Section IX of Ordinance Number 118 Amended.* Section IX of Ordinance Number 118, dated 24 August 1946, is hereby amended by adding thereto the following:

Until such election ordinance is enacted, vacancies caused by the death, resignation, disqualification or expulsion of any member shall, upon notification of the presiding chairman (Speaker), be filled in the same manner by which such member was originally chosen.

Each member of the Assembly shall register during the three days prior to the opening of the session thereof, by presenting his credentials and personal history, and shall report at the opening session. A member who fails to register within seven days after the opening of the session shall not be seated unless the Assembly determines that there was adequate reason for such failure. The Assembly may refuse to seat or may disqualify or expel a member for failure to register within fifteen days after the opening of the session, or for grave and serious violations of its discipline, or for violations of law.

SECTION III. *Effective Date.* This ordinance shall be effective on the tenth day after the date appearing hereon.

ARCHER L LERCH
Major General United States Army
Military Governor in Korea

軍政廳　官報　法令　第一三一號　一九四七年三月六日

336

HEADQUARTERS
UNITED STATES ARMY MILITARY
GOVERNMENT IN KOREA
Office of the Military Governor
Seoul, Korea

ORDINANCE
NUMBER 132

24 February 1947

TEMPORARY SUSPENSION OF ORDINANCE NUMBER 112

Pending action on the subject matters hereof by the Korean Interim Legislative Assembly and the enactment of an ordinance pursuant thereto, the following shall govern:

SECTION I. *Ordinance Number 112 Suspended in Part.* Ordinance Number 112, dated 18 September 1946 (*Regulation of Child Labor*) is hereby suspended until 1 June 1947; *provided, however,* that the hiring or employment of children under the age of 14 full years under contracts or agreements for hire or employment entered into on and after the effective date hereof is nevertheless prohibited.

SECTION II. *Effective Date.* This ordinance shall be effective on the date appearing hereon.

ARCHER L LERCH
Major General United States Army
Military Governor in Korea

HEADQUARTERS
UNITED STATES ARMY MILITARY
GOVERNMENT IN KOREA
Office of the Military Governor
Seoul, Korea

ORDINANCE
NUMBER 133

8 March 1947

RAW SILK INSPECTION LAW

Pending action on the subject matters hereof by the Korean Interim Legislative Assembly and the enactment of an ordinance pursuant thereto, the following shall govern:

SECTION I. *Purpose.* The purpose of this ordinance is to provide for the inspection and grading of raw silk produced in licensed filatures, thereby insuring the production of an improved product.

SECTION II. *Laws Repealed.* Article 34 and Section I of Article 13 of Governor General Ordinance Number. 24, dated 25 March 1942 (*Sericulture Control Ordinance*), are hereby repealed.

SECTION III. *Inspection of Raw Silk Required.* All raw silk produced in licensed filatures shall be inspected and tested for quality, condition, weight, and grade by or under the supervision of the Department of Agriculture. No raw silk required to be inspected and tested shall, prior to such inspection and testing, be exported or used domestically.

SECTION IV. *Preparation and Publication of Regulations.* The Director of the Department of Agriculture shall prepare regulations for the inspection and testing of raw silk produced in licensed filatures. Such regulations, upon approval by the Military Governor and upon publication in the *Official Gazette,* shall have the force of law.

SECTION V. *Transportation of Uninspected Raw Silk Prohibited* No raw silk required to be inspected and tested shall be transported by any means whatsoever unless and until it is inspected and tested; *provided, however,* that prior to inspection and testing, raw silk may be transported to an inspection

軍政廳　官報　法令　第一三三號　　　　一九四七年三月八日

point or may be transported to a warehouse for storage, where such warehouse is owned by or leased under written lease to the licensed filature producing the raw silk.

SECTION VI. *Records.* Each licensed filature shall maintain such written records as may be required by the Director of the Department of Agriculture. Such records shall be kept at the office of the licensed filature and shall be open for inspection at all times during regular business hours by employees and representatives of the Department of Agriculture.

SECTION VII. *Penalties.* Any raw silk which is transported in violation of this ordinance or any materials produced therefrom and any private ship, boat, motor vehicle, cart, animal, or other conveyance or instrumentality used for illegal transportation thereof shall be forfeited, as provided by law, to the Government of Korea. Any person violating the provisions of this ordinance or any regulation issued hereunder shall, upon conviction by a Military Occupation Court, suffer such punishment as the court shall determine.

SECTION VIII. *Definitions.* As used herein:

.. *Raw silk* means all silk reeled from the cocoon by or in the filature, but not further processed. It does not include waste silk.

b. Licensed filatures means raw silk production factories which are duly licensed pursuant to law.

SECTION IX. *Effective Date.* This ordinance shall be effective on the tenth day after the date appearing hereon.

ARCHER L LERCH
Major General, United States Army
Military Governor in Korea

HEADQUARTERS
UNITED STATES ARMY MILITARY
GOVERNMENT IN KOREA
Office of the Military Governor
Seoul, Korea

ORDINANCE
NUMBER 134

14   March 1947

*AMENDING THE PATENT LAW, 1946**

Pending action on the subject matters hereof by the Korean Interim Legislative Assembly and the enactment of an ordinance pursuant thereto, the following shall govern:

SECTION I. *Purpose.* The purpose of this ordinance is to amend *The Patent Law, 1946,* so as to provide for charges for publications sold by the Bureau of Patents.

SECTION II. *Art. 14 Amended. Art. 14* of *The Patent Law, 1946,* is hereby amended to read as follows:

Art. 14. The Commissioner shall publish a Patent Gazette in which shall be printed particulars provided in the law, and an Annual Report in which shall be printed matters deemed by him to be necessary.

SECTION III. *Art. 16. Amended. Art. 16* of *The Patent Law, 1946,* is hereby amended so as to delete therefrom the following:

4. Annual reports of the Commissioner of Patents.

SECTION IV *Art. 251 Amended. Art. 251* of *The Patent Law, 1946,* is hereby deleted in its entirety, and the following is substituted therefor:

Art. 251. The Commissioner shall, subject to the approval of the Director of the Department of Commerce, prescribe fees for other services rendered to persons having business before the Bureau of Patents, and for the cost of publications sold by the Bureau of Patents, including the Patent Gazette and the Annual Report, on a scale commensurate with the cost of such services and publications to the Bureau of Patents.

See Ordinance No 91, dated 5 October 1946; and *The Patent Law, 1946.*

軍政廳　　官報　　法令　第一三四號　　一九四七年三月十四日

The Commissioner may, under the provisions of the foregoing paragraph, make charges for the preparation and copying of drawings, translations, for assistance to patent agents and other persons in the use of the Bureau of Patents library and for similar services.

SECTION V *Effective Date.* This ordinance shall be effective on the date appearing hereon.

ARCHER L LERCH
Major General United States Army
Military Governor in Korea

HEADQUARTERS
UNITED STATES ARMY MILITARY
GOVERNMENT IN KOREA
Office of the Military Governor
Seoul, Korea

ORDINANCE
NUMBER 135

15 March 1947

APPOINTMENT AND REMOVAL OF GOVERNMENT OFFICIALS

SECTION I. Pending enactment of the Korean Interim Legislative Assembly of a general franchise law which will prescribe the procedure to be followed in the election of officials under the provisions of Ordinance Number 126, dated 15 November 1946, and in the election of officials generally, the following instructions shall govern appointments:

a. Appointment of administrative officials of the rank of Civil Administrator and above shall be made by the Military Governor subject to confirmation by the Korean Interim Legislative Assembly as provided in Section V, Ordinance Number 118, dated 24 August 1946.

b. Appointment of Directors of Departments and Offices, Provincial Governors, and the Mayor of the Independent City of Seoul, shall be made by the Military Governor upon recommendation of the Civil Administrator, who will submit the recommendation of the majority of the Korean Directors. The Civil Administrator will indicate his approval or disapproval of the recommendation of the majority of the Directors with such reasons as he may see fit to state. Such appointments by the Military Governor shall be subject to confirmation by the Korean Interim Legislative Assembly as provided in Section V, Ordinance Number 118, dated 24 August 1946.

c. All appointments within the several Provinces shall be made by the respective Provincial Governors, subject to the approval of the Office of Korea Civil Service in those cases in which existing law requires such approval. This provision shall apply to the Mayor of the Independent City of Seoul with reference to appointments within that municipality.

軍政廳　官報　法令　第一三五號　　　一九四七年三月十五日

342

d. Appointment of officials within the several departments shall be made by the Director of the Department concerned, subject to approval by the Office of Korea/Civil Service in those cases in which existing law requires such approval.

SECTION II. Nothing in this Ordinance shall be held to interfere in any may with the duties and powers of the Korean Interim Legislative Assembly to review, confirm and consent to certain appointments set forth in Section V of Ordinance Number 118, dated 24 August 1946; or to change the procedure in filling or vacating offices in those cases where such procedure is otherwise prescribed by law.

SECTION III. The provisions of this Ordinance shall continue in effect only until the Korean Interim Legislative Assembly shall have provided General Election laws or shall have otherwise enacted legislation governing election to the offices concerned.

SECTION IV. Removals from office in each case will be made by the same authority and under procedures corresponding to that herein prescribed for appointments to office.

SECTION V. Provincial Governors may, on thirty days notice to the Military Governor of Korea, substitute popular election for appointments as provided herein.

SECTION VI. The Military Governor reserves the right to revoke any authority granted under the provisions of this Ordinance.

Effective Date: 15 March 1947.

APPROVED:

ARCHER L LERCH
Major General United States Army
Military Governor in Korea

RECOMMENDED:

AHN CHAI HONG
Civil Administrator

HEADQUARTERS
UNITED STATES ARMY MILITARY
GOVERNMENT IN KOREA
Office of the Military Governor
Seoul, Korea

ORDINANCE

NUMBER 138

20-March 1947

AMENDING ORDINANCE NUMBER 88

Pending action on the subject matters hereof by the Korean Interim Legislative Assembly and the enactment of an ordinance pursuant thereto, the following shall govern:

SECTION I. *Amendments.* Ordinance Number 88, dated 29 May 1946 (*Licensing of Newspapers and Other Periodicals*) is hereby amended as follows:

a. *Paragraph b of Section II* thereof to read:

The application shall be filed with the Department of Public Information of the Government of Korea, Seoul. The licensing authority shall be the Director of the Department of Public Information.

b. In *Section V* thereof the phrase *Director of the Department of Commerce* is deleted, and the phrase *Director of the Department of Public Information* is substituted in place thereof.

SECTION II. *Effective Date.* This ordinance shall be effective on the date appearing hereon.

APPROVED:

ARCHER L LERCH
Major General United States Army
Military Governor in Korea

RECOMMENDED:

AHN CHAI HONG
Civil Administrator

HEADQUARTERS
UNITED STATES ARMY MILITARY
GOVERNMENT IN KOREA
Office of the Military Governor
Seoul, Korea

ORDINANCE
NUMBER 137   12 June 1947

REVISION OF INHERITANCE TAX RATES

Pending action on the subject matters hereof by the Korean Interim Legislative Assembly and the enactment of an ordinance pursuant thereto, the following shall govern:

SECTION I. *Purpose.*   The purpose of this ordinance is to revise inheritance tax rates.

SECTION II. *Inheritance Tax Law Amended.*   Ordinance Number 19, dated June 1934 (*Korean Inheritance Tax Ordinance*), is hereby amended as follows:

a. Article 5-2 is changed to read:

"*a. .Where the inheritance is by a person who becomes the new head of the deceased's household, and the taxable value of the inherited property is ¥ 100,000 or less, a deduction of ¥ 3,000 per person shall be allowed for dependents living with the new head of the household who are under 18 or over 60 years of age or who are disabled.*

"*b. Where the inheritance is by a person who does not become the new head of the deceased's household and the taxable value of the inherited property is ¥ 60,000 or less, a deduction of ¥ 3,000 per person shall be allowed for each child of the person receiving the inheritance who lives with him and is under his parental control and who is under 18 years of age, or who is disabled.*"

b. Article 6 is changed to read:

"*a. Where the inheritance is by a person who becomes the new head of the deceased's household and the taxable value of the inherited property, after deductions for dependents, is below ¥ 20,000, no tax shall be imposed.*

"*b. Where the inheritance is by a person who does not become the new head of the deceased's household and the taxable value of the inherited property, after deductions for dependents, is below ¥ 5,000, no tax shall be imposed.*"

— 1 —

c. Article 8 is changed to read:

"Inheritance taxes shall be imposed at the following rates and in accordance with the following classifications:

Inheritance by Person Becoming New Head of Household

1	2	3
Taxable Amount	Inheritance by Person Who is Lineal Descendant and a Member of Deceased's Household.	Inheritance by Person other than set forth in Column 2.
First ¥ 20,000	18/1000	24/1000
¥ 20,001 to ¥ 30,000	24/1000	36/1000
¥ 30,001 to ¥ 40,000	30/1000	50/1000
¥ 40,001 to ¥ 50,000	45/1000	70/1000
¥ 50,001 to ¥ 70,000	65/1000	95/1000
¥ 70,001 to ¥ 100,000	90/1000	120/1000
¥ 100,001 to ¥ 150,000	120/1000	145/1000
¥ 150,001 to ¥ 200,000	150/1000	175/1000
¥ 200,001 to ¥ 300,000	180/1000	200/1000
¥ 300,001 to ¥ 400,000	200/1000	245/1000
¥ 400,001 to ¥ 500,000	240/1000	280/1000
¥ 500,001 to ¥ 700,000	270/1000	320/1000
¥ 700,001 to ¥ 1,000,000	310/1000	370/1000
¥ 1,000,001 to ¥ 2,000,000	350/1000	420/1000
¥ 2,000,001 to ¥ 3,000,000	400/1000	470/1000
¥ 3,000,001 to ¥ 5,000,000	450/1000	530/1000
Above ¥ 5,000,000	500/1000	600/1000

— 2 —

Inheritance by Person Not Becoming New Head of Household

1	2	3	4
Taxable Amount	Inheritance by Person who is Lineal Descendant.	Inheritance by Spouse or Lineal Ascendant	Inheritance by Person other than set forth in Columns 2 or 3.
First ¥ 5,000	24/1000	36/1000	48/1000
¥ 5,001 to ¥ 10,000	36/1000	48/1000	72/1000
¥ 10,001 to ¥ 20,000	48/1000	65/1000	100/1000
¥ 20,001 to ¥ 30,000	60/1000	90/1000	130/1000
¥ 30,001 to ¥ 40,000	80/1000	120/1000	160/1000
¥ 40,001 to ¥ 50,000	105/1000	150/1000	190/1000
¥ 50,001 to ¥ 70,000	130/1000	180/1000	220/1000
¥ 70,001 to ¥ 100,000	160/1000	200/1000	250/1000
¥ 100,001 to ¥ 150,000	200/1000	245/1000	280/1000
¥ 150,001 to ¥ 200,000	240/1000	285/1000	320/1000
¥ 200,001 to ¥ 300,000	280/1000	330/1000	360/1000
¥ 300,001 to ¥ 400,000	320/1000	375/1000	405/1000
¥ 400,001 to ¥ 500,000	360/1000	420/1000	450/1000
¥ 500,001 to ¥ 700,000	400/1000	465/1000	500/1000
¥ 700,001 to ¥ 1,000,000	460/1000	500/1000	550/1000
¥ 1,000,001 to ¥ 2,000,000	500/1000	560/1000	600/1000
¥ 2,000,001 to ¥ 3,000,000	560/1000	600/1000	670/1000
¥ 3,000,001 to ¥ 5,000,000	620/1000	670/1000	730/1000
Above ¥ 5,000,000	680/1000	730/1000	800/1000

— 3 —

SECTION III. *Effective Date.* This ordinance shall be effective on 1 May 1947.

APPROVED:

ARCHER L LERCH
Major General United States Army
Military Governor in Korea

RECOMMENDED:

AHN CHAI HONG
Civil Administrator

HEADQUARTERS
UNITED STATES ARMY MILITARY
GOVERNMENT IN KOREA
Office of the Military Governor
Seoul, Korea

ORDINANCE
NUMBER 138

1 April 1947

AMENDING HORSE-RACING ORDINANCE

Pending action on the subject matters hereof by the Korean Interim Legislative Assembly and the enactment of an ordinance pursuant thereto, the following shall govern:

SECTION I. *Purpose.* The purpose of this ordinance is to amend the existing law regulating horse-racing and to provide a means of revenue which may be used to improve the breed of domestic horses.

SECTION II. *Amendments.* Governor General Ordinance Number 3, dated October 1922, *Korean Horse-Racing Ordinance,* is hereby amended as follows:

a. In *Section 4* thereof, the words *not less than two won and not more than twenty won* are deleted, and the words *not less than twenty won and not more than 100 won* are substituted therefor.

b. In paragraph 1 of *Section 6* thereof, the words *provided that the amounts shall not exceed ten times the face value of the race ticket* are deleted.

SECTION III. *Effective Date.* This ordinance shall be effective on the date appearing hereon.

<table>
<tr><td>APPROVED:</td><td>RECOMMENDED:</td></tr>
<tr><td>ARCHER L LERCH
Major General United States Army
Military Governor in Korea</td><td>AHN CHAI HONG
Civil Administrator</td></tr>
</table>

軍政廳　官報　法令　第一三八號　　一九四七年四月一日

HEADQUARTERS
UNITED STATES ARMY MILITARY
GOVERNMENT IN KOREA
Office of the Military Governor
Seoul, Korea

ORDINANCE
NUMBER 139

19 April 1947

TAX DELINQUENCY PENALTIES REVISED

Pending action on the subject matters hereof by the Korean Interim Legislative Assembly and the enactment of an ordinance pursuant thereto, the following shall govern:

SECTION I. *Purpose.* The purpose of this ordinance is to modify tax collection procedures and to increase penalties for failure to pay national or provincial taxes within the time prescribed by law.

SECTION II. *Penalty for Non-Payment of Taxes.* A penalty of five percent of the amount of tax due is hereby imposed and added to such tax on the date upon which the tax becomes delinquent in payment. A further penalty of five percent of the amount of tax due is hereby imposed and added to each tax not paid within thirty (30) days after the tax becomes delinquent in payment. For each further period of thirty (30) days any tax is delinquent in payment, there is hereby imposed and added to the tax due a further penalty of ten percent of the amount thereof. In no case shall the total penalty imposed hereunder exceed thirty percent of the amount of the unpaid tax. Penalties hereunder shall be imposed on the amount of tax delinquent in payment, exclusive of penalties previously imposed.

SECTION III. *Tax Delinquency Notices.* Article 9 of National Tax Collection Law Number 21, dated 26 March 1897, is hereby repealed.

SECTION IV. *Definitions.* As used herein:

a. *Delinquent in payment* means unpaid or not paid in full within the time allowed by the law pursuant to which the tax is assessed.

軍政廳　　官報·　法令　　第一三九號　　一九四七年四月十九日

b. *Tax* means a national or provincial x, whether periodical or occasional.

SECTION V. *Repeal of Inconsistent Laws.* All laws and regulatory measures and parts thereof, which are inconsistent herewith or in conflict with the provisions hereof, are to the extent that they are inconsistent or in conflict herewith, hereby repealed.

SECTION VI. *Application.* This ordinance shall apply to taxes delinquent in payment on the effective date hereof and to taxes which may thereafter become delinquent in payment.

SECTION VII. *Effective Date.* This ordinance shall be effective on the thirtieth day after the date appearing hereon.

APPROVED:

ARCHER L LERCH
Major General United States Army
Military Governor in Korea

RECOMMENDED:

AHN CHAI HONG
Civil Administrator

HEADQUARTERS
UNITED STATES ARMY MILITARY
GOVERNMENT IN KOREA
Office of the Military Governor
Seoul, Korea

ORDINANCE
NUMBER 140

9 June 1947

LIMITING SLAUGHTER OF CATTLE

Pending action on the subject matters hereof by the Korean Interim Legislative Assembly and the enactment of an ordinance pursuant thereto, the following shall govern:

SECTION I. *Purpose.* The purpose of this ordinance is to conserve and make possible the restoration of a normal stock of cattle in Korea. A considerable reduction in the number of cattle has occurred. This depletion has correspondingly affected the draft power available for the tilling of lands in the production of rice and other grains. The unrestricted slaughter of cattle threatens the continued existence of the species. Therefore, effective measures to limit the slaughter of the existing stock are deemed necessary.

SECTION II. *Repeal of Existing Law.* Government General Instruction No. 9, dated 11 March 1912 '*Encouraging Improvement and Conservation of Cattle*' is hereby repealed.

SECTION III. *Cattle Slaughter Prohibited.* The slaughter of cattle under the age of ten (full) years is hereby prohibited, *provided however* that this prohibition shall not extend to dairy bulls. It is provided further that pregnant cows shall not be slaughtered, regardless of age.

SECTION IV. *Sanitary Controls Not Altered.* Existing provisions of law relating to sanitary and inspection controls in the slaughter of cattle and to the regulation of slaughterhouses are not intended to be altered or amended by the provisions of this ordinance. Inspections such as are now provided by law shall include inspection and inquiry concerning age.

SECTION V. *Excepted Areas.* The Director of the Department of

— 1 —

352

Agriculture may except any province, island, or other area, from the application of this ordinance, subject to the approval of the Military Governor.

SECTION VI. *Presumption As To Age.*   Cattle slaughtered at any place not licensed as a slaughterhouse, or slaughtered without prior inspection as required by law shall be presumed to be below the age of ten (full) years. Upon a prosecution for violation of this ordinance, evidence contrary to such presumption may be presented.

SECTION VII. *Penalties.*

a.   Any person who knowingly acts or conspires to violate the provisions of this ordinance shall, upon conviction by a Military Occupation Court, suffer such punishment as the Court shall determine.

b.   A government official charged with the duty of the inspection of cattle intended for slaughter who, through wilful or gross negligence, fails to enforce the provisions of this ordinance shall, upon conviction by a Military Occupation Court, be dismissed from public service, and shall suffer such punishment as the Court shall determine.

c.   The owner or manager of a slaughterhouse who knowingly permits, or an employee of a slaughterhouse who knowingly participates in, an act or conspiracy in violation of the provisions of this ordinance shall, upon conviction by a Military Occupation Court, suffer such punishment as the Court shall determine.

d.   In case the owner, manager, or employee of a slaughterhouse is duly convicted of an act or conspiracy in violation of the provisions of this ordinance, the license authorizing the operation of such slaughterhouse may be revoked in the discretion of the licensing authority.

e.   Any cattle, live or slaughtered, which is found to be the subject of an act or conspiracy in violation of the provisions of this ordinance shall be seized. In the event of a conviction adjudged by a Military Occupation Court for such act or conspiracy, the cattle seized shall be forfeited to the Government of Korea, in the manner provided by law. A permanent record of such forfeiture and of the disposition of the cattle and proceeds of the sale thereof shall be made in detail and retained as part of the court records.

— 2 —

SECTION VIII. *Cattle Defined.* "Cattle" as used herein means domesticated bovine animals.

SECTION IX. *Effective Date.* This ordinance shall be effective on the date appearing hereon.

APPROVED:

ARCHER L LERCH

Major General United States Army

Military Governor in Korea

RECOMMENDED:

AHN CHAI HONG

Civil Administrator

SOUTH KOREAN INTERIM GOVERNMENT
·Seoul, Korea·

ORDINANCE

NUMBER 141   ·17 May 1947

DESIGNATION OF SOUTH KOREAN INTERIM GOVERNMENT

Pending action on the subject matters hereof by the Korean Interim Legislative Assembly and the enactment of legislation pursuant thereto, the following· shall govern:

SECTION I. *Designation of Elements of Government.* The Korean elements of United States Army Military Government in Korea concerned with governing Korea south of 38° north latitude ·including the Legislative, Executive, and Judiciary branches, are hereby designated as *South Korean Interim Government.*

SECTION II. *Effective Date.* This ordinance shall be effective on the date appearing hereon.

APPROVED: RECOMMENDED:

   ARCHER L LERCH   AHN CHAI HONG

Major General United States Army   Civil Administrator

. Military Governor in Korea

SOUTH KOREAN INTERIM GOVERNMENT
Seoul, Korea

ORDINANCE
NUMBER 142   21 June 1947

AMENDING THE PERSONAL INCOME TAX ORDINANCE

Pending action on the subject matters hereof by the Korean Interim Legislative Assembly and the enactment of legislation pursuant thereto, the following shall govern:

SECTION I. *Purpose.* The purpose of this ordinance is to increase revenue and to adjust personal income tax rates and exemptions.

SECTION II. *Income Tax Ordinance Amended.* Governor General Ordinance Number 6, dated 30 April 1934 (*Korean Income Tax Ordinance*) (as amended), is hereby amended as follows:

a. Article 16 is hereby amended to increase the figure *ten thousand won* (¥ *10,000*) to *one hundred thousand won* (¥ *100,000*), to increase the figures *five thousand won* (¥ *5,000*) to *fifty thousand won* (¥ *50,000*), to increase the figures *twenty percent* (*20%*) to *forty percent* (*40%*), and to increase the figures *ten percent* (*10%*) to *twenty percent* (*20%*), wherever such figures appear in this Article.

b. Article 17 is hereby amended to increase the supporting family allowance from *fifty won* (¥ *50*) to *one thousand won* (¥ *1,000*) for each person therein named, and by adding the following at the end of said Article:

"The maximum total supporting family allowance authorized under this Article shall not exceed five thousand won (¥ 5,000)."

c. Articles 18 and 18-2 are hereby repealed.

d. The cumulative income tax schedule set forth in Article 36 is hereby revised to read as follows:

Amount of Income (in Won)	*Tax Rate*
The first 12,000	3%
The next 8,000 (12,001 to 20,000)	5%
The next 10,000 (20,001 to 30,000)	8%

Amount of Income (in Won)	Tax Rate
The next 20,000 (30,001 to 50,000)	12%
The next 30,000 (50,001 to 80,000)	16%
The next 40,000 (80,001 to 120,000)	20%
The next 80,000 (120,001 to 200,000)	25%
The next 100,000 (200,001 to 300,000)	30%
The next 200,000 (300,001 to 500,000)	35%
The next 200,000 (500,001 to 700,000)	40%
The next 300,000 (700,001 to 1,000,000)	·50%
The next 500,000 (1,000,001 to 1,500,000)	60%
The next 500,000 (1,500,001 to 2,000,000)	70%
The next 500,000 (2,000,001 to 2,500,000)	80%
The amount over 2,500,000	90%

Any person with a net annual income under this Article, of less than twelve thousand won (¥ 12,000) shall not be subject to the above tax.

  e. The minimum personal income tax ("third income tax") authorized to be paid in two installments, as provided in Article 57, is hereby changed from *fifteen won* (¥ 15) to *five hundred won* (¥ 500).

  SECTION III. *Effective Date.* This ordinance shall be effective on the date appearing hereon and shall be applicable to income for the years commencing with 1946.

APPROVED:

    ARCHER L LERCH

Major General United States Army

Military Governor in Korea

RECOMMENDED:

    AHN CHAI HONG

Civil Administrator

軍政廳　官報　法令　第一四二號　一九四七年六月二十一日

SOUTH KOREAN INTERIM GOVERNMENT
Seoul, Korea

ORDINANCE
NUMBER 143

21 June 1947

CROP REPORTING

Pending action on the subject matters hereof by the Korean Interim Legislative Assembly and the enactment of legislation pursuant thereto, the following shall govern:

SECTION I. *Purpose.* The purpose of this ordinance is to establish an organization and prescribe policy for crop reporting for the government, so that there shall be one source of authoritative estimates relating to agricultural, fishery and forest products for the information of public officials and the public generally.

SECTION II. *Crop Reporting Board.* *a. Establishment and Organization.* The Crop Reporting Board is hereby established in the Department of Agriculture. It shall consist of nine members, including a Chairman and Executive Secretary. The Chief of the Bureau of Agricultural Economics shall serve as Chairman of the Board, and the Chief of the Agricultural Statistics Section shall serve as Executive Secretary. The seven remaining members of the Board shall be designated by the Director of the Department of Agriculture from the personnel of the following units: one statistician from the Bureau of Fisheries, one statistician from the Bureau of Forestry and five statisticians from the Bureau of Agricultural Economics. Paid officials of the government serving on the Board shall receive no additional compensation for such service.

b. Provincial Agricultural Statistician. The positions of Provincial Agricultural Statisticians (one for each province) and Agricultural Statistician for Seoul City are hereby established in the national Department of Agriculture. The Provincial Agricultural Statistician shall assist the Board in carrying out its duties and functions. He shall be responsible directly to the Chairman of the Board. Each provincial Bureau of Agriculture shall transfer to the

— 1 —

appropriate Provincial Agricultural Statistician such statistical records and personnel as he requires to establish a provincial agricultural statistics sub-section.

    c. County and City Agricultural Statistician. (1) The positions of County Agricultural Statistician and City Agricultural Statistician (one for each county and city respectively) are hereby established in the national Department of Agriculture. The County and City Agricultural Statisticians shall assist the Board in carrying out its duties and functions. They shall be responsible directly to their respective Provincial Agricultural Statisticians.

    (2) Each County Agricultural Statistician shall have one assistant statistician except that the County Agricultural Statistician for Sosan County in South Chungchong Province shall have two assistant statisticians.

    d. Village and Township Agricultural Statisticians. The positions of Village Agricultural Statistician and Township Agricultural Statistician (one for each village and township respectively) are hereby established in the national Department of Agriculture. The Village and Township Agricultural Statisticians shall assist the Board in carrying out its duties and functions. They shall be responsible directly to their respective County Agricultural Statisticians.

    SECTION III. *Duties and Functions.* The Board shall:

    a. Develop a system of statistical estimates for agricultural, fishery and forest products.

    b. In collaboration with the Bureau of Agricultural Economics, the Bureau of Fisheries and the Bureau of Forestry, collect and collate such data as are required for the preparation of statistical estimates relating to agricultural, fishery and forest products, including but not limited to estimates of past and future production and inventories.

    c. Publish monthly and other reports of its findings and make reports and estimates to other departments, offices, agencies and instrumentalities of the government upon written request of the heads of such departments, offices, agencies or instrumentalities. Public reports shall be signed by the Chairman of the Board and the Director of the Department of Agriculture.

    SECTION IV. *Duty to Furnish Information.* Every farmer, warehouse-man, processor, distributor or other person dealing in agricultural, fishery or forest products shall furnish to the Board or its agents such information relating to such products and in such form as the Board or its agents may require.

— 2 —

SECTION V. *Other Agencies.* *a.* The National Food Administration, Controller of Commodities, Provincial Governors, Provincial Food Services, Korean Agricultural Association, Federation of Financial Associations, and all other departments, offices, agencies, instrumentalities and political subdivisions of the government shall furnish to the Board or its agents such data, office space, facilities and other assistance as they may from time to time require.

b. No department, office, agency or instrumentality of the government shall publish statistical estimates relating to agricultural, fishery, or forest products except with the prior approval of the Board.

SECTION VI. *Limitation of Use of Information.* The information furnished under Section IV of this ordinance shall be used only for the statistical purposes for which it is supplied. No publication or report shall be made by the Board whereby the data furnished by any particular person can be identified; nor shall the Board permit anyone other than its employees or members to examine the individual reports.

SECTION VII. *Rules and Regulations.* The Director of the Department of Agriculture may issue such rules and regulations as he deems necessary or appropriate to the administration of this ordinance. Such rules and regulations shall, when published in the *Official Gazette* upon approval of the Military Governor, have the force and effect of law.

SECTION VIII. *Penalties.* Any person (natural or juridical), required by this ordinance or by rules and regulations issued hereunder to furnish reports or information, who refuses to provide such reports or information, or wilfully provides false or misleading reports or information, shall, upon conviction by a Military Occupation Court, suffer such punishment as the court shall determine.

SECTION IX. *Effective Date.* This ordinance shall be effective on the tenth day after the date appearing hereon.

APPROVED:

    ARCHER L LERCH
Major General United States Army
  Military Governor in Korea

RECOMMENDED:

  AHN CHAI HONG
  Civil Administrator

軍政廳 官報　法令　第一四三號　一九四七年六月二十一日

SOUTH KOREAN INTERIM GOVERNMENT
Seoul, Korea

ORDINANCE

NUMBER 144   21 June 1947

ADMINISTRATIVE SUPERVISION OF FINANCIAL
ASSOCIATIONS TRANSFERRED

Pending action on the subject matters hereof by the Korean Interim Legislative Assembly and the enactment of legislation pursuant thereto, the following shall govern:

SECTION I. *Purpose.* The purpose of this ordinance is to transfer supervision of Financial Associations from the provincial governors to the Federation of Financial Associations.

SECTION II. *Financial Association Ordinance and Regulations Amended.* In the following ordinances, order and instructions, and in all amendments thereto, the words *Provincial Governor* are deleted wherever they appear and the words *Federation of Financial Associations* substituted therefor.

a. Governor General Ordinance Number 22, dated 22 May 1914 (*Financial Association Ordinance*).

b. Government General Order Number 38, dated 30 April 1929 (*Supervision Regulations on the Operation of Financial Associations*).

c. Instruction Number 50, dated 5 December 1925 (*Regulation of Occupational Duties and Discipline of Directors and Vice Directors of Financial Associations*).

d. Instruction Number 2, dated 14 February 1927 (*Regulation on Personal Bonds of Directors and Vice Directors of Financial Associations*).

e. Governor General Ordinance Number 12, dated 30 August 1935 (*The Industrial Guild (Shiksan Kae) Ordinance*).

f. Government General Order Number 145, dated 13 December 1935 (*Regulation on the Enforcement of Industrial Guild Ordinance*).

SECTION III. *Financial Association Ordinance Further Amended.* In Sections 65, 66 and 67 of Governor General Ordinance Number 22, dated 22 May 1914 (*Financial Association Ordinance*), the words *the supervising government offices* are hereby deleted wherever they appear and the words *the supervising authority* substituted therefor.

軍政廳　官報　法令　第一四四號　・一九四七年六月二十一日

SECTION IV. *Governor General Ordinance Number 7 Amended.* Governor General Ordinance Number 7, dated 17 August 1933 (*Korean Federation of Financial Associations Ordinance*) is hereby amended as follows:

a. Section I thereof is amended to read:

The Korean Federation of Financial Associations (hereinafter referred to as the Federation) is hereby established as a juridical person with the purpose of furnishing funds to its members, furnishing guidance to them on business matters, acting for the advancement of their mutual interests and supervising their activities.

The main office shall be in the City of Seoul.

b. Section XV, Paragraph 1 thereof, is amended by:

(1.) Substituting the sentence, *The Federation shall manage or have the following business, functions and duties,* for the sentence, *The Federation shall manage the following business,* and

(2) Adding thereto as item 8, *To supervise its members and take such action with respect to them as may be directed or authorized by law.*

SECTION V. *Effective Date.* This ordinance shall be effective on the tenth day after the date appearing hereon.

APPROVED:

ARCHER L LERCH

Major General United States Army

Military Governor in Korea

RECOMMENDED:

AHN CHAI HONG

Civil Administrator

SOUTH KOREAN INTERIM GOVERNMENT
Seoul, Korea

ORDINANCE
NUMBER 145   16 June 1947

CREATION OF THE KOREAN FOREIGN EXCHANGE BANK, LTD

Pending action on the subject matters hereof by the Korean Interim Legislative Assembly and the enactment of legislation pursuant thereto, the following shall govern:

SECTION I. *Purpose.* The purpose of this ordinance is to establish banking and foreign exchange facilities for the promotion of foreign trade with Korea.

SECTION II. *Creation.* The Korean Foreign Exchange Bank, Ltd. (hereinafter referred to as "the Bank") is hereby created as a corporate body.

SECTION III. *Management.* The officers of the Bank shall be a President, a Vice President, and a Managing Director. The Board of Directors shall consist of the officers and four other directors. Officers and directors shall be appointed and may be removed by the Military Governor. The Bank shall be operated under rules and by-laws to be prescribed by its Board of Directors. The principal office of the Bank shall be in Seoul, with such branch banks and offices as the Board of Directors may determine.

SECTION IV. *Advisory Board.* The Department of Finance, with the approval of the Military Governor, shall appoint and may remove members of an Advisory Board consisting of the President of the Bank, who will act as Chairman, and five or more members, at least two of whom shall be representatives of other Korean Banks and at least one of whom shall represent Korean commerce. The Advisory Board shall meet at the call of the Chairman. The Advisory Board will make recommendations to the Board of Directors with respect to operation of the Bank and the Board of Directors shall consult the Advisory Board on major questions of policy.

SECTION V. *Capital Stock.* The Bank shall have a capitalization of five hundred million won (￥500,000,000), divided into shares of ten thousand won

363

(¥ 10,000) each. The United States Army Military Government in Korea shall initially subscribe to shares totalling two hundred million won (¥ 200,000,000) in paid-up capital stock. Military Government may subscribe to additional amounts of the Bank's unissued capital stock upon recommendation of the Board of Directors, with the approval of the Military Governor. Liability of Military Government shall be limited to its capital stock subscriptions. The Board of Directors may, subject to the approval of the Military Governor, authorize the sale of unissued shares or of any or all of the shares owned by Military Government to other Korean financial institutions and/or the general public.

SECTION VI. *Powers and Duties.* The Bank shall have the following powers, functions and duties:

a. To transact a business in foreign exchange.

b. To maintain accounts, in United States or other foreign currency, in the names of authorized individuals, organizations, or juridical persons, subject to foreign exchange regulations of the Military Government in Korea.

c. To maintain accounts in won with Korean financial institutions.

d. To appoint banks in foreign countries to act as the Bank's corre spondents and to maintain accounts with them in any currencies.

e. To transmit to, and receive from, the Bank's appointed corre spondents and other approved agents, bills of exchange and other documents in connection with exports from and imports into Korea, subject to such regulations as may from time to time be issued by the Military Government.

f. To transact such other business in connection with, or to facilitate, foreign trade and/or foreign exchange, as the Board of Directors may deem necessary or advisable, subject to the approval of the Military Government.

g. To rent, purchase or otherwise provide suitable premises necessary to carry on its business.

h. To appoint and compensate such officers and employees as the Board of Directors may determine, but no person already receiving compensation from the Military Government in Korea shall receive additional compensation from the Bank.

SECTION VII. *Effective Date.* This ordinance shall be effective as of the date of approval hereof.

APPROVED:16 June 1947   RECOMMENDED:

ARCHER L LERCH   AHN CHAI HONG
Major General United States Army   Civil Administrator
Military Governor in Korea

SOUTH KOREAN INTERIM GOVERNMENT
Seoul, Korea

.ORDINANCE

NUMBER 146

30 June 1947

PROVINCIAL AND LOCAL TAX EXEMPTIONS REVISED

Pending action on the subject matters hereof by the Korean Interim Legislative Assembly and the enactment of legislation pursuant thereto. the following shall govern :

SECTION I. *Purpose.* The purpose of this ordinance is to adjust personal income exemptions under Ordinance No 109, dated 15 October 1946 (*Revision of Provincial and Local Taxes*).

SECTION II. *Exemptions.* Section II-c of Ordinance No 109, dated 15 October 1946, is hereby amended to read as follows:

" *c. Exemptions.* The following earned-income exemptions shall be the only permissible deductions from taxable income under the provincial and local household income tax :

(*1*) One thousand won (¥ 1,000) for the head of the household.

(*2*) One thousand won (¥ 1,000) for each member of the household under eighteen (18) or over sixty (60) years of age, but in no case to exceed a total of five thousand won (¥ 5,000).

(*3*) An allowance of forty percent (40%) of that portion of the total salary and wages received by the taxpayer which is fifty thousand won (¥ 50,000) or less, and an allowance of twenty percent (20%) of that portion of the total salary and wages received by the taxpayer which is in excess of fifty thousand won (¥ 50,000) but not in excess of one hundred thousand won (¥ 100,000)."

— 1 —

SECTION III. *Effective Date.* This ordinance shall be effective on the date appearing hereon and shall be applicable to income for the years commencing with 1946.

APPROVED: RECOMMENDED:

    C G HELMICK   AHN CHAI HONG

Brigadier General United States Army   Civil Administrator

    Acting Military Governor

SOUTH KOREAN INTERIM GOVERNMENT
Seoul, Korea

ORDINANCE
NUMBER 147

30 July 1947

FISHERY PRODUCTS INSPECTION REGULATION REVISED

Pending action on the subject matters hereof by the Korean Interim Legisla-tive Assembly and the enactment of legislation pursuant thereto, the following shall govern :

SECTION I. *Purpose.* The purpose of this ordinance is to revise the fees set forth in Section 5 of Government General Order Number 26, dated 25 February 1943 *(Fishery Products Inspection Regulation)*.

SECTION II. *New Inspection Fees Established.* The fishery products inspection fees. set forth in Section 5 of Government General Order Number 26. dated 25 February 1943, are hereby revised to read as follows :

Class 1-Fishery Products. Dried	*Quantity*	*Fee*
Shark fins, abalone, sea slugs (sea cucumbers), oysters and shellfish ligaments	60 90 kg.	¥ 5.00
Sand eels, codfish, stiegel musseis (shellfish family), razor clam, mactra sachalinensis (clam) and processed sardine meal	60-90 kg.	2.50
Cuttle fish, mussels (sea mussels), heartshell clams, and lobsters (without shell)	60-90 kg.	3.00
Anchovies	3.75 kg.	0.50
Laver	100 sheets	0.10
Lobsters (with shell)	60-90 kg.	2.00
Tang (myuk)	60-90 kg.	1.50
Other dried fishery products	60-90 kg.	1.50
Class 2-Fishery Products. Canned		
Abalone, crab, and top shellfish	1 dozen cans	1.00
Mackerel, spanish mackerel, shellfish ligament. mactra sachalinensis (clam), eel, sardines. and lobster	1 dozen cans	0.50
Other canned fishery products	1 dozen cans	0.50

Class 3-Fishery Products, Salted.	*Quantity*	*Fee*
Sardines, Alaska pollock, herring, Spanish mackerel, hairtail, skipper (mackerel pike), and other salted fish	60-90 kg.	¥ 1.50

Class 4-Fertilizers

Sardine cake, herring cake, hairtail cake, codfish meal, sandfish cake, dried sardine, dried herring, and other fishery fertilizers	60-90 kg.	1.00

Class 5-Seaweed

Japanese jelly plant, glue weed, and iridaea laminarioides.	42-78 kg.	2.50
Grateloupia flabellata, ceramium rubrum, campylaephora hypnaeoids, gracilaria greville, gloiopeltis tenax, and turbinaria fusiformis	42-78 kg.	2.00
Agar-agar	60 kg.	10.00

Class 6-Fish Oils

Sardine oil	16.5 kg.	1.00

Class 7-Fish Meal

Sardine meal	50 kg.	1.00
Sardine powder	18.75 kg.	0.50
Other fish meals	30 kg.	1.00

Section III. *Effective Date.* This ordinance shall be effective on the tenth day after the date appearing hereon.

APPROVED: RECOMMENDED:

    C G HELMICK   AHN CHAI HONG

Brigadier General United States Army   Civil Administrator

    Acting Military Governor

軍政廳　官報　法令　第一四七號　　一九四七年七月三十日

SOUTH KOREAN INTERIM GOVERNMENT
Seoul, Korea

ORDINANCE
NUMBER 148

22 August 1947

PETROLEUM PRODUCTS SALES TAX LAW

Pending action on the subject matters hereof by the Korean Interim Legislative Assembly and the enactment of legislation pursuant thereto, the following shall govern:

SECTION I. *Purpose.* The purpose of this ordinance is to establish a sales tax on petroleum products at the distribution level.

SECTION II. *Sales Tax Levied.* Each purchaser of petroleum products from the Petroleum Distributing Agency shall pay to the Petroleum Distributing Agency at the time of payment of the purchase price a tax of twenty-five percent (25%) of the Petroleum Distributing Agency's Published Retail Price.

SECTION III. *Transmittal of Tax.* The Petroleum Distributing Agency shall, not later than the tenth day of each month, deliver all petroleum products sales taxes received by it on sales during the preceding month, less authorized returns of tax, to the Treasury Bureau, Department of Finance, SKIG, for the account of *General Revenue Account of SKIG.*

SECTION IV. *Authorized Returns of Tax.* In those cases where, after payment of purchase price and sales tax, the Petroleum Distributing Agency refunds the purchase price to the purchaser because of the cancellation of the sale or loss of the goods in transit, the Petroleum Distributing Agency is authorized to refund the sales tax paid by the purchaser.

SECTION V. *Records.* The Petroleum Distributing Agency and all purchasers of petroleum products from it shall maintain such written records with respect to the payment of petroleum products sales taxes as may be required by the Director of the Department of Finance. Such records shall be open at all times for inspection by employees and representatives of the Department of Finance.

軍政廳　官報　法令　第一四八號　一九四七年八月二十二日

SECTION VI. *Definitions.* As used herein :

a. *Petroleum products* includes gasoline, kerosene, fuel oil, Diesel oil, lubricating oil and lubricating greases.

b. *Purchaser* means any buyer, whether private or governmental.

SECTION VII. *Effective Date.* This ordinance shall be effective on the tenth day after the date appearing hereon.

APPROVED :

C G HELMICK

Brigadier General United States Army

Acting Military Governor

RECOMMENDED :

AHN CHAI HONG

Civil Administrator

SOUTH KOREAN INTERIM GOVERNMENT
Seoul, Korea

ORDINANCE
NUMBER 149

25 August 1947

REGULATING FOREIGN COMMERCE

Pending action on the subject matters hereof by the Korean Interim Legislative Assembly and the enactment of legislation pursuant thereto, the following shall govern:

SECTION I. *Purpose.* The purpose of this ordinance is to prepare for a wider scope of foreign commerce, to coordinate all existing commerce laws, and to repeal or revise those which do not conform hereto.

SECTION II. *Repealer.* Ordinance No. 39, dated 3 January 1946 (*Regulations of Foreign Trade*) is hereby repealed.

SECTION III. *Ordinance No. 82 Amended.** Ordinance No. 82, dated 17 May 1946 (*Regulation of Foreign Trade*), is amended as follows: In Section I, the words "in accordance with Ordinance No. 39, dated 3 January 1946" are hereby deleted. In Section II, the words "in accordance with Sections I to IV of Ordinance No. 39" are hereby deleted.

SECTION IV. *Ordinance No. 98 Amended.** Ordinance No. 98, dated 11 July 1946 (*Bureaus of Foreign Commerce and Domestic Commerce Created in Department of Commerce; Ordinance No. 82 Amended*), is amended as follows: In Section II *b*, the words, "in accordance with Sections I to IV of Ordinance No. 39, dated 3 January 1946" are hereby deleted.

SECTION V. *Foreign Commerce Regulated.* No foreign trade or commerce, of any description or to any extent whatsoever, shall be conducted, and no interzonal or coastwise trade shall take place, except in full compliance with rules and regulations of the Department of Commerce and with all orders, directives and instructions issued by competent authority.

---

* This amendment is not intended to change the scope or effect of Ordinance 82 or 98; it is a formal amendment only.

SECTION VI. *Transportation of Property*. All transportation of property of any description into or from southern Korea is hereby prohibited, except with the knowledge and consent (by license or otherwise) of competent authority.

SECTION VII. *Ports*. All transportation of property of any description into or from southern Korea is hereby prohibited except by way of the following ports hereby designated as ports of entry: Inchon, Kunsan, Mukho, Mokpo, Pusan, Cheju, and Kimpo Airport (near Seoul).

SECTION VIII. *Interzonal Vessels*. Masters of vessels engaged in interzonal trade shall be required to report to the Customs authorities for a thorough check of papers and cargo at one of the following ports: Chumunjin, Yongho-do, or Yosu.

SECTION IX. *Contraband*. All property of any description whatsoever, transported into or in the course of transportation from southern Korea contrary to any existing law, order or regulation pertaining thereto, shall be considered contraband and disposed of according to law.

SECTION X. *Seizure of Conveyances*. Any vessel, airplane, cart, animal, motor vehicle or other conveyance or instrumentality, not owned by any government, which is utilized in violation of this ordinance, may be impounded pending the trial of the violator or violators and, upon conviction of such violators, shall be forfeited to the South Korean Interim Government and disposed of according to law.

SECTION XI. *Inadvertent violations.* Any person appearing in good faith with any property at any point on the border or coastline with respect to foreign commerce of southern Korea, but who inadvertently violates any laws or regulations, shall have an opportunity to ascertain the legal procedures and requirements set forth herein and elsewhere and shall be given reasonable time and opportunity to comply therewith. If such person then complies with said procedures and requirements he shall not suffer the penalties or forfeitures of Sections IX and X hereof.

SECTION XII. *Fees*. The Department of Commerce shall have authority to fix and determine fees to the extent necessary to cover the cost of granting of import and export licenses and of handling matters incidental or pursuant thereto.

SECTION XIII. *Penalties.* Any person violating the provisions of this ordinance or of any order or regulation issued hereunder shall, upon conviction by a duly constituted court, suffer such punishment as the Court shall determine.

SECTION XIV. *Excepted Persons and Property.* This ordinance does not apply to the transportation of the personal effects of any officer or employee of the Military Government or of the United States. It shall not be construed to apply to the transportation of the personal effects, goods or valuables of any persons subject to the military or naval laws of the United States or any of the United Nations nor shall it be construed as a prohibition of the transportation to or from Korea of the personal effects, goods, or valuables of persons duly authorized to enter or depart from Korea under repatriation programs approved by the Military Government of Korea or of any persons in the course of performance of official business for such government.

SECTION XV. *Definitions.* As used herein :

"Foreign commerce" means trade between southern Korea and countries other than Korea.

"Domestic commerce" means trade wholly within southern Korea.

"Interzonal commerce" means trade (by land, air or water) between southern and northern Korea.

"Southern Korea" means that portion of Korea which is south of 38 degrees north latitude.

"Coastwise trade" means domestic or interzonal commerce via coastal ports.

SECTION XVI. *Effective Date.* This ordinance shall be effective as of the date of approval hereof.

APPROVED : 25 August 1947
   ARCHER L LERCH
Major General United States Army
 Military Governor in Korea

RECOMMENDED :
  AHN CHAI HONG
  Civil Administrator

SOUTH KOREAN INTERIM GOVERNMENT

Seoul, Korea

**ORDINANCE
NUMBER 150**

4 October 1947

ORDINANCE NUMBER 93 AMENDED

Pending action on the subject matters hereof by the Korean Interim Legislative Assembly and the enactment of legislation pursuant thereto, the following shall govern:

SECTION I. *Purpose.* The purpose of this ordinance is to provide more effective means of enforcing:

a. The prohibition on transactions in gold and other precious metals without authorization of the Director of the Department of Finance.

b. The registration of gold and other precious metals, including coin or bullion.

SECTION II. *Foreign Exchange Controls Revised.* Section II of Ordinance No. 93, dated 4 July 1946 (*Foreign Exchange Control*) is hereby amended by adding the following:

"In the case of gold, silver, platinum and other precious metals in Korea reported pursuant to the provisions of this ordinance, the Bank of Chosun shall, upon receipt of a properly executed report, issue a nontransferable certificate of registration to the holder of such metals indicating the type and quantity of such precious metals of which he has possession. It shall be unlawful for any person in Korea, other than duly licensed dentists, artisans, mine operators and/or ore-processing industries, to hold, possess, or own any gold, silver, platinum or other precious metal (except personal jewelry and objects of art) without such a certificate of registration from the Bank of Chosun."

SECTION III. *Penalties.* Any person violating the provisions of this ordinance shall, upon conviction by a duly constituted court, forfeit all unre-

gistered gold or other precious metal to the South Korean Interim Governme
and suffer such other punishment as the court shall determine.

SECTION IV. *Effective Date.* This ordinance shall be effective on t
date appearing hereon.

APPROVED :

C G HELMICK

Brigadier General United States Army

Acting Military Governor

RECOMMENDED :

AHN CHAI HONG

Civil Administrator

SOUTH KOREAN INTERIM GOVERNMENT
Seoul, Korea

ORDINANCE
NUMBER 151

30 September 1947

APPOINTMENT OF JUDGES AND PROSECUTORS

Pending action on the subject matters hereof by the Korean Interim Legislative Assembly and the enactment of legislation pursuant thereto, the following shall govern:

SECTION I. *Judges and Prosecutors, Method of Appointment.* Appointments of all judges and prosecutors shall be made by the Military Governor.

SECTION II. *Ordinance No. 85 Amended.* Ordinance No. 85 (*Amending Ordinance Number 67*), dated 16 May 1946, is amended to conform herewith, by deleting from Section II of said Ordinance No. 85 so much as reads: "He shall appoint, with the approval of the Military Governor and from nominations submitted by the Central Council of the National Bar Association, judges of courts other than the Supreme Court and Courts of Review and all prosecutors."

SECTION III. *Effective Date.* This ordinance shall be effective on the appearing hereon.

APPROVED:

  C. G. HELMICK

Brigadier General, United States Army

  Acting Military Governor

RECOMMENDED:

  AHN CHAI HONG

  Civil Administrator

軍政廳　官報　法令　第一五一號　　一九四七年九月三十日

SOUTH KOREAN INTERIM GOVERNMENT
Seoul, Korea

ORDINANCE
NUMBER 152

29 October 1947

AREA OF CITY OF IRI INCREASED

Pending action on the subject matters hereof by the Korean Interim Legislative Assembly and the enactment of legislation pursuant thereto, the following shall govern:

SECTION I. *Purpose.* The purpose of this ordinance is to increase the area of the City of Iri.

SECTION II. *Areas Added to City of Iri.* The following areas are hereby transferred to and made a part of the City of Iri in North Cholla Province:

a. Moin Ri located in Bukil Myun in Iksan Gun.

b. All of Songhak Ri located in Osan Myun in Iksan Gun, except the following lot numbers: Numbers 290 to 320, inclusive; Numbers 551-1 to 607, inclusive; Numbers 611 to 857, inclusive; Number 860; Numbers 863-1 to 874-2, inclusive; and Numbers 873 to 884, inclusive.

c. All of Mokchun Ri located in Osan Myun in Iksan Gun, except the following lot numbers: Numbers 461-2 to 493-2, inclusive; Numbers 640 to 653-2, inclusive: Numbers 926 to 933, inclusive; Numbers 937 to 957-1, inclusive; Numbers 967-1 to 976-1, inclusive; Numbers 1005-1 to 1089, inclusive; and Numbers 1128 to 1176-3, inclusive.

SECTION III. *Effective Date.* This ordinance shall be effective on the tenth day after the date appearing hereon.

APPROVED:

    C G HELMICK

Brigadier General United States Army

    Acting Military Governor

RECOMMENDED:

    AHN CHAI HONG

    Civil Administrator

SOUTH KOREAN INTERIM GOVERNMENT
Seoul, Korea

ORDINANCE
NUMBER 153

3 November 1947

AIDING AMERICAN PERSONNEL TO VIOLATE MILITARY DIRECTIVES PROHIBITED

Pending action on the subject matters hereof by the Korean Interim Legislative Assemby and the enactment of legislation pursuant thereto, the following shall govern:

SECTION I. *Purpose.* The purpose of this ordinance is to clarify the obligations of Koreans to refrain from aiding in the violation of military directives by military and civilian personnel under the jurisdiction of United States Army Forces in Korea.

SECTION II. *"Off Limits" Establishments.* No person who owns or operates an establishment declared "off limits" by any authority under United States Army Forces in Korea shall permit or suffer any military or civilian personnel under the jurisdiction of United States Army Forces in Korea to enter or remain in such establishment.

SECTION III. *Prostitution.* No person shall procure or aid in procuring a prostitute for any military or civilian personnel under the jurisdiction of United States Army Forces in Korea.

SECTION IV. *Penalties.* Any person who violates the provisions of this ordinance shall, upon conviction by a duly constituted court, suffer such punishment as the court shall determine.

SECTION V. *Effective Date.* This ordinance shall be effective on the tenth day after the date appearing hereon.

APPROVED:

    C G HELMICK

Brigadier General United States Army

    Acting Military Governor

RECOMMENDED:

    AHN CHAI HONG

Civil Administrator

軍政廳　官報　法令　第一五三號　　一九四七年十一月三日

SOUTH KOREAN INTERIM GOVERNMENT

Seoul, Korea

ORDINANCE
NUMBER 154

1 November 1947

BEVERAGE TAX LAW

Pending action on the subject matters hereof by the Korean Interim Legislative Assembly and the enactment of legislation pursuant thereto, the following shall govern:

SECTION I. *Purpose.* The purpose of this ordinance is to promote the general welfare of Korea, raise needed tax revenues, prevent inflation, improve administrative procedure, develop a tax administrative staff, improve collection, and to establish a review procedure for all national taxes.

SECTION II. *Internal revenue tax; levy and rate.*

a. Distilled spirits and fermented liquors generally; rate. There shall be levied, collected and paid on all distilled spirits and fermented liquors (alcoholic beverages) produced or manufactured in or imported into Korea a tax at the following rates:

A	B	C	D
Kind of beverage	Alcoholic content	Rate in won per suk	Rate per suk where alcoholic content is greater than the established amount.
Takju	10% or less	W 800	W 300 per percent in excess of 10%
Yakju	15% or less	5,000	600 per percent in excess of 15%
Beer	3.5% & up*	15,000	
Fruit syrup wine	15% or less	13,000	1,200 per percent in excess of 15%
Whangju	15% or less	11,000	1,000 per percent in excess of 15%
Japanese wine (including sake)	18% or less	47,000	3,000 per percent in excess of 18%
Other fermented liquors	15% or less	22,500	2,000 per percent in excess of 15%

*Beer under 3.5% is to be taxed at the rate established.

— 1 —

A	B	C	D
Kind of beverage	*Alcoholic content*	*Rate in won per suk*	*Rate per suk where alcoholic content is greater than the established amount.*
Soju	25% or less	W 8,000	W 400 per percent in excess of 25%
Koliang Koaryangju	50% or less	20,000	500 per percent in excess of 50%
Other distilled wine	40% or less	47,500	1,500 per percent in excess of 40%
Kwahaju	20% or less	15,000	800 per percent in excess of 20%
Mixed Japanese wine	18% or less	25,000	1,500 per percent in excess of 18%
Mixed liquor	40% or less	47,500	1,500 per percent in excess of 40%
Alcohol	85%	30,000	400 per percent in excess of 85%

(1) Where the alcoholic content of any of the foregoing beverages (except alcohol, the last item in the above list) exceeds fifty percent (50%), each percent in excess of fifty percent (50%) shall be taxed at twice the rate in effect for the excess over the established amount (i. e., at twice the rate in column D).

(2) There shall be levied, collected and paid a proportionate tax on all multiple parts or fractional parts of suks.

b. Products containing distilled spirits or fermented liquors. All products of distillation or fermentation, by whatever name known, which contain spirits or alcohol, on which no specific tax is imposed herein, shall be considered and taxable as mixed liquor generally.

c. Nonalcoholic beverages; rate. There shall be levied, collected and paid. on all nonalcoholic beverages produced in or imported into Korea, a tax at the following rates:

First Class (i. e., bottled)...................   W   3,000 per suk

Second Class (i. e., other than bottled)   1,000 for each kilogram, prorated to fraction thereof, of carbonic gas contained therein.

(1) A proportionate tax at a like rate shall apply on all multiple parts or fractional parts of such containers.

— 2 —

軍政廳　官報　法令　第一五四號　一九四七年十一月　一日

d. *Tax on stored liquors.* There shall be levied, collected and paid, on all nonalcoholic beverages and distilled spirits and fermented liquors, on hand or in storage, for sale, or in excess of the amount excluded by law, a tax at the rates specified in Sections II *a, b* and *c* hereof, such tax to be due and payable on the effective date of this ordinance and in the manner prescribed herein.

SECTION III. *Payment of tax.* All taxes imposed by this ordinance shall be paid by the purchase of Internal Revenue Stamps at the offices of the Bureau of Internal Revenue or its authorized agent immediately prior to the time of completion of the producing, manufacturing or importing of the taxable items; such revenue stamps shall be affixed over the cap or cork of the container containing taxable distilled spirits, fermented liquors, or beverages in such a manner that the stamp will be destroyed at the time of opening such container, or affixed as provided by regulations of the Department of Finance.

SECTION IV *Licenses.*

a. General. Every person, corporate body, government agency, association or group, or officer (or official) or group of officers (officials) of such corporate body, government agency, association or group, engaging in or intending to engage in the business of producing, importing, distributing, serving, or handling in any manner for sale or storage, beverages described in Section II hereof, is required to procure a license from or pursuant to the directives of the Bureau of Internal Revenue for the privilege of engaging in such business.

b. License fees, distilled spirits and fermented liquors generally There shall annually be levied on, collected from and paid by all manufacturers, producers, distillers, and importers, license fees for each fiscal period 1 September to 30 August, inclusive, based on the following schedule :

A	B	C	D	E	F	G
Annual Production or Importation for Each Class of Beverage	License Fees for Various Classes of Beverages					
	Takju	Yakju	Wine-Fermented	Remaden Wine	Beer	Wine-Distilled
1. 10 suk or less	W 5,000	W 5,000	W 30,000	W 20,000	W 30,000	W 5,000
2. Not more than 20 but more than 10	5,000	5,000	30,000	20,000	30,000	10,000
3. Not more than 50 but more than 20	5,000	10,000	30,000	30,000	30,000	20,000
4. Not more than 100 but more than 50	10,000	20,000	50,000	50,000	50,000	30,000
5. Not more than 200 but more than 100	20,000	30,000	100,000	100,000	100,000	50,000
6. Not more than 300 but more than 200	30,000	50,000	15,000	150,000	150,000	100,000
7. Not more than 500 but more than 300	50,000	100,000	200,000	200,000	200,000	150,000
8. Not more than 1000 but more than 500	100,000	150,000	250,000	250,000	250,000	200,000
9. Not more than 1500 but more than 1000	150,000	200,000	300,000	300,000	300,000	250,000
10. More than 1500	200,000	200,000	300,000	300,000	300,000	300,000

These fees apply when the alcoholic content does not exceed the maximum percentage set forth in Section II *a*, column B. If the alcoholic content is in excess of such maximum percentage, the beverage will be taxed as though the number of suk indicated in column A, were proportionately greater. The minimum license fee shall be that attributable to a production of ten suk or less.

(1) No single license fee shall apply to more than one place of business or more than one class of beverage. However, subnumbered licenses of the master license may be issued upon payment of the regular license fee applicable to alcoholic and nonalcoholic beverages.

c. *License fees; nonalcoholic beverages.* There shall annually be levied on, collected from and paid by all manufacturers, producers, distillers, and importers of nonalcoholic beverages, the following license fees for each fiscal period 1 September to 30 August, inclusive:

— 4 —

Annual Production for Each Class of Beverage – Suk	License Fee
1. 50 suk or less	W 5,000
2. Not more than 100 but more than 50	10,000
3. Not more than 200 but more than 100	20,000
4. Not more than 300 but more than 200	40,000
5. Not more than 500 but more than 300	80,000
6. Not more than 1,000 but more than 500	200,000
7. More than 1,000	300,000

d. Retail license fees. There shall annually be levied on, collected from and paid by all wholesalers and retailers of alcoholic and nonalcoholic beverages covered in this ordinance the following license fees for the fiscal period 1 September to 30 August, inclusive:

Class of Dealer	License Fee
1. Retailer of Korean alcoholic beverages	W 2,000
2. Retailer of imported alcoholic beverages	5,000
3. Wholesaler of Korean alcoholic beverages	30,000
4. Wholesaler or importer of imported alcoholic beverages	50,000
5. General dealer in nonalcoholic beverages	2,000
6. Operators of restaurants and other consumption-on-premises dealers	2,000

e. License fees: yeast and yeast products. There shall annually be levied on, collected from and paid by all wholesalers and retailers and manufacturers of yeast and yeast products, for the fiscal period 1 September to 30 August, inclusive, a license fee of an amount equivalent to the license fee for manufacturers, producers and processors of Takju as provided in Section IV *b* hereof, column B.

f. Acquisition of taxable products. Licensed liquor, beverage and yeast dealers are prohibited from acquiring liquor, beverages or yeast from sources other than other licensed liquor, beverage or yeast dealers under this ordinance, or through importation, except by written permission of the district tax offices.

g. Application for license.

(1) General. Every person engaged in the business of a distiller, producer, or manufacturer of alcoholic or nonalcoholic beverages, (2)

manufacture or distribution of yeast, and yeast products, or (*c*) importer of alcoholic or nonalcoholic beverages, who is distributing or handling any of such _beverages in any manner, for sale or exchange, or intending to engage therein, or who wishes to continue in such business on or after the effective date of this ordinance, or on or after the first day of September in any year, shall give notice of such intention to the Bureau of Internal Revenue on forms prescribed by the Bureau of Internal Revenue, not less than thirty (30) days before engaging in such business. Such notice is not and shall not be considered as the license to engage in such business. No person shall be permitted to engage in the beverage business under this ordinance without first securing a written license from the Bureau of Internal Revenue.

(2) The license application shall include all of the information required by laws in effect relating to the description of premises, condition of title to premises, and apparatus and equipment.

(3) The minimum license fee shall accompany the application for license.

SECTION V. *Reports, Labels, Obliteration, Bonds and Records.*

a. Reports. Every person engaged in the business of distiller, producer, manufacturer or importer of alcoholic or nonalcoholic beverages shall submit a monthly report to the Bureau of Internal Revenue within fifteen days following the close of each month. Such report shall set forth the beginning inventories, production and purchases, and closing inventories, by alcoholic content, kind of commodity, total commodities, and such other information as may be prescribed by the Bureau of Internal Revenue, for the preceding period and cumulative from the beginning of the taxable year. There shall be submitted therewith payment for all taxes due and at the close of each fiscal year payment shall be submitted therewith for license fees due on the production, sale or distribution in excess of the amount on which license fee was previously paid.

b. Labels. Every bottle or container containing alcoholic or nonalcoholic beverages must have affixed thereto, by the producer, manufacturer or importer thereof, a label showing the name and address of such producer, manufacturer or importer, the kind of beverage, the alcoholic content or

— 6 —

carbonic gas content, and the total content by volume of the bottle or container and the license number under which such beverage was produced or imported. The producer, manufacturer or importer may add such other identifying data as may be desired as long as it does not obliterate the required information.

c. *Obliteration of stamps, marks and brands on bottles, packages and containers.* When bottles, packages or containers of alcoholic or nonalcoholic beverages are emptied, all stamps, marks and brands required to be placed thereon shall be completely effaced and obliterated in such manner that they cannot be reused.

d. *Performance bond.* Every person intending to commence or to continue the business of distiller, manufacturer, processor or importer shall, upon filing application for license and before proceeding with such business, execute a performance bond in accordance with the existing law which provides for a bond equal to least one-fourth (1/4) times the estimated annual taxes due, or a guarantee by the Brewers' Association.

e *Records.* Every person liable to any tax imposed by this ordinance, or for the collection thereof, shall keep such records as will clearly reflect all of the transactions involved for the determination of correct tax liability of such taxpayer, render such statements and make such returns based upon such records as are necessary and prescribed by the Bureau of Internal Revenue. Such records shall be open for inspection at all times.

SECTION VI. *Penalties and forfeitures.* Failure to comply with any of the requirements of this ordinance shall cause such violator to be subject to the following penalties.

a. *Creation of fictitious proof.* (1) *Penalty.* Every person who adds or causes to be added any ingredient or substance to any alcoholic beverage, before the tax is paid thereon, for the purpose of creating a fictitious proof, shall be fined and shall suffer the penalties provided by law with respect to false reports.

(2) *Forfeiture.* Every such cask or package, with its contents, shall be forfeited to the Korean government.

— 7 —

386

b. Unlawful affixing, cancelling, issue or failure to issue of stamps by officer; penalty. Whenever any Internal Revenue officer (1) fails to issue or cancel any stamps required by law, relating to alcoholic or nonalcoholic beverages, or (2) affixes or cancels, or causes or permits to be affixed or cancelled, any such stamp in any other manner or in any other place or issues such stamp to any other person than as provided by law or by regulation pursuant thereto, or (3) permits any such stamp to be affixed to any cask or package or container of beverages, the whole or any part of which has been distilled, rectified, compounded, removed or sold in violation of law or which has in any manner escaped payment of beverage tax due thereon, such Internal Revenue Officer shall, upon conviction for every such offense, be fined not less than fifty thousand won (W 50,000) nor more than three hundred thousand won (W 300,000) and be imprisoned for not less than one year nor more than three (3) years.

c. Evasion of tax; penalty. Whenever any person evades or attempts to evade the payment of tax on any alcoholic or nonalcoholic beverage, in any manner whatever, he shall forfeit such beverage and pay at least five times the amount of the tax so evaded or attempted to be evaded, and suffer other penalties as provided by law.

d. Fraud. Whenever any person, engaged in carrying on any business defined and covered by this ordinance, defrauds or attempts to defraud the Korean government of any tax on any products or any part thereof made by him, he shall be subject to the penalties provided by law.

e. Offenses not specifically covered. If any manufacturer, distiller, processor or importer, or any other person subject to this ordinance, omits, neglects or refuses to do or cause to be done any of the things required by law in the carrying on or conducting of his business, or does anything prohibited by this ordinance, and there happens to be no specific penalty or punishment therefor elsewhere in this ordinance, such person shall suffer the penalties provided by existing laws which include imprisonment as well as advalorem penalties.

f. Cancellation of license. If any manufacturer, distiller, processor, or importer, or any other person subject to this ordinance, omits, neglects, or refuses to do or cause to be done any of the things required by law in the carrying on or conducting of his business or does anything by this ordinance prohibited, he shall suffer a penalty consisting of the cancellation of all his licenses, in addition to any other penalty provided by law.

SECTION VII. *Penalty for breaking locks or gaining unlawful access to cistern room or buildings.* Every person who destroys, breaks, damages or tampers with any lock or seal which may be placed on any cistern, cistern room or building by the duly authorized officer or officers of the Internal Revenue Bureau, or who opens such lock or seal or the doors to such cistern room or building. or in any other manner gains unlawful access to the contents therein, in the absence of a proper Internal Revenue Bureau officer, shall be fined and suffer the penalties provided by law, consisting of imprisonment as well as payment of advalorem penalties.

SECTION VIII. *Entry and examination of distillery, brewery, plant; power of revenue officers.* In addition to the powers under existing law, it shall be lawful for any revenue officer to enter, at all times by day or night, any distillery or building or place used for the business of distilling, or used in connection therewith for storage or for other purposes, and to examine, gauge, measure and take an account of every still or other vessel, utensil or container of any kind, and of all low wines. and the quantity and gravity of all mash, wort and beer, and of all yeast and other compositions for exciting or producing fermentation in any mash or spirits which may be in such distillery or premises or in possession of the distiller.

SECTION IX. *Refund of tax.*

a. *Industrial alcohol; alcohol for export purposes.* The tax paid in accordance with the provisions of this ordinance shall be refunded to the user of alcohol for industrial purposes and to the exporter of alcohol from Korea, subject to the limitations and qualifications set forth in Section IX *b* hereof.

b. *Limitations and qualifications.* (1) A person who acquires alcohol for industrial purposes and uses it for industrial purposes, or who acquires alcohol for export and exports such alcohol, shall be paid a refund of the tax paid on such alcohol.

— 9 —

(2) Refund under above Section IX *b (1)* may be obtained by submitting properly executed copies of the invoices from the dealer selling the' alcohol intended to be used for industrial purposes or exported. such invoice to indicate the proof spirit and total suks. Each such invoice shall be certified by the importer, manufacturer, distiller or processor that the beverage tax has been paid thereon, and certified by a member of the Bureau of Internal Revenue or Customs that the beverage was used for industrial purposes or exported. Upon receipt of such invoices and claim for refund and upon approval thereof, the Bureau of Internal Revenue shall remit to claimant the tax due and claimed. Such claim must be filed not later than six months from the date of such industrial use or export of such alcohol subject to refund.

(3) Existing laws with reference to refund or exemption of beverage taxes are not hereby repealed but are modified to the extent of the provisions of this Section IX.

SECTION X. *Penalty for producing in excess of established quotas.* There shall be levied on. collected from and paid by each person who produces more alcoholic beverages than permitted under his individual established quota, a penalty equal to three times the total tax due and payable on such excess production.

SECTION XI. *Definitions.* Applicable definitions in existing laws shall apply to this ordinance. in addition to the following:

a. Korean liquors are deemed to. be all beverages manufactured in Korea. Any definitions in other laws, to the contrary, are hereby superseded.

b. Korea means that portion of Korea which is south of 38 degrees north latitude.

c. Bureau of Internal Revenue. The term *Bureau of Internal Revenue* as used herein refers to the Bureau of the Treasury of the Department of Finance and is so used unless otherwise indicated in this ordinance or suusequent ordinances.

SECTION XII. *Statute of limitations.*

a. *General rule..* The amount of taxes imposed by this ordinance shall be assessed within five years after the filing of the return, except in the case of fraudulent returns, and no proceeding in court for the collection of beverage taxes shall be commenced after the expiration of such period or without assessment having been made.

b. *Fraudulent returns.* Where there has been an evasion or attempted evasion of the tax or fees, in whole or in part, there shall be no statutory limit as to the time of instituting proceedings in court for the collection of taxes or license fees imposed by this ordinance.

SECTION XIII. *Interest.* Taxes and license fees hereunder which are not paid on or before the date due are considered delinquent. Interest at the rate of five percent (5%) of the amount due is hereby imposed and added to such tax or license fee on the date upon which the tax or license fee becomes delinquent. Additional interest of five percent (5%) of the amount of tax or license fee due is hereby imposed on taxes or license fees which are not paid within thirty (30) days after becoming delinquent. For each further period of thirty (30) days of delinquency, there is hereby imposed and added to the tax or license fee due further interest charges of ten percent (10%) of the delinquent tax or license fee. In no case shall the total interest attributable to any single tax or license fee hereunder exceed thirty percent (30%). "Interest" hereunder is separate from and in addition to "penalties" hereunder. This section supersedes Section II of Ordinance No. 139 dated 19 April 1947 (*Tax Delinquency Penalties Revised*), for beverage-tax purposes.

SECTION XIV. *Informers. a.* Where a person other than the taxpayer concerned, and other than a member of the Bureau of Internal Revenue, discloses to the Bureau of Internal Revenue information which is instrumental in establishing a deficiency of tax and assessment of penalties under this ordinance, such informer shall receive twenty-five percent (25%) of such penalties assessed, collected and paid, but shall not receive more than fifty thousand won (W 50,000). In determining informer's reward hereunder, neither the deficiency assessment nor any interest charges attributable thereto shall be included in the basis; the reward shall be computed solely on the basis of penalties.

Where information resulted in the disclosure of a deficiency but other investigations or disclosures also contributed to the uncovering of such deficiency, the payment to the informer shall be determined by the Director of the Department of Finance and his determination shall be final.

b. In accordance with the purpose of this ordinance as set forth in Section I hereof, and to develop an independent investigation staff, it is provided that twenty-five percent (25%) of all penalties levied, assessed and paid because of deficiencies under this ordinance shall be paid to the Korean Cooperative Welfare Association of the Bureau of Internal Revenue. No such payment, however, shall be made in any case where payment is made under Section XIV *a* hereof.

SECTION XV. *Collection and enforcement procedure.* In addition to the existing applicable law, the following shall govern: The Bureau of Customs shall be charged with and accountable for revenue stamps to be sold to importers, and such stamps shall be affixed to bottles and containers of beverages in accordance with the provisions of Section III hereof. The Customs tariff provisions shall additionally apply.

SECTION XVI. *Grievance procedure.* The Director of the Department of Finance shall cause to be created a board, panel or group, as he shall elect, the duties and functions of which shall be to hear and decide the merit of alleged deficiencies and overassessments of taxes, licenses and penalties pertaining to all national taxes and to review, as requested by such Director or upon written request of the taxpayer, determinations made by the Bureau of Internal Revenue.

SECTION XVII. *Bureau of Internal Revenue; jurisdiction. a.* The Bureau of Internal Revenue shall have full and complete jurisdiction over all tax and license matters hereunder and of the enforcement of this ordinance.

b. The Bureau of Internal Revenue shall establish, subject to the approval of the Director of the Department of Finance, individual production quotas (subject to overall production quotas established by law) in accordance with existing laws and shall control production and distribution of liquors and beverages in accordance with the provisions of existing laws.

c. The Department of Finance shall prescribe such regulations as the Bureau may deem necessary in order that the tax liability of any taxpayer hereunder may be determined, computed, assessed, collected or adjusted in such manner as clearly to reflect the correct tax liability, and to prevent avoidance of tax liability hereunder.

SECTION XVIII. *Inconsistent laws.*

a. Section XI b (9) of Ordinance No. 114, dated 23 October 1946 (*Reorganization of Provincial Governments*), is hereby repealed. All "laws, ordinances, orders, regulations, directives and instructions, or parts thereof", to the extent they were repealed by Section XIX of said Ordinance No. 114 as inconsistent or in conflict with said Section XI b (9), are hereby reinstated and reenacted.

b. All other existing laws, statutes and ordinances are hereby amended, abolished or superseded to the extent they are inconsistent with the provisions of this ordinance; provided, however, that no taxpayer shall thereby be relieved of any liability, penalty or interest which may have been incurred prior to such amending, abolishing or superseding.

SECTION XIX. *Effective date.* This ordinance shall be effective on the date of approval hereof and shall be applicable to all taxpayers and all commodities on hand or produced on or after the effective date hereof.

APPROVED:

    C G HELMICK

Brigadier General United States Army

    Acting Military Governor

RECOMMENDED:

    AHN CHAI HONG

Civil Administrator

SOUTH KOREAN INTERIM GOVERNMENT
Seoul, Korea

ORDINANCE
NUMBER 155

30 September 1947

DISSOLUTION OF MATERIALS CONTROL CORPORATION

Pending action on the subject matters hereof by the Korean Interim Legislative Assembly and the enactment of legislation pursuant thereto, the following shall govern:

SECTION I. *Purpose.* The purpose of this ordinance is to dissolve the Materials Control Corporation.

SECTION II. *Dissolution.* The Materials Control Corporation (formerly the Korean Important Materials Control Corporation) created by Ordinance No. 24 dated 5 November 1945 (*Materials Control Corporation*) is hereby dissolved, and the Office of Property Custody is appointed its liquidator.

SECTION III. *Transfer to Office of Property Custody.* All functions, property, assets, records and personnel of the Materials Control Corporation are hereby transferred to the Office of Property Custody.

SECTION IV. *Effective date.* This ordinance shall be effective on the date appearing hereon.

APPROVED:

  C G HELMICK

Brigadier General United States Army

  Acting Military Governor

RECOMMENDED:

  AHN CHAI HONG

Civil Administrator

軍政廳　官報　法令　第一五五號　一九四七年九月三十日

SOUTH KOREAN INTERIM GOVERNMENT
Seoul, Korea

ORDINANCE
NUMBER 156   21 November 1947

PROPERTY CUSTODIAN AUTHORIZED TO CALL STOCKHOLDERS' MEETINGS

Pending action on the subject matters hereof by the Korean Interim Legislative Assembly and the enactment of legislation pursuant thereto, the following shall govern:

SECTION I. *Purpose.* The purpose of this ordinance is to facilitate the calling of stockholders' meetings for corporations in which the Property Custodian has a stock interest.

SECTION II. *Power to call meetings.* The Property Custodian shall have the authority, as to any company in which he has a stock interest, to call a meeting of the stockholders of such company for the purpose of electing directors and conducting such other business as may be properly carried on at stockholders' meetings.

SECTION III. *Notice.* Thirty days notice of any meeting called under Section II hereof shall be given to stockholders, by ordinary mail, at their addresses as registered in the company's records, to the extent that such records are available, and copies of such notice shall be publicly posted at each Provincial Property Custody office for not less than fifteen days before such meeting.

SECTION IV. *Effect.* The effect of this ordinance is to facilitate the calling of stockholders' meetings by the Property Custodian. It is not intended to modify, eliminate or add to any other provisions of the corporation's charter, articles of incorporation, or its rules and by-laws, or any other provisions of law.

SECTION V. *Inconsistent laws.* All laws and regulatory measures and parts thereof, which are inconsistent herewith or in conflict with the provisions hereof, are to such extent hereby repealed.

SECTION VI. *Effective date.* This ordinance shall be effective on the date of approval hereof.

APPROVED: 30 October 1947
    C G HELMICK
Brigadier General United States Army
Acting Military Governor

RECOMMENDED:
    AHN CHAI HONG
Civil Administrator

軍政廳　官報　法令　第一五六號　　一九四七年十一月二十一日

SOUTH KOREAN INTERIM GOVERNMENT
Seoul, Korea

ORDINANCE
NUMBER 157

25 November 1947

ESTABLISHMENT OF NATIONAL POLICE BOARD

Pending action on the subject matters hereof by the Korean Interim Legislative Assembly and the enactment of legislation pursuant thereto, the following shall govern:

SECTION I. *Purpose.* The purpose of this ordinance is to establish a National Police Board.

SECTION II. *Establishment of Board.* The National Police Board is hereby established in the South Korean Interim Government. It shall be directly responsible to the Military Governor.

SECTION III. *Functions and powers.* The National Police Board shall have the following functions and powers:

a. To initiate major matters of police policy and pass upon matters of police policy and practice referred to it by the Director of the National Police.

b. To compel the attendance of police officials for questioning with reference to police action.

c. To pass upon appointments and dismissals proposed by the Director of the Department of Police of police officials of the Civil Service status of Class 5 or better; *provided, however,* that such appointments and removals shall remain subject to existing law.

d. To hear appeals from decisions of Police Trial Boards.

e. To act on such other matters concerning the Department of Police as may be referred to it by the Military Governor.

SECTION IV. *Other agencies.* The Department of Police and all other departments, offices, agencies, instrumentalities and political subdivisions of the government shall furnish to the National Police Board or its agents such data, facilities and other assistance as they may from time to time require.

軍政廳 官報 法令 第一五七號　　一九四七年十一月二十五日

SECTION V. *Organization of Board.* The National Police Board shall consist of six members appointed by the Military Governor, two of whom shall be appointed on recommendation of the Civil Administrator from among the Directors of Departments and two upon recommendation of the Director of the Department of Justice from among the judiciary or prosecutors. The Director of the Department of Police shall be an additional member of the Board but shall have no vote. Paid officials of the government serving on or for the National Police Board shall receive no additional compensation for such service.

SECTION VI. *Officers.* The officers of the National Police Board shall consist of a chairman, secretary and such other officers as may be provided by the Board.

    a. The chairman of the Board shall be designated from among its members by the Military Governor. The chairman shall preside over meetings of the Board and perform such other duties as are provided for in this ordinance or as the Board shall decide.

    b. The secretary shall be elected by the Board. The secretary shall keep minutes of the meetings and all other records of the Board.

SECTION VII. *Meetings.* The National Police Board shall meet at the call of the chairman. The chairman shall call a meeting upon written request of two or more members no later than six days after such request. The secretary shall send notices of all meetings, including tentative agenda, to the members of the Board.

SECTION VIII. *Quorum.* Four members shall constitute a quorum.

SECTION IX. *Effective date.* This ordinance shall be effective on the tenth day after the date appearing hereon.

APPROVED:

    WILLIAM F DEAN

Major General, United States Army

    Military Governor in Korea

RECOMMENDED:

    AHN CHAI HONG

    Civil Administrator

— 2 —

SOUTH KOREAN INTERIM GOVERNMENT
Seoul, Korea

ORDINANCE
NUMBER 158   30 December 1947

TRANSFER AND ABOLITION OF CERTAIN LICENSING FUNCTIONS

Pending action on the subject matters hereof by the Korean Interim Legislative Assembly and the enactment of legislation pursuant thereto, the following shall govern :

SECTION I. *Purpose.* The purpose of this ordinance is to transfer and abolish certain licensing functions.

SECTION II. *Transfer of licensing functions from Department of Police.* All licensing functions hitherto exercised by the Department of Police, except those listed in Section III, are hereby transferred as provided in Section IV of this ordinance.

SECTION III. *Licensing functions remaining in Department of Police.* The following licensing functions hitherto exercised by the Department of Police, and to the extent that they are so exercised, shall continue to be within the jurisdiction of the Department of Police :

a. Licensing functions concerned with guns, ammunition, and explosives and the storage thereof as provided in Governor General Ordinance No. 3, 1912 (*Order Concerning Guns, Ammunition, and Explosives*).

b. Licensing of investigating companies as provided in Government General Ordinance No. 82, 1911 (*Order Concerning Credit Investigation*).

c. Licensing of shipping as provided in provincial governors' orders concerned with shipping.

d. Licensing of pawnshops as provided in Governor General Ordinance No. 3, 1912 (*Order Concerning Pawnshops*).

e. Licensing of second-hand shops as provided in Governor General Ordinance No. 2, 1912 (*Order Concerning Second Hand Shops*).

— 1 —

f. Licensing of seal engraving as provided in provincial governors' orders concerned with seal engraving.

g. Licensing of motor vehicles and operators thereof as provided in Government General Ordinance No. 131, 1934 (*Order Concerning Motor Vehicles*).

SECTION IV. *Transfer to provincial governors.* All other licensing functions hitherto exercised by the Department of Police are hereby transferred to the appropriate provincial governors and the Mayor of the City of Seoul.

Such provincial governors and the Mayor of the City of Seoul shall exercise such functions through the appropriate provincial bureau, which shall before issuance of such license consult all agenies concerned.

SECTION V. *Abolition of licensing functions.* Licensing functions relating to the following subject matters, to the extents that they have hitherto been exercised by the Police, are abolished and the following laws insofar as they relate to such licensing functions are repealed:

a. Messenger services as provided in provincial governors' orders concerned with messenger services.

b. Drayage and carts as provided in Government General Ordinance No. 54, 1913, concerning drayage, and provincial governors' orders concerned with carts.

c. Manufacturing companies as provided in provincial governors' orders concerned with manufacturing companies.

d. Advertising as provided in provincial governors' orders concerned with advertising.

e. Engines as provided in Government General Ordinance 83, 1925 (*Order Concerning Engines*).

f. Store stalls as provided in Government General Ordinance No 231, 1938 (*Order Concerning Roads*).

SECTION VI. *Transfer of records and property.* All records and property used in the administration of licensing functions transferred by Section IV shall be turned over to the appropriate bureau designated by the provincial governor or the Mayor of the City of Seoul within thirty days from the date appearing hereon.

SECTION VII. *Effective date.* This ordinance shall be effective thirty (30) days from the date appearing hereon.

APPROVED: RECOMMENDED:

WILLIAM F DEAN   AHN CHAI HONG
Major General, United States Army   Civil Administrator
Military Governor in Korea

SOUTH KOREAN INTERIM GOVERNMENT
Seoul, Korea

ORDINANCE
NUMBER 159   30 December 1947

AMENDING ORDINANCE NO. 135: KOREA CIVIL SERVICE APPROVAL OF APPOINTMENT AND REMOVAL OF CERTAIN PROVINCIAL OFFICIALS ELIMINATED

Pending action on the subject matters hereof by the Korean Interim Legislative Assembly and the enactment of legislation pursuant thereto, the following shall govern:

SECTION I. *Purpose.* The purpose of this ordinance is to eliminate the requirement that the Office of Korea Civil Service approve the appointments and removals of certain officials.

SECTION II. *Amending Section I c of Ordinance No. 135.* Section I c of Ordinance No. 135 dated 15 March 1947 (*Appointment and Removal of Government Officials*) is amended to read as follows:

"All appointments within the several Provinces and City of Seoul shall be made by the respective Provincial Governors and Mayor of the City of Seoul, subject to the approval of the Office of Korea Civil Service in those cases in which existing law requires such approval; *provided, however,* that appointments to and removals from the positions of county, town, village or island headman shall not require the approval of the Office of Korea Civil Service. Except as herein provided such positions shall remain subject to rules and regulations of the Office of Korea Civil Service."

SECTION III. *Effective date.* This ordinance shall be effective on the tenth day after the date appearing hereon.

APPROVED: RECOMMENDED:

   WILLIAM F DEAN   AHN CHAI HONG

Major General, United States Army   Civil Administrator

   Military Governor in Korea

軍政廳 官報 法令 第一五九號   一九四七年十二月三十日

SOUTH KOREAN INTERIM GOVERNMENT
Seoul, Korea

ORDINANCE
NUMBER 160

15 December 1947

AGRICULTURAL RESEARCH AND EDUCATION

Pending action on the subject matters hereof by the Korean Interim Legislative Assembly and the enactment of legislation pursuant thereto, the following shall govern:

SECTION I. *Purpose.* The purpose of this ordinance is to establish the Agricultural Improvement Service and to define functions and relationships of government elements engaged in agricultural research and agricultural education.

SECTION II. *Agricultural Improvement Service*

a. *Establishment and Organization.* The Agricultural Improvement Service is hereby established. It shall consist of the National Agricultural Extension Service and the Agricultural Experimental Stations and Colleges hereafter mentioned.

b. *Administrative Direction.* The Agricultural Improvement Service shall be under the administrative direction of a Coordinator appointed by the Military Governor upon recommendation by the Board for Agricultural Improvement Service.

c. *Duties and Functions.* The Agricultural Service shall:

(1) Through the Central Agricultural Experiment Station and the branch experiment stations, develop and administer programs of research in agriculture and farm homemaking designed to increase agricultural production and improve conditions on farms and in farm homes.

(2) Through the Suwon Agricultural College and the Taegu Agricultural College and through assistance to other agricultural colleges and the agricultural middle schools, develop and administer programs of resident

— 1 —

teaching in agriculture and farm homemaking designed to increase agricultural production, train agricultural extension personnel, and improve conditions on farms and in farm homes.

(3) Through the National Agricultural Extension Service, develop and administer programs of agricultural extension education designed to disseminate agricultural information to farmers and farm homemakers which shall increase agricultural production and improve conditions on farms and in farm homes.

SECTION III. *Board for Agricultural Improvement Service.*

a. Establishment and Organization. The Board for the Agricultural Improvement Service is hereby established. It shall consist of six voting members and four non-voting members. The six voting members shall be: The Director of the Department of Education, the American Advisor to Director of Department of Education, the Director of the Department of Agriculture, the American Advisor to Director of the Department of Agriculture, the President of Seoul National University, and an agricultural economist from the staff of the National Economic Board. The four non-voting members shall be: The Coordinater of the Agricultural Improvement Service, the Director of the Central Agricultural Experiment Station, the Dean of the Suwon Agricultural College, and the Director of the Agricultural Extension Service. The voting members of the board shall elect a chairman and a secretary.

b. Duties and Functions. The Board shall:

(1) Establish policy governing the administration of research in all fields relating to agriculture and farm homemaking, resident instruction in agriculture and related fields, and agricultural extension education.

(2) Prepare and submit budgets for the Agricultural Improvement Service and its constituent agencies.

(3) Subject to Civil Service Rules and Regulations, appoint, compensate, suspend and remove persons on the staffs of the Agricultural Improvement Service and its constituent agencies, except the Coordinator of the Agricultural Improvement Service. The Board may by written authorization delegate to the Coordinator of the Agricultural Improvement Service such of its

compensation, appointment, suspension and removal powers as it deems appropriate.

SECTION IV. *Central Agricultural Experiment Station and Branch Stations.*

a. *Transfer of Experiment Stations.* All duties, functions, personnel, physical facilities (including lands, buildings, laboratories, laboratory equipment, office fixtures and furniture, livestock, feed and seed, farm supplies, farm equipment and tools), other property including scientific data both published and unpublished, records, publications and unexpended funds from the 1947—48 fiscal year budget allocation of the following experiment stations, are hereby transferred from the Department of Agriculture to the Agricultural Improvement Service:

(*1*) Central Agricultural Experiment Station, Suwon.
(*2*) Whasan Livestock Farm, Suwon
(*3*) Sungwhan Livestock Farm, Sungwhan.
(*4*) Kyongju Sheep Station, Kyongju
(*5*) Kaipung Sheep Station, Kaipung
(*6*) Sosa Branch Station, Sosa
(*7*) Chun Chon Branch Station, Chun Chon
(*8*) Kangnung Sub-Branch Station, Kangnung
(*9*) Taegu Branch Station, Taegu
(*10*) Taijon Branch Station, Taijon
(*11*) Iri Branch Station, Iri
(*12*) Chongju Branch Station, Chongju
(*13*) Kwangju Branch Station, Kwangju
(*14*) Chinju Branch Station, Chinju
(*15*) Mokpo Branch Station, Mokpo

SECTION V. *Suwon Agricultural College and Taegu Agricultural College.*

a. *Transfer of Colleges.* All duties, functions, personnel, unexpended funds from the 1947-1948 fiscal year budget allocations, and administrative responsibility for the operation, use and maintenance of all physical facilities including lands, buildings, laboratories, scientific and other equipment, livestock, farm equipment and tools, feed and seed, farm supplies, office fixtures and furniture), scientific data both published and unpublished, records, publications, and other property of the Suwon Agricultural College (except the

School of Veterinary Medicine) aud the Taegu Agricultural College, are hereby transferred from the Department of Education and Seoul National University to the Agricultural Improvement Service.

b. Personnel actions with respect to the teaching staff of Suwon Agricultural College shall be subject to the prior approval of the Board of Regents of Seoul National University.

c. Standards of Instruction. Entrance requirements, requirements for graduation, and the content of the curriculum of Suwon Agricultural College shall be in accordance with standards prescribed by Seoul National University.

SECTION VI. *National Agricultural Extension Service.*

a. Establishment and Organization. The National Agricultural Extension Service is hereby established. It shall consist of a Director and Assistant Director, nine (9) District Agents, and a Gun Agent and Asstistant Gun Agent in each Gun primarily agricultural. The chain of responsibility shall be from the Gun Agent to the District Agent, from the District Agent to the National Director, from the National Director to the Coordinator of the Agricultural Improvement Service.

b. Duties and Functions. The National Agricultural Extension Service shall disseminate agricultural information to the rural population of South Korea. It shall assist farmers and farm homemakers to increase agricultural production and to improve conditions on farms and in farm homes. It shall have no part whatsoever in any compulsory production or collection programs, and shall engage in no political activities whatsoever. The District Agents and Gun Agents will maintain close liaison with the provincial bureaus and the gun agencies, respectively.

c. Transfer of Agricultural Extension Activities. All duties, functions, necessary personnel, physical facilities (including lands, buildings, laboratories, laboratory equipment, livestock, farm equipment and tools, seed and feed, farm supplies, scientific equipment, office furniture and fixtures), scientific data both published and unpublished, records, publications, other property and unexpended funds in the 1947-48 fiscal year national budget allocations pertaining to agricultural extension education of the Korea Agriculture Association, the Bureau of Agricultural Economics, the Bureau of Agricultural Production, the Suwon College of Agriculture, the Taegu College of Agriculture, the Central

Agricultural Experiment Station and the Branch experiment stations, are hereby transferred from the respective agencies to the National Agricultural Extension Service.

SECTION VII. *Publication of Policy Directives.* The Coordinator of the Agricultural Improvement Service is authorized to issue policy directives, conforming with the policies established by the Board for Agricultural Improvement Service, governing the administration of the Agricultural Improvement Service and its constituent agencies.

SECTION VIII. *Other Aricultural Colleges and Schools.* National policies pertaining to agricultural instruction in primary and middle schools and pertaining to colleges of agriculture not specified in Section V hereof shall be determined by the Agricultural Improvement Service, and shall be made effective through the Provincial Bureaus of Education.

SECTION IX. *Inconsistent Laws.* All existing laws, statutes, ordinances, directives and regulations which are inconsistent with the provisions hereof, are, to the extent that they are so inconsistent, hereby repealed or amended.

SECTION X. *Effective Date.* This ordinance shall be effective on the date appearing hereon.

APPROVED:

    WILLIAM F DEAN

Major General, United States Army

Military Governor in Korea

RECOMMENDED:

    AHN CHAI HONG

Civil Administrator

SOUTH KOREAN INTERIM GOVERNMENT
Seoul, Korea

ORDINANCE
NUMBER 161   15 January 1948

REVISION OF KOREAN AUTOMOBILE REGULATION

Pending action on the subject matters hereof by the Korean Interim Legislative Assembly and the enactment of legislation pursuant thereto, the following shall govern:

SECTION I. *Purpose.* The purpose of this ordinance is to increase the fees provided for in Section 113 of Government General Ordinance Number 131, dated 29 December 1933 (*Korean Automobile Regulation*).

SECTION II. *New Fees Established.* The fees set forth in Section 113 of Government General Ordinance Number 131, dated 29 December 1933 (*Korean Automobile Regulation*), are hereby changed to read as follows:

a. Fee for certificate of registration, as provided for in paragraph 32:

    Common automobile............................80 won
    Special automobile ....80 won
    Small automobile ....40 won

b. Fee for certificate of registration, as provided for in paragraph 1 of Article 42:

    Common automobile............................40 won
    Special automobile...........................40 won
    Small automobile ....25 won

c. Fee for renewal of certificate of registration, as provided for in paragraph 1 of Article 36 or paragraph 1 of Article 40: 25 won.

d. Fee for replacement of certificate of registration: 25 won.

e. Fee for driver's license:

    Common automobile......... ....120 won
    Special automobile.........................120 won

— 1 —

407

f. Fee for driver's license for small automobile: 40 won.

g. Fee for replacement of driver's license: 25 won.

SECTION III. *Effective Date.* This ordinance shall be effective on the tenth day after the date appearing hereon.

APPROVED: RECOMMENDED:

WILLIAM F DEAN   AHN CHAI HONG
Major General, United States Army   Civil Administrator
Military Governor in Korea

軍政廳　官報　法令　第一六一號　　一九四八年一月十五日

SOUTH KOREAN INTERIM GOVERNMENT
Seoul, Korea

**ORDINANCE
NUMBER 162**

15 January 1948

PROPERTY FORMERLY UNDER CONTROL OF JAPANESE
ENEMY PROPERTY CUSTODIAN

Pending action on the subject matters hereof by the Korean Interim Legislative Assembly and the enactment of legislation pursuant thereto, the following shall govern:

SECTION I. *Purpose.* The purpose of this ordinance is to clarify the status of titles to certain real property formerly controlled by the Japanese Enemy Property Custodian and to provide a simple procedure for correct registration of such titles.

SECTION II. *Property Controlled by Japanese Enemy Property Custodian.* Registration is hereby cancelled of all real property of American, Allied Nationals or Korean juridical persons, presently entered in the name of the Japanese Enemy Property Custodian for custodial purposes, pursuant to Law Number 99, dated 23 December 1941 (*Enemy Property Custody Law*).

SECTION. III. *Application for Change of Registration.* An American, Allied National or Korean juridical person whose property is presently entered for custodial purposes, in the name of the Japanese Enemy Property Custodian may have such entry stricken from the real property records by presenting to the appropriate registrar an application, verified under oath, giving the following information:

    a. The name of the applicant.

    b. The location and description of the real property owned by the applicant, and presently registered in the name of the Japanese Enemy Property Custodian.

    c. A statement that the applicant requests, pursuant to the provisions of this ordinance, that the entry in the name of the Japanese Enemy Property

— 1 —

Custodian be stricken from the records.

SECTION IV. *Effective Date.* This ordinance shall be effective ten days from the date appearing hereon.

APPROVED:

    WILLIAM F DEAN

Major General, United States Army

Military Governor in Korea

RECOMMENDED:

    AHN CHAI HONG

Civil Administrator

SOUTH KOREAN INTERIM GOVERNMENT
Seoul, Korea

ORDINANCE
NUMBER 163   **23 January 1948**

SANITARY STANDARDS OF SEAFOOD AND SEAFOOD ESTABLISHMENTS

Pending action on the subject matters hereof by the Korean Interim Legislative Assembly and the enactment of legislation pursuant thereto, **the following shall govern:**

SECTION I. *Purpose.* The purpose of this ordinance is to provide a system of sanitary inspection to maintain the sanitation and wholesomeness of the edible seafood supply in Korea.

_SECTION II. *Definitions.* a. "Seafood" as used herein means all fish, shellfish, and any marine animal used for food, whether fresh, frozen, canned, salted, dried or otherwise processed and all food products processed in whole or substantial part from such seafood.

b. "Fresh seafood" means seafood which has not been frozen, canned, dried, or salted.

c. "Seafood establishment" as used herein means any fishing boat, building, carrier, shed, yard, or other place where seafood is handled, sold, stored or processed. It does not include a restaurant or other institutional user.

SECTION III. *Sanitary Rules and Regulations.* The Department of Public Health and Welfare may issue rules and regulations published in the Official Gazette providing for (*a*) standards of sanitation for any class of seafood or seafood establishment, (*b*) sanitary permits issued by the Provincial Bureau of Public Health and Welfare, and (*c*) enforcement of sanitary standards.

SECTION IV. *Prohibitions* On and after the effective date of any rules or regulations issued by the Department of Public Health and Welfare pursuant to Section III of this ordinance. no operator of a seafood establishment of the class regulated shall handle. sell. store or process any seafood which does not conform to sanitary standards published in such rules or regulations, or without such sanitary. permit as may be required, or otherwise in violation of such rules or regulations.

SECTION V. *Applications.* Applications for such sanitary permits shall be submitted in writing to the Provincial Bureau of Public Health and Welfare for the province in which the regulated seafood establishment is located by the operator thereof and shall set forth the following:

    a. The name and address of the owner and operator.

    b. The name and address of the seafood establishment.

    c. A statement that the applicant has read the applicable rules and regulations and is in compliance therewith.

    d. Such other information as the applicable rules and regulations may require.

    SECTION VI. *Refusal or Suspension of Sanitary Permits.* *a.* The Provincial Bureau of Public Health and Welfare, after due notice and hearing, may refuse to grant or may suspend a sanitary permit if such seafood establishment does not comply with the applicable sanitary rules or regulations. The applicant shall be notified in writing of any such refusal or suspension and may appeal to the Provincial Governor. No suspension under this section shall exceed a period of thirty (30) days.

    b. Notification of the refusal or suspension of any sanitary permit for any regulated seafood establishment licensed by the National Department of Agriculture shall be given to such Department.

    SECTION VII. *Inspections.* *a.* Each operator of a seafood establishment shall keep his premises. facilities and merchandise available for inspection by an inspector of the Provincial Bureau of Public Health and Welfare during usual business hours. Such inspectors shall make periodic inspection: and shall certify on the official form provided by the Department of Public Health and Welfare whether or not the regulated seafood establishment complies with applicable sanitary rules and regulations. One copy of such form shall be filed with the Provincial Bureau of Public Health and Welfare, and one copy shall be posted and kept posted in a conspicuous place in such seafood establishment by the operator thereof.

    b. An operator of a regulated seafood establishment shall, upon request of an inspector, deliver to him a sample of his seafood if such inspector leaves a receipt for such sample.

    SECTION VIII. *Condemnation of Unwholesome Seafood.* An inspector of the Provincial Bureau of Public Health and Welfare may condemn and cause to be destroyed any fresh seafood which does not meet the sanitary standards established by applicable rules and regulations. He shall immediately report such condemnations to the Provincial Bureau of Public Health and Welfare.

— 2 —

412

SECTION IX. *Penalties.* a. Any person who violates the provisions of this ordinance or any rule or regulation issued hereunder shall, upon conviction by a duly constituted court, suffer such punishment as the court may determine.

b. Upon conviction by a court of a violation of this ordinance or any rule or regulation issued hereunder the Provincial Bureau of Public Health and Welfare may revoke or suspend for any period any sanitary permit issued pursuant to this ordinance. The notification and appeal provisions of Section VI hereof shall apply to any such revocation or suspension.

c. In addition to the power to condemn fresh seafood provided in Section VIII hereof, the Provincial Bureau of Public Health and Welfare may seize any processed seafood not conforming to the sanitary standards published in rules or regulations issued pursuant to this ordinance. In event of a conviction by a judge of a duly constituted court for possession of such seafood, the seafood seized shall be forfeited to the Government of Korea in the manner provided by law. A permanent record of such forfeiture and of the disposition of the seafood shall be made and retained as part of the court records.

SECTION X. *Effective Date.* This ordinance shall be effective on the tenth day after the date appearing hereon.

APPROVED:

    WILLIAM F DEAN
    Major General, United States Army
    Military Governor in Korea

RECOMMENDED:

    AHN CHAI HONG
    Civil Administrator

SEAL
USAMGIK

SOUTH KOREAN INTERIM GOVERNMENT
Seoul, Korea

ORDINANCE

NUMBER 164   23 January 1948

RAW HIDES AND LEATHER CONTROL

Pending act on on the subject matters hereof by the Korean Interim Legislative Assembly and the enactment of legislation pursuant thereto. the following shall govern:

SECTION I. *Purpose.* The purpose of this ordinance is to regulate the distribution of hides and leather in the best interests of the economy of Korea.

SECTION II. *Prohibitions.* a. No slaughterer shall sell or deliver raw hides except to a local dealer licensed pursuant to Section III of this ordinance. and no person shall buy or receive raw hides on his own behalf from a slaughterer unless he is a licensed local dealer.

b. No licensed local dealer shall sell or deliver raw hides except to a national agent designated pursuant to Section IV of this ordinance. and no person shall buy or receive raw hide on his own behalf from a licensed local dealer unless he is a designated national agent.

c. No designated national agent shall sell or deliver raw hides except to a tanner who has received an allocation of such raw hides from the Department of Commerce, and no person shall buy or receive raw hides on his own behalf from a designated national agent unless he is authorized to receive such raw hides on the basis of an allocation by the Department of Commerce.

d. A tanner shall use or dispose of all raw hides he receives in accordance with the terms of his allocation from the Department of Commerce

e. No person except a designated national agent may transfer raw hides from the province in which the cattle were slaughtered. and no person except a slaughterer. veterinary inspector in the course of his official duties. licensed local dealer. designated national agent or tanner who has received an allocation from the Department of Commerce. shall sell or deliver. buy or receive. or have possession of on his own behalf any raw hides.

SECTION III. *License of local raw hide dealers.* The provincial Bureau of Agriculture. and the Agriculture Section, Bureau of Commerce. Seoul City, shall, upon application, license such local dealers as have facilities adequate to handle raw hides. At least one shall be licensed for each *gun* or *pu* in which a slaughter house is in operation.

SECTION IV. *National raw hide agents.* The Director of the Department of Agriculture shall designate one or more national raw hide agents to operate in such areas as he may decide.

SECTION V. *Records and reports. a.* Each designated national agent shall submit to the National Department of Agriculture on the 5th and 2th of each month following the effective date of this ordinance a report of his inventory of raw hides and the province from which he received them.

b. Each licensed local dealer shall submit such reports required in paragraph *a* of this section to the head of the *gun* or *pu* for which he is licensed for transmission to the Provincial Governor. Licensed local dealers in the City of Seoul shall submit such reports to the Mayor. The Provincial Governors and the Mayor of the City of Seoul shall forward such reports to the National Department of Agriculture.

c. The National Department of Agriculture may require such other reports as it deems necessary.

SECTION VI. *Definition. Raw hides* as used herein means the hides of domesticated bovine animals prior to tanning.

SECTION VII. *Prices.* Prices for sales of raw hides shall be fixed by the National Price Administration.

SECTION VIII. *Authority to allocate.* The Department of Commerce is hereby authorized to allocate in the interest of the Korean economy raw hides to tanners and to allocate leather to manufacturers, processors or industrial users.

SECTION IX. *Repeal of inconsistent laws.* All national, provincial or other laws, ordinances, orders, regulations, directives and instructions, or parts thereof, which are inconsistent herewith or in conflict with the provisions hereof are, to the extent they are inconsistent or in conflict herewith, hereby repealed.

SECTION X. *Penalties. a.* Any person who violates the provisions of this ordinance or the terms of any allocation by the Department of Commerce shall, upon conviction by a duly constituted court, suffer such punishment as the court shall determine.

— 2 —

b. Upon conviction by a court of a violation of this ordinance or the terms of any allocation by the Department of Commerce, the Provincial Governor, the Director of the Department of Agriculture or Director of Department of Commerce may revoke any license designation or allocation issued pursuant to this ordinance.

c. All raw hides or leather found to be the subject of an act or conspiracy in violation of the provisions of this ordinance shall be seized.   In the event of a conviction adjudged by a duly constituted court for such act, the goods seized shall be forfeited to the Government of Korea in the manner provided by law. A permanent record of such forfeiture and of the disposition of the goods and proceeds of the sale thereof shall be made in detail and retained as part of the court records.

SECTION XI. *Effective date.* This ordinance shall be effective on the thirtieth day after the date appearing hereon.

APPROVED: RECOMMENDED:

    WILLIAM F DEAN   AHN CHAI HONG
Major General, United States Army   Civil Administrator
Military Governor in Korea

SEAL
USAMGIK

軍政廳　官報　法令　第一六四號　　一九四八年一月二三日

SOUTH KOREAN INTERIM GOVERNMENT
Seoul, Korea.

ORDINANCE
NUMBER 165   31 January 1948

ORGANIZATION OF THE KOREA AGRICULTURAL
ASSOCIATION REVISED

Pending action on the subject matters hereof by the Korean Interim Legislative Assembly and the enactment of legislation pursuant thereto, the following shall govern:

SECTION I. *Purpose.* The purpose of this ordinance is to provide for certain administrative changes in the organization of the Korea Agricultural Association (hereinafter referred to as the Association).

SECTION II. *Relief from office.* Provincial Governors, Chiefs and Assistant Chiefs of the Bureaus of Agriculture, *gun, eup* and *myun* heads and any other provincial, *gun, eup* and *myun* government officials who are serving as officials of the Association are, upon the appointment of their successors as hereinafter set forth, relieved from their duties and responsibilities as such officials of the Association. They shall turn over to their designated successors, when notified to do so, all the records, accounts, moneys, supplies, land, buildings and other tangible assets of the Association under their jurisdiction.

SECTION III. *Appointment of Association officials and employees.* The National office of the Association shall appoint the heads of Association activities on the provincial level. Such heads shall, subject to the prior approval of the National Office of the Association, appoint and remove persons in charge of the functions of the Association on the *gun, eup* and *myun* level. The National Office of the Association is empowered for cause to remove officials and employees of the Association on the provincial level and below.

SECTION IV. *Line of authority.* The line of authority in the Association shall be from its National Office to its provincial branches, and from the provincial branches to the branches in the *guns, eups* and *myuns*, with communication to be direct, only through Association channels.

SECTION V. *Relationship with the Bureaus of Agriculture.* The provincial Bureaus of Agriculture and their personnel and the Association officials and employees

軍政廳　官報　法令　第一六五號　一九四八年一月三十一日

at the provincial level and below shall maintain close liaison with each other, so as to avoid conflicting policies and actions between the provincial Bureaus and the Association.

SECTION VI. *Continuation of duties.* Employees of the provinces, *guns*, *eups* and *myuns* who are presently exclusively engaged in handling Association matters shall, irrespective of the source of their salary funds, continue to perform such duties, but shall be responsible to the Association heads for the prompt and efficient handling of their work.

SECTION VII. *Definition. Province* or *provincial* as used herein includes the City of Seoul.

SECTION VIII. *Repeal of inconsistent laws.* All laws, ordinances, orders, regulations, directives and instructions, or parts thereof, which are inconsistent herewith or in conflict with the provisions hereof are, to the extent they are inconsistent or in conflict herewith, hereby repealed.

SECTION IX. *Effective date.* This ordinance shall be effective on the date appearing hereon.

APPROVED:

    WILLIAM F DEAN
    Major General, United States Army
    Military Governor in Korea

RECOMMENDED:

    AHN CHAI HONG
    Civil Administrator

SEAL
USAMGIK

SOUTH KOREAN INTERIM GOVERNMENT
Seoul, Korea

ORDINANCE
NUMBER 166   24 January 1948

DISCIPLINARY PUNISHMENT FOR PROSECUTORS

Pending action on the subject matters hereof by the Korean Interim Legislative Assembly and the enactment of legislation pursuant thereto, the following shall govern:

SECTION I. The purpose of this ordinance is to provide for disciplinary punishment for failure to discharge official duties for acts detrimental to official dignity of prosecutors.

SECTION II. No prosecutor shall be subject to disciplinary punishment except in accordance with the provisions of this ordinance.

SECTION III. A disciplinary punishment shall be by decision of the Committee of Disciplinary Punishment for Prosecutors (hereinafter referred as the Disciplinary Committee).

SECTION IV. Disciplinary punishments shall be of the following four categories:

   1. Reprimand

   2. Forfeiture of pay

   3. Suspension without pay

   4. Dismissal

SECTION V. A forfeiture of pay shall be not more than one-third of the monthly salary for a period of not less than one month and not more than one year.

A suspension without pay shall be for a period of not less than one month and not more than six months, during which the person suspended from office shall be prohibited from carrying out his official duties.

A dismissal, suspension, or forfeiture of pay shall be enforced, upon report thereof to the appointing authority, by the chief prosecutor of the Supreme Court, who shall attach the original text of the decision of the Disciplinary Committee.

A reprimand shall be conducted by the appropriate superior.

— 1 —

SECTION VI. No person who is punished by dismissal shall be eligible for official position for two years after his dismissal.

SECTION VII. When a case to be referred to the Disciplinary Committee is pending before a criminal court, such case shall not be referred to the Disciplinary Committee during such pendency.

When a public prosecution is lodged in a case, before the decision of the Disciplinary Committee is made in such case, the consideration thereof shall be suspended until the final judgment of the court on it.

SECTION VIII. The Disciplinary Committee shall be within the Department of Justice of the South Korean Interim Government.

It shall consist of a chairman and four other members thereof, and shall have four provisional members, to serve in the absence of regular members.

SECTION IX. The Director of the Department of Justice shall be, ex officio, the chairman of the committee.

The members and provisional members shall be appointed by the Military Governor from among the prosecutors and bureau chiefs of the Department of Justice.

The term of office of the members and provisional members shall be three years.

The tenure of a member filling a vacancy shall be the remainder of his predecessor's term.

SECTION X. The quorum of the Committee shall consist of five, including the chairman, and all business of the Committee shall be transacted by a majority of the members present and voting. The chairman shall not vote, except in case of a tie.

SECTION XI. In absence of the chairman, the senior member shall be the acting chairman.

In case of absence or vacation of office by a member, the chairman shall appoint an acting member or fill the vacancy from among the provisional members.

SECTION XII. In any of the following cases the chairman or a member of the Committee shall be disqualified from office on the Committee:

1. When he no longer holds his official position;

2. When he is transferred to a place other than where the Committee is located.

SECTION XIII. The Disciplinary Committee shall have the necessary clerks.

The clerks shall be designated by the chairman from among the clerks of the Department of Justice.

SECTION XIV. The meetings of the Disciplinary Committee shall be initiated by the order of the Military Governor, or by the request of the chief prose-cutor of the Supreme Court.

SECTION XV. No chairman or member of the Committee shall take part in a meeting of the Disciplinary Committee which is considering his case or a case of a relative in the paternal line within the eighth degree, or in the maternal line, or by marriage within the fourth degree.

SECTION XVI. Should the Committee deem it necessary, it may order the person under charges to present himself or submit a written answer to the charges to the Committee on or by a date fixed by the Committee.

SECTION XVII. Should the Committee deem it necessary it may order a member or members to collect evidence for the case.

SECTION XVIII. A member of the Committee or the designated members may examine witnesses in accordance with the principles of the provisions of the Criminal Code.

SECTION XIX. When the hearing of the case is finished, the Committee shall decide the case by vote, after taking the views of the Chief Prosecutor of the Supreme Court.

SECTION XX. The Committee shall determine the disciplinary punishment for the person in charge, taking into consideration his past behavior and prior discharge of official duty

SECTION XXI. When a case is referred to the Disciplinary Committee the Director of the Department of Justice may suspend from office the person under charges.

SECTION XXII. When a disciplinary punishment is decided upon it shall be reported to the Military Governor, together with a copy of the decision; and the original copy of the decision shall be served on the appropriate head.

— 3 —

SECTION XXIII. When a case referred to the Committee is not triable as a disciplinary case, or the charges not proven, decision to such effect shall be given to the person under charges and to his appropriate head.

Additional Rules

This ordinance shall be effective on the date appearing hereon.

This ordinance shall apply to the cases committed before as well as after the effective date of this ordinance.

APPROVED:

    WILLIAM F DEAN
Major General. United States Army
Military Governor in Korea

RECOMMENDED:

    AHN CHAI HONG
Civil Administrator

SEAL
USAMGIK

SOUTH KOREAN INTERIM GOVERNMENT
Seoul, Korea

ORDINANCE
NUMBER 168   21 February 1948

PENALTIES ON VIOLATIONS OF RATION SYSTEM OF
FOOD AND GOODS

Pending action on the subject matters hereof by the Korean Interim Legislative Assembly and the enactment of legislation pursuant thereto, the following shall govern:

SECTION I. The purpose of this ordinance is to provide penalties for violations of the ration system for food and goods.

SECTION II. The distribution of food and goods shall be made by agencies concerned in accordance with the system established by the provinces and Shi (City of Seoul). Provided that, as to the distribution of food, directives on the controlled food ration issued by the central government shall be followed.

SECTION III. No more than one ration card for food and goods shall be issued to any one person or, where applicable, to any one household.

The quota for each person shall be equal except as specially provided by the directives on the controlled food ration.

SECTION IV. Ration card for food and goods shall not be issued to a person until he has finished his registration of temporary residence in accordance with Korean Temporary Residence Act.

When a person moves from a district where his ration card was issued, he shall take the procedure to correct the residence record within ten days.

Provided that, when he moves to a place where a ration card is not necessary or when the ration card becomes invalid, he shall without delay turn in the ration card to the authorities concerned, or have his household ration card corrected accordingly.

SECTION V. When a person who has been issued a ration card is absent over one month from that place, for traveling or other reasons, he shall request the issuing authorities to suspend distribution during that period.

— 1 —

SECTION VI. Application for the ration card for food and goods shall be required to accompany the certificate of residence by the Dong (Ri) chief, Tong (Ku) chief, and Ban chief of the district where the applicant is residing.

When there is misrepresentation on the application for ration card as to the number of the household or the number of persons in a household, the person who certified their residence shall be responsible.

SECTION VII. The Dong (Ri) chief, Tong (Ku) chief and Ban chief shall pay attention to the movement of residents under their respective jurisdictions, shall examine the number of households and number of residents in the appropriate districts as of January 1, April 1, July 1, and October 1, every year, and shall report by the fifteenth day of each such month to the chiefs of Pu, Ku office, Eup and Myun.

Provided that, when they find any misrepresentation in the number of residents in a household, they will without delay make reports.

SECTION VIII. a. The chiefs of Pu, Ku office, Eup and Myun shall examine the number of households and number of residents who are receiving the distribution of food or goods as of April 1 and October 1 every year, and check with the numbers of houses on which the provincial tax and house tax are imposed, and with the record of population statistics in accordance with the Population Statistics Regulations.

b. When anyone is negligent in his duties set forth in this Section, or by willful act or negligence causes any facts within the scope of this Section to be misrepresented, he shall be punished by imprisonment for not more than one year or fine not to exceed fifty thousand won (W50,000).

SECTION IX. Any person who violates any of the following provisions shall be punished with penal servitude for not less than three months but not more than 10 years or a fine of fifty thousand won (W50,000). Provided that both may be imposed when circumstances require.

1. When one acquires or possesses a ration card by an unlawful measure.

2. When one misuses a ration card or when he uses an unlawful ration card.

3. When one issues an unlawful ration card or treats it as a proper ration card.

4. When one unlawfully allocates or distributes food or rationed goods or receives such unlawful allocation or distribution.

5. When one receives or gives or requests or offers or promises bribery in connection with the allocation or distribution of food or rationed goods.

6. Any attempt of crime described in any of the foregoing numbers.

SECTION X. When any official or person who certified the residence as prescribed in Section VI causes any facts described in the second paragraph of the same Section to occur by willful act or negligence. he shall be punished by imprisonment not more than one year or a fine of not more than twenty thousand won (W20,000).

The same shall be the case when anyone, by willful act or negligence causes to occur any misrepresentation to a report in accordance with Section VII.

SECTION XI. Anyone who violates Section IV or Section V shall be punished by a fine of not more than ten thousand won (W10,000).

SECTION XII. Anyone who does not respond to the investigation by the authorities concerned on the record concerning ration card distribution or allocation, and the facts concerning allocation or distribution shall be punished by a fine of not more than ten thousand won (W10,000).

SECTION XIII. The "ration card" mentioned in this ordinance means any paper, coupon or any other evidence which makes it possible for the possessor to receive the food or goods which are issued to the residents by the authorities distributing food or goods.

SECTION XIV. In case of violations concerning the receiving or distributing of food or goods both the violator and the head of the household to which he belongs shall be punished.

Provided. however, that if the head of the household did not know of the criminal act at the time of such commitment he shall not be punished

SECTION XV. Any laws orders or any other rules of law, or any parts thereof which are in conflict with this ordinance are hereby repealed to such extent.

— 3 —

SECTION XVI. This ordinance shall be effective on and after the date appearing hereon.

APPROVED:

WILLIAM F DEAN
Major General, United States Army
Military Governor in Korea

RECOMMENDED:

AHN CHAI HONG
Civil Administrator

SOUTH KOREAN INTERIM GOVERNMENT
Seoul, Korea

ORDINANCE
NUMBER 169

1 March 1948

REORGANIZATION OF BUREAU OF INTERNAL REVENUE

Pending action on the subject matters hereof by the Korean Interim Legislative Assembly and the enactment of legislation pursuant thereto, the following shall govern:

SECTION I. *Purpose.* The purpose of this ordinance is to reorganize and improve existing procedures covering the administration, control, and enforcement of national tax laws and collection of national taxes, and to transfer the direct supervision of national tax offices to the national government.

SECTION II. *Definitions.* As used herein:

    a. *Director* means the Director of the Department of Finance.

    b. *Commissioner* means the Commissioner of Internal Revenue.

    c. The Bureau of the Treasury of the Department of Finance shall hereafter be known as the *Bureau of Internal Revenue.*

SECTION III. *Commissioner of Internal Revenue.* There shall be in the Department of Finance a Commissioner of Internal Revenue, who shall be appointed by the Director.

SECTION IV. *Organization.* The Internal Revenue Service shall consist of the Office of the Commissioner and of the following:

BUREAU OF INTERNAL REVENUE:

    (*a*) Divisions of the Bureau of Internal Revenue:

        1. Accounts and Collections Division

        2. Direct Tax Division

        3. Indirect Tax Division

        4. Field Investigations Division:

        Branch Offices (Offices of the Chief National Tax Investigator)

    (*b*) The field service, comprised of the following offices:

    National Tax Supervision Offices. (These are regional offices. All these offices report to the Bureau of Internal Revenue.)

        Accounts and Collections Section

        Direct Tax Section

— 1 —

Indirect Tax Section

National Tax Offices.   These are district offices.   All these offices report to their respective National Tax Supervision Offices.)

Accounts and Collections Section

Direct Tax Section

Indirect Tax Section

SECTION V.   *Duties and functions of the Commissioner.*   The Commissioner shall have complete jurisdiction over the Bureau of Internal Revenue.   (Reference to *Bureau of Internal Revenue* throughout this ordinance shall include the field service.)   The Commissioner, through such officers and employees of the Bureau as he shall designate, shall have general superintendence over all matters which are assigned to or may be placed under the jurisdiction of the Bureau. In addition, the Commissioner shall have the following duties and functions:

a.   *Provincial and local taxes.*   The Commissioner, through such officers and employees as he may designate, shall prescribe and enforce standard procedures, and issue regulations for the implementation of provincial and local tax laws; conduct periodic inspection and investigation of the assessment and collection of, and the records pertaining to, provincial and local taxes; maintain records of the provincial and local taxes assessed and collected; recommend the enactment, abolition or modification of laws pertaining to provincial and local taxes; and be charged with such other powers, duties and functions pertaining to provincial and local taxes as may be assigned by law or by the Director, with the approval of the Military Governor.

b.   *Instructions, forms, etc.*   The Commissioner shall prepare and distribute all the instructions, directives, forms, blanks, stamps and the like pertaining to (1) the assessment and or collection of national taxes, (2) the enforcement of national tax laws, and (3) such other matters as are, or may be hereafter, placed under his jurisdiction.

c.   *Personnel.   1) Appointment.*   All personnel of the Bureau shall be appointed by the Commissioner; provided, however, that the Chiefs of the National Tax Supervision Offices and of the National Tax Offices and the Chief National Tax Investigators shall be appointed by the Commissioner with the approval of the Director.   The provisions of Section V of Ordinance Number 118 dated 21 August 1946 (*Establishment of Korean Interim Legislative Assembly*)

- 2 -

shall not apply to such appointments.

(2) *Transfer, detail and suspension.* The Commissioner (a) shall detail officers or employees of the Bureau for such special duty as he may desire: (b) shall make any transfers within the Bureau as he sees fit; and (c) may. with the approval of the Director, immediately suspend, pending investigation, any officer or employee of the Bureau.

d. *Field service.* The Commissioner may by departmental order, published in the Official Gazette. order the establishment, abolition or consolidation of any office or offices in the field service.

SECTION VI. *Duties and functions of the Bureau.* a. The Bureau shall be charged with all matters relating to the assessment and collection of national taxes and the enforcement of all national tax laws; provided, however, that except as otherwise prescribed by the Commissioner, with the approval of the Director. the actual collection of all national taxes, penalties, and interest shall be performed by the National Tax Offices, and no other office in the Bureau of Internal Revenue except the National Tax Offices shall have the power or authority to make actual collection of any national taxes, penalties and/or interest, unless authorized to do so by the Commissioner with the approval of the Director. The foregoing provisions of this Section notwithstanding, the existing procedures with regard to the collection of taxes, penalties and interest by or through the various *pus, eups, myuns,* post offices and other agencies shall remain in effect, unless specifically changed or modified by the Commissioner with the approval of the Director. All powers, duties and functions of the Bureau of Banking and Taxation (hereinafter abolished) relating to national taxes and properties are hereby transferred to and vested in the Bureau of Internal Revenue, to be exercised by the National Tax Supervision Offices.

b. The Bureau shall further be charged with such other matters transferred to it in this ordinance.

c. The Bureau shall further be charged with such matters as may be assigned to it from time to time.

d. All powers and authority transferred (1) from the District Tax Offices to the National Tax Offices under Section XI hereof and (2) from the provincial Bureaus of Banking and Taxation to the National Tax Supervision Offices under Section VI a hereof, may be similarly exercised by the Bureau

— 3 —

of Internal Revenue.

e. All jurisdiction, powers. duties, functions. property and personnel of the Bureau of the Treasury are hereby transferred to the Bureau of Internal Revenue.

SECTION VII. *Deputy Commissioners. a. Appointments.* There shall be in the Bureau of Internal Revenue four Deputy Commissioners, who shall be appointed by the Commissioner. with the approval of the Director.

b. Duties and functions. Each of the Deputy Commissioners, under the direction of the Commissioner. shall have general superintendence over one of the four Divisions of the Bureau and such additional duties and functions as the Commissioner, with the approval of the Director, may prescribe.

c. To act as Commissioner. The Director may designate any Deputy Commissioner to act as Commissioner during the Commissioner's absence.

SECTION VIII. *Duties and functions of the Divisions.* The four Divisions of the Bureau shall have, respectively, the following general duties and functions:

a. The Accounts and Collections Division shall be primarily charged with all matters relating to the collection and accountability of taxes and revenues, and all matters relating to budgets and the disbursement of funds.

b. The Direct Tax Division shall be primarily charged with all matters relating to the assessment and determination of national direct taxes, and all matters under the jurisdiction of the Bureau relating to national properties.

c. The Indirect Tax Division shall be primarily charged with all matters relating to the assessment and determination of national indirect taxes, including beverage taxes and licenses.

d. The Field Investigation Division, notwithstanding the foregoing provisions of this Section, shall be charged primarily with national tax investigations, and also with such other matters as may be assigned by the Commissioner.

SECTION IX. *Branch Offices of the Field Investigations Division.*

a. Organization and establishment. Branch offices of the Field Investigations Division shall be established in such cities or provinces and shall be responsible for such territories as the Commissioner, with the approval of the

— 4 —

Director. may prescribe. Each branch office shall be designated as an "Office of the Chief National Tax Investigator" and shall be headed by a Chief National Tax Investigator who shall be responsible directly to the Deputy Commissioner in charge of the Division.

b. Duties and Functions. The Branch Offices shall be charged with such duties and functions as the Deputy Commissioner in charge of the Division may assign.

SECTION X. *National Tax Supervision Offices. a. Establishment.* There are hereby established nine National Tax Supervision Offices. which shall respectively inherit the jurisdiction of the nine provincial Bureaus of Banking and Taxation (abolished hereinafter). Each of these offices shall have jurisdiction over such National Tax Offices as the Commissioner with the approval of the Director. shall prescribe.

b. Organization. Each National Tax Supervision Office shall consist of the Office of the Chief and the following three sections: (1 Accounts and Collections Section. (2) Direct Tax Section, (3) Indirect Tax Section. The duties and functions of each Section shall correspond generally to those assigned to the corresponding Divisions of the Bureau of Internal Revenue.

c. Duties and Functions. Each National Tax Supervision Office shall be charged with (1) the supervision. control and efficient operation of the National Tax Offices over which it is assigned jurisdiction. (2) such duties and functions transferred to it in this ordinance, and (3) such other duties and functions as the Commissioner may prescribe.

SECTION XI. *Bureau of Banking and Taxation abolished. a.* The Bureau of Banking and Taxation of each provincial government is hereby abolished.

b. Ordinance Number 114 amended. Ordinance Number 114 dated 23 October 1946 (*Reorganization of Provincial Governments*) is hereby amended as follows:

 (1) Section VI *c* is deleted.

 (2) Section X is repealed.

c. Transfer. Certain powers. duties and functions of the Bureau of Banking and Taxation have been transferred under Section VI *a* of this Ordinance. All other duties. powers and functions are hereby transferred to the

— 5 —

provincial Bureaus of Home Affairs. The records, property and personnel of the Financial Section and of the Local Tax Subsection of the Bureau of Banking and Taxation are hereby transferred to the provincial Bureaus of Home Affairs. All other records, property and personnel of the Bureau of Banking and Taxation are hereby transferred to the Bureau of Internal Revenue.

SECTION XII. *National Tax Offices: District Tax Offices.* The District Tax Offices shall hereinafter be known as the National Tax Offices. All existing powers, duties, functions, personnel and property of the District Tax Offices are hereby transferred to the National Tax Offices, except that the powers, duties, functions, personnel and property of the Local Tax Subsections of the District Tax Offices are hereby transferred to the provincial Bureaus of Home Affairs. The transfers made by this Section shall be effected as soon as practicable, in any event to be completed not later than 31 May 1948.

SECTION XIII. *Examination of books and witnesses.* The Commissioner, (a) for the purpose of ascertaining the correctness of any return, (b) for the purpose of making a return or assessment where none has been made, or (c) for the purpose of any tax determination, is authorized through any officer or employee of the Bureau of Internal Revenue designated by him for that purpose, (d) to enter any building or place in which any books, papers, records, or memoranda bearing upon the matters required to be included in the return, or required for the tax determination, are kept, and (e) to examine such books, papers, records or memoranda, and (f) to require the attendance of the person rendering the return or of any officer or employee of such person, or the attendance of any other person having knowledge of the foregoing, and may take his testimony with reference to any matter relating to the tax liability of any person or any matter required by law to be included in such return, or required for the tax determination.

SECTION XIV. *Penalties.* If any person shall (a) deliberately obstruct or hinder any officer or employee of the Bureau of Internal Revenue in the execution of any power or authority lawfully vested in him, or (b) forcibly rescue or attempt to rescue, or cause to be rescued, any property, articles or objects after the same shall have been seized by an officer or employer of the Bureau - such offender shall (except in cases otherwise provided for), for every such offense, forfeit and pay the sum of twenty thousand won (W 20,000) or double the value of the property concerned, or be imprisoned for a term not exceeding two years, at the discretion of the court.

SECTION XV. *Jeopardy assessments. a. Assessment and collection.* If the Commissioner or any Deputy Commissioner or the Chief of any National Tax Supervision Office or any Chief National Tax Investigator, believes that the collection of any tax imposed by any of the national tax laws will be jeopardized unless immediate assessment thereof and demand for payment be made (whether or not the time otherwise prescribed by law for making the return and paying the tax has expired), such tax, together with all penalties and interest the assessment of which is provided by law, shall be immediately assessed. Such tax, penalties and interest shall thereupon become immediately due and payable, and immediate notice and demand shall be made by the Chief, National Tax Office, for payment thereof. Upon failure or refusal to pay such tax, penalty and interest collection thereof by such means as are authorized for the collection of delinquent taxes shall be lawful. An alternative to immediate payment under this subsection *a* is provided in the following subsection *b*.

b. Deferment of payment. The collecion of the whole or any part of the amount of an assessment made under Section XV *a* hereof may be stayed until the normal due date by filing with the Chief, National Tax Office, a bond at least equal to but not more than double the amount desired to be deferred, with such sureties as the Chief deems necessary.

SECTION XVI. *Informers.* Where a person other than the taxpayer concerned, and other than an officer or employee of the Bureau of Internal Revenue, discloses to the Bureau of Internal Revenue information which is primarily instrumental in establishing a liability for national tax or a deficiency of a national tax and assessment of penalties, such informer shall receive twenty-five percent (25%) of such penalties assessed, collected and paid, but shall not receive more than fifty thousand won (W 50,000). In determining informer's reward hereunder, neither the deficiency assessment nor any interest charges attributable thereto shall be included in the basis; the reward shall be computed solely on the basis of penalties collected.

SECTION XVII. *Rules and regulations.* The Commissioner, with the approval of the Director, shall prescribe all rules and regulations deemed necessary or desirable for the enforcement of this ordinance. Such rules and regulations shall be in the form of Orders of the Office of the Commissioner and shall be published in **the Official Gazette.**

SECTION XVIII. *Inconsistent laws.* All existing laws, ordinances, orders, regulations, directives and instructions, or parts thereof, which are inconsistent or in conflict herewith are, to the extent of such inconsistency or conflict, hereby repealed; provided, however, that no taxpayer shall thereby be relieved of any liability, penalty or interest which may have been incurred prior to the effective date hereof.

SECTION XIX *Effective date.* This ordinance shall be effective on the date appearing hereon.

APPROVED:

WILLIAM F DEAN
Major General, United States Army
Military Governor in Korea

RECOMMENDED:

AHN CHAI HONG
Civil Administrator

軍政廳　官報　法令　第一六九號　　　　一九四八年　月一日

SOUTH KOREAN INTERIM GOVERNMENT
Seoul, Korea

ORDINANCE
NUMBER 170   20 March 1948

RADIO TRANSMISSION OF PRESS ITEMS

Pending action on the subject matters hereof by the Korean Interim Legislative Assembly and the enactment of legislation pursuant thereto, the following shall govern:

SECTION I. *Purpose.* The purpose of this ordinance is to provide for the licensing of radio transmission of press items.

SECTION II. *Repealer.* Governor General Ordinance Number 251, dated 21 November 1940 (*Regulation of Radio-Telegram*) is hereby repealed.

SECTION III. *Licensing procedures.* Press agencies or offices desiring to transmit or receive press items must submit a written request to the Director of the Department of Communications, (hereinafter referred to as the "Director") inclosing, in case of receiving agencies, the proposed sender's written consent, and giving the following information:

a. The purpose of the intended transmission and type of items to be transmitted.

b. Time of transmission and reception and estimated number of words to be transmitted.

c. Name and address of sender and receiver.

d. The name of telegraph office which will handle the collection of transmittal and receiving fees.

SECTION IV. *Changes after license.* Licensees who desire to change any of the items set forth in Section III. shall submit written application for such change to the Director.

SECTION V. *Changes in transmittal.* The Director may, as necessity requires, direct the change of times of transmission or reduce the number of words to be transmitted. Any agency or office desiring to cancel press transmission must give one week's notice to the Director of the Department of Communications.

SECTION VI. *Fees.* Fees for press transmission shall be those fixed by

— 1 —

the Director, by Department order.

SECTION VII. *Billing procedures.* a. Statements for the above-mentioned communications services shall be rendered by the appropriate local telegraph office not later than the 10th of the month following the period in which the services were rendered.

b. An agency or office shall make payment of its accounts by the 20th of the month in which such statement was rendered.

c. An agency or office whose account is in arrears for 30 days, or which has been in arrears in payment three times within a year, may, in the discretion of the Director. lose the privilege of transmitting or receiving radio press items.

SECTION VIII. *Language of transmission.* All items shall ordinarily be in the Korean language. Transmission of items in English may be made upon authorization of the Director.

SECTION IX. *General regulations.* a. Requests for receiving services shall be made to the Department of Communications at least seven days in advance of the period for which service is desired.

b. Advertisements or personal items shall not be transmitted.

c. Press items shall not be authorized the privileges set forth in Section 7 of Governor General Ordinance Number 93, dated 10 October 1925 (*Regulation of Telegraph*).

d. The Director may suspend the privilege of transmission for non-payment of fees or suspend or cancel for violation of radio broadcast regulations.

SECTION X. *Effective date.* This ordinance shall be effective on the tenth day after the date appearing hereon.

APPROVED: RECOMMENDED:

  WILLIAM F DEAN   **AHN CHAI HONG**
 Major General, United States Army Civil Administrator
  Military Governor in Korea

SOUTH KOREAN INTERIM GOVERNMENT
Seoul, Korea

ORDINANCE
NUMBER 171   11 March 1948

AMENDING ORDINANCE NUMBER 110; ISSUE
OF
KOREAN REVENUE STAMPS

Pending action on the subject matters hereof by the Korean Interim Legislative Assembly and the enactment of legislation pursuant thereto, the following shall govern:

SECTION I. *Purpose.* The purpose of this ordinance is to authorize the issuance of additional denominations of Korean revenue stamps.

SECTION II. *Ordinance Number 110 amended.* Section I of Ordinance Number 110, dated 14 September 1946 (*Issue of Korean Revenue Stamps; Discontinuance of Former Issues*) is hereby amended by revising the table therein to read as follows:

Denomination		Color
1	won	Orange
2	"	Purple
3	"	Lavender
5	"	Green
10	"	Dark red
50	"	Light red
100	"	Blue
500	"	Yellow

SECTION III. *Effective date.* This ordinance shall be effective on the date appearing hereon.

APPROVED: RECOMMENDED:

WILLIAM F DEAN   AHN CHAI HONG
Major General, United States Army   Civil Administrator
Military Governor in Korea

— 1 —

軍政廳　官報　法令　第一七一號　一九四八年三月一一日

437

SOUTH KOREAN INTERIM GOVERNMENT
Seoul, Korea

ORDINANCE
NUMBER 172

11 March 1948

TIME ALLOWANCE TO PRISONERS FOR GOOD BEHAVIOR

Pending action on the subject matters hereof by the Korean Interim Legislative Assembly and the enactment of legislation pursuant thereto, the following shall govern:

SECTION I. The purpose of this ordinance is to encourage repentance of prisoners. thereby making them fit for life in society, so that they can stand on their own feet.

SECTION II. Each prisoner, who has been or who is sentenced to confinement or penal servitude and whose record of conduct shows that he has faithfully observed all of the rules and has not been subjected to punishment, shall be discharged at the expiration of his term of sentence less the time deducted as follows. The provisions of Article 174 of the *Enforcing Regulation for Korean Prison Ordinance*, dated March 1912, shall apply *mutatis mutandis:*

1. Upon a sentence of not less than six months nor more than one year, five days for each month.

2. Upon a sentence of more than one year and not more than three years the deduction computed as in item *1* for the first one year, and six days for each month for the remainder of the term.

3. Upon a sentence of more than three years and not more than five years, the deduction computed as in item *2* for the first three years, and seven days for each month for the remainder of the term.

4. Upon a sentence of more than five years and not more than ten years the deduction computed as in item *3* for the first five years, and eight days for each month for the remainder of the term.

5. Upon a sentence of more than ten years, the deduction computed as in item *4* for the first ten years, and nine days for each month for the remainder of the term. Confinement or penal servitude for life shall be regarded as confinement or penal servitude for thirty years, for the purpose of this computation.

When a prisoner has two or more sentences, the aggregate of his several sentence shall be the basis upon which his deduction shall be estimated.

SECTION III.   In case any prisoner violates any existing rule or misconducts himself, the whole or a part of the deduction for good behavior which has been estimated may be forfeited.

Such forfeiture shall be decided by the Director of the Department of Justice upon recommendation of the appropriate prison warden.

The provisions of Article 173 of the *Enforcing Regulation of the Korean Prison Ordinance*, dated March 1912, shall apply *mutatis mutandis* in the case of the recommendation set forth in the preceding paragraph.

SECTION IV.   In case there exist circumstances meriting sympathy, a prisoner who has forfeited his right to a deduction for good behavior may be allowed suspension of decision as to forfeiture, for a definite period.

In case the prisoner under suspension of decision commits no violation of rules or misconduct during suspension of decision, no forfeiture shall be imposed on such prisoner.

SECTION V.   In case a prisoner who has forfeited his deduction for good behavior indicates penitence through subsequent good behavior, the whole or a portion of his forfeiture may be revoked, as may be proper.

The provisions of paragraphs 2 and 3 of Section III hereof shall apply *mutatis mutandis* in such cases.

SECTION VI.   This ordinance shall be effective on the date appearing hereon.

APPROVED: RECOMMENDED:

WILLIAM F DEAN   AHN CHAI HONG
Major General, United States Army   Civil Administrator
Military Governor in Korea

軍政廳　官報　　　　　　　　　　　　　　　　　一九四八年三月一一日

SOUTH KOREAN INTERIM GOVERNMENT
Seoul, Korea

ORDINANCE
NUMBER 173

22 March 1948

CREATION OF THE NATIONAL LAND ADMINISTRATION

Pending action on the subject matters hereof by the Korean Interim Legislative Assembly and the enactment of legislation pursuant thereto, the following shall govern:

SECTION I. *Purpose.* The purpose of this ordinance is to assist landless tenant farmers to become independent farm owners so as to strengthen the agriculture of Korea by fostering wider ownership of the land, through sale of farm land which was formerly Japanese owned.

SECTION II. *National Land Administration created.* The National Land Administration (hereinafter referred to as the "Administration") is hereby created. It shall have the authority to dispose of land transferred to it, to issue rules and regulations and to perform other functions to accomplish the purposes of this ordinance. Expenses shall be paid from the proceeds of sales and operations. It shall operate on a budget to be approved by the Military Governor.

SECTION III. *Appointment of Directors and Administrators.* a. The Administration shall have a Board of Directors of ten members, appointed by the Military Governor. The Board of Directors shall establish, consistently with the purposes of this ordinance, the operating policies of the Administration. The Board of Directors shall adopt bylaws, subject to the approval of the Military Governor, to govern its operations.

b. An Administrator and Deputy Administrator shall be appointed by the Military Governor, and they shall direct the activities of the Administration in accordance with the policies established by the Board of Directors.

SECTION IV. *Duties and responsibilities of the Administration.* The Administration shall:

a. Dispose of the farm lands transferred to it, and manage them until such disposition. It shall have the authority to dispose of any other properties transferred to it.

b. Examine into all problems connected with effectively carrying out its program and make recommendations thereon to the Military Governor.

c. With respect to land purchased from a a long term payment plan, under regulations approved by the Military Governor, make loans to farmers and farmers' cooperatives for farm production and family needs, land repair, irrigation, reclamation and improvement, processing and marketing of agricultural products and other purposes, until such time as an adequate system of farm credit is established.

d. Establish and direct subordinate agencies of the Administration and delegate duties and responsibilities to them.

e. Collect the proceeds of the claims and obligations that may become due the Administration in its carrying out of this ordinance.

f. Account for the proceeds of the sale of its lands and other property in accordance with regulations issued by the Military Governor.

g. Provide, in cooperation with other agencies of the government, supervision and direction to purchasers in their farming operations, to insure the completion of their payments and fulfillment of their contracts.

h. Perform all other acts necessary to carry out the purposes of this ordinance.

SECTION V. *Transfers to the Administration,* *a.* There is hereby transferred to the Administration title to all farm land, classified as such in registry records, ownership or control of which vested in Military Government by reason of the provisions of Ordinance Number 33, dated 6 December 1945 (*Vesting Title to Japanese Property Within Korea*). There is hereby excepted from such transfer farm land, control of which has been heretofore transferred to the government for scientific educational or experimental work. If any question arises hereunder as to whether a particular plot or piece of land has been transferred to the Administration hereunder, the matter shall be submitted to the Military Governor for his decision.

b. The assets, managed property and personnel of all government agencies performing agricultural or land management functions in connection with the land transferred to the Administration, shall, to the extent they are needed by the Administration to perform such functions, hereafter be transferred to the Administration as ordered by the Military Governor.

c. The personnel of the Administration shall hereafter become subject to the rules and regulations of the Korean Civil Service, at such time as the Military Governor may direct.

SECTION VI. *Distribution of dry fields and paddy lands.* a. Dry fields and paddy lands sold by the Administration under the provisions of this ordinance shall be distributed fairly, and insofar as possible, in economic units. No land shall be sold to a farmer which, when added to the area of land that, on 1 March 1948 or at the time land is bought from the Administration, he owned or rented from other landlords, would increase his total of dry fields and paddy land to more than two chungbo.

b. Preference shall be given to the purchasers in the following order:

(1) Present tenants of the land to be sold, who have satisfactorily cultivated their farms and complied with the terms of tenancy shall have the highest priority to buy their farms.

(2) Other farmers, farm laborers, including refugee farmers from north of the 38° parallel and repatriated farmers, living in the neighborhood and having experience in farm operation, shall have second priority. Those farmers mentioned in this paragraph who have lost the ownership of their land within the past ten years as a result of irrigation and reclamation projects recognized by the government shall be given special consideration.

c. If the Administration offers the land to the present tenant, stating the price, and he does not, within thirty days after such offer, accept in writing, the land may thereafter be sold to a person other than the present tenant, subject to the right of the present tenant to harvest his then current crop.

SECTION VII. *Listing of land sold.* The Administration shall, every three months, submit for public inspection, in the area where located, a list of farm lands sold and the names of the purchasers.

SECTION VIII. *Sale to priority-two applicants.* a. As to land not sold to the present tenant thereof, the Administration shall submit for public inspection a list of farm lands to be sold, a description thereof and the terms of sale. There shall be stated the period of time during which applications for purchase by priority-two applicants may be submitted.

b. The Administration shall set up, as needed, consultive committees at various lower levels of government, for the purpose of assisting the Administration in selecting which priority-two applicant shall be given a right of purchase. The duties and composition of each consultive committee shall be provided for in regulations issued by the Administration At least half of each local consultive committee (myun or village) shall be farmers who are tenants or who rent at least one half of the land farmed by them.

c. Any priority-two applicant who asserts that he should have been preferred above a priority-two applicant given a right of purchase by the Administration shall have the right to appeal directly to the Administrator.

SECTION IX. *Price.* A farmer who purchases dry fields and paddy land under the provisions of this ordinance shall pay in kind an amount equal to three times the annual production of the principal crop on the land he purchases. The annual production is to be determined on the basis of the land's production classification and previous production records.

SECTION X. *Payment for dry fields and paddy land.* The purchaser of dry fields and paddy land shall deliver in kind twenty percent of the annual production of the principal crop, as agreed upon by the purchaser and the Administration, for a period of fifteen years. However, payments may be made in advance of their due dates, or when justified by extraordinary conditions, payment of annual installments may be extended by the Administration.

SECTION XI. *Sale of other property.* The Administration shall, subject to the approval of the Military Governor, issue regulations setting forth policies, plans, procedures and terms of sale for all property other than dry fields and paddy land to be sold by it.

SECTION XII. *Land whose ownership is disputed.* *a.* Non-Japanese juridical persons, whose assets or controlling stock interests have not vested in Military Government under Ordinance Number 33, dated 6 December 1945, or non-Japanese natural persons, who claim ownership of land transferred to the Administration may, not later than 1 July 1948, file suit against the Administration in the district court where such land is located. Such court may, if title is found in the claimant, award the land to the claimant if it has not already been sold by the Administration. If the land has been sold, the court shall award monetary damages in lieu of the land. In no case where the land has been sold by the Administration may the judgment of the district court affect the validity of the title thereto.

The suits above referred to may be filed, even though a claim has theretofore been filed with the Property Claims Commission.

b. The Administration shall make every reasonable effort not to sell land to which its title is disputed, but all titles given by it shall be valid as against all persons who claim any adverse property rights therein.

— 4 —

SECTION XIII. *Prohibited transactions.* a. Regarding dry fields and paddy land purchased from the Administration, the following transactions are prohibited until full payment of the purchase price or ten years from date of purchase, whichever is the later, unless they are otherwise authorized by this ordinance or by other laws.

(1) Free sale donation and other kinds of disposition;

(2) Tenancy contracts and other rental contracts, including all transactions aiming at such contracts;

(3) Mortgages, superficies and establishment of preferential rights.

b. In case it is needed by any special or unavoidable reason to conduct any of the following transactions, they may be authorized by the Administration.

(1) Exchange, adjustment and annexation of dry fields and paddy land or change of farm classification;

(2) Transactions to entrust the custody of dry fields and paddy land temporarily to other persons on behalf of the owner, in cases such as illness, attending school or assuming public office.

SECTION XIV. *Special provisions on transfer.* a. In case a purchaser of dry fields and paddy land under the provisions of this ordinance dies at a time when the restrictions of Section XIII hereof still apply, an heir designated by law or by will may take over the land, provided the new owner cultivates the land.

b. In case a purchaser of land under the provisions of this ordinance or a successor to his agreement is unable to continue cultivation of the land or wishes to discontinue completion of his agreement the land may be returned to the Administration, with the owner to be compensated under the provisions of Section XVI hereof; or the land may, with the consent of the Administration, be transferred to a son or a son-in-law, provided that the total holdings of agricultural farm land of the transferee do not after transfer exceed two chungbo, and further provided that the new owner assumes the mortgage and cultivates the land.

SECTION XV. *Limitations on seizure for debt.* During the period of "Prohibited Transactions" set forth in Section XIII hereof, farm land purchased from the Administration shall be exempt from seizure by creditors except as specified herein Implements, animals, fertilizer and manure, indispensable for carrying on husbandry, shall not be subject to seizure for debt. The owner's share in the crop may serve as security for seeds, seedlings, fertilizer, tools, animals or other similar farm needs, provided that the credit is granted by a recognized financial institution. Agricultural products indispensable for husbandry of the farm until the next harvest shall not be subject to removal or levy of judgment debt.

SECTION XVI. *Foreclosure powers.* In the event that any purchaser of land from the Administration defaults in his payment or other obligations to the Administration or in payment of taxes or irrigation fees, the Administration shall have power to institute action of mortgage foreclosure or any other action warranted by the agreement of the parties. The Administration may pay past due taxes and irrigation fees, and add such payments to the principal of the mortgage. Application for mortgage foreclosure shall not be filed if the non-payment is due to an act of Providence and not until every reasonable means have been exhausted to collect the claim against the owner. In the event of foreclosure, the Administration shall pay the owner of the land seventy-five percent of the value of the agricultural products already received from him, less the expenses of foreclosure, moneys due the Administration, and taxes and irrigation fees outstanding. Unusual depletion caused by poor husbandry, neglect, or depreciation beyond ordinary wear and tear shall also be deducted by the Administration from the amount to be repaid. The reasonable value of unexhausted improvements made by the owner subsequent to purchase will be adjudicated and paid him separately.

SECTION XVII. *Re-sale of land.* All land sold by the Administration, and thereafter reconveyed to it, either voluntarily or through foreclosure, shall be resold by the Administration under the provisions of this ordinance.

SECTION XVIII. *Encumbrances to be cleared.* Except for irrigation charges maturing after title passes to the new owner, title in all land to be sold by the Administration in accordance with the provisions of this ordinance shall be given free of all debts and encumbrances, other than a purchase money mortgage to the Administration. Where debts and encumbrances cannot be cleared prior to transfer of title, the Administration shall remove them thereafter as promptly as possible. There shall be no power, except by the Administration, to foreclose for failure to pay irrigation charges.

SECTION XIX. *Registration of sales. a.* A deed given by the Administration shall be registered by the officials charged with such duty, notwithstanding that land to which a deed is given does not appear on the land registry records in the name of the Administration.

b. Registration of land by a buyer from the Administration shall serve as notice that such land is subject to the restrictions set forth in Section XIII hereof.

c. Registrars shall effect registration of mortgages to the Administration, even though payable in crops and not in money.

SECTION XX. *Registration fees and procedures. a.* No transfer tax shall be collected on sales of land by the Administration or on mortgages to the Administration.

b. No fees or other payments, whether by revenue stamp or otherwise, shall be collected from the Administration or purchaser therefrom for the registration of deed or mortgage on property sold by the Administration.

c. When the Administration commits for registration to the competent registration office a deed given by it or a mortgage to it, no certificate of seal impression shall be required, and the personal appearance of the buyer of the land and mortgagor shall not be required before the registrar. The registrar shall record such deeds and mortgages without delay.

d. When land transferred to the Administration pursuant to this ordinance is registered under the Korean Real Property Certificate Order, and the Administration sells such land, the initial registration of such land under the Korean Real Property Registration Order shall be in the name of the purchaser from the Administration. The application for registration and personal appearance of the owner appearing of record under the Korean Real Property Certificate Order shall not be required. Upon registration of land, pursuant hereto, under the Korean Real Property Registration Order, the former registration under the Korean Real Property Certificate Order shall become void.

SECTION XXI. *Penalties.* Those who intentionally violate Section XIII of this ordinance, with a view to circumventing the provisions thereof, shall, upon conviction, be punished with penal servitude not to exceed two years or a fine not to exceed 20,000 won.

SECTION XXII. *Violations by juridical persons.* In case a legal representative of a juridical person or an agent or employee of a juridical or natural person, with the knowledge and approval of said person, violates Section XIII hereof in the performance of the business of the juridical or natural person, not only the direct violator shall be punished but also the juridical person shall be fined as provided in Section XXI.

SECTION XXIII. *Inconsistent laws repealed.* All laws, ordinances, orders, regulations, directives or instructions, or parts thereof, which are inconsistent herewith, or in conflict with the provisions hereof, to the extent that they are in conflict or inconsistent with this ordinance, are hereby repealed.

SECTION XXIV. *Effective date.* This ordinance shall be effective on the date appearing hereon.

APPROVED:

    WILLIAM F DEAN
  Major General, United States Army
   Military Governor in Korea

RECOMMENDED:

   AHN CHAI HONG
   Civil Administrator

SOUTH KOREAN INTERIM GOVERNMENT
Seoul, Korea

ORDINANCE

NUMBER 174   22 March 1948

DISSOLUTION OF THE NEW KOREA COMPANY, LIMITED

Pending action on the subject matters hereof by the Korean Interim Legislative Assembly and the enactment of legislation pursuant thereto, the following shall govern:-

SECTION I. *Purpose.* The purpose of this ordinance is to dissolve the New Korea Company, Limited and provide for its liquidation.

SECTION II. *Dissolution.* The New Korea Company, Limited, created by Ordinance No. 52, dated 21 February 1946 (*Creation of the New Korea Company, limited*), as amended by Ordinance No. 80, dated 7 May 1946, is hereby dissolved, and the Office of Property Custody is appointed its liquidator.

SECTION III. *Transfer to Office of Property Custody.* The following are hereby transferred from the New Korea Company, Limited to the Office of Property Custody:

a. All property other than the property transferred to the National Land Administration by the provisions of Ordinance Number 173, dated 22 March 1948 (*Creation of the National Land Administration*).

b. All records connected with the property transferred above.

c. All funds of the company.

SECTION IV. *Effective date.* This ordinance shall be effective on the date appearing hereon.

APPROVED: RECOMMENDED:

WILLIAM F DEAN   AHN CHAI HONG
Major General, United States Army   Civil Administrator
Military Governor in Korea

448

SOUTH KOREAN INTERIM GOVERNMENT
Seoul, Kor

ORDINANCE
NUMBER 175   17 March 1948

LAW FOR THE ELECTION OF REPRESENTATIVES OF THE KOREAN PEOPLE

Pending action on the subject matters hereof by the Korean Interim Legislative Assembly and the enactment of legislation pursuant thereto, the following shall govern:

CHAPTER I. GENERAL RULES

Section 1. Any citizen twenty-one or more years of age has the right to vote for representative to the National Assembly regardless of sex, property, education or religion.

Any citizen twenty-five or more years of age has the right to be elected as a representative to the National Assembly regardless of sex, property, education or religion.

Age shall be counted as of the date the election is held.

Section 2. Persons to whom any of the following categories are applicable are not eligible to vote:

(1) persons who have been declared incompetent by a decision of a court of justice;

(2) weakminded persons who have been placed as quasi-incompetent under guardianship by a decision of a court of justice;

(3) persons who are serving prison sentences, or who are under suspended sentence or under a sentence not yet executed;

(4) persons who accepted peerages from the Japanese government;

(5) persons who were members of the Japanese Imperial Diet.

Section 3. A person is not eligible to be elected:

(1) if he is ineligible to vote in accordance with Section 2 of this law; provided, however, Section 2, Number (3) does not apply if the sentence was imposed for a political offense;

(2) persons who have received a sentence of one or more years penal servitude or imprisonment; provided, however, that if three or more years have elapsed since the completion of sentence, or since the time when final

— 1 —

decision was made not to execute the sentence or if the sentence was imposed for a political offense, the person concerned will not be included in this category;

(*3*) persons who under the Japanese regime held ranks of "HANNINKAN" or higher in the Civilian Police Force, or who served as "KEMPEI" or "KEMPEI-HO" in the Japanese Military Police, Force, or persons who held positions in the police in charge of "thought control," or those who acted as a spy for the police in charge of "thought control" under the Japanese regime:

(*4*) persons who were advisors, members, or vice chairman of the Central Advisory Council under the Japanese regime;

(*5*) persons who were members of an advisory or deciding council of PU or DO (Province) under the Japanese regime;

(*6*) persons who held positions of the third class or higher of the "KOTOKAN" or who received a medal (KUN) of the seventh class or higher; provided, however, that educators and technical officials are not included in this category.

Section 4   Government officials shall not, while in office, hold membership in the National Assembly in addition to such position except *Jungmukwan*

Section 5.   The members of local self-governing bodies shall not, while in office, hold membership in the National Assembly in addition to such position.

Section 6.   The members of election committees shall not be eligible to be elected representatives to the National Assembly within the districts concerned.

Section 7.   Expenses incurred in the election of National Assembly members shall be defrayed by the national treasury.

CHAPTER II. ELECTORAL DISTRICTS AND REPRESENTATION

Section 8.   Each electoral district shall be represented by one delegate.

Section 9.   Electoral districts in the meaning of this law are:

(1) GUN, PU and KU of the City of Seoul; provided that their population does not exceed 150,000.

(2) Those sectors of GUN, PU and KU of the City of Seoul with a population exceeding 150,000 which for the purpose of this election have been established by the Chief Executive as electoral districts in accordance with Section 10 of this law.

(3) DO (island) which form administrative units on the level of GUN

and PU. NOTE: wherever the term "GUN" is used in the following sections of this law, it shall include DO (island) in the meaning of Section *9 (3)* of this law

Section 10. GUN, PU and KU of the City of Seoul with a population between 150.000 and 250,000 will be divided into two sectors of approximately equal population; PU with a population between 250,000 and 350,000 will be divided into three sectors of approximately equal population; PU with a population between 350.000 and 450,000 will be divided into four sectors of approximately equal population.

Section 11. Names, boundaries and population of the electoral districts established by the Chief Executive in accordance with Section *10* of this law are set forth in Appendix I which forms a part of this law.

Section 12, Each electoral district shall be divided into voting districts; each voting district shall have a population of not more than 2,000. Each voting district shall form a part of one EUP, MYUN or DONG respectively.

Voting districts shall be established by the electoral district election committees on the basis of detailed recommendations to be prepared by the chiefs of the administrative units (GUN, PU and KU of the City of Seoul) concerned.

The establishment of voting districts shall be published before the beginning of voters registration

Section 13. The National Election Committee is authorized to order the establishment of election committees on the EUP, MYUN and DONG level wherever such election committees are deemed necessary to assist, instruct, and supervise the election committees of the voting districts.

Section 14. Electoral districts and voting districts will be established on the basis of the census of 24 August 1946.

CHAPTER III. POLL REGISTERS

Section 15. Registration shall be accomplished by signing the registration paper, or by making thumbprint upon the registration paper before two witnesses who are literate, in the registration place designated by the chairman of the electoral district election committee, sometime during a period of 10 days beginning 40 days before the date of election.

Each voter shall register at only one registration place.

Section 16. The election committee of the voting district shall prepare poll registers indicating the name, address, date of birth, sex, date of registration, and other pertinent data with respect to all voters who heve been in residence in that district since the date sixty days prior to the date of the election concerned. The regiters shall be open to the public at the registration places for a period of 7 days beginning 25 days prior to the date of the election. The poll registers shall become final on the 2nd day before the date of election.

Section 17. When there are incorrect recordings or omissions, or when a person who has no right to vote is registered in the poll register, each voter may object to the election committee concerned during the Period of public notification.

Examination and decision in regard to these objections shall be made within 3 days by the election committee concerned.

In case objections to the foregoing decision be raised, the voters may request, within 5 days, the electoral district election committee to review this decision.

The election committee of the electoral district shall on a request for review in accordance with the preceding paragraph review and decide each case within 3 days and shall notify in writing the person who asked for review and other persons concerned.

CHAPTER IV. ELECTION COMMITTEES

Section 18. The National Election Committee organized by the Chief Executive in accordance with Public Act No. 5, Section 13, shall be constituted as National Election Committee.

The National Election Committee consists of 15 members and is under the chairmanship of a Justice of the Supreme Court.

Section 19. The National Election Committee shall organize election committees in each province and in the City of Seoul and appoint its members. Each provincial election committee and the City of Seoul election committee shall consist of one chairman, ten members and an alternate chairman and ten alternate members. The chairman and the alternate chairman shall be judges of an appellate or district court located in the province (City of Seoul) concerned.

The Chief Judge of the District Court in each province and the City of Seoul shall recommend to the National Election Committee the names of three judges who are residents of such provinces (City of Seoul). The National Elec-

— 4 —

tion Committee shall select and appoint from among the judges recommended the chairman of each provincial election committee and his alternate (City of Seoul Committee).

The Chief Judge of the District Court in each Province (City of Seoul) and the governor of each province (Mayor, City of Seoul)—each acting inpependently— shall submit to the National Election committee a list of twenty respectable citizens who are residents of such province (City of Seoul) and indicate the party affiliation – if any – of each of them. Not more than one-third of the persons recommended on each list shall belong to the same political party. The National Election Committee shall select from the persons recommended 20 persons and appoint 10 persons members of the provincial (City of Seoul) election committee and 10 persons their alternates; provided, however, that not more than one-third of the members and not more than one-third of the alternates of each provincial (City of Seoul) election committee shall belong to the same political party

Section 20. There shall be established in each electoral district an electoral district election committee. Each electoral district election committee shall consist of one chairman, 8 members, one alternate chairman and 8 alternate members The chairman and the alternate chairman of each electoral district election committee shall be appointed by the Chief Judge of the District Court in the province (City of Seoul) to which the electoral district belongs.

The highest ranking Judge of the District Court of the Branch Court which has jurisdiction over, and the highest administrative official of the administrative district (GUN PU, KU of the City of Seoul) to which such electoral district belongs shall each appoint four residents of such a district, members, and four residents of such a district, alternates, of the electoral district election committee concerned. Not more than one-third of the members and one-third of the alternates of each electoral district election committee shall belong to the same political party.

Section 21 The administrative head of each GUN, PU and KU of the City of Seoul shal without delay submit to the provincial (City of Seoul) election committee concerned and the National Election Committee the names of all persons within his district who have been appointed chairman, alternate chairman, member and alternate member of an electoral district election committee and indicate their party affiliation if any and also give a brief personal history of each.

The provincial (City of Seoul) election committee concerned shall without delay review such appointments and within ten days from the day of appointment may at its own discretion decide to recall the chairman, alternate chairman, members and alternate members of such electoral district election committees and to replace them by appointees of its own choice. All appointments not recalled within ten days shall be final.

Section 22. There shall be established in each voting district a voting district election committee.

Each voting district election committee shall consist of one chairman, one alternate chairman, eight members and eight alternates. The chairman, alternate chairman, the members and the alternate members of each voting district election committee shall be appointed by the head of the administrative district (EUP, MYUN, DONG) to which the voting district belongs. Not more than one-third of the members and not more than one-third of the alternates shall belong to the same political party.

The administrative head of each EUP, MYUN and DONG shall without delay submit to the electoral district election committee concerned, to the provincial (City of Seoul) election committee concerned and to the National Election committee the names of all persons within his district who have been appointed chairman, alternate chairman, member or alternate member of a voting district election committee and indicate their party affiliation, if any, and also give a brief personal history of each.

The electoral district election committee concerned shall without delay review such appointments and within ten days from the day of appointment may at its own discretion, decide to recall the chairman, alternate chairman, members and alternate members of such voting district election committees and to replace them by appointees of their own choice. All appointments not recalled within ten days shall become final.

In case the head of a MYUN, EUP, or DONG fails to appoint within reasonable time one or several election officials (chairman, alternate chairman, member or alternate member) of a voting district election committee, the election committee of the electoral district to which the voting district belongs shall appoint without delay such election officials.

Section 23. Section 22 applies *mutatis mutandis* to the establishment of election committees on the MYUN, EUP and DONG level wherever such election committees have been organized in acc dance with Section *13* of this law.

Section 24. A candidate for election is not eligible for appointment to the election committee in the electoral district in which he is a candidate.

Each type of election committee may employ several secretaries and/or clerks.

Section 25. Each election committee shall act in accordance with laws and regulations and in accordance with the orders and instructions of *higher* election committees and shall render various reports on its activities to the higher election committees and shall supervise the activities of the lower election committees

Upon request, each election committee shall show its various records and documents to a court

Any chairman, alternate chairman, member or alternate member of any election committee, other than the National Election Committee, who fails to act in accordance with laws and regulations and in accordance with the orders and instructions of higher election committees may be removed by the next higher election committee, which shall replace any person so removed with a person of its own choice. Any chairman or member of the National Election Committee who fails to act in accordance with the laws and regulations may be removed and replaced by the Chief Executive.

Section 26. The quorum of each election committee shall be a majority of its members, and all the actions of the committee shall be determined by majority vote of the members present. In case of a tie in voting, the chairman of the committee shall decide.

CHAPTER V. CANDIDATES AND ELECTION CAMPAIGNS

Section 27. A person who wishes to stand for election to the National Assembly shall register with the election committee of the electoral district sometime between the date of the beginning of voters registration and a date to be determined by the National Election Committee and submit at the same time, a recommendation signed by two hundred or more registered voters.

In case of recommendation of one person by another to be a candidate, the person making the recommendation must secure the signature of two hundred or more registered voters and the written consent of the person being

軍政廳　官報　法令　第一七五號　　一九四八年三月一七日

recommended and register with the election committee of the electoral district

In case an electoral district election committee reaches the conclusion that a person who applied for registration as a candidate is ineligible to be elected for one of the reasons set forth in section 3 of this law, it shall without delay refer the case to the National Election Committee with a report stating the reasons. Simultaneously, the electoral district election committee shall transmit to the applicant a copy of this report.

The National Election Committee shall decide the case referred to above as a matter of first priority. The National Election Committee shall hear the applicant on his request at a public meeting. The applicant may introduce documentary evidence and summon witnesses.

The chairman of the National Election Committee shall notify the Chief Executive, the chairman of the electoral district election committee and the applicant of its decision.

The registration of a candidate who is registered in two or more electoral districts shall be invalid.

Section 28. Election committees of electoral districts shall give public notice of the name, address, age, occupation, and party or organization affiliation, if any, of every candidate within three days after each candidate's registration by publishing his name in newspapers, announcing it over the radio if possible, and by message to all voting district election committees which form part of the electoral district concerned.

Election committees of electoral districts shall give public notice immediately of the withdrawal or death of any candidate in the same manner.

Section 29. Registered candidates may freely conduct campaigns for election

Members of each election committee, public officials connected with election matters and any other public official may not take part in campaigns.

CHAPTER VI. ELECTION PROCEDURE AND
SUCCESSFUL CANDIDATES

Section 30. The election of representatives to the National Assembly shall take place on the same day throughout the area in which this law is applicable.

The election committees of the electoral districts shall make official announcements of the date time of the election and the location of the polling places

40 days before election by publishing such facts in newspapers, posters in each voting district, and announcement, if possible, over the radio.

Polling places shall be the same places in which the voters register; however, when any polling place must be changed on account of a calamity or other unavoidable emergency, the electoral district election committee concerned shall immediately give public notice in the same manner as prescribed in the preceding paragraph.

Section 31. Balloting shall begin at 7:00 A. M. The polling places shall be closed at 7:00 P. M; provided however, that voters waiting for admission to the polling place at that time shall be admitted to the polling place until 8:00 P. M.

Voters who were present at the voting place at 7:00 P. M. and voters admitted to the polling place until 8:00 P. M., in accordance with the preceding paragraph shall be entitled to vote.

The ballot boxes shall be closed after the last voter entitled to vote has cast his ballots. No ballots shall be cast after the boxes are closed.

Section 32, The voting shall be accomplished by the casting of secret, unsigned ballots each of which shall designate one candidate only.

Section 33 Only uniform ballots to be prepared by each electoral district election committee and to be printed under the supervision of the National Election Committee, and uniform envelopes to be prepared by and to be printed and officially stamped under the supervision of the National Election Committee, shall be used at each polling place.

The ballots to be used at each polling place must conform to a standard form to be prepared by the National Election Committee. Each ballot shall consist of the printed names of the duly nominated candidates of the electoral district for which the ballot is prepared, a symbol before the name of each candidate and a suitable space after the name of each candidate in which the voter may indicate by a mark the candidate of his choice. The names of the candidates shall be printed in Korean, as well as in Chinese letters. The ballots and envelopes shall not be numbered or otherwise marked.

The electoral district election committee shall determine by lot in a meeting to be held not later than 2 days after the time registration of candidates has elapsed at which the candidates and their representatives may be present, the order of the names of the candidates on the ballot and shall determine in the same meeting the symbols assigned to each candidate. Such symbols shall be for instance, *one stroke, two strokes, three strokes,* etc.

Section 24. Upon presenting himself at the polling place and upon identification as a registered voter, the voter shall sign or thumb-print the poll register in the presence of at least one member of the voting district election committee.

In case of doubt as to the identity of any voter, the election committee of the voting district shall decide the question. In such case, the head of the DONG or PAN in which the voter resides may be called upon as a witness. Persons whose names are not registered on the poll register connot vote; provided, however, that a person may vote who brings with him the notification of a decision rendered by the election committee of the electoral district or voting district certify ng that the person shall be registered on the poll register.

The chairman of the voting district election committee shall hand to each voter one envelope and one ballot after the chairman has put his seal on the ballot form in the presence of the voter.

In a separate room provided for this purpose, in which the voter shall be alone, he shall mark the ballot and put the marked ballot into the envelope. The voter shall then, in front of the chairman and members of the election committee, place the envelope containing the marked ballot into the ballot box. In the case of a voter who spoiled a ballot form, the chairman shall hand to such voter only one other ballot form; provided that the spoiled ballot form is returned to the chairman.

A blind voter may be accompanied to the ballot room by a member of his family or another person of his choice who may assist him in marking the ballot and putting the ballot into the envelope. The chairman may invite one of the members of the election committee to be present during the marking of the ballot.

In the absence of the chairman, his alternate or a member of the election committee specifically designated by the chairman for this purpose shall perform the functions as assigned to the chairman of the voting district election committee in this section.

Section 35. Each electoral district election committee and each voting district election committee shall give suitable publicity to the official ballot to be used in such electoral district for the general information of the voters by

軍政廳　官報　法令　第一七五號　　一九四八年三月一七日

reproducing it on posters and in newspapers generally.

At the entrance of each polling place and in each room where the marking of the ballot takes place, photographs of the candidates, to be furnished by such candidates, shall be placed in the same order as the candidates are placed on the official ballot. The photograph of each candidate shall bear in Korean, as well as in Chinese letters, the candidate's name and the symbol assigned to him.

The secrecy of voting is guaranteed. The voter shall be under no obligation to disclose the identity of the candidate for whom he has voted.

No legislative, executive or administrative agency and no court shall ever question a voter as to the candidate for whom he has voted.

Section 36 One representative of each candidate duly nominated in the electoral district to which such voting district belongs shall be authorized to stay in the polling place during the whole time of voting; provided that the candidate submitted three days before the date of voting, the name of such representative, to the chairman of the voting election committee concerned. Not more than five representatives of duly nominated candidates shall be admitted to each polling place. In case that more than five duly nominated candidates submitted the names of representatives to the chairman of a voting district election committee, such chairman shall in a special meeting, determine by lot the names of the five representatives who are admitted to the polling place. Duly admitted representatives of candidates may watch the election procedure at the place designated by the voting district election committee; they are not authorized to interfere in any way with the election procedure, render speeches or otherwise take any steps which may influence the voters.

Section 37. Voters shall stay in the polling place no longer than is required to perform the acts prescribed in Section 34 of this law. They shall leave the polling place immediately after they cast the envelope in the ballot box. No voter and no duly admitted representative of candidate shall carry a weapon while on the premises of the polling place.

Section 38. Except when entering the polling premises as voters, members of the police force shall not be authorized to enter the polling place unless they are called by the chairman of the voting district election committee to preserve the order of the polling place. When preserving the order of the polling place, the police shall act under the direction of the chairman of the voting district

— 11 —

electio committee concerned and shall leave the polling place immediately after bei g dismissed by the chairman.

Section 39. The chairman of the voting district election committee concerned may restrain speech, discussion, electioneering, clamor at the polling place, the waiting hall and the area immediately adjacent to the polling place. He may expel from the polling place any person who disobeys such restraining orders.

A voter who has been expelled from the voting place in accordance with the preceding paragraph, may vote at the end of the voting day. However, when the chairman of the voting district election committee concerned finds that there is no longer any danger of disturbance of the polling place, he may, at his discretion, admit such person to the polling place.

Section 40. The ballot boxes shall be locked after the last voter who is entitled to vote has cast his vote.

The voting district election committees shall convey the ballot boxes and the records to the election committee of the electoral district to which such voting district belongs. The voting district election committee shall proceed with conveying the ballot boxes and records immediately after the voting has been closed.

Section 41. Immediately after the election committee of a voting district has conveyed its ballot box to the headquarters of the election committee of the electoral district, the chairman of the election committee of the electoral district shall take the ballot box in his custody. The chairman of the election committee of the electoral district shall order that the ballot boxes shall be opened and shall ascertain whether the number of ballots counted corresponds with the number of registered voters who received ballots according to the poll records.

The candidates or their representatives may be present at the opening of the ballot boxes. Applications for presence at the opening of the ballot boxes shall be made at least two days before the date of opening.

At the time of the opening of ballot boxes, the chairman of the election committee of the electoral district shall announce the opening. In the presence of more than half of the members of the election committee of the electoral district, the chairman of the election committee of the electoral district shall clearly show that the ballot boxes are closed and sealed; then he shall open the boxes and examine the ballots.

Section 42. The following types of ballots shall be null and void:

(*1*) Ballot other than the regular official ballot;

(*2*) Ballot on which the name of no candidate is marked;

(*3*) Ballot on which the names of more than one candidate are marked;

(*4*) Ballot on which it cannot be told, which name is marked;

(*5*) Ballot on which the voter records other matters than a mark behind the name of the candidate of his choice;

(*6*) If any envelope shall be found containing two or more ballots, both of them shall be null and void;

(7) Ballot not placed in an official envelope.

Section 43. The candidate who receives the greatest number of valid votes shall be elected. In case of tie, the chairman of the electoral district election committee concerned shall decide the winner by lot, at a public meeting at which the candidates or their representatives may be present.

In case only one person was duly registered as a candidate in an electoral district, such candidate will be elected automatically, without election vote.

After the winner has been determined, the election committee of the electoral district shall, without delay, inform the successful candidate of his election, and give public notice of the results of the election.

Section 44. Within a period of two weeks after the date of the election, the Military Governor may, after consultation with the National Election Committee, declare the election in any of the electoral districts null and void if:

(*1*) the election could not be held in all voting districts of such an electoral district on account of a calamity or violence; or

(*2*) If the ballot boxes of one or more voting districts of such an electoral district were opened illegally or were lost; or

(*3*) If there is clear evidence that the election results within such electoral district have been substantially affected by irregularities, fraud or improper actions of election officials.

Any group of voters has the right to bring to the attention of the National Election Committee and the Military Governor all facts prescribed in paragraph *1* of this section.

If the Military Governor, after consultation with the National Election Committee, decided to nullify the elections in any electoral district for one of the reasons set forth in this section or in the event any election in any electoral district becomes null and void, he shall order that a new election take place

軍政廳　官報　法令　第一七五號　　一九四八年三月一七日

within a period of not more than 30 days.

Section 45. After the voting has been closed, the election committees of the voting districts shall deliver the poll register and other documents pertaining to the election to the heads of MYUN, EUP, PU and KU of the City of Seoul to which such voting districts belong. The heads of MYUN, EUP, PU and KU of the City of Seoul shall have those records in their safe-keeping during the term of the elected representatives of the National Assembly.

Section 46. After completion of the ballot tabulation, the election committees of the electoral districts shall deliver their election records without delay to the appropriate provincial or Seoul City election committee.

The election committees of the electoral districts shall separate and identify valid and invalid ballots after the completion of the election, and shall deliver them together with all pertinent documents, to the heads of PU, GUN or KU of the City of Seoul, who shall keep them during the term of the elected representative of the National Assembly.

As soon as the provincial and Seoul City election committees receive all the election records from their respective electoral districts, they shall submit the reports on the election without delay to the National Election Committee.

CHAPTER VII. TERM OF OFFICE AND BY-ELECTIONS

Section 47. The term of office of members of the National Assembly is hereby limited to two years from its opening session, unless a general dissolution of the National Assembly is decided upon before such time, by a competent authority.

Section 48. When a vacancy occurs, a by-election shall be held to fill it.

The Chief Executive shall hold the by-election within seventy days after receipt of notification from the chairman of the Assembly that the vacancy exists.

The date of the by-election shall be announced at least fifty days before it is to be held.

If the by-election is held within six months after general election day, the poll registers which were used for the general election shall be used for the by-election also.

Section 49. Besides the regulations stated in this chapter all provisions of this law are applicable to by-elections.

CHAPTER VIII. LITIGATION CONCERNING ELECTIONS

Section 50.   Without the prejudice to the powers of the Military Governor acting in consultation with the National Election Committee to invalidate an election in accordance with Section 44 of this law, questions relating to the validity of election shall be submitted to an Election Review Board of five members: two justices of the Supreme Court, appointed by the Chief Executive; two members elected by the National Assembly; with a chairman designated by the Chief Justice of the Supreme Court.

Section 51.   Any defeated candidate may raise the question of the validity of an election before the Election Review Board within fourteen days after the Election Review Board has been established.

The Election Review Board shall declare the election void if such election is found by the Election Review Board to involve violation of the provisions of this law and the provisions of the regulations promulgated in pursuance thereof and such violations may have caused a change in the result of the election.

In addition to the regulations contained in this chapter, pertinent provisions of ordinary civil procedure shall be applicable to election cases.

Section 52..   The chairman of the Election Review Board shall inform the Chief Executive, National Election Committee and the electoral district election committees concerned of the election cases filed with it.   The chairman of the Election Review Board shall send copies of its decisions to the Chief Executive, the National Election Committee, the electoral district election committees concerned and the chairman of the National Assembly.

CHAPTER IX. PENAL REGULATIONS

Section 53.   Any of the following violators of the law shall be punished by penal servitude for not more than five years, or a fine, of not more than 100.000 won provided that both penal servitude and fine may be imposed in case that special circumstances require it:

1) Any person who either registers on the poll register or votes by fraudulent means;

(2) Any person who gives or receives or promises to give or receive money, goods, entertainment, or any other property gain, or gives or promises to give position of honor on favorable terms for the promise of votes or the abandonment of right to vote as a condition;

(3) Any person who tries to prevent anyone from voting or being a candidate, or forces anyone not to vote by the use of violence, threats, arrest or confinement, or any other method;

(4) Any head of a DONG or PAN or any other person who deliberately makes false statements when heard as a witness in accordance with Section 34 of this law;

(5) Any person who, for the purpose of hindering the election, uses violence or threats against the election committee members or public officials or captures or destroys the ballot boxes or the election records;

(6) Any person who interferes with the free exercise of the right to vote or with the election process in general, by mass disturbances or demonstrations at or near the polling place;

(7) Any person who forces entrance into the polling place carrying firearms, a sword, cudgel, or any other weapon;

(8) Any member of an election committee or public official who violates any laws or regulations pertaining to the election.

Section 54. Any person who has been punished for the crimes described in this chapter shall be deprived of the rights to vote and to be elected for a period of three years from the date on which he completes the serving of his sentence.

Section 55. The statute of limitations for public prosecution with respect to the offenses described in this chapter shall be one year.

Supplementary Rules.

Section 56. The Chief Executive may enact such detailed regulations as may be necessary to carry out this law.

Section 57. This law shall be effective on the 17th day of March 1948.

WILLIAM. F. DEAN
Major General, United States Army
Military Governor in Korea

Name of Province and Guns	Population	Sub-division Figures		Names of Koos and Dongs, Myuns and Up's comprising Sub-division Group I
		Group I	Group II	

Seoul City

Chong No Koo 199,028 101,441 97,587

I. Chong Un, Shin Kyo, Kung Jong, Hyo Ja, Ru Sang, Ru Ha, Ok In, Tong In, Chang Sung, Pil Un, Sa Juk, ChuBu, Ton Ii, Ne Ja, Chuk Sun, Ne Su, Do Rium, Kang Ju, Suode Moon 1, Suode Moon 2, Se Jong Ro, Chong Ro 1, Chong Ro 2, Su Rin, Chung Jin, Su Son, Chung Hak, Kwal Hoon, In Sa, Kyun Ji, Cong Pyung, Kwan Chul, Song Hyun, Sa Ryun, An Kuk, So Kak, Hwa, Pal Pan, Sam Chung, Chai, Ka Hai,

II. Won Suh, Kwon Nong, Wa Ryong, Wun Ni, Soo Un, Dong Eui, Kyung Woon, Ik Sun, Rak Won, Chong Ro 3, Kwan Soo, Chang Sa, Chong Ro 4, Hook Chung, Bong Ik, Yae Chi, In Eui, Won Nam, Yun Chi, Chong Ro 5, Chong Ro 6, Choon Shin, Hyo Jae, Dong Sung, Ih Wha, Hoi Wha, Myung Yoon 1, Myung Yoon 2, Myung Yoon 3, Myung Yoon 4, Yun Koon, Kae Dong,

Dong Dai Moon Koo 178,072 86,951 91,121

I. Don Am, Sung Puk, Chang Shin, Sung In,

II. Shin Sul, An Am, Yong Doo, Jae Ki, Chong An, Chung Nong, Tap Ship Ree, Chung Yang Ree, Hoi Ki, Mee Kyong, Ih Moon,

Sung Dong Koo 117,287

Sah Kun, Mah Chang, Un Dang, An Bong, Kim Ho, Ha Wang Ship Ree, Sang Wang Ship Ree, Sin Dang, Ok Soo,

軍政廳 官報 法令 第一七五號附錄의一   一九四八年三月一七日

Name of Province and Guns	Population	Sub-division Figures Group I	Group II	Names of Koos and Dongs, Myuns and Ups comprising Sub-division Group I
Choong Koo	141,563	·		Tae Pyung Ro 1, Tae Pyung Ro 2, Nam Tai Mun Ro 4, Nam Tai Mun Ro 5, Yang, Do Dong 2, Do Dong 1, Dong Ja, Bong Nai Dong 2, Bong Nai Dong 1, Bang San, Chu Kyo, San Rim Ip Chong, Soo Pyo, Chang Kyo, Soo Ha, San Kwk, Nam Dai Mun Ro 1, Da, Moo Kyo, Ul Chi Ro 1, Ul Chi Ro 2, Ul Chi Ro 3, Ul Chi Ro 4, Ul Chi Ro 5, Ul Chi Ro 6, Ul Chi Ro 7, Kwang Hee Dong 2, Kwang Hee Dong 1, O Chang, Yae Kan, In Hyun Dong 2, In Hyun Dong 1, Cho, Ja Dong 2, Ja Dong 1, Myung Dong 2, Myung Dong 1, Nam Dai Mun Ro 2, So Cong, Nam Dai Mun Ro 3, Choong Mo Roo 1, Choong Mo Ro 2, Choong Mo Ro 3, Choong Mo Ro 4, Choong Mo Ro 5, Sang Rim, Chang Chung Dong 1, Chang Chung Dong 2, Huk Chong, Pul Dong 3, Pul Dong 2, Pul Dong 1, Nam Hak, Yae Chang, Chu Ja, Nam San Dong 3, Nam San Dong 2, Nam San Dong 1, Hai Hyun Dong 3, Hai Hyun Dong 2, Hai Hyun Dong 1, Puk Chang, Nam Chang,
Suh Dai Moon Koo	131,509			Chung, Suh So Moon, Man Rhee 2, Man Rhee 1, Soon Wha, Eui Choo Ro 2, Choong Rim, Hap, Choong Chung Ro 3, Dai Hyun, Ro Ko San, Chang Chun, Yun Hee, Shin Chon, Bong Won, Puk A Hyun, Choong Chun Ro 2, Mi Pyun, Eui Choo Ro 1, Choong Chung Ro 1, Pyung, Song Wol, Kyo Nam, Naing Chun, Chun Yun, Ok Chun, Yung Chun, Kyo Puk, Hong Pa, On

Name of Province and Guns	Population	Sub division Figures		Names of Koos and Dongs, Myuns and Up's comprising Sub division Group I
		Group I	Group II	
				Chon, Hyun Cha, Hang Hai, Hong Chi, Chin Young, Boo Am,
Ma Po Koo	130,334			Ah Hyun, Kong Duk, Shin Kong Duk, Do Wha, Ma Po, To Chung, Yong Kang, Yum Ree, Dai Hung, Shin Soo, Koo Soo, Hyun Suk, Shin Chung, Ko Choong, Chang Chun, Dong Kyo, Suh Kyo, Sang Soo Ik, Ha Soo Ik, Dang Chin, Hap Chun, Mang Won, Yool Do,
Yong San Koo	125,291			Suh Hai, Hoo Am, Hae Wol, Nam Yung, Han Kang Ro 1, Han Kang Ro 2, Han Kang Ro 3, Yong San Dong 1, Yong San Dong 2, Yong San Dong 3, Yong San Dong 4, Yong San Dong 5, Yong San Dong 6, Ree Tai Won, Han Nam, Bo Kwang, Choo Sung, Dong Bing Ko, Suh Bing Ko, Ih Chon, Chung Pa 1, Chung Pa 2, Chung Pa 3, Moon Bai, Shin Sul, Won Hyo Ro 1, Won Hyo Ro 2, Won Hyu Ro 3, Won Hyo Ro 4, Jai Am, San Chun, Shin Chang, Do Won, Yong Mun, Hyu Chang,
Young Dung Po Koo	118,682			You Eui, Dong Jak, Hook Suk, Bun Dong, Ro Ryang Chin, Sang Do, Dai Bang, Shin Kil, Do Rim, Yung Dung Po, Dan San, Yang Pyung, Yang Wha,
Kae Sung City	87,962			
In Chon City	215,784	112,207	103,577	I. Choong Ang, Kwan, Song Hak, Hai An Dong, Hang, Sa, Dap, Shin Hung Dong 2, Shin Hung Dong 1, Shin Hung Dong 3, Shin Saing, Shin Po, Sun Wha, Yu, Do

軍政廳　官報　法令　第一七五號附錄의一　　一九四八年三月一七日

Name of Province and Guns	Population	Sub-division Figures Group I	Group II	Names of Koos and Dongs. Myuns and Up's comprising Sub-division Group I
				Won, Yul Mok, Sun Rim, Puk Sung, Song Wol, In Hyun, Chun, Nai, Kyung, Yong, Soon Ui, Yong Hyun, Hak Ik, Ok Ryun, Within the jurisdiction of Choo An Branch, Within the jurisdiction of Nam Dong Branch, Within the jurisdiction of Moon Hak Branch,
				II. Man Suk, Wha Mok, Wha Pyung, Chang Young, Kim Kok, Song Rim, Song Hyun, Within the jurisdiction of Boo Pyung Branch, Within the jurisdiction of Suh Kwan Branch,
Koh Yang Gun	156,198	75,125	81,073	I. Eun Pyong, Soong In,
				II. Sin Do, Won Dang, Chi Do, Choong, Song Po, Byuk Je, Duk Do,
Kwang Ju Gun	102,944			Kwang Ju, Oh Po, Cho Wol, Do Chuk, Sil Chon, Toe Chon, Nam Chong, Choong Boo, Dong Boo, Seu Boo, Koo Chun, Choong De, Dae Wang, Awn Choo, Lak Saeng, Dol Ma,
Yang Ju Gun	151,378	77,063	74,315	I. Ee Chung Boo, Chee Nae, Hoe Chon, Ee Dan, In Hyun, Kwang Jok, Baek Suk, Chang Heung, Byul Nae, Nam,
				II. Chin Chub, Chin Kun, Wha Do, Wa Bu, Mi Kim, Koo Ri, Ro Hae,
Po Chun Gun	58,382			Po Chun, Gun Nae, Nae Chon, Kah San, So Henl, Sin Puk, Chung Sam, Tchang Soo, Su Choong, Ill Tong,
Kah Pyung Gun	57,780			Kah Pyung, Sul Ak, Oe Seu, Sang, Ha, Puk,

Name of Province and Guns	Population	Sub-division Figures Group I	Group II	Names of Koos and Dongs, Myuns and Up's comprising Sub-division Group I
Yang Pyung Gun	83,506			Yang Pyung, Kang Sang, Kang Ha, Yang Su, Ok Chon, Su Chong, Dan Wol, Chung Woon, Yang Dong, Chi Je, Yong Moon,
Yeu Ju Gun	79,070			Yeu Ju Up, Chum Dong, Ka Nam, Lyung Su, Henng Chun, Kim Sa, Kae Gun, Dae Sin, Puk Nae, Kang Chun,
Ri Chun Gun	85,167			Ri Chun Up, Chang Ho Won Up, Sin Don, Back Sa, Boo Bal, Ho Bob, Ma Chang, Dae Wol, Mo Ka, Sul Sung, Yul,
Yong In Gun	91,711			Yong In, Po Kok, Mo Hyun, Ku Sung, Soo Chi, Ki Heung, Nam Sa, Ee Tong, Ko Sam, Won Sam, Oe Sa, Nae Sa,
An Sung Gun	104,825			An Sung Up, Bo Ge, Kim Kwang, Su Woon, Mi Yang, Dae Uuk, Yung Sung, Kung Do, Yen Kok, Ile Chook, Ee Chook, Sam Chook,
Pyung Tack Gun	92,751			Pyung Tack Up, Song Tan, Puk, Su Tan, Ko Duk, O Sung, Chung Puk, Po Seung, Hyun Duk, Peng Sung,
Soo Won Gun	206,784	99,645	107,139	I. Soo Won Up, Ill Wang, Ban Wol, Mae Song, Bong Dam, Bi Bong, Eum Duk, Ma Do,
				II. Song San, Su Sin, Pal Tan, Chang An, Woo Chung, Hyang Nam, Yang Kam, Chung Nam, Cho San, Dong Tan, Dae Chang, An Yong,
Si Heung Gun	86,241			Dong, Seu, Sin Dong, Kwa Chun, An Yang, Nam, Seu Am, Joon Ja,

軍政廳 官報 法令 第一七五號附錄의一 一九四八年三月一七日

Name of Province and Guns	Population	Sub-division Figures Group I	Group II	Names of Koos and Dongs, Myuns and Up's comprising Sub-division Group I
Boo Chun Gun	85,455			So Sa Up, So Rao, O Chung, Ke Yang, Yung Chong, Puk Do, Yong Yu, Duk Chuk, Yung Heung, Dae By,
Kim Po Gun	71,618			Kim Po, Koh Chon, Yang Su, Yang Dong, Kim Dan, Yang Chon, Dae Got, Wol Got, Ha Sung,
Kang Wha Gun	98,414			Kang Wha, Sun Won, Bul Ern, Kil Sang Wha Do, Yang Do, Nae Ka, Ha Chub, Yang Sa, Song Hae, Kyo Dong, Sam San, Su Do,
Pa Ju Gun	86,842			Im Chin, Wol Nong, Tan Hyun, Kyo Ha, A Dong, Cho Ri, Kwang Tan, Chu Nae, Chun Hyun, Pa Pyung, Chuk Sung,
Chang Dan Gun	40,863			Chang Dan, Gun Nae, Chin Dong, Chang Do, Chin Seu,
Kae Poong Gun	79,933			To Sung, Nam, Seu, Chung Kyo, Bong Dong, Choong, Sang Do, Rim Han, Heung Kyo, Dae Sung, Kwang Duko,
Yun Back Gun	211,678	106,005	105,673	I. Yun An Up, Song Bong, Bong Seu, Hae Ryong, Yong Do, Ke Koong, Chy Wha, Chung Yung, Ill Sin, Nae Sung,
				II. He Dong, Ho Nam, Hae Sung, Ern Chon, Yu Kok, Hae Wol, On Jong, Do Chon, Suk San, Bong Wha,
Ong Jin Gun	151,183	75,687	75,496	I. Ong Jin Up, Puk, Sou, Yong Chun, Baek Nyung,
				II. Boo Min, Yong Yun, Whan Myung, Heung Mi, Song Rim, Hae Nam, Dong Kang, Dong Nam,

Name of Province and Guns	Population	Sub-division Figures Group I	Group II	Names of Koos and Dongs, Myuns and Up's comprising Sub-division Group I
Choong Chun Pukto				
Chung Uu City	51,522			
Chung Won Gun	175,534	89,480	86,054	I. Sa Ju, Lang Sung, Mi Won, Ka Duk, Nam Ill, Nam Ee, Moon Ee, Hyun To, Bo Yung,
				II. Kang Su, Kang Nae, Kang Oe, Ok San, O Chang, Puk Ee, Puk Ill,
Bo Ern Gun	82,109			Bo Ern, Sok Lee, Ma Ro, Tan Bu, Sam Sung, Soo Han, Hoe Nam, Hoe Puk, Nae Puk, San Oe,
Ok Chon Gun	85,799			Ok Chon, Dong Ee, An Nam, An Nae, Chung Sung, Chung San, Ee Won, Gun Su, Gun Puk,
Yong Dong Gun	110,830			Yung Dong Up, Youg San, Whang Gan, Whang Kum, Mae Kok, Sang Chen, Yang Kang, Yong Hwa, Hak San, Yang Sah, Sim Chun,
Thin Chun Gun	65,995			Chin Chun, Duk San, Cho Pyung, Moon Back, Bak Kok, Ri Wol, Man Seung,
Kwe San Gun	134,732			Kwe San, Kam Mul, Chang Yun, Sang Doon, Yun Poong, Chil Sung, Moon Kwang, Chung Chun, Chung An, Hoe Pyung, Do An, Sa Ri, Teho Soo. Bul Chung,
Eum Sung Gun	96,326			Eum Sung, So Ee, Won Nam, Naeng Dong, Dae Se, Sam Sung, Kim Wang, Sumg Kuk, Kam Kok,
Choong Ju Gun	144,182			Choong Ju Up, Sa Mi, Ri Ryu, Ju Duk, Sin Ni, Ro In, Ang Sung, Ka Kim, Kim Ka, Dong Yang, San Chuck, Erm Chung, So Dae,
Je Chon Gun	105,120			Je Chon Up, Kim Sung, Chung

Name of Province and Guns	Population	Sub-division Figures Group I	Group II	Names of Koos and Dongs, Myuns and Up's comprising Sub-division Group I
				Poong. Soo San, Duk San, Han Soo, Back Woon, Bong Yang, Song Hak,
Dan Yang Gun	60,745			Dan Yang, Dae Kang, Ka Kok, Yung Choong, Eu San Chon, Mae Po, Chuk Sung,
Choong Chung Nam Do				
Tae Jon City	96,207			
Dae Duk Gun	93,932			San Nae, Dong, Hoe Duk, Puk, Koo Chuk, Tang Dong, Yu Sung, Chin Cham, Ki Sung, Yu Chun,
Yun Ki Gun	82,456			Cho Chi Won Up, Dong, Su, Nam, Gin Nam, Chon Ee, Chon Dong,
Kong Ju Gun	154,926	71,791	83,135	I. Kong Ju Up, Ri Chun, Tan Chon, Ke Ryong, Ban Po,
				II. Chang, Ee, Ee Dang, Chung An, Woo Sung Su Kok, Sin Poong, I Koo,
Non San Gun	184,722	95,885	88,836	I. Non San Up, Kang Kyung Up, Sung Dong, Kwang Suk, Ro Sung, Sang Wol, Boo Chuk,
				II. Yun San. Doo Ma, Bul Kok, Yang Chon, Ka Ya Kok, Koo Ja Kok, Ern Jin, Tche Woor,
Boo Yu Gun	151,869	71,338	80,531	I. Boo Yu, Koong Am, Ern San, Oe San, Nae San, Koo Ryong, Hong San,
				II. Ok San, Nam, Choong Wha, Yang Wha, Rim Chun, Chang Am, Se Do, Suk Sung, Cho Chun,
Chaw Tchun Gun	130,345			Chang Hang Up, Chaw Tchun, Ma Su, Wha Yang, Ki San, Han San, Ma San, Si Cho, Moon San. Pam Kyo, Chong Chun, Bi In, Su,

Name of Province and Guns	Population	Sub-division Figures Group I	Group II	Names of Koos and Dongs, Myuns and Up's comprising Sub-division Group I
Bo Hyung Gun	104,312			Dae Chun, Choo Po, O-Chon, Chun Puk, Chung So, Chung La, Lam Po, Oon Chon, Chu San, Mi San,
Chung Yang Gun	85,432			Chung Yang, Woon Kok, Dae Tchi, Chung San, Mok, Chung Nam, Chuk Kok, Sa Yang, Wha Sung, Bi Bong,
Hong Sung Gun	120,376			Hong Sung Up, Kwang Chun Up, Hong Puk, Kim Ma, Hong Dong, Chang Kok, Ern Ha, Kyul Sung, Su Boo, Kal San, Ku Hong,
Ye San Gun	135,761			Ye San Up, Dae Sool, Sin Yang, Kwang Si, Dae Heung, Ern Bong, Sab Kyo, Duk San, Bong San, Ko Duk, Sin Am, O Ka,
Su San Gun	192,651	87,184	105,467	I. Su San Up, In Chi, Pal Bong, Dae San, Sung Yun, On Am, Dae Ho Ji Chung Mi, Woon San,
				II. Hae Mi, Bu Suk, Chi Kok, Ko Puk, An Mim, Nam, Tae An, Keun Heung, So Won, Won Puk,
Dang Jin Gun	118,680			Dang Jin, Ko Dae, Suk Moon, Myun Tchun, Soon Sung, Hab Deuk, Oo Kang, Sin Pyung, Song Ak, Song San,
Ah San Gun	118,684			On Yang Up, Song Byung, Be Bang, On Jung, Yum Tchi, Eum Bong, Don Po, Yung In, In Ju, Sun Chang, Do Ko, Sin Chang,
Chun An Gun	139,052			Chun An Up, Hwan Sung, Poong Se, Kwang Duk, Mok Chun, Puk, Sung Nam, Soo Sin, Byung Chon, Dong, Sung, Hwan, Chik San, Sung Ku, Ib Chang,

Name of Province and Guns	Population	Sub-division Figures		Names of Koos and Dongs, Myuns and Up's comprising Sub-division Group I
		Group I	Group II	
Cholla Puk To				
Gun San City	66,715			
Chon Ju City	83,333			
Wan Ju Gun	172,323	82,928	89,395	I. Cho Po, Yong Jin, Cho Chun, Bong Dong. Sam Ye, Bi Bong,
				II. Sang Kean, Woo Jon, Ee Su, So Yang, Koo Ee, Ko San, Woon Ju, Wha San, Dong Sang,
Chin An Gun	72,590			Chin An, Sang Jon, Back Woon, Sung Soo. Ma Ryung, Boo Ki, Yong Dam, Choo Chun, Dong Hyang, An Chun, Chung Chun,
Kum San Gun	89,255			Kum San Up, Kum Sang, Che Won, Boo Ri, Gun Puk, Nam Ill, Nam Ee, Chin San, Bok Soo, Choo Boo,
Moo Chu Gun	63,238			Moo Chu, Moo Poong, Sul Chun, Chuk Sang, An Sung, Boo Nam,
Chang Soo Gun	64,430			Chang Soo, San Su, Ban Am, Ke Nae, Chun Chun, Ke Nam, Ke Nam, Ke Puk,
Im Sil Gun	91,241			Im Sil, Chung Oong, Woon Am, Sin Pyung. Sung Soo, Dong Nam, Sin Duk, Sam Ke, Kwan Chon, Kang Jin, Duk Tchi, Ji Sa,
Nam Won Gun	137,475			Nam Won Up, Ee Back. Ju Chon, Song Dong, Chu Saeng, Dae San, Dae Kang, Sa Me, Duk Kwa, Bo Jul, Wang Tchi, Kim Chi, San Dong, Soo Chi. Oon Bong, A Yung, San Nae, Dong,
Soon Tchang Gun	87,849			Soon Tchang, In Ke, Dong Ke, Poong San. Kim Gwa, Pal Duk, Sang Chi, Bok Heung. Jok Sung, Yu Deung, Ku Choo,

— 10 —

Name of Province and Guns	Population	Sub-division Figures Group I	Group II	Names of Koos and Dongs, Myuns and Up's comprising Sub-division Group I
Chung Up Gun	237,837	120,354	117,483	I. Chung Ju Up, Puk, Nae Chang, I, Am, So Sung, Ko Bu, Yung Won, Duk Chun, Ri Pyong,
				II. Sin Tae In Up, Chung Woo, Tae In, Kam Kok, Ong Dong, Chil Bo, San Nae, San Oe,
Koh Chang Gun	153,155	81,202	71,953	I. Koh Chang, Ko Soo, Moo Chang, Kong Eum, Sang Ha, Sung Song, Dae San,
				II. Ah San, Hae Ri, Sin Won, Heung Dok, Sung Nae, Sin Rim, Boo Ah,
Boo Ahn Gun	131,163			Boo An Up, Choo San, Ha Su, San Nae, Sang Su, Back San, Bo An, Haeng An, Chool Po, Dong Jin,
Kim Je Gun	219,455	102,791	116,664	I. Chook San, Back San, Bak Koo, Man Kyung, Kong Duk, Chung Ha, Sung Duk, Chin Bong,
				II. Kim Je Up, Wol Chon, Yong Chi, Boo Yang, Kim Koo, Bong Nam, Bong San, Kim San,
Ok Koo Gun	120,013			Mi, Ok Koo, Ok San, Hoe Hyun, Rim Pa, Su Soo, Dae Ya, Kae Chung, Sung San, Ra Po,
Ik San Gun	189,839	93,541	96,298	I. O San, Pook Ill, Whang Dong, Kin Ma, Wang Goong, Choong Po, Pal Bong, Sam Ki,
Ri Ri City	36,502			II. Ham Ra, Woong Po, Sung Dang, Yong An, Ham Yul, Nang San,
Cholla Nam Do				Mang Sung, Whang Ha, Lyu San,
Mok Po City	103,081			
Kwang Ju City	100,451			Song Jung Up, Hyo Chi, Suk Kok, Chi San, Bi Ah, Ha Nam, Rim Kok, Dong Kok, Dae Chon, Su Chang, Kuk Nak, Su Bang,
Kwang San Gun	121,045			

Name of Province and Guns	Population	Sub-division Figures Group I	Group II	Names of Koos and Dongs, Myuns and Up's comprising Sub-division Group I
Dam Yang Gun	108,155			Dam Yang Up, Bong San, Wol San, Kim Sung, Yong, Moo Chung, Chang Pyung, Ko Su, Nam, Dae Duk, Soo Puk, Dae Jon,
Kok Sung Gun	82,329			Kok Sung, O Kok, Sam Chi, Suk Kok, Mok Sa Dong, Chook Kok, Ko Tal, Ok Kwa, Rip, Kyum, Wha,
Koo Re Gun	63,384			Koo Re, Moon Chuk, Yung Jon, To Chi, Ma San, Kwang I, Yong Bang, San Dong,
Kwang Yang Gun	77,832			Kwang Yang, Bong Kang, Ok Ryong, Kol Yak, Ok Kok, Chin Sang, Chin Sang, Chin Wol, Da Ab,
Lyu Soo Gun	168,136	88,623	79,513	I. Lyu Soo Up, Sang Bong, Sam Ill, So Ra.
				II. Yul Chon, Wha Yang, Dol San, Nam, Wha Chung, Sam San,
Soon Tchon Gun	172,955	84,250	88,705	I. Soon Tchon Up, Hae Ryong, Su, Whang Jon,
				II. Wol Dung, Sung Am, Choo Am, Song Kwang, Oe Su, Nak An, Byul Ryang, Do Sa, Sang Sa,
Ko Heung Gun	168,192	88,764	79,428	I. Ko Heung, Poong Yang, Do Yang, Gum San, Do Wha, Po Doo.
				II Bong Nae, chum Am, Kwa Yuk, Nam Yang, Dong Kang, Dae Su, Doo Wun,
Bo Sung Gun	136,202			Bo Sung Up, Byul Kyo Up, Ro Dong, Mi Lyuk Kyum Baek, Yul Aw, Bok Nae, Moon Duk, Che Sung, Derk Lyang, Hoe Chun, Woong Sa,

Name of Province and Guns	Population	Sub-division Figures Group I	Group II	Names of Koos and Dongs, Myuns and Up's comprising Sub-division Group I
Wha Soon Gun	120,676			Wha Soon, Han Chun, Choon Yang, Chung Poong, Ri Yang, Nerng Ju, Do Kok Do Am Ee Seu, Puk, Dong Pok, Nam, Dong,
Chang Heung Gun	114,155			Chang Heung Up, Yong San Kwan San, Dae Duk, An Ryang, Chang Dong, Chang Pyung, Yu Chi, Boo San,
Kang Jin Gun	103,220			Kang Jin Up, Gun Dong, Chil Ryang, Dae Koo Do Am, Sung Jon, Chak Chen, Byung Yung, Am Chon,
Hae Nam Gun	174,084	91,725	82,359	I. Hae Nam Sam San, Wha San, Hyun San, Song Chi, Puk Pyung, Ok Chun, II. Ke Kok, Ma San Whang San, San I, Moon Nae Wha Won,
Yung Am Gun	118,195			Yuny Am, Duk Jin, Kim Chung, Sin Puk, Si Chong, To Po Gun Su, Seu Ho, Hak San, Mi Am, Sam Ho,
Moo An Gun	244,127	116,805	127,322	I. I Ro, Sam Hyang, Ill Ro, Mong Tan, Chung Ke, Kum Sung, Hyun Kyong, Mang Oon, Hae Che, II. Chi Do, Im Cha, Cha Ern, Bi Kerm, To Cho, Herk San, Ha I, Chang San, An Cha, Am Tae Ab Hae,
Ra Ju Gun	227,380	114,034	113,346	I. Yong San Po Up, Da Do, Wang Kok, Ban Nam, Kong San, Dong, Kang, Da Si, Bong Hwang, Se Chi, II. Ra Ju Up, Moon Pyung, Sam Do, Bon Rang, Pyung Dong, Ro An, Kim Chon, San Po, Nam Pyung,

軍政廳　官報　法令　第一七五號附錄의一　　一九四八年三月一七日

Name of Province and Guns	Population	Sub-division Figures Group I	Group II	Names of Koos and Dongs, Myuns and Up's comprising Sub-division Group I
Ham Pyung Gun	113, 464			Ham Pyung, Son Bool, Sin Kwang, Hak Kyo, Erm Da, Dae Dong, Ra San, Hae Bo, Wol Ya,
Yung Kwang Gun	128,630			Yung Kwang, Dae Ma, Se Pang, Bool Kab, Gun Su, Gun Nam, Yun San, Pak Soo, Eub Sung, Hons Nong, Wi Don, Nak Wol,
Chang Sung Gun	117,999			Chang Sung Up, Chin Won, Nam, Dong Wha, Sam Su, Sam Ke, Whang Yong, Su Sam, Puk Ill, Puk Ee, Puk Sang, Puk Ha,
Wan Do Gun	106,521			Wan Do Up, Gun Oe, Sin Chim Ko Kum, Kim Ill, Chung San, So An, Ro An,
Chin Do Gun	74,629			Chin Do, Gun Nae, Ko Gun, Ee Sin, Kim Hoe, Chi San, Che To,

Kyung Sang Puk to

Tae Gu City	269,113	93,281	83,195	92,637

I. Suh Nai, Puk Nai, Poo Chung, Suh Moon Ro 1, Suh Sung Ro 2, Tai Pyung Ro 2, Chun, Nyang Chon, Kae San Dong 2, Chong no 1, Chong no 2, Moon Wan Sang, Duk, Yong Duk, Kong Pyung, Wan Chun, Long In, Kyo, Dong Sung Ro 1, Jong Sung Ro 2, Dong Sung Ro 3, Nam Il, Yun San, Nam Sung, Dong Il, Chang Kwan, Puk Sung Ro 1, Puk Sung Ro 2, Dai An, Sam Duk, Sa Il, Wha Chun, Tai Pyung Ro 1, Chil Sung, Dong, Moon, Shin Am, San Kyuk, Bok Hyun, Kym Dan, Soo, Sang Su, Ha Suh,

II. Dong San, Shi Chang Puk, Suh Moon Ro 2, Do Won, Suh Ya, In

Name of Province and Guns	Population	Sub-division Figures Group I	Group II	Names of Koos and Dongs, Myuns and Up's comprising Sub-division Group I
				Kyo, Dal Sung, Kae San Dong 1, Soo Chang, Dai Shin, Suh Sung Ro 1, Won Dai, Tai, Pyung Ro 3, Bi San, Sang Rhee, Choong Rhee Pyung Rhee, Ree Hyun, Ro Ui, Cho Ya, Nai Tang, Sung Dang,
				III. Nam San, Bong San, Dai Bong, Duk San, Dai Myung, Choong, Ha, Sang, Shin Chun, Hyo Mook, Man Chon, Bum Uh, Whang Chung, Chi San, Bum Mol, Dee San, Bong Duk,
Tai Sung Gun	135,470			Dong Chon, Kong San, Ka Chang, Sung Su, Da Sa, Ha Bin, Wol Bye, Wha Won, Ok Po, Non Sam, Hyun Poong, Yu Ka, Ku Chi,
Gun Ee Gun	71,588			So Bo, Gun Ee, Hyo Lyung, Bu Ke, Oe Bo, Ee Heung, San Sung, Ko Lo,
I Sung Gun	179,653	88,988	90,665	I. I Sung Up, Dan Chon, Jum Kok, Ok San, Sa Kok, Choon San, Kah Erm, Kim Sung, Bong Yang,
				II. Bi An, Ku Chun, Dan Mil, Dan Puk, An Ke, Da In, Sin Pyung, An Pyung,
An Dong Gun	199,750	103,551	96,199	I. An Dong Up, Wa Ryong, Pook Hoo, Su Ho?, Poong San, Poong Chun,
				II. Ill Chik, Nam Hoo, Nam Sun, Rim Ha, Kil An, Rim Tong, Wol Kok, Ye An, Di San, Nok Chun,
Chung Song Gun	62,678			Chung Song, Boo Dong, Boo Nam, Hyun Dong, Hyun Su, An Duk, Pa Chun, Chin Bo,

— 15 —

Name of Province and Guns	Population	Sub-division Figures Group I	Group II	Names of Koos and Dongs, Myuns and Up's comprising Sub-division Group I
Yung Yang Gun	48,138			Yung Yang, Rip Am, Chung Ki, Il Wol, Soo Bi, Suk Bo,
Yung Duk Gun	93,444			Yung Duk, Kang Koo, Nam Ching, Tal Son, Chi Pum, Dal San, Nyung Hae Byong Kok, Tchang Soo,
Yung Ill Gun	220,997	118,701	102,296	I. Po Hang Up, Tal Jon, Heung Hae, Kok Kang, Sin Kwang, Ki Ke, Chuk Chang,
				II. Koo Ryong Po, Chung Ha, Song Ra,
				Yun Ill, Dae Song, O Chon, Dong Hae, Chi Haeng,
Kyong Zu Gun	219,883	109,798	110,085	I. Kyong Zu Up, Kam Po, Nae Dong, Yany Puk, Yang Nam, Oe Dong,
				II. Nae Nam, San Nae, Su, Kyun Kok, Kang, Su, Kang Dong, Chun Puk,
Yung Chun Gun	155,129	79,947	75,182	I. Yung Chun Up, Chung Tong, Sin Nyung, Wha San, Wha Puk,
				II. Cha Yang, Im Ko, Ko Kyung, Puk An, Dae Chang, Kum Ho,
Kyong San Gun	126,210			Kyong San, Ko San, An Sim, Ha Yang, Wa Chon, Chin Ryang, Cha, In, Yong Sung, Nam San, Ab Yang, Nam Chon,
Chung Do Gun	109,529			Chung Do, Wha Yang, Kak Nam, Poong Kak, Kak Puk, I Su, Woon Moon, Gum Chon, Mae Jon,
Ko Lyong Gun	68,356			Ko Lyong, Duk Kok, Woon Soo, Sung San, Da San, Kae Chin, Woo Kok, Sang Rim,

Name of Province and Guns	Population	Sub-division Figures Group I	Group II	Names of Koos and Dongs, Myuns and Up's comprising Sub-division Group I
Sung Zu Gun	103,681			Sung Zu, Sun Nam, Yong Am, Soo Rin, Ka Chon, Kim Soo, Dae Ka, Byuk Chin, Cho Zon, Wol Hyang,
Tchil Kok Gun	97,986			We Kwan, Chi Chon, Chil Kok, Dong Myung, Ka San, In Dong, Suk Chuk, Puk Sam, Yak Mok,
Kim Chon Gun	184,158	99,461	84,697	I. Kim Chon Up, Nong So, Nam Er Whai, An Po, Kae Nyung, Gam Moon.
				II. Er Whae, Bong San, Dae Hang, Gam Chon, Cho Ma, Ku Sung, Chi Re, Bu Hang, Dae Duk, So San,
Sun San Gun	101,360			Sun San, Moo Eul, Ok Sung, Do Kae, Bae Pyung, San Dong, Chang Chon, Ku Mi, Ko Ah,
Sang Ju Gun	221,703	111,794	109,909	I. Sang Ju Up, Sa Bul, Choong Dong, Nak Dong, Chung Ri, Kong Sung, Oe Nam,
				II. Ne Su, Mo Dong, Mo Su, Wha Dong Wha Su, Wha Puk, Oe Su, Eun Chuk, Kong Kerm, Ham Chang, Ri An,
Moon Kyung Gun	129,523			Moon Kyung, Ma Sung, Ka Eun, Yong Am, Yong Ke, Ho Su Nam, Yung Soon, San Yang, San Puk, Dong Ro,
Yawe Tchon Gun	139,889			Yawe Tchon Up, Yong Moon, Sang Ri, Ha Ri, Gam Chon, Bo Moon, Ho Myung, Lyu Chon, Yong Goong, Kae Po, Chi Bo, Poong Yong,
Yung Ju Gun	115,832			Yung Ju Up, I San, Pyung Ern, Moon Ju, Chan Soo, An Chung Bong Hyun, Poong Ki, Soon Heung, Dan San, Bo Suk,

Name of Province and Guns	Population	Sub-division Figures		Names of Koos and Dongs, Myuns and Ups comprising Sub-division
		Group I	Group II	Group I
Bong Wha Gun	101,436			Nae Sung, Mool Ya, Bong Sung, Bub Jon, Choon Yang, So Chen, Chae San, Myung Ho, Sang Woohn,
Wool Lyong Do	13,244			Nam, Su, Puk,

Kyung Sang Nam Do

Pu San City   400,156   92,073   104,855   99,656   103,572

I. Su An, Bok Chon, Chil San, Rk Min, Myong Nyun, On Chon, Sa Chik, Ku Jae, Ryon San, An Rak, Ban Nyo, Myong Chang, Su Kum Sa, Whae Dong, Suk To, Ban Song, O Ryun, Pu Kok, Chang Chon Su Yong Mang Pu Kwang An, Min Lak, Nam Chon, Jae Song, U, Chung, Cha, Pu Chon, Bum Chon, Cho Up, Chon Po, Mun Hyun, Pu Am, Dang Kam, Ryong Ho, Ryon Dang, Ram Man, U Am, Yang Chong, Ryon Chi, Tae Yun, Kae Kum, Ka Yo,

II. Bum Il, Cha Chon, Su Chong, Chor Yang, Yong Chu,

III. Tong Tae Shin, Su Tae Shin, Pu Yong, Bu Min, A Mi, Cho Chang, Wan Wol, Nam Bu Min, Am Nam, Kwe Jong, Ha Tan, Kam Chon, Ku Pyong, Shin Pyong, Chang Lim, Dang Li, Ta Te, Hong Ji,

IV. Bo Su, Tae Chong, Tong Kwang, Chung Ang, Kwang Pok, Shin Chang, Chang Sun, Bu Pyong, To Sung, Chang Mu, Nam Po, Tae Kyo, Tae Pyong, Nam Hang, Yong Sun, Shin Sun, Bong Re, Chong Hak, Tong Sam,

Ma San City   82,175

— 18 —

Name of Province and Guns	Population	Sub-division Figures		Names of Koos and Dongs, Myuns and Up's comprising Sub-division Group I
		Group I	Group II	
Chin Ju City	86,852			
Chin Yang Gun	133,606			Na Dong, Chung Chon, Kim Kok, Moon San, Chim Sung, Il Ban Sung, Ee Ban Sung, Sa Bong, Chi Soo, Dae Kok, Kum San, Chip Hyun, Mi Chun, Myung, Suk, Dae Pyung, Soo Kok,
I Yung Gun	126,305			I Yung, Ka Rye, Chil Kok, Dae I, Wha Chung, Yong Duk, Chung Kok, Boo Chung, Nak Su, Boo Rim, Bong Soo, Koong Ryu, Ryu Kok,
Ham An Gun	131,745			Ham An, Ka Ya, Gun puk, Bub Soo, Dae San, Chil Sul, Chil Puk, Chil Won Sam In, Yaw Hang,
Chang Nyung Gun	141,138			Chang Nyung, Koh Am, Sung San, Dae Hap, Ri Bang, Yu Aw, Dae Chi, Chang Nak, Kee Sung, Yung San, Dae Ma, Nam Chi, Gun Chun, Kil Kok, Boo Kok,
Mil Yang Gun	188,771	89,070	99,701,	I. Mil Yang Up, Boo puk, Sang Dong, San Oe, San Nae, Dan Yang,
				II. Sam Nang Jin, Sang Nam, Ha Nam, Cho Dong, Moo An, Chung Do,
Yang San Gun	55,448			Yang San, Dong, Mool Kum, Won Dong, Sang Puk, Ha Puk, Oong Sang,
Wool San Gun	194,399	99,978	94,421	I. Wool San Up, Bang Aw Jin Up, Ha Sang, On San, Su Sang, Chung Ryang, Dae Hyun,
				II. Nong So, Kang Tong, On Yang Oong Tchon, Boum Su, Doo Dong, Doo Su, Ern Yang, Sang Puk, Sam Nam,

軍政廳　官報　法令　第一七五號附錄의一　一九四八年三月一七日

Name of Province and Guns	Population	Sub-division Figures Group I	Group II	Names of Koos and Dongs, Myuns and Up's comprising Sub-division Group I
Dong Nae Gun	70,354			Ku Po, Puk, Sa Sang, Ki Chang, Ill Kwang, Chang An, Dung Kwan, Tchaul Ma,
Kim Hae Gun	188,402	95,204	93,198	I. Kim Hae Up, Ka Rak, Dae Su, Myong Ji, Sang Dong, Ha Dong,
				II. Chin Yung, Nok San, Chang Yu, Choo Tchon, Chin Re, Ee Puk, Seng Rim,
Chang Won Gun	207,163	104,289	102,874	I. Chin Hae Up, Dong, Puk, Dae San, Sang Nam,
				II. Nae Su, Chang Won, Woong Nam, Ku San, Chin Dong, Chin Buk, Chin Jon, Oong Tchon, Oong Tong, Oong Ka,
Tong Yung Gun	208,393	100,340	108,053	I. Tong Yung Up, Yong Nam, San Yang, Do San, Kwang Do, Han San, Don Duk,
				II. Chang Seung Po Up, Won Yang, Chang Mok, Geu Chae, Sa Deung, Ill Oon, Dong Boo, Ha Chung, Yun Cho,
Ko Sung Gun	117,271			Ko Sung, Sam San, Ha Ill, Ha Ee, Sang Ri, Dae Ka, Yung Hyun, Yung O, Kae Tchon, Koo Man, Hoe Wha, Ma Am, Dong Hae, Keu Ryu,
Sa Tchon Gun	129,735			Sam Tchon Po Up, Sa Tchon Up, Chung Dong, Sa Nam, Yong Kyun, Nam Yang, Tchoo Dong, Kon Yong, Kon Myung, Su Po,
Nam Hae Gun	111,513			Nam Hae, Ee Dong, Sam Dong, Nam, Su, Ko Hyun, Woon Tchon, Chang Sun,

Name of Province and Guns	Population	Sub-division Figures		Names of Koos and Dongs, Myuns and Up's comprising Sub-division Group I
		Group I	Group II	
Ha Dong Gun	121,214			Ha Dong Up, Hwa Kae, Ak Yang, Chuk Yang, Hoeng Chon, Ko Ill, Kim Nam, Jin Kyo, Yang Po, Puk Chon, Chung Am, Ok Chong,
San Tchong Gun	103,904			San Tchung, Cha Whang, O Bu, Seng Cho, Kum Su, Sam Chang, Sil Chon, Dan Sung, Sin An, Saeng Bi Yang, Sin Dung,
Ham Yang Gun	102,109			Ham Yang, Suk Ha, Jo Tchon, Hyoe Tchon, Yu Rim, Soo Dong, Chi Kok, An I, Su Ha, Su Sang, Back Jon, Byung Kok,
Kaw Chang Gun	118,842			Kaw Chang Up, Wol Tchon, Chu Sang, Oong Yang, Ko Che, Book Sang, Ooi Chon, Ma Ri, Nam Ha, Sin Won, Ka Cho, Ka Puk,
Hyub Tchon Gun	166,337	83,136	83,201	I. Hyub Tchon, Bong San, Myo San, Ka Ya, Ya So, Yul Kok, Tcho Ke, Sang Tchek,
				II. Duk Kok, Chung Duk, Chuk Choong, De Yang, Sang Back, Sam Ka, Ka Hoe, Dae Byung, Yong Ju,

Kang Won Do

Name of Province and Guns	Population	Sub-division Figures		Names of Koos and Dongs, Myuns and Up's comprising Sub-division Group I
		Group I	Group II	
Choon Chon City	46,809			
Choon Sung Gun	63,982			Dong, Dong San, Sin Dong, Nam, Su, Sa Puk, Sin Puk, Puk San,
Kang Nyung Gun	175,489,	83,524	91,965	I. Kang Nyung Up, Sung Duk, Kang Dong, Ok Ke, Mook Ho Up,
				II. Koo Chung, Sung San, Wang San, Kyung Po, Sa Chom, Yun Kok, Choo Moon Jin Up, Hyun Nam, Sin Su,

Name of Province and Guns	Population	Sub-division Figures		Names of Koos and Dongs. Myuns and Up's comprising Sub-division Group I
		Group I	Group II	
Sam Tchuk Gun	147,816			Sam Chuk Up, Kun Duk, Won Dok, Sang Chang, Ha Chang, So Won, Ro Kok, Mi Ro, Puk Pyong,
Wool Chin Gun	94,490			Wool Chin, Puk, Su, Kun Nam, Won Nam, Ki Sung, Pyung Hae, On Chung,
Jung Syn Gun	59,675			Jung Sun, Dong Nam, Puk, Rim Ke, Sin Dong,
Pyung Chang Gun	91,186			Pyong Chang, Mi Tan, Bang Rim, De Wha, Bong Pyung, Chin Boo, Do Am,
Nyung Wol Gun	91,566			Nyung Wol, Ha Dong, Sang Dong, Puk, Nam, Su, Chu Chun, Soo Chu,
Won Ju Gun	101,977			Won Ju Up, So Cho, Ho Chaw, Chi Chung, Moon Mak, Boo Ron, Kwi Rae, Heung Up, Pan Boo, Sin Rim,
Hoeng Sung Gun	93,904			Hoeng Sung, Woo Chun, An Heung, Don Nae, Kap Chun, Chung Ill, Kong Keun, Su Won,
Hong Chun Gun	144,942			Hong Chun, Hwa Chon, Du Chon, Ne Chon, Su Suk, Dong, Nam, Su, Puk Bang, Ki Rin Nae, Sin Nam,

Che Ju Do

Name of Province and Guns	Population	Sub-division Figures		Names of Koos and Dongs. Myuns and Up's comprising Sub-division Group I
		Group I	Group II	
Puk Che Ju Gun	178,393	103,960	74,433	I. Che Ju Up, Cho Chon, Ku Jwa, II. Ae Wol, Ha Nlim, Chu Za,
Nam Che Ju Gun	97,755			Te Jon, An Duk, Choong Moon, Su Gwi, Nam Won, Pyo Son, Song San,

軍政廳 官報 法令 第一七五號附錄의一   一九四八年三月一七日

SOUTH KOREAN INTERIM GOVERNMENT
Seoul, Korea

ORDINANCE
NUMBER 176   20 March 1948

CHANGES IN CRIMINAL PROCEDURES

Pending action on the subject matters hereof by the Korean Interim Legislative Assembly and the enactment of legislation pursuant thereto, the following shall govern:

SECTION I. The purpose of this ordinance is to provide for changes in the Criminal Procedure so that rights of the people to freedom from unlawful arrest and detention may be more adequately secured.

SECTION II. As used herein, the following terms shall be interpreted as indicated:

a. All words used in this ordinance to refer to restraint, restraint of body, or restrained person, shall be deemed to include all forms of restraint such as koo in, koo ryu, oochi, chepo kum sok, etc.

b. The competent court for hearing applications under Section XVII hereof shall be the District Court or the Branch Court having Jurisdiction over the place where the person named in the application is actually restrained. Any judge of such court may receive and make decision on such application. Special Judicial Officers shall not have the power to receive or make decision on such application.

c. The competent court for the issuance of warrants of arrest (koo sok yung jang) shall be the District Court or Branch Court. Special Judicial Officers shall not have the power to issue warrants of arrest.

d. In any case which is lawfully before a court, the question of what was reasonable fear, reasonable ground or necessary measures shall be determined by the court from all the evidence.

SECTION III. No person shall suffer restraint of body except pursuant to a warrant of arrest (koo sok yung jang) issued by a court, stating the name of the person to be arrested and the offense with which he is charged; *provided,* however, that no warrant of arrest shall be required in the following cases where prompt action is necessary:

a. When the suspect has no fixed residence.

b. In case of *flagrante delicto* as defined in Article 130 of the Code of

— 1 —

Criminal Procedure, whether the suspect is on the scene of the offense or not, except that such arrest without warrant must be made within forty-eight (48) hours of the time the offense was committed.

c. In the case of an accomplice who has been discovered in the course of investigation of a flagrant offense, except that such arrest without warrant must be made within forty-eight (48) hours of the time the offense was committed.

d. When a convicted prisoner or a lawfully restrained person awaiting trial has escaped.

e. When an offender has been discovered as a result of examination of a dead body.

f. When there is reasonable fear of destruction of evidence by the suspect.

g. When there is reasonable fear that the suspect will flee.

h. When there is reasonable ground to believe that the suspect has committed an offense punishable by imprisonment for one (1) year or more or by graver penalty.

SECTION IV. In addition to other powers given by law, prosecutors or police shall have the power to enter any building at any time, day or night, without a warrant where there is reasonable ground to believe that an offense punishable by imprisonment for one (1) year or a graver penalty has been or is being or is about to be committed therein, to take all necessary measures for the effectual prevention or detection of such offense and may then and there restrain all persons suspected of being concerned in such crime, and also may seize all property which there is reasonable ground to suspect has been stolen or used in the commission of such offense.

SECTION V. No prosecutor, judicial police or other constituted authority shall make search or seizure without a search warrant (soo sak yung jang) issued by a court, except as provided in Section IV hereof; and *provided,* further, that property owned by, held by, or in custody of, persons subject to arrest without warrant, in accordance with the provisions of Section III, hereof, shall be subject to search and seizure without warrant within the time allowed by Section VI hereof for obtaining a warrant of arrest. The warrant of search (soo sak yung jang) herein provided for shall describe the place to be searched and the thing to be seized in accordance with the provisions of the Code of Criminal Procedure.

SECTION VI. In case a prosecutor, a judicial police or other constituted authority has restrained the body of a person without a warrant of arrest, under provisions of Sections III and IV hereof, he shall obtain from a court a warrant of arrest (koo sok yung jang) for such person within forty-eight (48) hours of such restraint in the city of Seoul and in a city, county or island where a court is located or within five (5) days in a city, county or island where a court is not located. If he fails to obtain such a warrant of arrest within the prescribed time, the person restrained shall be immediately released. In such case, articles seized without a warrant of search (soo sak yung jang) shall be returned to the persons from whom they were seized. No person released under provisions of this Section shall be subject to restraint of body on the same facts except pursuant to a warrant issued by a court.

SECTION VII. The provisions pertaining to warrants of arrest (koo in jang) and warrants of detention (koo ryoo jang) in the Code of Criminal Procedure shall apply *mutatis mutandis* to the warrant of arrest (koo sok yung jang) hereof, and the provisions pertaining to arrest (koo in) and detention (koo ryoo) in the Code of Criminal Procedure shall apply *mutatis mutandis* to arrest (koo sok) under warrant of arrest (koo sok yung jang) hereof. The provisions pertaining to warrant of search (soo sak yung jang) and warrant of seizure (ap soo yung jang) shall apply *mutatis mutandis* to the warrant of search (soo sak yung jang) hereof, and the provisions pertaining to search (soo sak) or seizure (ap soo) in the Code of Criminal Procedure shall apply *mutatis mutandis* to the search (soo sak) or seizure (ap soo) under warrant of search (soo sak yung jang) hereof.

SECTION VIII. When the body of a person has been restrained by the police, the judicial police officer shall complete his investigation and deliver the suspect to the prosecutor within ten (10) days from the date of actual restraint or release him unless prior to the expiration of such period of ten (10) days an order of extension has been obtained from a competent court. But if the completion of the investigation shall require more than ten (10) days, the judicial police officer may apply to the competent court, through the prosecutor, for an extension of the period of detention, giving reasons therefor. The court, if convinced of the necessity therefor, may order an extension of not more than ten (10) days. Not more than one extension may be granted. At the expiration of the time fixed by the court, the suspect must be delivered to the prosecutor or be released.

— 3 —

SECTION IX   The prosecutor shall institute public action against the suspect or release him within ten (10) days from the date he actually restrains the suspect or receives the suspect from the police, unless within that time he shall obtain an order from a competent court for an extension. On application by the prosecutor, giving reasons therefor, the court, if convinced of the necessity, may order an extension of the detention for a period not exceeding ten (10) days. Not more than one extension shall be granted. At the expiration of the extension, the prosecutor shall institute public action against the suspect or release him.

SECTION X.   An order of extension under Sections VIII and IX hereof, shall state the date of expiration of the extension granted and shall be effective from the time the judge signs the same.

SECTION XI.   When a suspect or an accused is arrested, he shall immediately be informed of the charge against him and the particulars thereof and shall be informed that he may provide legal counsel for himself, subject to the provisions of Section XIV hereof. If the arrest is made in the presence of a member of his family, or if a member of his family makes inquiry, such member of his family shall likewise be notified of the charge and the particulars thereof and of the right of the suspect or the accused to have legal counsel.

SECTION XII.   When a person is under restraint of body, he may provide legal counsel for himself; or his legal representative, supervisor, lineal ascendants or descendants, spouse or family head may provide legal counsel for him independently. The designation of legal counsel shall be effective from the date of designation and shall continue to be effective through the first instance.

SECTION XIII.   Upon inquiry, the legal counsel for the suspect or the accused shall be informed of the charges against the suspect or the accused and the particulars thereof. He shall have the right to present evidence on behalf of the suspect or the accused to the judicial police officer, the prosecutor or the court.

SECTION XIV   Prior to institution of public action, interviews and written communications between the suspect and his legal counsel shall be permitted; *provided*, however, that the judicial police or the prosecutor, if there should be reasonable ground to fear that interviews and communications between the suspect and his legal counsel would result in destruction or fabrication of evidence or the escape of the suspect, may prohibit such interviews

— 4 —

軍政廳　官報　法令　　第一七六號　　一九四八年三月二十日

or communications. In the event such a prohibition is made, report thereof shall be made to the competent court, together with the ground of the prohibition. Legal counsel may object to the prohibition and apply to the court for an order of rescission of the prohibition. When the court receives such application, it shall decide upon the application, within a period of two (2) days.

SECTION XV. After the case has been committed for public trial, legal counsel and the accused in detention shall not be forbidden to correspond with or see each other.

SECTION XVI. Should legal counsel not appear, or should no legal counsel have been appointed, legal counsel may be designated by the presiding judge after hearing the opinion of the public prosecutor in the following cases:

a. If the accused is under 20 or is at least 70 years of age.

b. If the accused is female,

c. If the accused is a deaf or dumb person.

d. If the accused is suspected of being a person of unsound or weak mind.

e. If such a course is otherwise deemed necessary

SECTION XVII. a. When a person is restrained in his body by constituted authority or otherwise, he, his legal counsel or persons specified in Section XII hereof, may apply to the competent court for inquiry into the legality of the restraint. The application shall state all the pertinent facts and shall set out the following:

(1) The grounds on which it is claimed that the restraint is illegal.

(2) Whether or not a prior application has been made for release from the same restraint.

(3) If a prior application has been made, in what court the prior application was made and the action of the court thereon.

(4) If a prior application has been made, the reasons for the new application.

(5) If known, whether or not the person named in the application is restrained pursuant to a warrant issued by a court.

(6) If the person is restrained pursuant to a warrant issued by a court, the name of the court which issued the warrant and the alleged defects in the warrant.

b. The court which receives the application shall examine it at once.

If the application shows on its face that the restraint is legal, the application shall be denied. If the application shows on its face that the restraint is prima facie illegal, the court shall fix a day for the hearing thereof, not more than seven (7) days from the date the application is received, and shall order the person who is restraining the person named in the application to bring such person before the court and show cause for continuance of the restraint. If before the day set for the hearing, the person who is restraining the person named in the application shall exhibit to the court a valid warrant issued by a competent court within the time limits provided in Sections VIII and IX hereof, the court may deny the application before the day set. If, after hearing, the reasons for the restraint and the objections of the restrained person thereto and such evidence bearing on the legality of the restraint as the court deems necessary, the court determines the restraint to be unlawful, it shall order the release of the restrained person. No person released under provisions of this Section shall thereafter be restrained in his body on the same facts except pursuant to a warrant issued by a court.

SECTION XVIII. If either party to a hearing under Section XVII hereof is dissatisfied with the decision of the court, he may within three (3) days make application to the next higher court, which court shall proceed as provided in Section XVII hereof; *provided*, however, that if the lower court has ordered the release of the restrained person, he shall be released immediately.

SECTION XIX. a. Courts shall be liberal in releasing persons against whom warrants of arrest or detention have been executed on their giving reasonable bail, either before or after institution of public action. In fixing the amount of bail, the court shall take into consideration the circumstances of the case and shall fix an amount high enough to give the reasonable assurance that the suspect or accused will not attempt to escape.

b. Persons accused of offenses punishable by fine or by penal servitude or imprisonment for not more than six (6) months, shall be released on giving reasonable bail to be fixed by the court.

c. When a person has been detained for thirty (30) days after institution of public action without trial, the court, on application by or for him, must fix reasonable bail for his release unless the prosecutor shall show to the court good reason for denying bail. When the accused has been detained without trial for sixty (60) days after institution of public action, the court, on application by or for him, must fix reasonable bail for his release.

d. The provisions of paragraph *c.* of this Section will not apply to cases where the offense charged is punishable by death or by penal servitude or imprisonment for a term of fifteen (15) years or more.

e. Except as provided in this Section, the provisions of the Code of Criminal Procedure as to release on bail shall apply.

f. In cases in which the accused has been declared innocent (moo choi), has been acquitted (myun so), or the prosecution has been dismissed (kong so ee kak) in the lower court, and the prosecutor has appealed to a higher court, liberation on bail previously granted shall not be revoked except for causes provided in the Code of Criminal Procedure, and, if the liberation has not been previously granted, the court shall fix reasonable bail for release of the accused.

g. In cases in which the accused has appealed to a higher court from a conviction by a lower court, liberation on bail or revocation of bail previously granted shall be governed by the following:

(1) If the punishment fixed by the sentence is a minor fine, fine, detention or penal servitude or imprisonment for not more than six (6) months, liberation on bail shall not be revoked except for causes provided in the Code of Criminal Procedure, and, if liberation on bail has not been previously granted, the court shall fix reasonable bail for release of the accused.

(2) In cases where the punishment fixed by the sentence is penal servitude or imprisonment for more than six (6) months but less than fifteen (15) years, revocation of previous liberation on bail or approval of application for liberation on bail shall be discretionary with the court.

(3) In cases where the punishment fixed by the sentence is for death, penal servitude or imprisonment for life or for fifteen (15) years or more, liberation on bail shall be at once revoked and not thereafter allowed unless the higher court changes the sentence.

(4) If the prosecutor, suspect or accused is dissatisfied with the order of the court fixing or denying bail, he may appeal to the next higher court within three (3) days.

SECTION XX. The second paragraph of Article 471 of the Code of Criminal Procedure is amended so as to read as follows:

"Any person dissatisfied with a measure adopted by a judicial

軍政廳　官報　法令　第一七六號　　一九四八年三月二十日

police officer concerning detention, seizure or restoration of articles under seizure may demand from the court having jurisdiction over the district in which such judicial police officer exercises his functions the cancellation or alteration of such measure."

The provisions of Article 471 of the Code of Criminal Procedure shall apply not only to cases already covered thereby but also to cases of illegal seizure or illegal detention of articles and to cases of articles legally seized but not required as evidence in the trial of a case.

SECTION XXI. The Chief Prosecutor of each District Court shall assign one or more of the prosecutors attached to his office under his supervision to inspect the police jails and police stations and police substations in his province for the purpose of determining whether persons are illegally restrained therein. It is his responsibility that each police jail or police station or substation be inspected at least once each month. The prosecutor making the inspection shall question each prisoner who is restrained and inspect the records of restraint. If he shall find that any person has been restrained otherwise than as provided by law, he shall at once take jurisdiction of the case in the same manner as any other case which is turned over to him by the police. If he shall find that any law has been violated with respect to the restraint or treatment of persons, he shall investigate and prosecute in accordance with law. Any person who hinders the prosecutor in the performance of the duties imposed on him by this Section, shall be punished with penal servitude for not less than six (6) months nor more than seven (7) years.

SECTION XXII. *a.* Every person who shall unlawfully restrain a person's body shall be liable civilly to the person unlawfully restrained, in damages computed at one thousand won (W 1,000) for each day of illegal restraint. Any person who does not obey an order of a court made under this Ordinance and does not comply with provisions of Sections III, V, VI. VIII, and IX hereof, shall be punished with penal servitude for not less than six (6) months nor more than seven (7) years.

b. If any chief prosecutor of a district court, senior prosecutor of a branch court, chief of a police division, or chief of a police station fails to take appropriate action against subordinates directly under his control for violation of the provisions of this Ordinance, he shall be discharged immediately and shall not be eligible to any of the positions in the Department of Justice

— 3 —

or in the Police Department for a period of two (2) years thereafter.

SECTION XXIII. The period of restraint referred to in Article 113 of the Code of Criminal Procedure shall run from the date of the warrant of arrest (koo sok yung jang) provided in this Ordinance. **The provisions of Article 113 of the Code of Criminal Procedure shall not authorize the restraint of a person beyond the periods specified in Sections VIII and IX hereof, unless public action has been instituted.**

SECTION XXIV. The following laws and provisions of law are hereby repealed:

a. The Administrative Execution Act of July, 1914 (Governor General Act No. 23, July, 1914).

b. Articles 3, 12, 13, 14, 15, 16, 27, 38-2 and paragraph 10-2 of Article 1 of the Korean Criminal Act (Governor General Act No. 11, 18 March 1912) as amended.

c. All provisions of laws and regulations inconsistent with this Ordinance to the extent that they are so inconsistent.

SECTION XXV. This Ordinance shall be effective on 1 April 1948.

APPROVED: RECOMMENDED:

    WILLIAM F DEAN   AHN CHAI HONG
Major General, United States Army   Civil Administrator
Military Governor in Korea

軍政廳　官報　法令　第一七六號　　一九四八年三月二十日

SOUTH KOREAN INTERIM GOVERNMENT
Seoul, Korea

ORDINANCE
NUMBER 177

10 April 1948

REVISION OF CUSTOMS TARIFF

Pending action on the subject matters hereof by the Korean Interim Legislative assembly and the enactment of legislation pursuant thereto, the following shall govern:

SECTION I. *Purpose.* The purpose of this ordinance is to augment the national revenue by increasing customs on certain articles, and to encourage importation of food stuffs by abolition of customs on certain other articles.

SECTION II. *Revised customs tariff.* Section II of Ordinance Number 116 dated 8 October 1946 (*Revision of Customs Laws*) is hereby amended to read as follows:

"*Ad valorem customs duty.* Customs duties levied and collected upon all dutiable articles imported from foreign countries shall be uniformly ten percent (10%) *ad valorem.*"

SECTION III. *Exemptions.* In addition to the exemptions provided under existing laws, including said Ordinance Number 116, the following articles of food shall be exempt from import duties:

Tariff Number	Article
a. 12	Rice and paddy
b. 13	Barley
c. 14	Pearl barley
d. 15	Malt
e. 16	Wheat
f. 17	Oats
g. 17-2	Millet
h. 18	Kaoliang
i. 19	Maize, or Indian corn
j. 20	Buckwheat
k. 21-1	Soja beans
l. 21-2	Beans, red and white (small)
m.	Food and food supplies for private household use (not for sale)

— 1 —

SECTION IV. *Registration of private households.* Persons claiming the private-household exemption under Section III m her of shall register with district customs authorities the following information: (*a*) Name and specimen signature of person claiming exemption, (*b*) address of his residence, (*c*) number of persons in his household for whom duty-exempted food is intended, and (*d*) such other pertinent information as the customs authorities may require, in the form in which they may require it.

SECTION V. *Effective date.* This ordinance shall be effective thirty days after the date appearing hereon; except that, as to contracts already in force on the date appearing hereon, the increases under this ordinance shall not apply provided that copies of such contracts or evidence thereof are filed with the district customs authorities, by the party claiming the waiver, prior to the effective date hereof.

APPROVED:

WILLIAM F DEAN
Major General, United States Army
Military Governor in Korea

RECOMMENDED:

AHN CHAI HONG
Civil Administrator

SOUTH KOREAN INTERIM GOVERNMENT
Seoul, Korea

ORDINANCE
NUMBER 178

10 April 1948

AMENDING KOREAN HORSE ASSOCIATION ORDINANCE

Pending action on the subject matters hereof by the Korean Interim Legislative Assembly and the enactment of legislation pursuant thereto, the following shall govern:

SECTION I. *Purpose.* The purpose of this ordinance is to amend the existing law regulating horse racing and to provide additional revenue to encourage the development of the horse industry.

SECTION II. *Amendments.* Law Number 47 dated 10 April 1923 (*Horse Racing Act*), referred to in Governor General Ordinance Number 1 dated 14 February 1942 (*Korean Horse Association Ordinance*), is hereby amended as follows:

a. In paragraph 1 of Section 4 of said Law, the words "*not less than five won and not more than twenty won*" are deleted and the words "*not less than twenty won and not more than one hundred won*" are substituted therefor.

b. In paragraph 1 of Section 6 of said Law, the words "*provided that the amounts shall not exceed ten times the face value of the race ticket*" are deleted.

SECTION III. *Repealer.* Ordinance Number 138 dated 1 April 1947 (*Amending Horse-Racing Ordinance*) is hereby repealed.

SECTION IV. *Effective date.* This ordinance shall be effective on the date appearing hereon.

APPROVED:
WILLIAM F DEAN
Major General, United States Army
Military Governor in Korea

RECOMMENDED:
AHN CHAI HONG
Civil Administrator

SOUTH KOREAN INTERIM GOVERNMENT
Seoul, Korea

ORDINANCE
NUMBER 179   1 April 1948

PROVISIONS FOR TEMPORARY FAMILY REGISTERS

Pending action on the subject matters hereof by the Korean Interim Legislative Assembly and the enactment of legislation pursuant thereto, the following shall govern:

SECTION I. The purpose of this ordinance is to provide, until North and South Korea shall be unified, measures for temporary family registers for persons now residing in Korea, South of Thirty-Eight Degrees North Latitude (hereinafter referred to as *South Korea*) whose family registers are presently in Korea, North of Thirty-Eight Degrees North Latitude (hereinafter referred to as *North Korea*).

SECTION II. *a.* Any person who on 15 August 1945 had his family register in North Korea and who now or hereafter resides in South Korea, may designate his existing place of residence in South Korea as his provisional permanent domicile, and apply to the competent authority for initial entry in the family registration books.

b. The application mentioned above shall state expressly the permanent domicile in North Korea, and a copy or extract of material facts from the family registration book and the matters set forth in Section 11 of Government General Order Number 154, dated December, 1922, (*Korean Family Registration Order*) shall be submitted therewith. If the copy or extract required cannot be supplied, a certificate of the facts made by two persons over twenty years of age having knowledge of the facts, shall be submitted with the application.

c. Heads of pu, ku, eup, or myun, upon the receipt of applications pursuant to the above, shall compile the temporary family registers in conformity with Sections *3* to *6*, inclusive, of the Korean Family Registration Order.

SECTION III. *a.* Heads of pu, eup, and myun having jurisdiction thereof, shall compile temporary family registers in conformity with Section 29 of Governor General Instruction Number 15, dated March, 1923 (*Enforcement Procedure for the Korean Family Registration Order*), in behalf of those persons

— 1 —

having permanent domiciles in South Korea on 15 August 1945, and whose family registers are located in North Korea and not available to them for use.

b. Any person coming under the foregoing subsection, may apply for a family registration, with the provisions of Section *II* hereof to apply *mutatis mutandis.*

SECTION IV. *a.* The family registers compiled in accordance with Section III and IV hereof, shall have legal effect as family registers under the Korean Family Registration Order until family registers in North Korea shall be available again, after North and South Korea are reunited.

b. The acts notified in domestic relations, based upon the temporary family registers created pursuant to Sections *III* and *IV* hereof, shall have the same legal effect as though based upon the original family registers, insofar as not inconsistent with the original family register.

SECTION V. Any person who makes a false statement, application, or certificate under Sections *II* and *III* hereof, shall be punished upon conviction by not more than ten (*10*) years' penal servitude or fine of not more than two hundred thousand won (*W 200,000*).

SECTION VI. This ordinance shall be effective on the date appearing hereon.

APPROVED:

    WILLIAM F DEAN
Major General, United States Army
Military Governor in Korea

RECOMMENDED:

    AHN CHAI HONG
Civil Administrator

SOUTH KOREAN INTERIM GOVERNMENT
Seoul, Korea

ORDINANCE

NUMBER 180   31 March 1948

ORDINANCE NUMBER 176 SUPPLEMENTED

Pending legislative action on the subject matters hereof and the enactment
f legislation pursuant thereto, the following shall govern:

SECTION I. *Purpose.* The purpose of this ordinance is to provide for
integrating present criminal procedures into those provided by Ordinance
Number 176, dated 20 March 1948 (*Changes in Criminal Procedure*).

SECTION II. *The effect of warrants issued prior to Ordinance Number 176.*
A warrant issued by a court, public prosecutor, or judicial police pursuant to
the Korean Criminal Act (*Governor General Act Number 11, dated 18 March
1912*), or the Criminal Procedure Code. prior to the effective date of Ordinance
Number 176, shall continue to have legal effect until expiration, pursuant to
such Act or Code.

SECTION III. *Computation of time.* The periods *thirty days* and *sixty
days* appearing in Section XIX c of Ordinance Number 176, shall be construed,
as to cases pending on the effective date of Ordinance Number 176, to run
from such effective date.

SECTION IV. *The effect of records made pursuant to provisions of repealed
law.* Records heretofore made pursuant to provisions of law repealed by Ordinance Number 176, shall continue to be legally admissible in evidence to the
same extent provided in such repealed provisions of law.

SECTION V. *Procedure to secure warrants of restraint or search.* Applications for warrants of restraint or search shall be made as follows:

a. A public prosecutor shall apply to the court to which he is assigned.

b. A judicial policeman or other constituted authority shall apply to the
public prosecutor. who shall submit the application to the court.

SECTION VI. *Authority of special judicial officers.* The chief judge of the
district court may authorize the special judicial officer in any city, gun. or island
where there is no court. to issue warrants of arrest and search under Ordinance
Number 176. as representative of the appropriate judge. The chief judge shall
give notice to the appropriate official of the names and location of the special

景刑題　官報　法令　第一八〇號　　一九四八年三月三一日

judicial officers who are given such authority. In such places, the police officials or other constituted authority may apply for a warrant directly to the special judicial officer without the necessity of going through the prosecutor. In such places, the time within which a warrant must be obtained under Section VI of Ordinance Number 176 shall be forty-eight (48) hours.

SECTION VII. *Effective date.* This ordinance shall be effective on 1 April 1948.

APPROVED: RECOMMENDED:

W. F. DEAN
Major General, United States Army
Military Governor in Korea

AHN CHAI HONG
Civil Administrator

— 2 —

SOUTH KOREAN INTERIM GOVERNMENT
Seoul, Korea

ORDINANCE
NUMBER 181   1 April 1948

SPECIAL WAR TIME LEGAL PROCEDURE ORDINANCE REPEALED

Pending action on the subject matters hereof by the Korean Interim Legislative Assembly and the enactment of legislation pursuant thereto, the following shall govern:

SECTION I. *Purpose.* The purpose of this ordinance is to restore the three instance trial system, to repeal the Special Regulations for the Korea Courts Ordinance, for the Korea Civil Ordinance and for the Korea Criminal Ordinance, which were emergency measures in war time, and to amend other ordinances connected therewith, and thereby to return the judicial system to normal procedures.

SECTION II. *The War Time Special Regulation for Korea Courts Ordinance repealed and the Korea Courts Ordinance amended.*

a. Governor General Ordinance Number *2*, dated 15 February 1944, (*War Time Special Regulation for Korea Courts Ordinance*), is hereby repealed; provided, however, that in cases in which the oral pleading of first instance has been concluded and the appeal against a rule of first instance taken prior to the effective date of this ordinance, the former provisions shall apply.

b. The provisions of Section VI, (*Jurisdiction over Civil Cases*) of Appointment Order Number *36*, dated 19 November 1945, are hereby repealed.

c. Imperial Ordinance Number *236*, dated 18 October 1909, (*Korea Courts Ordinance*) is hereby amended by substituting in Section IV thereof the words *one hundred thousand won* for the words *one thousand yen*.

d. The provisions of Section *IV* of Imperial Ordinance Number *236*, dated 18 October 1909 (*Korea Courts Ordinance*) shall apply to cases of violation of Military Government Ordinances which provide that "Any person violating the provisions of this Ordinance shall, upon conviction by a Military Occupation Court, (or "by a duly constituted court") suffer such punishment as the court shall determine", and which do not specifically limit the kinds or terms of punishment, or the amounts of fine. The single judge mentioned in Section *IV* of the Korea Courts Ordinance shall in all such cases examine the record presented by the prosecutor, and consider all pertinent circumstances;

— 1 —

503

and if it appears to him that if there were a conviction a case would presumably be punishable by death or by hard labor or imprisonment for more than ten (*10*) years or for life, the single judge shall, pursuant to the provisions of the Korean Criminal Ordinance, transmit the papers in the case to the appropriate three-judge court.

e. If the trial of a principal is transferred to a three-judge court, pursuant to sub-section "d" above, the cases of all those triable jointly with the principal shall also be transferred to such three-judge court.

SECTION III. *Korea War Time Civil Special Regulation repealed and Korea Civil Ordinance amended.*

a. Governor General Ordinance Number *3*, dated 15 February 1944 (*Korea War Time Civil Special Ordinance*) is hereby repealed; provided, however, that Sections *10-2* and *10-3* of Public Act Number *63*, dated 24 February 1942 (*War Time Civil Special Law*), applied in Korea by paragraph *1*, Section *I* of the said Governor General Ordinance Number *3*, shall apply to cases in which the oral pleadings of first instance have been concluded prior to the effective date of this ordinance; and the provisions of Sections XI to XXI of the Korea War Time Civil Special Ordinance shall apply to arbitration cases pending prior to the effective date of this ordinance.

b. Governor General Ordinance Number 7, dated 19 March 1912 (*Korea Civil Ordinance*) is hereby amended by substituting in Sections *44* and *45* thereof the words *five thousand won* for the words *fifty yen*.

c. In Section *398* of Public Act Number *29*, dated 21 April 1890 (*Code of Civil Procedure*), the words *twenty days* are substituted for the words *thirty days*.

SECTION IV. *Korea War Time Criminal Special Regulation repealed and Korea Criminal Ordinance amended.*

a. Governor General Ordinance Number *4*, dated 15 February 1944 (*Korea War Time Criminal Special Regulation*), is hereby repealed, provided, however, the provisions of Sections *27* to *29* of Public Act Number *64*, dated 24 February 1942 (*War Time Criminal Special Law*), applied in Korea by paragraph *1*, Section *1* of the said Governor General Ordinance Number *4*, shall apply to cases in which oral pleadings have been concluded prior to the effective date of this ordinance, and the provisions of Sections *29-3* to *29-5* of the said law shall apply to a case in which a summary order has been served, but the case has not been finally decided.

— 2 —

b. Sections *412* and *414* of Public Act Number *75*, dated 1922 (*Code of Criminal Procedure*) are hereby repealed; provided, however, that these sections shall continue to apply to cases presently on appeal before the Appeal Court.

c. Subsection *I* of Section *422* of the Code of **Criminal** Procedure is hereby amended to read as follows:

*Upon the receipt of the record of the case, the **Supreme Court** shall promptly notify the appellant and the respondent of such fact.*

d. Section *423* of the Code of Criminal Procedure is hereby amended to read as follows:

The appellant shall file a statement of reasons for second appeal (Sanko) to the Supreme Court within twenty days from the date on which he has received the foregoing notification.

e. Subsection *I* of Section *424* of the Code of Criminal Procedure is hereby amended to read as follows:

The respondent may file notice of cross-appeal within twenty days after he has received the notification mentioned in subsection 1 of Section 422.

f. Section *442* of the Code of Criminal Procedure is hereby amended to read as follows:

Should it consider that ground for appeal clearly does not exist, due to the statement of the reasons (Sanko) or any record before it, the Supreme Court may enter judgment of dismissal of the second appeal, without hearing.

SECTION V. *Effective date.* This ordinance shall be effective on the thirtieth day after the date appearing hereon.

APPROVED:

W. F. DEAN
Major General, United States Army
Military Governor in Korea

RECOMMENDED:

AHN CHAI HONG
Civil Administrator

SOUTH KOREAN INTERIM GOVERNMENT
Seoul, Korea

ORDINANCE
NUMBER 182

7 April 1948

EXCHANGE OF 100-WON BANKNOTES WITH JAPANESE FEATURES

Pending action on the subject matters hereof by the Korean Interim Legislative Assembly and the enactment of legislation pursuant thereto, the following shall govern:

SECTION I. *Purpose.* The purpose of this ordinance is to provide for the removal from circulation of certain 100-won banknotes, bearing Japanese features.

SECTION II. *Circulation prohibited.* a. There shall be removed from circulation all 100-won-Bank-of-Chosun banknotes which bear:

(1) The Paulowina Crest (*Japanese Imperial Household flower*);

(2) The statement that they may be exchanged for Bank-of-Japan notes; and

(3) The statement that they are printed by the Imperial Japanese Government Printing Bureau.

b. The above-described notes are named officially *Cho Series, Cho Kap Series, Cho Kai Series, Cho Ul Series, and Cho Byung Series.* They will hereinafter be referred to as "*Japanese-feature banknotes.*"

c. After 11 April 1948, Japanese-feature-100-won banknotes shall not constitute legal tender in Korea.

SECTION III. *Exchange of banknotes.* a. During the period beginning 2 April 1948 and ending 24 April 1948, Japanese-feature-100-won banknotes may be exchanged for notes bearing the Korean everlasting flower and the statement that they are printed in the Chosun Book Printing Plant in Seoul.

b. All banks, branch banks, financial associations, and federations of financial associations (*hereinafter collectively referred to as "banks"*) shall during this period effect this exchange or accept Japanese-feature-100-won banknotes for deposit.

SECTION IV. *Disposition of banknotes.* a. On 26 April 1948, all banks, by telegraph or messenger, shall notify the Seoul Head Office of the Bank of Chosun of the amount of Japanese-feature-100-won banknotes on hand at the close of business on 24 April 1948.

    b. Between 26 April 1948 and 1 May 1948, inclusive, all banks shall transfer all Japanese-feature-100-won banknotes to the nearest branch of the Bank of Chosun for credit or exchange for legal tender.

    SECTION V. *Illegal possession.* After 24 April 1948, no person, natural, juridical, or governmental, shall possess Japanese-feature-100-won banknotes, unless authorized by this ordinance or by the Department of Finance.

    SECTION VI. *Banknotes illegal'y imported.* If any person, natural, juridical or governmental, submits for exchange Japanese-feature-100-won banknotes which the bank where submitted believes may have been imported in violation of existing laws and directives limiting the amount of cash repatriates are permitted (*1,000 won per repatriate*) to bring into South Korea, the bank shall:

    a. Accept the currency, but pay out no legal tender in exchange;

    b. Deposit the funds in a blocked account to the credit of the depositor;

    c. Notify the Department of Finance through the Seoul Head Office of the Bank of Chosun, or, in the case of Financial Associations through the Federation of Financial Associations;

    d. Permit no withdrawals from the blocked account without permission of the Department of Finance.

    SECTION VII. *Penalties.* *a.* Any person who after 11 April 1948 circulates or attempts to circulate as legal tender Japanese-feature-100-won banknotes, or

    b. Any person who exchanges or attempts to exchange Japanese-feature-100-won banknotes which have been imported into South Korea in violation of law or directive, or

    c. Any bank employee who accepts Japanese-feature-100-won banknotes for exchange after 24 April 1948, or

    d. Any person who, after 24 April 1948, possesses Japanese-feature-100-won banknotes in violation of Section V hereof,

shall, upon conviction by a duly constituted court, be fined not less than one thousand won (*W 1,000*) or more than fifty thousand won (*W 50,000*), or be sentenced to not less than thirty (*30*) days' or more than one (*1*) year's penal servitude, or both.

    SECTION VIII. *Banknotes not redeemed.* The Bank of Chosun shall pay to the South Korean Interim Government the difference between the amount of outstanding Japanese-feature-100-won banknotes and the amount presented for exchange pursuant to the provisions hereof.

SECTION IX. *Effective date.* This ordinance shall be effective on the date appearing hereon.

APPROVED: RECOMMENDED:

W. F DEAN   AHN CHAI HONG
Major General, United States Army   Civil Administrator
Military Governor in Korea

軍政廳　官報　法令　第一八二號　　　一九四八年四月七日

SOUTH KOREAN INTERIM GOVERNMENT
Seoul, Korea

ORDINANCE
NUMBER 184   16 April 1948

INCREASING PORT FEES

Pending action on the subject matters hereof by the Korean Interim Legislative Assembly and the enactment of legislation pursuant thereto, the following shall govern:

SECTION I. *Purpose.* The purpose of this ordinance is to increase the governmental revenues of southern Korea.

SECTION II. *Port fees increased.* All existing port fees, including those set forth in the following laws and regulations, are hereby increased to fifty times the amounts prevailing immediately prior to the effective date of this ordinance:

a. Law Number 88; dated 24 march 1899 (*Tonnage Tax Law*).

b. Governor General Ordinance Number 2, dated 25 March 1911 (*Custom Broker Ordinance*).

c. Government General Ordinance Number 29, dated 25 March 1911 (*Enforcement Regulations for Customs Broker Ordinance*).

d. Government General Ordinance Number 86, dated 13 July 1911 (*Customs License Fee Ordinance*).

e. Government General Ordinance Number 115, dated 28 August 1920 (*Enforcement Regulations for Customs Law and Law No. 53, dated 1920*).

f. Government General Ordinance Number 118, dated 28 August 1920 (*Enforcement Regulations for Korea Tonnage Tax Ordinance*).

g. Government General Ordinance Number 47, dated 1 April 1922 (*Regulation for Use of Customs Pier Moorings and Locks*).

h. Government General Ordinance Number 85, dated 31 August 1927 (*Enforcement Regulations for Bonded Factory Law*).

i. Government General Ordinance Number 86, dated 31 August 1927 (*Enforcement Regulations for Bonded Warehouse Law.*)

j. Government General Ordinance Number 114, dated 3 December 1927 (*Enforcement Regulatons for provisions Relative to Customs in Aviation Law*).

軍政廳　官報　法令　第一八四號　一九四八年四月一六日

SECTION III. *Effective date.* This ordinance shall be effective on the date appearing hereon.

APPROVED:

W. F. DEAN
Major General, United States Army
Military Governor in Korea

RECOMMENDED:

AHN CHAI HONG
Civil Administrator

SOUTH KOREAN INTERIM GOVERNMENT
Seoul, Korea

ORDINANCE

NUMBER 185   24 April 1948

CORPORATION PROCEDURES SIMPLIFIED FOR VESTED JURIDICAL PERSONS

Pending action on the subject matters hereof by the Korean Interim Legislative Assembly and the enactment of legislation pursuant thereto, the following shall govern:

SECTION I. *Purpose.* The purpose of this ordinance is to simplify procedures for the reorganization of juridical persons in which interests are vested under Ordinance Number 33, dated 6 December 1945 (*Vesting Title to Japanese Property Within Korea*).

SECTION II. *Transfer of vested shares.* Upon the delivery of a certificate. signed by the Property Custodian, showing the names and holdings of those persons whose shares he has ascertained are vested in the Military Government by virtue of the provisions of Ordinance No. 33, the juridical person to whom such certificate is delivered shall forthwith cause to be cancelled on its register of shareholders the registration in the names of such persons and shall effect registration of such shares in the name of Military Government. It shall thereupon issue to the Military Government in Korea a new share certificate dated 9 August 1945 for the total number of shares certified to by the Property Custodian as being vested. The new share certificate shall be delivered to the Property Custodian. All restrictive provisions concerning the transfer of shares shall be ineffective with respect to vested shares.

SECTION III. *Additional certificates.* The filing of a certificate by the Property Custodian showing the names and holdings of those persons whose shares he has ascertained are vested shall be no bar to his filing of an additional or amended certificate with the same juridical person.

SECTION IV. *Claims of title by non-Japanese.* Non-Japanese individuals and juridical persons who claim ownership of shares declared vested by the Property Custodian shall be limited to presentation of such claims to the Property Claims Commission, established by Ordinance Number 103, dated 31 August 1946. All such claims must be filed within six months from the date hereof or within three months after registration of shares in the name of Military

— 1 —

Government, whichever date is later.

SECTION V. *Directors and auditors.* Directors and auditors nominated and elected by persons voting the vested shares, need not be shareholders in the company and need not make any deposit of company shares.

SECTION VI. *Continuation of legal entity.* In those cases where all of the shares of a joint stock company have vested in Military Government under Ordinance Number 33, the joint stock company shall continue as a separate legal entity, notwithstanding any principle of law or legal enactment to the contrary.

SECTION VII. *Failure to give proper notice.* The failure to publicly post notice at each Provincial Property Custody Office not less than fifteen (15) days before the date set for a general meeting of the shareholders, as provided for in Ordinance No. 156, dated 21 November 1947, shall not invalidate any proceedings had at such meeting; provided, however, that more than 50% of the number of shares outstanding are represented at such meeting and that notice of such meeting has been mailed to each person registered as a shareholder in the company's register of shareholders (*other than those shareholders whose shares have been certified as being vested*) addressed to the address furnished by such shareholder to the company.

SECTION VIII. *Inconsistent provisions.* Notwithstanding the provisions of any general or special law, or the articles of incorporation, by-laws or resolutions of a juridical person, the foregoing shall apply to all juridical persons in which interests are vested under Ordinance No. 33.

SECTION IX. *Effective date.* This ordinance shall be effective on the date appearing hereon.

APPROVED:

    W. F. DEAN
    Major General, United States Army
    Military Governor in Korea

RECOMMENDED:

    AHN CHAI HONG
    Civil Administrator

SOUTH KOREAN INTERIM GOVERNMENT
Seoul, Korea

ORDINANCE
NUMBER 186

1 April 1948

AMENDING THE KOREAN INCOME TAX ORDINANCE

Pending legislative action on the subject matters hereof and the enactment of legislation pursuant thereto, the following shall govern:

SECTION I. *Purpose.* The purpose of this ordinance is to increase revenues and revise the bases and rates of existing income tax laws.

SECTION II. *Income tax ordinance amended.* Governor General Ordinance Number 6 dated 30 April 1934 (*Korean Income Tax Ordinance*) as amended, is further amended as follows:

a. *Section 3.* Paragraphs *d* and *e* under the *2nd Class* are repealed.

b. *Section 14.* (1) The figure *three thousand won* (*W 3,000*) in Section 14 is increased to *one hundred thousand won* (*W 100,000*).

(2) Paragraph *1*, items *4* and *5*, and paragraphs 2 and 3 of Section 14 are repealed.

c. *Section 15.* (1) The figure *three thousand won* (*W 3,000*) in paragraph 1, item *1-3*, of Section 15 is increased to *one hundred thousand won* (*W 100,0 00*).

(2) Paragraph *1*, item *5*, of Section 15 is amended by adding at the end thereof the following: *The "amount received" shall include the fair market value. as of the date of receipt of any property, goods or assets received in lieu of, or in addition to, any of the foregoing.*

(3) Paragraph *1*, item *6*, of Section 15 is amended by adding at the end thereof the following: *However, the gain from the sale, exchange or transfer of immovable property, mining rights and registered vessels in the case of a person other than a dealer or broker in such properties, shall be included in income only to the extent hereinafter provided in Section 15-5 hereof.*

d. *Section 15-2* is amended as follows:

(1) By deleting therefrom the following: *Area of the Income Tax Jurisdiction, Formosa, Kwantung Chu, South Ocean Islands or Saghalien and notwithstanding Section 23, No. 6 and No. 7.*

(2) The figure *thirty thousand won* (*W 30,000*) appearing therein is

— 1 —

increased to *one hundred twenty six thousand won* (W 126,000).

    e. Section 15-2. The figure *five thousand won* (W 5,000) is increased to *fifty thousand won* (W 50,000).

    f. Sections *15-5* and *15-6* are repealed and a new Section *15-5* is substituted in lieu thereof to read as follows:

"SECTION 15-5. GAIN FROM SALE, EXCHANGE OR TRANSFER OF IMMOVABLE PROPERTY, MINING RIGHTS OR REGISTERED VESSELS.

"*a. Applicability of this section.* The provisions of this section shall apply only to those cases where the total *gross gains* (exclusive of losses) realized during the preceding year from the sale, exchange or transfer of immovable property, mining rights or registered vessels *were* one hundred thousand won (W 100,000) or more. If such total gross gains are less than one hundred thousand won (W 100,000), they shall not be included in third-class incomes.

"*b. Gains and losses recognized.* Gains and losses under this section shall be recognized for purposes of third-class incomes only to the following extent:

                                                                   Extent

    (I) Property acquired on or before 8 August 1945...................20%

    (II) Property acquired on or after 9 August 1945

        and on or before 31 December 1946............................60%

    (III) Property acquired on or after 1 January 1947.................80%

"*c. Basis of property.* (*I*) *Property acquired on or before 8 August 1945.* The 'basis' of the property shall be the cost thereof or the fair market value as of 8 August 1945, whichever is the greater.

"(*II*) *Property acquired on or after 9 August 1945.* The 'basis' of the property shall be the cost thereof, except that:

"(a) If the property was acquired by exchange, bequest, devise or inheritance, the basis shall be the fair market value as of the date of acquisition.

"(b) If the property was acquired by gift, the basis shall be the cost to the donor or the fair market value as of the date of the gift, whichever is the lesser.

"(c) If the property was acquired by distribution in liquidation of a juridical person, the basis shall be the fair market value as of the date of distribution or the amount used for the determination of taxable income to the distributor, whichever is the lesser.

— 2 —

"d *Computation of gain or loss; Definitions.* (I) The gain realized is the excess of the amount realized from the sale or other disposition of the property during the preceding year over the basis thereof.

"(II) The *loss incurred* is the excess of the basis over the amount realized during the preceding year.

"(III) The *amount realized* is the sum of money received, plus the fair market value as of the date of receipt of any property (other than money), less necessary expenses incurred in connection with the sale, exchange, or transfer of the property.

"(IV) The *gain recognized* is the amount determined by applying the percentages set forth in paragraph *b* of this section to the 'gains realized.'

· "(V) The *loss recognized* is the amount determined by applying the percentages set forth in paragraph *b* of this section to the 'losses incurred.'

"(VI) A *net gain* is the excess of the total 'gains recognized' over the total 'losses recognized.'

"(VII) A *net loss* is the excess of the total 'losses recognized' over the total 'gains recognized.'

"e. *Amount taxable.* The amount to be included in taxable third-class income is the 'net gain.' A 'net loss' is not deductible.

"f. *Gain on sale, exchange or transfer of personal residence.* The gain realized on the sale, exchange or transfer of the taxpayer's personal residence shall be exempt from income tax; provided the taxpayer actually had occupied the residence as his personal abode for at least three of the four years immediately preceding the date of sale, exchange or transfer; or, if the residence had been owned for less than four years, then for at least seventy-five percent (75%) of the period during which owned; provided, however, that no exemption shall be allowed for periods of occupancy of less than one year; and provided further that this exemption may be taken for not more than one residence during any one period of time."

g. *Section 17.*. The maximum total supporting family allowance of *five thousand won* (W 5.000), authorized in Section 17, is increased to *ten thousand won* (W 10.000)

h. *Section 20* is amended by adding at the end thereof the following: *Provided, however, that this exemption shall not extend to any foundational juridical person whose profits, or any part thereof, inure directly or indirectly to the benefit of*

— 3 —

any private contributor or individual.

    i. Section 22-2. The figure *five thousand won* (W 5 000) is increased to *fifty thousand won* (W 50 000).

    j. Section 23 is amended as follows:

    (1) Items *(5)*, *(6)*, and *(7)* are repealed and the following are substituted in lieu thereof:

    "*(5)* Property acquired by inheritance.

    "*(6)* Amounts received under a life insurance contract by reason of the death of the insured.

    "*(7)* Amounts received as insurance or damages for personal injuries sustained."

    (2) Item *(8)* is amended by substituting *Korean* for *Japanese* and by deleting therefrom the following: *The areas of the Income Tax jurisdiction, Formosa, Kwantung Chu, Saghalien and South Ocean Islands.*

    k. Section 32. The figure *twelve thousand won* (W 12,000) is increased to *twenty-five thousand won* (W 25,000)

    l. Section 33 is amended as follows:

    (1) Paragraph *1* is repealed and the following is substituted in lieu thereof:

    "The tax on the first class incomes shall be as follows:

    "Ordinary income..40%

    "Income from liquidation:

    "Composed of reserve funds and nontaxable income..............20%

    "Other..40%"

    (2) Paragraphs *5* and *6* are repealed.

    m. Section 35 is amended as follows:

    (1) The table of withholding rates appearing in paragraph 1 A is repealed and the following shall be substituted in lieu thereof:

    "Interest on national bonds 10%

    "Interest on public bonds and debentures guaranteed by the government..25%

    "Others... .... 27%"

    (2) The table of withholding rates appearing in paragraph 1 C is revised to read as follows:

    "Not more than W 100,000..15%

"More than W 100,000 but not more than W 1,000,000... . ...25%

"More than W 1,000,000 but not more than W 5,000,000......... 35%

"More than W 5,000,000................. ....45%"

(3) Paragraphs *1- D* and *1-E* are repealed.

(4) Paragraphs *2, 3, 4, 7* and *8* are repealed.

   n. *Section 36.* (*1*) The cumulative income tax schedule in paragraph *1*
is revised to read as follows:

"*Amount of Income (in Won)*

Less than W 25,000

25,000 and over but less than	30,000
30,000 and over but less than	40,000
40,000 and over but less than	60,000
60,000 and over but less than	100,000
100,000 and over but less than	150,000
150,000 and over but less than	200,000
200,000 and over but less than	400,000
400,000 and over but less than	700,000
700,000 and over but less than	1,000,000
1,000,000 and over but less than	5,000,000
5,000,000 and over but less than	10,000,000

10,000,000 and over

   Tax (in Won)

No tax

1,000 plus 10% in excess over	25,000
1,500 plus 13% in excess over	30,000
2,800 plus 16% in excess over	40,000
6,000 plus 20% in excess over	60,000
14,000 plus 29% in excess over	100,000
28,500 plus 39% in excess over	150,000
48,000 plus 48% in excess over	200,000
144,000 plus 59% in excess over	400,000
321,000 plus 63% in excess over	700,000
510,000 plus 65% in excess over	1,000,000
3,110,000 plus 68% in excess over	5,000,000
6,510,000 plus 70% in excess over	10,000,000"

(2) Paragraphs *4* and *5* are repealed.

o. Section 39. Paragraph 2, Section 39, is hereby repealed and the following is substituted in lieu thereof:

"In any case where it is determined by the Government that the amount of tax due under this ordinance for any taxable year is in excess of the amount previously assessed for that year, the additional tax may be assessed immediately."

p. Section 45. The words *Governor General of Korea* are changed to read *Director of the Department of Finance.*

q. Section 50-2. The following section is added:

"SECTION 50-2. ADVANCE ASSESSMENTS ON CURRENT YEAR INCOME

"*a. Assessment.* The amounts of all third-class income tax assessments made on and after the effective date of this ordinance covering income years ending after 31 December 1946, shall be increased for 'advance assessments' according to the following schedule, to be amended from time to time:

Assessments for taxable year 1948 based on incomes received during the income year 1947...25%

"*b. Definitions* For purposes of this Section *50-2:*

"(I) *Income year* is the year preceding the 'taxable year' and refers to the year during which the income was received.

"(II) *Taxable year* is the year following the 'income year' and refers to the year during which the tax, except for this section, would be normally assessed.

"(III) *Advance assessment* is the amount of income tax which, in accordance with subsection *a* of this section, is assessed in advance of the date otherwise prescribed for the assessment of the tax.

"*c. Credits against assessments.* Third-class income tax assessments made on or after 1 April 1949, based on income received during any 'income year' shall be credited with all 'advance assessments' which may have been made with respect to the income received during that 'income year.'

"*d. Refunds and cancellations.* In any case where the aggregate 'advance assessments' with respect to a particular 'income year' exceed the total corrected assessments relating to said income year, the excess assessment, if collected, shall be refunded or credited in accordance with regulations prescribed by the

— 6 —

Commissioner with the approval of the Director; or, if not collected, the excess assessment shall be abated in accordance with regulations prescribed by the Commissioner with the approval of the Director.

"*e. Applicability of other provisions.* All provisions of existing laws, which are not inconsistent with the provisions of this section 50-2, and which relate to the assessment and collection of third-class income taxes, including penalty provisions, shall apply with equal force and effect to the 'advance assessments' provided in subsection *a.* hereof."

r. Section 53. The word *province* is changed to read *National Tax Supervision Office.*

s. Section 54. The words *Governor General of Korea* are changed to read *Director of the Department of Finance.*

t. Section 57. The minimum personal income tax (third income tax) authorized to be paid in two installments is changed from *five hundred won* (W 500) to *twelve hundred won* (W 1,200).

u. Section 66 is amended by adding at the end thereof the following: *The monetary penalty imposed by this section shall be assessed, collected and paid at the same time and in the same manner as the tax.*

v. Section 66-2. The following section is added:

"SECTION 66-2. FRAUD

"If any person evades or attempts to evade, directly or indirectly, any tax imposed by this ordinance, he shall, upon conviction thereof, and in addition to other penalties provided by law, be fined not more than one hundred thousand won (W 100,000) or imprisoned for not more than two years, or both."

w. Section 69. The penalty imposed is increased to *fifty thousand won* (W 50,000).

x. Sections repealed. The following Sections are hereby repealed in their entirety: 6-2, 7, 13, 21, 22, 22-3, 22-4, 24, 25, 26, 27, 27-2, 27-3, 27-4, 27-5, 27-6, 27-7, 27-8, 27-9, 27-10, 27-11, 28, 29 30, 30-2, 31-3, 31-5, 31-6, 35-2, 35-4, 36-2, 36-3, 36-4, and 46-3.

SECTION III. *Informers. a.* The provisions of Section XVI of Ordinance Number 169, dated 1 March 1948 (*Reorganization of Bureau of Internal Revenue*), providing for payment to certain informers of a part of the penalties collected, shall apply to penalties collected under the Korean Income Tax Ordinance, as amended by this ordinance.

b. Twenty-five percent (25%) of all penalties levied, assessed, collected and paid because of income tax deficiencies shall be paid to the Korean Cooperative Welfare Association of the Bureau of Internal Revenue; provided however, that no such payment shall be made in any case where payment is made under subsection *a* of this section.

SECTION IV. *Specific repealer.* a. The following ordinances are hereby repealed:

(1) Ordinance Number 5 dated 31 March 1937 (*The Korean Capital Interest Tax Law*), as amended.

(2) Ordinance number 8 dated 31 March 1940 (*The Public Loan and Debenture Interest Tax Law*), as amended.

(3) Ordinance Number 4 dated 31 March 1937 (*The Korean Foreign Bond Special Tax Law*), as amended.

(4) Ordinance Number 17 dated 5 October 1937 (*Ordinance for Reduction or Remission of Taxes and Postponement of Collection of Taxes for Army and Navy*).

(5) Ordinance Number 24 dated 24 March 1942 (*War-Time Damage National Tax Remission Law*).

(6) Ordinance Number 23 dated 24 March 1942 (*Ordinance for Prevention of Double Assessment of Taxes Between Japan and Manchuria*).

b. The effective date of the repeal under subsection *a* of this section shall be the first day of the month following the date appearing hereon; and no taxpayer shall be relieved of any tax, penalty or interest which may have been incurred prior to the effective date of this section, under the provisions of the ordinances repealed by this section.

SECTION V. *Effective date.* Except as otherwise provided herein, the provisions of this ordinance shall be effective on the date appearing hereon and shall be applicable to the computation of income for the following years:

First Class (juridical persons):

Ordinary income.   For business years ending after 31 December 1946.

Liquidation income.   For dissolutions and amalgamations completed after 31 December 1946.

Second Class (withholding): The provisions relating to withholding shall become effective on the first day of the month following the date appearing

— 3 —

hereon.

Third Class (individuals):   For income years ending after 31 December 1946.

APPROVED: RECOMMENDED:

W. F. DEAN   AHN CHAI HONG
Major General, United States Army   Civil Administrator
Military Governor in Korea

SOUTH KOREAN INTERIM GOVERNMENT
Seoul, Korea

ORDINANCE

NUMBER 167   1 April 1948

REVISION OF HOUSEHOLD INCOME TAX

Pending legislative action on the subject matters hereof and the enactment of legislation pursuant thereto, the following shall govern:

SECTION I. *Purpose.* The purpose of this ordinance is to coordinate household income tax rates with the national income tax rates.

SECTION II. *Units.* The table of income grades and units for computation of the household income tax, which is uniformly in effect in the provinces, is hereby amended to read as follows:

Grade	Amount of Income							Number of Units
1	Less than W 5,000							2
2	W 5,000	and	over	but	less	than	W 7,000	3
3	7,000	"	"	"	"	"	10,000	5
4	10,000	"	"	"	"	"	15,000	10
5	15,000	"	"	"	"	"	20,000	20
6	20,000	"	"	"	"	"	25,000	35
7	25,000	"	"	"	"	"	30,000	53
8	30,000	"	"	"	"	"	35,000	73
9	35,000	"	"	"	"	"	40,000	101
10	40,000	"	"	"	"	"	45,000	131
11	45,000	"	"	"	"	"	50,000	161
12	50,000	"	"	"	"	"	60,000	202
13	60,000	"	"	"	"	"	70,000	254
14	70,000	"	"	"	"	"	80,000	310
15	80,000	"	"	"	"	"	90,000	369
16	90,000	"	"	"	"	"	100,000	432

軍政廳

Grade	Amount of Income		
17	W 100,000 and over but less	W 120,000	[illegible]
18	120,000 ″ ″ ″ ″ ″	140,000	[illegible]
19	140,000 ″ ″ ″ ″ ″	160,000	673
20	160,000 ″ ″ ″ ″ ″	180,000	[illegible]
21	180,000 ″ ″ ″ ″ ″	200,000	865
22	200,000 ″ ″ ″ ″ ″	250,000	962
23	250,000 ″ ″ ″ ″ ″	300,000	1,200
24	300,000 ″ ″ ″ ″ ″	350,000	1,440
25	350,000 ″ ″ ″ ″ ″	400,000	1,680
26	400,000 ″ ″ ″ ″ ″	450,000	1,920
27	450,000 ″ ″ ″ ″ ″	500,000	2,160
28	500,000 ″ ″ ″ ″ ″	600,000	2,400
29	600,000 ″ ″ ″ ″ ″	700,000	2,880
30	700,000 ″ ″ ″ ″ ″	800,000	3,370
31	800,000 ″ ″ ″ ″ ″	900,000	3,850
32	900,000 ″ ″ ″ ″ ″	1,000,000	4,330
33	1,000,000 ″ ″ ″ ″ ″	1,200,000	4,810
34	1,200,000 ″ ″ ″ ″ ″	1,400,000	5,770
35	1,400,000 ″ ″ ″ ″ ″	1,600,000	6,730
36	1,600,000 ″ ″ ″ ″ ″	1,800,000	7,690
37	1,800,000 ″ ″ ″ ″ ″	2,000,000	8,650
38	2,000,000 and over		9,620 units,

plus 960 units for every additional W200,000 of income.

SECTION III. *Installment payment of tax.* Annual liabilities under the table set forth in Section II hereof, exceeding one hundred and fifty won (W 150) shall be collected semi-annually; similar liabilities of one hundred and fifty won (W 150) or less shall be collected annually at the time otherwise provided for payment of the first installment.

SECTION IV. *Repealer.* Section II b of Ordinance Number 109 dated

15 October 1946 (*Revision of Provincial and Local Taxes*) is hereby repealed.

SECTION 5.   *Effective date.*   The provisions of this ordinance shall be effective on the date appearing hereon and shall be applicable for the computation of tax on incomes for the year 1947 and thereafter.

APPROVED: RECOMMENDED:

W. F. DEAN   AHN CHAI HONG
Major General, United States Army   Civil Administrator
Military Governor in Korea

SOUTH KOREAN INTERIM GOVERNMENT
Seoul, Korea

ORDINANCE
NUMBER 188

10 May 1948

PETROLEUM PRODUCTS RATIONING VIOLATIONS

Pending legislative action on the subject matters hereof and the enactment of legislation pursuant thereto, the following shall govern:

SECTION I. *Purpose.* The purpose of this ordinance is to provide penalties for violations of the petroleum products rationing system.

SECTION II. *Illegal use.* Petroleum products supplied by the government or secured from or allotted by a governmental source shall, where secured through governmental rationing procedures for a particular use or purpose, be used only for such use or purpose, and shall not be sold, bartered, loaned, or given away, in violation of such use or purpose.

SECTION III. *Records of petroleum products consumers.* a. Each governmental and non-governmental consumer of petroleum products, except consumers of kerosene and wax for domestic lighting, shall keep written records showing:

(1) The source of all petroleum products on hand.

(2) The date upon which secured.

(3) The name of the person from whom secured.

(4) The amount secured.

(5) The place at which secured.

(6) The purpose for which secured.

(7) Persons to whom petroleum products received have been transferred.

(8) Reasons for and date of transfer.

(9) Purposes for which petroleum products received have been used.

b. All records shall be kept for one (1) year.

SECTION IV. *Records of motor vehicle drivers.* Each driver, governmental and non-governmental, of a motor vehicle and each person in charge of the operation of any other internal-combustion equipment shall at all times while engaged in its operation, have in his possession a written record, showing for the preceding thirty (30) days:

— 1 —

a. The amount, date, and place any petroleum product was placed in the fuel tank of the motor vehicle or equipment.

b. The name of the person supplying such petroleum product.

SECTION V. *Transportation of petroleum products.* The captain of each ship and the person operating or driving any motor vehicle, wagon, cart, or other conveyance shall, as to all petroleum products being transported, have in his possession a written record as to such petroleum products, showing:

a. The person from whom, and place where, secured.

b. The date when secured.

c. The name of the owner, if known.

d. The consignee.

e. The kind and amount being transported.

SECTION VI. *Motor vehicle operation.* An owner of a motor vehicle shall not operate or permit it to be operated through the use of gasoline at any time later than forty-five (45) days after gasoline was last lawfully secured for its operation. Any such operation of the motor vehicle after said period of forty-five (45) days, shall be presumed to be through use of stolen or illegally procured gasoline.

SECTION VII. *Violations of petroleum products rationing.* The following acts or omissions shall be illegal, with respect to rationing of petroleum products:

a. A false material statement in an application.

b. Failure to state all material facts.

c. Issuance of a ration authorization where the issuing authority knows that misrepresentation has been made or material information withheld.

d. Knowingly to sell rationed petroleum products to a person not entitled to purchase.

e. Altering, forging or counterfeiting a ration authorization.

f. Knowingly to honor an altered, forged, or counterfeited ration authorization.

g. Sale, barter, loan, or transfer of a ration authorization except in connection with a legitimate sale of the business to which the ration authorization was issued.

h. Issuance by an employee of the government to a person known to be not entitled to receive.

i. Use of a ration authorization knowing that it is altered, forged, or counterfeited.

j. Securing petroleum products from any source not legally entitled to engage in the sale or distribution thereof.

k. Possession of petroleum products secured from any source not legally entitled to engage in the sale or distribution thereof.

l. Failing to keep or keeping falsely any record required hereunder.

m. Knowingly to transport stolen or illegally acquired petroleum products.

SECTION VIII. *Definitions* – As used herein:

a. *Governmental source* includes the Petroleum Distributing Agency and all wholesale and retail sellers of rationed petroleum products legally entitled to engage in the sale thereof.

b. *Petroleum products* includes gasoline, kerosene, fuel oil, diesel oil, lubricating oil, lubricating greases, and petroleum wax.

SECTION IX. *Penalties.* Any person who violates this ordinance shall, upon conviction by a duly constituted court, be punished as follows:

a. For violation of Section *II* hereof, he shall be sentenced to not less than thirty (*30*) days' or more than five (*5*) years' penal servitude or be fined not less than five thousand won (*W 5,000*) or more than two hundred fifty thousand won (*W 250,000*), or both.

b. For violation of Section *III* hereof, he shall:

(*1*) For a first offense, be fined not less than five hundred won (*W 500*) or more than twenty-five thousand won (*W 25,000*).

(*2*) For a second offense, be fined not less than two thousand five hundred won (*W 2,500*) or more than fifty thousand won (*W 50,000*).

(*3*) For a third or further offense, be sentenced to not less than thirty (*30*) days' or more than one (*1*) year's penal servitude and be fined not less than five thousand won (*W 5,000*) or more than one hundred thousand won (*W 100,000*).

c. For violation of Section *IV* hereof, he shall:

(*1*) For a first offense, be fined not less than two hundred fifty won (*W 250*) or more than five thousand won (*W 5,000*).

(*2*) For a second offense, be fined not less than one thousand won (*W 1,000*) or more than ten thousand won (*W 10,000*).

(3) For a third or further offense, be fined not lessthan two thousand five hundred won (W 2,500) or more than twenty thousand won (W 20,000). In addition, where the violation is by the operator of a motor vehicle, his license to drive a motor vehicle shall be suspended for ninety (90) days.

d. For violation of Section V hereof, he shall:

(1) For a first offense, be fined not less than five hundred won (W 500) or more than twenty-five thousand won (W 25,000).

(2) For a second offense, be fined not less than two thousand five hundred won (W 2,500) or more than fifty thousand won (W 50,000).

(3) For a third or further offense, be sentenced to not less than thirty (30) days' or more than one (1) year's penal servitude and be fined not less than five thousand won (W 5,000) or more than one hundred thousand won (W 100,000).

e. For violation of Section VI hereof, he shall:

(1) For a first offense, be fined not less than two thousand five hundred won (W 2,500) or more than ten thousand won (W 10,000). In addition, the registration of the motor vehicle involved in the offense shall be suspended for ninety (90) days.

(2) For a second offense, be sentenced to not less than thirty (30) days' or more than one (1) year's penal servitude. In addition, the motor vehicle involved in the offense shall be forfeited to the South Korean Interim Government.

f. Upon conviction of any act or omission set forth in Section VII hereof, where any specific penalty with respect thereto is fixed in this ordinance or any other applicable law, such penalty shall govern. Where no specific penalty is so fixed by applicable law, the offender shall:

(1) For a first offense, be sentenced to not less than thirty (30) days' or more than one (1) year's penal servitude or be fined not less than one thousand won (W 1,000) or more than ten thousand won (W 10,000), or both.

(2) For a second or further offense, be sentenced to not less than thirty (30) days' or more than one (1) year's penal servitude and be fined not

less than two thousand five hundred won ($W\ 2,500$) or more than twenty-five thousand won ($W\ 25,000$).

SECTION X. *Effective date.* This ordinance shall be effective twenty (20) days after the date appearing hereon.

APPROVED: RECOMMENDED:

W. F. DEAN   AHN CHAI HONG
Major General, United States Army   Civil Administrator
Military Governor in Korea

軍政廳　官報　法令　第一八八號　　一九四八年五月一○日

SOUTH KOREAN INTERIM GOVERNMENT
Seoul, Korea

ORDINANCE
NUMBER 189

10 May 1948

FUNCTIONS OF KOREAN COAST GUARD

Pending legislative action on the subject matters hereof and the enactment of legislation pursuant thereto, the following shall govern:

SECTION I. *Purpose.* The purpose of this ordinance is to clarify the scope of powers and functions of the Korean Coast Guard Bureau established by Ordinance Number 86, dated 15 June 1946 (*Korean Constabulary and Korean Coast Guard*).

SECTION II. *Inspection of ships; transfer of functions. etc.* All duties functions, records, property, funds and personnel concerned with or pertaining to the inspection of ships, of the Bureau of Marine Transportation of the Department of Transportation, are hereby transferred to the Bureau of Coast Guard of the Department of Internal Security. All other authority with respect to the inspection of ships shall also be placed in the Korean Coast Guard Bureau.

SECTION III. *Jurisdiction of Coast Guard.* The provision in Section III of said Ordinance Number 86, reading "*The Korean Coast Guard is hereby established as a marine unit for the Government of Korea for maintaining inshore and interisland patrol of Korean coastal waters*" is hereby amplified as follows:

a. *Coastal waters.* The term *coastal waters* means *territorial waters*, in the terminology of international law. The following is hereby adopted as defining the territorial waters of Korea, south of 38 degrees north latitude.

b. *Territorial waters* include the ports, harbors, bays and other enclosed arms of the sea along the coast of Korea and a marginal belt of the sea extending from the coastline outward one marine league, or three geographic miles.

SECTION IV. *Licensing and certificating of Merchant Marine personnel.* All duties, functions, records, property, funds, and personnel of the Inspection of Ships Division of the Bureau of Marine Transportation of the Department of Transportation pertaining to licensing and certificating of officers, and seamen; suspension and revocation of licenses and certificates; investigation of marine

— 1 —

530

casualties; enforcement of manning requirements, citizenship requirements of crew and requirements for the mustering and drilling of crews; control of log books; shipment. discharge, protection and welfare of merchant seamen; enforcement of duties of owners and officers after accidents; licensing of motor boat operators, are hereby transferred to the Bureau of Coast Guard of the Department of Internal Security.

SECTION V. *Inconsistent laws repealed.* All laws, ordinances, orders regulations, directives and instructions, or parts thereof, which are inconsistent herewith or in conflict with the provisions hereof, are to the extent of such inconsistency or conflict hereby repealed. Included in this category are Governor General Ordinance Number 7, dated April 1914 (*Korean Shipping Act*) and Governor General Ordinance Number 2, dated January, 1935 (*Korean Ship Safety Act*).

SECTION VI. *Effective date.* This ordinance shall be effective ten days after the date appearing hereon.

RECOMMENDED:

W. F. DEAN
Major General, United States Army
Military Governor in Korea

AHN CHAI HONG
Civil Administrator

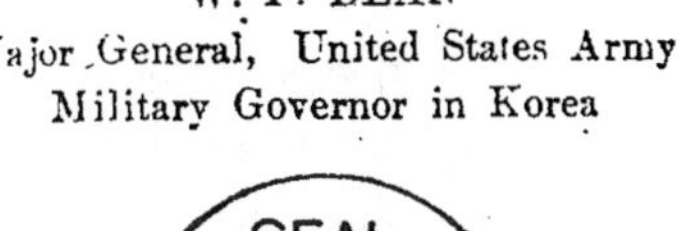

SOUTH KOREAN INTERIM GOVERNMENT
Seoul, Korea

ORDINANCE
NUMBER 190

23 April 1948

RADIO LICENSING AND FEES AMENDED

Pending legislative action on the subject matters hereof and the enactment of legislation pursuant thereto, the following shall govern:

SECTION I. *Purpose.* The purpose of this ordinance is to make provision for license fees for certain privately-owned radio equipment.

SECTION II. *Amendments.* Government General Order Number 39, dated 13 March 1935 (*Personal Wireless Telephone and Telegraph Regulations*) is hereby amended as follows:

a. Section *V* shall be renumbered as Section *V-I.*

b. An additional section shall be added, numbered *V-II*, to read as follows:

"(*a*) *Application under V-I, hereof, shall be accompanied by a license fee of fifty von (W 50) in postage stamps.*

"(*b*) *If application under Section V-I hereof is refused, the accompanying fee shall be refunded.*

"(*c*) *Section V-II (a) hereof, shall not apply to any government radio (Transmitting and/or receiving) operated in government service.*

"(*d*) *Standard Broadcast Receivers (550 kc to 1500 kc only) are exempt from the requirements of this section and shall be licensed and fee paid according to Section XVII (a) of Private Radio Broadcasting Regulations, Communications Department Order Number 98, dated 21 December 1923, Government of Japan, as revised, applied in Korea by Government General Order Number 48, dated 6 September 1924, as revised.*"

SECTION III. *Amendments.* Ministry of Communications Order Number 98, dated 21 December 1923 (*Private Radio Broadcasting Regulations*), as applied in Korea by Government General Order Number 48, dated 6 September 1924, as revised, is hereby amended as follows:

— 1 —

532

Section *XVII*, (*a*), is amended by the deletion of the words "*fifty chun (YO. 50)*" and the substitution of the words "fifty won (W 50)" therefor.

SECTION IV. *Effective date.* This ordinance shall become effective on 15 April 1948.

APPROVED: RECOMMENDED:

          W. F. DEAN   AHN CHAI HONG
Major General, United States Army   Civil Administrator
   Military Governor in Korea

SOUTH KOREAN INTERIM GOVERNMENT
Seoul, Korea

ORDINANCE
NUMBER 191

12 May 1948

ORDINANCE NUMBER 33 CLARIFIED

SECTION I. *Purpose.* The purpose of this Ordinance is to clarify certain provisions of Ordinance Number 33, dated 6 December 1945 (*Vesting Title to Japanese Property Within Korea*).

SECTION II. *Relinquishment of Japanese nationality.* Property previously vested in Military Government by Ordinance Number 33, shall continue to be vested in cases of former owners who after the effective date hereof relinquish Japanese nationality and retroactively regain Korean nationality.

SECTION III. *Effective date.* This Ordinance shall be effective on the date appearing hereon.

APPROVED: RECOMMENDED:

W. F. DEAN   AHN CHAI HONG
Major General, United States Army   Civil Administrator
Military Governor in Korea

軍政廳　官報　法令　第一九一號　　一九四八年五月一二日

534

SOUTH KOREAN INTERIM GOVERNMENT
Seoul, Korea

ORDINANCE

NUMBER 192   4 May 1948

COURT ORGANIZATION LAW

Pending legislative action on the subject matters hereof and the enactment of legislation pursuant thereto, the following shall govern:

BOOK 1.- ORGANIZATION AND POWER OF THE COURT

Chapter I. General Provisions.

SECTION 1. *Purpose.*. The purpose of this ordinance is to transfer the administration of the courts from the Department of Justice to the supreme court and to provide for the reorganization of the court system.

SECTION 2. *Jurisdiction.* The court shall decide civil, criminal, administrative, and the election cases and all of the legal disputes. It shall control non-litigation cases and other matters provided by statute.

SECTION 3. *Courts* There shall be four kinds of courts as follows:

    a. supreme court.

    b. high court.

    c. district court and

    d. summary court.

SECTION 4. *Creation of courts.*. The establishment, abolition and territorial jurisdiction of courts and their change shall be determined as provided by statute.

SECTION 5. *Branch courts.* The supreme court may establish a branch of the district court within its jurisdiction in order to dispose of a part of the work of the district court; provided, however, that the number of branches of district courts shall not be more than forty (40).

SECTION 6. *Justices and judges,* There shall be justices (*dai bop kwan*) in the supreme court and judges (*pansa*) in each other court. The number of justices shall not be more than eleven (11). The number of judges shall be fixed by statute.

SECTION 7. *Temporary duty.* The chief justice of the supreme court upon the resolution of the council of justices may assign a judge of a high court to act temporarily as a justice of the supreme court

軍政廳　官報　法令　第一九二號　　　　一九四八年五月四日

In like manner, the chief justice of the supreme court may assign a judge of a high court to act temporarily as a judge of another high court or of another district court. In like manner, the chief justice of the supreme court may assign a judge from a district court to act temporarily as a judge of a high court or of another district court. In like manner, the chief justice of the supreme court may assign a judge of a high court or of a district court to act temporarily as a judge of a summary court. In like manner, the chief judge of a high court, within his jurisdiction, or the chief judge of a district court may order such temporary duty as described in this section.

SECTION 8. *Benches.* The power of judgment of the supreme court shall be exercised in a bench composed of five justices; provided, however, that when a grand-bench judgment is required, the grand-bench composed of all the justices shall exercise the power of the court.

The power of judgment of a high court shall be exercised by a bench composed of three judges.

The power of judgment of a district court or a branch thereof or a summary court shall be exercised by a single judge; provided, however, that when a collegiate judgment is required at a district court or at a branch court thereof which is a collegiate court, a bench composed of three judges shall exercise the power.

SECTION 9. *Clerks.* There shall be clerks at the courts. The clerks shall be present at trials, prepare and keep records and other court documents, and perform such other duties as are prescribed by statute. They shall dispose of the secretarial work under the order of a higher official.

SECTION 10. *Secretariats.* There shall be a bureau of secretariat in the supreme court and a section of secretariat in the high courts, district courts, branch district courts, and summary courts. There shall be a chief of bureau of secretariat in the bureau of secretariat of the supreme court and a chief of section in each section of secretariat of the high courts and district courts. There shall be a chief clerk in each section of secretariat of branch district courts and summary courts.

Chief of bureau of the secretariat and chief of section of secretariat and chief clerk, shall under the order of the chief justice or appropriate chief judge, manage the work of his bureau or section and direct and supervise the personnel under him.

SECTION 11. *Interpreters.* There may be interpreters at the courts. The interpreters shall interpret and translate under the order of higher officials.

SECTION 12. *Marshals.* There shall be marshals at the district courts. The marshals shall be engaged in the execution of judgments, service of documents and other duties prescribed by statute.

SECTION 13. *Number of clerks, etc.* The number of clerks, interpreters and marshals shall be decided by the supreme court.

Chapter II. Supreme Court.

SECTION 14. *Supreme court.* The supreme court shall be the highest court.

SECTION 15. *Location of supreme court.* The supreme court shall be located in the capital city; provided however, that in case of national emergency, it may move to another place designated by resolution of the council of justices.

SECTION 16. *Chief justice.* There shall be a chief justice in the supreme court. The chief justice shall control the general affairs of the supreme court and direct and supervise the judicial administration in all the courts.

SECTION 17. *Jurisdiction of supreme court.* In order to unify and establish the interpretation and application of laws, the supreme court shall have jurisdiction as a court of last resort, over the following cases:

a. *Sang go* appeals from judgments rendered in the first instance by a collegiate court of a district court, when the parties to civil cases agree to waive *Kong so* appeal.

b. *Sang go* appeals from judgments rendered by a high court as a court of first instance or as a court of second instance.

c. *Hang go* appeals which are within the jurisdiction of the supreme court by statute.

SECTION 18. *Interpretation.* The interpretation of laws made by the supreme court in its judgment shall bind the lower courts in reference to the respective case.

SECTION 19. *Grand bench.* In the following cases, a grand bench of the supreme court shall render the judgments:

a. When it considers that judgment made previously by the supreme court on the same point of law was not proper.

b. When it considers that a certain case is especially important.

SECTION 20. *Opinions.* In the statement of decision of the supreme court, the different opinions of the justices who have participated in the case may be expressed.

SECTION 21. *Divisions of supreme court.* There shall be civil divisions, criminal divisions, and special divisions in the supreme court.

SECTION 22. *Regulations.* The supreme court shall by resolution by the council of justices enact regulations concerning the appointment, training, and education of court officials and regulations for the general affairs of the court.

SECTION 23. *Judicial administration.* There shall be an office of judicial administration in the supreme court which shall handle administrative affairs for all the courts.

SECTION 24. *Judicial training institute.* There shall be a judicial training institute in the supreme court which shall supervise the education and training of court officials.

Chapter III. High Court

SECTION 25. *Chief judge.* There shall be a chief judge in each high court. The chief judge of the high court shall control the general administrative affairs of his court and direct and supervise the administrative affairs of the courts under his jurisdiction.

SECTION 26. *Jurisdiction* The high court shall have jurisdiction over the following cases:

a. *Sang go* appeals from judgments rendered in the first instance by a single judge of a district court or by a summary court.

b. *Kong so* appeals from judgments rendered by a collegiate court of a district court.

c. *Hang go* appeal cases which belong to the high court by statute.

d. Other cases which belong to the high court by statute.

SECTION 27. *National government administrative legislation.* The Seoul High Court shall have exclusive original jurisdiction of all administrative litigation concerning the orders and actions of the executive branch of the national government.

SECTION 28. *Divisions.* There shall be civil divisions, criminal divisions and special divisions in each high court. There shall be a division chief in each division. A division chief shall, under the direction of the chief judge, supervise the work of his division.

Chapter IV. District Court.

SECTION 29. *Chief judge.* There shall be a chief judge in each district court. The chief judge of the district court shall control the general administrative affairs of that court and direct and supervise the administrative affairs of the branch courts and summary courts in his district.

SECTION 30. *Administrative affairs.* There shall be a chief judge in each branch district court which has more than one judge. The chief judge of the branch district court or the judge of a branch district court which has only one judge shall, under the direction and supervision of the chief judge of the district court, conduct the administrative affairs of that branch court.

SECTION 31. *Jurisdiction* The district court shall handle cases through a single judge, except that the following cases shall be handled by a collegiate court:

a. *Kong so* appeals from judgements rendered by a single judge of a district court, in which case the judge who rendered the decision shall not take part, provided such appeal shall not be handled by a branch court.

b. *Kong so* appeals from judgments rendered by summary courts.

c. Civil cases, the object of which exceeds the value of W 500 000.

d. Litigation concerning personal matters (*in sa*).

e. Cases in which the maximum penalty is death or in which the minimum penalty is imprisonment or penal servitude for life or for not less than one (1) year.

f. Cases of accomplices in connection with cases described in paragraph e hereof, when they are tried at the same time with such cases

g. Administrative litigation, except that reserved for other courts.

h. Such election litigation as may by statute be placed under jurisdiction of the district court.

i. Cases which the council of judges by resolution decide should be tried by a collegiate court.

SECTION 32. *Divisions* There shall be civil divisions, criminal divisions and special divisions in each district court. There shall be a division chief in each division who shall, under the direction of the chief judge, supervise the work of his division.

SECTION 33. *Juvenile jurisdiction.* The district court shall have juris-

— 5 —

diction over all matters now under the juvenile division. Such matters shall be handled in accordance with the laws relating to protective disposal of juveniles.

SECTION 34. *Juvenile divisions.* The council of justices may direct the establishment in each district of a branch of the district court of that district to be known as the juvenile division of the district court of that district Such juvenile division of the district court shall exercise jurisdiction over the protective disposal of juveniles in that district under the supervision of the chief judge of the district court. Such juvenile division shall have its office in the same city as the district court

The chief judge of the juvenile division shall be a judge of the rank of division chief judge of the district court. The chief judge of the district court may assign other judges of the district court or of a branch thereof to handle cases involving protective disposal of juveniles, either in addition to their other duties or as their sole duty. Such judges, when they are engaged in the handling of cases involving the protective disposal of juveniles, shall be under the supervision of the chief judge of the juvenile division.

SECTION 35. *Administrative affairs.* The chief judge of the juvenile division shall, under the direction of the chief judge of the district court, direct and supervise the general administrative affairs of the juvenile division.

SECTION 36. *Districts without juvenile division.* In districts where there is no juvenile division established as provided in Section 34 hereof, the chief judge of the district court shall designate judges of the district court or of the branch courts thereof who shall exercise jurisdiction over the protective disposal of juveniles in accordance with the laws relating to the protective disposal of juveniles.

SECTION 37. *Decision* The decision of the juvenile division shall be made by a single judge.

There shall be probation officers in each juvenile division. The probation officer shall assist the judge by investigations and gathering of evidence. He shall also perform all the duties imposed by law on probation officers.

The chief judge of the district court shall appoint the probation officers from among persons who have had experience in protection and education of juveniles and other proper persons.

Chapter V. Summary Court

SECTION 38. *Summary Court.* The summary court shall be a lower court of the first instance.

SECTION 39. *Chief judge.* There shall be a chief judge in each summary court which has more than one judge. The chief judge of the summary court or the judge of a summary court which has only one judge shall, under the direction and supervision of the chief judge of the appropriate district court, conduct the administrative affairs of that summary court

SECTION 40. *Jurisdiction.* The summary court shall have jurisdiction over the following cases:

a. Matters concerning arbitration, conciliation (*wha hae*) rule *nisi* on monetary claims (*tok chuk*), tenant farming, and leases of land and houses.

b. Civil cases the object of which does not exceed the value of W 100,000.

c. Summary cases in which the maximum penalty is a fine or detention or confinement of less than three months.

d. Criminal cases in which the maximum penalty is penal servitude or prison for not more than one year.

e. Summary cases involving trespassing in dwellings, habitual gambling, theft and violation of price, food and other administrative regulations where the penalty given is not greater than penal servitude or imprisonment for one year.

f. Cases in which the summary court is given jurisdiction by statute.

g. Cases over which the special judicial officer has jurisdiction at the time this ordinance becomes effective.

When a judge of a summary court considers that a case is not within the jurisdiction of his court in accordance with the provisions of this section, he shall by decision transfer the case to the appropriate district court or branch district court.

SECTION 41. *Contempt of court.* The judge of the summary court has jurisdiction of cases of contempt of his court provided in section 72, irrespective of the provisions in the preceding section.

SECTION 42. *Time for appeal.* The accused, the prosecutor or a party to a civil case may, within seven days from the sentence of the summary court, take a *Kong so* appeal from the sentence to the appropriate district court.

BOOK 2 - COURT OFFICIALS

Chapter I. - Justices and Judges.

SECTION 43. *Appointment of chief justice and justices.* The chief justice and justices shall be appointed from among those who have one of the following qualifications:

a. Persons who have been justices, judges or prosecutors for 10 years or more.

b. Persons who have been lawyers or professors or assistant professors of law in recognized law colleges for 15 years or more.

c Persons who have held two or more positions mentioned above for a total period of 15 years or more.

SECTION 44. *Appointment of chief judges of high courts.* The chief judges of high courts shall be appointed from among persons who have the qualifications described in the preceding section.

SECTION 45. *Qualifications of judges of high courts, etc.* Judges of the high courts, chief judges of the district courts, division chief judges of the district courts and chief judge of the juvenile divisions shall be appointed from among those who have one of the following qualifications:

a. Persons who have been justices, judges or prosecutors for 5 years or more.

b. Persons who have been lawyers or professors or assistant professors of law in recognized law colleges for 8 years or more.

c. Persons who have held 2 or more of the positions described above for a total of 9 years or more.

SECTION 46. *Qualifications of judges of district court, etc.* Judges of the district court and of the branches thereof including the juvenile divisions thereof and judges of a summary court shall be appointed from among those who have one of the following qualifications:

a. Judicial apprentices who have completed 2 years or more of the required training and have passed the final examination

b. Persons who have been prosecutors, lawyers or professors or assistant professors of law in recognized law colleges 2 years or more.

c. Persons who have held 2 or more of the positions mentioned above for a total period of not less than 3 years.

SECTION 47. *Appointment of chief justice and justices.* The chief justice and the justices of the supreme court shall be appointed by the Military Governor or by the chief executive succeeding him in authority from among those recommended by a committee which shall consist the chief justice, the justices of the supreme court, the chief judges of the high courts, the chief judges of the district courts, the director and deputy directors of the Department of Justice, the chief prosecutor of the supreme court, the chief prosecutors of the high courts and district courts and the chairmen of the officially recognized bar associations.

A majority of the members of the Committee shall constitute a quorum. Voting on persons to be recommended shall be by secret ballot. The Committee shall recommend to the appointing authority a panel for appointment equal to three times the number of justices to be appointed. Each of those nominated shall receive a majority of the votes of those present. The committee shall itself determine the rules and means of its own voting, so that the final result shall be that each of those persons nominated shall receive the vote of a majority of the committee members present at the meeting.

SECTION 48. *Appointment of chief judge of the high courts etc.* The chief judge of the high courts, the chief judges of the district courts and other judges shall be appointed by the Military Governor (or by the chief executive succeeding him in authority) upon recommendation of the council of justices. A majority of the members of the council shall constitute a quorum. Voting on persons to be recommended shall be by secret ballot. The council shall recommend to the appointing authority, a panel for appointment equal to three times the number of appointments to be made. Each of those nominated shall receive a majority of the votes of those present. The council shall itself determine the rules and means of its own voting, so that the final result shall be that each of those persons nominated shall receive the vote of a majority of the council members present at the meeting.

For appointments to positions below the rank of Division Chief Judge, the requirement that three times the number of appointments be recommended may be waived.

SECTION 49. *Term of office.* The justices and judges shall hold their office during good behavior and shall not be discharged, removed, demoted, suspended nor have their salaries reduced unless they shall be sentenced to penal servitude or imprisonment or heavier penalty or shall be found guilty of improper conduct by the Korean judicial disciplinary committee.

SECTION 50. *Korean judicial disciplinary committee.* The supreme court shall appoint a judicial disciplinary committee, as provided by the Korean Judicial Disciplinary Ordinance (*Governor General Ordinance No. 5. May 1911*) which committee shall have the powers provided in said ordinance. The supreme court may make regulations for the work of the committee not inconsistent with said ordinance.

SECTION 51. *Retirement.* Justices and judges shall retire when they reach the age of retirement. The age of retirement for chief justice and justices shall be 70 and for judges shall be 65.

SECTION 52. *Retirement on account of health.* When a justice or a judge becomes mentally ill and has no hopes of recovery and is unable to perform his duties he shall retire upon request of the council of justices.

SECTION 53. *Restrictions on justices and judges.* Justices and judges shall not while in office engage in the following affairs:

    a. become a member of a legislative body or hold any office in the executive branch of the government.

    b. participate in political activities.

    c. accept a paid position without permission of the appropriate chief judge.

    d. operate any business for the purpose of pecuniary profit.

SECTION 54. *Compensation.* The compensation of the chief justice, the justices and the judges shall be determined by statute.

Chapter II. - Judicial Apprentices

SECTION 55. *Judicial apprentices.* Judicial apprentices shall be appointed by the chief justice upon resolution of the council of justices from among those who have passed the examination prescribed by the supreme court. All matters concerning the study and examination of judicial apprentices shall be determined by the supreme court.

SECTION 56. *Salary and discharge of judicial apprentices.* The judicial apprentices shall receive a fixed salary from the national treasury during the period of study. When a judicial apprentice is guilty of conduct which discredits the dignity of his office, or when there appears to be no hope of his successfully completing his course of study, the chief justice upon resolution of the council of justices may discharge him.

Chapter III. - Clerks, Assistant Clerks and Technicians.

SECTION 57. *Clerks.* Clerks shall be appointed and removed by the

軍政廳　官報　法令　　一九二號　　　一九四八年五月四日

chief of the office of judicial administration ···on the recommendation of the chief justice or appropriate chief judge or o. the judge of courts that have only one judge from among those who have the required qualifications. The regulations concerning the qualifications of clerks shall be determined by the supreme court.

SECTION 58. *Duties of clerks.* Clerks shall obey the orders of the justice or judge while executing their duties.

When a clerk receives an order of a justice or a judge concerning the recording of oral statements or the preparation and change of other documents but such order is contrary to his own opinion, the clerk may file his own opinion in the record.

SECTION 59. *Assistant clerks.* There shall be assistant clerks in the courts to assist the work of clerks. Assistant clerks shall be appointed and removed by the chief of the office of Judicial Administration upon recommendation of *chief justice* or appropriate chief judge or of the judge of courts that have only one judge.

SECTION 60. *Court technicians.* There may be court technicians in the courts. Technicians shall do technical work under the order of the higher officials.

SECTION 61. *Interpreters.* There may be interpreters in the courts whenever necessary.

SECTION 62. *Appointment of interpreters and technicians.* Interpreters and technicians shall be appointed and removed by the chief of the office of judicial administration upon recommendation by the *chief justice* or appropriate chief judge or of the judge of courts that have only one judge.

Chapter IV. – Marshals

SECTION 63. *Marshals.* Marshals shall be appointed and removed by the appropriate chief judge of the district court.

SECTION 64. *Fees of marshals.* Marshals shall receive fees for their services. The duties and fees of marshals shall be as fixed by statute.

SECTION 65. *Deposit of security.* Marshals shall deposit in the appropriate district court security for faithful and proper performance of their duties.

Chapter V – Bailiffs

SECTION 66. *Bailiffs.* Bailiffs shall be appointed and removed by the appropriate chief judge; provided, however, that bailiffs of the supreme court shall be appointed and removed by the chief of the office of judicial administration.

SECTION 67. *Duty of bailiffs*. Bailiffs shall under the order of justices or judges, guide the parties and witnesses and perform other duties determined by the supreme court.

SECTION 68. *Service of legal papers*. When a marshal is not available, a court may have a bailiff serve legal papers.

BOOK 3 - CONDUCT OF TRIAL

Chapter I. – Court Sessions.

SECTION 69. *Public trials*. Public trials shall be held at the courthouse. The chief judge may, however, in accordance with regulations provided by the supreme court order trials to be held at other places.

SECTION 70. *Exclusion of public from public trials*. All public trials shall be open to the public; provided, however, that the public may be excluded when there is danger of disturbance of public order or good morals. The decision to exclude the public from a public trial shall be announced and clearly explain the reason for the decision. In cases when the public is excluded from the public trial, the presiding judge may permit the presence of persons whom he considers should be present.

SECTION 71. *Order in the court*. The presiding justice or judge or the judge who opens a court session, shall maintain order in the court. The presiding justice or judge or the judge who has opened a court session, may prohibit the entrance of any person who there is reason to believe may disturb the dignity or order of the court or may order such person to leave the court or issue such other orders as may be necessary to maintain order in the court.

SECTION 72. *Contempt of court*. The presiding justice or judge or a judge who has opened a court session, may detain until the close of the court session any person who violates the orders mentioned in the preceding section. any person who disturbs the dignity or order of the court, any person who interferes with the trial or any person who is guilty of any misconduct in the court. At the close of the session, the court may order his release or may impose by judgement penal servitude or imprisonment for not more than one year or fine of not more than ₩ 50,000; provided, however, when such an act is also a violation of other provisions of the criminal law, the person may

— 12 —

also be punished criminally f r such an act. The person sentenced may appeal from the judgement to the next higher court within seven (7) days.

SECTION 73. *Court at places other than courthouse.* The provisions of the preceding four sections shall be applied *mutatis mutandis* to cases when justices or judges hold court at places other than at the courthouse.

SECTION 74. *Improper conduct.* In case any prosecutor or lawyer uses improper language in court, the presiding justice or judge or the judge who opened the court session may prohibit further participation in the case by the prosecutor or the lawyer.

Such prohibition is no b r to punishment for such acts. In cases where the further participation of a lawyer is prohibited, the presiding judge or the judge who has opened the c urt session shall give the parties concerned opportunity to appoint another lawyer if they so desire; or upon their failure to do so, may on his own authority appoint a lawyer if required for trial of case. In case the further participation of a prosecut r is prohibited, the presiding judge or the judge who has opened the court, shall request the appropriate chief prosecutor to have another prosecutor attend the trial.

SECTION 75. *Refusal of prosecutor to attend trial.* In case the presence of a prosecutor at a public trial is necessary, the court shall give the prosecutor notice of the time and place of the opening of the first session of the court in the case, not less than one full week prior to such opening. At the close of each session, the court shall give the prosecutor notice of the time and place of the next session in that case.

In case the prosecutor who has been notified fails to attend, the court shall postpone the session and give the prosecutor a second notice. In case the prosecutor fails to attend the session of the court on the second notice, or in case no prosecutor attends a trial in spite of the request mentioned in the last paragraph of section 74, the presiding judge or a judge who has opened the session of the court, may designate a judicial apprentice, a chief of section of clerks or a chief of a police station to act as a special prosecutor and have him attend the public trial.

The special prosecutor shall have the power of a prosecutor only so far as the particular instance is concerned.

In case a special prosecutor is designated as provided above, the judge shall immediately report the matter to the appropriate chief judge.

— 13 —

SECTION 76. *Refusal of court to act.* In case the judge or judges fail to attend the session of court at the time set, or in case there is unreasonable delay in setting the date for public trial after request by the prosecutor, the prosecutor shall report the matter to the chief justice or appropriate chief judge, who shall make all necessary orders for the prompt handling of the business of the court.

SECTION 77 *Setting cases for trial.* The supreme court shall make regulations to implement existing law concerning the manner in which cases are set for public trial and in which prosecutors and lawyers are notified.

Chapter II—Language of the Court.

SECTION 78. *National language.* The national language shall be used in the court. When a party concerned does not speak the national language, he may use an interpreter appointed or approved by the court.

Chapter III—Deliberation.

SECTION 79. *Deliberation.* The deliberation of a collegiate court shall not take place in public, but the presence of judicial apprentices may be permitted by the court.

SECTION 80. *Court decisions.* Collegiate court decisions shall be made by the majority opinion of the court; provided, however, that if there are three or more different opinions, the decision shall be made in accordance with the procedures set out below; in civil cases the number of opinions in favor of the largest amount shall be added to the number of opinions in favor of the next largest amount and so on, until an absolute majority is attained. The amount of the majority opinion shall be that of the opinion in favor of the smallest amount which is held within the majority group.

In criminal cases, the number of opinions most unfavorable to the accused shall be added to the number of opinions next most unfavorable and so on, until an absolute majority is attained. The majority opinion is then that of the opinion most favorable to the accused which is held within the majority group.

SECTION 81. *Supplementary judge.* When a collegiate court expects to continue in session for a long time, a supplementary justice or a supplementary judge may be present at the trial. When a member of the collegiate court becomes unable to participate in the trial because of unavoidable circumstances occurring during the trial, the supplementary justice or judge shall be substituted for such member and perform all of his duties.

Chapter IV. Cooperation of Courts.

SECTION 82. *Cooperation.* The courts s... render necessary mutual assistance in the conduct of trials.

BOOK 4 - JUDICIAL ADMINISTRATION

Chapter I. Supervision of Judicial Administration.

SECTION 83. *Supervision.* The powers of the Chief Justice or the appropriate chief judge to supervise the judicial administration shall include the following:

    a. when the performance of duties is improper or insufficient, attention may be called and instructions may be issued.

    b. when there is behavior unfit for the position, charges may be made to the Korean Judicial Disciplinary Committee

The power of supervision provided in this section shall not limit or influence the power of decision of justices and judges.

CHAPTER II. Council of Justices.

SECTION 84. *Council of justices.* There shall be a council of justices as the highest deliberative organ of judicial administration.

SECTION 85. *Members of council of justices.* The council of justices shall consist of the chief justice and all the justices. A quorum of the council shall be two-thirds of all the members. Chief judges of the high courts may attend sessions of the council of justices and express their opinions but may not vote.

SECTION 86. *Chairman.* The chief justice of the supreme court shall be the chairman of the council of justices. In case the chief justice is absent, the justice among those present who has been a justice the longest shall preside over the meeting.

SECTION 87. *Powers of council of justices.* The council of justices shall have power over the following:

    a. Matters concerning the recommendation of persons to be appointed as judge or to be transferred.

    b. Matters concerning the retirement of justices and judges before they reach the age of retirement.

    c. Matters concerning the temporary duty of judges.

    d. Enactment of regulations concerning court procedure and administration to implement existing law.

    e. Matters concerning the appointment, removal and promotion of chief and deputy chief of the office of judicial administration, chief of the judical training institute, chiefs of section of the office of judicial administration.

 f. Matters concerning research, collection and publication of precedents.

 g. Matters concerning the preparation of the budget and the approval of accounts

 h. Matters concerning the establishment, change and promotion of branch courts, including juvenile divisions, and assignment of officials thereof.

 i Matters concerning various examinations of court officials and concerning the appointment and removal of clerks, interpreters and technicians of the supreme court.

 j. Important matters concerning the judicial administration and other matters which belong to the power of the council of justices by statute.

SECTION 88. *General affairs.* General affairs concerning the council of justices shall be handled by the chief of the supreme court office of judicial administration.

Chapter III. Council of Judges.

SECTION 89. *Council of Judges.* There shall be a council of judges as the deliberative organ of judicial administration in each high court and in each district court As to the council of judges, the provisions in Section 85 and 86, shall be applied *mutatis mutandis.*

SECTION 90. *Power of council of judges.* The council of judges shall have power over the following:

 a. Matters concerning personnel.

 b. Matters concerning temporary duty of judges.

 c. Matters coming under Section 81. paragraph *i.*

 d Important matters concerning judicial administration and other matters which belong to such council of judges by statute.

Chapter IV. Supreme Court Office of Judicial Administration.

SECTION 91. *Judicial administration.* There shall be a supreme court office of judicial administration, hereinafter referred to as office of judicial administration, directly under the chief justice of the supreme court which shall manage judicial administrative affairs in all the courts.

SECTION 92. *Chief and deputy chief.* There shall be one chief and one deputy chief in the office of judicial administration. The chief of the office of judicial administration, shall, under the supervision of the chief justices of the

supreme court, have charge of the affairs of the office. The deputy chief shall
assist the chief and act on his behalf when he is absent or unable to perform
his duties.

SECTION 93. *Sections.* In the office of judicial administration, there
shall be the following sections: section of secretariat, section of accounts, section
of administration, which shall have the powers and duties as set up in the
following three sections unless otherwise provided by the council of justices.

SECTION 94. *Section of Secretariat.* The section of secretariat shall have
charge of confidential matters, personnel matters and all other matters which
are not assigned to other sections.

SECTION 95. *Section of accounts* The section of accounts shall have
charge of the preparation of the budgets, accounts and financial matters.

SECTION 96. *Section of administration.* The section of administration
shall have charge of matters concerning decisions, registration, family registers,
statistics deposits and marshals.

SECTION 97. *Section chief.* There shall be a section chief in each section.
The section chief shall, under the direction of the chief of the office of judicial
administration, handle the business of his section.

SECTION 98. *Appointment and removal of personnel.* The chief justice of the
supreme court shall upon the resolution of the council of justices appoint and remove
the chief of office, the deputy chief, and the section chiefs of the office of judicial
administration. The chief of the supreme court office of judicial administration shall
appoint and remove other employees of the office of judicial administration.

Chapter V. Judicial Training Institute.

SECTION 99. *Judicial training institute.* There shall be a judicial training
institute directly under the chief justice which shall have charge of matters
concerning the training and education of judges, judicial apprentices and other
court officials.

SECTION 100. *Chief, judicial training institute,* There shall be a chief of
the judicial training institute and judicial training attorneys in the judicial
training institute. The chief of the institute shall, under the supervision of the
chief justice. manage the affairs of the institute and the judicial training.
Attorneys shall, under the order of their superiors, handle the work of the
institute. The chief justice. shall, upon resolution of the council of justices,
appoint and remove the chief of the judicial training institute. The chief of

the institute shall appoint and remove the other officials and employees of the institute.

SECTION *101. Regulations, judicial training institute.* The supreme court shall make regulations concerning the judicial training institute.

BOOK 5 - BUDGET OF THE COURTS - TRANSITION PROVISIONS

SECTION *102. Budget.* The budget of the courts shall be prepared independently in the office of judicial administration.

SECTION *103. Certain officials to continue to hold offices* Those who are holding positions of chief justice, justices and judges at the time this ordinance becomes effective shall be regarded as appointed to such positions, irrespective of the provisions of this ordinance.

SECTION *104. Credit for time served to be given.* As to those persons who held the position of justice, judge or prosecutor, or, lawyer, or director or deputy director of the Department of Justice and bureau chiefs of the Department of Justice before the effective date of this law, the period for such positions held shall be counted as a similar period in the position of judge in the application of section 43 to 46, inclusive, of this ordinance.

SECTION *105. Certain officials to continue to hold office.* Those who hold the position of judge before the effective date of this ordinance, may be appointed as judges of district courts, branch district courts, including juvenile divisions and summary courts, irrespective of provisions in section 46. Persons who have held the position of special judicial officer and have performed the duties of their office satisfactorily, may be appointed to the position of judge of summary court. The supreme court may provide for a special examination for persons to be appointed as judges of the summary courts at the time this ordinance becomes effective, irrespective of the provisions of section 46.

SECTION *106. Power of assignment to two positions.* The chief justice of the supreme court may upon the resolution of the council of justices, assign judges to hold, in addition to their position as judges, the position of chief or deputy chief of the supreme court office of judicial administration, chief of the judicial training institute, or judicial training attorneys; provided, however, such person shall receive only one salary which shall be the highest salary of the two positions which he holds.

SECTION *107. Temporary continuance of special judicial officers.* The special judicial officers shall continue to perform the duties now provided by law until the establishment of summary courts to take their place. At the time of the establishment of each summary court, the supreme court shall order the abolition of the office of special judicial officer at that place.

SECTION 108. *Establishment of summary courts.* The supreme court shall establish summary courts as soon as personnel is available therefor. There shall be as early as possible a summary court at each place where there is a police station.

SECTION 109. *Jurisdiction of summary court.* The summary court shall have jurisdiction to issue warrants of arrest and warrants of search provided for in Ordinance No. 176, dated 20 March 1948 (*Changes in Criminal Procedures*). In places where there is a summary court but no prosecutor's office, the police may make applications for warrants direct to the summary courts. The summary court shall not have jurisdiction over applications for inquiry filed under section XVII of ordinance 176.

SECTION 110. *Continuance of Korean judicial disciplinary committee.* The present Korean judicial disciplinary committee shall continue to serve until the supreme court appoints a new committee.

SECTION 111. *Transference of certain personnel.* The personnel of the courts bureau of the Department of Justice at the time this ordinance becomes effective, except the chief thereof, shall be transferred to the supreme court office of judicial administration. Clerical and accounting personnel of the administrative bureau of the Department of Justice, who, at the time this ordinance becomes effective, are engaged in work relating to personnel or financial affairs of the courts, shall be transferred to the supreme court office of judicial administration. Personnel who at the time this ordinance becomes effective are engaged partly on work relating to functions which are transferred to the courts and partly on work relating to functions which remain in the Department of Justice, shall be equitably divided between the Department of Justice and the supreme court office of judicial administration.

SECTION 112. *Transference of certain buildings.* Buildings which at the time this ordinance becomes effective are entirely used by functions which are transferred to the courts, shall be controlled by the courts. Buildings which at the said time are entirely used for functions which remain in the Department of Justice, shall be controlled by the Department of Justice. Buildings which

— 19 —

at such time are jointly used, shall continue to be so used and shall remain under the control of the agency which controls them at such time.

SECTION 113. *Judicial adjustment board.* There is hereby created a judicial adjustment board. The board shall consist of one member appointed by the chief justice of the supreme court, one member appointed by the director of Department of Justice and one member appointed by the director of the Office of Administration and two members appointed by the director of the Department of Finance. The appointments shall be made and certified to the Military Governor within five days after this ordinance is signed. The board shall have the power to settle all disputes concerning and to make all necessary adjustments of personnel, budgeted funds, buildings, grounds, automobiles, and other property between the courts and the Department of Justice.

SECTION 114. *Budgeted funds.* The director of the Department of Finance shall make an equitable division of budgeted funds for the current year between the courts and the Department of Justice, subject to review by the judicial adjustment board.

SECTION 115. *Succession of certain courts and records.*

a. The supreme court in existence on the effective date of this ordinance, shall become the supreme court under this ordinance.

b. The Seoul court of review and the Taegu court of review in existence on the effective date of this ordinance, shall become the Seoul high court and the Taegu high court, respectively, under this ordinance.

c. The district courts and the branches thereof in existence on the effective date of this ordinance, shall become the district courts and the branches thereof under this ordinance.

d. The juvenile courts of Seoul, Taegu, Pusan and Kwangju shall become branches of the district courts of Seoul, Taegu, Pusan, and Kwangju, respectively, without the necessity for the establishment thereof under Section 34 hereof, and shall be governed by sections 25 to 37, inclusive, hereof.

e. The records of the above-mentioned courts in existence at the effective date of this ordinance shall become records of the courts which succeed them as provided in the foregoing four paragraphs.

f. All cases or other matters which have been accepted by or are otherwise pending in each of the above-mentioned courts at the effective date of this ordinance under the law existing on said date, shall be deemed to be

— 20 —

accepted or pending in the court which succe   it as provided in the first four paragraphs of this section.

g. All papers relating to reappeal, appeal, complaint, or other proceeding, which, at the effective date of this ordinance have been sent to any court and not accepted, shall be deemed to have been sent to the sucessor to that court as provided in the first four paragraphs of this section.

h. In all cases in which *Kong So Appeals, San Go Appeals* or *Han Go Appeals* have been made at the time this ordinance becomes effective by sending the papers to the proper court under the law in effect on said date, whether the papers have been accepted by the higher court or not, such appeals shall be governed by the law as to appellate jurisdiction in effect on said date.

i. Each court mentioned in the first four paragraphs of this section shall have the jurisdiction, territorial and otherwise, which, at the time this ordinance becomes effective, vested in the court which it succeeded as provided in said four paragraphs, except as changed by this ordinance.

SECTION 116. *Permissible references to summary courts.* In appropriate cases, the chief judge of a district court may refer to a summary court for trial, civil actions which have been filed with the district court or a branch thereof. In appropriate cases, the chief judge of a district court, with concurrence of the chief prosecutor of the district court, may refer to a summary court for trial, criminal actions which have been started in the district court or branch thereof.

SECTION 117 *Excepted positions.* Justices. judges, chief and deputy chief of the supreme court office of judicial administration, chief of bureau of secretariat of the supreme court, chief of the judicial training institute, and judicial training attorneys shall be excepted positions within the meaning of the regulations of the Korean Civil Service.

Pending revision of the regulations of Korean Civil Service, the employment, compensation and removal of all other personnel of the courts shall be in accordance with the law and regulations in effect at the time this ordinance becomes effective.

SECTION 118. *Transfer of certain functions.* On the effective date of this ordinance, the administration of the courts (*including the juvenile divisions and the special judicial officers*), examination and training of judicial (*court*) apprentices, the registration offices, the supervision of the family register offices, the deposit offices, and all other matters referred to in sections *94, 95*, and *96* hereof, insofar as such other matters relate to the courts, shall be transferred from the Department of Justice to the supreme court.

The director of the Department of Justice and the chief justice shall effectuate the transfer as rapidly as possible under the provisions of this ordinance.

SECTION 119. *Inconsistent laws repealed.* All laws, ordinances, orders, regulations, directives or instructions, or parts thereof, which are inconsistent herewith, or in conflict with the provisions hereof, to the extent that they are in conflict or inconsistent with this ordinance, are hereby repealed.

SECTION 120. *Effective date.* This ordinance shall be effective on 1 June 1948.

APPROVED: RECOMMENDED:

    W. F. DEAN   AHN CHAI HONG
Major General, United States Army   Civil Administrator
  Military Governor in Korea

SOUTH KOREAN INTERIM GOVERNMENT
Seoul, Korea

ORDINANCE
NUMBER 193   1 May 1948

AMENDMENT OF AMUSEMENT AND ADMISSIONS TAX LAWS

Pending legislative action on the subject matters hereof and the enactment of legislation pursuant thereto, the following shall govern:

SECTION I. *Purpose.* The purpose of this ordinance is to increase revenue and to revise rates and exemption provisions of existing laws in order to facilitate more effective enforcement.

• SECTION II. *Amusement, drinking, and eating taxes revised.* Section I c of Ordinance Number 101 dated 31 August 1946 (*Amendment of Sales Tax Laws*) is hereby amended to read as follows:

"*c. Amusement drinking, and eating tax revised.* The tax rates provided in Tax Ordinance Number 19, dated 31 March 1940, as revised, are hereby further revised and shall be as follows:

(*1*) *Class A:* Restaurant with *Kisang* License:
on charges of over W 300 per person, *30%* of entire bill.

(2) *Class B:* Restaurant without *Kisang* License:
on charges of over W 300 per person 20 % of entire bill.

(3) *Class C:* Hotel charges per person per day:

Without meals:

Charge. W 300 and under	*No tax*
Charges over W 300 but not over W 400	10%
Charges, over W 400	25%

With meals:

Charge W 550 and under	*No tax*
Charges, over W 550 but not over W 700	10%
Charges, over W 700	25%"

SECTION III. *Admission tax ordinance amended.* Section I d of Ordinance Number 101 dated 31 August 1946 (*Amendment of Sales Tax Laws*) is hereby amended to read as follows:

"*d. Admission tax revised:* The tax rates provided in Tax Ordinance Number 22, dated 31 march 1940, as revised, are hereby further revised and shall be as follows:

— 1 —

(1) *1st Class:* Admission charges of over W 10. 100% of entire charge.

(2) *2nd Class:* Admission charge to billiard games, *mahjong* club, and dance halls, 100% of entire charge. Taxes on admission charges to skating rinks, ping pong games, bow and arrow and shooting practice, are hereby repealed.

(3) *3d Class:* Special admission tax on events given to the public without profit (non-professional) is hereby repealed."

SECTION IV. *Exemption provision repealed.* Section 5 of Tax Ordinance Number 22, dated 31 March 1940, as amended, is hereby repealed.

SECTION V. *Card and dice tax revised.* Section I *g* of Ordinance Number 14 dated 31 August 1946 (*Amendment of Sales Tax Laws*) is hereby amended to read as follows:

"*g. Rates.* The tax rates provided in Tax Ordinance Number 1, dated 15 April 1931, as revised, are hereby further revised and shall be as follows:

(1) Manufacture of playing cards, W 50 per deck.

(2) Manufacture of *kolpai* (dice), W 100 per set.

(3) Manufacture of *mahjong*, W 500 per set.

"*Payment of the tax - manufacturer or importer.* The tax provided in this subsection shall be paid by the manufacturer or importer of the articles by the purchase of revenue stamps at the offices of the Bureau of Internal Revenue or its authorized agent immediately prior to the time of completion of the manufacture or importation. Such stamps shall be affixed to the package or the article and cancelled in the manner provided by existing law.

"*Payment of the tax - articles held for sale.* Upon the items enumerated in this subsection, which on the effective date of this ordinance are held by any person for sale, there shall be levied, assessed, collected, and paid by the purchase and affixation of revenue stamps as above provided, a tax equal to the difference between the rate effective immediately prior to the issuance of this ordinance and the rate provided by this ordinance upon each of the articles as follows:

	Old Rate	*New Rate*	*Additional Rate*
(1) Playing cards	W 5	W 50	W 45

— 2 —

		Old Rate	New Rate	Additional Rate
(2)	*Kolpai*	W 15	W 100	W 85
(3)	*Mahjong*	100	500	400"

SECTION VI. *Informers awards.* a. The provisions of Section XVI of Ordinance Number 169, dated 1 March 1948 (*Reorganization of Bureau of Internal Revenue*), providing for payment to certain informers of a part of the penalties collected, shall apply to penalties collected under this ordinance.

b. Twenty-five per cent (25%) of all penalties levied, assessed, collected, and paid because of Amusement Tax, Admissions Tax, or Card and Dice Tax deficiencies shall be paid to the Korean Cooperative Welfare Association of the Bureau of Internal Revenue; provided, however, that no such payment shall be made in any case where payment is made under subsection *a* of this section.

SECTION VII. *Repeal of inconsistent laws.* All laws and regulating measures or parts thereof which are inconsistent herewith or in conflict with the provisions hereof, are, to the extent that they are inconsistent or in conflict herewith, hereby repealed.

SECTION VIII. *Effective date.* This ordinance shall become effective ten days after the date appearing hereon.

APPROVED: RECOMMENDED:

    W. F. DEAN   AHN CHAI HONG
Major General, United States Army   Civil Administrator
  Military Governor in Korea

SOUTH KOREAN INTERIM GOVERNMENT
Seoul, Korea

ORDINANCE

NUMBER 194   17 May 1948

MANAGEMENT OF CONFUCIAN-SHRINE PROPERTY

Pending legislative action on the subject matters hereof and the enactment of legislation pursuant thereto, the following shall govern:

SECTION I. The purpose of this ordinance is to make regulations for the most effective and proper management of Confucian-shrine property.

SECTION II. *"Confucian-shrine property"* as used in this ordinance means the movable, immovable and all other property which has been constituted for the purpose of supporting the Confucian shrines in South Korea.

SECTION III. Confucian-shrine property shall not be permitted to be bought, sold, transferred, exchanged, mortgaged or otherwise disposed of except pursuant to the provisions of this ordinance.

SECTION IV· A Confucian-Shrine Foundation shall be established in each province for ·Confucian-shrine property. Immovable things, such as land and buildings and ritual articles for religious ceremonies of the Confucian Shrine, shall constitute the basic property. Cash and other movable things shall constitute the floating property.

SECTION V. Each Confucian-Shrine Foundation shall maintain the Confucian shrines, conduct educational work. and promote the development of Confucianism and other culture in the province.

SECTION VI. The income of each Confucian-Shrine Foundation shall not be spent for other than the following objects:

  a. Maintenance of Sung Kyun Kwan Institute and Sung Kyun Kwan College.

  b. Maintenance of the Confucian shrines in each province.

  c. Education and operation of other cultural work.

SECTION VII. Ten per cent of the total income each year in each Confucian-Shrine Foundation shall be delivered to the Sung Kyun Kwan Institute for its upkeep and twenty per cent to the Sung Kyun Kwan College for its upkeep.

SECTION VIII. Each year each provincial Confucian-Shrine Foundation shall appropriate not less than eighty per cent of its total income less the amount mentioned in the preceding Section and other reserves, for education and other cultural work.

SECTION IX. The manager of each provincial Confucian-Shrine property shall present an inventory of all property, cash, and other movable things as of the effective date of this ordinance, to the competent provincial governor, within twenty days after the effective date hereof.

SECTION X. The provincial governor receiving the inventory of all property, cash, and other movable property from the manager of the Confucian-shrine property, within ten days after receipt, shall report the details to the Director of the Department of Education.

SECTION XI. Each provincial governor shall take all proceedings for the formation of the Confucian-Shrine Foundation in accordance with this ordinance and other applicable laws, within forty days after the effective date hereof.

SECTION XII. The number of the directors of each Confucian-Shrine Foundation shall be from seven to fifteen, and they shall be elected by the representatives of each Confucian shrine. But one of the directors shall be designated by the provincial governor out of the personnel of the provincial government.

SECTION XIII. All lawful rights and obligations that have accrued with respect to Confucian-shrine property before the effective date hereof, shall be succeeded to by the Confucian-Shrine Foundation established by this ordinance.

SECTION XIV. Government General Order Number 91, dated 29 June 1919 (*Regulations for Control of the Confucian-Shrine Property*) and Government General Order Number 110, dated May 14, 1945 (*Regulations on the Provincial Confucian Shrines*) are repealed, effective twenty days after the date appearing hereon.

SECTION XV. This ordinance shall be effective on the date appearing hereon.

APPROVED:

W. F. DEAN
Major General, United States Army
Military Governor in Korea

RECOMMENDED:

AHN CHAI HONG
Civil Administrator

軍政廳　官報　法令　第一九四號　　一九四八年五月一七日

SOUTH KOREAN INTERIM GOVERNMENT
Seoul, Korea

ORDINANCE

NUMBER 195

18 May 1948

LICENSING OF JUDICIAL SCRIVENERS

Pending legislative action on the subject matters hereof and the enactment of legislation pursuant thereto, the following shall govern:

SECTION I. The purpose of this ordinance is to provide for the transference of jurisdiction over judicial scriveners from the courts to the Director of the Department of Justice, in accordance with the enactment of Court Reorganization (*Ordinance Number 192, dated 4 May 1948*) and to provide for the regulation of judicial scriveners.

SECTION II. Judicial scriveners shall engage only in the drawing up of documents upon the request of another person for submission to a court, public prosecutor's office, or other judicial agency.

SECTION III. The Director of the Department of Justice shall have the power to license the judicial scrivener business, to determine fees authorized to be received by judicial scriveners and to exercise supervision over such businesses. The Director of the Department of Justice, may, however, delegate to the chief public prosecutor of a district or branch court authority to handle the functions of such licensing and supervision.

SECTION IV. No judicial scrivener shall in the conduct of his business:

a. Refuse to draw up a document as requested by another person.

b. In a case he has handled upon the request of one party, draw up a document on behalf of an adverse party.

c. Take over a financial interest in the result of litigation where he acts as scrivener for one of the parties.

d. Appear in court as a representative of any party to litigation.

e. Reveal secrets of another person which he has received in the course of his scrivener business.

SECTION V. The Director of the Department of Justice shall have power to cancel or suspend a judicial scrivener business if a judicial scrivener shall violate his business obligation or degrade himself.

SECTION VI. Governor General Ordinance Number 5, dated December,

— 1 —

192: (*Korean Judicial Scrivener Ordinance*) is hereby repealed. Government General Ordinance Number 13, dated 16 March 1925 (*Enforcement Regulation of Korean Judicial Scrivener Ordinance*), issued pursuant to the Korean Judicial Scrivener Ordinance, shall continue to be effective as enforcement regulation hereof, provided that for "*Chief Judge of District Court,*" as used in such ordinance, there shall be substituted "*Director of the Department of Justice.*"

SECTION VII. This ordinance shall be effective on the date appearing hereon.

APPROVED:

      W. F. DEAN
    Major General, United States Army
      Military Governor in Korea

RECOMMENDED:

    AHN CHAI HONG
    Civil Administrator

SOUTH KOREAN INTERIM GOVERNMENT
Seoul, Korea

ORDINANCE
NUMBER 196   24 May 1948

REVISION OF MINING FEES

Pending legislative action on the subject matters hereof and the enactment of legislation pursuant thereto, the following shall govern:

SECTION I. *Purpose.* The purpose of this ordinance is to readjust various fees pertaining to mining activities, and to amend existing provisions of law relating to such fees.

SECTION II. *Mining ordinance amended.* The first paragraph of Section 43 of Government General Ordinance Number 9, dated 29 February 1916 (*Enforcement Regulation of Chosun Mining Ordinance*), is hereby amended to read as follows:

There shall be levied, assessed, collected and paid upon the following applications, requests or reports of every person a fee equal to the following:

		Fee Per Application
(1)	Application for mining claims	*Won* 1,000
(2)	Application to change name of claimant	500
(3)	Application for succession of title to mining claim	500
(3-2)	Application for seceding joint title to mining claim	500
(4)	Application for increasing, or increasing and decreasing, area of claim and mine lot	500
(5)	Application for decreasing area of claim and mine lot	500
(6)	Application for correcting area of claim in accordance with Sections 12, 13 or 22 of Chosun Mining Ordinance	500
(7)	Application for union or division of mine lot	500
(8)	Application for changing kind of mineral	500

— i —

Fee Per Application

(9) Application for actual inspection of
mine in accordance with Section 26
hereof   *Won 3,000*

(10) Request for using or expropriating
another's mining land in accordance
with Sections 32 or 40 of Chosun mining
Ordinance   *500*

(11) Application for trespass or removal of
obstacles in accordance with Section 39
hereof   *500*

(12) Request for reading claim record per
kun, per hour   *100*

(13) Request to examine mining claim map
per hour   *50*

SECTION III. *Mining Registration Ordinance amended.* Paragraph 3 of
Section 8 of Government General Ordinance Number 10, dated 29 February
1916 (*Regulations of Chosun Mining Registration*), is hereby amended to read as
follows: .

Rate of Fee. Fees set forth in the first paragraph hereof shall be as
follows:

(1) Application for copy or abstract of mining   *Won*
registration book – per page   *50*

(2) Application for copy of mining map
(Per. 100,000 pyong or per 2 Ri)   *50*

(3) Request for reading of mining registration book and examining of mine lot –
per one mine-lot, per hour   *50*

SECTION IV. *Payment of fee.* The fees imposed by this ordinance shall
be paid by the purchase of revenue stamps, which shall be affixed to the
application and cancelled, in accordance with the revenue-stamp laws in effect
when the fees are due and payable.

SECTION V. *Applicability of revenue-stamp laws.* All matters relating to
any fee or penalty to be levied, assessed, collected or paid under the provisions
of this ordinance shall be determined in accordance with the revenue-stamp

— 2 —

laws in effect as of the date the fees are due or payable; or, in the case of penalty, as of the date the act or acts giving rise to the penalty are committed.

SECTION VI. *Effective date.* This ordinance shall be effective on the tenth day after the date appearing hereon.

APPROVED:

W. F. DEAN
Major General, United States Army
Military Governor in Korea

RECOMMENDED:

AHN CHAI HONG
Civil Administrator

SOUTH KOREAN INTERIM GOVERNMENT
Seoul, Korea

ORDINANCE
NUMBER 197   25 May 1948

KOREAN COAST GUARD EMPOWERED TO SEARCH VESSELS

Pending legislative action on the subject matters hereof and the enactment of legislation pursuant thereto, the following shall govern:

SECTION I. *Purpose.* The purpose of this ordinance is to give the Korean Coast Guard police powers sufficiently broad to insure better compliance with the law as pertaining to vessels and crews thereof.

SECTION II. *General powers to board, search and seize vessels.* Commiss'oned, warrant, and petty officers of the Coast Guard are hereby empowered, and it shall be their duty, to make inquiries, examinations, inspections, searches, seizures, and arrests upon the high seas and the navigable waters of Korea south of 38 degrees north latitude. For such purposes, such officers are authorized to go on board any vessel at any time, subject to the jurisdiction or to the operation of any law of the South Korean Interim Government, to address inquiries to those on board, to examine the ship's documents and papers, and to examine, inspect, and search the vessel and use all necessary force to compel compliance with the law. When it shall appear that a breach of law is being or has been committed, the offender shall be arrested; or if escaping to shore, he shall be immediately pursued and arrested on shore, or other lawful and appropriate action shall be taken. Or if it shall appear that a breach of the law has been committed (*a*) so as to render such vessel or the merchandise or any part thereof (whether on board or brought into southern Korea by such vessel) liable to forfeiture, or (*b*) so as to render such vessel liable to a fine or penalty and if necessary to secure such fine or penalty, such vesssel shall be seized.

SECTION III. *Effective date.* This ordinance shall be effective on the date appearing hereon.

APPROVED: RECOMMENDED:

    W. F. DEAN —   AHN CHAI HONG
Major General, United States Army   Civil Administrator
Military Governor in Korea

— 1 —

SOUTH KOREAN INTERIM GOVERNMENT
Seoul, Korea

ORDINANCE

NUMBER 198   26 May 1948

PENALTIES FOR VIOLATION OF FOOD COLLECTION PROGRAMS

Pending legislative action on the subject matters hereof and the enactment of legislation pursuant thereto, the following shall govern:

SECTION I.   The purpose of this ordinance is to provide penalties for violations of governmental food collection programs.

SECTION II   As used herein:

a.   *Food* includes summer grains, rice or any other item of human food.

b.   *Food collection program* means a program of the National Government, or any agency thereof, duly promulgated, for the collection of food.

SECTION III.   Persons who during any food collection program illegally store or conceal or wrongfully fail to deliver food required to be delivered by governmental authority, shall, upon conviction by a duly constituted court, be punished by penal servitude for not exceeding six (6) months or a fine not exceeding fifty thousand won (W 50,000), or both.   In addition, any food involved in the violation shall be forfeited to the government.

SECTION IV.   Persons who during any food collection program sell, offer for sale, transfer or offer to transfer, transport or offer to transport, buy or offer to buy, any food included in a food collection program except by permission or authority of the government, or who otherwise illegally deal in such food, shall, upon conviction by a duly constituted court, be punished by penal servitude for not exceeding six (6) months or a fine not exceeding one hundred thousand won (W 100,000), or both.   In addition, any food involved in the violation shall be forfeited to the government.

SECTION V.   Persons who fail or refuse to receive, collect, inspect, properly assign quotas, store, grade, safeguard, transport, process, issue proper receipts, keep accurate records and accounts, or make payments for, or reports concerning, any food included in a food collection program, or who are guilty of other official misconduct, failure or willful neglect of duty in connection with such program, shall, upon conviction by a duly constituted court, be punished by penal servitude for not exceeding six (*6*) months.

SECTION VI. Persons who make or attempt any transaction or act with the intent of evading or avoiding the provisions of any law setting forth a food collection program, shall, upon conviction by a duly constituted court, be punished by penal servitude for not exceeding six (*6*) months or by a fine not exceeding fifty thousand won (*W 50,000*), or both.

SECTION VII. Persons who during any food collection program hamper or attempt to prevent any authorized government official from investigating the amount of food grown or produced, or who knowingly spread false propaganda, or who threaten farmers with the intention of lessening the farmers' willingness to deliver food to the authorized collection agency, shall, upon conviction by a duly constituted court, be punished by penal servitude for not exceeding three (*3*) months or by a fine not exceeding fifty thousand won (*W 50,000*), or both.

SECTION VIII. Persons who in a manner not heretofore stated in this ordinance violate the provisions of law which contain a food collection program, shall, upon conviction by a duly constituted court, be punished by penal servitude for not exceeding three (*3*) months or a fine not exceeding fifty thousand won (*W 50,000*), or both.

SECTION IX. Where specific provisions of the penal or other law fix or permit heavier punishments for acts heretofore stated in this ordinance than those set forth herein, persons found guilty shall be subject to such heavier punishments.

SECTION X. The provisions of Section VIII of Ordinance Number 90, dated *28* May 1946 (*Economic Controls*), shall be deemed limited by the provisions of this ordinance.

SECTION XI. This ordinance shall be effective ten days after the date appearing hereon.

APPROVED: RECOMMENDED:

W. F. DEAN   AHN CHAI HONG
Major General, United States Army   Civil Administrator
Military Governor in Korea

— 2 —

SOUTH KOREAN INTERIM GOVERNMENT
Seoul, Korea

ORDINANCE

NUMBER 190

29 May 1948

REGULATIONS FOR CENTRAL RESEARCH LABORATORY AUTHORIZED

Pending legislative action on the subject matters hereof and the enactment of legislation pursuant thereto, the following shall govern:

SECTION I. *Purpose.* The purpose of this Ordinance is to authorize the promulgation of rules and regulations for the Central Research Laboratory.

SECTION II. *Authorization.* The Department of Commerce is hereby authorized by Department Order, to make rules and regulations for the Central Research Laboratory, including the fixing of fees.

SECTION III. *Effective date.* This ordinance shall be effective on the date appearing hereon.

APPROVED: RECOMMENDED:

W. F. DEAN   AHN CHAI HONG
Major General, United States Army   Civil Administrator
Military Governor in Korea

57C

SOUTH KOREAN INTERIM GOVERNMET
Seoul, Korea

ORDINANCE
NUMBER 200   21 June 1948

TRANSFER OF FUNCTIONS TO CUSTOMS BUREAU

Pending legislative action on the subject matters hereof and the enactment of legislation pursuant thereto, the following shall govern:

SECTION I. *Purpose.* The purpose of this ordinance is to transfer certain functions which more properly belong to the Bureau of Customs.

SECTION II. *Documentation of vessels.* All duties, functions, records, property, funds and personnel concerned with or pertaining to the documentation, licensing, and registration of vessels including those now belonging to the Department of Transportation, shall belong to the Bureau of Customs, Department of Finance.

SECTION III. *Search of vessels in port* All functions heretofore exercised by the Department of National Police in connection with control, inspection, and searching of ships and crews at ports of entry shall hereafter be exercised exclusively by the Bureau of Customs, Department of Finance; provided that the provisions of this Section are to be considered as supplemental to, but not inconsistent with or contravening, the provisions of Ordinance Number 197, dated 25 May 1948 (*Korean Coast Guard Empowered to Search Vessels*).

SECTION IV. *Harbor control.* Authority is hereby given to the Director of the Department of Finance to issue, amend and abolish such rules and regulations as may be deemed necessary and/or advisable to carry out the provisions of this Section.

The Bureau of Customs of the Department of Finance shall have full control over all matters pertaining to the movement of shipping within the open ports of southern Korea. The Director of the Department of Finance shall appoint for each port a Harbor master who shall act under the direct supervision of the local Collector of Customs, and whose duties and powers shall pertain to the enforcement of the harbor rules and regulations.

SECTION V. *Repealer.* a. Section III c of Ordinance Number 158 dated 30 December 1947 (*Transfer and Abolition of Certain Licensing Functions*) is hereby repealed.

— 1 —

軍政廳　官報　法令　第二〇〇號　　　一九四八年六月二一日

6. All other laws, ordinances, orders, regulations, directives and instructions, or parts thereof, including national, provincial and local laws which are inconsistent herewith or in conflict with the provisions hereof, are to the extent of such inconsistency or conflict hereby repealed. Included in this category is Public Act Number 46 dated March, 1899 (*Shipping Law*).

SECTION VI. *Effective date.* This ordinance shall be effective ten days after the date appearing hereon.

W. F. DEAN
Major General, United States Army
Military Governor in Korea

SOUTH KOREAN INTERIM GOVERNMENT
Seoul, Korea

ORDINANCE
NUMBER 201   21 June 1948

REVISION OF PROVINCIAL AND LOCAL TAXES

Pending legislative action on the subject matters hereof and the enactment of legislation pursuant thereto, the following shall govern:

SECTION I. *Purpose.* The purpose of this ordinance is to increase the provincial and local revenues by revision of tax rates and simplification of the tax structure, so as to promote the more efficient functioning of provincial and local governments and the economy of Korea.

SECTION II. *Building Tax.*

a. *Rates.* Commencing with the calendar year 1948, the provincial tax and local surtax rates on buildings shall be as follows:

TAXING AUTHORITY	ANNUAL TAX RATE
Province (Do)	W 7.00 per unit
City (Bu)	240% surtax on the provincial tax
Town (Eup)	200% surtax on the provincial tax
Village (Myun)	160% surtax on the provincial tax

b. *Units.* The units to which the above rates apply are set forth in the following table;

Table of Number of Units for Imposing Building Tax Classified by Grades

Type of Construction

Type of Use	A Class			B Class			C Class		
	1st Group	2nd Group	3rd Group	1st Group	2nd Group	3rd Group	1st Group	2nd Group	3rd Group
Grade of Sites 50.00 1 or above	10.00	7.00	6.00	7.00	4.90	4.20	5.00	3.50	3.00
2 45.00	9.09	6.37	5.46	6.37	4.46	3.82	4.55	3.19	2.73

軍政廳　官報　法令　第二〇一號　　　一九四八年六月二一日

573

Type of Use	A Class			B Class			C Class		
	1st Group	2nd Group	3rd Group	1st Group	2nd Group	3rd Group	1st Group	2nd Group	3rd Group
3 40.00	8.18	5.73	4.91	5.73	4.01	3.44	4.09	2.87	2.46
4 35.00	7.27	5.09	4.37	5.09	3.57	3.06	3.64	2.55	2.19
5 30.00	6.36	4.46	3.82	4.46	3.12	2.68	3.18	2.23	1.91
6 25.00	5.45	3.82	3.27	3.82	2.68	2.29	2.73	1.91	1 64
7 20.00	4.54	3.18	2.73	3.18	2.23	1.91	2.27	1.59	1.37
8 16.00	3.80	2.66	2.28	2.66	1.87	1.60	1.90	1.33	1.14
9 13.00	3.23	2.27	1.94	2.27	1.59	1.36	1.62	1.14	.97
10 10.00	2.66	1.87	1.60	1.87	1.31	1.12	1.33	.94	.80
11 7.00	2.09	1.47	1.26	1.47	1.03	.88	1.05	.74	.63
12 5.00	1.73	1.22	1.04	1.22	.85	.73	.87	.61	.52
13 4.00	1.55	1.09	.93	1.09	.76	.66	.78	.55	.47
14 3.00	1.37	.96	.83	.96	.68	.58	.69	.48	.42
15 2.50	1.27	.89	.77	.89	.63	.54	.64	.45	.39
16 2.00	1.17	.82	.71	.82	.58	.50	.59	.41	.36
17 1.50	1.07	.75	.65	.75	.53	.45	.54	.38	.33
18 1.00	.97	.68	.59	.68	.48	.41	.49	.34	.30
19 .70	.93	.66	.56	.66	.46	.40	.47	.33	.29
20 .50	.89	.63	.54	.63	.44	.38	.45	.32	.28
21 .35	.87	.61	.53	.61	.43	.37	.44	.31	.27
.20 22 or above	.85	.60	.51	.60	.42	.36	.43	.30	.26
23 under .20	.83	.59	.50	.59	.41	.35	.42	.29	.25

c. *Payment of tax.* The building tax shall be due and payable semi-annually, as follows:

First term – 1 May – 31 May

Second term – 1 October – 31 October

In case the annual building tax for one person in one bu eup, or myun is *W 300* or less, the entire tax shall be paid or collected during the month of May.

SECTION III. *Forestry tax.*

a. *Rates.* Effective for the fiscal year beginning 1 April 1948, and annually thereafter, the provincial tax on forest areas shall be determined according to the following table of grades and rates:

Forest Tax Rates Per One Chungbo (3000 Pyung)
Classified by Grades

Area Grade	Forest Field Grade			
	Special Grade	Higher Grade	Middle Grade	Lower Grade
1	W 17.50	W 15.00	W 13.00	W 11.00
2	13.50	11.50	10.00	8.50
3	10.50	9.00	8.00	6.50
4	8.00	7.00	6.00	5.00

Area grades and forest field grades shall be those established by laws in effect prior to the fiscal year 1945. The grades of lands newly devoted to forestry use shall be established in the same manner as those of forestry lands now in existence.

b. Date of payment. The entire tax is due and payable annually in the month of December.

SECTION IV. *Slaughter tax.* Commencing with the effective date of this ordinance, the provincial tax rate on the slaughter of cows and pigs shall be *W 300* per cow and *W 30* per pig.

SECTION V. *Fisheries and fishing taxes.* Commencing with the fiscal year 1948, the provincial tax rates on fisheries and fishing shall be as set forth in the following subsections:

a. Fish culture.

Type of Area	Tax Rate per 1,000 Square Meters
(1) Area enclosed by stationary bamboo poles—1,000 sq. Meters	(1) W 17 per year
(2) Area enclosed by single bamboo net – 1,000 sq. Meters	(2) W 10 per year
(3) Oyster culture area	(3) W 10 per year
(4) Culture area for other forms of marine life	(4) W 10 per year

b. Trap net fishing.

Two per cent of the value of the total yearly catch.

c. Exclusive fishing right.

— 3 —

Two per cent of the value of the total yearly catch of fish, and one and one-half per cent of the value of the total yearly catch of other forms of marine life.

d.   *Whale fishing.*

Species	Tax Rate
(1) Ballaena	W 15,000 per whale
(2) Humpback, sperm and fin	W 6,000 per whale
(3) California Grey (devilfish)	W 1,000 per whale

e.   *Bottom dragnet fishing (by motor driven boat):*

Value of Yearly Catch	Tax Rate per Boat
(1) Under W 1,000,000	(1) .20% per year
(2) W 1,000,000 to W 2,000,000	(2) .30% per year
(3) W 2,000,000 to W 3,000,000	(3) .50% per year
(4) W 3,000,000 to W 4,000,000	(4) 1.00% per year
(5) W 4,000,000 to W 5,000,000	(5) 1.50% per year
(6) W 5,000,000 to W 7,000,000	(6) 2.00% per year
(7) W 7,000,000 to W 10,000,000	(7) 2.50% per year
(8) W 10,000,000 and over	(8) 3.00% per year

f.   *Purse Seine Fishing.*

Value of Yearly Catch	Tax Rate per Boat Equipped with Purse Seine
(1) Under W 2,000,000	(1) .3% per year
(2) W 2,000,000 to W 5,000,000	(2) .5% per year
(3) W 5,000,000 to W 10,000,000	(3) 1.0% per year
(4) W 10,000,000 to W 15,000,000	(4) 1.5% per year
(5) W 15,000,000 to W 25,000,000	(5) 2.0% per year
(6) W 25,000,000 to W 30,000,000	(6) 2.5% per year
(7) W 30,000,000 and over	(7) 3.0% per year

g. *Submarine armor fishing.*

W 1,750 per year for each submarine armor diving suit.

h. *Shore dragnet fishing (including shore seine and running tuck net).*

Length of Floating Net	Tax Rate per Net
(1) Under 300 Meters	(1) W 140 per year
(2) 300 meters to 500 meters	(2) W 280 per year
(3) 500 meters to 800 meters	(3) W 560 per year
(4) 800 meters and over	(4) W 840 per year

i. *Dragnet fishing (by boat without motor).*

Length of Boat	Tax Rate Per Net
(1) Under 15 meters	(1) W 490 per year
(2) 15 metres and over	(2) W 700 per year

j. *Scoop net fishing.*

Method	Tax Rate Per Net
(1) One boat and light	(1) W 210 per year
(2) More than 4 boats	(2) W 560 per year
(3) Other	(3) W 100 per year

k. *Roe net Fishing.*

Method	Tax Rate Per Net
(1) With boat	(1) W 210 per year
(2) Without boat	(2) W 100 per year

l. *Swig net fishing (angler type net).*

Beam of Boat	Tax Rate Per Boat
(1) Under 6 meters	(1) W 140 per year
(2) 6 to 8 meters	(2) W 210 per year
(3) over 8 meters	(3) W 280 per year

— 5 —

m. *Set net fishing.*

Length of Main Net Floater	Tax Rate Per Net
(1) Under 90 meters	(1) W 140 per year
(2) 90 to 180 meters	(2) W 350 per year
(3) Over 180 meters	(3) W 700 per year

n. *Dam net fishing.*

Length of Net	Tax Rate Per Net
(1) Under 200 meters	(1) W 140 per year
(2) 200 to 400 meters	(2) W 280 per year
(3) over 400 meters	(3) W 490 per year

o. *Haul net fishing. (reeled by hand or hauled by rowing),*

Length of Boat	Tax Rate Per Boat
(1) 8 meters or under	(1) W 100 per year
(2) Over 8 meters	(2) W 140 per year

p. *Dredge net fishing.*

Length of Boat	Tax Rate Per Boat
(1) 9 meters or under	(1) W 100 per year
(2) Over 9 meters	(2) W 140 per year

q. *Ring net fishing.*

Species	Tax Rate Per Net
(1) Sardine, horse mackeral, mackeral, gray mullet, herring	(1) W 1,750 per year
(2) Others	(2) W 560 per year

r. *Defensive net. (including bind net and scoop net into which fish are driven by stones).*

W 700 per year per net.

s. Drift net. (with sailboat).

Species	Tax Rate per Boat
(1) Sardine. Alaska pollock, crab, mackeral, spanish mackeral	(1) W 420 per year
(2) Other	(2) W 140 per year

t. Drift net. (with motor boat).

Species	Type of Motor	Tax Rate Per Boat
(1) Mackeralor Spanish mackeral	(a) Under 20 h. p. (b) 20 to 40 h. p. (c) Over 40 h. p.	(a) W 1050 per year (b) W 1400 per year (c) W 2100 per year
(2) Other	(a) Under 20 h. p. (b) 20 to 40 h. p. (c) Over 40 h. p.	(a) W 560 per year (b) W 1050 per year (c) W 1400 per year

u. Trawling line. (artificial bait).

Length of Boat	Tax Rate Per Boat
(1) 9 meters and under	(1) W 100 per year
(2) Over 9 meters	(2) W 210 per year

v. Diving. (without submarine armor).

W 100 per year per diver.

w. Harpooning. (by harpoon gun).

W 700 per year per gun.

x. Angling. (including spearing or trawling from boat with natural bait).

Method	Length of Boat	Type of Motor	Tax Rate
(1) Angling with single rod and line	Any	Any	W 100 per year per rod
(2) Trawling with natural bait or spearing	(a) 9 meters or under (b) over 9 meters	(a) None (b) None	(a) W 140 per year per boat (b) W 280 per year per boat

— 7 —

Method	Length of Boat	Type of Motor	Tax Rate
(3) Trawling with natural bait or spearing	(a) Any	(a) Under 10 h. p.	(a) W 350 per year per boat
	(b) Any	(b) 10-20 h. p.	(b) W 700 per year per boat
	(c) Any	(c) 21-30 h. p.	(c) W 1050 per year per boat
	(d) Any	(d) Over 30 h. p.	(d) W 1400 per year per boat

y.   Other boat fishing. (not specified above).

Length of Boat	Tax Rate Per Boat
(1) 6 meters or under	(1) W 100 per year
(2) Over 6 meters	(2) W 140 per year

z.   Period and payment.   The taxes are based on the fiscal year beginning 1 April and the entire tax is due and payable annually in the month of June.

SECTION VI.   *Provincial vehicle tax.*

a.   Taxes repealed.   Effective 1 April 1948, the following provincial vehicle taxes are repealed:

(1) Trolley car tax, (2) bicycle tax, (3) trailer tax, (4) carriage tax, (5) Rikisha tax, (6) horse cart tax, and (7) cart tax.

b   Self-propelled vehicles tax.   On and after 1 April 1948, the provincial tax on self-propelled vehicles shall be determined annually by the use of the following table:

Type of Vehicle	Passenger capacity or pay load in kilograms	Tax per Vehicle	
		Business Use	Private Use
(1) Automobile Bantam type	(a) Any	W 5,400	W 8,100
Automobile	(b) 3 passenger or under	W 10,800	W 16,200
Automobile	(c) 4 passenger or over	W 16,200	W 24,300
(2) Passenger Bus	(a) 15 passenger or under	W 21,600	W 32,400
	(b) 16 to 23 passenger	W 27,000	W 40,500
	(c) 24 passenger or over	W 32,100	W 54,000
(3) Trucks	(a) Under 1,000 kilograms	W 10,800	W 16,200
	(b) 1,000 to 2,000 kilograms	W 16,200	W 24,300
	(c) 2,001 to 3,000 "	W 21,600	W 32,400
	(d) 3,001 to 4,000 "	W 27,000	W 40,500
	(e) Over 4,000 kilograms	W 35,100	W 54,000
(4) Motorcycle		W 2,700	W 13,500

c. No "bu", "up" or "myun" surtaxes shall be levied on self-propelled vehicles for the fiscal year 1948 and thereafter.

d. *Date of payment.* Provincial vehicle taxes shall be paid or collected semi-annually, as follows:

The first term from 1 May to 31 May.

The second term from 1 October to 31 October.

SECTION VII. *Bu, eup and myun vehicle taxes.*

a. *Effective date.* Effective on and after 1 April 1948, annual *shi* (Seoul City), bu, eup and myun vehicle taxes shall be assessed under the provincial vehicle assessment rule and shall be paid or collected annually in the month of May at the rates shown in the following table:

Type of Vehicle	Class	Tax per Vehicle per year	
		Business Use	Private Use
(1) Bicycle		W 130	W 130
(2) Trailer		W 70	W 70
(3) Carriage	(a) 1 horse	W 1,400	W 4,200
"	(b) more than 1 horse	W 2,700	W 5,500
(4) Rikisha		W 1,350	W 1,350
(5) Horse (or cow)-drawn cart	(a) 2-wheeled	W 550	W 550
"	(b) 4-wheeled	W 900	W 900
(6) Hand-drawn cart		W 270	W 270

b. *Provincial vehicle tax assessment rule.* In applying the provincial vehicle assessment rule, the words *"within or without the province"*, *"the provincial governor or chief of the national tax office"* and *"the national tax office"* in the provincial vehicle tax rule, shall be considered respectively as *"within or without the city, town, village"*, *"mayor or chief of city, town, village"* and *"city, town or village"*.

SECTION VIII. *Dog tax.* Effective on and after 1 April 1948, the shi, bu, eup and myun dog tax rates shall be *W 100* per year per dog.

SECTION IX. *Boat tax.* Effective on and after 1 April 1948, the shi, bu, eup and myun boat tax rates shall be as shown in the following table:

Capacity of Boat	Tax Rate per Boat
(1) 1 to 5 tons	(1) W 150
(2) 5 to 10 tons	(2) W 500
(3) Over 10 tons	(3) W 500, plus W 40 for each ton

SECTION X. *Restaurant attendance tax.* Effective thirty (30) days after the date appearing hereon, the rate of provincial tax on attendance in restaurants shall be *W 30* per person, of which two-sevenths shall be paid by the province to the bu, eup or myun in which the tax is collected.

SECTION XI. *Specific repealer.* The Residence **Tax,** provided in Section X of Ordinance Number 109, dated 15 October 1946 (*Revision of Provincial and Local Taxes*) is hereby repealed and shall not be levied, collected or paid on and after the effective date of this ordinance.

SECTION XII. *Repeal of inconsistent laws.* All laws, ordinances, orders, regulations, directives and instructions, or parts thereof, which are inconsistent herewith or in conflict with the provisions hereof are, to the extent they are in conflict herewith, hereby repealed.

SECTION XIII. *Effective date.* Except as otherwise specifically provided herein, this ordinance shall be effective on the date appearing hereon,

SEAL
USAMGIK

W. F. DEAN
Major General, United States Army
Military Governor in Korea

SOUTH KOREAN INTERIM GOVERNMENT
Seoul, Korea

ORDINANCE
NUMBER 202

21 June 1948

LAND TAXES REVISED

Pending legislative action on the subject matters hereof and the enactment of legislation pursuant thereto, the following shall govern:

SECTION I. *Purpose.* The purpose of this ordinance is to provide the provincial and local governments and the schools with more adequate revenues, to repeal provisions requiring mandatory payments to the Korean Agricultural Association, to provide exemption for small land owners, and to combat inflation in Korea.

SECTION II. *Ordinance Number 128 amended.* Section II of Ordinance Number 128, dated 31 December 1946 (*Revision of Land Taxes and Service Fees*, is hereby amend to read as follows:

"Section II. *Land Taxes.* Land taxes shall be levied at the rate of 6%, established by Imperial Ordinance Number 6, dated 31 March 1943 (*Korea land Tax Law*), of an assessed rental value equal to fifty times the assessed rental value determined in accordance with Imperial Ordinance Number 37, dated 4 December 1940 (*Concerning Investigation of Rental Values of Land in Korea*)."

SECTION III. *Imperial ordinance Number 6 amended.* Sections 60, 61, and 87 of Imperial Ordinance Number 6, dated 31 March 1943 (*Koreans Land Tax law*) are hereby repealed.

SECTION IV. *Government-General Ordinance Number 3 Amended.* Section 24 of Government-General Ordinance Number 3, dated 25 January 1926 (*Regulation of Korean Agricultural Association*), authorizing a surtax of 25% or less on the land tax, is hereby repealed.

SECTION V. *Exemption.* Any taxpayer whose total annual land tax in any one bu, eup or myun does not exceed twenty-five won (W 25) shall be exempt from payment of such tax on land in such bu, eup or myun.

SECTION VI. *Allocation of land tax.* The land tax collected under the provisions of this ordinance shall be allocated as follows:

a. Ninety per cent of the national land taxes collected in each gun,

— 1 —

583

eup or myun shall be paid quarterly, in accordance with regulations prescribed by the Director, Department of Finance, for the authorized use of the schools.

b. Ten per cent of the land taxes shall be retained by the national government.

SECTION VII. *Liability in case of the sale or transfer of lands, leases, or superficies.* In the case of lands, superficies of more than twenty (20) years' duration, or leases registered on the land ledger, sold or transferred to a successor, without the full payment of land taxes and accrued interest thereon due at the time of transfer, the land or rights therein may be seized from a successor and sold at public auction for the payment of the unpaid taxes, penalties, and interest. The procedures set forth in the National Tax Collection Law (as amended), Law Number 21, dated 26 March 1897, shall apply *mutatis mutandis.*

SECTION VIII. *Repeal of inconsistent laws.* All laws, ordinances, orders, regulations, directives and instructions, or parts thereof, which are inconsistent herewith or in conflict with the provisions hereof are, to the extent they are in conflict with the provisions hereof, hereby repealed.

SECTION IX. *Effective date.* This ordinance shall be effective 1 January 1948.

W. F. DEAN
Major General, United States Army
Military Governor in Korea

SOUTH KOREAN INTERIM GOVERNMENT
Seoul, Korea

ORDINANCE

NUMBER 203   1 July 1948

LITIGATION COSTS AND FEES AMENDED

Pending legislative action on the subject matters hereof and the enactment of legislation pursuant thereto, the following shall govern:

SECTION I. *Purpose.* The purpose of this ordinance is to increase the statutory litigation fees and costs.

SECTION II. *Amendment to Civil Litigation Cost Act.* Act Number 64, dated August, 1890 (*Civil Litigation Cost Act*), as amended, is further amended as follows:

a. Costs for writing fixed in Section *2* and costs of translation fixed in Section *3* shall be increased by *200* times.

b. Allowances for parties or witnesses fixed in Section *9* shall be increased by *150* times.

c. The maximum allowance for an expert witness or interpreter fixed in Section *11* shall be increased by *150* times.

d. The maximum lodging cost fixed in Section *12* shall be increased by *150* times.

e. Transportation costs fixed in Section *13* shall be increased to *W 5* for one mile on water or for one kilometer on land.

SECTION III. *Amendment to Civil Litigation Stamp Act.* Act Number 65, dated August, 1890 (*Civil Litigation Stamp Act*), as amended, is further amended to read as follows:

a. *Section 3, Subsection 1.* "In litigation in which the subject matter is non-property, a *W 300* revenue stamp shall be affixed."

b. *The first sentence of Section 6-2.* "On the following applications, a *W 20* revenue stamp shall be affixed when the value of subject matter does not exceed *W 100,000*, or a *W 40* revenue stamp shall be affixed when the value of the subject matter exceeds *W 100,000*."

c. *The first sentence of Section 6-3.* "On the following applications, a *W 50* revenue stamp shall be affixed when the value of the subject matter does not

— 1 —

exceed *W 100,000*, or a *W 100* revenue stamp shall be affixed when the value of the subject matter exceeds *W 100,000*."

d. *Section 10.* "On the answer or other application not particularized in the foregoing sections, a *W 20* stamp shall be affixed when the value of subject matter does not exceed *W 100,000*. A *W 30* revenue stamp shall be affixed when the value of subject matter exceeds *W 100,000*."

e. *Section 16, Subsection 1.* "On applications in non-litigation cases, a *W 20* stamp shall be affixed when the value of the subject matter does not exceed *W 100,000*, or a *W 30* stamp shall be affixed when the value of the subject matter exceeds *W 100,000*."

f. *The first sentence of Section 16 Subsection 2.* "On the following applications, a *W 50* stamp shall be affixed when the value of the subject matter does not exceed *W 100,000*, or a *W 100* stamp shall be affixed when the value of the subject matter exceeds *W 100,000*."

g. *Section 16, Subsection 3.* "An application in a non-litigation case in which the subject matter has no monetary value, shall be deemed not to exceed *W 100,000* in value."

SECTION IV. *Amendments to Marshal Fee Regulation.* Act Number 52, dated July 1890 (*Marshal Fee Regulation*), as amended, is hereby further amended as follows:

a. *Section 2.* The fee for service of documents shall be *W 20*.

b. *Section 3.* The fees for attachment or provisional attachment shall be *W 100* when the value of the subject matter to be attached does not exceed *W 100,000*, or shall be *W 500* when such value exceeds *W 100,000*.

c. *Section 6.* The fee for delivery up of specific movable property or a specific amount of goods, shall be *W 400* when the value of subject matter does not exceed *W 100,000*, or shall be *W 600* when such value exceeds *W 100,000*, provided, however, that after two hours, *W 50* shall be paid for each further hour required to make delivery.

d. *Section 7, Subsection 1.* For executions in accordance with Section 731, Subsection 1 or Section 733, Subsection 1 of the Civil Procedure Code, the fee for execution shall be *W 400* if three hours or less is required, and *W 100* for each hour additional.

e. *Section 9, Subsection 1.* Fees for public auction shall be *W 400* where the value of the subject matter does not exceed *W 100,000*, or *W 600* where the value of subject matter is more than W 100,000 but less than W 200,000; and where the value of the subject matter is W 200,000 or more, W 600 plus W 2 for each W 1,000 or value in excess of W 200,000.

f. Section 10 and Section 11. Fees laid down in each proviso shall be W 200.

g. Section 14, Subsection 2. The fee for writing shall be W 10 for one full page.

h. Section 15. The fee shall be W 30.

i. Section 16. The fee for composing a protest shall be W 150, provided, however, that if it takes more than one hour, the fee shall be increased W 50 for each hour in excess.

j. Section 16-2. The fee for inspection shall be W 10 one time.

k. Section 17. The allowance for a witness shall not exceed W 150, and the allowance for an expert witness shall not exceed W 250.

l. Section 18, Subsection 1. Transportation costs shall be W 5 per one kilometer.

m. Section 18, Subsection 2. Lodging costs shall not exceed W 600.

SECTION V. *Amendment to Fee for Personal Mediation.* Government General Order Number 121, dated 2 August 1939 (*Fees for Personal Mediation*) is hereby amended as follows:

a. Section 1. The fee for application of mediation shall be W 100 per case.

b. Section 2. The fee for inspection and copy of records shall be W 10.

c. Section 4. The allowance for mediation committee shall not exceed W 500 per day. The maximum lodging cost shall not exceed W 1,000 per day.

d. Section 5. Transportation cost shall be W 7 per mile on water routes which do not allow steamboats and W 7 per kilometer on land.

SECTION VI. *Amendment to lease mediation fees.* Government General Order Number 300, dated 23 December 1940 (*Fees for Lease Mediation*) is hereby amended to read as follows:

Section 1. "The fees for application of lease mediation shall be as follows:

a. The fee shall be W 100 when the value of subject matter does not exceed W 50,000; W 200 when it exceeds W 50,000.

"*b.* The value shall be deemed to exceed W 50,000 when it cannot be valued in money."

— 3 —

SECTION VII. The fee fixed in Section 31 of Governor General Order Number 5, dated 10 December 1932 (*Korean Tenancy Mediation Order*) shall be *W 20* per case.

SECTION VIII. *Amendment to fee for delivery of copy or extract of criminal decision.* Government General Order Number 3, dated 5 August 1912 (Fees for Copy and Extract of Criminal Judgment) is hereby amended so that the fees fixed in this order shall be *W 10* per page.

SECTION IX. *Amendment to fee for copies or extracts from land registry book.* Government General Order Number 206 (*Fee Regulation for Copy or Extract from Registry Books*) dated 14 August 1942, is hereby amended as follows:

    a. *Section 2.* The fee shall be *W 20* per sheet.

    b. *Section 3.* The fee for inspection shall be *W 10* per time.

    c. *Section 4.* The fee for certification shall be *W 20* per case.

    d. *Section 5.* The fee for certification of registration shall be *W 10* per case.

    e. *Section 6* Fee for certification of registration of ship shall be *W 200* per case.

SECTION X. *Regulations of fee for family registration and temporary residence registration.* Government General Order Number 156, dated 27 December 1922 (*Korean Fee Regulation for Family Registration and Temporary Residence Registration*) is hereby amended as follows:

    a. *Section 1.* The fee for inspection shall be *W 10* per time.

    b. *Section 2, Subsection 1.* The fee for copy and extract of original copy shall be *W 15* per sheet.

    c. *Section 2, Subsection 2; and Section 3.* The fee for certification shall be *W 10* per case.

SECTION XI. *Amendment of fee for certification of date.* Government General Order Number 208, dated 14 August 1941 (*Fee for Application for Certification of Date on Agreement*), is hereby amended to read as follows:

"In the case of applications to the registration office for certification of the date on an agreement, the fee shall be *W 50*."

SECTION XII. *Amendment of fee for change of name.* Government General Order Number 222, dated 26 December 1939 (*Change of Korean Name*), is hereby amended as follows:

Section 3. The fee shall be *W 50*.

軍政廳　官報　法令　第二〇三號　　　一九四八年七月一日

SECTION XIII. *Amendment of fee for notarial action* Government General Order Number 21, dated 17 March 1913 (*Korean Notarial fee Regulation*) is hereby amended as follows:

_a. *Section 1,, Subsection 1 is amended to read:* "The fee for drawing a legal document shall be W 150 when the value of subject matter does not exceed W 100,000 and W 20 shall be added for every W 50,000 in excess of W 100,000."

b. *Section 11.* The figure "Y 500" is changed to "W 100,000."

c, *Section 13.* The figure "sen 20" is changed to "W 20."

d. *Section 14.* The fee for drawing a non-legal document shall be W 100 per hour, and W 50 shall be added for each hour in excess of the first hour.

e. *Section 18.* The fee for drawing a will as a secret document shall be W 200.

f. *Section 19.* The fee shall be W 100, and W 50 shall be added for each hour in excess of the first hour.

g. *Section 20-2.* The fee for certification of company or other by-laws shall be W 500, and W 30 shall be added for each W 100,000 that the capital exceeds W 1,000,000.

h. *Section 21,* The fee for certification of date on a private agreement shall be W 50.

i. *Section 22.* The fee for executory certification on the document shall be W 50 when the value of subject matter does not exceed W 100,000, and W 100 when it exceeds W 100,000.

j. *Section 23.* The fee for copy or extract particularized in this section shall be W 30 per sheet, provided that in the cases set forth in Subsection 1. of Section 55 of the Notary Act (*including the cases applied mutatis mutandis Section 62-6 of the same act*), it shall be W 10 per sheet.

k. *Section 24.* The fee for inspection particularized in this section shall be W 20.

l. *Section 26* is amended as follows:

Allowance	W 300
Cost for train and ship	*Passenger fare of 2nd class*
Cost for other vehicles	W 7 *per kilometer*
Water route	W 7 *per mile*

Cost for lodging   *W 750*

SECTION XIV. *Effective date.* This Ordinance shall be effective on the
tenth day after the date appearing hereon.

W. F. DEAN
Major General, United States Army
Military Governor in Korea

SOUTH KOREAN INTERIM GOVERNMENT
Seoul, Korea

ORDINANCE
NUMBER 204   14 June 1948

REVISION OF ORDINANCE NUMBER 81
TRANSFER OF FUNCTIONS TO NATIONAL ECONOMIC BOARD

Pending legislative action on the subject matters hereof and enactment of legislation pursuant thereto, the following shall govern:

SECTION I. *Purpose.* The purpose of this Ordinance is to transfer the functions of the Research and Statistics Divisions from the Office of Administration to the National Economic Board.

SECTION II. *Transfer.* The Research and Statistics Divisions of the Office of Administration are hereby abolished and all duties, functions, records and property thereof, together with such personnel as are administratively deemed necessary for efficient operation, are hereby transferred to the National Economic Board.

SECTION III. *Effective date.* This ordinance shall be effective as of 17 May 1948.

W. F. DEAN
Major General, United States Army
Military Governor in Korea

SOUTH KOREAN INTERIM GOVERNMENT
Seoul, Korea

ORDINANCE
NUMBER 207

1 July 1948

LAW RELATING TO THE LEGAL PROFESSION

Pending legislative action on the subject matters hereof and the enactment of legislation pursuant thereto, the following shall govern:

SECTION I. *Purpose.* The purpose of this ordinance is to provide for the supervision and regulation of the practice of law and to further provide for the creation of the Korean Federal Bar Association and District Bar Associations.

SECTION II. *Eligibility to practice law.* a. No person shall handle legal proceedings in court or perform any legal functions for another unless he has been duly admitted to the practice of law.

b. Any Korean person twenty (20) or more years of age who comes within any of the following categories, shall be admitted to the practice of law:

1. A person qualified to be admitted to the practice of law in accordance with the laws effective on 15 August 1945.

2. A person who has successfuly passed the legal examination after training in the practice of law as a legal apprentice or probationer.

3. A person who has served as an attorney or research specialist in the Department of Justice for two (2) years, after having been qualified as a legal apprentice or probationer.

4. A person who does not come within any of the foregoing subsections and who has been a judge, including a judge of the juvenile court or a public prosecutor, prior to effective date of this ordinance.

5. Any of the following persons who do not come within Subsection b 4 of this section, with the prior consent of the Director:

(a) One who has become a judge or a public prosecutor after the effective date of this ordinance.

(b) One who has taught law for more than seven (7) years as a full-time professor in a recognized law college.

SECTION III. *Admission to practice.* Before commencing the practice of law, a lawyer shall apply to the Director, through the bar association of which he is a member, to be entered on the Registry of Practicing Lawyers.

SECTION IV. *Admission of foreign lawyers.* A foreigner authorized to

— 1 —

practice law in his own country may, with the prior permission of the Director, be admitted to the practice of law in Korea, subject, however, to the principles of international reciprocity.

SECTION V. *Disqualification.* Persons who come within any of the following categories shall not be qualified to practice law:

a. One who has been declared to be incompetent or quasi-incompetent by a court.

b. One who has been sentenced to imprisonment or penal servitude.

c. Any public official who has been discharged from his position by disciplinary disposition, provided, however, that any such person may be admitted to the practice of law by the Director, after the expiration of two (2) years from such discharge.

SECTION VI. *Duties and obligations of a practicing lawyer.* a. A practicing lawyer shall perform his duties conscientiously and shall commit no act or acts unbecoming a member of the legal profession or such as would impair the administration of justice.

b. A practicing lawyer shall not betray the confidential relationship existing between himself and his clients and shall not, without reasonable grounds, refuse to handle a matter assigned to him pursuant to law by the governmental authorities or by a bar association.

c. A practicing lawyer shall not accept the following cases:

1. Cases wherein he is called upon to represent the adversary of one to whom he has previously given advice.

2. Cases in which he was involved, directly or indirectly, as a judge, prosecutor, or arbitrator.

d. A practicing lawyer shall not charge exorbitant fees or otherwise take any unfair advantage of his clients, nor shall he knowingly conceal the truth or make a false statement.

e. A practicing lawyer shall not maintain, directly or indirectly, more than one office for the practice of law. Notice of the establishment or the removal of an office to another location shall be filed with the Director.

f. The Director, with the advice of the Central Council of the Korean Federal Bar Association, shall prescribe, from time to time, such other rules of legal ethics as are deemed necessary, which rules shall become part hereof, violations thereof to be punished as provided in Section VII hereof.

— 2 —

SECTION VII. *Punishment. a.* The Director shall appoint a disciplinary committee, with himself as *ex officio* chairman, said committee to consist of four (*4*) regular members and four (*4*) alternate members. The members thereof shall serve for a term of one (*1*) year, and shall be appointed from the Registry of Practicing Lawyers and from among those who occupy the positions of bureau chief, or higher, in the Department of Justice, in equal numbers.

b. When a practicing lawyer violates this ordinance, the by-laws of his bar association, rules of legal ethics, or otherwise engages in conduct detrimental to the dignity of the profession, the Chief Prosecutor of the Supreme Court, or the President of the bar association to which the lawyer belongs, shall initiate action by filing a complaint with the disciplinary committee.

c. Disciplinary punishment shall consist of one or more of the following:

1. A reprimand.

2. A fine not to exceed ten thousand won (*W 10,000*).

3. Suspension from the practice of law for not more than one (*1*) year.

4. Disbarment from the practice of law.

d. An appeal may be taken directly to the Supreme Court, said appeal to be made within three (*3*) days after the receipt of the decision of the disciplinary committee.

e. The provisions of Ordinance Number 166, dated 24 January 1948 (*Disciplinary Punishment for Prosecutors*), shall apply *mutatis mutandis*, except as otherwise provided herein.

f. Any person who has not been admitted to the practice of law, or any person who has been suspended or barred therefrom, shall not engage in the practice thereof, appear on behalf of any party to any judicial proceeding other than in his own behalf, or perform any professional act or service for which a practicing lawyer could collect a fee. A violation of this provision shall be punished by the courts by a fine of not more than ten thousand won (*W 10,000*), or imprisonment for not more than one (*1*) year. In addition to any other penalty or punishment, an additional fine will be levied equal to at least three (*3*) times all the fees collected in violation of this provision.

SECTION VIII. *District bar associations. a.* The practicing lawyers residing within the territorial jurisdiction of each district court shall constitute

軍政廳　官報・法令　第二〇七號　　一九四八年七月一日

themselves a bar association, provided, however, that no bar association shall be formed in a district having less than five (5) practicing lawyers, such lawyers to become members of the bar association in an adjoining district, as designated by the Director. Each said bar association is hereby created as a juridical person, the stated purposes thereof being to uphold the honor and dignity of practicing lawyers, the maintenance of a law library, the exchange of professional information, and to improve the services of practicing lawyers to clients and the public.

  b. The temporary chairman of each said bar association shall be the lawyer who has practiced the greatest number of years in the district. Within thirty (30) days after the effective date of this ordinance, he shall call a general meeting to elect a president, a vice-president, a secretary and an accounter, to hold office for one (1) year. The general meeting of each said bar association shall also draft by-laws and submit them to the Director for approval, the by-laws to include the following:

  1. Name and location of the association.

  2. Offices created and the functions thereof.

  3. Provisions concerning meetings.

  4. Provisions concerning minimum and maximum fees to be charged clients.

  5. Provisions concerning membership fees. Amendments from time to time made to by-laws shall likewise be submitted to the Director for approval.

  c. Each said association shall, in accordance with its by-laws, hold a general meeting at least once a year to elect officers, determine and allocate appropriations, approve accounts, and to resolve other matters properly brought before the meeting. Special meetings shall be held as often as necessary upon call of the President, or upon call of the number of members fixed by the by-laws. The general meeting may create a number of standing committees to deal with special matters, the chairman and members thereof to be appointed by the President. Five (5) days' notice of all meetings shall be given the Director and a copy of the minutes thereof shall be filed with him.

  d. The Director shall exercise supervision over practicing lawyers and district bar associations. The Director may revoke resolutions and suspend the holding of meetings when in his opinion there is involved a violation of law, by-laws or an injury to the public interest.

— 4 —

e. Those district bar associations wherein high courts are located, may elect as honorary members persons of distinction in judicial circles, with the consent of not less than two-thirds of the members of the association, and subject to the approval of the Director.

SECTION IX. *The Korean Federal Bar Association* a. The Korean Federal Bar Association is hereby created as a juridical person, with its principal office in Seoul, Korea. It shall be composed of all district bar associations, and its objects shall be to advance the science of jurisprudence, promote the administration of justice, uphold the honor of the legal profession, and to further international friendship among lawyers. Each district bar association shall contribute such funds as are necessary to operate the Korean Federal Bar Association.

b. A Central Council of the Korean Federal Bar Association is likewise hereby created. It shall be composed of one representative elected from each district bar association, together with such additional representatives as are elected at the ratio of one representative for each fifty (50) members of each district bar association. The Council shall elect a president, a vice-president, a secretary and an accounter from among its members, by unsigned, secret ballot of the representatives present. The Council shall meet at least once each year in Seoul. A quorum of the Council shall be a majority of its members present, and all actions of the Council shall be resolved by a majority vote of the members present. The provisions of Paragraph *d* of Section VIII hereof, shall apply to the meetings and resolutions of the Korean Federal Bar Association.

SECTION X. *Orders.* a . Bureau of Justice Order Number 4, dated 19 November 1945, as amended, creating the National Bar Association of Korea and district chapters thereof, is hereby superseded, and all their rights, duties, functions, funds, fixtures and equipment are hereby transferred to the Korean Federal Bar Association and the district bar associations created herein, respectively; provided, however, that the officers of the National Bar Association, and of the district chapters thereof, shall continue to hold their respective offices until the organization of the Korean Federal Bar Association and the district bar associations shall have been completed in accordance with this ordinance.

b. All other Orders, except where inconsistent with the provisions hereof, shall remain *mutatis mutandis* in full force and effect.

— 5 —

SECTION XI. *Definitions.* As used herein, the term "Director" shall mean the Director of the Department of Justice.

SECTION XII. *Effective date.* This ordinance shall be effective on the date appearing hereon.

W. F. DEAN
Major General, United States Army
Military Governor in Korea

SOUTH KOREAN INTERIM GOVERNMENT
Seoul, Korea

ORDINANCE

NUMBER 209   3 July 1948

SECTION XII OF ORDINANCE NUMBER 173 AMENDED

Pending legislative action on the subject matters hereof and the enactment of legislation pursuant thereto, the following shall govern:

SECTION I. *Purpose.* The purpose of this ordinance is to amend Section XII of Ordinance Number 173, dated 22 March 1948 (*Creation of the National Land Administration*).

SECTION II. *Amendment.* Section XII of Ordinance Number 173 is hereby amended to read as follows:

"SECTION XII. *Land whose ownership is disputed; 31 August 1948 time limitation.* a. Non-Japanese juridical persons, whose assets or controlling stock interests have not vested in Military Government under Ordinance Number 33, dated 6 December 1945; or non-Japanese natural persons who claim ownership of land transferred to the Administration may, not later than 31 August 1948, file suit against the Administration in the district court or branch thereof having jurisdiction where such land is located. Such court shall have jurisdiction to determine title to such land and to all incidental building sites, farm buildings and dwellings, ownership of which has vested in Military Government, where claim to incidental building sites, farm buildings and dwellings arises from the identical transaction on which claim to the land is based. Where incidental building sites, farm buildings and dwellings are involved, the Administration shall represent the interests of the Property Custodian, and the Property Custodian shall not be separately named as a party to the action. Such court will, if title is found in the claimant, award to him the property claimed, if it has not already been sold by the Administration or Property Custodian. If the property has been sold, the court shall award monetary damages against the Administration in lieu of the property. In no case where the property has been sold by the Administration or the Property Custodian, may the judgment of the court affect the validity of the title thereto.

"b. Suits which may be filed hereunder by a person described above shall be forever barred if not filed on or before 31 August 1948. Claims on which suit could be brought hereunder, may not be filed with the Property

— 1 —

Claims Commission after the effective date hereof.

"c. If claim has been filed with the Property Claims Commission, the claimant may not file suit against the Administration. The Property Claims Commission shall, as to claims filed with it before the effective date hereof for property described in subsection '*a*' hereof transfer such claims to the appropriate district or branch court for determination, with the Administration thereupon to become the defendant. Transfer of such a claim shall have the effect of filing of suit by the claimant. No filing fee shall be required to be paid in such cases. Such transfers shall be made as quickly as possible, but without requirement that all transfers be made by 31 August 1948. Upon transfer of a case to the court, copies of the petition need not be served on the parties, but the court shall notify the parties that suit has been filed, with notice to the Administration to be given at the Regional Office. Thereafter the court shall handle the suit pursuant to the ordinary provisions of law, as modified by the provisions of this ordinance.

"d. When a new suit is filed, the plaintiff must (*1*) state in his petition that no claim has been filed with the Property Claims Commission; (*2*) state the location of the Regional Office of the Administration which is located nearest to the land; (*3*) file with his petition copies in duplicate of the documents, if any, on which his claim is based; and (*4*) give a brief statement of the facts supporting his claim. The court shall serve a copy of the petition on the Administration in Seoul. The court shall also serve copies of the petition and the documents, if any, on which the claim is based on the Offices of the Administration in accordance with the location of the land, as follows:

'Land located in Seoul, Kyunggi Do, and Kangwon Do on the Seoul Regional Office (*Chi Bang Chung*); land located in Chungchong Pukto and Chungchong Namdo on Taejon Regional Office (*Chi Bang Chung*); land located in Cholla Pukto on the Iri Regional Office (*Chi Bang Chung*); land located in Cholla Namdo on the Mokpo Regional Office (*Chi Bang Chung*); land located in Kyongsang Namdo on Pusan Regional Office (*Chi Bang Chung*); land located in Kyongsang Pukto on Taegu Regional Office (*Chi Bang Chung*); and land located in Cheju Do on Seoul National Office (*Bon Chang*).'

"e. In the cases referred to in subsections '*c*' and '*d*' hereof, the court shall fix a date for hearing not less than thirty (*30*) days after service upon the Regional Office of the Administration of the petition and accompanying papers or the notice of filing. Within such time or at the hearing, the Administration shall answer, stating one of the following: (1) that the Administration does not desire to defend the suit; (2) that the Administration is ready to defend the suit, in which case its defense shall be stated. If the Administration needs more time to investigate in order to answer, it shall ask for a postponement of the trial and give reason therefor, and the court shall give the Administration reasonable time in which to complete its investigation.

"*f.* In every suit filed against the Administration claiming title to property described in subsection 'a' hereof, if the Administration fails to appear or states that it does not desire to defend the suit, it shall be the duty of the court to examine the plaintiff and his proofs, and the court shall not award judgment to the plaintiff unless it appears, without regard to the admissions of the Administration, that on the facts and law the plaintiff is entitled to judgment.

"*g.* Any default judgment entered against the Administration prior to 31 August 1948 shall be opened by the court upon the application of the Administration, made not later than 31 October 1948.

"*h.* In all suits involving claims previously filed with the Property Claims Commission, expenses which the claimant has incurred in preparing documents for the Commission shall not be allowed as court costs.

"*i.* In case a suit heretofore filed sets forth a claim previously filed with the Property Claims Commission, the court may request the Commission to send it the papers relating to the claim. When any suit on a claim previously filed with the Property Claims Commission involves property not the subject matter of the suit, the court shall, after using the papers, return them to the Commission.

"*j.* The Administration shall make every reasonable effort not to sell land to which its title is disputed, but all title given by it shall be valid as against all persons who claim any adverse property rights therein."

SECTION III. *Effective date.* This Ordinance shall be effective on the date appearing hereon.

W. F. DEAN
Major General, United States Army
Military Governor in Korea

SOUTH KOREAN INTERIM GOVERNMENT
Seoul, Korea

ORDINANCE
NUMBER 210

12 July 1948

RELEASE OF PROPERTY SEIZED BY JAPANESE AUTHORITIES AS ENEMY PROPERTY

Pending legislative action on the subject matters hereof and the enactment of legislation pursuant thereto, the following shall govern:

SECTION I. *Purpose.* The purpose of this ordinance is to supplement Ordinance Number 162, dated 15 January 1948 (*Property Formerly Under Control of Japanese Enemy Property Custodian*), and to provide a simple procedure for the release or return of property which was seized by Japanese authorities as enemy property.

SECTION II. *Release of certain property.* The property described below is released from the operation of Ordinance Number 33 of 6 December 1945 (*Vesting Title to Japanese Property within Korea*), and will be returned to the natural or juridical persons, or their successors or legal representatives, owning such property at the time it was seized by or taken under the control of the former Government General of Korea or any agency thereof, including the Japanese Enemy Property Custodian, or required to be transferred:

a. Property which on 7 December 1941 and at the time it was seized or was taken under control was owned by an American or an allied national, or by a Korean *zaidan* or other Korean juridical person determined by General Order Number 774 of 22 May 1942 or by other order or administrative decision, to be an enemy organization, and which property was later sold and on 9 August 1945 was owned by a Japanese national or by the Government General of Korea or any provincial or municipal subdivision thereof or by any instrumentality thereof.

b. Property which on 7 December 1941 was owned by an American or an allied national or by a Korean *zaidan* or other juridical person determined by General Order Number 774 of 22 May 1942 or by other order or administrative decision to be an enemy organization and which property was sold or otherwise transferred at the request or by order of any official of the former Government General of Korea by any person acting as agent or representative of such owner

軍政廳　官報　法令　第二一〇號　　一九四八年七月一二日

and which on 9 August 1945 was owned by a Japanese national or by the Government General of Korea or any provincial or municipal subdivision thereof or by any instrumentality thereof

SECTION III. *Return of certain property.* The property described below will be returned to the natural or juridical persons, or their successors or legal representatives, owning such property at the time it was seized by or taken under the control of the former Government General of Korea or any agency thereof, including the Japanese Enemy Property Custodian, or required to be transferred:

a. Property which on 7 December 1941 and at the time it was seized or was taken under control was owned by an American or an allied national or by a Korean zaidan or other Korean juridical person determined by General Order Number 774 of 22 May 1942 or by other order or administrative decision to be an enemy organization, and which property was later sold but on 9 August 1945 did not have the character of Japanese-owned property.

b. Property which on 7 December 1941 was owned by an American or an allied national or by a Korean zaidan or other juridical person determined by General Order Number 774 of 22 May 1942 or by other order or administrative decision to be an enemy organization, and which property was sold or otherwise transferred at the request or by order of any official of the former Japanese Government General of Korea by any person acting as agent or representative of such owner but on 9 August 1945 did not have the character of Japanese-owned property.

SECTION IV *Procedure for transfer of title.* An American or allied national who, or an American, allied or Korean juridical person who, was deprived of his property as described above, may upon application have the entries showing such transfer and all subsequent transfers cancelled from the Land Registry upon compliance with the conditions hereinafter set forth:

a. If the immovable property is still registered in the name of the Japanese Enemy Property Custodian upon compliance with Ordinance 162, dated 15 January 1948 (*Property Formerly Under Control of Japanese Enemy Property Custodian*).

b. If the immovable property was sold or transferred to a Japanese national or juridical person, and such immovable property is now vested in the Military Government of Korea, under Ordinance Number 33, dated 6 December

1945 (*Vesting Title to Japanese Property Within Korea*) an application will be made to the Land Registry Office of the district court having jurisdiction to cancel the entry turning the immovable property over to the Japanese Enemy Property Custodian, or any other entry made under any law or legal proceeding or by an official act or order by which alien enemies of the Japanese Empire or natural or juridical persons adjudged to be alien enemies or under the control of alien enemies were deprived of immovable property, and all subsequent entries on the Land Register. The application will list the entries to be cancelled and will be approved and countersigned by the Office of Property Custody and will show that the funds, if any, referred to in Section V hereof, have been transferred to the Office of Property Custodian. Upon production of the approved and countersigned application by the applicant or the Office of Property Custody, the Land Registry Office will immediately expunge and delete the listed items from the Land Register and will make an entry on the Land Register that the cancellation was made under authority of this ordinance, which entry will show the date of the cancellation.

c. If the immovable property was sold or otherwise transferred to the Japanese Government or the Government General of Korea or the Government of any of the Provinces of South Korea, or any political subdivision, or department, bureau, or agency thereof, and is now vested in the Military Government of Korea under Ordinance Number 33, the procedure set forth in Subsection *b*, above, will be followed, but the funds transferred to the Property Custodian will be held for the account of the agency designated in Section V hereof, and the indorsement of the Property Custodian on the application will state that fact.

d. If the immovable property was sold or transferred to a Korean national or a Korean juridical person, or to a Japanese national or juridical person who thereafter sold or transferred it to a Korean national or juridical person, whether or not such juridical person is controlled in whole or in part by the Property Custodian, United States Army Military Government in Korea, on or prior to 8 August 1945, and title to the said immovable property is now held in the name of a Korean national or juridical person, in addition to the application to the Land Registry Office, an application will be made to the Property Custodian for an appraisal of the current value of the immovable property. The Property Custodian will appraise the immovable property and serve a notice in writing on the present owner, stating the amount of the appraisal and that he or it shall accept the sum for which the immovable property has been appraised, which sum has been set aside in a special account to his or its credit, or he or it may accept other vested property, if available, of an equal appraised value in lieu thereof, or property and money. The application to the Land Registry Office of the district court to cancel the listed registrations referred to in Subsection *b*, above, will be submitted to the Office of Property Custody, together with the application for the appraisal of the current value of the immovable property. After immovable property has been appraised, the application for cancellation will be approved and countersigned by the Property Custodian, and will show that the current value of the immovable property has been appraised, and that the amount of the appraised value has been set aside in a special account to the credit of the said present owner. Upon production of the approved and countersigned application for cancellation before the Land Registry Office, the entries listed on the application will be immediately expunged and deleted from the Land Register.

SECTION V. *Disposition of funds.* Any funds or other thing of value deposited for the account of the former American owner or other former owner described in Section IV hereof in payment or consideration for the transfer of such property sold or transferred, shall be disposed of as follows:

a. In case the owner on 9 August 1945 was a Japanese national, such funds or other thing of value deposited or held for the account of the former owner, or the equivalent thereof, shall be delivered to the Office of the Property Custodian to be held as vested property.

b. In case the owner on 9 August 1945 was the former Government General of Korea or any provincial or municipal subdivision thereof or any instrumentality thereof, such funds or other thing of value deposited or held for the account of the former owner, or the equivalent thereof, shall be delivered to the successor in the South Korean Interim Government to the agency of government which made payment for the property in funds or other things of value or shall be delivered to the Finance Department of the South Korean Interim Government.

c. Any funds or other thing of value deposited for account of the former American owner or other former owner described in Section IV hereof, shall be delivered to the present owner of such property as partial or complete

軍政廳　　官報　　法令　　第二一〇號.　　一九四八年七月一二日

payment therefor.

SECTION VI. *Closing of cases.* Upon completion of any transfers made pursuant to this ordinance, the Property Claims Commission shall close the cases pending before it for return of such property.

SECTION VII. *Effective date.* This ordinance shall be effective on the date appearing hereon.

W. F. DEAN
Major General United States Army
Military Governor in Korea

SOUTH KOREAN INTERIM GOVERNMENT
Seoul, Korea

ORDINANCE
NUMBER 211

July 1948

CHANGES IN THE BONDED WAREHOUSE AND CUSTOMS LAWS

Pending legislative action on the subject matters hereof and the enactment of legislation pursuant thereto, the following shall govern:

SECTION I. *Purpose.* The purpose of this ordinance is to provide for the reduction of the permissible period in which goods may be stored in bonded warehouses, or may be retained in custody by the Bureau of Customs after being removed from bonded zones, and to further provide for the disposal of goods which are stored, or retained in custody. beyond the permitted period.

SECTION II *Period of storage.* a. Goods imported into Korea south of thirty-eight degrees north latitude shall be withdrawn from storage in bonded warehouses within six (6) months from the date of original storage.

b The Director of the Department of Finance, at his discretion, is hereby authorized to extend the above period upon due application made by the owner of the goods.

c. Paragraphs *a* and *b* above, shall likewise apply. where, pursuant to law. goods have been taken into custody by the Bureau of Customs from bonded zones.

SECTION III. *Disposition of goods.* a. Goods stored or retained in custody beyond the permitted period shall be regarded as abandoned to the South Korean Interim Government and shall be seized by the Collector of Customs, the former owners thereof to be compensated therefor at prices established by the National Price Administration or by Customs appraisal, or from proceeds of public auctions, as the case may be, less unpaid customs duties, warehouse fees and all other charges and expenses incurred in connection with the goods.

b. Goods so abandoned may be utilized for public use by the South Korean Interim Government or may be sold under such regulations as the Director, Department of Finance shall prescribe.

SECTION IV. *Repeal of inconsistent laws.* All laws, ordinances, orders, regulations, directives. and instructions, or parts thereof, which are inconsistent herewith or in conflict with the provisions hereof, are, to the extent they are inconsistent or in conflict herewith, hereby repealed.

SECTION V. *Effective date.* This ordinance shall be effective on the date appearing hereon.

W. F. DEAN

Major General, United States Army

Military Governor in Korea

HEADQUARTERS
UNITED STATES ARMY MILITARY
GOVERNMENT IN KOREA
Office of the Military Governor
Seoul, Korea

EXECUTIVE ORDER
NUMBER 1   22   April 1947

DISSOLUTION OF THE GREAT KOREAN DEMOCRATIC
YOUNG MEN'S ASSOCIATION (TAI HAN MIN CHUNG)

WHEREAS, members of The Great Korean Democratic Young Men's Association, commonly known as Tai Han Min Chung, headquarters at Seoul, Korea, openly identifying themselves as such, have committed acts of violence and intimidation, engaging in terrorism and gangsterism;

WHEREAS, such organization has adopted terrorism, gangsterism and violence as a matter of policy in exerting political influence;

WHEREAS, such activities have been secretly conducted from the head-quarters of such organization; and

WHEREAS, such activities are an immediate threat to orderly government and the development of true democracy in Korea:

NOW, THEREFORE, IT IS ORDERED:

1. THAT The Great Korean Democratic Young Men's Association (Tai Han Min Chung) be and it is hereby dissolved;

2. THAT its registration pursuant to Ordinance No 55 be revoked;

3. THAT no member of this organization shall join any other organization with similar policies;

4. THAT no group or organizaton shall hereafter use the name *The Great Korean Democratic Young Men's Association (Tai Han Min Chung)*;

5. THAT the Department of Police immediately padlock the headquarters of this organization, seize and hold all its records, and impound its funds and other property;

軍政廳　官報　行政命令　第一號　　　一九四七年四月二十二日

CHAPTER 4   EXECUTIVE ORDERS (USAMGIK)

6. THAT a complete investigation and report be made for the Military Governor of all financial and other support such organization has received and of all its affiliates;

7. THAT the Department of Police take immediate steps to apprehend and bring to justice all members of this association connected with any specific act of terrorism or violence;

8. THAT any person violating or evading the provisions of this order shall, upon conviction by a Military Occupation Court, suffer such punishment as the court shall determine

9. THAT this order be effective on the date appearing hereon.

REQUESTED:

CHOUGH PYUNG O

Director, Department of Police

APPROVED:

RECOMMENDED:

ARCHER L LERCH

Major General United States Army

Military Governor in Korea

AHN CHAI HONG

Civil Administrator

HEADQUARTERS
UNITED STATES ARMY MILITARY
GOVERNMENT IN KOREA
Office of the Military Governor
Seoul, Korea

EXECUTIVE ORDER
NUMBER 2

16 May 1947

DISSOLUTION OF THE CHOSUN DEMOCRATIC
YOUTH ALLIANCE (CHOSUN MIN CHUNG)

WHEREAS, members of Chosun Democratic Youth-Alliance, commonly known as Chosun Min Chung, headquarters at Seoul, Korea, openly identifying themselves as such, have committed acts of violence and intimidation, engaging in terrorism and gangsterism;

WHEREAS, such organization has adopted terrorism, gangsterism and violence as a matter of policy in exerting political influence;

WHEREAS, such activities have been secretly conducted from the headquarters of such organization; and

WHEREAS, such activities are an immediate threat to orderly government and the development of true democracy in Korea;

NOW, THEREFORE, IT IS ORDERED:

1. THAT The Chosun Democratic Youth Alliance (Chosun Min Chung) be and it is hereby dissolved;

2. THAT its registration pursuant to Ordinance No 55 be revoked;

3. THAT no member of this organization shall join any other organization with similar policies;

4. THAT no group or organization shall hereafter use the name Chosun Democratic Youth Alliance (Chosun Min Chung);

5. THAT the Department of Police immediately padlock the headquarters of this organization, seize and hold all its records, and impound its funds and other property;

軍政廳　官報　行政命令　第二號　　一九四七年五月十六日

6. THAT a complete investigation and report be made for the Military Governor of all financial and other support such organization has received and of all its affiliates;

7. THAT the Department of Police take immediate steps to apprehend and bring to justice all members of this association connected with any specific act of terrorism or violence;

8. THAT any person violating or evading the provisions of this order shall, upon conviction by a Military Occupation Court, suffer such punishment as the court shall determine;

9. THAT this order be effective on the date appearing hereon.

REQUESTED :

CHOUGH PYUNG OK

Director, Department of Police

APPROVED :   RECOMMENDED :

ARCHER L LERCH   AHN CHAI HONG

Major General United States Army   Civil Administrator

Military Governor in Korea

SOUTH KOREAN INTERIM GOVERNMENT
Seoul, Korea

EXECUTIVE ORDER
NUMBER 3
          30 June 1947

CODE DRAFTING COMMISSION

IN ORDER to consolidate, coordinate and expedite the work of drafting proposed basic codes of law which has been in progress for many months in the Department of Justice, among the judges and among the prosecutors in the various courts of the South Korean Interim Government, and

. IN ORDER to insure the development of a modern and democratic system of administering justice which is purely Korean in origin :

1. There is hereby established a CODE DRAFTING COMMISSION.

2. The CODE DRAFTING COMMISSION is hereby charged with the duty of preparing complete drafts of a set of basic codes to be adopted, after necessary approval, as substitutes for codes of law now in force regulating civil rights, property rights, family relations, commercial relations and punishment of crimes and minor offenses and prescribing procedures to be followed in the enforcement of law and the administration of justice.

The drafts shall be prepared in such form as to reflect democratic principles and sound, modern trends, with due regard to the customary law and traditions of Korea.

3. The CODE DRAFTING COMMISSION shall make periodic reports to the Military Governor regarding the progress of its work, and upon completing the several codes it shall submit them to the Military Governor for reference to the Korean Interim Legislative Assembly and for his approval.

4. The CODE DRAFTING COMMISSION is authorized to appoint committees of persons learned in the law, composed of officials of the South Korean Interim Government or public spirited private citizens of Korea, to assist the Commission in its task. Such committees shall perform such duties as the Commission may prescribe, and will receive appropriate recognition for their patriotic work ; *provided, however,* that no official of the South Korean

Interim Government shall receive compensation in addition to his official salary for work performed on any such committee, and *provided further*, that any necessary expenses incurred in connection with the work of the Commission or its committees shall be paid out of the funds appropriated for the budget of the Department of Justice and the courts and in accordance with the laws regulating expenditure of public funds.

5. The CODE DRAFTING COMMISSION is authorized to request assistance and information from the various departments, offices, agencies and instrumentalities of the South Korean Interim Government and shall have free access to the National Library, the Capitol Library, the National Law Library, the libraries of the various courts and the library of Seoul National University.

6. The following persons are appointed to the positions indicated as members of the CODE DRAFTING COMMISSION, and shall receive no compensation therefor in addition to their official salaries:

> KIM YOUNG MOO (Chief Justice of the Supreme Court),
> Member and Chairman
>
> KIM BYONG RO   (Director of the Department of Justice),
> Member
>
> LEE IN   (Chief Prosecutor of the Supreme Court),
> Member

7. Other appointments to membership on the CODE DRAFTING COMMISSION may be made from time to time as may be deemed necessary.

8. This Order shall be effective on the date appearing hereon

APPROVED: RECOMMENDED:

   C G HELMICK   AHN CHAI HONG
Brigadier General United States Army   Civil Administrator
Acting Military Governor

SOUTH KOREAN INTERIM GOVERNMENT
Seoul, Korea

EXECUTIVE ORDER
NUMBER 4

28 June 1947

ESTABLISHMENT OF KOREAN AS OFFICIAL LANGUAGE
OF THE SOUTH KOREAN INTERIM GOVERNMENT

Pursuant to authority granted by General of the Army of the United States, Douglas MacArthur, Commanding the Far East Command, the following amendment of paragraph 5 of Proclamation No 1 (*To the People of Korea*), issued by General MacArthur on 7 September 1945, is announced:

"For all purposes during the military control, English will be the official language, and in the event of any ambiguity or diversity of interpretation or definition between any English and Korean or Japanese text, the English text shall prevail, *provided that*, effective 1 July 1947, the official language of the South Korean Interim Government will be Korean."

C G HELMICK

Brigadier General United States Army

Acting Military Governor

軍政廳　官報　行政命令　第四號　　　一九四七年六月二十八日

SOUTH KOREAN INTERIM GOVERNMENT
Seoul, Korea

EXECUTIVE ORDER
NUMBER 5

4 August 1947

MEMORIAL CEREMONIES FOR EMANCIPATION

SECTION I. The purpose of this order is to regulate 1947 memorial ceremonies for emancipation, in order to maintain public peace and order.

SECTION II. Memorial ceremonies for emancipation to be held on August 15, 1947, in the City of Seoul and all other localities shall be held only under the sponsorship of the appropriate administrative heads of the localities.

SECTION III. No political party, social organization or other group shall separately hold an open-air memorial ceremony or hold an open-air ceremonial demonstration in any form. However, memorial ceremonies within buildings are permitted.

SECTION IV. Any person violating any of the provisions of this order shall be subject to conviction for violation of Proclamation Number 2, Headquarters, United States Army Forces, Pacific.

SECTION V. This order shall be effective on the date appearing hereon.

APPROVED

    C G HELMICK

Brigadier General United States Army

.Acting Military Governor.

RECOMMENDED :

AHN CHAI HONG

Civil Administrator

군정청　官報　行政命令　第五號　　　一九四七年八月四日

SOUTH KOREAN INTERIM GOVERNMENT
Seoul, Korea

EXECUTIVE ORDER
NUMBER 6

6 October 1947

INVESTIGATIONS, ARRESTS AND PROSECUTIONS INVOLVING BUSINESSES CONTROLLED PURSUANT TO ORDINANCE NO. 33

WHEREAS, it is necessary to the economy of Korea that businesses controlled pursuant to Ordinance No. 33, dated 6 December 1945 (*Vesting Title to Japanese Property Within Korea*) remain in continuous operation free from interruptions caused by the investigation, arrest or prosecution of managers or assistant managers of such businesses; and

WHEREAS, it is necessary that the public officials charged with the immediate supervision of the management of such businesses be notified of investigations, arrests and prosecutions of such managers and assistant managers so that such supervising officials may arrange to continue the operation of such businesses without interruption;

NOW, THEREFORE. IT IS ORDERED:

1. THAT no police official, prosecutor, or other law enforcement officer of the South Korean Interim Government or any of its subdivisions shall investigate the affairs or books of any business controlled pursuant to Ordinance No. 33 unless or until such law enforcement officer has notified the public official charged with the immediate supervision of the management of such business. Upon such notice the law enforcement officer may forthwith impound and examine the books and records of the company in accordance with law regardless of any objection by the official notified.

2. THAT no police official, prosecutor, or other law enforcement officer of the South Korean Interim Government or any of its subdivisions shall arrest or prosecute any manager or assistant manager of such business unless or until he has notified the public official charged with the immediate supervision of the management of such business. Such arrest may be made on the third day after notice is given regardless of any objection by the official notified;

— 1 —

provided, however, that the arrest may be made earlier with the consent of such notified official.

3. THAT law enforcement officers shall inform the National Property Custodian by letter sent through the property custodian of the appropriate province or the City of Seoul of action taken in accordance with the foregoing paragraph.

4. THAT this order shall be effective on the date appearing hereon.

APPROVED: RECOMMENDED :

C G HELMICK   AHN CHAI HONG
Brigadier General United States Army   Civil Administrator
Acting Military Governor

SOUTH KOREAN INTERIM GOVERNMENT
Seoul, Korea

EXECUTIVE ORDER
NUMBER 7

18 October 1947

CLOSING OF THE HAIBANG NEWS AGENCY

WHEREAS, The Haibang News Agency has violated regulations and orders concerning the use of radio equipment, and committed acts to the prejudice of good order and hostile and prejudicial to the life, safety and security of the persons and property of the United States, calculated to disturb public peace and order and prevent the administration of justice, in violation of Proclamation Number 2, General Headquarters, United States Army Forces, Pacific, 7 September 1945;

NOW, THEREFORE, IT IS ORDERED:

1. That The Haibang News Agency be and it is hereby closed and all its business activities discontinued.

2. That all officers, agents, servants, employees, and representatives of any kind, employed by The Haibang News Agency, be and they are hereby restrained and enjoined from performing or doing, directly or indirectly, any further acts in the course of said employment or in any way calculated to continue the business or operations of The Haibang News Agency.

3. That all persons, firms, associations and corporations be and they are hereby restrained and enjoined from entering into the employment of or doing any business of any kind with The Haibang News Agency, either directly or indirectly.

4. That any person violating or evading the provisions of this order shall, upon conviction, suffer such punishment as the court shall determine.

— 1 —

5. That this order shall be effective on the date appearing hereon.

APPROVED: RECOMMENDED:

    C G HELMICK   AHN CHAI HONG

Brigadier General United States Army   Civil Administrator

   Acting Military Governor

SOUTH KOREAN INTERIM GOVERNMENT
Seoul, Korea

EXECUTIVE ORDER
NUMBER 8

20 October 1947

COLLECTION AND TRANSPORTATION OF RICE

SECTION I. *Purpose.* The purpose of this Executive Order is to implement and clarify provisions of Public Act No. 6 and National Food Regulation No. 6 and to prevent the unauthorized and illegal transportation and sale of rice.

SECTION II. *Provisions of National Food Regulation No. 6.* The following provisions appearing in Section 6 of National Food Regulation No. 6 are hereby specifically confirmed and will be strictly enforced:

"All transportation of rice shall be controlled by the government. Except for the transportation of rice to collection ports and to and from retail ration stores, no rice shall be transported by other than an authorized governmental agency. Rice in unauthorized transit or on sale in other than legal distribution channels shall be seized without payment and shall be delivered to the appropriate Provincial Food Service for sale through the regular legal distribution channels."

SECTION III. *Other authorization to transport or sell rice.* Any authorization heretofore given for the transportation or sale of rice in any quantity or under any conditions which is inconsistent with provisions of National Food Regulation No. 6 quoted above is hereby revoked.

SECTION IV. *Effective date.* This order shall be effective on the date appearing hereon.

APPROVED:

C G HELMICK

Brigadier General United States Army

Acting Military Governor

RECOMMENDED:

AHN CHAI HONG

Civil Administrator

軍政廳　官報　行政命令　第八號　一九四七年十月二十日

SOUTH KOREAN INTERIM GOVERNMENT
Seoul, Korea

EXECUTIVE ORDER
NUMBER 9   15 December 1947

EMERGENCY ELECTRIC POWER BOARD

The Commanding General, United States Army Forces in Korea, recognizing that an emergency shortage of electrical energy presently exists in South Korea and is likely to continue to exist, and further recognizing the necessity of assuring equitable and proper distribution of electrical energy so as to provide for the maintenance of vital needs of the people of South Korea, has directed the establishment of an Emergency Electric Power Board with authority to control the use and distribution of electrical energy during the emergency period.

In compliance with the above directive

AN EMERGENCY ELECTRIC POWER BOARD IS HEREBY ESTABLISHED

1. The Emergency Electric Power Board shall consist of seven members, one of whom shall be designated by the Commanding General, United States Army Forces in Korea and the others being the following six persons:

> Kim, Do Yun, Member, Korean Interim Legislative Assembly
> Chun, Hang Shup, Chamber of Commerce and Industry of Korea
> Chief, Bureau of Industry
> Assistant Adviser, Department of Public Works
> Adviser, Bureau of Utilities, who will serve as executive officer of
> the Board
> Adviser on Utilities, National Economic Board, who will serve as
> Chairman of the Board

Meetings of the Board will be on the call of the Chairman.

2. The Emergency Electric Power Board shall have the power to issue orders, instructions, priorities and restrictions, both general and specific,

---

軍政廳　官報　行政命令　　第九號　　一九四七年十二月十五日

dealing with the production, distribution and use of electrical power, directed to all persons in South Korea, individual, corporate and governmental. The orders of the Board shall have the effect of law.

3. The Board shall have authority to call upon and receive the services of any department or agency of the South Korean Interim Government for assistance deemed necessary in carrying out the duties of the Board and assistance required for those engaged in the production, transmission or distribution of electrical power.

4. The following uses of electrical power are prohibited unless specifically authorized by the Board:

a. As a source of heat in offices, public buildings, homes, restaurants or other structures.

b. For cooking purposes.

c. For display and advertising.

5. The following general priorities for essential use of electrical power for limited operations are hereby established subject to further action by the Board:

Priority I. Use by waterworks, communications, coal mines, briquette factories, hospitals, railways, prisons and jails.

Priority II. Use by rice and cereal mills, gas plants and essential military, police, coastguard and constabulary installations.

Priority III. Use by essential industries, plants, businesses and establishments and for essential house and billet lighting.

6. Among its powers, the Board shall have the power to:

a. Change, alter, add to or eliminate the above priorities.

b. Fix the kind, extent and time of use of electrical power by any person, plant, installation or activity given a priority

c. Define "limited operations" generally or in specific cases.

d. Define "essential" use, generally or in specific cases.

e. Define, fix and limit "priority" use, as set forth in paragraph 5 above, either generally or in specific cases.

— 2 —

f. Require information and reports, written or oral, on the use of electric power, from any user of electric power.

g. Order the physical disconnecting of electric power lines to any user of electric power.

h. Limit or prohibit the distribution, sale or lease of electrical appliances, devices or instruments.

i. Define violations of this order, and of orders and directives issued by it.

7. The Board shall inform the Commanding General, USAFIK, of each directive issued by it and shall request issuance of appropriate directives to the military establishment, when or where applicable.

8. The electric power companies are hereby empowered to discontinue service, including the physical disconnection of lines, to any person who violates this order or any order or directive of the Board.

9. The Board shall have the power to issue its orders in writing, verbally, by telephone, telegraph, radio or by any other form of communication.

10. Any person violating the provisions of this order or of any order or directive of the Board shall, upon conviction by a duly constituted court, suffer such punishment as the court shall determine. The Board shall issue schedules of suggested punishments for violations of this order or of any order or directive issued by it.

11. All laws, ordinances, orders, regulations, directives and instructions, or parts thereof, which are inconsistent herewith or in conflict with the provisions hereof are, to the extent they are inconsistent or in conflict herewith, hereby suspended during the period of emergency.

12. All decisions, directives or orders of the Board shall be subject to review by the Military Governor and to further review by the Commanding General, USAFIK, but shall have full force and effect until such time as changed, modified or rescinded on such review.

13. The provisions of this order, and the powers of the Board thereunder,

shall continue in effect until such time as a public declaration of the end of the period of emergency is issued.

14. This order shall be effective on the date appearing hereon.

APPROVED :

WILLIAM F DEAN
Major General, United States Army
Military Governor in Korea

RECOMMENDED :

AHN CHAI HONG
Civil Administrator

SOUTH KOREAN INTERIM GOVERNMENT
Seoul, Korea

EXECUTIVE ORDER

NUMBER 10

24 December 1947

TEMPORARY SUSPENSION OF CERTAIN PROVISIONS WITH RESPECT TO LABOR

SECTION I.   The shortage of electrical energy in South Korea has resulted in an emergency which requires a temporary rearrangement of working hour during the period of such emergency.   Therefore, until such time as a public declaration of the end of the emergency is issued, as provided for in Executive Order Number 9, dated 15 December 1947, the following provisions of law are suspended:

(a) The provision of Section II (b) of Ordinance Number 121, dated 7 November 1946, which provides that employees subject to said ordinance shall be entitled to not less than twenty-four consecutive hours of rest in each work week.

(b) The provision of Section IV of Public Act Number 4, dated 16 May 1947 which provides that children shall not engage in night work.

(c) The provision of Section V of Public Act Number 4, dated 16 May 1947, which provides that children shall not work more than six days in any one week.

(d) The provisions of Section 2 of Department of Labor Order Number 1 dated 15 September 1947.

(e) The provisions of Section 3 of Department of Labor Order Number 1 dated 15 September 1947, provided that if the hours at the work place exceed four at least one-half hour shall be given for meals at the end of the fourth hour.

SECTION II. This order shall be effective on the date appearing hereon.

APPROVED:

WILLIAM F DEAN
Major General, United States Army
Military Governor in Korea

RECOMMENDED:

AHN CHAI HONG
Civil Administrator

(SEAL USAMGIK)

— 1 —

SOUTH KOREAN INTERIM GOVERNMENT
Seoul, Korea

EXECUTIVE ORDER

NUMBER 11

31 December 1947

PERMISSIBLE TRANSPORTATION OF RICE

SECTION I. *Purpose.* The purpose of this Executive Order is to allow non-governmental transportation of rice as to provinces which have delivered their rice collection quotas for 1947.

SECTION II. *Permission to transport rice.* a. The City of Seoul and Kyunggi Do, Chung-hong Pukto, Chungchong Namdo, Cholla Namdo, Kyongsang Namdo. and Kangwon Do having fully delivered their rice collection quotas for the year 1947, nongovernmental transportation of rice within and between them is authorized. No transportation of rice between them shall be through a province which has not delivered its full 1947 rice collection quota.

b. On the public announcement by the National Food Administration that any other province has delivered its full rice collection quota for the year 1947, there is hereby authorized nongovernmental transportation of rice within such province and between it and other provinces which have also delivered their full rice collection quotas for 1947, provided that no part of such transportation is through a province which has not delivered its full 1947 rice collection quota.

SECTION III. *Inconsistent laws.* The provisions of Executive Order No. 8, dated 20 October 1947 (*Collection and Transportation of Rice*) and of National Food Regulation No. 6, dated 18 August 1947 (*Collection of Rice*) and of directives and instructions in conflict with the provisions hereof, are modified hereby to the extent that they are in conflict.

SECTION IV. *Effect on prosecution and offenses.* This Executive Order is not intended and shall not be construed to extinguish offenses, under any of the measures herein modified or abrogated, committed prior to the effective date hereof, nor to abate or bar existing or future prosecutions based on such offenses.

SECTION V. *Effective date.* This order shall be effective on the date appearing hereon.

APPROVED:

    WILLIAM F DEAN
Major General, United States Army
Military Governor in Korea

RECOMMENDED:

    AHN CHAI HONG
    Civil Administrator

— 1 —

軍政廳　官報　行政命令　第一一號　一九四七年一二月三一日

SOUTH KOREAN INTERIM GOVERNMENT
Seoul, Korea

EXECUTIVE ORDER

NUMBER 12   12 January 1948

ESTABLISHMENT OF KOREAN GOVERNMENT EMPLOYEES WELFARE OFFICE

SECTION 1. *Purpose.* The purpose of this Executive Order is to establish the Korean Government Employees Welfare Office and to define responsibilities for the allocation and distribution of consumer goods to government and vested industry employees.

SECTION II. *Korean Government Employees Welfare Office.*

a. Establishment and Organization. The Korean Government Employees Welfare Office is hereby established as a subsidiary agency of the National Economic Board. It shall consist of a director, deputy director, and an adviser appointed by the Chairman of the National Economic Board. The office shall consist of a government division, a vested industry division, and a general affairs unit, each of which shall consist of a chief and an adviser, together with such other employees as may be necessary to carry out the duties and functions outlined in this order.

b. Duties and Functions. The Korean Government Employees Welfare Office shall:

(1) Prepare recommendations to the Consumer Goods Allocation Board for the allocation of consumer goods to government and vested industry employees on both the national and provincial levels.

(2) On the basis of approved allocations to it, allocate and distribute such goods equitably to government and vested industry employees.

(3) Establish priorities of distribution for such consumer goods as are received in quantities insufficient to insure even distribution to each employee.

(4) Maintain current records of employees eligible to receive consumer goods.

(5) Maintain records of goods received and goods distributed.

(6) Cause to be published periodic statements showing goods received

— 1 —

軍政廳　官報　行政命令　第一二號　—　一九四八年一月一二日

and to whom distributed; such statements to indicate what each employee receives from each allocation.

(7) Assist government and vested industry employees in strengthening existing cooperative organizations and in establishing other such organizations to aid them in solving for themselves their problems of living.

SECTION III. *Consumer Goods Allocation Board.*

a. *Establishment and Organization.* The Consumer Goods Allocation Board is hereby established. It shall consist of the following seven (7) members: the Director and Adviser to the Director, Department of Commerce; the Director and Adviser to the Director, Korean Civil Service; the Director of the USAMGIK Services of Civilian Supply; the Director and Adviser to the Director of the Korean Government Employees Welfare Office. The Adviser to the Director of the Korean Government Employees Welfare Office will serve as chairman of the Board. The meetings of the Board will be on the call of the chairman.

b. *Duties and Functions.* The Board shall:

(1) Make periodic allocations of consumer goods to the Department of Public Health and Welfare for relief distribution and to the operating agencies of the South Korean Interim Government for the carrying out of their assigned functions.

(2) Make periodic allocations of consumer goods to the Korean Government Employees Welfare Office.

(3) Prepare recommendations to the National Economic Board on the percentage of domestic production of consumer goods which should be allocated to the Korean Government Employees Welfare Office.

(4) Perform such other duties and functions as relate to the allocation and distribution of consumer goods.

SECTION IV. *Abolition of the Korean Civil Service Relief Committee.* The Korean Civil Service Relief Committee, established by letter directive of the Military Governor dated 24 March 1947, is hereby abolished and its employees, funds, records, and property are hereby transferred to the Korean Government Employees Welfare Office.

SECTION V. *Administrative Direction.* Any administrative action taken by the director or deputy director of the Korean Government Employees Welfare

Office or by the division and section chiefs thereof relating to the allocation and distribution of consumer goods, shall have the concurrence of the appropriate adviser prior to implementation.

SECTION VI. *Effective Date.* This Executive Order shall be effective on the date appearing hereon.

APPROVED.

RECOMMENDED:

WILLIAM F DEAN
Major General, United States Army
Military Governor

AHN CHAI HONG
Civil Administrator

SEAL
USAMGIK

軍政廳　官報　行政命令　第一二號　一九四八年一月一二日

SOUTH KOREAN INTERIM GOVERNMENT
Seoul, Korea

EXECUTIVE ORDER
NUMBER 14   3 March 1948

NATIONAL ELECTION COMMITTEE

SECTION I.   There is hereby created a National Election Committee, and the persons named below are hereby appointed as members of that Committee:

NAME:	NAME:
CHANG. Myon	LEE. Sung Bok
KIM. Bub Nin	PAIK. Lin Jai
PAHK. Seung Ho	HYUN. Sang Yun
YEE. Kap Sung	RO. Chin Sul
YUN, Ki Sup	CHOI. Kyu Dong
KIM. Chi Whan	CHOI. Too Sun
KIM. Dong Sung	CHYUN. Kyu Hong
OH. Sang Hyung	

Appointments to membership on the Central Election Committee, heretofore made in accordance with provisions of Public Act Number 5, dated 3 September 1947, Law for the Election of Members of the Korean Interim Legislative Assembly, are hereby confirmed as appointments to membership on the National Election Committee.

SECTION II.   There are hereby conferred upon the National Election Committee, and the members thereof, all powers and duties conferred upon the Central Election Committee, and the members thereof, by Public Act No. 5, dated 3 September 1947, Law for the Election of Members of the Korean Interim Legislative Assembly.   The National Election Committee will exercise such powers and perform such duties in accordance with the provisions of that Act, incident to the election to be held on 9 May 1948, pursuant to the Proclamation of Election of Representatives of the Korean People, issued on 1 March 1948, by the Commanding General, United States Army Forces in Korea.

SECTION III.   The National Election Committee hereby established is authorized to employ a staff of assistants and to obtain through the appropriate agencies of the South Korean Interim Government necessary supplies and funds

— i —

for the discharge of its functions, subject to the approval of the Military Governor and such regulations as he may prescribe.

SECTION IV. This Order shall be effective on the date appearing hereon.

WILLIAM F DEAN
Major-General, United States Army
Military Governor in Korea

軍政廳　官報　行政命令　第一四號　　一九四八年三月三日

SOUTH KOREAN INTERIM GOVERNMENT
Seoul, Korea

EXECUTIVE ORDER
NUMBER 15

5 March 1948

AMENDMENT OF EXECUTIVE ORDER NUMBER 12

SECTION I. Executive Order Number 12, dated 12 January 1948 (*Establishment of Korean Government Employees Welfare Office*), is hereby amended as follows:

a. Wherever the words "Korean Government Employees Welfare Office" appear, there shall be substituted in place thereof the words "National Consumers Goods Office".

b. The last sentence of Section II *a* is changed to read: "The office shall consist of a *requirements division, an operations division* and a general affairs unit, each of which shall consist of a chief and an adviser, together with such other employees as may be necessary to carry out the duties and functions outlined in this order."

c. Section III *a* is changed to read: "*Establishment and Organization.* The Consumer Goods Allocation Board is hereby established. It shall consist of the following *eight* (8) members: the Director and Adviser to the Director, Department of Commerce; the Director and Adviser to the Director, Korea Civil Service; the Director of the USAMGIK Services of Civilian Supply; the Director and Adviser to the Director of the National Consumers Goods Office *and the Director of the Department of Labor.* The Adviser to the Director of the National Consumers Goods Office will serve as Chairman of the Board. The meetings of the Board will be on the call of the chairman.

SECTION II. This Executive Order shall be effective on the date appearing hereon.

APPROVED:

WILLIAM F DEAN
Major General, United States Army
Military Governor in Korea

RECOMMENDED:

AHN CHAI HONG
Civil Administrator

SOUTH KOREAN INTERIM GOVERNMENT
Seoul, Korea

EXECUTIVE ORDER
NUMBER 16   19 March 1948

ABOLISHMENT OF PUBLIC PROSTITUTION LAW

SECTION I. The purpose of this order is to implement Public Act Number 7 dated 14 November 1947 (*Abolishment of the Public Prostitution Law*), as amended by public Act Number 9 dated 12 February 1948 (*Public Act Number 7 Amended*).

SECTION II. A person who was conducting a business of prostitution as of 13 February 1948 or thereafter shall evict all prostitutes and touters from his brothels and other places under his control, within ten days after the effective date of this order. Any such person who fails so to evict shall be deemed to be within category *a* of Section III of Public Act Number 7 and punishable as such.

SECTION III. All persons engaged in public prostitution as of 13 February 1948 or thereafter shall vacate the brothels and other places where they engaged in such business, within ten days after the effective date of this order. Any prostitute failing so to vacate shall be deemed to be within category *a* of Section III of Public Act Number 7 and punishable as such.

SECTION IV. The police shall procure a list of all brothels and premises licensed as of 13 November 1947 and shall investigate each such place to determine compliance with Sections II and III hereof and furnish a written report on each to the Civil Administrator, giving the facts as found to exist and the disposition made of each case. Such report shall be made within fifteen days from the effective date of this order.

SECTION V. The police shall have the authority to padlock all brothels and premises referred to in Sections II, III and IV hereof and to take any other measures relative to such brothels and premises which may be necessary for the enforcement of this order.

SECTION VI. Any person with any interest in any premises padlocked or blocked under the provisions of this order may apply to the Chief of Police or the Director of the Police Department for an order suspending or revoking the action taken with regard to such premises, by filing a written statement of

官報　行政命令　第一六號　　一九四八年三月一九日

he facts and the grounds on which suspension or revocation is asked.

SECTION VII. This order shall be effective on the date appearing hereon.

WILLIAM F DEAN
Major General, United States Army
Military Governor in Korea

SOUTH KOREAN INTERIM GOVERNMENT
Seoul, Korea

EXECUTIVE ORDER

NUMBER 18   30 March 1948

APPOINTMENTS TO NATIONAL ELECTION COMMITTEE

SECTION I. The following persons, appointed as members of the National Election Committee by Executive Order Number 14, have resigned:

NAME:	*NAME:*
CHANG, Myon	KIM, Bub Nin
YUN, Ki Sup	LEE, Sung Bok

SECTION II. The persons named below are hereby appointed as members of the National Election Committee, in place of those whose resignation is stated above. The appointments shall be effective as of the dates appearing opposite their names:

NAME:	*DATE:*
PYUN, Sung Ok	23 March 1948
PAK, Hyun Sook	23 March 1948
LEE, Chong Sun	23 March 1948
KANG, Ki Duk	26 March 1948

WILLIAM F DEAN
Major General, United States Army
Military Governor in Korea

軍政廳　官報　行政命令　第一八號　一九四八年三月三十日

SOUTH KOREAN INTERIM GOVERNMENT
Seoul, Korea

EXECUTIVE ORDER
NUMBER 19   **31 March 1948**

TEMPORARY SUSPENSION OF PENALTIES IN RESIDENCE REGISTRATION ORDINANCE

SECTION I. *Purpose.* The purpose of this Executive Order is to suspend temporarily the provisions of law which provide a penalty for persons failing to register a new residence.

SECTION II. *Suspension of penalty.* Section 5 of Governor General Ordinance Number 32, dated 26 September 1942 (*Korean Temporary Residence Registration Ordinance*) is hereby suspended. Such suspension shall remain in effect until the holding of the election pursuant to Ordinance Number 175, dated 17 March 1948 (*Law for the Election of Representatives of the Korean People*).

SECTION III. *Effective date.* This order shall be effective on the date appearing hereon.

APPROVED: RECOMMENDED:

W. F. DEAN
Major General, United States Army
Military Governor in Korea

AHN CHAI HONG
Civil Administrator

— 1 —

SOUTH KOREAN INTERIM GOVERNMENT
Seoul, Korea

EXECUTIVE ORDER

NUMBER 20   5 April 1948

CHANGE OF DATE FOR ELECTION OF THE
REPRESENTATIVES TO THE NATIONAL ASSEMBLY

SECTION I. The day for election of representatives to the National Assembly has been changed to Monday, 10 May 1948. Therefore, in order to avoid the necessity of re-registration of persons who registered on 30 March 1948, their registration shall be regarded as having been made within the period of registration prescribed by Section *15* of the *Law for the Election of Representatives of the Korean People* (*Ordinance Number 175, dated 17 March 1948*).

SECTION II. This Executive Order shall be effective on the date appearing hereon.

W. F. DEAN
Major General, United States Army
Military Governor in Korea

軍政廳　公報　行政命令　第二〇號　一九四八年四月五日

SOUTH KOREAN INTERIM GOVERNMENT
Seoul, Korea

EXECUTIVE ORDER

NUMBER 21   28 April 1948

REGULATION OF DEMONSTRATIONS AND SALE
OF LIQUOR DURING THE ELECTION PERIOD

SECTION I. *Purpose.* The purpose of this Order is to maintain public peace and order during the election period.

SECTION II. *Prohibition on sale of intoxicating liquor.* It is hereby declared unlawful to sell intoxicating liquor on 9 and 10 May 1948 in South Korea.

SECTION III. *Prohibition of parades and demonstrations on election date.* No political party, social organization, youth organization, or other group, shall hold a parade or a demonstration on 9 and 10 May 1948.

SECTION IV. *Penalties.* Any person who violates the "provisions of this Executive Order shall, upon conviction by a duly constituted court, suffer such punishment as the court shall determine.

SECTION V. *Effective date.* This order shall be effective on the date appearing hereon.

APPROVED: RECOMMENDED:

W. F. DEAN   AHN CHAI HONG
Major General, United States Army   Civil Administrator
Military Governor in Korea

軍政廳　官報　行政命令　第二一號　一九四八年四月二八日

688

SOUTH KOREAN INTERIM GOVERNMENT
Seoul, Korea

EXECUTIVE ORDER
NUMBER 22   10 June 1948

BY-ELECTION ON ISLAND OF CHEJU-DO
INDEFINITELY POSTPONED

SECTION I. By virtue of the power vested in me by Section 44, Law for the Election of Representatives of the Korean People. of 17 March 1948, acting on recommendation of the National Election Committee, I declared on 24 May 1948 the election held on 10 May 1948 in Electoral District A and Electoral District B of North CHEJU DO null and void because voting was held in less than 50 per cent of the voting districts of these two electoral districts due to activities and violence of subversive elements. Simultaneously, I ordered that a new election take place in these two electoral districts on 23 June 1948.

In view of the continuing efforts of subversive elements to disturb public peace and order on CHE JU DO Island, motivated by the desire to guarantee to the voting population of Electoral District A and Electoral District B of North CHE JU DO a peaceful and undisturbed election which truly represents the will of the people of these districts, I herewith order that the by-election in these two electoral districts is postponed for an indefinite period.

SECTION II. This Executive Order shall be effective on the date appearing hereon.

W. F. DEAN
Major General, United States Army
Military Governor in Korea

SOUTH KOREAN INTERIM GOVERNMENT
Seoul, Korea

EXECUTIVE ORDER
NUMBER 23   22 June 1948

ELECTION REVIEW BOARD ESTABLISHED

Pursuant to Section 56, Ordinance Number 175 (*Law for the Election of Representatives of the Korean People*), dated 17 March 1948, this Executive Order is promulgated in order to carry out the Law for the Election of Representatives of the Korean People:

SECTION I. The Election Review Board, prescribed in Chapter VIII, Law for the Election of Representatives of the Korean People, is hereby established, effective 25 June 1948.

SECTION II. The Election Review Board will be composed of the following persons:

KIM, Young Moo	Chairman, Chief Justice of the Supreme Court
KIM, Chan Yong	Justice of the Supreme Court, appointed by the Military Governor
YANG, Dai-Kyung	Justice of the Supreme Court, appointed by the Military Governor
CHANG, Myun	Member, designated by the National Assembly
SUH, Soon Yung	Member, designated by the National Assembly

SECTION III. Membership of the Election Review Board is honorary, with appropriate allowances and travel expenses only. Allowances and travel expenses are determined by the Regulations issued by the National Election Committee in accordance with Section 25, Regulations for Implementing the Law for the Election of Representatives of the Korean People, dated 22 March 1948.

SECTION IV. The Election Review Board enacts its own rules and regulations, subject to Sections V and VI of this Executive Order.

SECTION V. The meetings of the Election Review Board are open to the public.

SECTION VI. Any defeated candidate may file a motion for review of the validity of the election in the electoral district in which he was duly

— 1 —

nominated as a candidate, by submitting to the Elect. Review Board (*Seoul, Capitol Building*) not later than 9 July 1948.a written application stating:

a.   The name of the electoral district concerned

b.   His own name and address and the name of the elected candidate

c.   A brief summary of the alleged violations of the Election Law and /or the Election Regulations.

The application may contain the name and address of witnesses and other evidence tendered.

W. F. DEAN
Major General, United States Army
Military Governor in Korea.

SOUTH KOREAN INTERIM GOVERNMENT
Seoul, Korea

EXECUTIVE ORDER

NUMBER 24   26 July 1948

EMERGENCY WATER CONSERVATION BOARD

The Commanding General, United States Army Forces in Korea, recognizing that an acute shortage of potable water presently exists in South Korea, and will continue to exist, and further recognizing the necessity for a better and more equitable distribution of the available water for the essential needs of the people of South Korea, has directed the establishment of an Emergency Water Conservation Board with authority to regulate the use and distribution of potable water until such time that a more adequate supply of such potable water becomes available.

In compliance with such directive, there is hereby created and established a Board to be known as the "Emergency Water Conservation Board," generally referred to hereinafter as the "Board."

1. The Board shall consist of eleven (11) members. Seven (7) of these members shall represent the National interest, and four (4) shall represent local interests. Two (2) local members shall represent the Seoul-Inchon Water District and two (2) shall represent the Kyongsang Namdo-Pusan Water District.

The National members shall be three (3) persons named by the Commanding General, United States Army Forces in Korea, and four (4) persons to be named by the Military Governor. The local members shall be named by the Military Governor.

2. Meetings of the Board will be on the call of the Chairman. A quorum shall consist of four (4) of the National members, together with one (1) member each from the Local water districts specified when dealing with general matters affecting the whole of South Korea; and of four (4) of the National members and one (1) member from a specified water district when dealing with matters solely affecting such specified water district; and of four (4) National members when dealing with matters affecting water districts other than those specified. A member of the Board, in case of unavoidable absence, may be represented by an alternate named by him.

3. The Board shall inform the Commanding General, United States Army

軍政廳　官報　行政命令　第二四號　　一九四八年七月二六日

Forces in Korea, of each directive issued by it, and shall request issuance of appropriate directives to the military establishment when and where applicable.

4. The Board is empowered to, and may:

a. Issue orders, instructions, priorities, and restrictions, both general and specific, dealing with the distribution and use of potable water, directed to all persons in South Korea, individual, corporate and governmental. Any such action by the Board shall have the effect of law.

b. Call upon and receive the services of any department or agency of the South Korean Interim Government, for such assistance as may be deemed necessary in carrying out its duties.

c. Ration the use of potable water for all domestic purposes, for limited but necessary business and industrial purposes, and with the consent of the Commanding General of United States Army Forces in Korea, for the essential needs of the military forces.

d. To the extent deemed necessary, prohibit the use of potable water for stated purposes, other than for use against fires.

e. Inspect or order the inspection of, any water supply and distribution system, including all taps and outlets wherever found, and peremptorily to order whatever remedial measures may be necessary or desirable to stop the voluntary or involuntary wastage of water, or to order repair of leaky water mains, water lines, water taps, or outlets.

f. Inspect all Korean warehouses and storehouses, whether governmental, private or otherwise, for the purpose of locating equipment, materials, supplies and whatever other maintenance items may be necessary to repair pumping plants, treatment plants, water mains and distribution system for highest operating efficiency and to requisition such equipment, materials, supplies, etc., for immediate delivery to the several water systems of South Korea, unless held for military use of equal or greater importance or priority.

g. Hold hearings, administer oaths, take testimony, admit evidence, and make whatever investigations as may be necessary or desirable in all matters affecting the use or wastage of potable water, and to prepare and promulgate rules and regulations based on the findings of such hearings and investigations.

h. Call upon the water companies or agencies to deliver promptly to it any operating and financial reports.

i. Define violations of any order or directive issued by it

— 2 —

j. Issue its orders in writing. verbally, by telephone, telegraph. radio, or by any other form of communication.

k. Make studies and recommendations for enlargement of existing water plants and facilities or for new and additional water facilities. if and where needed.

5. The Board shall, as soon after completing its organization as possible. determine the availability of potable water in each water supply and distribution area, and proceed to establish ration quotas for the civil population; the military establishment as stipulated by the Commanding General, United States Army Forces in Korea; the welfare institutions. such as hospitals, encampments. prisons, jails, etc.; business and industrial entities; and other public purposes; and to establish such prohibitions as may be warranted on all unnecessary use of water, or on the use of potable water for irrigating purposes; but the use of such water for fire protection and extinguishment shall not be impaired. The Board may vary such quotas, from time to time, as it may determine to be necessary or desirable.

6. Any person who may be convicted in any duly constituted court of violating this order, or of any order or directive the Board may issue hereunder shall suffer such punishment as the court shall determine. The Board may issue schedules of suggested punishments for violations of this order or of any order or directive issued by it.

7 All decisions. directives, or orders of the Board, shall be subject to the review of the Military Governor, and to the further review of the Commanding General, United States Army Forces in Korea, but shall have full force and effect until such time as changed, modified or voided on such review.

8. All laws, ordinances, orders, regulations, directives, and instructions, or parts thereof, which are inconsistent herewith, or in conflict with the provisions hereof, are, to the extent to which they may be so inconsistent or in conflict, hereby suspended during the period of this emergency

9. The provisions of this order, and the powers of the Board hereunder, shall continue in effect until such time as a permanent water supply and conservation board shall be brought into being, or a public proclamation declaring

— 3 —

the end of the emergency period be issued by the Mi——y Governor.

10. This order shall be effective on the date appearing hereon.

W. F. DEAN
Major General, United States Army
Military Governor in Korea

軍政廳　官報　行政命令　第二四號　一九四八年七月二六日

HEADQUARTERS
UNITED STATES ARMY MILITARY
GOVERNMENT IN KOREA
Department of Agriculture
Seoul, Korea

DEPARTMENT ORDER
NUMBER 1

15 June 1946

KOREAN AGRICULTURAL ASSOCIATION FEES

1. *Increased Fees Authorized.* City, county, and island Agricultural Associations are hereby authorized to increase Association fees. New fees established upon the basis of this order shall not exceed the following amounts:

    Per farm family _______________________ 5 yen

    Per cow or horse _______________________ 10 yen

    Per calf _______________________ 5 yen

2. *Effective Date.* This order shall be effective on the tenth day after the date appearing hereon.

BY DIRECTION OF THE MILITARY GOVERNOR:

JAMES I MARTIN
Lt Col CAC
Director
Department of Agriculture

재조선미국육군사령부 군정청 법령집

(전2권)

인쇄일: 2025년 6월 01일
발행일: 2025년 6월 16일
지은이: 재조선 미육군 사령부
발행인: 윤영수
발행처: 한국학자료원
서울시 구로구 개봉본동 170-30
전화: 02-3159-8050 팩스: 02-3159-8051
문의: 010-4799-9729
등록번호: 제312-1999-074호

잘못된 책은 교환해 드립니다.

정가   350,000원

在朝鮮美國陸軍司令部

軍政廳法令集 英文版

（下）

韓國學資料院

刊行辞

우리民族의 피는 数十世紀에 걸쳐 保族을 하기 為하여 経済와 같이 再生産되어져 連綿하게 오늘에 이어졌고 또한 未来에도 傳하여져 永久히 그 피의 保族을 하기為한 經濟生活의 營爲는 來日도 繼續되어 火食의 斷絶을 除去할 것이다.

이에는 必然的으로 어떠한 某一種의 規範이 定하여져 온 것은 否認할 수 없는 事實로 우리 國家制度에 外來의 文物制度가 強制로 加工되어 아무리 많은 變遷을 가져왔다고는 하지만 固有의 各種制度上 줄기는 連綿하게 名脈을 維持하여 傳하여져 왔고 또 그 줄기는 이어주고 있어 우리는 이를 다시 찾아 보아 不足한 点을 補充하고 長点은 再善用하여 未來 法生活의 發展을 期하고자 旧代의 法令을 再問世시켜 斯道에 이바지 하여 보겠다는 信念에서 이 大全을 出刊하는 바이오니 江湖 諸賢諸位의 叱正과 鞭撻을 바라마지 않는바이다.

在朝鮮美國陸軍司令部

軍 政 廳 法 令 集

이 法令集은 元來 太平洋美陸軍總司令部가 布告한 布告文과 在朝鮮美陸軍司令部軍政廳이 그

官報에 揭載 公布하였던 法令을 編輯한 것이다。

이 布告文과 法令은 一九四五年 八月 十日 밤 八時에 日本이 聯合國에게 無條件 降伏을 宣言한

지 四週가 되는 九月 七日에 太平洋美陸軍總司令部가 「朝鮮住民에게 布告함」이라는 布告 第一

號外 三件을 飛行機로 撒布한 布告文을 비롯하여 滿二年 十一個月 六日間인 一九四八年 八月

二日까지 즉 全年 八月 十五日 大韓民國政府가 樹立되기 前에 在朝鮮美陸軍總司令部 軍政廳이

布告六件、法令二一九件、行政命令二四件、部令 및 指令(處·局令 및 規則包含)一一五件、在朝

鮮過渡政府法律一四件、立法議院立法決議案四件、其他 一一件、計 三九七件을 法令으로서 公布

하였던 것이다。(當時의 官報는 紙類의 求得難으로 後期에는 馬糞紙로까지 發行 되었음。)

그리고 이러한 法令은 우리 民族이 異民族의 굴레에서 벗어난 時代에 豫想하지 못했던 遺物

이라고 하겠으나 한 나라의 經綸에 있어 間斷이 있을 수 없기 때문에 近三年間이나 이 땅위에

서 施行되었던 制度로서 法史學研究에 있어 重要한 資料가 된다。

그런데 이러한 法令은 美軍政法令集이란 名稱으로나 軍政廳官報綴로서 法制處、內務部、國會、

韓國銀行 및 韓國産業銀行 등의 各 圖書舘과 圖書室에 또는 資料로서 故 鄭光鉉 博士의 書

齋에 散在해 있었다。 이와같이 散藏되어 있던 美軍政 當時의 法令을 拾遺集録하여 在朝鮮

美國 陸軍司令部 軍政廳法令集이라고 題名한 것이다.

이번에 平素 民族文化遺産의 普及에 至大한 關心을 보여왔던 民族文化의 金容和 代表가

法史學과 더불어 斯道에 이바지하고자 하는 뜻에서 이 法令集을 影印 出刊하게 되것이라고

보아져 諸賢諸位의 鞭撻과 以諒을 乞하는 바이다。

東亞大學校 法政大學 教授（法制史）

尹　載　秀　識

TABLE OF CONTENTS

CHAPTER I PROCLAMATION

(G.H.Q.U.S. ARMY FORCES, PACIFIC)

CHAPTER 2 KILA PUBLIC ACTS.

CHAPTER 3 ORDINANCES (USAMGIK)

CHAPTER 4 EXECUTIVE ORDERS
(USAMGIK)

CHAPTER 5 USAMGIK DEPARTMENT ORDERS AND DIRECTIVES (INCLUDING THOSE ISSUED BY BUREAUS, BOARDS, OFFICES OR SIMILAR AUTHORITIES)

I. DEPARTMENT OF AGRICULTURE ORDERS

- END -

CHAPTER 5 USAMGIK DEPARTMENT ORDERS AND DIRECTIVES

DEPARTMENT ORDER
NUMBER 2   15 June 1946

RICE RENTAL COLLECTION
BY NEW KOREA COMPANY LIMITED

1. *Tenants in Default.* Every tenant on agricultural land owned or managed by the New Korea Company, Limited, who has failed to pay his 1945 rice rental in full, shall within thirty days after the effective date of this order:

a. Pay such rental in kind, in full, to the New Korea Company, Limited; *or*

b. Pay such rental in cash, in full, to the New Korea Company, Limited, and at the same time present written evidence that he has surrendered his rice to Military Government or designated competent authorities.* In such case, cash payment shall be computed at the rate shown to have been received by him when he surrendered such rice, or, if such rate cannot be shown, at the rate of ¥ 150 per straw bag of 54 kg; *or,*

c. Produce written evidence acceptable to the local representative of the New Korea Company, Limited, that his 1945 rice crop failed or was destroyed by an Act of God (flood, fire, wind, hail weather, blight or other inevitable accident or crop destructive disease or pest not reasonably controllable by man). In such case he may be excused from any and all payment of rice rental for 1945 in kind or in cash, or from payment representing the portion of such crop destroyed.

2. *Penalty for Noncompliance.* Every such tenant who has failed to comply with the requirements of Section 1 hereof, shall be required to pay the full amount of rice rental in kind due and owing for the year 1945, plus a penalty of ten percentum (10%) thereof in kind, together with the full amount of the 1946 rental in kind, on or before the due date of such 1946 rice rental. Such payments of defaulted 1945 rentals and penalties shall be a first charge against 1946 rice crops.

3. *Effect upon Ordinances 9 and 45.* Legal payments under this order shall not be deemed to be violations of Ordinance Number 9, dated 5 October 1945, fixing maximum farm rents, nor of Ordinance Number 45, dated 25 January 1946, concerning national rice collection.

BY DIRECTION OF THE MILITARY GOVERNOR:

JAMES I MARTIN
Lt Col CAC
Director
Department of Agriculture

---

*See Ordinance Number 45, dated 25 January 1946.

HEADQUARTERS
UNITED STATES ARMY MILITARY GOVERNMENT IN KOREA
Department of Agriculture
Seoul, Korea

DEPARTMENT ORDER
NUMBER 3

26 September 1946

GRAIN INSPECTION REGULATIONS

1. *Purpose.* The purpose of this order is to implement and effectuate the provisions of Ordinance Number 111, dated 18 September 1946.

2. *Grain Inspections.* The following varieties of grain shall neither be transported from one of the areas which shall be fixed by or under the authority of the respective provincial governors, nor be used within such area for any commercial transaction or for any economic benefit, unless first submitted for inspection to authorized grain inspectors at the appropriate Grain Inspection Office, collection points fixed by law, or other appropriate place, in containers conforming to the standards established by *Section 6* hereof, and unless such containers have been appropriately stamped to indicate the contents thereof conform to one of the grades established by *Section 8* hereof:

Unhulled rice, brown rice, polished rice, barley, bare barley, cleaned barley, pressed barley, wheat, rye, unhulled Italian millet, cleaned Italian millet, unhulled Indian millet, cleaned Indian millet, corn, soya beans, kidney beans, red beans and peas.

3. *Waiver of Inspections.* Inspection, unless requested, may be waived upon presentation of proof to the Grain Inspection Office and issuance of a certificate by such office that the grain is within one of the following classes:

    a. A sample not exceeding one bag;

    b. Required for household consumption;

    c. Requisitioned by or in the possession of a governmental agency;

    d. Intended to be utilized for research, seed, exhibits, competitive exhibitions or prize shows;

    e. Waste or broken rice;

    f. Not intended for commercial use or economic benefit.

4. *Procedure on Inspection.* The following procedure shall be followed in the inspection of grain: -

a. Application for inspection shall be made to the local Grain Inspection Office on the appropriate form, obtainable at such office, with the appropriate fee in revenue stamps affixed thereto.

b. The applicant or his authorized agent shall be present at the inspection. Authorized agents shall present appropriate forms, obtainable at the aforementioned office, duly executed by the owner of such grain.

c. The appropriate inspection form shall be attached to the container of the grain examined.

d. Expenses incident to inspection shall be paid by the applicant. The office of the Controller of Commodities and the Provincial Food Service shall be exempt from the payment of inspection fees.

e. Inspection shall be made at collection points or such other assembly points as may be designated by the Grain Inspection Office.

f. Each local Grain Inspection Office having jurisdiction may in its discretion direct or make a re-inspection of the grain. No fee shall be charged for such re-inspection.

g. Determinations made by Grain Inspectors may be appealed to the appropriate provincial Grain Inspection Office and may be further appealed to the Chief of the provincial Bureau of Agriculture.

h. If, after inspection of the contents and the stamping of the grade upon the container thereof, any container is opened for any purpose, except by authority of the Chief Grain Inspector of a Grain Inspection Office, it shall be unlawful to transport such container or its contents outside of the areas fixed by or under the authority of the respective provincial governors, or to use such container or its contents in any commercial transaction or for any economic benefit unless duly submitted for re-inspection.

— 2 —

5. *Fees.* The following fees are established for the inspection of grains:

a. *Type of Grain*	*Fee*
(1) Brown rice, cleaned rice, cleaned barley, pressed barley, cleaned Italian millet, cleaned Indian millet	30 sen per straw bag
(2) Cleaned rice	15 sen per cloth bag

6. *Containers of Grain.* Grain containers shall conform to the following standards:

a. *Type of Grain*	*Container and Weight*
(1) Unhulled rice, cleaned Italian millet, rye	One straw bag to contain 54 kg net weight plus 600 grams as a reserve
(2) Brown rice, bare barley, wheat, cleaned barley, unhulled Indian millet, cleaned Indian millet, red beans, kidney beans, peas	One straw bag to contain 60 kg net weight plus 600 grams as a reserve
(3) Polished rice	One straw bag to contain 60 kg net weight plus 600 grams as a reserve. One cloth bag to contain 30 kg net weight plus 300 grams as a reserve
(4) Unhulled barley, pressed barley	One straw bag to contain 42 kg net weight plus 600 grams as a reserve

a. Type of Grain	Container and Weight
(5) Unhulled Italian millet ~	One straw bag to contain 45 kg net weight plus 600 grams as a reserve
(6) Soya beans	One straw bag to contain 60 kg net weight plus 1.5 kg as a reserve
(7) Corn	One straw bag to contain 60 kg net weight plus 900 grams as a reserve; one hemp bag to contain 90 kg net weight plus 1.5 kg as a reserve

b. Type of bag	Description
(1) Straw bag	With the exception of straw bags containing unhulled rice and barley, straw bags shall be new, never having previously been used, and shall have been certified as conforming to Straw Bag Inspection Regulation, Department of Agriculture Order Number 4 dated September 1946. Binding rope must be smooth and pliable, measuring 4 or 4.5 *bu* in diameter. Vertical binding: straw bags containing unhulled rice. barley,

Type of Bag	Description
(1) Straw bag	cleaned barley, pressed barley, bare barley, wheat, rye, unhulled Italian millet, cleaned Italian millet, unhulled Indian millet, cleaned Indian millet, or corn shall be encircled vertically by a double rope in three places. For other grains, the bag shall be encircled in four places. See Illustration Number 1, attached. Horizontal binding: straw bags containing unhulled rice or barley shall be encircled once by a double rope; bags containing other grains by a double rope in two places. See Illustration Number 1, attached.
(2) Cloth bag	Cloth bags containing clean rice shall be sound, durable and new, never having previously been used. The mouth of the bag shall be securely sewn with strong cotton thread. Binding rope shall be smooth and pliable, measuring 3.5 or 4 *bu* in diameter.

Type of Bag	Description
(2) Cloth bag	Binding: the rope, doubled, shall encircle the bag once vertically, and in two places horizontally. See Illustration Number 2, attached.
(3) Hemp bag	Hemp bags containing corn shall be sound, durable and new, never having previously been used. The mouth shall be securely sewn with strong cotton thread. Binding rope shall be smooth and pliable, measuring 4 or 4.5 *bu* in diameter. Binding: same as that prescribed for cloth bags. See Illustration Number 2, attached.
(4) In case of emergency, upon the request of the Provincial Governor or the Office of the Controller of Commodities, the Director of the Department of Agriculture may issue instructions permitting the use of satisfactory bags with three vertical ropes rather than the required four, provided such bags are not to be exported from Korea.	

7. *Pre-Grading Qualifications.* Upon preliminary examination, bags containing grains, other than unhulled rice, shall be stamped as containing grain "unfit" without grading of the contents, and the transportation thereof, outside of the area which shall be fixed by or under the authority of the respective provincial governors, or use in commercial transactions or for any economic benefit, is prohibited in the following cases:

a. If the bag contains grains harvested in more than one crop year;

b. If the grain is mixed with degenerate rice :

c. If the amount of stone in the brown rice, cleaned rice or pressed barley contained in the bag exceeds 3 granules per *toi*;

d. If the amount of stone in cleaned barley or cleaned Indian millet contained in the bag exceeds 6 granules per *toi*;

e. If the amount of stone in unhulled Italian millet contained in the bag exceeds 10 granules per *hop.*

3. *Grades.* In the event that preliminary examination reveals that the contents are not "unfit" within the provisions of *Section 7* hereof, grains contained shall be graded according to the standards of quality and moisture content established annually for each grade by the Director of the Department of Agriculture and in accordance with the following specifications:

Type of Grain	*Grade*	*Description of Grade*
(1) Unhulled rice	1st grade	Amount of red rice not exceeding 20 granules per hop; mixture of stone, earth, barn grass, unripened rice and other impurities not exceeding 3%
	2nd grade	Amount of red rice not exceeding 50 granules per *hop* ; mixture of stone, earth, barn grass, unripened rice and other impurities not exceeding 6%
	Substandard, out of grade, or exception	Grain which does not meet the above specifications
(2) Brown rice	1st grade	Amount of red rice not exceeding 20 granules per *hop* ; mixture of earth, barn grass,

Type of Grain	Grade	Description of Grade
(2) Brown rice	1st grade	unhulled rice, dead rice, broken rice and other impurities not exceeding 4%
	2nd grade	Amount of red rice not exceeding 50 granules per *hop*; mixture of earth, barn grass, unhulled rice, dead rice, broken rice and other impurities not exceeding 5%
	3d grade	Amount of red rice not exceeding 80 granules per *hop*; mixture of earth, barn grass, unhulled rice, dead rice, broken rice and other impurities not exceeding 6%
(3) White rice (cleaned or polished)	1st grade	Containing no barn grass, unhulled rice or other impurities; amount of broken rice not exceeding 10%
	2nd grade	Containing no barn grass, unhulled rice or other impurities; amount of broken rice not exceeding 12%
(4) Unhulled barley, bare barley, wheat, rye	1st grade	Mixture of stone, earth, barn grass, unripe, wormeaten, damaged or diseased grain and other impurities not exceeding 5%, if unhulled barley; not exceeding 3%, if bare barley; not exceeding 4%, if wheat or rye

Type of Grain	Grade	Description of Grade
	2nd grade	Mixture of stone, earth, unripe, worm-eaten. damaged or diseased grain and other impurties not exceeding 8%, if unhulled barley ; not exceeding 5% if bare barley; not exceeding 6%, if wheat or rye
(5) Cleaned barley	1st grade	Containing no rough barley, barn grass or other impurities : amount of broken barley not exceeding 4%
	2nd grade	Containing no rough barley, barn grass or other impurities ; amount of broken barley not exceeding 7%
(6) Pressed barley	Acceptable	Containing no rough barley, barn grass or other impurities ; amount of broken barley not exceeding 4%
(7). Unhulled Italian millet	1st grade	Mixture of stone, earth, barn grass, unripe grain and other impurities not exceeding 2%
	2nd grade	Mixture of stone, earth, barn grass, unripe grain and other impurities not exceeding 4%

Type of Grain	Grade	Description of Grade
(8) Cleaned Italian millet	Acceptable	Mixture of rough Italian millet, barn grass, broken Italian millet and other impurities not exceeding 2%
(9) Unhulled Indian millet	1st grade	Mixture of stone, earth, barn grass, unripe grain and other impurities not not exceeding 5%
	2nd grade	Mixture of stone, earth, barn grass, unripe grain and other impurities not exceeding 8%
(10) Cleaned Indian millet	Acceptable	Mixture of barn grass, broken Indian millet and other impurities not exceeding 6%
(11) Corn	1st grade	Mixture of stone, earth, unripe, broken, worm-eaten, damaged or degenerate corn and other impurities not exceeding 1%
	2nd grade	Mixture of stone, earth, unripe, broken, worm-eaten, damaged or degenerate corn and other impurities not exceeding 4%
(12) Soya beans	1st grade	Containing no soya beans of varied colored skins; mixture of impurities not exceeding 0.5%

Type of Grain	Grade	Description of Grade
	2nd grade	Containing no soya beans of varied colored skins; mixture of impurities not exceeding 1%
	3rd grade	Amount of soya beans of varied colored skins not exceeding 2 granules per *hop*; mixture of impurities not exceeding 3%
	4th grade	Amount of soya beans of varied colored skins not exceeding 5 granules per *hop*; mixture of impurities not exceeding 4%
(*13*) Red cleaned kidney beans, peas	1st grade	Mixture of different varieties not exceeding 3%; mixture of impurities not exceeding 2%
	2nd grade	Mixture of different varieties not exceeding 5%; mixture of impurities not exceeding 3%

9. *Disposition of Substandard Grains.* Bags containing grains other than unhulled rice, which do not meet the standards prescribed in *Section 8* hereof shall be stamped "unfit" and the transportation thereof outside of the area which shall be fixed by or under the authority of the respective provincial governors, or use in commercial transactions or for any economic benefit. is prohibited. In the event of crop failure, emergency or poor crops. grains which do not meet the specifications prescribed for the lowest grade provided in *Section 8* hereof may be further graded as determined by the Director of

the Department of Agriculture and transportation of such substandard grades outside the area which shall be fixed by or under the authority of the respective provincial governors, or use in commercial transactions or for any economic benefit, may thereupon be authorized.

10. *Markings.* After inspection by the Grain Inspection Office, the appropriate grade, seals and marks of the Department of Agriculture shall be affixed and prominently displayed on the surface of the container. Such containers, as hereinabove prescribed, shall not be inclosed in any other bag, sack, box or other container.

11. *Exportation.* Prior to exportation from Korea south of 38° north latitude, grain shall be submitted for further inspection at the port of exit.

12. *Importation.* The receipt of grain at any port of entry shall be reported immediately to the appropriate Grain Inspection Office, except in any of the following cases:

    a. If the grain is imported by or under direction of a governmental agency;

    b. If the grain is a specimen or sample, weighing not more than 10 kg;

    c. If the grain is hand-carried, weighing not than 60 kg.

13. *Trade-Marks.* Trade-marks validly granted shall be submitted for approval to the Director of the Department of Agriculture prior to use on grain containers. If so approved, such approval shall be reported to the appropriate Grain Inspection Office within five days. Such trade-marks shall be placed on the side of the container other than that on which seals and other official marks are affixed.

14. *Credentials of Inspectors.*—Personnel appointed by the Department of Agriculture as Grain Inspectors shall at all times carry an official identification card issued by the Department of Agriculture and shall display such card upon request.

15. *Assistance to Grain Inspectors.* The Korea Agricultural Association, the Office of the Controller of Commodities and the Provincial Food Service, upon the request of the Chief of the provincial Bureau of Agriculture, shall render such assistance to the grain inspectors as may be required.

16. *Delegation of Authority.* Each provincial governor shall take appropriate measures to enforce these requirements.

— 12 —

17. *Repeal of Inconsistent Laws*. In accordance with Section I of Ordinance Number 111, dated 18 September 1946, all laws, ordinances, orders, regulations, directives and instructions, and parts thereof, which are inconsistent herewith or in conflict with the provisions hereof are hereby repealed.

18. *Effective Date*. This order shall be effective on the date appearing hereon.

  BY DIRECTION OF THE MILITARY GOVERNOR:

SEAL

JAMES I MARTIN
Lt Col CAC
Director

穀叺證印記號及封緘의結付個所及捺印個所

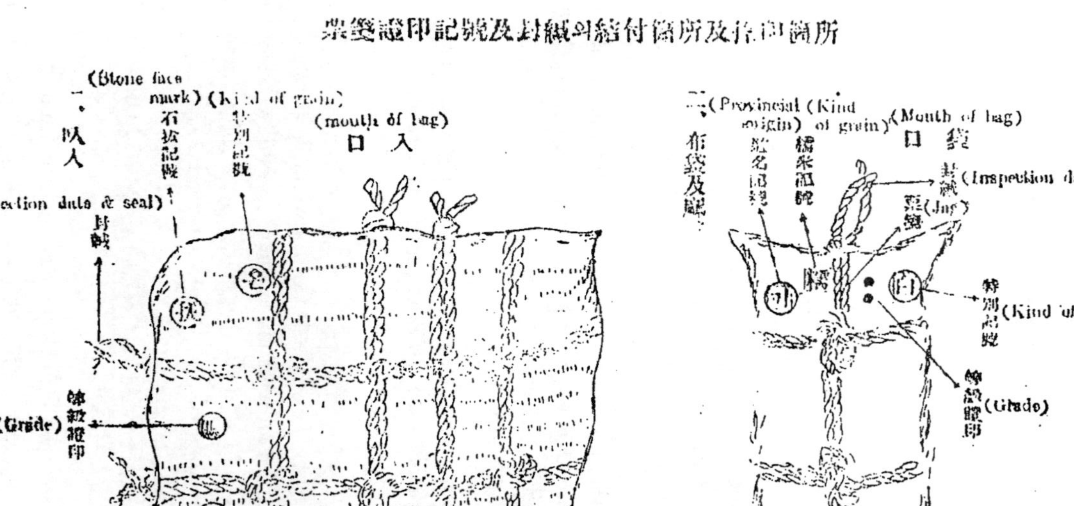

HEADQUARTERS
UNITED STATES ARMY MILITARY
GOVERNMENT IN KOREA
Department of Agriculture
Seoul, Korea

DEPARTMENT ORDER
NUMBER 4   5 October 1946

STRAW BAG INSPECTION REGULATIONS

1. *Purpose.* The purpose of this order is to implement and effectuate the provisions of Ordinance Number 111, dated 18 September 1946.

2. *Straw Bag Inspections.* Straw bags and bundles thereof, shall neither be transported from one of the areas which shall be fixed by or under the authority of the respective provincial governors, nor used within such area for any commercial transaction or for any economic gain, unless first submitted for examination to authorized government inspectors at the appropriate Grain Inspection Office or other appropriate place, and unless such bags and bundles have been appropriately stamped to indicate conformity with the standards established by *Sections 5* and *6* hereof.

3. *Waiver of Inspections.* Such inspection, unless requested, may be waived upon presentation of proof to the Grain Inspection Office, and issuance of a certificate by such office, that the straw bags are within one of the following classes:

   a. Requisitioned by or in the possession of a governmental agency;

   b. Intended to be utilized for research, exhibits, competitive exhibitions or prize shows, in which event the grades may differ from the standard grades established;

   c. Not intended for sale or gain;

   d. Not intended to be utilized as containers for grain, salt or fertilizer.

4. *Fees.* A fee of one yen per bundle of bags shall be paid prior to inspection. Application for inspection shall be made to the local Grain Inspection Office on the appropriate form, obtainable at such office, with the appropriate fees in revenue stamps affixed thereto.

— 1 —

5. *Grading of Individual Bags.* Bags meeting the specifications listed in Table 1, attached, shall be graded as " passed"; bags which do not shall be graded as " unfit".

6. *Standards for Bundles.* The transportation, or use for any commercial transaction or economic benefit, of bundles of straw bags which have not been inspected and which have not been found to meet the following standards and requirements, is hereby prohibited :

a. Binding rope, not less than 5 *bu* in diameter, shall encircle such bundle in four places; triple strands at both ends; double strands at two places in the center.

b. If intended for the use of grain, each such bundle shall consist of 30 bags, folded in half, not more than 3 *cha* in height.

c. If intended for the use of fertilizer, each such bundle shall consist of 50 bags, not more than 3.3 *cha* in height.

d. If intended for use of salt, each such bundle shall consist of 30 bags folded in half, not more than 2.7 *cha* in height.

7. *Disposition of Substandard Bundles.* Bundles of straw bags which do not conform to the standards provided in *Section 6* hereof, or which contain uninspected or ineligible straw bags, straw bags of different grades, or damaged or discolored straw bags, shall be. stamped "unfit", and the transportation or use thereof in commercial transactions or for any economic benefit is prohibited. If transported, used in any commercial transaction or for any economic benefit in violation of this order, such bundles of straw bags may be confiscated or ordered returned to point of origin.

8. *Procedure.* The following procedure shall be followed in the inspection of straw bags:

a. Straw bags qualified for use as containers of grain shall be stamped with the appropriate inspection marks on the lower right corner; those for salt or fertilizer shall be stamped on the upper portion of such bag;

b. Bundles of straw bags passed on inspection shall be appropriately stamped;

c. No markings resembling or simulating the official stamps and seals shall be placed upon a straw bag or bundle of such bags;

d. Each local Grain Inspection Office having jurisdiction may, in its discretion, direct re-inspection of any straw bag or bundle of such bags;

e. Determinations of grain inspectors may be appealed to the appropriate Provincial Grain Inspection Office, and may be further appealed to the Chief of the provincial Bureau of Agriculture;

f. Each applicant or his authorized agent shall be present at each inspection.

9. *Care in Transportation and Storage.* Transportation of straw bags or bundles thereof, not adequately protected against the elements or the hazards of transportation and storage, is prohibited.

10. *Credentials of Inspectors.* Personnel appointed by the Department of Agriculture as Grain Inspectors shall at all times carry an official identification card issued by the Department of Agriculture and shall display such card upon request.

11. *Delegation of Authority.* Each provincial Military Governor shall take appropriate measures to enforce these requirements.

12. *Repeal of Inconsistent Laws.* In accordance with *Section 1* of Ordinance Number 111, dated 18 September 1946, all laws, ordinances, orders, regulations, directives and instructions, and parts thereof, which are inconsistent herewith or in conflict with the provisions hereof, are hereby repealed.

13. *Effective Date.* This order shall be effective on the date appearing hereon.

BY DIRECTION OF THE MILITARY GOVERNOR.

JAMES I MARTIN
Lt Col   CAC
Director, Department of Agriculture

TABLE NO 1   (To Accompany Department Order No 4, Department of Agriculture, USAMGIK)

	For use of Grain	For use of Salt	For use of Fertilizer
Texture.	19 or 20 vertical pairs of straw strings per bag	18 vertical pairs of straw strings per bag	17 vertical pairs of straw strings per bag
Vertical string	1. Rubbing—made with beaten straw (i.e. the string is rubbed to smooth it) 2. Measures 2 *bu* in diameter. 3. 18 loops per *cha*	1. Same as left 2. Measures 1.5 *bu* in diameter 3. Same as left	1. Same as left 2. ,, ,, ,, 3. ,, ,, ,,
Weave	Beaten straw is woven one by one, inserting eight from one side, next eight from the other side, and so on. It is closely woven.	Beaten straw is woven one by one, inserting twenty from one side, next twenty from the other side, and so on. It is closely woven.	Same as for grain
Salvage (a weave at the side of the bag)	A round selvages with 5 pieces (length of selvage board, 3 *bu*).	Same as left	Same as left
End of weave	Vertical rope is braided with three other vertical ropes and its end must be outside the straw bag.	Same as left	Same as left
Depth of fold in the bottom	15 *bu*	13 *bu*	10 *bu*
Width of straw bag	2.7 *cha* (selvage board excepted)	2.65 (selvage board excepted)	2.4 (selvage board excepted)
Length of straw bag (to the center of the fold at the bottom)	Front, 3.1 *cha* Back, 2.9 *cha*	Front, 3.15 *cha* Back, 2.85 *cha*	Front, 2.75 *cha* Back, 2.55 *cha*
Parental rope (rope which combines the halves of the fold over straw mat making the bag)	1. Rubbing—made with beaten straw (see vertical string) 2. Measures 4 *bu* in diameter 3. Rubbing knot, 16 per *cha*	1. Same as left 2. Measures 3.5 *bu* in diameter 3. Same as left	1. Same as left 2. ,, ,, ,, 3. ,, ,, ,,

— 4 —

	For use of Grain	For use of Salt	For use of Fertilizer
Rope at the mouth, fastens corners inside straw bag)	10 *cha*	Same as left	Same as left
Sewing rope (attaches parental rope to bag)	1. Rubbing—made with beaten straw (see vertical string). 2. Measures 2.5 *bu* in diameter 3. Rubbing knot 19 per *cha*	1. Same as left 2. Measures 2 *bu* in diameter 3. Same as left	1. Same as left 2. ., ,, ,, 3. ,, ,, ,,
Sewing method	Parental rope is placed between second and third vertical ropes and is sewn with sewing rope, opposite to the twist of the parental rope.	Same as left	Same as left
Intervals of sewing rope ('bu)	(Width of sewing rope excepted)	Same as left	Same as left
Weight	750-900 me	500-700 me	500-650 me

Reference : 1 bu – about 1/10 inch.
1 cha – about 1 foot (.994 foot)
1 me – 1 mommei (.00826 lb)

SOUTH KOREAN INTERIM GOVERNMENT
Department of Agriculture
Seoul, Korea

DEPARTMENT ORDER
NUMBER 5   18   September 1947

KOREAN AGRICULTURAL ASSOCIATION FEES

1. *Increased Fees Authorized.* City, county, and island Agricultural Associations are hereby authorized to increase Association fees. New fees established upon the basis of this order shall not exceed the following amounts:

Per farm family	15 yen
Per cow or horse	50 yen
Per calf	25 yen

2. Department Order No. 1 (*Korean Agricultural Association Fees*), dated 15 June 1946 issued by the Department of Agriculture, is hereby revoked.

3. *Effective Date.* This order shall be effective on the tenth day after the date appearing hereon.

BY DIRECTION OF THE MILITARY GOVERNOR:

LEE HOON KOO
Director
Department of Agriculture

SOUTH KOREAN INTERIM GOVERNMENT
Department of Commerce
Seoul, Korea

DEPARTMENT ORDER
NUMBER 1

25 August 1947

FOREIGN COMMERCE REGULATIONS

1. PURPOSE. The purpose of this order is to implement Ordinance No. 149, dated 25 August 1947 (*Regulating Foreign Commerce*), and to facilitate the conduct of merchant-to-merchant trade in privately-owned goods so as to effect a balance of imports and exports.

2. REPEALER. The following Bureau of Foreign Commerce Regulations are hereby repealed: No. 1, dated 12 July 1946 (*Import and Export Licenses and Permits*); No. 2, dated 20 June 1947 (*Fees for Licenses and Permits*).

3. IMPORT AND EXPORT LICENSES. In order to exercise governmental control over trade with foreign countries all goods and commodities exported from southern Korea or imported into southern Korea will be licensed. A separate license for each specific transaction will be issued by the Bureau of Foreign Commerce, Department of Commerce, South Korean Interim Government.

4. LICENSE REQUIREMENTS. Licenses will be issued, provided that:

(*a*) Goods or commodities are on the published list of approved imports and exports prepared by the Bureau of Foreign Commerce.

(*b*) All goods or commodities imported into southern Korea will be (1) bartered for a comparable value of goods or commodities to be exported from southern Korea within a reasonable time after the said importation, or (2) sold for won, said won to be used only for the purchase of goods or commodities to be exported from southern Korea within a reasonable time. The determination of "a reasonable time" shall be made by the Bureau of Foreign Commerce.

— 1 —

軍政廳　官報　商務部令　第一號　一九四七年八月二十五日

(c) Goods or commodities exported from southern Korea may be sold for foreign exchange but the foreign exchange derived from the sale of these exports will be used exclusively for the purchase of goods or commodities to be imported into southern Korea. All foreign exchange transactions will be handled and controlled by the Korean Foreign Exchange Bank, Ltd.

(d) Prices of goods and commodities imported into southern Korea or exported from southern Korea will not be outside the control prices set by the Department of Commerce.

5. CUSTOMS CLEARANCE. No Customs clearance will be allowed on any goods or commodities prior to issuance of export or import license covering those specific goods or commodities.

6. PENALTIES. A violation or attempted violation of any of the terms or conditions of these regulations shall be sufficient cause for the Department of Commerce to suspend the party or parties who committed the violation, or any other party to the transaction, from future dealing in foreign commerce. Also, the party or parties violating or attempting to violate these regulations may be subject to criminal prosecution.

7. EFFECTIVE DATE. This order shall be effective on the date appearing hereon.

CHUNGSOO OH
Director
Department of Commerce

SOUTH KOREAN INTERIM GOVERNMENT
Department of Commerce
Seoul, Korea

DEPARTMENT ORDER
NUMBER 2

28 October 1947

DEPOSIT OF WON PROCEEDS FROM SALE OF IMPORTS

1. PURPOSE. The purpose of this order is to assure that won proceeds from an initial sale of imports are utilized for the purchase of goods or merchandise for export, so that imports and exports may be balanced.

2. WON DEPOSITS. All won proceeds received by an importer on the sale of goods or commodities imported under the provisions of Section 4 *(b) (2)** of Department of Commerce Order Number 1, dated 25 August 1947 (*Foreign Commerce Regulations*), shall be deposited in the Korean Foreign Exchange Bank, Ltd., to the credit of the importer or his nominee. Withdrawals shall be permitted only with approval of the Chief of the Bureau of Foreign Commerce.

3. PENALTIES. The penalties provided in paragraph 6 of said Department Order Number 1 shall also apply to any violation of the provisions hereof.

4. EFFECTIVE DATE. This order shall be effective on the date appearing hereon.

CHUNGSOO OH

Director
Department of Commerce

---

* This section reads: "(b) All goods or commodities imported into southern Korea will be ***(2) sold for won, said won to be used only for the purchase of goods or commodities to be exported from southern Korea within a reasonable time." It refers to situations where goods or commodities are imported into Korea without compensatory exports having first been made or arranged.

軍政廳　官報　商務部令　第二號　一九四七年十月二十八日

SOUTH KOREAN INTERIM GOVERNMENT
Department of Commerce
Seoul, Korea

DEPARTMENT ORDER
NUMBER 3
7 June 1948

STANDARD OPERATING PROCEDURE AND FEES
FOR CENTRAL RESEARCH LABORATORY

Pursuant to Section II of Ordinance Number 199 dated 29 May 1948 (*Regulations for Central Research Laboratory Authorized*), the following shall govern:

1. *Purpose.* The purpose of this order is to establish a standard operating procedure and scale of fees for technical experiments, analysis and the use of the equipment of the Central Research Laboratory.

2. *Applications.* Applicants for technical experiments, industrial judging special work, or analysis, must submit requests in writing on forms which are required by the Central Research Laboratory. A sample of the material sufficient for the analysis or experiment shall be submitted in the amount required by the Laboratory.

3. *Use of facilities.* Any person or corporation who desires to use the facilities of the Central Research Laboratory shall submit a request in writing on form required by the Central Research Laboratory, and such use will be granted subject to the discretion of the Laboratory.

4. *Special equipment, materials, etc.* In all cases where special equipment, material, or labor is required to perform the experiment or analysis requested, the Central Research Laboratory may require the applicant to furnish such equipment, material or labor.

5. *Fees.* The following fees shall be the fees payable in advance by applicant:

		Won
a. *Qualitative analysis*		
(1) Appointed components (for one component)		50
(2) Total of chief components		150—500
b *Quantitative analysis*		
(1) Appointed components (for one component)		100
(2) Total of chief components		250—750
c. *Experiment on clays, stones and its products.*		
(1) To determine refractoriness,		

— 1 —

	contraction and expansion of clays, stones and its products (for one item)	50—500
(2)	mechanical analysis of clay	50—500
(3)	Test on applying raw materials used in the manufacture of pottery, glass, enamel, brick, tile and cement	150—2500
(4)	To determine specific gravity, setting time, degrees of fineness, expansion and cracking of cement (for one item)	50
(5)	To determine ductility, compressive strength, and permeability of cement (for one item)	50
(6)	Test on stones and bricks for suction activity, ductility, compressive strength and refrigeration (for one item)	100
(7)	To determine compressive strength of earthen pipe	100—1500

d. *Experiment on fats, waxes, essential oils and mineral oils.*

(1)	Test for specific gravity, viscosity, solidifying point, melting point, boiling point, flash point, combustion point, and luminosity (for one item)	50
(2)	Test for iodine value, acid value, saponification value, and the reaction to metals, acids and alkalies (for one item)	100
(3)	Test on applying fats, waxes, essential oils and mineral oils	150—2500

— 2 —

e. *Test on paints and pigments.*

(1) To determine transparency, lustre, dryness, viscosity, and specific gravity of paints

(2) To determine fineness and covering power of pigments (for one item)   50

(3) Tests on applying paints or pigments   150—2500

f. *Test on paper.*

(1) To determine bursting strength and internal tearing strength   (for one item)   50

(2) Test on applying pulp or other raw materials used in paper making   150—2500

g. *Test of rubber.*

(1) Vulcanization   500

(2) Tensile strength and elongation   100

(3) Hardness and elasticity (for one item)   50

(4) Aging   250

h. *Testing of fibre, yarn and cloth.*

(1) Tensile strength, elongation, and friction factor   (for one item)   50

(2) Loss on boiling   100

(3) Refining and bleaching (for one item)   50—250

(4) Finishing of yarn and cloth   50

(5) Materials, system and density of cloth   (for one item)   50

(6) Plan of weaving   100—1000

(7) Weaving   100—1000

Testing of dyes and mordants.

(1) Application or quality (for one item)   50—500

(2) Test on dying of yarn or cloth   50—500

— 3 —

    (3) Fastness of dyed material   50

j. *Copy of report, when requested*   50

k. Translation into English language   100

l. Applicants desiring the completion of their analyais or experiment in advance of the normal lengh of time required for such work, shall be required to pay *five* times the stipulated fee.

For all items not listed above, and for the use of facilities, industrial judging, etc, Director of the Central Research Laboratory shall determine the fee to be charged for such work. The fees shall be paid by attaching revenue stamps for the amount of the fee to the application for the work, and the revenue stamps shall be cancelled by the Director.

6. *Sealing of sample.* When the applicant wants the sample or experiment to be sealed at the time it is submitted, a request shall be submitted in writing on the form required by the Central Research Laboratory and the fee required paid.

7. *Reports.* The results of an experiment or analysis will be reported to the person who desired the experiment or analysis.

8. *Return of unused samples.* The unused portion of any sample which is submitted for analysis or experiment shall not be returned unless requested in advance. If the nature of the sample is such that it can be returned, the expense of returning it shall be paid by the person requesting it.

9. *Exemption of fees for government.* No fees or rent for any of the above services shall be charged to the government or any agency thereof.

10. *Effective date.* This Order shall become effective upon the date appearing hereon.

LAH, KI HO
Acting Director
Department of Commerce

SOUTH KOREAN INTERI GOVERNMENT
Department of Commerce
Seoul, Korea

DEPARTMENT ORDER

NUMBER 4   2 August 1948

OVERLAND INTERZONAL TRADE

1. Purpose. The purpose of this Order is to implement Ordinance Number 149, dated 25 August 1947 (*Regulating Foreign Commerce*), by providing for the control and supervision of private overland interzonal trade.

2. *Permits.* a. Traders who desire to engage in barter with persons in Korea north of thirty-eight degrees, shall, for each barter arrangement, apply for a permit to the Bureau of Domestic Commerce, Department of Commerce. or to such agencies as may be authorized to issue permits, from time to time. by the Department of Commerce.

b. As a condition to the issuance of a permit, the Bureau may designate the terms of the barter arrangement, including the manner of disposition of goods and commodities brought into South Korea by traders.

c. As a further condition to the issuance of a permit, an applicant may be required to deposit with the Bureau a cash bond in such amount as is deemed necessary to secure the proper and faithful performance of the barter arrangement.

3. Goods and commodities. The following goods and commodities may, generally, be the subject matter of barter:

a. Goods and commodities that are on the List of Approved Export Items issued, from time to time, by the Bureau of Foreign Commerce, Department of Commerce.

b. Such kinds and quantities of "nationally controlled" goods and commodities as may be permitted by the Bureau of Domestic Commerce, Department of Commerce.

c. Such other goods and commodities as may be specifically authorized.

4. Controls. a. Cargoes destined for, or coming from, Korea north of thirty-eight degrees, must be routed through any one of the following places, or such other place or places as may, from time to time, be designated by the Bureau of Domestic Commerce:

— 1 —

軍政廳　官報　商務部令　第四號　　　一九四八年八月二日

Name of Place	Approximate Location
CHUNG DAN (*Chun Dan*)	N 37°-59′, E 125°-50′
PAEK CHON (*Bak Chun*)	N 37°-59′, E 126°20′
YOUHION JIN (*Yu Hyon*) (*Yo Hyon*)	N 37°-59′, E 126°-25′
CHUK AM (*Juk Am*)	N 37°-59′, E 126°-58′
YANG MUN NI	N 37′-59′, E 127°-15′

b. Cargoes will not be permitted to cross the thirty-eight-degree parallel in vehicles or railroad cars.

c. Pursuant to interdepartmental agreement, the Department of Police is hereby delegated the authority and duty of administering controls at or as near the thirty-eight-degree parallel as possible. The Department of Police shall check cargoes against issued permits; shall maintain records of movements of cargoes; andshall submit by the 10th day of each month a report for the month preceding, revealing names and addresses of traders, permit numbers, kinds and quantities of goods transported, the dates thereof, and such additional information as, and when, requested by the Bureau of Domestic Commerce.

d. When inspections reveal failure to observe any of the provisions of this Order, including variances between contents of cargoes checked and contents as listed in the issued permits, traders shall be taken into custody, cargoes and vehicles detained, and prompt notification thereof given to the Bureau of Domestic Commerce. 'Disposition in each case shall be pursuant to law.

5. Violations. Any private trade with Korea north of the thirty-eight-degree parallel shall be unlawful unless carried on by barter arrangements pursuant to the provisions hereof. Any person violating or attempting to violate any of the provisions of this Order shall be punished as provided in Ordinance Number 149.

6. Effective date. This Order shall be effective on the date appearing hereon.

BY DIRECTION OF THE MILITARY GOVERNRO:

OH, CHUNG SOO
Director .
Department of Commerce

HEADQUARTERS
UNITED STATES ARMY MILITARY
GOVERNMENT IN KOREA
Department of Communications
Seoul, Korea

DEPARTMENT ORDER
NUMBER 1

29 July 1947

KOREAN BROADCASTING CORPORATION; RULES
CONCERNING LISTENING TO RADIO BROADCASTS

1. In accordance with the provisions of Section II of *Private Broadcasting Regulations**, the following changes are approved in *Korean Broadcasting Corporation Rules Concerning Listening to Radio Broadcasts*:

a. Article 6 is amended to read as follows:

The fee for listening to radio broadcasting is ten yen per month for each individual listening contract. The fee is payable as of the first day of the month in which the contract is signed, and is payable for the full month in which the contract is terminated, regardless of the day of the month on which it is signed or terminated.

b. Article 7 is amended to read as follows:

The fee shall be payable on the first day of each month or on the first day of every other month at the option of the Corporation.

However, when the fee is paid through a book transfer savings account at the post office, it shall be payable quarterly, on the first day of January, April, July and October of each year.

A discount of 10% shall be allowed in the event that fees for an aggregate of six months or one year are paid in advance.

2. This order shall be effective on the date appearing hereon.

BY DIRECTION OF THE MILITARY GOVERNOR:

LLOYD C PARSONS
Colonel   Sig C
Director
Department of Communications

---

*Communications Department Order Number 93, dated 20 December 1923, Government of Japan, as revised, applied in Korea by Government General Order Number 43, dated 6 September 1924, as revised.

677

HEADQUARTERS
UNITED STATES ARMY MILITARY
GOVERNMENT IN KOREA
Department of Communications
Seoul, Korea

DEPARTMENT ORDER
NUMBER 2

2 August 1946

AMENDING POSTAL, TELEPHONE AND TELEGRAPH RATES

1. *Change in Postal Rates.*

   a. The schedule of postal rates fixed by Law Number 54,* dated 13 March 1900 (*Postal Law*), as amended, is hereby amended and shall be as follows:

Class of Mail	Type of Mail	Weight	Rate
1st	Letters	20 grams or fraction thereof	50 cheun
1st	Ordinary post cards	—	25 cheun
2d	Return post cards	—	50 cheun
2d	Letter-style post cards	—	50 cheun
3d	Periodicals	100 grams or fraction thereof	25 cheun
3d	Daily newspapers	100 grams or fraction thereof	15 cheun
4th	Printed letters and circulars	100 grams or fraction thereof	50 cheun
4th	Printed matter for the blind	100 grams or fraction thereof	15 cheun
4th	Printed booklets, commercial papers and samples of merchandise	100 grams or fraction thereof	50 cheun

* Applied to Korea by Japanese Imperial Order Number 412, dated 30 September 1910.

— 1 —

Class of Mail	Type of Mail	Weight	Rate
5th	Agricultural seeds	100 grams or fraction thereof	12 cheun

b. The schedule of postal rates fixed by Government General Order Number 86, dated 25 April 1938 (*Postal Regulations*), as amended, is hereby amended and shall be as follows:

(1) Parcel post rates:

Weight	Rate
Not exceeding 2,000 grams	5.00 weun
Exceeding 2,000 grams but less than 4,000 grams	6.00 weun

(2) Fees for special services:

Type of Service	Rate
(a) Registering of mail	1.50 weun
(b) Insuring of mail Merchandise, for each 20 weun of valuation	50 cheun
(c) Return receipt of delivery	1.00 weun
(d) Return receipt of delivery of court papers	2.00 weun

c. The schedule of money order fees fixed by Government General Order Number 27, Dated 23 March 1911, as amended, is hereby amended and shall be as follows:

Type of Money Order	Amount of Money Order	Rate
(a) Ordinary money order (redeemable only at post office designated thereon)	Up to 100 weun	2.50 weun
	From 101 to 300 weun	5.00 weun
	From 301 to 500 weun	7.50 weun
	From 501 to 1000 weun	10.00 weun
(b) Money order (redeemable at any postoffice)	Up to 100 weun	1.50 weun
	From 101 to 200 weun	3.00 weun
	From 201 to 300 weun	4.50 weun
(c) Telegraph money order	Up to 100 weun	6.00 weun
	From 101 to 300 weun	12.00 weun
	From 301 to 500 weun	18.00 weun

2. *Change in Telegraph Rates.* The schedule of telegraph rates fixed by Government General Order Number 93, dated 10 October 1925 (*Telegraph Regulation*). as amended, is hereby amended and shall be as follows:

    a. Ordinary telegrams

(1) Within Korea

Type of Telegram		*Rate*
Korean telegrams	7 characters (syllables) or less	2.00 yen
	For each additional two characters	40 sen
Japanese telegrams	10 characters (*kana*) or less	2.00 yen
	For each additional five characters	40 sen
Telegrams in European languages	5 words or less	2.00 yen
	For each additional word	40 sen

(2) Telegrams between Korea and Japan

Type of Telegram		*Rate*
Japanese telegrams	10 characters (*kana*) or less	2.00 yen
	For each additional 5 characters	40 sen
Telegrams in European languages	5 words or less	2.00 yen
	For each additional word	40 sen

(3) If more than one addressee in the Korean and Japanese type telegrams, 20 sen will be charged for each additional addressee.

b. Press telegrams

(1) Within Korea

Type of Telegram		Rate
Korean telegrams	25 characters (syllables) or less	1.60 yen
	For each additional 25 characters	1.20 yen
Telegrams in European languages	10 words or less	1.60 yen
	For each additional 10 words	1.20 yen

(2) Telegrams between Korea and Japan

Type of Telegram		Rate
Japanese telegrams	50 characters (*kana*) or less	1.60 yen
	For each additional 50 characters	1.20 yen
Telegrams in European languages	10 words or less	1.60 yen
	For each additional 10 words	1.20 yen

c. Subscription press telegram fee (appointment calls for limited time and words) within Korea or between Korea and Japan

Type of Telegram		Rate
Korean telegrams	250 characters (syllables) or less	2,800.00 yen
	From 251 to 500 characters (syllables)	5,200.00 yen
	From 501 to 750 characters (syllables)	8,000.00 yen
Japanese telegrams	500 characters (*kana*) or less	2,800.00 yen
	From 500 to 1,000 characters (*kana*)	5,200.00 yen
	From 1,001 to 1,500 characters (*kana*)	8,000.00 yen

Telegrams in European languages	100 words or less	2,800.00 yen
	From 101 to 200 words	5,200.00 yen
	From 201 to 300 words	8,000.00 yen

d. Special handling fees

		Rate
(1)	Urgent telegram fee—3 times the ordinary telegram fee	
(2)	Collated telegram fee (guarantees delivery)—1-1/2 times the ordinary telegram fee	
(3)	Express (special delivery) fee	
	Distance of 8 Km or less	4.00 yen
	For each additional 4 Km	4.00 yen
	Delivery to any island	4.00 yen
	(If actual cost of delivery exceeds the fee, difference must be paid by the addressee)	
(4)	Ship delivery fee	4.00 yen
	(If actual cost of delivery exceeds the fee, the difference must be paid by the addressee)	
(5)	Fee for cancellation before transmission	40 sen
(6)	Sending telegrams by phone	40 sen
(7)	Sending telegrams on credit basis	40 sen

e. The following special services which had been suspended for the
duration of the war shall be resumed at the following rates:

	Type of Service	Rate
(1)	Acknowledgement of receipt by telegram (Korean language) (Japanese language) (European languages)	Equal to basic charge of telegram
(2)	Acknowledgement of receipt by mail	50 sen
(3)	Multiple address telegram (Korean language) (Japanese language) (European languages)	1/2 of charge paid for text letters 1/2 of charge paid for text letters 1/2 of charge paid for text and signature
(4)	Press multiple address	1/2 of charge paid for text letters of words
(5)	Subscription multiple address	1/2 of charge paid for subscription fee
(6)	Receipt fee	40 sen
(7)	Special designation of delivery Ordinary—per year Special—per month	120.00 yen 12.00 yen
(8)	Perusal fee (message taken from files and contents received by customer)	40 sen
(9)	Certified copy : Korean language, 50 characters (syllables) or less Japanese language, 100 characters (*kana*) or less European language, 25 words or less	80 sen 80 sen 80 sen
(10)	Ship information registration--per year	12.00 yen

(11)	Telegraphic report of passing vessels
	  a. If registration fee paid
	  b. If registration fee not paid
		1.60 yen
		2.00 yen

(12) Semaphore telegram (ship to shore to lighthouse) *a.* Signal fee	6.00 yen
b. Telegram or Mail fee	Actual expense of delivery
(13) Report of casualties at sea (SOS from ship to shore to owner of ship)	Actual expense of delivery

3. *Change in Radio-telegram Rates*

The schedule of radio-telegram rates fixed by Government General Order Number 97, dated 10 October 1925 (*Radio-telegram Regulation*), as amended, is hereby amended and shall be as follows:

a. Radio telegrams to Korean ships and aircraft

Type of Telegram		*Rate*
Korean telegrams	7 characters (syllables) or less	4.00 yen
	For each additional 2 characters	80 sen
	Medical telegrams of 7 characters or less	2.40 yen
	For each additional two characters	40 sen
Telegrams in European languages	5 words or less	4.00 yen
	For each additional word	80 sen
	Medical telegram of 5 words or less	2.40 yen
	For each additional word	40 sen

b. Rates between land stations shall be the same as for ordinary telegrams.

c. Press radio-telegram

— 7 —

Type of Telegram		Rate
Korean language	25 characters (syllables) or less	3.60 yen
	For each additional 25 characters	3.00 yen
Japanese language	50 characters (*kana*) or less	3.60 yen
	For each additional 50 characters	3.00 yen
European languages	10 words or less	3.60 yen
	For each additional 10 words	3.00 yen

d. Special handling fees

Service	Rate
(1) Urgent radio-telegram	Equal to radio or press telegram charge
(2) Collated radio-telegram (delivery guaranteed)	1-1/2 times radio-telegram charge

(3) The following special handling services which had been suspended for the duration of the war shall be resumed at the following fees:

Service	Rate
Acknowledgement of receipt by radio-telegram	Equal to basic charge for radio-telegram
Multiple address	1/2 of charge paid for text, letters or words
Press multiple address radio-telegram	1/2 of charge paid for text, letters or words.

   c. Telegraph money order fees
       See section 1, subsection c, of this order.
   4. *Change in Meteorological Telegram Rates.* The schedule of telegraph rates for meteorological telegrams fixed by Government General Order Number 102, dated 25 December 1916, as amended, is hereby amended and shall be as follows:

Type of Telegram	Monthly Charge	
	Abbreviated Rate	Interpreted Sentence Rate
General weather forecast	24.00 yen	32.00 yen
Special weather announcement	10.00 yen	12.00 yen
General storm warning	10.00 yen	8.00 yen
Local weather forecast	18.00 yen	24.00 yen
Local weather announcement	8.00 yen	10.00 yen
Local storm warning	8.00 yen	6.00 yen
Present detailed weather announcement		
Once a day	—	180.00 yen
Twice a day	—	300.00 yen

   5. *Change in Telephone Rates.* Schedule of telephone rates fixed by Government General Order Number 116. (*Telephone Regulations*) dated 19 June 1919, as amended, is hereby amended and shall be as follows:
     a. Subscription charges

Type of Service	Basis	Class of Area	Rate
(1) Message rate system	Yen per year	1	280.00 yen
		2	240.00 yen
	Sen per message in same exchange area	1	40 sen
		2	40 sen

Type of Service	Basis	Class of area	Rate
(2) Uniform rate system	Yen per year	3 4 5 6	680.00 yen 640.00 yen 560.00 yen 480.00 yen
(3) Special telephone for trunk call (suburban party line service)	Yen per year	—	400.00 yen
(4) Private telephone line—local	Yen per year	1 2 3 4 5 6	800.00 yen 760.00 yen 680.00 yen 640.00 yen 560.00 yen 480.00 yen
(5) Private telephone line toll	Charge determined in each case by Director, Department of Communications		

b. Supplementary charges

Type of Service	Basis	Rate
(1) Subscriber out of exchange area	Yen per year per 100 meters of line from exchange area	64.00 yen
(2) Availability of special long distance trunks	Yen per year per trunk	24.00 yen
(3) (a) Telephone handset instrument (Dept. owned) (b) Desk set (Dept. owned)	Yen per year Yen per year	120.00 yen 80.00 yen
(4) Extra telephone (a) Wall (Dept. owned) (Subscriber owned) (b) Handset (Dept. owned) (Subscriber owned)	Yen per year Yen per year	120.00 yen 80.00 yen 240.00 yen 80.00 yen

— 1 0 —

Type of Service	Basis	Rate
(c) Desk set (Dept. owned) (Subscriber owned) (d) For each extra telephone not for private use (e) Desk set available to two exchange lines—for each additional telephone set	Yen per year Yen per year Additional to above, yen per year	200.00 yen 80.00 yen 20.00 yen 80.00 yen
(5) Private switchboard (Dept. owned)	Charge determined in each case by Director, Department of Communications	
(6) Extra bell and receiver	Yen per year	40.00 yen
(7) Extra telephone on private switchboard	Yen per year	80.00 yen
(8) Each telephone on a private line	Charge determined in each case by Director, Department of Communications	

c. Registration fee

Type of Service	Rate
(1) 1st to 4th class areas	80.00 yen
(2) 5th and 6th class areas	40.00 yen

d. Transferral fee

Changing name of subscriber (within immediate family only permissible) all areas	300.00 yen

e. Installation charge

(1) 1st and 2nd class areas	600.00 yen
(2) 3rd class area	450.00 yen
(3) 4th and 5th class areas	350.00 yen
(4) 6th class area	200.00 yen

f. Charge for moving telephone instruments

(1) Moving instrument on premises	40.00 yen
(2) Temporary withdrawal	40.00 yen

Type of Service	Basis	Rate
(3) (a) Moving instrument to new premises		200.00 yen
(b) If on a subscriber construction line		80.00 yen
(4) Moving or temporary withdrawal of extra receiver and bells		40.00 yen
(5) Moving or temporary withdrawal of accessories		40.00 yen
(6) Moving or temporary withdrawal of switchboard		No charge

g. Cost of directory listing

Type of Service	Basis	Rate
(1) 1st and 2d class areas	Yen per listing	40.00 yen
(2) 3d to 6th class areas	Yen per listing	20.00 yen

h. Message rate, message withdrawal rate, calling fee and advice fee

Type of Service	Rate
(1) Message rate-local, per message	40.00 yen
(2) Withdrawal of message	
(a) Before completion of message	No charge
(b) Reduced time of message	No charge
(c) After party has answered	
(1) Appointment call	No charge
(2) Other calls	No charge
(d) Calling party not present when call is completed	No charge
(3) Calling fee in same exchange area (calling non-subscriber to telephone) per call	80 sen
(4) Fee of advice (change a call covered by (3) above)	80 sen
(5) Regular conversation (appointment call)	No charge

(6) Message rate and calling fee outside the same exchange area

Distance	Basis	Ordinary Cal:	Calling Rate
Within the same pu, eup, or myun	sen per three minute period or fraction thereof	80 sen	80 sen
Outside area-less than 8 km		160 sen	120 sen
8—16 km		240 sen	120 sen
16—32 km		320 sen	120 sen
32—64 km		400 sen	120 sen
64—104 km		560 sen	120 sen
104—152 km		720 sen	160 sen
152—200 km		800 sen	160 sen
200—248 km		880 sen	160 sen
248—296 km		960 sen	160 sen
296—344 km		1120 sen	160 sen
344—440 km		1360 sen	160 sen
440—536 km		1600 sen	160 sen

Distance	Basis	Ordinary Call	Calling Rate
536—632 km	sen per three minute period or fraction thereof	1840 sen	160 sen
632—824 km		2160 sen	200 sen
824-1016 km		2480 sen	200 sen
1016-1200 km		2800 sen	200 sen
1200-1400 km		3120 sen	200 sen

Rates of this section shall be doubled for each 3 minute period over 9 minutes.

i. Special Service Rates

Type	Rate
(1) Urgent Message	Twice the rate of ordinary message
(2) Special urgent message (military and government)	Three times the rate of ordinary message
(3) Regular message (appointment call)	Four times the rate of an ordinary message
(4) Regular press calls	4/5 the rate of one ordinary message, times 360 for annual contract
(5) Urgent call (public)	Twice the rate of an ordinary call

6. *Definition.* As used herein, *Korea* means that portion of Korea which is south of 38 degrees north latitude.

7. This order is issued pursuant to Ordinance No 99, dated 14 July 1946.

8. *Effective Date.* This order shall be effective on the tenth day after the date appearing hereon.

LLOYD C PARSONS
Colonel Sig C
Director
Department of Communications

HEADQUARTERS
UNITED STATES ARMY MILITARY
GOVERNMENT IN KOREA
Department of Communications
Seoul, Korea

DEPARTMENT ORDER
NUMBER 3

4 July 1946

RESUMPTION OF MAIL SERVICE FOR LETTERS, POST CARDS
AND PRINTS (INCLUDING PRINTED MATTER IN GENERAL,
PERIODICALS AND BOOKS) MAILED IN KOREA TO THE
UNITED STATES

1. Effective 4 July 1946, letters, post cards and prints, (including periodicals and books) will be accepted in Korea for mailing to the United States of America and its territories and insular possessions, subject to the regulations governing Universal Postal Union mails. No registered, special delivery or collect on delivery matter will be accepted. The following is a description of the articles, units of weight, rates and dimensions:

LETTERS — Unit of weight, 20 grams: rate, first unit of weight or fraction thereof, 10 yen; each additional unit or fraction thereof, 6 yen; total weight not to exceed 2 kilograms. Dimensions: Length, breadth and thickness combined, 90 centimeters: greatest length, 60 centimeters. When mailed in the form of a roll, length plus twice the diameter not to exceed 100 centimeters; greatest length, 80 centimeters.

POST CARDS — Rate: 6 yen for single card and 12 yen for card with reply paid. Dimensions: Maximum, 15 by 10.5 centimeters: minimum, 10 by 7 centimeters.

PRINTS — Unit of weight, 50 grams; rate, 10 yen for each unit or fraction thereof; limit of weight, 2 kilograms (books, 3 kilograms for single volumes); dimensions, same as for letters.

2. All matter excluded from the mails in either country will be excluded from the mails from Korea to the United States.

3. All mail matter addressed to the United States, its territiories and insular possessions must be addressed in English.

BY DIRECTION OF THE MILITARY GOVERNOR:

LLOYD C PARSONS
Colonel, Sig C
Director
Department of Communications

693

HEADQUARTERS
UNITED STATES ARMY MILITARY
GOVERNMENT IN KOREA
Department of Communications
Seoul, Korea

DEPARTMENT ORDER
NUMBER 4   14 October 1946

RESUMPTION OF MAIL SERVICE FOR LETTERS, POST CARDS
AND PRINTS (INCLUDING PRINTED MATTER IN GENERAL,
PERIODICALS AND BOOKS) MAILED IN KOREA TO THE
COMMONWEALTH OF THE PHILIPPINES.

1. Effective 20 October 1946, letters, post cards and prints (including periodicals and books) will be accepted in Korea for mailing to the Commonwealth of the Philippines subject to the regulations governing Universal Postal Union Mails. No registered, special delivery, or collect on delivery matter will be accepted. The following is a description of the articles, units of weight, rates and dimensions:

LETTERS — *Unit of weight*: 20 grams. *Rate*: first unit of weight or fraction thereof, 10 yen; each additional unit or fraction thereof, 6 yen; *total weight* not to exceed 2 kilograms. *Dimensions*: length, breadth and thickness combined, 90 centimeters; greatest length, 60 centimeters. When mailed in the form of a roll, length plus twice the diameter not to exceed 100 centimeters; greatest length, 80 centimeters.

POST CARDS — *Rate*: 6 yen for single card and 12 yen for card with reply paid. *Dimensions*: maximum, 15 by 10.5 centimeters; minimum, 10 by 7 centimeters.

PRINTS — *Unit of weight*: 50 grams. *Rate*: 10 yen for each unit or fraction thereof. *Limit of weight*: 2 kilograms (books, 3 kilograms for single volumes). *Dimensions*: same as for letters.

2. All matter excluded from the mails in either country will be excluded from the mails from Korea to the Commonwealth of the Philippines.

3. All mail for the Commonwealth of the Philippines must be addressed in Roman characters.

BY DIRECTION OF THE MILITARY GOVERNOR:

LLOYD C PARSONS
Colonel Sig C
Director

HEADQUARTERS
UNITED STATES ARMY MILITARY
GOVERNMENT IN KOREA
Department of Communications
Seoul, Korea

DEPARTMENT ORDER
NUMBER 5   1 November 1946

RESUMPTION OF MAIL SERVICE BETWEEN KOREA AND JAPAN

1. Mail service between Korea and Japan is authorized, effective 20 November 1946, subject to the following provisions:

a. The service will be restricted to postcards to and from Korea; and the mailing to Korea of such scientific and professional publications as may be approved by the Supreme Commander for the Allied Powers.

b. Communications on postcards must be of a personal or family nature, written in Korean, Chinese, English, French, Russian, Spanish or Japanese. Letters (other than official mail pertaining to repatriation) and commercial and financial communications are prohibited.

2. Regular mail service will be governed by the provisions of the Universal Postal Union Convention of Buenos Aires, of 23 May 1939.

3. Mails will be dispatched from Korea to Japan in Korean or United States vessels

4. All incoming and outgoing mail will be subject to censorship to the extent deemed advisable by military authorities.

5. Repatriation mail service authorized 28 October 1945 (SCAPIN 202) has been discontinued. Hereafter all repatriation mail, including official correspondence pertaining only to repatriation of Japanese forces and Japanese nationals, will be handled as International Mail.

— 1 —

軍政廳　官報　遞信部令　第五號　　一九四六年十一月一日

6. The following is a description of the articles (rates of postage and dimensions):

POST CARDS *Rate:* 1 yen for single card and 2 yen for card with reply paid. *Dimensions:* maximum, 15 by 10.5 centimeters ; minimum, 10 by 7 centimeters.

7. All matter excluded from the mails in either country will be excluded from the mails from Korea to Japan.

BY DIRECTION OF THE MILITARY GOVERNOR:

LLOYD C PARSONS
Colonel Sig C
Advisor to Director

KIEL WON BONG
Director

軍政廳　官報　遞信部令　第五號　　一九四六年十一月一日

HEADQUARTERS
UNITED STATES ARMY MILITARY
GOVERNMENT IN KOREA
Department of Communications
Seoul, Korea

DEPARTMENT ORDER
NUMBER 6

30 November 1946

RESUMPTION OF MAIL SERVICE FOR LETTERS, POST CARDS AND PRINTS (INCLUDING PRINTED MATTER IN GENERAL, PERIODICALS AND BOOKS) BETWEEN KOREA AND CHINA.

1. Effective 10 December 1946, letters, post cards and prints (including periodicals and books) will be accepted in Korea for mailing to China, subject to the regulations governing Universal Postal Union Mails. No registered, special delivery, or collect on delivery matter will be accepted for mailing to China, but one-way gift parcels from China will be accepted; such parcels to be limited to 5 kilograms in weight and contents to be restricted to essential relief items such as non-perishable food, clothing, soap, and mailable medicines; each parcel and customs declaration to be endorsed "Gift Parcel". The following is a description of the articles, units of weight, rates and dimensions:

LETTERS *Unit of weight:* 20 grams. *Rate:* first unit of weight or fraction thereof, 2 yen; each additional unit or fraction thereof, 2 yen; total weight not to exceed 2 kilograms. *Dimensions:* length, breadth and thickness combined, 90 centimeters; greatest length, 60 centimeters. When mailed in the form of a roll, length plus twice the diameter not to exceed 100 centimeters; greatest length, 80 centimeters.

POST CARDS *Rate:* 1 yen for single card and 2 yen for card with reply paid. *Dimensions:* maximum 15 by 10.5 centimeters; minimum, 10 by 7 centimeters.

— 1 —

PRINTS *Unit of weight:* 50 grams. *Rate:* 2 yen for each unit or fraction thereof. Limit of weight, 2 kilograms (books, 3 kilograms for single volumes). *Dimensions*, same as for letters.

2. Correspondence, until further notice, will be restricted to messages written in Korean, Chinese, English, French, Japanese, Portuguese, Russian or Spanish.

3. All mail from Korea to China will be transported in Korean or United States vessels.

4. All matter excluded from the mails in either country will be excluded from the mails from Korea to China.

BY DIRECTION OF THE MILITARY GOVERNOR :

LLOYD C PARSONS
Colonel Sig C
Adviser to Director

KIEL WON BONG
Director

HEADQUARTERS
UNITED STATES ARMY MILITARY
GOVERNMENT IN KOREA
Department of Communications
Seoul, Korea

DEPARTMENT ORDER
NUMBER 7

14 February 1947

RESUMPTION OF INTERNATIONAL POSTAL SERVICE

1. *Purpose.* The purpose of this order is to provide generally for the resumption of postal service between Korea and other countries, and to prescribe the conditions therefor. Department Orders Nos. 3, 4, and 6, of the Department of Communications, respectively dated 4 July 1946, 14 October 1946, and 30 November 1946, are hereby revoked, and this order is substituted therefor.

2. *Universal Service.* International postal service between Korea and any other country is hereby authorized, as provided hereinafter.

3. *Regular Mails.* Letters, postcards (single and with reply paid), commercial papers, prints (including periodicals, books, and printed matter for the blind), samples of merchandise, and small packets, as defined in Articles 112 to 125, inclusive, of the Univeral Postal Union Convention of Buenos Aires, of 23 May 1939, will be accepted in Korea, for mailing to any country, and if mailed to Korea from any country, subject to the provisions of the said Convention, and to the conditions outlined hereinafter. Further restrictions will be provided in the postal service between Korea and Japan.

4. *Parcel Post.* Gift parcels will be accepted in Korea from any country which may agree to allow Korea a parcel post terminal delivery credit of 20 centimes (in accordance with the monetary standard referred to in Article 29 of the Universal Postal Union Convention of Buenos Aires, of 23 May 1939) per pound, or 40 centimes per kilogram, or fraction thereof, whichever weight may be used. Such parcels will be limited to 11 pounds or 5 kilograms in weight. The contents thereof will be restricted to essential relief items such

軍政廳　　官報　　遞信部令　　第七號　　一九四七年二月十四日

as non-perishable food, clothing. soap, and mailable medicines.   Each parcel and related customs declaration will be indorsed "Gift Parcel".   The sender will prepare a customs declaration for each parcel on a special form provided for this purpose, by the Administration of origin.

5.   *Rates and General Conditions.*   Except as otherwise provided in separate orders, the following is a description of the mailable articles, units of weight, dimensions, and rates of postage:

LETTERS.
Unit of weight: 20 grams.   *Rate:* first unit of weight or fraction thereof, 10 weun; each additional unit or fraction thereof, 6 weun; total weight not to exceed 2 kilograms.   *Dimensions:* length, breadth and thickness combined, 90 centimeters; greatest length, 60 centimeters.   When mailed in the form of a roll, length plus twice the diameter not to exceed 100 centimeters; greatest length, 80 centimeters.

POST CARDS
Rate: 6 weun for single card and 12 weun for card with reply paid.   *Dimensions:* maximum, 15 by 10.5 centimeters; minimum, 10 by 7 centimeters.

COMMERCIAL PAPERS
Unit of weight: 50 grams. *Rate:* 3 weun for each unit or fraction thereof; minimum charge, 10 weun; limit of weight, 2 kilograms. *Dimensions:* same as for letters.

PRINTS
Unit of weight: 50 grams.   *Rate:* 3 weun for each unit or fraction thereof; limit of weight, 2 kilograms (books, 3 kilograms for single volume).   *Dimensions:* same as for letters.

RAISED PRINTS FOR THE BLIND
Each 1000 grams, 2 weun; limit of weight, 7 kilograms. Prints sent in the form of folded or unfolded cards will be subject to the same minimum limits as post cards.

SMALL PACKETS
Unit of weight: 50 grams.   *Rate:* 5 weun for each unit or fraction thereof; minimum charge, 30 weun; limit of weight, 1 kilogram. *Dimensions:* same as for letters.   (Small packets should not be accepted for countries which have expressed an unwillingness to accept them).

— 2 —

SAMPLES OF
MER-
CHANDISE

Unit of weight: 50 grams. *Rate:* 3 weun for each unit or fraction thereof; limit of weight, 500 grams. *Dimensions :* same as for letters; minimum charge, 5 weun.

6. *Language.* Correspondence will be restricted to messages written in the Korean, Chinese, English, French, Japanese, Portuguese, Russian, or Spanish language.

7. *Non-Acceptable Matter.* Insured, special delivery, or collect on delivery matter will not be accepted in Korea for, or from, other countries. No registered mail will be accepted in Korea for other countries.

8. *Excluded Matter.* All matter required to be excluded from the mail of any country will be excluded from the mail from Korea to that country.

9. *Obligations.* Korean international postal obligations, including charges for transportation of international mail from Korea on vessels of other than Korean registry, incurred hereunder, will be settled in United States dollars, upon approval by the United States Army Military Government in Korea.

10. *Billing.* International mail will be billed to the Seoul Central Post Office, Seoul, Korea, as the designated international exchange office.

11. *Dispatch.* Mail will be dispatched from Korea about twice a month, via Inchon and Pusan, Korea, depending upon the availabilability of the vessels sailing.

12. *Routing.* All international mail will be routed via Inchon or Pusan, Korea.

13. *Trade Correspondence.* Business, financial, and commercial correspondence will be limited to ascertainment of facts and exchange of information, except that transactional communications will be permitted concerning transactions licensed by the United States Army Military Government in Korea.

— 3 —

14. *Effective Date*. This order shall be effective on the date appearing hereon.

BY DIRECTION OF THE MILITARY GOVERNOR:

LLOYD C PARSONS
Colonel Sig C
Adviser to Director

KIEL WON BONG
Director
Department of Communications

軍政廳　官報　遞信部令　第七號　一九四七年二月十四日.

HEADQUARTERS
UNITED STATES ARMY MILITARY
GOVERNMENT IN KOREA
Department of Communications
Seoul, Korea

DEPARTMENT ORDER
NUMBER 8

14 February 1947

RATES AND GENERAL CONDITIONS OF POSTAL SERVICE
BETWEEN KOREA AND CHINA

1. *Purpose.* The purpose of this order is to provide specifically. for the rates and general conditions of postal service between Korea and China. Except as otherwise provided herein, all provisions of Department Order No. 7, of the Department of Communications. dated 14 February 1947 (*Resumption of International Postal Service*), will be applied with full force and effect to postal service between Korea and China.

2. *Rates and General Conditions.* The following is a description of the mailable articles, units of weight, dimensions, and rates of postage in the postal service between Korea and China:

LETTERS   *Unit of weight:* 20 grams.   *Rate:* first unit of weight or fraction thereof, 2 weun; each additional unit or fraction thereof. 2 weun; total weight not to exceed 2 kilograms. *Dimensions:* length, breadth and thickness combined, 90 centimeters; greatest length, 60 centimeters. When mailed in the form of a roll, length plus twice the diameter not to exceed 100 centimeters; greatest length, 80 centimeters.

POST CARDS   *Rate:* 1 weun for single card and 2 weun for card with reply paid. *Dimensions:* maximum 15 by 10.5 centimeters; minimum. 10 by 7 centimeters.

COMMERCIAL PAPERS   *Unit of weight:* 50 grams.   *Rate:* 3 weun for each unit or fraction thereof; minimum charge. 10 weun; limit of weight. 2 kilograms. *Dimensions:* same as for letters.

PRINTS	*Unit of weight.* 50 grams. *Rate:* 3 weun for each unit or fraction thereof; limit of weight, 2 kilograms (books, 3 kilograms for single volume). *Dimensions:* same as for letters.
RAISED PRINTS FOR THE BLIND	Each 1000 grams, 2 weun ; limit of weight, 7 kilograms. Prints sent in the form of folded or unfolded cards will be subject to the same minimum limits as post cards.
SMALL PACKETS	*Unit of weight:* 50 grams. *Rate:* 5 weun for each unit or fraction thereof; minimum charge, 30 weun; limit of weight, 1 kilogram. *Dimensions:* same as for letters. (Small packets should not be accepted for countries which have expressed an unwillingness to accept them).
SAMPLES OF MERCHANDISE	*Unit of weight:* 50 grams. *Rate:* 3 weun for each unit or fraction thereof; limit of weight, 500 grams. *Dimensions:* same as for letters; minimum charge, 5 weun.

3. *Effective Date.* This order shall be effective on the date appearing hereon.

BY DIRECTION OF THE MILITARY GOVERNOR:

LLOYD C PARSONS
Colonel Sig C
Adviser to Director

S E A L

KIEL WON-LONG
Director
Department of Communi ations

HEADQUARTERS
UNITED STATES ARMY MILITARY
GOVERNMENT IN KOREA
Department of Communications
Seoul, Korea

DEPARTMENT ORDER
NUMBER 9   14 February 1947

POSTAL SERVICE BETWEEN KOREA AND JAPAN

1. *Purpose.* The purpose of this order is to provide specific conditions of postal service between Korea and Japan. Department Order No. 5, of the Department of Communications, dated 1 November 1946 (*Resumption of Mail Service Between Korea and Japan*), is hereby revoked, and this order is substituted therefor.

2. *Regular Mails.* Letters and post cards, as defined in Articles 112 to 125, inclusive, of the Universal Postal Union Convention of Buenos Aires, of 23 May 1939, will be accepted in Korea, for mailing to Japan, and if mailed to Korea from Japan, subject to the provisions of the said Convention, to the conditions outlined hereinafter, and to such further restrictions as may now or later be applicable to mail exchanged between Japan and other countries.

3. *Rates and General Conditions.* The following is a description of the mailable articles, units of weight, rates of postage and dimensions in the postal service between Korea and Japan:

LETTERS   *Unit of weight:* 20 grams. *Rate:* 2 weun for each unit or fraction thereof; each additional unit or fraction thereof, 2 weun; total weight not to exceed 2 kilograms. *Dimensions:* length, breadth and thickness combined, 90 centimeters; greatest length, 60 centimeters. When mailed in the form of a roll, length plus twice the diameter not to exceed 100 centimeters; greatest length, 80 centimeters.

POST CARDS   *Rate:* 1 weun for single card and 2 weun for card with reply paid. *Dimensions:* maximum, 15 by 10.5 centimeters; minimum, 10 by 7 centimeters.

4. *Other Conditions.* The provisions of paragraphs 6 to 13, inclusive, of Department Order No. 7, of the Department of Communications, dated 14 February 1947 (*Resumption of International Postal Service*), will be applied, insofar as applicable, to postal service between Korea and Japan.

5. *Effective Date.* This order shall be effective on the date appearing hereon.

BY DIRECTION OF THE MILITARY GOVERNOR:

LLOYD C PARSONS
Colonel Sig C
Adviser to Director

S E A L

KIEL WON BONG
Director
Department of Communications

HEADQUARTERS
UNITED STATES ARMY MILITARY GOVERNMENT IN KOREA
Department of Communications
Seoul, Korea

DEPARTMENT ORDER
NUMBER 10

31 March 1947

REVISION OF POSTAL RATES

1. *Department Order No 2 Amended.* The schedule of postal rates fixed by paragraph 1 of Department Order No 2, Department of Communications, dated 2 August 1946 (*Amending Postal, Telephone and Telegraph Rates*) is hereby revised, and shall be as follows:

REGULAR MAILS

Class of Mail	Type of Mail	Weight	Rate
1st	Letters	20 grams or fraction thereof	1.00 won
2d	Ordinary post cards	—	50 chon
2d	Return post cards	—	1.00 won
2d	Letter-style post cards	—	1.00 won
3d	Periodicals	100 grams or fraction thereof	50 chon
3d	Daily newspapers	100 grams or fraction thereof	30 chon
4th	Printed letters and circulars	100 grams or fraction thereof	1.00 won

— 1 —

Class of Mail	Type of Mail	Weight	Rate
4th	Printed matter for the blind	100 grams or fraction thereof	30 chon
4th	Printed booklets, commercial papers, and samples of merchandise	100 grams or fraction thereof	1.00 won
5th	Agricultural seeds	100 grams or fraction thereof	30 chon

PARCEL POST

Weight	Rate
Not exceeding 2,000 grams	20.00 won
Exceeding 2,000 grams, but not exceeding 4,000 grams	30.00 won

SPECIAL SERVICES

Type of Service	Rate
Registering of mail	4.00 won
Insuring of mail, or of merchandise, for each 100 weun of valuation	5.00 won
Return receipt of delivery	5.00 won
Return receipt of delivery of court papers	5.00 won

MONEY ORDERS

Type of Money Order	Amount of Money Order	Rate
Ordinary money order (redeemable only at post office designated thereon)	Up to 300.00 won	10.00 won
	From 300.01 to 500.00 won	15.00 won
	From 500.01 to 1,000.00 won	20.00 won
	From 1,000.01 to 3,000.00 won	25.00 won
	From 3,000.01 to 5,000.00 won	30.00 won
	From 5,000.01 to 8,000.00 won	35.00 won
	From 8,000.01 to 10,000.00 won	40.00 won

軍政廳　官報　遞信部令　第十號　一九四七年三月三十一日

Type of Money Order	Amount of Money Order	Rate
Money order (redeemable at any post office)	Up to 100.00 won From 100.01 to 200.00 won From 200.01 to 300.00 won From 300.01 to 500.00 won	3.00 won 6.00 won 9.00 won 12.00 won
Telegraph money order	Up to 300.00 won From 300.01 to 500.00 won From 500.01 to 1,000.00 won From 1,000.01 to 3,000.00 won From 3,000.01 to 5,000.00 won From 5,000.01 to 8,000.00 won From 8,000.01 to 10,000.00 won	25.00 won 35.00 won 50.00 won 60.00 won 70.00 won 85.00 won 100.00 won

2. *Effective Date.* This order shall be effective on and after 1 April 1947.

BY DIRECTION OF THE MILITARY GOVERNOR:

LLOYD C PARSONS
Colonel Sig C
Adviser to Director

KIEL WON BONG
Director
Department of Communications

軍政廳　官報　遞信部令　第十號　一九四七年三月三十一日

HEADQUARTERS
UNITED STATES ARMY MILITARY
GOVERNMENT IN KOREA
Department of Communications
Seoul, Korea

DEPARTMENT ORDER

NUMBER 11   31 March 1947

REVISION OF TELEGRAPH AND TELEPHONE RATES

1. *Telegraph Rates.* The schedule of telegraph rates fixed by paragraph 2 of Department Order No 2, Department of Communications, dated 2 August 1946 (*Amending Postal, Telephone and Telegraph Rates*) is hereby revised, and shall be as follows:

a. ORDINARY TELEGRAMS

(1) Within Korea

Type of Telegram	Unit	Rate
In Korean language	10 characters (syllables) or less For each additional character	10.00 won 1.00 won
In a European language	5 words or less For each additional word	10.00 won 1.00 won

— 1 —

軍政廳　官報　遞信部令　第一一號　一九四七年三月三十一日

711

(2) Telegrams between Korea and Japan

Type of Telegram	Unit	Rate
In Japanese language	10 characters (*kana*) or less	10.00 won
	For each additional character	1.00 won
In a European language	5 words or less	10.00 won
	For each additional word	1.00 won

(3) If more than one addressee in telegrams in Korean or Japanese language, 2.00 won for each additional addressee

b. PRESS TELEGRAMS

Within Korea

Type of Telegram	Unit	Rate
In Korean language	For each 30 characters (syllables) or fraction	4.00 won
In a European language	For each 10 words or fraction	4.00 won

c. SUBSCRIPTION PRESS TELEGRAM FEE (appointment calls for limited time and words) within Korea or between Korea and Japan

Type of Telegram	Unit	Rate
In Korean language	250 characters (syllables) or less	6,000.00 won
	From 251 to 500 characters (syllables)	10,000.00 won
	From 501 to 750 characters (syllables)	15,000.00 won
In a European language	100 words or less	6,000.00 won
	From 101 to 200 words	10,000.00 won
	From 201 to 300 words	15,000.00 won

軍政廳　官報　遞信部令　第一一號　一九四七年三月三十一日

d. SPECIAL HANDLING FEES

(1) Urgent telegram fee—3 times ordinary telegram fee	
(2) Collated telegram fee (delivery guaranteed)—1-1/2 times ordinary telegram fee	

(3) Express (special delivery) fee :	Rate
Distance of 8 km or less	20.00 won
For each additional 4 km	20.00 won
Delivery to any island	20.00 won
(Note : If actual cost of delivery exceeds the fee, difference to be paid by addressee)	

(4) Ship delivery fee	20.00 won
(Note : If actual cost delivery exceeds the fee, difference to be paid by addressee)	

	Rate
(5) Fee for cancellation before transmission	2.00 won
(6) Sending telegram by telephone	2.00 won
(7) Sending telegram on credit basis	2.00 won

e. SPECIAL SERVICES

Type of Service	Rate
(1) Acknowledgment of receipt by telegram—in Korean, Japanese, or a European language	Equal to basic charge for telegram
(2) Acknowledgment of receipt by mail	1.00 won
(3) Multiple address telegram—in Korean, Japanese or a European language	1/2 of charge for telegram

軍政廳 官報 遞信部令 第一一號 一九四七年三月三十一日

Type of Service	Rate
(4) Press multiple address telegram	1/2 of charge for telegram
(5) Subscription multiple address telegram	1/2 of charge paid for subscription fee
(6) Receipt fee	2.00 won
(7) Special designation of delivery Ordinary (per year) Special (per month)	500.00 won 50.00 won
(8) Perusal fee (message taken from files and contents received by customer)	2.00 won
(9) Certified copy: In Korean language, 50 characters (syllables) or less In a European language, 25 words or less	5.00 won 5.00 won
(10) Ship information registration—per year	50.00 won
(11) Telegraphic report of passing vessels: a. If registration fee paid b. If registration fee not paid	5.00 won 10.00 won
(12) Semaphore telegram (ship to shore to lighthouse): a. Signal fee b. Telegram or mail fee	25.00 won Actual expense of delivery
(13) Report of casualties at sea (SOS from ship to shore to owner of ship)	Actual expense of delivery

2. *Radio—Telegram Rates.* The schedule of radio telegram rates fixed by paragraph 3 of Department Order No 2, Department of Communications, dated 2 August 1946 (*Amending Postal, Telephone and Telegraph Rates*) is hereby revised, and shall be as follows:

— 4 —

a. RADIO-TELEGRAMS TO KOREAN SHIPS AND AIRCRAFT

Type of Telegram	Unit	Rate
In Korean language	10 characters (syllables), or less	20.00 won
	For each additional character	2.00 won
	Medical telegram of 10 characters or less	10.00 won
	For each additional character	1.00 won
In a European language	5 words or less	2.00 won
	For each additional word	2.00 won
	Medical telegram of 5 words or less	10.00 won
	For each additional word	1.00 won

b. RATES BETWEEN LAND STATIONS: same as for ordinary telegram.

c. PRESS RADIO-TELEGRAMS

Type of Telegram	Unit	Rate
In Korean language	30 characters (syllables) or less	10.00 won
In a European language	10 words or less	10.00 won

d. SPECIAL HANDLING FEES

Service	Rate
(1) Urgent radio-telegram	Equal to radio or press telegram charge
(2) Collated radio-telegram (delivery guaranteed)	1-1/2 times radio-telegram charge
(3) Acknowledgment of receipt of telegram	Equal to basic charge for radio telegram
(4) Multiple address	1/2 of charge paid for text, letters or words
(5) Press multiple address radio-telegram	1/2 of charge paid for text, letters or words

— 5 —

3. *Meteorological Telegram Rates.* The schedule of telegraph rates for meteorological telegrams fixed by paragraph 4 of Department Order No 2, Department of Communications, dated 2 August 1946 (*Amending Postal, Telephone and Telegraph Rates*) is hereby revised, and shall be as follows:

Type of Telegram		Monthly Rate	
		Abbreviated	*Interpreted Sentence*
General weather forecast		80.00 won	120.00 won
Special weather announcement.		40.00 won	48.00 won
General storm warning		—	32.00 won
Local weather forecast		72.00 won	96.00 won
Local weather announcement		32.00 won	40.00 won
Local storm warning		—	24.00 won
Detailed weather announcement	Once a day	—	720.00 won
	Twice a day	—	1,440.00 won

4. *Telephone Rates.* The schedule of telephone rates fixed by paragraph 5 of Department Order No 2, dated 2 August 1946 (*Amending Postal, Telephone and Telegraph Rates*) is hereby revised, and shall be as follows:

a. SUBSCRIPTION CHARGES

Type of Service	Basis	Class of Area	Rate
(1) Message rate system	Per year	1	1,200.00 won
	Per message in same exchange area	1	2.00 won

Type of Service	Basis	Class of Area	Rate
(2) Uniform rate system	Per year	1 2 3 4 5 6	3,400.00 won 3,100.00 won 2,800.00 won 2,600.00 won 2,300.00 won 2,000.00 won
(3) Special telephone for trunk call (suburban party-line service)	Per year	—	1,600.00 won
(4) Private telephone line-local	Per year	1 2 3 4 5 6	3,400.00 won 3,100.00 won 2,800.00 won 2,600.00 won 2,300.00 won 2,000.00 won
(5) Private telephone line toll	Charge determined in each case by Director, Department of Communications		

(6) The class of an exchange area is to be determined by the number of telephone subscribers in the area, as follows:

Class	Number of Subscribers
1	10,000 or more
2	1,000 to 9,999
3	500 to 999
4	300 to 499
5	100 to 299
6	Less than 100

b. SUPPLEMENTARY CHARGES

Type of Service	Basis	Rate (per year)
(1) Subscriber outside of exchange area	Per 100 meters of line from exchange area	260.00 won

軍政廳 官報、遞信部令 第一一號 一九四七年三月三十一日

Type of Service	Basis	Rate (per year)
(2) Availability of special long distance trunks	Per trunk	100.00 won
(3) (a) Telephone handset instrument (Dept-owned)	—	480.00 won
(b) Desk set (Dept-owned)	—	320.00 won
(4) Extra telephone		
(a) Wall: Dept-owned	—	480.00 won
Subscriber-owned	—	320.00 won
(b) Hand set: Dept-owned	—	960.00 won
Subscriber-owned	—	320.00 won
(c) Desk set: Dept-owned	—	800.00 won
Subscriber-owned	—	320.00 won
(d) For each extra telephone not for private use	—	80.00 won
(e) Desk set available to 2 exchange lines-for each additional telephone set	—	320.00 won
(5) Private switchboard (Dept-owned)	Charge determined in each case by Director, Department of Communications	
(6) Extra bell and receiver	—	160.00 won
(7) Extra telephone on private switchboard	—	320.00 won
(8) Each telephone on a private line	Charge determined in each case by Director, Department of Communications	

    c. REGISTRATION FEE (all areas)   300.00 won

    d. TRANSFERRAL FEE

Changing name of subscriber (permissible only within immediate family)-all areas	2,000.00 won

軍政廳　官報　遞信部令　第一一號　一九四七年三月三十一日

e. INSTALLATION CHARGE

Type of Service	Rate
(1) 1st class area	8,000.00 won
(2) 2nd class area	7,000.00 won
(3) 3rd class area	6,000.00 won
(4) 4th, 5th and 6th class areas	4,000.00 won

f. CHARGE FOR MOVING TELEPHONE INSTRUMENTS

Service	Rate
(1) Moving instrument on premises	200.00 won
(2) Temporary withdrawal	200.00 won
(3) (a) Moving instrument to new-premises	1,000.00 won
(b) If on a subscriber-constructed line	400.00 won
(4) Moving or temporary withdrawal of extra receiver and bells	200.00 won
(5) Moving or temporary withdrawal of accessories	200.00 won
(6) Moving or temporary withdrawal of switchboard	Actual cost.

g. DIRECTORY LISTINGS

Type of Service	Basis	Rate
(1) 1st and 2nd class areas	Per listing	160.00 won
(2) 3rd to 6th class areas	Per listing	80.00 won

h. MESSAGE RATE, MESSAGE WITHDRAWAL RATE, CALLING FEE AND ADVICE FEE

Type of Service	Rate
(1) Message rate-local, per message	2.00 won
(2) Withdrawal of message	
(a) Before completion of message	1/2 ordinary rate
(b) Reduced time of message	1/2 ordinary rate
(c) After party has answered, on appointment or other calls	Full charge
(d) Calling party not present when call is completed	Full charge
(3) Calling fee in same exchange area (calling non-subscriber to telephone), per call	4.00 won

— 9 —

Type of Service	Rate
(4) Fee of advice (change a call covered by (3) above)	4.00 won
(5) Regular conversation (appointment call)	Four times the rate of ordinary message

(6) Message rate and calling rate outside the exchange area:

Distance	Basis	Message Rate	Calling Rate
Within the same city, town or township (*pu, eup* or *myun*)	Per 3-minute period, or fraction thereof, up to 9 minutes.	2.00 won	2.00 won
Outside area-less than 8 km		4.00 won	3.00 won
8 to 16 km	For each 3-minute period	5.00 won	3.00 won
16 to 32 km	over 9 minutes	7.00 won	3.00 won
32 to 64 km	rate shall be	8.00 won	3.00 won
64 to 104 km	doubled.	12.00 won	3.00 won
104 to 152 km		15.00 won	4.00 won
152 to 200 km		16.00 won	4.00 won
200 to 248 km		18.00 won	4.00 won
248 to 296 km		20.00 won	4.00 won
296 to 344 km		23.00 won	4.00 won
344 to 440 km		28.00 won	4.00 won
440 to 536 km		32.00 won	4.00 won
536 to 632 km		37.00 won	4.00 won
632 to 824 km		44.00 won	5.00 won
824 to 1016 km		50.00 won	5.00 won
1016 to 1200 km		56.00 won	5.00 won
1200 to 1400 km		63.00 won	5.00 won

SPECIAL SERVICE RATES

Type	Rate
(1) Urgent message	Two times the rate of ordinary message
(2) Regular message (appointment call)	Four times the rate of ordinary message

軍政廳　官報　遞信部令　第一一號　　一九四七年三月三十一日

Type	Rate
(3) Regular press call	4/5 the rate of ordinary message (multiplied by 360, for annual contract)
(4) Urgent call (public)	Two times the rate of ordinary message

5. *Effective Date.* This order shall be effective on and after 1 April 1947.

BY DIRECTION OF THE MILITARY GOVERNOR

LLOYD C PARSONS
Colonel Sig C
Adviser to Director

KIEL WON BONG
Director
Department of Communications

HEADQUARTERS
UNITED STATES ARMY MILITARY
GOVERNMENT IN KOREA
Department of Communications
Seoul, Korea

DEPARTMENT ORDER
NUMBER 12   11 April 1947

KOREAN BROADCASTING CORPORATION; RULES
CONCERNING LISTENING TO RADIO BROADCASTS

1. In accordance with the provision of Section II of *Private Broadcasting Regulations**, the following change is approved in *Korean Broadcasting Corporation Rules Concerning Listening to Radio Broadcasts :* Article 6 as amended in Department Order Number 1, dated 29 July 1946, is further amended to read as follows :

> *The fee for listening to radio broadcasting is forty won per month for each individual listening contract. The fee is payable as of the first day of the month in which the contract is signed, and is payable for the full month in which the contract is terminated, regardless of the day of the month on which it is signed or terminated.*

2. This order shall be effective on 1 May 1947.

BY DIRECTION OF THE MILITARY GOVERNOR :

LLOYD C PARSONS   KIEL WON BONG

Colonel Sig C   Director

Adviser to Director   Department of Communications

---

*Communications Department Order Number 93, dated 20 December 1923, Government of Japan, as revised, applied in Korea by Government General Order Number 48, dated 6 September 1924, as revised.

軍政廳　官報　遞信部令　第一二號　　一九四七年四月十一日

SOUTH KOREAN INTERIM GOVERNMENT
Department of Communications
Seoul, Korea

DEPARTMENT ORDER   22 May 1947
NUMBER ·13

CHANGE OF NAME OF KOREAN SIMPLE LIFE INSURANCE

1. PURPOSE. The purpose of this order is to confirm and make of record the changes made by Bureau of Communications Order No 10, dated 11 March 1946, which changes were not heretofore officially published.

2. CHANGE OF DESIGNATION. The content of Bureau of Communications Order No 10, dated 11 March 1946, is as follows:

"Effective 1 April 1946 the title of the Korean Simple Life Insurance is temporarily amended as follows, pending revision or amendment of all laws and orders:

(1) Korean Simple Life Insurance Order is changed to
NATIONAL LIFE INSURANCE ORDER.

(2) The phrase, Korean Simple Life Insurance, is changed wherever it may appear in Korean Simple Life Insurance Order, Simple Life Insurance Act where applicable to same order, Korean Simple Insurance Regulations, Bureau Orders, Notifications or other directives, to the phrase NATIONAL LIFE INSURANCE."

3. EFFECTIVE DATE. This order confirms and makes effective as of 1 April 1946 the changes above set forth.

BY DIRECTION OF THE MILITARY GOVERNOR:

KIEL WON BONG

Director

Department of Communications

SOUTH KOREAN INTERIM GOVERNMENT
Department of Communications
Seoul, Korea

DEPARTMENT ORDER
NUMBER 14

22 May 1947

LIFE INSURANCE PAYMENTS CHANGED

1. PURPOSE. The purpose of this order is to confirm and make of record the changes made by Bureau of Communications Order No 14, dated 19 March 1946, which changes were not heretofore officially published.

2. INCREASE IN RATES. The content of Bureau of Communications Order No 14, dated 19 March 1946, is as follows:

"Effective 1 April 1946 the terms of the National Life Insurance Act, National Life Insurance Order and National Life Insurance Regulations are amended by the following:

(1) The term 2,000 won (¥2,000) appearing in Paragraph 1, Section 4 of the National Life Insurance Act is changed to 5,000 won (¥5,000) and in the Excepted Clause the term 1,000 won (¥1,000) is changed to 3,000 won (¥3,000).

(2) Part 2, Section 7 of the National Life Insurance Order is amended by changes stated below:

Under 4 years old	75 won	Is	225 won
5	105 won	Amended	315 won
6	135 won	to	405 won
7	165 won		495 won
8	195 won		585 won
9	225 won		675 won
10	240 won		720 won

(3) The terms '50 chon (¥.50), 1 won (¥1.), 1 won 50 chon (¥1.50)' appearing in paragraph Pertinent to Infantile Insurance Monthly Premiums, Separated List No. 1 as provided for by Section 2 of National Life Insurance Regulations, are amended to '1 won 50 chon

軍政廳　官報　遞信部令　第一四號　　一九四七年五月二十二日

72

(¥1.50), 3 won (¥3.), 4 won 50 chon (¥4.50)' respectively."

3. EFFECTIVE DATE. This order confirms and makes effective as of 1 April 1946 the changes above set forth.

.BY DIRECTION OF THE MILITARY GOVERNOR:

SEAL

KIEL WON BONG

Director

Department of Communications

SOUTH KOREAN INTERIM GOVERNMENT
Department of Communications
Seoul, Korea

DEPARTMENT ORDER
NUMBER 15   5 June 1947

NATIONAL LIFE INSURANCE AMENDMENTS

1. PURPOSE. The Purpose of this order is to further amend the National Life Insurance Law, Ordinance and Regulations.*

2. NATIONAL LIFE INSURANCE LAW AMENDED. Law Number 42, dated July 1916 (The *National Life Insurance Law* formerly the *Simple Life Insurance Law*) is hereby amended as follows:

In Paragraph 1, Section 4**, the words *5,000 won (¥5,000)* are deleted and the words *20,000 won (¥20,000)* substituted therefor, and in the Excepted Clause, the words *3,000 won (¥3,000)* are deleted and the words *10,000 won (¥10,000)* substituted therefor.

3. NATIONAL LIFE INSURANCE ORDINANCE AMENDED. Governor General Ordinance Number 5, dated 4 May 1929 (The *National Life Insurance Ordinance,* formerly the *Korean Simple Life Insurance Ordinance*) is hereby amended by changing the schedule set forth in Part 2, Section 7*** to read as follows:

Under 4 years old	750 won
Under 5 years old	1050 won
Under 6 years old	1350 won
Under 7 years old	1650 won
Under 8 years old	1950 won

---

* See Department of Communications Order Number 13, dated 22 May 1947, which changed "Korean Simple Life Insurance" to "National Life Insurance."
** See Section 2 (1) of Department of Communications Order Number 14, dated 22 May 1947.
*** See Section 2 (2) of Department of Communications Order Number 14, dated 22 May 1947.

軍政廳　官報　遞信部令　第一五號　　　一九四七年六月五日

Under 9 years old	2250 won
Under 10 years old	2400 won

4. NATIONAL LIFE INSURANCE REGULATIONS AMENDED. Government General Ordinance Number 78, dated 25 September 1929 (The *National Life Insurance Regulations*, formerly the *Korean Simple Life Insurance Regulations*) is hereby amended as follows:

a. Paragraph "A" of Schedule 1, provided for by Section 2 of the Regulations and pertaining to whole life insurance and endowment insurance, is amended by the addition of the following:

"5. Contracts concluded hereafter shall have:

(1). A minimum insurance amount of 2,000 won. (¥2,000)

(2) A minimum monthly premium of 10 won. (¥10)

(3) Monthly premiums in multiples of 1 won. (¥1)"

b. Paragraph "B" of Schedule 1, provided for by Section 2 of the Regulations and pertaining to insurance of infants, is amended by the addition of the following:

"4. Contracts concluded hereafter shall have monthly insurance premiums of 5 won (¥5), 1 won (¥10) or 15 won (¥15.)"

c. Paragraph 1, Section 17, is amended by the addition of the following:

"Where the amount of monthly premium on any one contract is less than 10 won (¥10), six (6) months premium shall be paid in advance."

5. EFFECTIVE DATE. This order shall be effective on the tenth day after the date appearing hereon.

BY DIRECTION OF THE MILITARY GOVERNOR:

.KIEL WON BONG

Director

Department of Communications

— 2 —

727

SOUTH KOREAN INTERIM GOVERNMENT
Department of Communications
Seoul, Korea

DEPARTMENT ORDER   5 June 1947
NUMBER 16.

AMENDMENT OF KOREAN POSTAL ANNUITY ORDINANCE

1. PURPOSE. The purpose of this order is to amend the Korean Postal Annuity Ordinance.

2. NAME OF KOREAN POSTAL ANNUITY ORDINANCE CHANGED The title of Governor General Ordinance Number 33, dated 18 June 1943, is hereby changed from *Korean Postal Annuity Ordinance* to *Postal Annuity Ordinance*. Wherever *Korean Postal Annuity Ordinance* appears in laws, acts, ordinances, orders, regulations, directives, instructions and notifications, *Postal Annuity Ordinance* in hereby substituted therefor.

3. POSTAL ANNUITY ACT AMENDED. In Law Number 39, dated 30 March 1926 (*Postal Annuity Law*)*, Section 3 is hereby amended by increasing the maximum permissible individual annuity for one beneficiary from 3,600 won (¥3,600) to 24,000 Won (¥24,000).

4. This order shall be effective on the tenth day after the date appearing hereon.

BY DIRECTION OF THE MILITARY GOVERNOR:

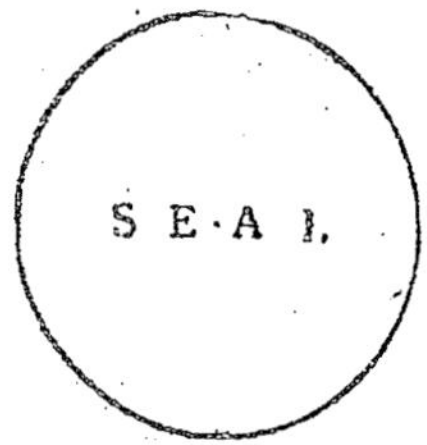

KIEL WON BONG

Director

Department of Communications

---

* Law Number 39, Government of Japan, was made applicable to Korea Governor General Ordinance Number 33, dated 18 June 1943 (*Postal Annuity Ordinance*).

軍政廳　官報　遞信部令　第一六號　　一九四七年六月五日

SOUTH KOREAN INTERIM GOVERNMENT
Department of Communications
Seoul, Korea

DEPARTMENT ORDER   19 May 1947
NUMBER 17

RESUMPTION OF REGISTRY SERVICE FOR INTERNATIONAL MAIL

1. PURPOSE. The purpose of this order is to provide for the registration of correspondence designated in Article 33 of the Universal Postal Union Convention of Buenos Aires, of 23 May 1939, and in accordance with the provisions of Chapter II, Articles 54 to 62, inclusive, of the Convention.

2. SERVICE. Registry service is authorized for the classes of mail, and in accordance with the conditions, mentioned in Department of Communications Orders Numbers 7, 8 and 9, dated 14 February 1947.

3. REGISTRY FEES. The fee for registering articles of correspondence will be 20 won, and for a return receipt requested at the time the article is mailed, 15 won additional. If a return receipt is requested after the article is mailed, the fee for a return receipt will be 20 won. All fees will be paid by the sender at the office of mailing by means of Korean postage stamps affixed to the articles registered. All fees will be in addition to the postage shown in Department of Communications Orders Numbers 7, 8 and 9.

4. IMDEMNITY. The Korean Postal Service will use all reasonable care in handling registered articles to prevent loss, damage or mishandling but, at the present time, is not in position to pay indemnity for loss, delay or failure of the addressee to receive registered articles in good condition.

5. EFFECTIVE DATE. This order shall be effective on and after 1 June 1947.

Kiel Won Bong
Director
Department of Communications

SOUTH KOREAN INTERIM GOVERNMENT
Department of Communications
Seoul, Korea

DEPARTMENT ORDER
NUMBER 18

5 August 1947

ESTABLISHMENT OF AIRMAIL SERVICE
FROM KOREA TO THE UNITED STATES

1. *Purpose.* The purpose of this order is to provide for airmail service from Korea to the United States of America for the following Articles of Correspondence mentioned in Article 33 of the Universal Postal Union Convention of Buenos Aires, of 23 May 1939, subject to the restrictions outlined hereinafter:

Letters, Post Cards, Commercial Papers, Prints including Printed Matter for the Blind and Samples of Merchandise, as described in Articles 112 to 123, inclusive, of the Universal Postal Union Convention of Buenos Aires, of 23 May 1939.

2. *Rates of Postage, Including Surcharge for Aerial Transportation, and General Conditions.* The following is a description of the articles, units of weight, dimensions, and rates of postage:

Letters, Commercial Papers, Prints and Samples of Merchandise	Unit of weight: 10 grams. Rate: 50 won for each unit or fraction thereof; total weight not to exceed 1000 grams. Dimensions: length, breadth and thickness combined, 90 centimeters; greatest length, 60 centimeters. When mailed in the form of a roll, length plus twice the diameter not to exceed 100 centimeters; greatest length 80 centimeters.
Post Cards	Rate: 50 won. Dimensions: maximum, 15 by 10.5 centimeters; minimum, 10 by 7 centimeters.

— 1 —

3. *Registration.* Articles may be registered on payment of a fee of 20 won, and for a return receipt, requested at the time the article is mailed, the fee will be 15 won additional. If a return receipt is requested after the article is mailed, the fee will be 20 won additional. All fees will be paid by the sender at the post office where the articles are mailed, by means of Korean postage stamps. All registry fees will be in addition to the postage required for sending articles by air mail. No indemnity for loss, delay or failure of the addressee to receive registered articles in good condition is authorized, but all reasonable care will be required in handling such articles while in custody of the Korean Postal Service.

4. *Languages.* Correspondence will be restricted to messages written in the Korean, Chinese, English, French, Japanese, Portuguese, Russian, or Spanish language.

5. *Non-Acceptable Matter.* Insured, special delivery, or collect on delivery matter will not be accepted in Korea for the United States of America.

6. *Excluded Matter.* All matter required to be excluded from the mails of either country will be excluded from the mails from Korea to the United States of America.

7. *Dispatch.* All air mail deposited in Korean post offices will be sent to the Seoul Central Post Office for dispatch to the United States of America.

8. *Obligations.* Korean postal obligations, including charges for transportation of International Air Mail from Korea, incurred hereunder, will be settled in United States dollars through such Korean dollar-balance trust fund as may be established, upon approval of the United States Army Military Government in Korea, in accordance with Article 14, paragraph 8, Provisions Concerning the Transportation of Regular Mails by Air, of the Universal Postal Union Convention of Buenos Aires, of 23 May 1939; or at the option of the transportation agency, such payments may be made in Korean currency at the rate authorized by the United States Army Military Government in Korea for converting military payment certificates into Korean currency.

— 2 —

9. ‧ *Effective Date.*   This order will become effective **on** the date appearing hereon.

BY DIRECTION OF THE MILITARY GOVERNOR：

KIEL WON-BONG
Director
Department of Communications

SOUTH KOREAN INTERIM GOVERNMENT
Department of Communications
Seoul, Korea

DEPARTMENT ORDER

NUMBER 19
1 October 1947

ESTABLISHMENT OF AIR MAIL SERVICE
TO VARIOUS COUNTRIES OF THE WORLD

1. *Purpose.* The purpose of this order is to provide for air mail service to various countries of the world, for the following articles of correspondence mentioned in Article 33 of the Universal Postal Union Convention of Buenos Aires, of 23 May 1939, subject to the restrictions outlined herein:

> Letters, Post cards, Commercial Papers, Prints, including Printed Matter for the Blind, and Samples of Merchandise, as described in Articles 112 to 123, inclusive, of the Universal Postal Union Convention of Buenos Aires, of 23 May 1939.

2. *General description and rates of postage, including surcharge for aerial transportation of mail.* Description of the articles, units of weight, and dimensions:

LETTERS, POST CARDS, COMMERCIAL PAPERS, PRINTS AND SAMPLES OF MERCHANDISE

Unit of weight: 10 grams. Total weight not to exceed 1000 grams. Dimensions: length, breadth and thickness combined, 90 centimeters; greatest length, 60 centimeters. When mailed in the form of a roll, length plus twice the diameter not to exceed 100 centimeters; greatest length 80 centimeters.

Post Cards. Dimensions: maximum, 15 by 10.5 centimeters; minimum, 10 by 7 centimeters.

— 1 —

Rates of Postage:

a. China, Japan and Rykyu Islands. Rate: 20 won for each 10 grams or fraction thereof.

b. Western Asia (Afghanistan, Iran, Iraq, Saudia Arabia, Syria, Turkey.) Rate: 60 won for each 10 grams or fraction thereof).

c. Borneo, Burma, Hongkong, India, Indo China, Java, Macao, Malaya, Nepal, Republic of the Philippines, Sumatra. Rate: 40 won for each 10 grams or fraction thereof.

d. Canada, Mexico, Newfoundland. United States of America (including its territories and island possessions). Rate: 50 won for each 10 grams or fraction thereof.

e. Cuba, Costa Rica, Dominican Republic, Gaudelupe, Guatemala, Leward Islands, Martinique, Nicaragua, Republic of Honduras, Salvador (other islands in what is known as the West Indies, and countries in what is known as Central America). Rate: 60 won for each 10 grams or fraction thereof.

f. Countries of South America (Argentina, Brazil, Chile, Colombia, Ecuador, British, French and Dutch Guiana, Panama, Paraguay, Peru, Uruguay, Venezuela). Rate: 70 won for each 10 grams or fraction thereof.

g. Australia, New Zealand, and islands adjacent thereto belonging to Great Britain. Rate: 50 won for each 10 grams or fraction thereof.

h. European countries (Albania, Austria, Belgium, Bulgaria, Czechoslovakia, Denmark, Eire (Ireland), Finland, France, Germany (subject to Army restrictions to that country), Great Britain and Northern Ireland, Greece, Holland, Hungary, Italy, Norway, Poland, Portugal, Rumania, Spain, Sweden, Switzerland, Union of Soviet Socialist Republics, Yugoslavia). Rate: 90 won for each 10 grams or fraction thereof.

i. Africa: All countries and possessions in Africa. Rate: 100 won for each 10 grams or fraction thereof.

j. Mail for other destinations within the Universal Postal Union will be accepted for transmission by air the distance regular air service is available;

and, if destination is not reached by air, such other regular means of transportation of mail available will be used.

3. *Registration.* Articles may be registered on payment of a fee of 20 won and, for a return receipt requested at the time the article is mailed, the fee will be 15 won additional. If a return receipt is requested after the article is mailed, the fee will be 20 won additional. All fees will be paid by the sender at the post office where the articles are mailed, by means of Korean postage stamps. All registry fees will be in addition to the postage required for sending articles by air mail. No indemnity for loss, delay or failure of the addressee to receive registered article in good condition is authorized, but all reasonable care will be required in handling such articles while in custody of the Korean Postal Service.

4. *Languages.* Correspondence will be restricted to messages written in the Korean, Chinese, English, French, Japanese, Portuguese, Russian, or Spanish languages.

5. *Nonacceptable matter.* Insured, special delivery, or collect-on-delivery matter will not be accepted in Korea for any country.

6. *Excluded matter.* All matter required to be excluded from the mails of any country will be excluded from the mails from Korea to that country.

7. *Dispatch.* All air mail deposited in Korean post offices will be sent to the Seoul Central Post Office for dispatch to destination.

8. *Obligations.* Korean postal obligations, including charges for transportation of International Air Mail from Korea, incurred hereunder, will be settled in United States dollars, upon approval of the United States Army Military Government in Korea, in accordance with Article 14, paragraph 8, *Provisions Concerning the Transportation of Regular Mails by Air,* of the Universal Postal Union Convention of Buenos Aires, of 23 May 1939; or, at the option of the transportation agency, such payments may be made in Korean currency at the rate authorized by the United States Army Military Government in Korea for converting military payment certificates into Korean currency.

— 3 —

9. *Effective date.* This order will become effective on the date appearing hereon.

BY DIRECTION OF THE MILITARY GOVERNOR:

KIEL WON PONG
Director
Department of Communications

SOUTH KOREAN INTERIM GOVERNMENT
Department of Communications
Seoul, Korea

DEPARTMENT ORDER
NUMBER 20   15 October 1947

AMENDMENT OF POSTAL DEPOSIT REGULATIONS

1. *Purpose.* The purpose of this order is to amend the Postal Deposit Regulations, issued March 1911, under Order No. 31, of the Government General of Chosen.

2. *Increase in fees.*

   a. The term *ten cheun (10 cheun)* appearing in Paragraph 2, Section 33 of the Postal Deposit Regulations is changed to *five won (W 5).*

   b. The term *five cheun (5 cheun)* appearing in Paragraph 2, Section 74 of the Postal Deposit Regulations is changed to *two won (W 2).*

3. *Increase in maximum withdrawal amounts.* The term *thirty won (W 30)* appearing in Section 80-2 of the Postal Deposit Regulations is changed to *five hundred won (W 500)* and the term *one hundred won (W 100)* appearing in the aforementioned paragraph is changed to *three thousand won (W 3,000).*

4. *Effective date.* This order shall be effective as of 15 October 1947.

KIEL WON BONG
Director
Department of Communications

軍政廳　官報　遞信部令　第二十號　·　一九四七年十月十五日

SOUTH KOREAN INTERIM GOVERNMENT
Department of Communications
Seoul, Korea

DEPARTMENT ORDER
NUMBER 21   15 October 1947

AMENDMENT OF POSTAL BOOK TRANSFER DEPOSIT REGULATIONS

1. *Purpose.* The purpose of this order is to amend the Postal Book Transfer Deposit Regulations, issued December 1909, under Order Number 69 of the Resident-General in Chosen.

2. *Increase in Amount of Basic Deposit.*

a. The term *five won* (W 5) appearing in Paragraph 1, Section 14 of the Postal Book Transfer Deposit Regulations is changed to *one hundred won* (W 100).

b. All existing accounts for which the basic deposit was five won (W 5) in accordance with former provisions of the Postal Book Transfer Deposit Regulations, are hereby required to be increased by an amount sufficient to result in a basic deposit of at least one hundred won (W 100).

3. *Increase in Charge for Voucher Book.*

a. The term *thirty cheun* (30 cheun) appearing in Paragraph 2, Section 22, of the Postal Book Transfer Deposit Regulations is changed to *thirty won* (W 30).

4. *Increase in Maximum Permissible Amount of a Single Payment Certificate.*

a. The term *three thousand won* (W 3,000) appearing in Section 27 of the Postal Book Transfer Deposit Regulations is changed to *ten thousand won* (W10,000).

5. *Increase in fees.*

a. Parts 1a, 2a, and 3a appearing in Paragraph 1, Section 28 of the

軍政廳　官報　遞信部令　第二十一號　　一九四七年十月十五日

Postal Book Transfer Deposit Regulations are amended by changing the schedules to the following :

(1) Fees for Ordinary Deposit :

Amount of Single Deposit			Fee
Cheun .50 to W	500.00		2 won
W 500.01	1,500.00		5 ″
1,500.01	2,000.00		10 ″
2,000.01	5,000.00		15 ″
5,000.01	8,000.00		20 ″
8,000.01	10,000.00		25 ″
10,000.01 and above			An additional 10 won will be charged for each unit of W 10,000 or any part thereof.

(2) Fee for Ordinary Transfer :

Fee

3 won

(3) Fees for Ordinary Payment and Direct Payment :

Amount of Single Payment		Fee
W .01 to W 500.00		5 won
500.01	1,000.00	10 ″
1,000.01	2,000.00	15 ″
2,000.01	3,000.00	20 ″
3,000.01	5,000.00	25 ″
5,000.01	8,000.00	30 ″
8,000.01	10,000.00	35 ″

b. The term *seven cheun (7 cheun)* appearing in Section 36, and in Paragraph 2, Section 49, of the Postal Book Transfer Deposit Regulations is changed to *two won (W 2)*.

The term *sixty cheun (60 cheun)* appearing in the above-mentioned paragraph and sections of the Postal Book Transfer Deposit Regulations is changed to *twenty-two won (W 22)*.

c. The term *twenty cheun (20 cheun)* appearing in Sections 56 and 57 of the Postal Book Transfer Deposit Regulations is changed to *two won (W 2)*.

d. The term *eighty cheun (80 cheun)* appearing in Paragraph 2, Section

— 2 —

61 of the Postal Book Transfer Deposit Regulations is changed to *twenty-six won* (*W 26*).

The term *ten cheun* (*10 cheun*) appearidg in the above-mentioned paragraph and section of the Postal Book Transfer Deposit Reguiations is changed to *two won* (*W 2*).

2. The term *one won, fifty cheun* (*W 1.50*) appearing in Section 76 of the Postal Book Transfer Deposit Regulations is changen to *eight won* (*W 8*).

6. *Effective Date.* This order shall be effective as of 15 October 1947.

KIEL WON BONG
Director
Department of Communications

SOUTH KOREAN INTERIM GOVERNMENT
Department of Communications
Seoul, Korea

DEPARTMENT ORDER
NUMBER 22

15 October 1947

RATES AND GENERAL CONDITIONS
OF POSTAL SERVICE BETWEEN KOREA AND CHINA.

1. *Purpose.* The purpose of this order is to amend the rates of postage from Korea to China in order to conform to the rates generally charged for mail service to other countries.

2. *Revocation; rates conformed.* Department of Communications Order No. 8, dated 14 February 1947 (*Rates and General Conditions of Postal Service Between Korea and China*), is hereby revoked and on and after 15 October 1947 the rates and general conditions of postal service between Korea and China will be the same as those mentioned in Department of Communications Order No. 7, dated 14 February 1947 (*Resumption of International Postal Service*).

3. *Effective date.* This order will become effective on the date appearing hereon.

BY DIRECTION OF THE MILITARY GOVERNOR:

PETER W SHUNK
Colonel Sig C
. Acting Adviser

KIEL   WON   BONG
Director
Department of Communications

軍政廳　官報　遞信部令　第二十二號　一九四七年十月十五日

SOUTH KOREAN INTERIM GOVERNMENT
Department of Communications
Seoul, Korea

DEPARTMENT ORDER
NUMBER 23   15 October 1947

POSTAL SERVICE BETWEEN KOREA AND JAPAN

1. *Purpose.* The purpose of this order is to further provide specific conditions of Postal Service from Korea to Japan. Department Order No. 9 dated 14 February 1947 (*Postal Service Between Korea and Japan*) is hereby revoked and this order is substituted therefor.

2. *Regular mails.* Letters, post cards, commercial papers, prints, raised prints for the blind, small packets, and samples of merchandise, as defined in Articles 112 to 125, inclusive, of the Universal Postal Union Convention of Buenos Aires, of 23 May 1939, will be accepted for mailing to Japan, subject to the provisions of said Convention and to such restrictions as the Supreme Commander for the Allied Powers may impose on articles for delivery in Japan.

3. *Letters and post cards.* Letters and post cards containing personal and family messages, and business, financial, commercial. and transactional correspondence, are mailable to Japan subject to the following prohibitions:

   a. Messages which transfer currency, checks, drafts. payment orders, or other credit or financial instruments.

   b. Mailing of instruments intended to defeat the purposes of USAMGIK's foreign exchange and foreign trade controls and/or those designed to defeat SCAP's regulations.

   c. Messages which grant or transfer translation, reproduction, performance, or other rights concerning books, articles, plays, music, motion pictures, or other media of information and expression, and messages relating in any way to patents or copyrights, except for description and

— 1 —

explanation of the authorized channels and procedures for handling such matters and except acknowledgments of rights ranged through the authorized channels.

4. *Commercial papers.* Commercial papers as defined by the Universal Postal Union Standards and Limitations are mailable to Japan with the following exceptions:

    a. Scores or sheets of music manuscript.

    b. Manuscripts of works or newspapers sent separately.

    c. All papers of legal procedure.

    d. Documents of all kinds drawn by Ministerial officers.

5. *Prints.* The mailing of prints, as defined by the Universal Postal Union Standards and Limitations, is restricted to the following specific categories:

    a. Photographs.

    b. Drawings.

    c. Plans.

    d. Maps.

    e. Patterns.

    f. Catalogues.

6. *Small packets and samples of merchandise.* These classifications of mail matters, as defined by the Universal Postal Union Convention, are mailable to Japan.

7. *Governing regulations.*

    a. The regular-mail service embracing letters, post cards, commercial papers, prints, samples of merchandise and small packets, as authorized, will be governed by the provisions of the Universal Postal Union Convention.

    b. Maintenance of records under the conditions of the Universal Postal Union Convention, the terms of the various bilateral agreements, and the preparation of the necessary accounts for settlement will be the responsibility of the Department of Communications, South Korean Interim Government.

8. *Language.* Correspondence will be restricted to messages written in the Korean, Chinese, English French. Japanese, Portuguese, Russian or Spanish language.

— 2 —

9. *Nonacceptable matter.* Insured, special delivery, or collect-on-delivery matter will not be accepted in Korea for Japan.

10. *Registered matter.* Articles may be registered on payment of a fee of 20 won and, for a return receipt requested at the time the article is mailed, the fee will be 15 won additional. If a return receipt is requested after the article is mailed, the fee will be 20 won additional. All fees will be paid by the sender at the post office where the articles are mailed, by means of Korean postage stamps. All registry fees will be in addition to the postage required for sending articles by ordinary mail. No indemnity for loss, delay or failure of the addressee to receive registered article in good-condition is authorized, but all reasonable care will be required in handling such articles while in the custody of the Korean Postal Service.

11. *Excluded matter.* All matter required to be excluded from the mail of either Korea or Japan will be excluded from the mail from Korea to Japan.

12. *Dispatch.* Mail will be dispatched from Korea to Japan about twice a month via Inchon or Pusan, Korea, depending upon the availability of vessels sailing.

13. *Billing.* Mail from Korea to Japan should be billed to the Tokyo Post Office. Mail from Japan to Korea should be to the Seoul Central Post Office as the designated international exchange office and dispatched through the ports of Pusan or Inchon, Korea.

14. *Rates and general conditions.* Except as otherwise provided in separate orders, the following is a description of the mailable articles, units of weight, dimensions, and rates of postage:

LETTERS   *Unit of weight:* 20 grams. *Rate:* first unit of weight or fraction thereof, 5 won; each additional unit or fraction thereof, 3 won; total weight not to exceed 2 kilograms. *Dimensions:* length, breadth and thickness combined, 90 centimeters; greatest length, 60 centimeters. When mailed in the form of a roll, length plus twice the diameter not to exceed 100 centimeters; greatest length, 80 centimeters.

POST CARDS   *Rate:* 3 won for single card and 6 won for card with reply paid. *Dimensions:* maximum, 15 by 10.5 centimeters; minimum, 10 by 7 centimeters.

— 3 —

COMMERCIAL
Unit of weight: 50 grams. *Rate:* 3 won for each unit or fraction thereof; minimum charge, 10 won; limit or weight, 2 kilograms. *Dimensions:* same as for letters.

PRINTS
Unit of weight: 50 grams. *Rate:* 3 won for each unit or fraction thereof; limit of weight, 2 kilograms (books, 3 kilograms for single volume). *Dimensions:* same as for letters.

RAISED
PRINTS FOR
THE BLIND
Each 1000 grams, 2 won; limit of weight, 7 kilograms. Prints sent in the form of folded or unfolded cards will be subject to the same minimum limits as post cards.

SMALL
PACKETS
Unit of weight: 50 grams. *Rate:* 5 won for each unit or fraction thereof; minimum charge, 30 won; limit of weight 1 kilogram. *Dimensions:* same as for letters. (Small packets should not be accepted for countries which have expressed an unwillingness to accept them).

SAMPLES OF
MERCHANDISE
Unit of weight: 50 grams. *Rate:* 3 won for each unit of fraction thereof; minimum charge, 5 won; limit of weight, 500 grams. *Dimensions:* same as for letters.

15. *Effective date.* This order will become effective on the date appearing hereon.

BY DIRECTION OF THE MILITARY GOVERNOR:

KIEL WON BONG
Director
Department of Communications

軍政廳　官報　遞信部令　第二三號　一九四七年十月十五日

SOUTH KOREAN INTERIM GOVERNMENT
Department of Communications
Seoul; Korea

DEPARTMENT ORDER
NUMBER 24

1 October 1947

AMENDING POSTAL, TELEPHONE AND TELEGRAPH RATES

1. *Change in Postal Rates.* The schedule of postal rates fixed by paragraph 1 of Department Order No. 10, Department of Communications, dated 31 March 1947 (*Revision of Postal Rates*) is hereby further revised, and shall be as follows :

REGULAR MAILS

Class of Mail	Type of Mail	Weight	Rate
1st	Letters	20 grams or fraction thereof	2.00 won
1st	Printed letters and circulars	100 grams or fraction thereof	2.00 won
2d	Ordinary post cards	20 grams or fraction thereof	1.00 won
2d	Return post cards .	—	2.00 won
2d	Letter-style post cards	—	2.00 won
3d	Periodicals	100 grams or fraction thereof	1.00 won
3d	Daily newspapers	100 grams or fraction thereof	50 chon

— 1 —

軍政廳　官報　遞信部令　第二十四號　一九四七年十月一日

REGULAR MAILS

Class of Mail	Type of Mail	Weight	Rate
4th	Printed matter for the blind	100 grams or fraction thereof	50 chon
4th	Printed booklets, commercial papers, and samples of merchandise	100 grams or fraction thereof	2.00 won
5th	Agricultural seeds	100 grams or fraction thereof	50 chon

PARCEL POST

Weight	Rate
Not exceeding 1000 grams	20.00 won
From 1001 to 2,000 grams	30.00 won
Exceeding 2,000 grams, but not exceeding 4.000 grams	50.00 won

SPECIAL SERVICES

Type of Service	Rate
Registering of mail	5.00 won
Insuring of mail or of merchandise, for each 100 won of valuation	5.00 won
Return receipt of delivery	10.00 won
Return receipt of delivery of court papers	10.00 won

— 2 —

軍政廳　官報　遞信部令　第二十四號　一九四七年十月一日

MONEY ORDERS

Type of Money Order	Amount of Money Order	Rate
Ordinary money order (redeemable only at post office designated thereon)	Up to and including 300.00 won	20.00 won
	From 300.01 to 500.00 won	30.00 won
	From 500.01 to 1,000.00 won	40.00 won
	From 1,000.01 to 3,000.00 won	50.00 won
	From 3,000.01 to 5,000.00 won	60.00 won
	From 5,000.01 to 8,000.00 won	70.00 won
	From 8,000.01 to 10,000.00 won	80.00 won
Money order (redeemable at any post office)	Up to and including 100.00 won	5.00 won
	From 100.01 to 200.00 won	10.00 won
	From 200.01 to 300.00 won	15.00 won
	From 300.01 to 500.00 won	25.00 won
Telegraph money order	Up to and including 300.00 won	50.00 won
	From 300.01 to 500.00 won	70.00 won
	From 500.01 to 1,000.00 won	100.00 won
	From 1,000.01 to 3,000.00 won	120.00 won
	From 3,000.01 to 5,000.00 won	140.00 won
	From 5,000.01 to 8,000.00 won	170.00 won
	From 8,000.01 to 10,000.00 won	200.00 won

REVISION OF TELEGRAPH AND TELEPHONE RATES

2. *Telegraph Rates.* The schedule of telegraph rates fixed by paragraph 1 of Department Order No. 11. Department of Communications. dated 31 March 1947 (*Revision of Telegraph and Telephone Rates*) is hereby revised. and shall be as follows :

— 3 —

a. ORDINARY TELEGRAMS

(1) Within Korea

Type of Telegram	Unit	Rate
In Korean language	10 characters (syllables) or less For each additional character	20.00 won 2.00 won
In a European language	5 words or less For each additional word	20.00 won 2.00 won

(2) Telegrams between Korea and Japan

Type of Telegram	Unit	Rate
In Japanese language	10 characters (kana) or less For each additional character	20.00 won 2.00 won
In a European language	5 words or less For each additional word	20.00 won 2.00 won

(3) If more than one addressee in telegrams in Korean or Japanese language, 2.00 won for each additional addressee

b. PRESS TELEGRAMS

Within Korea

Type of Telegram	Unit	Rate
In Korean language	For each 30 characters (syllables) or fraction	10.00 won
In a European language	For each 10 words or fraction	10.00 won

c. SUBSCRIPTION PRESS TELEGRAM FEE (appointment calls for limited time and words) within Korea or between Korea and Japan:

— 4 —

軍政廳　官報　遞信部令　第二十四號　一九四七年十月一日

Type of Telegram	Unit	Rate
In Korean language.	250 characters (syllables) or less	12,000.00 won
	From 251 to 500 characters (syllables)	20,000.00 won
	From 501 to 750 characters (syllables)	30,000.00 won
In a European language	100 words or less	12,000.00 won
	From 101 to 200 words	20,000.00 won
	From 201 to 300 words	30,000.00 won

d. SPECIAL HANDLING FEES

(1) Urgent telegram fee — 3 times ordinary telegram fee		
(2) Collated telegram fee (delivery guaranteed) — 1-1/2 times ordinary telegram fee		
(3) Express (special delivery) fee :		*Rate*
Distance of 8 km or less		100.00 won
For each additional 4 km		100.00 won
Delivery to any island		Actual cost
(4) Ship delivery fee		100.00 won
(Note : If actual cost of delivery exceeds the fee, difference is to be paid by addressee)		
(5) Fee for cancellation before transmission		5.00 won
(6) Sending telegrams by telephone		5.00 won
(7) Sending telegrams on credit basis		5.00 won

軍政廳　官報　遞信部令　第二十四號　一九四七年十月一日

c. SPECIAL SERVICES

Type of Service	Rate
(1) Acknowledgment of receipt by telegram—in Korean, Japanese or a European language	Equal to basic charge for telegram
(2) Acknowledgment of receipt by mail	2.00 won
(3) Multiple-address telegram—in Korean, Japanese or a European language	1/2 of charge for telegram
(4) Press multiple-address telegram	1/2 of charge for telegram
(5) Subscription multiple-address telegram	1/2 of charge paid for subscription fee
(6) Receipt fee	5.00 won
(7) Special designation of delivery: Ordinary (per year) Special (per month)	1,000.00 won 100.00 won
(8) Perusal fee (message taken from files and contents received by customer)-	5.00 won
(9) Certified copy : In Korean language, 50 characters(syllables) or less In a European language, 25 words or less	10.00 won 10.00 won
(10) Ship information registration — per year	100.00 won
(11) Telegraphic report of passing vessels: a. If registration fee paid b. If registration fee not paid	10.00 won 20.00 won

軍政廳　官報　遞信部令　第二十四號　一九四七年十月一日

Type of Service	Rate

(12) Semaphore telegram (ship to shore to lighthouse):
  a. Signal fee
  b. Telegram or mail fee
	50.00 won
	Actual expense of delivery

(13) Report of casualties at sea (SOS from ship to shore to owner of ship)	Actual expense of delivery

3. *Radio-Telegram Rates.* The schedule of radio-telegram rates fixed by paragraph 2 of Department Order No. 11, Department of Communications, dated 31 March 1947 (*Revision of Telegraph and Telephone Rates*) is hereby revised, and shall be as follows:

a. RADIO- TELEGRAMS TO KOREAN SHIPS AND AIRCRAFT

Type of Telegram	Unit	Rate
In Korean language	10 characters (syllables) or less	40.00 won
	For each additional character	4.00 won
	Medical telegram of 10 characters or less	20.00 won
	For each additional character	2.00 won
In a European language	5 words or less	40.00 won
	For each additional word	4.00 won

b. RATES BETWEEN LAND STATIONS: Same as for ordinary telegram.

c. PRESS RADIO-TELEGRAMS

Type of Telegram	Unit	Rate
In Korean language	30 characters (syllables) or less	20.00 won
In a European language	10 words or less	20.00 won

軍政廳　官報　遞信部令　第二十四號　一九四七年十月一日

PRESS BROADCAST		
Characters	*Transmitting Fee*	*Receiving Fee*
1000 characters per day	100 won	20 won
500 English characters per day	100 won	20 won

(1) When the Department of Communications furnishes transmitting services, the fee paid shall be that shown under the column headed "Transmitting Fee". When the Department of Communications furnishes receiving services only, the fee paid will be that shown under the column headed "Receiving Fee". In cases where the Department of Communications furnishes both services, fee paid will be a combination of both columns (120 won per 1000 Korean characters).

(2) If transmission of Press Broadcast is delayed beyond a period of four hours due to circuit and/or delivery failures, charges will be made at a rate of one fourth (1/4) of normal rate.

d. SPECIAL HANDLING FEES

Service	Rate
(1) Urgent radio-telegram	Equal to radio or press telegram charge
(2) Collated radio-telegram (delivery guaranteed)	1-1/2 times radio-telegram charge
(3) Acknowledgment of receipt of telegram	Equal to basic charge for radio telegram
(4) Multiple address	1/2 of charge paid for text, letters or words
(5) Press multiple-address radio-telegram	1/2 of charge paid for text, letters or words

4. *Meteorological Telegram Rates.* The schedule of telegraph rates for meteorological telegrams fixed by paragraph 3 of Department Order No. 11, Department of Communications, dated 31 March 1947 (*Revision of Telegraph and*

軍政廳　　官報　　遞信部令　　第二十四號　　一九四七年十月一日

Telephone Rates) is hereby revised, and shall be as follows :

Type of Telegram	Monthly Rate	
	Abbreviated	Interpreted Sentence
General weather forecast	160.00 won	240.00 won
Special weather announcement	80.00 won	100.00 won
General storm warning	—	70.00 won
Local weather forecast	150.00 won	200.00 won
Local weather announcement	70.00 won	80.00 won
Local storm warning	—	50.00 won

Detailed weather announcement:
  Once a day
  Twice a day
	—	1,500.00 won
		2,900.00 won

5. *Telephone Rates.* The schedule of telephone rates fixed by paragraph 4 of Department Order No. 11, dated 31 March 1947 (*Revision of Telegraph and Telephone Rates*) is hereby revised, and shall be as follows :

a. SUBSCRIPTION CHARGES

Type of Service	Basis	Class of Area	Rate
(1) Message-rate system	Per year	1	3,000.00 won
	Per message in same exchange area	1	5.00 won
(2) Uniform-rate system	Per year	1	12,000.00 won
		2	8,000.00 won
		3	7,000.00 won
		4	6,000.00 won
		5	5,000.00 won
		6	4,000.00 won

— 9 —

75

Type of Service	Basis	Class of Area	Rate
(3) Special telephone for trunk call (suburban party line service	Per year		3,000.00 won
(4) Private telephone line local	Per year	1	12,000.00 won
		2	8,000.00 won
		3	7,000.00 won
		4	6,000.00 won
		5	5,000.00 won
		6	4,000.00 won
(5) Private telephone line toll	Charge determined in each case by Director, Department of Communications		

(6) The class of an exchange area is to be determined by the number of telephone subscribers in the area, as follows:

Class	Number of Subscribers
1	10,000.00 or more
2	1,000.00 to 9,999
3	500 to 999
4	300 to 499
5	100 to 299
6	Less than 100

b. SUPPLEMENTARY CHARGES

Type of Service	Basis	Rate (per year)
(1) Subscriber outside of exchange area	Per 100 meters of line from exchange area	500.00 won
(2) Availability of special long distance trunks	Per trunk	300.00 won
(3) (a) Telephone hand-set instrument (Dept.-owned)	—	1000.00 won
(b) Desk set (Dept.-owned)	—	700.00 won

— 10 —

軍政廳　官報　遞信部令　第二十四號　一九四七年十月一日

Type of Service	Basis	Rate (per year)
(4) Extra telephone		
(a) Wall : Dept.-owned	—	1,000.00 won
Subscriber-owned	—	500.00 won
(b) Hand set : Dept.-owned	—	2,000.00 won
Subscriber-owned	—	500.00 won
(c) Desk set : Dept.-owned	—	1,500.00 won
Subscriber-owned	—	500.00 won
(d) For each extra telephone not for private use	—	500.00 won
(e) Desk set available to 2 exchange lines for each additional telephone set	—	500.00 won
(5) Private switchboard (Dept.-owned)	Charge determined in each case by Director, Department of Communications	
(6) Extra bell and receiver	—	500.00 won
(7) Extra telephone on private switchboard	—	500.00 won
(8) Each telephone on a private line	Charge determined in each case by Director, Department of Communications	

   c. REGISTRATION FEE (all areas)   500.00 won

   d. TRANSFERRAL FEE

Changing name of subscriber (permissible only within immediate family)-all areas	2,000.00 won

— 1 1 —

750

e. INSTALLATION CHARGE

Type of Service	*Rate*
(1) 1st class area (telephone)	13,000.00 won
(2) 2nd class area (telephone)	12,000.00 won
(3) 3rd class area (telephone)	10,000.00 won
(4) 4th, 5th and 6th class areas (telephone)	7,000.00 won
(5) Starting new exchange area, switchboard and telephone installation charge (according to Article No. 2 of Telephone Rule)	5,000.00 won
(6) Special telephone for trunk call (according to Article No. 40 of Telephone Rule):	
(a) Less than 300 meters	5,000.00 won
(b) More than 300 meters — Per 100 meters	1,500.00 won
(7) If subscriber supplies all materials for line construction on new installations the installation charges will be as follows :	
(a) For local telephones	2,000.00 won
(b) For special telephones for trunk call — 300 meters or less	2,000.00 won
NOTE : For each 100 meters above 300, on special telephones for trunk-call lines, add 500 won.	

f. CHARGE FOR MOVING SPECIAL TRUNK CALL TELEPHONE

(1) Line — per 100 meters	1,500.00 won
(2) Each telephone	500.00 won
(a) If telephone is on a subscriber-constructed line, the charge for each 100 meters is	500.00 won

g. CHARGE FOR MOVING TELEPHONE INSTRUMENTS

軍政廳　官報　遞信部令　第二十四號　一九四七年十月一日

Type of Service	Rate
(1) Moving instrument on premises	500.00 won
(2) Temporary withdrawal	500.00 won
(3) (a) Moving instrument to new premises	2,000.00 won
(b) If on a subscriber-constructed line	1.000.00 won
(4) Moving or temporary withdrawal of extra receiver and bells	500.00 won
(5) Moving or temporary withdrawal of accessories	500.00 won
(6) Moving or temporary withdrawal of switchboard	Actual cost

h. DIRECTORY LISTINGS

Type of Service	Basis	Rate
1) 1st and 2nd class areas	Per listing	300.00 won
(2) 3rd to 6th class areas	Per listing	150.00 won

i. MESSAGE RATE, MESSAGE WITHDRAWAL RATE, CALLING FEE AND ADVICE FEE

Type of Service	Rate
(1) Message rate, local, per message	5.00 won
(2) Withdrawal of message:	
(a) Before completion of message	1/2 ordinary rate
(b) Reduced time of message	1/2 ordinary rate
(c) After party has answered, on appointment or other calls	Full charge
(d) Calling party not present when call is completed	Full charge
(3) Calling fee in same exchange area (calling nonsubscriber to telephone), per call	20.00 won

Type of Service	Rate
(4) Fee of advice (change a call covered by (3) above)	20.00 won
(5) Regular conversation (appointment call)	Four times the rate of ordinary message

(6) Message rate and calling rate outside the exchange area :

Distance	Basis	Message Rate	Calling Rate
Within the same city, town or township (pu, eup or myun	Per 3-minute period, or fraction thereof, up to 9 minutes	10.00 won	20.00 won
Outside area - less than 8 km		12.00 won	30.00 won
8 to 16 km	For each 3-minute period over 9 minutes, rate shall be doubled	15.00 won	30.00 won
16 to 32 km		21.00 won	30.00 won
32 to 64 km		24.00 won	30.00 won
64 to 104 km		36.00 won	30.00 won
104 to 152 km		45.00 won	50.00 won
152 to 200 km		48.00 won	50.00 won
200 to 248 km		54.00 won	50.00 won
248 to 296 km		60.00 won	50.00 won
296 to 344 km		69.00 won	50.00 won
344 to 440 km		84.00 won	50.00 won
440 to 536 km		96.00 won	50.00 won
536 to 632 km		111.00 won	50.00 won
632 to 824 km		132.00 won	100.00 won
824 to 1016 km		150.00 won	100.00 won
1016 to 1200 km		168.00 won	100.00 won
1200 to 1400 km		189.00 won	100.00 won

j. SPECIAL SERVICE RATES

Type	Rate
(1) Urgent message	Two times the rate of ordinary message
(2) Urgent call (public)	Two times the ra etof ordinary message
(3) Urgent advice	Two times the rate of ordinary message
(4) Calling rate (information regarding toll call) sent by ship	30.00 won

k. MAINTENANCE CHARGES—PRIVATELY OWNED EQUIPMENT

Type	Rate
(1) Telephone and telegraph lines Per 100 meters per circuit Per year	500.00 won
(2) Telegraph equipment including wiring and equipment inside of building, per year	3,000.00 won
(3) Switchboard (5 local circuits or less) including wiring and accessories inside of building, per year *NOTE*: For each local circuit over 5 add 200 won per year.	3,000.00 won

Type	Rate
(4) Telephones, including dry cells, wiring and accessories inside of building.	
(a) Wall type, each, per year	700.00 won
(b) Hand set, each, per year	1,500.00 won
(c) Desk set, each, per year	1,000.00 won
NOTE : When the number of telephones exceeds 5, subtract 50 won for each telephone above 5; when the number of telephones exceeds 10 subtract 100 won for each telephone above 10	
(5) Storage batteries, each cell, per year *NOTE* : If number of calls exceeds 10, subtract 20 won for each cell above 10	200.00 won
(6) Extension bell, each, per year	200.00 won
(7) Extra receiver, Type "A" (Ordinary) Per year	200.00 won
(8) Extra Receiver, Type "B" (Watchcase) Per year	300.00 won

6. *Effective Date.* This order shall be effective on and after 1 October 1947.

BY DIRECTION OF THE MILITARY GOVERNOR :

KIEL WON BONG

Director

Department of Communications

軍政廳　官報　遞信部令　第二十四號　一九四七年十月一日

SOUTH KOREAN INTERIM GOVERNMENT
Department of Communications
Seoul, Korea

DEPARTMENT ORDER
NUMBER 25

1 May 1948

KOREAN BROADCASTING CORPORATION; RULES CONCERNING LISTENING TO RADIO BROADCASTS

1. In accordance with the provision of Section 11 of Private Broadcasting Regulations,* the following change is approved in Korean Broadcasting Corporation Rules Concerning Listening to Radio Broadcasts: Article 6 as amended in Department Order Number 1; dated 29 July 1946, is further amended to read as follows:

> *"The fee for listening to radio broadcasting is fifty won per month for each individual listening contract. The fee is payable as of the first day of the month in which the contract is signed, and is payable for the full month in which the contract is terminated, regardless of the day of the month on which it is signed or terminated."*

2. This order shall be effective on the date appearing hereon.

By DIRECTION OF THE MILITARY GOVERNOR:

KIEL, WON BONG
Director
Department of Communications

---

* Communications Department Order Number 98, dated 29 December 1923, Government of Japan, as revised, applied in Korea by Government General Order Number 48, dated 6 September 1924, as revised.

軍政廳　官報　遞信部令　第二五號　一九四八年五月一日

SOUTH KOREAN INTERIM GOVERNMENT
Department of Communications
Seoul, Korea

DEPARTMENT ORDER
NUMBER 26   25 May 1948

REVISION OF TELEGRAPH AND TELEPHONE RATES

1. Telegraph Rates. The schedule of telegraph rates fixed by paragraph 1 of Department Order No. 11, Department of Communications, dated 31 March 1947 (Revision of Telegraph and Telephone Rates), as amended in paragraph 2, Department Order No. 24, this Department, dated 1 October 1947 (Amending Postal, Telephone and Telegraph Rates), is hereby revised, and shall be as follows, effective 1 June 1948.

a. ORDINARY TELEGRAMS

(1) Within Korea

Type of Telegram	Unit	Rate
In Korean Language	10 characters (syllables) or less For each additional 2 characters	40.00 Won 5.00 Won
In a European Language	5 words or less For each additional word	40.00 Won 5.00 Won

(2) Telegrams between Korea and Japan

Type of telegram	Unit	Rate
In Japanese Language	10 characters (kana) or less For each additional character	40.00 Won 5.00 Won
In European Language	5 words or less For each additional word	40.00 Won 5.00 Won

(3) If more than one addressee in telegrams in Korean or Japanese language, 5.00 won for each additional addressee

b. PRESS TELEGRAMS

Within Korea

Type of Telegram	Unit	Rate
In Korean language	For each 30 characters (syllables) or fraction	20.00 Won
In a European language	For each 10 words or fraction	20.00 Won

c. SUBSCRIPTION PRESS TELEGRAM FEE (appointment calls for limited time and words) within Korea or between Korea and Japan:

Type of Telegram	Unit	Rate
In Korean language	250 characters (syllables) or less From 251 to 500 characters (syllables) From 501 to 750 characters (syllables)	38,400.00 Won 67,200.00 Won 96,000.00 Won
In a European language	100 words or less From 101 to 200 words From 201 to 300 words	38,400.00 Won 67,200.00 Won 96,000.00 Won

d. SPECIAL HANDLING FEES

(1) Urgent telegram fee – 3 times ordinary telegram fee	
(2) Collated telegram fee (delivery guaranteed) 1/2 times ordinary telegram fee.	
(3) Express (special delivery) fee: Distance of 8 km or less For each additional 4 km Delivery to any island	Rate 200.00 Won 200.00 Won Actual Cost
(4) Ship delivery fee (Note: If actual cost of delivery exceeds the fee, difference is to be paid by addressee)	200.00 Won
(5) Fee for cancellation before transmission	5.00 Won
(6) Sending telegrams by telephone	5.00 Won
(7) Sending telegrams on credit basis	5.00 Won

e. SPECIAL SERVICES

Type of Service	Rate
(1) Acknowledgement of receipt by telegram – in Korean, Japanese or a European language	Equal to basic charge for telegram
(2) Acknowledgement of receipt by mail	5.00 Won
(3) Multiple-address telegram – in Korean, Japanese or a European language	1/2 of charge for telegram

Type of Service	Rate
(4) Press multiple-address telegram	1/2 of charge for telegram
(5) Subscription multiple-address telegram	1/2 of charge paid for subscription fee
(6) Receipt fee	5.00 Won
(7) Special designation of delivery: Ordinary (per year) Special (per month)	1,000.00 Won 100.00 Won
(8) Perusal fee (message taken from files and contents received by customer)	5.00 Won
(9) Certified copy: In Korean language, 50 characters (syllables) or less In a European language, 25 words or less	10.00 Won 10.00 Won
(10) Ship information registration – per year	160.00 Won
(11) Telegraphic report of passing vessels: a. If registration fee paid b. If registration fee not paid	15.00 Won 25.00 Won
(12) Semaphore telegram (ship to shore to lighthouse): a. Signal fee b. Telegram or mail fee	80.00 Won Actual expense of delivery
(13) Report of casualties at sea (SOS from ship to shore to owner of ship)	Actual expense of delivery

2. Radio Telegram Rates. The schedule of radio-telegram rates fixed by paragraph 2 of Department Order No. 11, Department of Communications dated 31 March 1947 (Revision of Telegraph and Telephone Rates) as amended in paragraph 2, Department Order No. 24, this Department, dated 1 October 1947 (Amending Postal, Telephone and Telegraph Rates), is hereby revised, and shall be as follows, effective 1 June 1948.

a. RADIO TELEGRAMS TO KOREAN SHIPS AND AIRCRAFT

Type of Telegram	Unit	Rate
In Korean Language	10 characters (syllables) or less For each additional 2 characters Medical telegram of 10 characters or less For each additional 2 characters	80.00 Won 10.00 Won 40.00 Won 5.00 Won

— 3 —

Type of Telegram	Unit	Rate
In a European language	5 words or less For each additional word	80.00 Won 10.00 Won

    b. RATES BETWEEN LAND STATIONS: Same as for ordinary telegram.

    c. PRESS RADIO TELEGRAMS

Type of Telegram	Unit	Rate
In Korean Language	30 characters (syllables) or less	40.00 Won
In a European language	10 words or less	40.00 Won

PRESS BROADCAST		
Characters	Transmitting Fee	Receiving Fee
1000 characters per day 500 English characters per day	200 Won 200 Won	40 Won 40 Won

    (*1*) When the Department of Communications furnishes transmitting services, the fee paid shall be that shown under the column headed "Transmitting Fee". When the Department of Communications furnishes receiving services only, the fee paid will be that shown under the column headed "Receiving Fee". In cases where the Department of Communications furnishes both services, fee paid will be a combination of both columns (240 won per 1000 Korean characters).

    (*2*) If transmission of Press Broadcast is delayed beyond a period of four hours due to circuit and/or delivery failures, charges will be made at a rate of one fourth (1/4) of normal rate.

    d. SPECIAL HANDLING FEES

Service	Rate
(*1*) Urgent radio Telegram	Equal to radio or press telegram charge
(*2*) Collated radio-telegram (delivery guaranteed)	1/2 times radio-telegram charge

— 4 —

Service	Rate
(3) Acknowledgement of receipt of telegram	Equal to basic charge for radio telegram
(4) Multiple address	1/2 of charge paid for text, letters or words
(5) Press multiple-address radio-telegram	1/2 of charge paid for text, letters or words

3. Meteorological Telegram Rates. The schedule of telegraph rates for meteorological telegrams fixed by paragraph 3 of Department Order No. 11, Department of Communications. dated 31 March 1947 (Revision of Telegraph and Telephone Rates) as amended in paragraph 4, Department Order No. 24, this Department, dated 1 October 1947 (Amending Postal, Telephone and Telegraph Rates), is hereby revised, and shall be as follows, effective 1 June 1948.

Type of Telegram	Monthly Rate	
	Abbreviated	Interpreted Sentence
General Weather Forecast	960.00 Won	1200.00 Won
Special weather announcement	120.00 Won	160.00 Won
General storm warning	—	260.00 Won
Local weather forecast	200.00 Won	260.00 Won
Local weather announcement	80.00 Won	100.00 Won
Local storm warning	—	260.00 Won
Detailed weather announcement: Once a day Twice a day	—	2400.00 Won 4800.00 Won

4. Telephone Rates. The schedule of telephone rates fixed by paragraph 4 of Department Order No. 11, dated 31 March 1947 (Revision of Telegraph and Telephone Rates) as amended in paragraph 5, Department Order No. 24, this Department, dated 1 October 1947 (Amending Postal, Telephone and Telegraph Rates), is hereby revised, and shall be as follows, effective 1 June 1948.

— 5 —

a. SUBSCRIPTION CHARGES

Type of Service	Basis	Class of Area	Rate
(1) Message-rate system	Per year	1	4,000.00 Won
	Per message in same exchange area	1	5.00 Won
(2) Uniform-rate system	Per year	1 2 3 4 5 6	15,000.00 Won 14,000.00 Won 12,000.00 Won 10,000.00 Won 8,000.00 Won 7,000.00 Won
(3) Special telephone for trunk call suburban party line service)	Per year		3,000.00 Won
(4) Private telephone line local	per year	1 2 3 4 5 6	15,000.00 Won 14,000.00 Won 12,000.00 Won 10,000.00 Won 8,000.00 Won 7,000.00 Won
(5) Private telephone line toll			Charge determined in each case by Director, Department of Communications

(6) The class of an exchange area is to be determined by the number of telephone subscribers in the area, as follows:

Class	Number of Subscribers
1	10,000 or more
2	1,000 to 9,999
3	500 to 999
4	300 to 499
5	100 to 299
6	Less than 100

b. SUPPLEMENTARY CHARGES

Type of Service	Basis	Rate (per year)
(1) Subscriber outside of exchange area	Per 100 meters of line from exchange area	500.00 Won
(2) Availability of special long distance trunks	Per trunk	500.00 Won

Type of Service	Basis	Rate (per year)
(3) (a) Telephone hand-set instrument (Dept.-owned)	—	2000,00 Won
(b) Desk set (Dept.-owned)	—	1200.00 Won
(4) Extra telephone		
(a) Dept.-owned Wall	—	2000.00 Won
Subscriber owned	—	1200.00 Won
(b) Hand set:		
Dept.-owned	—	4000.00 Won
Subscriber owned	—	1200.00 Won
(c) Desk set:		
Dept. owned	—	3200.00 Won
Subscriber owned	—	1200.00 Won
(d) For each extra telephone not for private use	—	800.00 Won
(e) Desk set available to 2 exchange lines for each additional telephone set	—	800.00 Won
(5) Private switchboard (Dept. owned)	Charge determined in each case by Director, Department of Communications	
(6) Extra bell and receiver	—	800.00 Won
(7) Extra telephone on private switchboard	—	1,600.00 Won
(8) Each telephone on a private line	Charge determined in each case by Director, Department of Communications	

    c. REGISTRATION FEE (all areas)   500.00 Won

    d. TRANSFERRAL FEE

Changing name of subscriber (Permissible only within immediate family) all areas	2,000.00 Won

遞信部 官報 遞信部令 第二六號 · 一九四八年五月二五日

e. INSTALLATION CHARGE

Type of Service	Rate
(1) 1st class area (telephone)	20,000.00 Won
(2) 2nd class area (telephone)	17,000.00 Won
(3) 3rd class area (telephone)	15,000.00 Won
(4) 4th, 5th and 6th class area (telephone)	10,000.00 Won
(5) Starting new exchange area, switchboard and telephone installation charge (According to Article No. 2 of Telephone Rule)	10,000.00 Won
(6) Special telephone for trunk call (according to Article No. 40 of Telephone Rule):	
(a) Less than 300 meters	10,000.00 Won
(b) More than 300 meters-per 100 meters	2,000.00 Won
(7) If subscriber supplies all materials for line construction on new installations the installation charges will be as follows:	
(a) For local telephones	5,000.00 Won
(b) For special telephones for trunk call- 300 meters or less	5,000.00 Won
NOTE: For each 100 meters above 300, on special telephones for trunk-call lines, add 1,000.00 won.	

f. CHARGE FOR MOVING SPECIAL TRUNK CALL TELEPHONE

(1) Line-per 100 meters	4,000.00 Won
(2) Each telephone	800.00 Won
(a) If telephone is on a subscriber-constructed line, the charge for each 100 meters is	640.00 Won

g. CHARGE FOR MOVING TELEPHONE INSTRUMENTS

(1) Moving instrument on premises	640.00 Won
(2) Temporary withdrawal	640.00 Won
(3) (a) Moving instrument to new premises	4,000.00 Won
(b) If on a subscriber-constructed line	1,000.00 Won
(4) Moving or temporary withdrawal of extra receiver and bells	640.00 Won
(5) Moving or temporary withdrawal of accessories	500.00 Won
(6) Moving or temporary withdrawal of switchboard	Actual Cost

h. DIRECTORY LISTINGS

Type of Service	Basis	Rate
(1) 1st and 2nd class areas	per listing	800.00 Won
(2) 3rd to 6th class areas	per listing	400.00 Won

— 8 —

MESSAGE RATE, MESSAGE WITHDRAWAL RATE,
CALLING FEE AND ADVICE FEE

	Type of Service	Rate
(1)	Message rate, local, per message	5.00 Won
(2)	Withdrawal of message: (a) Before completion of message (b) Reduced time of message (c) After party has answered, on appointment or other calls (d) Calling party not present when call is completed.	1/2 ordinary rate 1/2 ordinary rate Full charge Full charge
(3)	Calling fee in same exchange area (calling nonsubscriber to telephone), per call	20.00 Won
(4)	Fee of advice (change a call covered by (3) above)	20.00 Won
(5)	Regular conversation (appointment call)	Four times the rate of ordinary message

(6) Message rate and calling rate outside the exchange area:

Distance	Basis	Message Rate	Calling Rate
Within the same city, town or township (pu, eup or myun)	Per 3 minute period, or fraction thereof, up to 9 minutes	10.00 Won	20.00 Won
Outside area-less than 8 km.		12.00 Won	30.00 Won
8 to 16 km	For each 3-minute period over 9 minutes, rate shall be doubled	24.00 Won	30.00 Won
16 to 32 km		32.00 Won	30.00 Won
32 to 64 km		40.00 Won	30.00 Won
64 to 104 km		56.00 Won	30.00 Won
104 to 152 km		64.00 Won	50.00 Won
152 to 200 km		72.00 Won	50.00 Won
200 to 248 km		80.00 Won	50.00 Won
248 to 296 km		88.00 Won	50.00 Won
296 to 344 km		105.00 Won	50.00 Won
344 to 440 km		120.00 Won	50.00 Won
440 to 536 km		144.00 Won	50.00 Won
536 to 632 km		170.00 Won	50.00 Won
632 to 824 km		200.00 Won	100.00 Won
824 to 1016 km		225.00 Won	100.00 Won
1016 to 1200 km		260.00 Won	100.00 Won
1200 to 1400 km		300.00 Won	100.00 Won

— 9 —

j. SPECIAL SERVICE RATES

Type	Rate
(1) Urgent message	Two times the rate of ordinary message
(2) Urgent call (public)	Two times the rate of ordinary call
(3) Urgent advice	Two times the rate of ordinary advice
(4) Calling rate (information regarding toll call) sent by ship	40.00 Won

k. MAINTENANCE CHARGES-PRIVATELY OWNED EQUIPMENT

Type	Rate
(1) Telephone and telegraph lines Per 100 meters per circuit Per year	500.00 Won
(2) Telegraph equipment including wiring and equipment inside of building, per year	4,400.00 Won
(3) Switchboard (5 local circuits or less) including wiring and accessories inside of building per year NOTE: For each local circuit over 5 add 360.00 won per year	4,400.00 Won
(4) Telephones, including dry cells, wiring and accessories inside of building. (a) Wall type, each, per year (b) Hand set. each, per year (c) Desk set. each, per year NOTE: When the number of telephones exceeds 5, subtract 160 won for each telephone above 5; when the number of telephones exceeds 10 subtract 320 won for each telephone above 10	1,400.00 Won 3,000.00 Won 2,000.00 Won
(5) Storage batteries, each cell, per year NOTE: If number of calls exceeds 10, subtract 56 won for each cell above 10.	240.00 Won
(6) Extension bell, each, per year	220.00 Won

Type	Rate
(7) Extra receiver, Type "A" (ordinary or watchcase) per year	220.00 Won
(8) Exra receiver, Type "B" (Headset) per year	640.00 Won

5. Effective Date. This order shall be effective on and after 1 June 1948.

BY DIRECTION OF THE MILITARY GOVERNOR:

KIEL WON BONG
Director
Department of Communications

S E A L

SOUTH KOREAN INTERIM GOVERNMENT
Department of Communications
Seoul, Korea

DEPARTMENT ORDER
NUMBER 27

15 July 1948

REVISION OF POSTAL RATES

1. *Postal Rates.* The schedule of postal rates fixed by paragraph 1 of Department Order Number 10 Department of Communications, dated 31 March 1947 (*Revision of Postal Rates*) as amended in paragraph 1, Department Order Number 24, this Department, dated 1 October 1947, (*Amending Postal, Telephone and Telegraph Rates*), is hereby revised and shall be as follows, effective 1 August 1948:

REGULAR MAILS

Class of Mail	Type of Mail	Weight	Rate
1st	Letters	20 grams or fraction thereof	4.00 won
1st	Printed letters and Circulars	100 grams or fraction thereof	3.00 won
2nd	Ordinary post cards	—	2.00 won
2nd	Return post cards	—	4.00 won
2nd	Letter style post cards	—	4.00 won
3rd	Periodicals	100 grams or fraction thereof	1.00 won
3rd	Daily newspapers	100 grams or fraction thereof	.50 chun
4th	Printed matter for the blind	1000 grams or fraction thereof	.50 chun
4th	Printed booklets, commercial papers, and samples of merchandise	100 grams or fraction thereof	2.00 won
5th	Agricultural seeds	100 grams or fraction thereof	.50 chun

— 1 —

PARCEL POST

Weight	Rate
Not exceeding 1,000 grams	30.00 won
Not exceeding 2,000 grams	40.00 won
Not exceeding 3,000 grams	50.00 won
Not exceeding 4,000 grams	80.00 won

Special Services

Type of Service	Rate
Registering of mail	10.00 won
Insuring of mail or of merchandise, for each 100 won of valuation	10.00 won
Return receipt of delivery	10.00 won
Return receipt of delivery of court papers	10.00 won

MONEY ORDERS

Type of Money Order	Amount of Money Order		Rate
Ordinary money order (redeemable only at post office designated thereon)	Up to and including	300.00 won	20.00 won
	From 300.01 to	500.00 won	30.00 won
	From 500.01 to	1,000.00 won	40.00 won
	From 1,000.01 to	3,000.00 won	50.00 won
	From 3,000.01 to	5,000.00 won	60.00 won
	From 5,000.01 to	8,000.00 won	70.00 won
	From 8,000.01 to	10,000.00 won	80.00 won
Money order (redeemable at any post office)	Up to and including	100.00 won	5.00 won
	From 100.01 to	200.00 won	10.00 won
	From 200.01 to	300.00 won	15.00 won
	From 300.01 to	500.00 won	25.00 won
Telegraph money order	Up to and including	300.00 won	50.00 won
	From 300.01 to	500.00 won	70.00 won
	From 500.01 to	1,000.00 won	100.00 won
	From 1,000.01 to	3,000.00 won	120.00 won
	From 3,000.01 to	5,000.00 won	140.00 won
	From 5,000.01 to	8,000.00 won	170.00 won
	From 8,000.01 to	10,000.00 won	200.00 won

2. Effective Date. This order shall be effective on and after 1 August 1948.

BY DIRECTION OF THE MILITARY GOVERNOR:

S E A L

PAK, SANG OK
Acting Director
Department of Communications

HEADQUARTERS
UNITED STATES ARMY FORCES IN KOREA
Office of the Military Governor
Bureau of Justice
Seoul, Korea

BUREAU ORDER
NUMBER   1   3 November 1945

PRISONS

1. The Bureau of Justice and the Penal Department of the Bureau of Justice are not apprehending agencies. Except with respect to the re-capture of escaped prisoners, prison officials have no authority to arrest or detain any person.

2. Arrest and detention of persons in Korea is the function of duly authorized military forces, the civilian police, officials of the Bureau of Police, and persons deputized by that bureau.

3. After the date of this instruction no person shall be received for detention without trial at the Westgate Prison or any other prison institution, who is not delivered with proper papers by duly constituted arresting authorities.

4. These instructions do not relate to or affect procedures for the imprisonment of persons convicted by the courts of Korea or by the military courts.

5. This order is effective immediately

EMERY J. WOODALL
Major AUS
Director Bureau of Justice

777

HEADQUARTERS
UNITED STATES ARMY FORCES IN KOREA
Office of the Military Governor
Bureau of Justice
Seoul, Korea

BUREAU ORDER
NUMBER 2   21 November 1945

    1. The name of the "Westgate Prison" is hereby changed to "Seoul Prison", and the prison formerly known as "Seoul Prison", will hereafter be a branch thereof without name.

    2. This order is effective at midnight 21 November 1945.

BY DIRECTION OF THE MILITARY GOVERNOR

MATT TAYLOR
Major FA
Director of the Bureau of Justice

HEADQUARTERS
UNITED STATES ARMY FORCES IN KOREA
Office of the Military Governor
Bureau of Justice
Seoul, Korea

BUREAU ORDER
NUMBER 3   22 November 1945

1. The supervision of the cell block and the physical property thereof at the Seoul Court~House, Seoul, Korea is hereby transferred from the Criminal Department of the Bureau of Justice to the Penal Department of the Bureau of Justice.

2. Effective 7 November 1945.

BY DIRECTION OF THE MILITARY GOVERNOR

MATT TAYLOR
Major FA
Director of the Bureau of Justice

BUREAU ORDER
NUMBER 4

19 November 1945

1. <u>Abolition of Existing Bar Associations</u>. All existing laws and regulations concerning the establishment, existence, and operations of Bar Associations in Korea and the juridical existence of Bar Associations are hereby repealed and abolished.

2. <u>Creation of National Bar Association of Korea</u>.

a. The National Bar Association of Korea is hereby created as a juridical person, with its principal office at the Bureau of Justice, Seoul, Korea.

The association is an organization of lawyers duly admitted to appear on behalf of parties to any judicial proceeding in any District Court, Court of Review, or the Supreme Court, of Korea by the order of admission to the practice of law of the Director of the Bureau of Justice after the successful passing of the required examination or examinations prescribed by the Director of the Bureau of Justice. No other person may become or remain a member. Any member holding judicial office, other than the office of the Chief Justice of the Supreme Court, will become an inactive member without paying dues while holding such office. Upon termination of his judicial office he will automatically become again an active member.

The association will have no capital fund and shall be a non-profit organization. No distribution of any funds accumulated by the association will be made to any member, except in payment of reasonable expenses incurred by such member, remuneration to such member for services actually rendered or as reimbursement for losses, injury or damages suffered as the direct result of services rendered in the performance of a duty to the association imposed by the Central Council of the association and recorded in the official minutes of the meetings of the Central Council signed and sealed by the members present at the meeting, except under the provisions of paragraph b of this Section. The association may incur and pay expenses of operation, not exceeding its current budget approved by the Director of the Bureau of Justice, from

— 1 —

membership fees and other income of the association. The association may accept bequests and gifts intervivos or testamentary, and may own personal, mixed or real property. All real property held by the association will be dedicated to the eleemosynary purposes of the association within twelve months after taking title thereto or will forthwith be sold to the highest of three or more bidders at a public auction held on at least thirty days' notice published in a newspaper having general circulation.

b. The income of the association in excess of its expenses within its current budget may be accumulated in one or more funds, as approved by the Central Council, for the education and improvement of its members in their profession, for the maintenance of a law library, for publicizing by books, journals or other publications professional information for the use and benefit of its members, and for such other public purposes as the Director of the Bureau of Justice approves. Such funds may be of two kinds : one in which the principal of the fund may be used, and the other in which the principal may not be expended but the income therefrom may alone be spent. All inter vivos or testamentary gifts will be funds of the association of which the income may alone be spent.

c. The original members of the Association will be those members of the legal profession of Korea duly admitted to the practice of law in Korea prior to the date of this order. New members will be all persons hereafter admitted to practice law before the Courts of Korea by the Director of the Bureau of Justice. Each such person will pay the prescribed entrance fee. No person shall be a member who fails to pay the annual dues of membership prescribed by the Central Council on or before 30 June each year, unless forgiven by the Central Council for good cause shown upon payment of all arrears of dues. A member will automatically cease to be a member upon his conviction by a District or higher Court for a crime for which the penalty provided by law exceeds 3 years imprisonment, with or without fine. If, however, such person is acquitted on an appeal or appeals taken, he shall automatically be reinstated as a member. Any member may be punished for a violation of the rules of legal ethics of the Association, as determined by the Disciplinary Committee on its own motion or upon a complaint filed in writing by any member or person claiming loss, injury or damage resulting from any professional act of omission or commission of the member, or by the Director of the Bureau of Justice. Such punishment may consist of:

— 2 —

(1) A fine not exceeding ten times the current annual dues, from which no appeal lies.

(2) The assessment of a money judgment or order for restitution in favor of an aggrieved person damaged as a direct result of any professional act of omission or commission, from which an appeal may be taken as from the decision of a District Court.

(3) A suspension from membership, which automatically for the duration of the suspension constitutes a suspension of the right to practice law or to receive any fee for an act or service performed during such suspension for which a lawyer could legally collect a professional fee. No such suspension shall exceed one year, but such suspension may be in addition to a fine, money judgment or order for restriction (1) or (2) above. An appeal will lie as from the decision of a District Court. The suspension will be effective at the expiration of the time for appeal or on the date of affirmance by the highest court to which appeal is actually taken.

(4) An irrovocable expulsion from membership, which automatically constitutes a disbarment from the legal profession for life. No such expulsion or disbarment by the Disciplinary Committee shall become effective until reviewed in extraordinary open session by the Supreme Court and affirmed.

No disciplinary penalty shall be assessed by the Disciplinary Committee except after a trial by the Committee with due process of law. Such trial shall be conducted under the Rules of the District Courts, for which purpose the Disciplinary Committee shall have the powers and exercise the processes of a District Court.

3. <u>District Chapters of National Bar Association.</u>

a. The members of the National Bar Association in good standing residing within the territorial jurisdiction of a District Court for a consecutive period of six months will constitute the District Chapter of the National Bar Association. On change of residence from one judicial district to another for other than temporary purposes the member will automatically, after six months' residence in the new district, become a member of that District Chapter. No member can be a member of two District Chapters at the same time. Until such residence requirement is met, he will retain membership in the former District Chapter and may participate in all business and elections of such District Chapter under the rules of election prescribed by

— 2 —

the Central Council with the approval of the Director of the Bureau of Justice.

b. Each District Chapter will elect by vote of the majority of the members thereof and will have the following officers: President, Vice-President, and Secretary, each of whom will serve without remuneration. No District Chapter will have any money transactions, receive any funds, incur any expenses, own any property or otherwise handle any funds for any purpose, except as the officers are directed to act as agent of the association in specific matters by the Central Council. A District Chapter has no juridical existence for any purposes and, except when acting as a disclosed and duly appointed agent of the association, cannot legally obligate or otherwise bind the association by any act of commission or omission. In law it is only a territorial committee of convenience for internal affairs of the association.

4. Central Council of National Bar Association.

a. The Central Council of the National Bar Association will be composed of one representative of each District Chapter elected for a term of one year. Such representatives will be elected within sixty days after the effective date of this order and annually thereafter in accordance with election procedures prescribed by the Director of the Bureau of Justice under the following principles:

Each member of a District Chapter in good standing, as shown by the membership record of the association, will have one vote. His vote will be cast by written secret ballot, and to be valid must be his voluntary act without promise or remuneration in any form from any person. Not more than five nor less than three nominees must have been selected as nominees to be voted upon at such election. Such selection will be from those members receiving the highest votes by similar written ballot. The representative elected will be the member among the nominees receiving a majority of the votes of all of the members of the District Chapter. No representative may serve more than five consecutive terms, after which he is not eligible as a nominee for three years.

b. The Central Council will meet in a court room of the Courts Building at Seoul on the second Monday of each January, April, July, and October for as long as the business of the Council requires. Reasonable expenses, as approved by the President of the association on recommendation of the Central Council, will be paid to Council members for actual travel to and from and for reasonable living expenses while attending the meetings of the

— 4 —

Council. No other remuneration will be paid to representatives as Council members.

c. The Council will take all action by majority vote of all the representatives then in office. Such a majority is required to constitute a quorum at any meeting. Any action of the Council which is contrary to this order or regulations thereunder, or contrary to the objects or purposes of the association may be annulled by the Director of the Bureau of Justice.

d. All Committees of the association will be appointed by the Central Council.

5. Officers of National Bar Association.

a. The association will have a President, who will always be the Chief Justice of the Supreme Court of Korea, a Vice President, a Treasurer, a Secretary, and such other officers and employees as are reasonably required to perform the work of the association as approved by the Director of the Bureau of Justice. The Vice-President, Treasurer and Secretary, and officers other than the President will be elected by the Central Council. The Vice-President will be elected from among the members of the Central Council and will serve without remuneration. The other officers will be elected from members of the association who are not members of the Central Council and will be paid salaries, commensurate with their duties, as recommended by the Central Council and approved by the Director of the Bureau of Justice. The salaries and wages of employees will be determined by the Central Council within the approved budget.

6. Rules of Legal Ethics.

a. The Rules of Legal Ethics will be prescribed by the Director of the Bureau of Justice, upon the advice of the Central Council.

b. The Director of the Bureau of Justice will make other regulations for the association and its members, prescribe such duties upon it in connection with the training, admission, and discipline of members, and approve such activities, including publications and participation in public matters, as are proper within the objects and purposes of the association.

c. No person who is not a member of the association in good standing will engage in the practice of law, appear on behalf of any party to any judicial proceeding other than in his own behalf, perform any professional act or service for which a member of the association in his profession as a lawyer could legally collect a professional fee, or collect any fee or remuneration from any person for any such act or service. A violation of this provision will be punished by the Courts as a crime not constituting felony. In every such case, in addition to any other penalty or punishment, a fine will be levied equal to at least three times all fees collected in violation of this provision. Any person other than a corporation may appear in any court on his own behalf.

7. **National Judicial Council.**

a. Every person holding office as a judge of any District or higher court of Korea south of 38° north latitude is automatically constituted a member of the National Judicial Council, which is hereby created. The Council will meet annually on the fourth Monday in January in a court room of the Courts Building at Seoul for the period necessary to transact its business. The President of the Council will be the Chief Justice of the Supreme Court of Korea. The Secretary will be the Chief Judge of the Courts of Appeals who has been longest in continuous service in such position. The Treasurer will be the Chief Judge of the District Courts of Korea (where more than one judge of a District Court in office) who has been longest in continuous service in such position. No other officer or employee will be appointed. No person will receive any remuneration for his services as an officer or otherwise as a member of the Council, except for necessary expenses actually incurred, with the approval of the Director of the Bureau of Justice, in connection with the work of the Council.

b. The National Judicial Council will perform such services for the Bureau of Justice as the Director may indicate by written request to the President, relating directly to the judicial processes of the administration of justice in Korea, including, without limitation, the promulgation and revision of Rules of the Courts, the Code of Civil Procedure, the Code of Criminal Procedure, and the administration of offices and functions supervised by the courts. The officials and employees of the courts will serve and assist the Council in its functions in accordance with instructions issued by the chief judicial officer of each court under directives of the President of the Council.

— 6 —

c. The National Judicial Council will have such standing committees as the President of the Council may appoint to perform services directed by the President of the Council at the written request of the Director of the Bureau of Justice. Such services will be consistent with the proper functions of the Council.

8. This order is effective at midnight 19 November 1945.

BY DIRECTION OF THE MILITARY GOVERNOR:

EMERY J. WOODALL
Major AUS
Director of the Bureau of Justice

HEADQUARTERS
UNITED STATES ARMY FORCES IN KOREA
Office of the Military Governor
Bureau of Justice
Seoul, Korea

BUREAU ORDER
NUMBER 5   4 December 1945

1. The Probationers Committee of the Bureau of Justice is hereby established for the purpose of examining and recommending the appointment of probationers to serve in the courts of Korea.

2. The following persons are appointed as members of the committee:

Kim Yong Mu, Chief Justice of the Supreme Court.

Kim Chang Yung, Chief Prosecutor of the Supreme Court.

Lee Sang Ki, Justice of the Supreme Court.

Choi Pyung Chu, Chief Secretary, Civil department.

Choi Chong Suk, Chief Secretary, Criminal department.

Choi Pyung Suk, Chief Secretary, Penal department.

Ex-Officio Members:

Matt Taylor, Major FA, Director Bureau of Justice.

George Anderson, Major AG, Executive Officer, Bureau of Justice.

Kim Yung Heui, Executive Assistant to the Director.

Wayne Geissinger, Major AG, Officer in Charge Bar Associations Department.

BY DIRECTION OF THE MILITARY GOVERNOR

MATT TAYLOR
Major FA
Director

HEADQUARTERS
UNITED STATES ARMY FORCES IN KOREA
Office of the Military Governor
Bureau of Justice
Seoul, Korea

BUREAU ORDER
NUMBER 7

17 December 1945

CHANNELS OF AUTHORITY

1. Effective this date greater responsibility will be placed upon the Korean officials and employees of this Bureau and the national institutions directed and supervised by it.

2. Military Government and Korean channels of authority will henceforth be maintained as separate channels.

3. Other than orders issued by the Director through the Associate Director, no orders will be issued by Military Government officers to Korean officials or employees of the Bureau or of the national institutions directed and supervised by it, unless expressly authorized by the Director.

4. Military Government officers of the Bureau will immediately assume the role of observers and advisors. Their duties will be:

a. To keep the Director advised of all developments within the scope of their assignment.

b. When their advice is sought in special situations or when they deem that the interests of Military Government require it, they will advise with the responsible officials of the national institutions. They will not substitute their opinions for the judgment of the responsible officials nor will they interrupt the normal established channels of authority.

c. To expedite the affairs of the Bureau in its relation with other Bureaus which remain under direct Military Government control.

5. The Associate Director will cause all officials and employees (other than Military personnel) through regular channels of authority to comply with the orders of the Director.

6. Orders issued by the Associate Director constitute orders of the Director. Korean officials and employees of the Bureau will promptly comply with all orders of the Associate Director.

BY DIRECTION OF THE MILITARY GOVERNOR

MATT TAYLOR
Major FA
Director

788

HEADQUARTERS
UNITED STATES ARMY FORCES IN KOREA
Office of the Military Governor
Seoul, Korea

BUREAU ORDER
NUMBER 8

17 December 1945

PROVINCIAL OBSERVERS AND ADVISORS

1. Selected representatives of the Bureau of Justice will forthwith proceed to assigned Provinces.

2. In the capacity of observers and advisors, Bureau representatives will perform the following functions:

    a. Establish liaison with Military Government Teams within their assigned Provinces.

    b. Survey and maintain contact with all institutions of the Bureau of Justice within their assigned Provinces.

    c. Comply with Bureau Order Number 7.

BY DIRECTION OF THE MILITARY GOVERNOR

MATT TAYLOR
Major FA
Director

HEADQUARTERS
UNITED STATES ARMY MILITARY
GOVERNMENT IN KOREA
BUREAU OF JUSTICE
Seoul, Korea

BUREAU ORDER
NUMBER 9

1 February 1946

Advisory Council

1. The Advisory Council of the Bureau of Justice is hereby reconstituted.

2. Members of the Council shall include the following Officials:
   a. Director of the Bureau of Justice, Chairman.
   b. Chief Justice of the Supreme Court.
   c. Chief Prosecutor of the Supreme Court.
   d. Chief Judges of the Courts of Review.
   e. Chief Prosecutors of the Courts of Review.
   f. Chief Warden of Seoul Prison.
   g. Chief Warden of Taegu Prison.
   h. Chief of the Administrative Department of the Bureau of Justice, Secretary.

3. The Advisory Council shall convene on call of the Director of the Bureau of Justice and shall render advice to the said Director as the same is required.

4. Except when expressly instructed, members of the Council who are officials in Taegu shall not meet with the Council, but shall receive for comment a report of all proceedings.

5. The Chief of the Administrative Department of the Bureau of Justice as Secretary thereof shall provide information for the Council, shall keep a record of all proceedings, shall prepare reports in final form and shall forward copies of the reports to absent members.

6. Members of the Advisory Council shall receive no additional compensation by virtue of services rendered pursuant to this Order.

BY DIRECTION OF THE MILITARY GOVERNOR

MATT TAYLOR
Major FA
Director
Bureau of Justice

**BUREAU ORDER
NUMBER 10**

28 February 1946

Whereas it is essential that there exist a means of proving family registrations and other official records which were destroyed in the fire at the Village Office of West Village in Kwangchoo County, Kyungi Province on 23 October 1945 and covers residents of Cho-Eeri, Cho-Ilri, Chun-Kungri, Hasa-Changri, Sangsa-Changri, Hangri, Kwang-Amri, Kam-Eeri, Kam-Ilri, Hak-Amri and Kam-Pukri in Kwangchoo County, Kyungi Province.

1. Now therefore: Any person who received any document concerning family registrations from the Chief of the Village Office of West Village in Kwangchoo County, Kyungi Province, from 1 September 1945 to 23 October 1945 will forthwith return the said documents before 31 March 1946.

2. Any person who reported or transmitted documents concerning family registrations to the Chief Clerk of the Village Office of West Village in Kwangchoo County, Kyungi Province, from 1 September 1945 to 23 October 1945 will prove the entries before 31 March 1946 in accordance with procedures which may be obtained from the Village Office aforesaid or the Seoul District Court.

BY DIRECTION OF THE MILITARY GOVERNOR

MATT TAYLOR
Major FA
Director
Bureau of Justice

HEADQUARTERS
UNITED STATES ARMY MILITARY
GOVERNMENT IN KOREA
Bureau of Justice
Seoul, Korea

BUREAU ORDER
NUMBER 11

27 February 1946

1. The second paragraph of Section 2a of Bureau Order Number 4 of the Bureau of Justice, dated 19 November 1945, is hereby amended to read as follows :

"The association is an organization of lawyers duly admitted to appear on behalf of parties to any judicial proceeding in any District Court, Court of Review, or the Supreme Court, of Korea by the order of admission to the practice of law of the Director of the Bureau of Justice after the successful passing of the required examination or examinations prescribed by the Director of the Bureau of Justice. Any person who has taught the subject of law in an accredited college or university for a period of not less than five years shall be entitled, upon application to and by permission of the Director of the Bureau of Justice, to practice law without being required to take the examination or examinations normally required, and shall likewise be entitled to membership in the association upon payment of the prescribed entrance fees. No other person may become or remain a member. Any member holding judicial office, other than the office of the Chief Justice of the Supreme Court, will become an inactive member without paying dues while holding such office. Upon termination of his judicial office he will automatically become again an active member."

2. This order is effective on the tenth day after date appearing hereon.

BY DIRECTION OF THE MILITARY GOVERNOR

MATT TAYLOR
Major, FA
Director
Bureau of Justice

792

HEADQUARTERS
UNITED STATES ARMY MILITARY
GOVERNMENT IN KOREA
Bureau of Justice
Seoul, Korea

BUREAU ORDER
NUMBER 12

28 March 1946

1. The name of the Seoul Branch Prison is hereby changed to Mapo Prison.

2. The following branch prisons are hereby abolished as branch prisons and established as independent prisons:

Prisons Abolished as Branch Prisons	Prisons Established as Independent Prisons	Location
Chunchon Branch Prisons of the Seoul Prison	Chunchon Prison	Kang Wondo
Chongju Branch Prison of the Taejon Prison	Chongju Prison	Chung Chong Pukto
Kunsan Branch Prison of the Chonju Prison	Kunsan Prison	Chol La Pukto
Andong Branch Prison of the Kumchon Prison	Andong Prison	Kyung Sang Pukto
Masan Branch Prison of the Pusan Prison	Masan Prison	Kyung Sang Wamdo
Chinju Branch Prison of the Pusan Prison	Chinju Prison	Kyung Sang Wamdo

3. This order shall be effective on the tenth day after the date appearing hereon.

BY DIRECTION OF THE MILITARY GOVERNOR:

MATT TAYLOR
Major   FA
Director of
Bureau of Justice

793

HEADQUARTERS
UNITED STATES ARMY MILITARY
GOVERNMENT IN KOREA
Department of Justice
Seoul, Korea

DEPARTMENT ORDER
NUMBER 2   14 September 1946

AMENDING PUBLIC NOTICES NUMBER 3

1. *Public Notices Number 3 Amended.* Public Notices Number 3, dated 15 June 1946, is hereby amended by deleting that part which is titled "Amended Ordinance of Civil Suits Costs Stamp", and such purported ordinance is hereby declared null and void.

2. *Effective Date.* This order shall be effective on the date appearing

BY DIRECTION OF THE MILITARY GOVERNOR

JOHN W CONNELLY JR
Major AUS
Director, Department of Justice

HEADQUARTERS
UNITED STATES ARMY MILITARY
GOVERNMENT IN KOREA
Department of Justice
Seoul, Korea

DEPARTMENT ORDER
NUMBER 3

29 March 1947

EXAMINATIONS FOR ADMISSION TO THE PRACTICE
OF LAW IN KOREA

SECTION I. *Purpose.* The purpose of this order is to establish temporary provisions for the examination of applicants for admission to the practice of law in Korea. The provisions contained herein are made pursuant to paragraph 2a of Bureau Order No. 4 of the Bureau of Justice, dated 19 November 1945, and shall not be deemed to revoke or impair the existing provisions of law regulating probationary periods and the examination of probationers.

SECTION II. *Law Examination Committee Established.*

a. A Law Examination Committee is hereby established to receive applications for admission to the practice of law in Korea, to examine candidates for such admission, to certify to the successful passing of the examination by such candidates, and to otherwise apply these provisions.

b. The Committee shall consist of the Chairman and not less than ten members, of which total number not less than one-third shall include general practitioners and teachers of law.

c. The Director of the Department of Justice shall be the Chairman of the Committee, *ex officio.* He shall appoint all other members of the Committee. The term of office of the members of the Committee shall be one year from the date of their appointments. No compensation shall be paid to any member of the Committee for services rendered as a member of the Committee.

d. Necessary clerical and administrative personnel shall be assigned from among the employees of the Department of Justice, by the Director of

— 1 —

軍政廳　官報　司法部令　第三號　　一九四七年三月二九日

795

the Department of Justice. No additional compensation shall be paid such personnel for services rendered in connection with such assignment.

e. The Chairman shall supervise the clerical and administrative functions of the Committee. In his absence, the senior member of the Committee shall act in his stead.

SECTION III. *Conduct of Examinations.* Examinations shall be prepared by the Committee, and shall be conducted in accordance with rules prepared by it. Examinations shall be held during the year 1947, in Seoul, Korea, at such particular times and places as shall be designated by the Committee. Public notification of the times and places of such examinations shall be published in the *Official Gazette,* sufficiently in advance of the dates thereof.

SECTION IV. *Applications. a.* An applicant for admission to the practice of law shall submit to the Committee an application in the form required by the Committee. The application shall show, among other matters, the record of his education, and shall be accompanied by a copy of his family register or other satisfactory proof of his age, and a recent photograph.

b. A fee of 100 yen shall be paid by the applicant, by the affixing of revenue stamps to the form of application. Such fees shall not be subject to refund.

SECTION V *Qualifications of Applicants.* The applicant must be of good moral character, and, at the time of examination, of the age of twenty two years or over. Upon proof of such qualifications, to the satisfaction of the Committee, the applicant shall submit to examination.

SECTION VI. *Form of Examination.*

a. Each examination shall consist of two related parts, to be designated as the preliminary examination and the main examination. Failure to pass the preliminary examination will bar entrance to the main examination.

b. A graduate of a university preparatory school, a college, or other school recognized by the Committee as of equivalent grade, shall not be subject to the preliminary examination. The purpose of the preliminary examination will be to determine whether or not the candidate is of sufficient scholastic and general abilities to qualify him to entrance to the main examination. The preliminary examination shall be in essay form. A person who has once successfully passed the preliminary examination shall not be again subject to such examination.

— 2 —

c. The purpose of the main examination will be to determine the knowledge and ability of the candidate, in law and in its practice. This examination shall be divided into two units, one of which is to be written and the other is be oral. The written unit shall embrace the following subjects: civil law, commercial law, criminal law; civil procedure, criminal procedure, conflict of laws, constitutional law; and economics. The oral unit shall consist of series of questions in the fields of civil law, commercial law, criminal law, civil procedure, and criminal procedure, and the candidate shall be examined in three of such fields. Failure to pass the written unit will bar entrance to examination in the related oral unit.

SECTION VII. *Successful Candidates.*

a. The Committee shall determine the passing grades of the parts of the examination.

b. A written document shall be issued by the Committee to each successful candidate, certifying to the successful completion of the examination.

c. The names of the successful candidates shall be published in the *Official Gazette.*

d. A successful candidate, to whom a certificate is issued as above provided, shall become eligible to serve as a *probationer* in accordance with the provisions of law pertaining thereto.

SECTION VIII. *Exemption.* An applicant who had successfully passed the written examination for admission to practice, given in August 1945, or the written examination for admission to the Law Refresher Course, given in 1946, shall not be subject to preliminary examination or to the written unit of the main examination.

SECTION IX. *Effective Date.* This order shall be effective on the date appearing hereon.

BY DIRECTION OF THE MILITARY GOVERNOR:

JOHN W CONNELLY JR
Major AUS
Adviser to Director
Department of Justice

KIM BYUNG NO
Director
Department of Justice

軍政廳　官報　司法部令　第三號　　一九四七年三月二九日

HEADQUARTERS
UNITED STATES ARMY MILITARY
GOVERNMENT IN KOREA
Department of Justice
Seoul, Korea

DEPARTMENT ORDER
NUMBER 4

31 March 1947

CONSTITUTING MAPO A SEPARATE PRISON*

1. *Separate Prison.* Mapo Prison, located in Mapo District, in the City of Seoul, and now a branch of Seoul Prison, is hereby constituted a separate prison.

2. *Effective Date.* This order shall be effective on the tenth day after the date appearing hereon.

BY DIRECTION OF THE MILITARY GOVERNOR

JOHN W CONNELLY JR
Major JAGD
Adviser to Director
Department of Justice

KIM BYUNG NO
Director
Department of Justice

軍政廳　官報　司法部令　第四號　一九四七年三月三十一日

*See Bureau Order No 12, Bureau of Justice, USAMGIK, dated 28 March 1947

SOUTH KOREAN INTERIM GOVERNMENT
Department of Justice
Seoul, Korea

DEPARTMENT ORDER

NUMBER 5   31 May 1947

REGISTRATION OF RELEASES OF MORTGAGES: INSTRUCTIONS TO JUDGES OF DISTRICT COURTS

1. *Purpose.* The purpose of this order is to provide for the registration of releases of mortgages held on property in which the government has a direct or indirect interest.

2. *Instructions.* All Judges of the District Courts will receive and register releases of mortgages, on property in which the government or any of its departments, agencies, elements or instrumentalities have a direct or indirect interest upon presentation by the Property Custodian or the appropriate Provincial Property Custodian of:

a. A deed of release, and

b. A request for cancellation of registration of mortgage.

Such documents shall be in the form set out in the Appendix hereto. They shall be duly completed and signed as provided in paragraph 3.

3. *Signature.* a. Where the mortgage to be released is owned by corporation or other juridical person organized or chartered in Korea in which the Property Custodian has an interest, the deed of release and the request for cancellation of registration of mortgage shall be signed by the appropriate officials of the corporation or other juridical person and countersigned by the Property Custodian or the appropriate Provincial Property Custodian.

b. Where such mortgage is held by the Property Custodian, he or the appropriate Provincial Property Custodian shall sign the deed of release and the request for cancellation of registration of mortgage.

c. Where such mortgage is held by any department, agency, instrumentality or element of the government other than the Property Custodian, the deed of release and the request for cancellation of registration of mortgage shall be signed by the head of such department, agency, instrumentality or

— 1 —

軍政廳　官報　司法部令　第五號　　一九四七年五月三十一日

799

element and countersigned by the Property Custodian or the appropriate Provincial Property Custodian.

d. Where so required by law, the deed of release and the request for cancellation of registration of mortgage shall be signed by such person, other than the Property Custodian or any department, agency, instrumentality or element of the government, as may have an interest in the mortgage.

4. *Applicability.* This order does not apply to pledges as described in Chapter IX, Section III, of the Civil Code.

5. *Effective Date.* This order shall be effective on the tenth day after the date appearing hereon.

BY DIRECTION OF THE MILITARY GOVERNOR:

KIM BYUNG NO

Director

Department of Justice

軍政廳　官報　司法部令　第五號　　一九四七年五月三十一日.

APPENDIX

Forms

DEED OF RELEASE

TO: ______________________________, Mortgagor

This letter certifies that the Mortgagee,___________________________
has received ¥___________________________as the interest and principal in full
satisfaction of the mortgage indicated as follows:

1. Date of Application for Registration of mortgage:

____

2. Receipt Number of mortgage at the Registration Office:

____

3. Indication of Immovable: (Brief description and location)

____

____

Date:_____________________________________

Mortgages:____

     Signed and sealed this the___________day of___________1947.

____

REQUEST FOR CANCELLATION OF REGISTRATION OF MORTGAGE

TO: The___________________________________District Court

Indication of Immovable: (Location and brief description)

Cause of Registration and its Date: (Deed of release dated___________.

Object of Registration: (Registration of release of mortgage)

Party Required to Release Mortgage: (Property Custodian or other creditor)

Party Entitled to Release of Mortgage: (Debtor)

Amount of Claim:_____________________________________

    It is requested that you register the above written matters.
    The deed of release is attached hereto.
    Signed and sealed this___________day of_____________, 1947.

SOUTH KOREAN INTERIM GOVERNMENT.
Department of Justice
Seoul, Korea

DEPARTMENT ORDER
NUMBER 6

31 July 1947

AMENDING DEPARTMENT ORDER NO 5; EXEMPTION OF BANKS AND FINANCIAL INSTITUTIONS FROM DEPARTMENT ORDER NO 5

1. Paragraph 4 of Department Order No 5, dated 31 May 1947, is amended to read as follows:

"4. *Applicability.* a. Department Order No 5 does not apply to pledges as described in Chapter IX, Section III, of the Civil Code.

b. Department Order No 5 does not apply to releases of mortgages by banks and financial institutions organized or chartered in Korea. Releases of mortgages by such banks and financial institutions and applications for registration of such releases do not require the signature of the Property Custodian and may be submitted directly by the bank or financial institution."

2. This order shall be effective as of June 10, 1947.

BY DIRECTION OF THE MILITARY GOVERNOR:

KIM BYONG RO
Director
Department of Justice

SOUTH KOREAN INTERIM GOVERNMENT
Department of Justice
Seoul, Korea

DEPARTMENT ORDER
NUMBER 7   12 February 1948

LEGAL APPRENTICESHIP REQUIREMENTS

SECTION I. *Purpose.* The purpose of this order is to provide for training of successful bar examination candidates and other qualified persons. as legal apprentices in practice of law and to provide for an examination necessary for admission to the bar.

SECTION II. *Training in practice of law.* a. A successful bar examination candidate or other person qualified to be a legal apprentice, desiring to be admitted to the bar, shall be subject to training in practice of law for a period of six full months at the office of a practicing lawyer.

b. Upon commencement of training in practice of law, a legal apprentice shall report, in writing to the Director of the Department of Justice, his commencement of training, together with a written consent of the lawyer from whom he will receive guidance.

c. A legal apprentice shall earnestly train himself in practice of law during his period of training by such means as observation of court proceedings, tentative preparation of documents for civil and criminal cases and other means necessary for practice of law.

d. The Korean Bar Association and its local branches shall have the duty to offer assistance and facilities in training of legal apprentices and to furnish assistance to them by available means. such as giving opportunity for collective observation of court proceedings.

e. A legal apprentice shall record in a diary in detail his training by his guiding lawyer. When he finishes his training he shall submit his diary together with an index to it to the Director of the Department of Justice.

f. A guiding lawyer shall directly submit to the Director of the Department of Justice a report of ratings of the legal apprentice, which shall indicate his diligence, behavior, performance and other available information.

SECTION III. *Examination of legal apprentices. a.* An examination shall

— 1 —

be held, upon application, for persons who are deemed to have sincerely finished training in practice of law.

b. A committee for examination of legal apprentices is hereby established in the Department of Justice. The committee shall consist of such members as the Director of the Department of Justice deems necessary. The Director of the Department of Justice shall be the chairman, ex officio. The members shall be named from time to time by the Director of the Department of Justice from among judges, prosecutors, bureau chiefs of the Department of Justice and practicing lawyers.

The committee shall receive applications for the examination, give the examination, certify the successful candidates, and otherwise enforce this ordinance.

The chairman shall supervise all the business of the committee.

In the case of the absence of the chairman, the senior member shall act as chairman.

c. The examination shall consist of a written and an oral examination in practical business of civil and criminal cases. The committee shall consider the moral character of the candidate, and shall have authority to investigate the background of the candidate. The committee shall choose the successful candidates upon the results of the examination, the ratings of training and upon consideration of moral character. Such decision shall be by a majority of the members.

d. A successful candidate in the examination shall be admitted to the practice of law.

e. Unsuccessful candidates may, on the expiration of three months from the date of the failed examination, apply to take another examination.

SECTION IV. *Effective date.* This order shall be effective on the date appearing hereon, and shall apply to those passing the bar examination of 1947, as well as those passing future examinations.

BY DIRECTION OF THE MILITARY GOVERNOR:

SEAL

KIM BYONG RO
Director
Department of Justice

軍政廳　　官報　　部令　　第七號　　　　一九四八年二月一二日

SOUTH KOREAN INTERIM GOVERNMENT
Department of Justice
Seoul, Korea

DEPARTMENT ORDER
NUMBER 8
1 April 1948

DEPARTMENT ORDER NUMBER 3 AMENDED

SECTION I. Department Order Number 3 dated 29 March 1947 (*Examinations For Admission to the Practice of Law in Korea*) is hereby amended as follows:

a. Section III is amended to read:

"*Conduct of Examinations.* Examinations shall be prepared by the Committee, and shall be conducted in accordance with rules prepared by it. *Examinations shall be held once each year in Seoul, Korea, at such particular times and places as shall be designated by the Committee.* Public notification of the times and places of such examinations shall be published in the Official Gazette, sufficiently in advance of the dates thereof."

b. Paragraph b of Section IV is amended to read:

"A fee of *500 won* shall be paid by the applicant, by the affixing of revenue stamps to the form of application. Such fees shall not be subject to refund."

c. Paragraph c of Section VI is amended by adding the following sentence at the end thereof:

"An applicant who has successfully passed the written examination, but failed the oral examination, is not required to take the written examination in the next ensuing year."

d. Section VIII is amended to read:

"*Exemption.* An applicant who had successfully passed the written examination for admission to practice, given in August, 1945, the written examination for admission to the Law Refresher Course, given in 1946, *or the written unit of the main examination, given in 1947, shall not be subject to the written unit of the main examination of 1948.*"

SECTION II. This order shall be effective on the date appearing hereon.

BY DIRECTION OF THE MILITARY GOVERNOR:

KIM BYONG RO
Director
Department of Justice

(S E A L)

BAR ADMISSION
ORDER NO. 2

28 November 1945

ADMISSIONS TO BAR OF KOREA

1. Within the analogy of the principle of reciprocity between states, the following officers of the United States Army Forces in Korea, who are qualified lawyers regularly admitted to practice before the courts of various State and Federal Courts of the United States of America, are hereby admitted to the Bar of Korea to practice law, to appear as attorneys at law before any court of Korea, and to hold any position in Korea eligible to be held by members of the Bar of Korea:

Lt. Colonel William G. Bray. 0-217086, INF
Lt. Colonel John R. Flynn, 0-207620, AC
Lt. Colonel Maurice Lutwack, 0-275727, CMP
Major Wayne T. Geissinger, 0-901182, AC
Major Walter B. Monagan, Jr., 0-344505, MI
Major Oswald H. Hilmore, 0-262617, CAC
Major Albert B. Mosebach,
Major Denny F. Scott, 0-305014, AC
Major William M. Sheehan, 0-903191, AC
Major Joseph E. Stermer, 0-274190, ORD DEPT
Major George E. Tobias, 0-317884, AC
Capt. Ralph Herrod, 0-432411, INF
Capt. John B. McKeel, 0-497586, AC
Capt. Arthur R. Murphy, 0-368643, QMC
Capt. Kenneth O'Leary, 0-276192, CMP
Capt. Bernard Peitzer, 0-499368, AGD
Capt. Hal J. Putman, 0-347910, MI
Capt. Leonard C. Rowe, 0-916043, AC
Capt. William L. Weiss, 0-391663, QMC
1st Lt. William S. Frates, 0-1058802, CAC
1st Lt. Tom I Gill, 0-1047133, CAC
1st Lt. Seymour Kaplan, 0-865217, AC

1st Lt. Donald MacDonald, 0-515964, IC
1st Lt. Bertram D. Sarafan, 0-1640912, SC
Lt. (JG) Karl F. Frisbie, 294317, D-V (S)

2. To meet the emergency scarcity of qualified lawyers in Korea caused by the restrictive and discriminatory administration of admissions to the Bar by the Japanese controlled Bureau of Justice, the following requirements having been met, the following persons are hereby admitted to the Bar of Korea to practice law, to appear as attorneys at law before any Court of Korea, and to hold any position in Korea eligible to be held by members of the Bar of Korea:

    a. <u>Requirements which have been met:</u>

      (1) Is of Korean nationality.

      (2) Has attained the age of twenty-four (24) years.

      (3) Has studied law and has ample education and experience for the practice of law.

      (4) Has appeared before a Board composed of membership selected by the Director of the Bureau of Justice and has been able to satisfy the Board, upon examination, of qualification both as to education and moral standards.

      (5) Is not prohibited from admission by any criminal bankruptcy, or insanity record.

    b. <u>Persons admitted:</u>

Eun Sung Yong, 102 Tai Wha St., Chonju
Yun Dong Sik, 17 An Kuk St., Seoul
Kim Yang, 530 Chun Yong St., Seoul
Kim Yong Dal, 121 Chung Ryang St., Seoul
Lee Chong Pak, 34 An Sung St., Seoul
Kim Eun Chai, 253 Chang Chum Lee, Sunchon
Kim Sang Keun, 224, 8 Don Am St., Seoul

BY DIRECTION OF THE MILITARY GOVERNOR

MATT TAYLOR
Major FA
Director
Bureau of Justice

UNITED STATES ARMY FORCES IN KOREA
Office of the Military Governor
Bureau of Justice
Seoul, Korea

BAR ADMISSION
ORDER NO. 3   4 December 1945

ADMISSIONS TO BAR OF KOREA

1. Par 1, Bar Admission Order No. 2, dtd 28 Nov 1945, is hereby amended to include the following additional qualified persons:

Major Matt Taylor, 0-274015, FA

Major Leal W. Reese, 0-117954, FA

Major John E. Deefe Jr, 0-252558, FA

Capt. Henry M. Leen, 0-1648359, SC

Capt. Patrick H. Walker, 0-260988, Inf

1st Lt Samuel A. Halpern, 0-1032713, CAV

1st Lt Ben Holstein, 0-1534759, MAC

Morton S. Jaffe, 0-1001813, AGD

Maury Maverick

Chairman, Small War Plants C

Washington, D. C., U. S. A.

BY DIRECTION OF THE MILITARY GOVERNOR

MATT TAYLOR
Major FA
Director

BAR ADMISSION
ORDER NO. 4 .   26 December 1945

ADMISSIONS TO BAR OF KOREA

1. The following named officers and enlisted men having been found
to be qualified lawyers regularly admitted to practice before the various State
and Federal Courts of the United States of America are hereby admitted,
under principles of reciprocity, to the Bar of Korea with all the privileges and
obligations attendant thereto :

  Lt Col George C Comstock, Jr. 0-276145, CAV
  Major Paul R Miller, 0-338840, AC
  Lt Commander H E Stevens, 166182, USNR
  1st Lt Gustus J Amrose, 0-1640461, SC
  1st Lt John B Edwards, 0-1104637, CE
  1st Lt Emanuel E Farber, 0-1584197, QMC
  1st Lt Sanford S Faunce, 0-1039208, TC
  1st Lt Donald D Moss, 0-1053743, CAC
  1st Lt Raymond L Perlman, 0-1036332, CWS
  1st Lt Harry G Travanti, 0-588676, AC
  1st Lt Arthur Weissman, 0-1544115, MAC
  T/4 John T Emerson, 36784195, SC
  Cpl. Walter W Hamill, 13137359, SC

2. The following named persons having been found qualified by previous
experience as judges in the Courts of Korea or as prosecutors in said courts
or having been previously engaged in the practice of law under the laws of
Korea then in effect, are hereby admitted to the Bar of Korea to practice law,
to appear as attorneys at law before any Court of Korea and to hold any
position in Korea which members of the Bar are eligible to occupy :

  Baak Uyn Wa, Seoul Mu Kyo Chon 13
  Kim Jyun Chong, Seoul Men Aun Cheon 162.1
  Iu Jin Yeong, Seoul Shing Sol Chon 175.15
  Han Chong Kun, Seoul Yon Sanku Sam Pan Thong 60.15
  Iu Hun Iyol, Seoul Ching Rim Chon 231
  Iyan Jin Ho, Seoul Seo Thae Mun Ku Chuk Chom Chon 20:
  Im Chol Ko, Seoul Sa Riong Chon 118

— 1 —

Kim Yun Su, Seoul Chong No Ku Uan Seo Cheon 1_ :
Kan Ko Book, Seoul Seo So Mun Cheon 33
Kan Jung In, Seoul Seo Thai Mun Ku Kyo Nam Cheon 111.3
Kim Oo Yeong, Seoul Miyong Chi Cheon 2,105
Kim Yaon Han, Choun Buok To Kun Sam Bu Chau Joun Cheon 57.7
Ree Byon Ring, Seoul Chong No Ku Chonjin Cheon 146
Sim San Bon, Seoul Ma Pc u Thae Huon Cheon 535
Lyan Tae Hiyang, Seoul Rai Pan Cheon 59
Kim Nam Ee, Chong No Ku Chun Hak Cheon 47
Chai Ta Won, Hwa Won Cheon 67
Han Kaak Man, Chuk Chom Cheon 3.3.102
Men Sun Keon, Sin Sol Cheon 425
Tin Kuk Nan, Ton Tai Mun Tong 3.101
Chai Chang Hong, Pil Un Cheon 174.1
Eaak Bong Je, Iyong Nak Cheon 2.86
Mun Hi Son, Su Cheon 17
Lim Han Kyong, Su Cheon 25.1
Chan Byol Chol, San Wang Shim Le Cheon 41.85
Bang Chai Ki, Sam Pan Tong 353.18
Chai Iyong Han, Won So Cheon 111.
Lee Mang Chung, Tong Tai Mun Ku Ton Am Cheon 53.3
Kim Ti Iyol, Tae Ku Tae Bong Cheon 163.10
Kim Hiyong Kun, Chong No Ku Hoa Cha Cheon 126
Oh Son Tok, Soh Tai Mun Ku Bok A Hjyong Chong 1.77
Kim Niyon Chai, Tong Tai Munku He Hwa Cheon 5.19
Chu Se, Iyong San Ku Kang Ki Cheon 98

3. The following named persons having been found qualified and who
have passed the prescribed Final Bar Examination are hereby admitted to the
Bar of Korea to practice law, to appear as attorneys at law before any Court
of Korea and to hold any position in Korea which members of the Bar are
eligible to occupy :

Lee Kyong Yong, Seoul Chom Chon 3.4
Kim Chun Jai, Seoul Tong Iin Chon 31.14
Kim Iyong Keyon, Seoul Chong No 4.186
Oh Kwan, Seoul Cho Um Chon 42
Iyun Chon Tai, Seoul Won Soh Chon 242.1

BY DIRECTION OF THE MILITARY GOVERNOR

MATT TAYLOR
Major FA
Director
Bureau of Justice

BAR ADMISSION
ORDER 5   10 January 1946

1. The following named officers having been found to be qualified lawyers regularly admitted to practice before the various State and Federal Courts of the United States of America are hereby admitted, under principles of reciprocity to the Bar of Korea with all the privileges and obligations attendant thereto:

Major-General Archer L Lerch, 0-6973, USA
Colonel Loren F Parmley, 0-9864, JAGD
Lt. Colonel Charles D Marsh, 0-324478, CAV
Lt. Comdr. Harold D Padgett, 59799, USNR
Lt. Comdr. Harry J Robinson, 159262 USNR
Capt. Murray M Halwer, 0-452362, CAC
Capt. Hershel L Seibelman, 0-1183514, FA
1st Lt. Norman H Abrahamson, 0-1894604, AUS
1st Lt. John R Crunelle, 0-1003704, AC
1st Lt. Harry E Dyer, Jr., 0-1080100, CAC
1st Lt. Charles R Gholston, 0-1797633, CMP
1st Lt. Samuel H Goldman, 0-1541468, MAC
1st Lt. Howard S Kennedy, 0-543635, TC
1st Lt. Murray Leicher, 0-156741, QMC
1st Lt. Julius L Malkin, 0-1544094, CMP
1st Lt. Edwin B Patterson, 0-1947797, TC
1st Lt. Martin Schlesinger, 0-1895990, AGD
1st Lt. Charles A Stuprich, 0-582976, AC
1st Lt. Simon Taub, 0-1798418, CMP
2nd Lt. Leonard M Bertsch, 0-1799818, CMP

FOR THE MILITARY GOVERNOR

MATT TAYLOR
Major FA
Director

BAR ADMISSION
ORDER No. 6

19 January 1946

1. The following named persons having been found to be qualified lawyers regularly admitted to practice before State and Federal Courts of the United States of America are hereby admitted, under principles of reciprocity, to the Bar of Korea, effective 14 January 1946, with all the privileges and obligations attendant thereto :

The Hon. Robert P Patterson,
The Secretary of War,
The United States of America.
Colonel Robert A. Ginsburgh.
Aide to the Secretary.
Capt. Barth P. Walker, 0-435109, AC
Capt. Robert B. Porter. 0-328336, CAC
Ist Lt Donald V. Bailey 0-1331362, Inf
Ist Lt Theodore C. Heyl, 0 585017, AC
Ist Lt Isadore I. Leiken, 0-1047635, CAC
Ist Lt Romolo F. Ottavi, 0-1302871, Inf
Ist Lt Martin J. Ross, 0-1588262, QMC

FOR THE MILITARY GOVERNOR

MATT TAYLOR
Major FA
Director
Bureau of Justice

HEADQUARTERS
UNITED STATES ARMY MILITARY
GOVERNMENT IN KOREA
Bureau of Justice
Seoul, Korea

BAR ADMISSION
ORDER NO. 7   24 January 1946

1. The following named persons having been found qualified and who have passed the prescribed Final Bar Examination are hereby admitted, effective 21 January 1946, to the Bar of Korea to practice law, to appear as attorneys at law before any Court of Korea and to hold any position in Korea which members of the Bar are eligible to occupy :

Oh Sok Pyo	Chul Ra Nam Do Kang Chu Pu Dai Chong Cheon 134.
Cho Keon Muk	Seoul Chong No Ku Tang Chu Cheon 14.1.
Chai Yun Tak	Seoul Chong No Ku Chung Hak Chion 47.
Chai Chon Mo	Seoul Chong No Ku Kei Ha Chon 22.28.
Han Mill	Seoul Chong No Ku Shin Kyo Chong 37.
Hong Nam Pyo	Seoul Chong No Ku Chong No 50. 227-7.

FOR THE MILITARY GOVERNOR

MATT TAYLOR
Major FA
Director
Bureau of Justice

813

BAR ADMISSION
ORDER NO. 8   16 February 1946

1.　The following named persons having been found to be qualified lawyers regularly admitted to practice before State and Federal Courts of the United States of America are hereby admitted, under principles of reciprocity, to the Bar of Korea, effective 3rd February 1946, with all the privileges and obligations attendant thereto:

Lt Colonel James C. Hamilton, 0-230242, Inf.

Major Robert E. Lee Counts, 0-382192, QMC.

Major Claude E. Broom, 0-529684, AUS.

Major Jay Kersbergen, 0-299123, FA.

Major James S. Killough, 0-298871, Inf.

Major Edward P. Machle, 0-163362, Inf.

Lt Cmdr. Draper W. Phillips, 183118, USNR.

Major Richard D. Reynolds, 0-439996, Ord.

Lt Cmdr. George A. Rupp, 242396, USNR.

Capt. Paul W. Bogikes, 0-496222, MAC.

Capt. Ralph L. Fusco, 0-1052349, GSC.

Capt. Charles A. Jens, 0-472494, CMP.

Capt. William J. Madden, 0-1795942, CMP.

Capt. John C. Martin, 0-427344, Ord.

Ist Lt. Philip J. Assiran, 0-1796018, CMP.

Ist Lt. Harry H. Dow, 0-179914, MI.

1st Lt. Stanley J. Jablonski, 0-1597687, QMC.

1st Lt. Edwin J. Kennedy, 0-1641460, SC.

1st Lt. Mack H. Austin, 0-575814, AC.

1st Lt. Malcolm B. Stark, 0-573466, AC.

2nd Lt. Sidney E. Kaner, 0-1560288, Ord.

2nd Lt. Dennis N. Key, 0-1799307, CMP.

2nd Lt. Richard C. Killin, 0-1598437, MI.

2nd Lt. Michael H. Lyons, 0-1799162, MI.

2nd Lt. Frank L. Manns, 0-2052345, JAGD.

WO. Ellis C. Duncan, W-2133232.

S/Sgt. John B. Coppinger, 36 690 697

Sgt. T. Ross Cissel Jr, 42 207 411.

Edward J. Sadek, Special Agent, Counter Intelligence Corps.

Ina D. Anderson, Assistant Field Director, ARC.

FOR THE MILITARY GOVERNOR:

MATT TAYLOR
Major FA
Director
Bureau of Justice

HEADQUARTERS
UNITED STATES ARMY MILITARY
GOVERNMENT IN KOREA
Bureau of Justice
Seoul, Korea

BAR ADMISSION
ORDER NO. 9

28 February 1946

1. The following named persons having been found qualified and who have passed the prescribed final Bar Examination are hereby admitted, effective 15 February 1946, to the Bar of Korea to practice law, to appear as attorneys-at-law before any Court of Korea and to hold any position in Korea which members of the Bar are eligible to occupy:

Name	Address
Chai Kyang Jin	Seoul Shin Kyong Cheong 17.12
Kim Iyeng Sang	Ka Bon Ko Lyo Cheong 183.21
Yang Tok Pyo	Seoul Chong No Ku Man Yung Cheong 2.166
Nam Ryung Su	Seoul Chong No Ku Sohoa Moon Cheong 1.191.9
Kang Kong Sun	Seoul Sang Yong Ku Shin Dang Cheong 251.66
Lee Cheng Su	Seoul Liyong San Ku Cheong Iyo Cheong 3.118.84
Kim Kwang Chui	Seoul San Lin Cheong 42
Kim Hong Shik	Seoul Song Tong Ku Shin Dang Cheong 340.68
Ok Seon Gin	Seoul Chong No Ku Ze Dong Cheong 6.2
Kim Yong Sik (Former judge, Seoul District Court).	Seoul Dai Hajung Samchungmok 27.

FOR THE MILITARY GOVERNOR

MATT TAYLOR
Major FA
Director
Bureau of Justice

BAR ADMISSION
ORDER NO. 10   16 March 1946

1. The following named persons having been found to be qualified lawyers regularly admitted to practice before the various State and Federal Courts of the United States of America are hereby admitted, under principles of reciprocity, to the Bar of Korea, effective 15 February 1946, with all the privileges and obligations attendant thereto:

Lt Colonel Teller Ammons, 0-167464, CMP

Lt Colonel Bernard A Brown, 0-256766, JAGD

Major Nelson Bohannan, 0-351166, TC

Major Joseph F Eichborn, 0-306811, INF

Capt. Della J Angst, L-306017, AC (WAC)

Capt. Walter M Bernhardt,   JAGD

Capt. Lawrence O Bilcovitch, 0-1100448, CE

Capt. Richard E Burton, 0-917294, AC

Capt. Colburn C Cherney, 0-1059119, QMC

Capt. Ira Kaye, 0-563478, AC

Capt. George F Meehan, Jr., 0-1796984, CMP

Capt. James O Monroe, Jr., 0-1846485, AC

Capt. Edward H Robinson, 0-414508, CAC

Capt. Myon M Ruby, 0-357101, INF

Capt. Robert A Snow, 0-265237, TC

1st Lt. Elics P Anderlin, 0-1796201, CMP

1st Lt. John M Bates, 0-1043014, CAC

1st Lt. Abraham H Belsky, 0-1104050, CE

1st Lt. George Cosson, Jr., 0-1796762, CMP

1st Lt. James F Dumas,   INF

1st Lt. Bernard J Freeman, 0-1797812, CMP

1st Lt. Richard S Hull, 0-1116671, CE

1st Lt. Ben F Mc Auley, 0-1281840, FD

1st Lt. Ira Monroe, 0-1796132, CMP

1st Lt. Anthony L Mueller, 0-327862, CE

1st Lt. Cregory A Mueller, 0-1290111, INF
1st Lt. Raymond D Munde, 0-557363, AC
1st Lt. Philip Plachinsky, 0-1308540, INF
1st Lt. Robert W Smart, 0-1178629, FA
1st Lt. Malcolm B Stark, 0-573466, AC
1st Lt. Robert B Stewart, 0-566777, AC
1st Lt. Joseph W Sullivan, 0-1895112, MI
1st Lt. Steven S Weisman,   -   AC
1st Lt. Jacob D Yaro, 0-1895902, TC
1st Lt. Dudley U Yoidicke,   CAC
2nd Lt. Venan J Alexsandroni, 0-1997023, CMP
2nd Lt. Oscar B Jones, 0-1331401, INF
2nd Lt. Frank L Manns, 0-2052545, JAGD
2nd Lt. Hiram L McDade, Jr., 0-1799064, CMP
2nd Lt. Melvin P Shapire, 0-1052884, CAC
T/Sgt. John H Marshall, 33964824
Sgt. Arthur Kulick, 32986654
Cpl. John F Mueller, 37717201
AGT. William H Kimbrough, Counter Intelligence Corps.

FOR THE MILITARY GOVERNOR

MATT TAYLOR
Major   FA
Director
Bureau of Justice

HEADQUARTERS
UNITED STATES ARMY MILITARY
GOVERNMENT IN KOREA
Bureau of Justice
Seoul, Korea

BAR ADMISSION
ORDER NO. 11   19 March 1946

1. The following named persons having been found qualified and who have passed the prescribed Final Bar Examination are hereby admitted, effective 5 March 1946, to the Bar of Korea to practice law, to appear as attorneys-at-law before any Court of Korea and to hold any position in Korea which members of the Bar are eligible to occupy:

Name	Address
Shin Tai Ik	Seoul, Iyong San Yeng St. 24.
Chai Ki Tai	Cholla Namdo Mokpo Dai Song St. 52.

BY DIRECTION OF THE MILITARY GOVERNOR:

MATT TAYLOR
Major   FA
Director
Bureau of Justice

819

HEADQUARTERS
UNITED STATES ARMY MILITARY
GOVERNMENT IN KOREA
Bureau of Justice
Seoul, Korea

BAR ADMISSION
ORDER NO 12

11 April 194

1. The following named persons having been found to be qualified lawyers regularly admitted to practice before the various State and Federal Courts of the United States of America are hereby admitted, under principles of reciprocity, to the Bar of Korea, effective 11 March 1945, with all the privileges and obligations attendant thereto:

Lt Colonel David M Bane, 0-359618, CAV.

Lt Colonel Victor H Kupferer, 0-291604, TC.

Lt Colonel Lee J Rutz, 0-277787, CAC.

Major William T Bennent, 0-218442, CMP.

Major James T Gaghan, 0-918739, AC.

Major Mitchell E Giblo, 0-913762, AC.

Major Leroy J Lillesand, 0-318283, Inf.

Major John W Prunty, 0-283101, Inf.

Capt. Thomas S Bown, 0-1796751, CMP.

Capt. Donald E Casper, 0-1797952, CMP.

Capt. John C Curd, 0-1040689, CAC.

Capt. Lawrence L Dail, 0-279086, CMP.

Capt. Charles S Delaney, 0-1176303, SC.

Capt. Paul Freeman, 0-1050260, CAC.

Capt. Robert H Gardner, 0-1113216, JAGD.

Capt. James A Harper, 0-1796273, CMP.

Capt. Jackson A Jordan, 0-1797301, CMP.

Capt. Jacob Kartman, 0-521388, AUS.

Capt. Edwin D Lasker, 0-1291656, Inf.

Capt. Thomas R Lea, 0-365930, CAC.

Capt. Gordon Lewis, 0-915195, AC.

Capt. Monroe F Marsh, 0-1280190, TC.

Capt. William A Messer, 0-1796129, CMP.

Capt. Luis A Rios, 0-215228, CMP.

Capt. Albert J Shults, 0-1044952, CAC.

Capt. Kenneth D Thomas, 0-1944829, TC.

1st Lt. Jean M Blum, 0-1057363, CAC.
1st Lt. Joseph F Ciccio, 0-1595678, AUS.
1st Lt. Jules L Druss, 0-1176316, FA.
1st Lt. John H Folks, 0-1172005, SC.
1st Lt. Julian A Gregory, 0-1051580, CAC,
1st Lt. George E Guinane, 0-1797120, CMP.
1st Lt. Harold L Martin, 0-1798691, CMP.
1st Lt. Seymour M Rowen, 0-1650563, SC.
1st Lt. Theodore D Schey, 0-1797539, CMP.
1st Lt. Howard M Schmidt, 0-1798067, CMP.
1st Lt. James P Simpson, 0-1046273, CAC.
1st Lt. Kenneth A Spitz, 0-1799076, CMP.
1st Lt. Carleton M Strouss, 0-152176, CAC.
1st Lt. Leo G Winzenburg, 0-1049401, CAC.
2nd Lt. Paul A Asch 0-1949348, TC.
2nd Lt. Harold A Manheim, 0-1949848, TC.
S/Sgt. Ernest F Ritter, 33 836 607.
Kurt M Falk (Civilian)
Ernst Fraenkel (Civilian)
John Kerr Rose (Civilian)

2. Bar Admission Order Number 2 dated 28 November 1945 is amended to show the following individual as being admitted to the Bar of Korea as of that date:

Major Oswald H Milmore, 0-262617, CAC.

FOR THE MILITARY GOVERNOR:

MATT TAYLOR
Major   FA
Director
Bureau of Justice

HEADQUARTERS
UNITED STATES ARMY MILITARY
GOVERNMENT IN KOREA
Department of Justice
Seoul, Korea

BAR ADMISSION
ORDER NO 13   20 May 1946

    1. The following named persons are hereby admitted to the Bar of Korea, effective 20 May 1946, with all the privileges and obligations attendant thereto:

    Lt. Colonel William A. Glass, 0-342729, Sig. C.

    Major Robert S. Gilliam, Jr., 0-314436, CAV.

    Captain H.O.H. Frelinghuysen, 0-565377, AC.

FOR THE MILITARY GOVERNOR:

RMERY J WOODALL
Lt Col AUS
Director
Department of Justice

BAR ADMISSION
ORDER NO 14   20 May 1946

1. The following named persons having been found to be qualified lawyers regularly admitted to practice before the various State and Federal Courts of the United States of America are hereby admitted, under principles of reciprocity, to the Bar of Korea, effective 20 May 1946, with all the privileges and obligations attendant thereto:

Major Gordon Flaherty, 0-1040711, CAC.

1st Lt. Charles J. Boettjer, 0-582358, AC.

1st Lt. Charlton E. Huntley, 0-2048343, CMP.

FOR THE MILITARY GOVERNOR:

EMERY J WOODALL
Lt Col AUS
Director
Department of Justice

SOUTH KOREAN INTERIM GOVERNMENT
Department of Justice
Seoul, Korea

BAR ADMISSION ORDER
NUMBER 17   25 May 1948

The following persons are hereby admitted, effective 18 March 1948, to the Bar of Korea, with all the privileges and obligations attendant thereon:

Dr. K. P. S. Menon

Dr. Marc Schreiber

The following person is hereby admitted, effective 30 April 1948, to the Bar of Korea, with all the privileges and obligations attendant thereon:

Custodio A. Villalva, Esq.

BY DIRECTION OF THE MILITARY GOVERNOR:

(S E A L)

KIM BYONG RO
Director
Department of Justice

SOUTH KOREAN INTERIM GOVERNMENT
Department of Justice
Seoul, Korea

BAR ADMISSION ORDER
NUMBER 18   25 May 1948

1. The following persons, qualified lawyers regularly admitted to practice before the various State and Federal Courts of the United States of America, under the principle of reciprocity, are hereby admitted, effective 21 October 1947, to the Bar of Korea, with all the privileges and obligations attendant thereon:

> Forbes E. Jordan, Esq.
> Paul J. Lacrosse, Esq.
> Wanda M. Lewandaski, Esq.

2. The following persons, qualified lawyers regularly admitted to practice before the various State and Federal Courts of the United States of America, under the principle of reciprocity, are hereby admitted, effective 60 April 1948, to the Bar of Korea, with all the privileges and obligations attendant thereon:

> Viva L. Anderson, Esq.
> Captain Casper V. Beimfohr
> Morris Berzon, Esq.
> Samuel S. Browning, Esq.
> Peter J. Carroll, Esq.
> Peter L. DeVita, Esq.
> Leonard M. Gardner, Esq.
> George Gurow, Esq.
> Hugo E. Hanser, Esq.
> Joseph W. Hardy, Esq.
> Marie Cavanaugh Helm, Esq.
> Bernard G. Lenske, Esq.
> Lt. Col. Marvin W. Ludington
> Alex E. MacDonald, Esq.
> Lloyd W. Maxwell, Esq.
> Cyril E. Morrison, Esq.

軍政廳　官報　法令　司法部指令　第一八號　一九四八年五月二五日

Major Robert P. Pike,
Henry N. Shabsin, Esq.
Fred I. Simon, Esq.
Major William D. Sommers, Jr.
Major Thomas H. Swan

BY DIRECTION OF THE MILITARY GOVERNOR:

KIM BYUNG RO
Director
Department of Justice

(S E A L)

INSTRUCTIONS TO JUDGES
NUMBER 1   19 November 1945

1. Special Duties. — Normal duties of all courts are augmented by special duties in connection with Military Occupation cases. All Justices of the Supreme Court and Judges of the Courts of Review, District Courts and Branches of District Courts will observe the following regulations.

2. Assignment and Supervision. — Consistent with their regular duties, Justices of the Supreme Court and other judges may be assigned special duties by the Director of the Bureau of Justice, through the Chief Justice of the Supreme Court. The Chief Justice will have general supervision and control of the work of all judges of all courts, reporting as required to the Director.

3. Assignment of Judges of Courts of Reveiw. — Judges of the Courts of Review will hear such cases in any court as are assigned to them by the Chief Judge of the Court of Review under direction of the Chief Justice of the Supreme Court. Assignment of such judges to District Court cases will generally be limited to serious crimes or difficult civil matters.

4. Assignment of Judges of District Courts. — Judges of the District Courts (including Branches, and the Juvenile and Special Property Courts) will be assigned by the Chief Judge of the District Court under direction of the Chief Justice of the Supreme Court. Any judge may be assigned to any case, including Special Property Court cases.

5. District Court Circuits: Special Prosecutors. -- District Court hearings will be held at the regular court buildings and at such other places within the District as may be required to expedite disposition of cases. To avoid repeated necessity of travel of over twenty miles for parties and witnesses, the Chief Judge of the District Court will arrange a regular circuit and set days for court hearings in the principal cities and towns of the District. Such arrangements will be reported to the Chief Justice of the Supreme Court for approval by the Director of the Bureau of Justice. The Chief Judge of any District Court may appoint one or more Special Prosecutors for the trial of criminal cases in emergencies, reporting such action immediately by

— 1 —

827

messenger cr registered mail to the Chief Justice of the Supreme Court, with reasons therefor.

6. Working Hours: Attendance Reports and Records. — The working hours of all judges, prosecutors and other court officials and personnel will be those prescribed for officers and employees of the Military Government at the National Capitol Buiding, Seoul, Korea.* All Korean and American National holidays will be observed. Sunday work will not be required except in emergencies. Attendance records will be kept for each official and employee of each court, on the form shown in Annex Number 1, attached. During the first week of each month a copy of the complete record for the preceding month will be sent to the Chief of the Administrative Department of the Bureau of Justice, with appropriate comments by the Chief Judge of the Court noted thereon for any absences reported. Promptness and regularity of attendance are required. Absence without valid reason approved by the Chief Judge will be cause for discharge. Judges, officials, and employees leaving their offices will sign in and out in a register, informing their superior officer of the purpose, destination, and duration of each departure.

7. Care of Buildings and Property. — The adequate maintenance and security of court records, buildings and property are the responsibility of the Chief Judge or other judge in charge of the court. He will make all necessary arrangements for hiring of building guards, and obtaining assistance of local police to prevent unlawful entry into court buildings and loss, damage or theft of records or property.

8. Court Reports. — On Saturday of each week, the Chief Judge of each Court of Review and District Court will forward one copy of every decision rendered by each judge during the preceding week, by messenger or by registered mail to the National Law Library, Court Building Annex #3, 38 Little West Gate Street, Seoul, Korea. Selected decisions on important matters of law will thereafter be printed in Court Reports distributed periodically to the courts in bound volumes for reference purposes. Court Reports will include memorandum decisions in less important cases and statistics of the work of the courts. Supreme Court decisions will be printed in Supreme Court Reports.

* 0800 to 1600 Mon–Fri, 1 May to 20 July
0800 to 1200 Sat, 1 May to 20 July
0800 to 1200 Mon–Sat, 21 July to 31 Aug
0830 to 1700 Mon–Sat, 1 Sept to 30 Apr

— 2 —

9.   Organization of Bureau of Justice and Courts. — All judges, prosecutors, and clerks of all courts will familiarize themselves with the present organization of the Bureau of Justice and of the cours. The Chief Judges will call this subject to the attention of all lawyers in their District by all available means. Copies of the chart of organization of the Bureau of Justice will be mailed by the Administrative Department of the Bureau of Justice upon written request.

10.   Effective Date. — This order is effective at midnight 19 November 1945.

BY DIRECTION OF THE MILITARY GOVERNOR

EMERY J. WOODALL
Major AUS
Director of the Bureau of Justice

ANNEX NUMBER I: ATTENDANCE RECORD

OFFICE _______________________

MONTH OF _____________ YR _____________

NAME	1	2	3	4	5	6	7	8	9	10	11	12	13	14	15	16	17	18	19	20	21	22	23	24	25	26	27	28	29	30	31
	*																														

* Place appropriate number or numbers in Box

1. Full Day Present
2. Half Day Absent
3. Full Day Absent
4. Sick
5. Government Trip

HEADQUARTERS
UNITED STATES ARMY MILITARY
GOVERNMENT IN KOREA
Department of Justice
Seoul, Korea

INSTRUCTIONS TO JUDGES
NUMBER 2
                                                    23 October 1946

VIOLATIONS OF PRICE REGULATIONS AND FOOD REGULATIONS

1. All cases involving alleged violations of price regulations or food regulations, where the accused is under arrest or imprisonment, shall be brought to trial and disposed of by judgment not later than 48 hours after the arrest of the accused, unless such accused requests additional time to prepare his defense, in which case adequate time may be granted in the discretion of the court. As used herein, *arrest* includes *detention.*

2. All cases involving alleged violations of price regulations or food regulations, where the accused is not under arrest or imprisonment, shall be brought to trial and disposed of by judgment not later than five days after the accused has been charged, unless such accused requests additional time to prepare his defense, in which case adequate time may be granted in the discretion of the court. As used herein, *arrest* includes *detention.*

3. In all cases involving alleged violations of price regulations or food regulations, where the accused desires to plead guilty and submit to the judgment of the court, he shall be given an opportunity to do so forthwith upon being charged, and shall not at any time while a court of competent jurisdiction is in session, be subjected to imprisonment or detention. He may, however, be continued in custody until judgment is pronounced and executed, as limited by Paragraph *1* hereof.

4. All cases involving alleged violations of price regulations or food regulations shall be given a trial preference on the trial date thereof, and shall be placed at the head of the trial calendar of each court.

5. These instructions shall also apply to Special Judicial Officers.

BY DIRECTION OF THE MILITARY GOVERNOR:

JOHN W CONNELLY JR   KIM BYONG NO

Major AUS   Director

Adviser to Director   Department of Justice

Department of Justice   (S E A L)

INSTRUCTIONS TO PROSECUTORS
NUMBER 1   18 October 1945

1. The Judicial power of the Police Chief was repealed on 9 October 1945 by Section 1 g of Ordinance Number 11, dated 9 Oct 1945 attached as Annex Number I, and the Judicial Authority for a Summary Court is abolished from and after that date. In order to have uniformity and early action by the police in the collection of information at the time of arrest the Bureau of Justice instructs as follows:

    a. When a person is apprehended by the police, they will turn over to a Prosecutor as soon as possible (in a simple case not longer than one day's delay) a report which contains the following facts:

        (1) The permanent address.
        (2) The present address.
        (3) The date of birth.
        (4) Occupation.
        (5) Former convictions.
        (6) Facts of the crime.
        (7) Remarks and opinions of the policeman.
        (8) The name of the chief of police.
        (9) The date of the report.

    b. A sample of a uniform report is attached as Annex Number II.

BY DIRECTION OF THE MILITARY GOVERNOR:

EMERY J WOODALL,
Major AUS.,
Director of the Bureau of Justice

Annex II

REPORT OF ARREST

Name

Permanent Address

Present Address

Date of Birth

Occupation

Former Convictions

Statement of Facts of the Crime

Remarks of Arresting Officer

Name of Police Chief

Date of Arrest

____
Arresting Officer

833

HEADQUARTERS
UNITED STATES ARMY FORCES IN KOREA
Office of the Military Governor
Bureau of Justice
Seoul, Korea

INSTRUCTIONS TO PROSECUTORS
NUMBER   2

19 November 1945

1. SPECIAL DUTIES : Normal duties of all prosecutors are augmented by special duties in connection with Military occupation cases. All prosecutors of all courts will observe the following regulations.

2. ASSIGNMENT AND SUPERVISION : Consistent with their regular duties, prosecutors may be assigned special duties by the Director of the Bureau of Justice, through the Chief prosecutor of the Supreme Court, who will have general supervision and control of the work of all prosecutors of all courts. All other Prosecutors of the Suprene Court will assist the Chief Prosecutor in such supervision. Two or more prosecutors may be assigned to the same case.

3. ASSIGNMENT OF PROSECUTORS OF COURTS OF REVIEW : Prosecutors of the Courts of Review may be assigned tony case in any court by the Chief prosecutor of the Court of Review. under direction of the Chief Prosecutor of the Supreme Court. Assignments to District Court cases will generally be limited to serious or difficult matters, and will not ordinarily be made where they interfere with the primary duties of the prosecutors.

4. ASSIGNMENT OF PROSECUTORS OF DISTRICT COURTS : Prosecutors of District Courts (including Branches) will be assigned by the Chief Prosecutor of the District Court, under direction of the chief prosecutor of the Supreme Court. Any prosecutor may be assigned to any criminal case, including cases being prepared by the Staff of the Special Criminal Investigating Committes of the Bureau of Justice. Cases will be presented at such places within the District as the Judges of the District Court determine.

5. DUTIES OF CHIEF PROSECUTORS : Chief Prosecutors will continue to supervise prosecutors in their courts, reporting as required to the Chief Prosecutor of the Supreme Court, who is responsible to the Chief of the Criminal Department of the Bureau of Justice for the efficient performance of duty by all prosecutors.

834

6. EFFECTIVE DATE : This order is effective at midnight 19 November 1945.

BY DIRECTION OF THE MILITARY GOVERNOR

EMERY J. WOODALL
Major   A.U.S,
Director of Bureau of Justice

INSTRUCTIONS TO PROSECUTORS
NUMBER 3
29 December 1945

1. The primary function of all prosecutors is the successful prosecution of cases before a Court of competent jurisdiction. The details of investigations are a burden, which Prosecutors should not be required to assume. The specialized training of prosecutors is vastly more important when directed to the legal aspects of their office.

2. In the interests of time saving and better efficiency, the prosecutors of all courts shall observe the following instructions:

    a. Concentrate on a program designed to expeditiously dispose of those cases entered for trial on Court Calendars.

    b. Request routine investigations be conducted by the Police Bureau. This is a function of the Police, not the Prosecutor's.

    c. Analyze the Police reports for legal sufficiency.

    d. Point out to police any deficiencies of evidence and request they be corrected, if possible.

    e. Engage, if necessary, only in that part of an investigation that actually requires legal analysis.

    f. Maintain liaison with Police chiefs on all matters incident to investigations. Especially, in regard to matters related to the evidence necessary to satisfactorily prosecute in a court of law. Prosecutors and police must work together to raise the standards of law enforcement in liberated Korea.

    g. It is the duty of chief prosecutors to assist in the administration of these instructions.

3. These instructions are not intended to modify, change or reduce the lawful authority, prerogatives, prestige or station of any prosecutor. They are intended to clarify the respective responsibilities of the Prosecutors and the Police. As a matter of policy for the Bureau of Justice, full compliance

is hereby directed. .

4. The aforementioned, being a matter of concern to both the Bureau of Justice and the Police Bureau, same was approved and coordinated by all parties concerned. Bureau of Police Memorandum No 1, Relations with Prosecutors, will be forwarded to prosecutors when it is made available to this Bureau.

BY DIRECTION OF THE MILITARY GOVERNOR:

MATT TAYLOR
Major FA
Director

INSTRUCTIONS TO PROSECUTORS
NUMBER 4

13 March 1946

1.   Instructions are hereby issued in seven (7) types of cases as follows:

a.   Criminal Offenses committed by members of the Armed Forces of the United States in contravention of the Military or Naval law of the United States or the criminal laws in force in Korea.

b.   Criminal Offenses committed by persons serving with the Occupying Forces and subject to the Military or Naval law of the United States in contravention of such laws or the criminal laws in force in Korea.

c.   Criminal Offenses committed by personnel of Nations Allied with the United States in contravention of the criminal laws in force in Korea.

d.   Criminal Offenses committed by persons in contravention of the criminal laws in force in Korea where such persons claim Diplomatic Immunity.

e.   Criminal Offenses committed by civilian personnel of the Axis Nations in contravention of the criminal laws in force in Korea.

f.   Criminal Offenses committed by any persons in contravention of the criminal laws in force in Korea where the prestige or security of the Armed Forces of the United States or its Allies is involved.

g.   Criminal Offenses committed by Koreans in contravention of the criminal laws in force in Korea.

2.   Prosecutors confronted with cases enumerated above will take action as follows :

a.   All cases coming under 1a, 1b, and 1d will be reported forthwith to the Director of the Bureau of Justice accompanied by all information available including the names, adresses and nationalities of the offenders, a list of the witnesses, a statement of the charges placed in each case and a summary of the facts therein.

b.   All Cases coming under 1c, 1e and 1f will be reported forthwith to the Provost Court of the area accompanied by all information available.

Such information will include the names, addresses and nationalities of the offenders, a list of the witnesses, a statement of the charges placed in each case and a summary of the facts therein.

c. In general the cases which will be accepted by the Provost Courts will be limited to those involving personnel of the Armed Forces of the United States as witnesses and those in which the property of the United States is involved. In some instances the Provost Courts may take jurisdiction in other types of cases, particularly those where the prestige or security of the Armed Forces of the United States is concerned. In the event that there is any doubt with regard to the proper category of a case Prosecutors are instructed to consult the Provost Courts.

d. Cases coming within 1g will be prosecuted by the Prosecutors Office under all established law including the Proclamations of the Commander in Chief, U. S. Army Forces, Pacific pertaining to Korea and all Ordinances, Orders and Notices of the Military Government of Korea.

3. The Chief Prosecutor of every Korean Court is hereby instructed to establish and maintain contact with the Provost Court of the area.

4. It is the stated policy of the Military Government of Korea to place the greatest amount of responsibility possible with Korean officials at the earliest practicable time. Accordingly Prosecutors are instructed to study the Proclamations of the Commander in Chief, U. S. Army Forces, Pacific pertaining to Korea and all Ordinances, Orders and Notices of the Military Government of Korea.

5. The Courts of Korea are a national institution supervised by and directed by the Director of the Bureau of Justice. Personnel of the Armed Forces of the United States, outside the Bureau of Justice may on occasion advise the personnel or observe the activities of the Courts but in all cases the responsibility for decisions pertaining to the operations of the courts is retained by the Director of the Bureau of Justice.

BY DIRECTION OF THE MILITARY GOVERNOR:

MATT TAYLOR
Major   FA
Director of Bureau of Justice

HEADQUARTERS
UNITED STATES ARMY MILITARY
GOVERNMENT IN KOREA
Bureau of Justice
Seoul, Korea

INSTRUCTIONS TO PROSECUTORS
NUMBER 5

11 February 1946

PROSECUTION OF PUBLIC OFFICIALS

In order to promote efficient functioning of the Government, and to avoid removals of public officials without justification, the following will be observed by all prosecutors:

1. Whenever, in the opinion of the Chief Prosecutor of any Court, it becomes necessary to prefer charges, arrest or otherwise interfere in the affairs of any public official the Chief Prosecutor will consult and advise with the superior of the official concerned before any action is taken.

2. If in any case a Chief Prosecutor believes that the procedure above outlined will prevent a proper administration of justice he will so report to the Director of the Bureau of Justice and await instructions from him.

MATT TAYLOR
Major   FA
Director
Bureau of Justice

HEADQUARTERS
UNITED STATES ARMY MILITARY
GOVERNMENT IN KOREA
Department of Justice
Seoul, Korea

INSTRUCTIONS TO PROSECUTORS
NUMBER 6   17 June 1946

PROSECUTIONS UNDER
ORDINANCE NUMBER 72

1. No prosecutions shall be undertaken or carried forward under Ordinance Number 72, dated 4 May 1946, without prior submission to and written approval of the Director of the Department of Justice.

2. This order is not intended to and shall have no effect whatsoever upon prosecutions for violations of Proclamation Number 2, GHQ USAFPAC, dated 7 September 1945, or for violations of any ordinances (other than Ordinance Number 72), laws or measures having the effect of law, whether such offenses are or are not enumerated in Ordinance Number 72.

3. This order shall be effective on the date appearing hereon.

BY DIRECTION OF THE MILITARY GOVERNOR:

JOHN W CONNELLY JR
Major AUS
Acting Director
Department of Justice

HEADQUARTERS
UNITED STATES ARMY MILITARY
GOVERNMENT IN KOREA
Department of Justice
Seoul, Korea

INSTRUCTIONS TO PROSECUTORS
NUMBER 7

23 October 1946

VIOLATIONS OF PRICE REGULATIONS
AND FOOD REGULATIONS

1. All cases involving alleged violations of price regulations or food regulations, where the accused is under arrest or imprisonment, shall be brought to trial and disposed of by judgment not later than 48 hours after the arrest of the accused, unless such accused requests additional time to prepare his defense, in which case adequate time may be granted in the discretion of the court. As used herein, *arrest* includes *detention*.

2. All cases involving alleged violations of price regulations or food regulations, where the accused is not under arrest or imprisonment, shall be brought to trial and disposed of by judgment not later than five days after the accused has been charged, unless such accused requests additional time to prepare his defense, in which case adequate time may be granted in the discretion of the court. As used herein, *arrest* include *detention*.

3. In all cases involving alleged violations of price regulations or food regulations, where the accused desires to plead guilty and submit to the judgment of the court, he shall be given an opportunity to do so forthwith upon being charged, and shall not at any time while a court of competent jurisdiction is in session, be subjected to imprisonment or detention. He may, however, be continued in custody until judgment is pronounced and executed, as limited by Paragraph *1* hereof.

4. All cases involving alleged violations of price regulations or food regulations shall be given a trial preference on the trial date thereof, and shall be placed at the head of the trial calendar of each court.

5. These instructions shall also apply to Special Judicial Officers.

BY DIRECTION OF THE MILITARY GOVERNOR:

JOHN W CONNELLY JR
Major   AUS
Adviser to Director
Department of Justice

KIM BYONG NO
Director
Department of Justice

SOUTH KOREAN INTERIM GOVERNMENT
Department of Justice
Seoul, Korea

INSTRUCTIONS TO PROSECUTORS
NUMBER 8

6 October 1947

INVESTIGATIONS, ARRESTS AND PROSECUTIONS INVOLVING BUSINESSES CONTROLLED PURSUANT TO ORDINANCE NO. 33: INSTRUCTIONS TO PROSECUTORS

1. Your attention is called to Executive Order No. 6 dated 6 October 1947 (*Investigations, Arrests and Prosecutions Involving Businesses Controlled Pursuant to Ordinance No. 33*). This order requires that police officials, prosecutors or any other law enforcement officers of the South Korean Interim Government shall, before investigating the affairs or books of any business controlled pursuant to Ordinance No. 33 dated 6 December 1945 (*Vesting Title to Japanese Property Within Korea*) or arresting or prosecuting any manager or assistant manager of such business, report to the official charged with immediate supervision of the management of such business.

2. The prosecutor is authorized to proceed with such investigation, arrest or prosecution in accordance with Executive Order No. 6 regardless of any objections by the official notified. The law enforcement officer shall, however, note any objections made by such notified official and transmit them to the Director of the Department of Justice.

3. To the extent that any previous instructions or orders are inconsistent with this order they are revoked.

4. This order shall be effective on the date appearing hereon.

BY DIRECTION OF THE MILITARY GOVERNOR:

KIM BYONG RO
Director
Department of Justice

— 1 —

軍政廳　官報　司法部訓令　第八號　一九四七年十月六日

HEADQUARTERS
UNITED STATES ARMY MILITARY
GOVERNMENT IN KOREA
Bureau of Public Health and Welfare
Seoul, Korea

BUREAU ORDER
NUMBER 1

23 April 1946

MAXIMUM SALES PRICES ON DRUGS, MEDICINES, PHARMACEUTICALS, AND RELATED ARTICLES

1. Pursuant to the provisions of Ordinance Number 62, dated 29 March 1946, prices shown on the attached list are hereby fixed as the maximum permissible sales prices for the drugs, medicines. pharmaceuticals, derivatives, compounds, substances, and other items of commerce used in the practice of medicine or public health activities, enumerated therein.

2. No drugs or other articles enumerated in the attached list may be sold, offered or exhibited for sale, or purchased or offered to be purchased, at prices in excess of those stated therein.

3. Any person violating the provisions of this order may be charged under Section III of Ordinance Number 19, dated 30 October 1945, and other applicable ordinances, proclamations, orders, and regulatory measures

4. This order is effective at 2400, 23 April 1946.

BY DIRECTION OF THE MILITARY GOVERNOR:

WILLIAM R. WILLARD
Major   USPHS
Acting Director

HEADQUARTERS
UNITED STATES ARMY MILITARY
GOVERNMENT IN KOREA
Department of Public Health and Welfare
Seoul, Korea

DEPARTMENT ORDER
NUMBER 2   25 August 1947

FOREIGN QUARANTINE REGULATION
SECTION I. GENERAL PROVISIONS

1. *Purpose.* The purpose of this regulation is to prevent the introduction of communicable diseases into Korea from foreign countries. For this purpose, quarantine procedures and requirements at ports of entry to, and departure from, Korea, for surface vessels and aircraft, are hereby established. *Tentative Port Quarantine Regulations* issued by the Department of Public Health and Welfare on 10 July 1946, are hereby revoked. All other regulatory measures and parts thereof which are inconsistent herewith or in conflict with the provisions hereof are, to the extent they are inconsistent or in conflict herewith, hereby revoked.

2. *Jurisdiction.* Any vessel or aircraft, military, naval or civil, together with its passengers, crew, and cargo, entering Korea from a foreign port, or leaving Korea destined for a foreign port, or which, during its current voyage or flight, has touched at any foreign port, shall be subject to this regulation. Entry and departure of subject vessels and aircraft shall be through designated seaports and airports only. INCHON, MOKPO, PUSAN and KUNSAN are hereby designated as such seaports. KIMPO is hereby designated as such an airport. Changes of, and additions to, such designations shall be made as authorized by the Military Governor.

3. *Administration.* The authority in charge of a designated port of entry and departure shall be responsible for the enforcement of this regulation. He shall be assisted by representatives of the Director of the Department of Public Health and Welfare. To such representatives, the

— 1 —

authority in charge of the port shall provide facilities for housing, transportation, and supplies necessary in the performance of their duties. The Director of the Department of Public Health and Welfare shall be responsible for the technical supervision of port quarantine activities and shall appoint as his representatives port quarantine officers and other technical personnel necessary to exercise such supervision. Direct communication between port quarantine officers and the Director of the Department of Public Health and Welfare is authorized.

4. *Definitions.* For the purpose of this regulation, the following terms are defined:

a. *Communicable Disease.* Any disease, the etiologic agent of which may pass or be carried, directly or indirectly, from one person to another.

b. *Contact.* Any person known to have been in such association with an infected person, animal, or vector, as to have been presumably exposed to infection

c. *Contamination.* Matter of undesirable substance, or material which may contain pathogenic micro-organisms, or the presence in an article of such matter.

d. *Disinfection.* The act of rendering anything free from the causal agents of disease.

e. *Disinfestation.* The act of destroying the vectors of a communicable disease.

f. *Fumigation.* The process by which the destruction of vermin and rodents is accomplished by the employment of gaseous agents.

g. *Immunity.* The condition of being protected against a particular disease either as a result of artificial immunization or through a previous attack of such disease.

h. *Incubation Period.* The period between the implanting of disease organisms in a susceptible person and the appearance of clinical manifestations of the disease.

i. *Infected Vessel or Aircraft.* A vessel or aircraft upon which a case of quarantinable disease exists or develops among persons or rodents aboard,

— 2 —

or upon which infected vectors of a quarantinable disease are found after embarkation.

j. Infestation. The condition of harboring insects or rodents capable of transmitting disease.

k. Isolation. The separation of human beings or animals from other human beings, animals, or vectors of disease, in such manner as to prevent the transmission of the disease.

l. Observation. The detention under medical supervision of a person in such place and for such period of time as may be specified in this regulation.

m. Pratique. A certificate issued by a quarantine officer releasing, or provisionally releasing, a vessel or aircraft from quarantine.

n. Quarantine. The detention of a person, vessel, aircraft or other conveyance, animal, or thing, in such place and for such period of time as may be specified in this regulation.

o. Quarantine Officer. A medical officer or other specially trained employee assigned to quarantine duty by the Director of the Department of Public Health and Welfare.

p. Quarantinable Diseases. The specific communicable diseases: Cholera, Plague, Smallpox, Louse-borne Typhus, and Yellow Fever.

q. Rodents. Gnawing mammals concerned in the transmission of quarantinable diseases.

r. Sanitary Log. A record of events and conditions of sanitary significance of the vessel.

s. Surveillance. The temporary supervision of a person who has been released from quarantine upon the condition that he will submit himself to further medical examination or inquiry.

t. Suspected Vessel or Aircraft. A vessel or aircraft arriving from a port infected or suspected of being infected with a quarantinable disease.

u. Typhus. Louse-borne Typhus.

v. Vector. An insect, animal, plant, or thing which conveys pathogenic organisms from a person or animal to another person or animal.

    w. Vermin. A species of insect capable of being a vector in the transmission of disease.

    5. *Incubation Period.* For purposes of this regulation, the incubation periods of quarantinable diseases shall be deemed to be:

Cholera	6 days
Plague	6 days
Smallpox	14 days
Typhus	12 days
Yellow Fever	6 days

SECTION II. INCOMING TRAFFIC.

    6. *Vaccination Requirements.* All persons arriving in Korea shall have in their possession certificates showing that the following vaccinations have been received:

Cholera, within the preceding four months;
Typhus, within the preceding two months;
Typhoid-paratyphoid, within the preceding twelve months;
Smallpox, within the preceding six months, except as to persons from an endemic smallpox area in which case vaccination is required within the preceding two months.

Persons arriving without the required certificates shall be given the necessary vaccinations, and placed under observation or surveillance for a sufficient period to insure their freedom from these diseases.

    7. *Measures at Foreign Ports.* Persons, vessels, and aircraft, arriving in Korea shall, prior to departure from foreign ports, have fulfilled the following requirements as a condition of entry into Korea:

    a. Persons.

      (1) Passengers and crew shall comply with the vaccination requirements of paragraph 6 above.

      (2) No person known or believed to be infected with a quarantinable disease or infested with the vector of a quarantinable disease shall be allowed on board.

— 4 —

(3) No person in recent contact with a quarantinable disease shall be allowed on board, unless known to be adequately protected by vaccination or by a previous attack of such disease.

(4) No person known or believed to be infected with a serious communicable disease shall be embarked, unless attended by a physician or-trained nurse or orderly, and then only if adequate arrangements for isolation are made available.

(5) Bodies may be embarked only if properly embalmed and contained within a hermetically-sealed metal container,

b. Vessels and Aircraft.

(1) Only water and food known to be safe for intended use shall be embarked.

(2) Sanitary conditions shall be good at time of departure.

(3) Cargo known or believed to be infected with the agent of, or infested with the vector of, a quarantinable disease shall not be embarked.

(4) Animals shall not be embarked for transport to Korea except in accordance with applicable veterinary quarantine regulations of the Government of Korea.

c. Vessels Only-Rat Control

(1) Immediately upon docking and during the entire time a vessel lies at a wharf, it shall be fended off at least six feet, wherever practicable; all connecting lines shall be properly fitted with rat guards; gangways and other means of access to the vessel shall be well lighted or separated from the shore at night.

(2) Prior to departure, the vessel shall be inspected for rats. If any are present, measures (trapping and/or fumigation) shall be taken for their destruction.

d. Aircaft Only. Disinsectization shall be performed immediately prior to departure for Korea.

軍政廳　官報　保健厚生部令　第二號　一九四七年八月二十五日

8. *Measures in Transit.* Vessels and aircraft arriving in Korea shall, while in transit, have fulfilled the following requirements as a condition of entry into Korea:

   a. Vessels and Aircraft.

      (1) Good sanitation shall be maintained.

      (2) Persons becoming ill with what is believed to be a dangerous communicable disease shall be isolated. Pertinent facts shall be reported by radio to the port of destination.

      (3) The authority in charge of the port of destination shall be notified by radio of expected arrival well in advance of the anticipated time of arrival.

   b. Vessels Only. When rats or signs of rats are found on board, trapping shall be undertaken.

   c. Aircraft Only. Prior to landing in Korea and subsequent to last take-off, all readily accessible compartments shall be disinsected. A potent insecticide, non-injurious to man, shall be employed in accordance with the manufacturer's recommendations. The aerosol bomb, as supplied to the United States Military Forces, is satisfactory. During the dispersion of the insecticide and for three minutes thereafter, the hatches, ports, exit doors, windows and ventilating apertures shall be kept closed.

9. *General Requirements upon Arrival.* Vessels subject to quarantine inspection shall fly a yellow flag (International Code Queen Flag) from the foremast upon arrival in a Korean port. anchor in the quarantine anchorage, and await inspection. After pratique has been issued, the yellow flag shall be lowered.

10. *Vessels or Aircraft Subject to Quarantine Inspection.* An incoming vessel or aircraft, subject to this regulation, shall not discharge passengers, crew, or cargo until after pratique has been received from the port quarantine officer.

11. *Vessels of Armed Services.* Vessels belonging to or operated by the armed services of the United States or any foreign nation may, in the discretion of the port quarantine officer, be exempted from quarantine inspection if a commissioned medical officer of such service certifies that:

— 6 —

851

a. Any person on board who is infected or suspected of being infected with a communicable disease shall be isolated until it is determined whether or not he is infected with a quarantinable disease, and that

b. The vessel is from a port where, at the time of departure, there was not present, or suspected of being present, cholera, plague, or yellow fever, or where there was not a significant increase in the prevalence of smallpox or typhus at the time the vessel touched there.

12. *Radio Pratique.* The port quarantine officer may grant pratique by radio to a vessel upon the basis of information regarding the vessel, its cargo and persons aboard, received prior to arrival of the vessel, when in his judgment, and in accordance with instructions from the Director of the Department of Public Health and Welfare, the entry of the vessel will not result in the introduction, transmission, or spread of communicable diseases.

13. *Intercoastal or Internal Traffic.* Quarantine inspection shall not be required of vessels or aircraft arriving from other ports in Korea, provided that during the current voyage or flight such vessels or aircraft have not touched at any foreign port, and provided further that there is no known epidemic of quarantinable disease at the point of departure. In case of epidemic at the point of departure, quarantine inspection shall be required, and such other measures shall be taken as may be ordered by the Director of the Department of Public Health and Welfare.

14. *Quarantine inspection.* Quarantine inspection of a vessel or aircraft shall include:

a. Inspection of the vessel or aircraft, its cargo, manifests, and sanitary log, to ascertain the sanitary history and condition of the vessel or aircraft.

b. Examination of the persons aboard the vessel or aircraft, and their personal effects, and medical records.

c. The determination of the measures necessary to prevent the introduction of quarantinable disease.

15. *Restrictions on Boarding Vessels.* Only the quarantine officer, quarantine employees, or pilots, shall be permitted to board any vessel subject to quarantine inspection until it has been inspected by the quarantine officer

— 7 —

and granted pratique, except with the permission of the quarantine officer. A person boarding such vessel, otherwise, shall be subject to the same restrictions as may be imposed on the persons on the vessel.

16. *Examination of Persons.* All persons on board shall be examined, except that, on an approved regular line vessel or aircraft which carries a ship or flight surgeon, such examination may be limited to persons designated by the port quarantine officer.

17. *Observation of Persons.* Persons may be held under observation at quarantine stations or on vessels or aircraft in quarantine pursuant to the provisions of paragraph 19 below.

18. *Expediting Quarantine Inspection.* Quarantine processing shall be performed as expeditiously as consistent with public health. Vessels arriving in the quarantine area between the hours of 0600 and 1800 shall be promptly boarded upon arrival. Those vessels arriving between 1800 and 0600 hours shall be boarded at 0600, unless otherwise arranged. Aircraft shall be boarded promptly, regardless of the hour. In the case of aircraft, if the arrival of the quarantine officer is delayed, passengers and crew may disembark, provided that they remain segregated within the quarantine area.

19. *Requirements Concerning Vessels or Aircraft Infected with Specific Diseases.*

   a. Cholera.

     (*1*) Vessels and Aircraft.

      (*a*) A cholera infected vessel or aircraft shall be detained in quarantine until disinfected.

      (*b*) The *dejecta* of all persons held under observation for cholera shall be disinfected before final disposition.

      (*c*) The personal effects of cholera cases and carriers shall be disinfected. Any other likely contaminated articles aboard shall be disinfected.

      (*d*) Fruits and vegetables aboard infected vessels and aircraft shall be destroyed or rendered noninfectious by proper cooking.

(e) The water supply of a cholera infected vessel or aircraft shall be disinfected.

(2) Persons.

(a) Persons ill with cholera, those suspected to be ill with cholera, known carriers, and all known and suspected contacts shall be isolated and kept under medical observation until known to be free of cholera.

(b) Satisfactory evidence of freedom from infection shall consist of 3 consecutive stool specimens taken not less than 48 hours apart, all of which fail to show the presence of cholera *vibrio* upon proper bacteriological examination.

b. Plague.

(1) Vessels and Aircraft.

(a) A plague infected vessel or aircraft shall be detained in quarantine, and immediate measures instituted for the destruction of rodents and vermin aboard.

(b) Disinfection of personal effects, baggage, bedding, quarters, kitchens, store rooms, and other areas, shall be accomplished as the quarantine officer may direct, to ensure that the vessel or aircraft is freed of infection.

(2) Persons.

(a) Persons infected or suspected to be infected with plague shall be isolated and kept under medical observation until known to be non-infectious.

(b) Contacts shall be held under observation or surveillance for a period of 6 days subsequent to last possible exposure.

c. Smallpox.

(1) Vessels and Aircraft.

An infected vessel or aircraft shall be detained in quarantine. until the personal effects of the sick, and the quarters

— 9 —

occupied by them, together with furnishings, shall have been disinfected.

(2) Persons.

(a) Persons ill with or suspected of being ill with smallpox shall be isolated and kept under medical observation until known to be non-infectious.

(b) All contacts shall be vaccinated and held under observation until the results of the vaccination indicate immunity. Persons refusing vaccination shall be held under observation until 14 days have elapsed from the date of their last contact.

d. Typhus.

(1) Vessels and Aircraft.

(a) Infected vessels and aircraft shall be detained in quarantine until vermin destruction has been completed.

(b) A louse-infested vessel or aircraft shall be disinfested.

(c) The personal effects and baggage of typhus cases, suspect typhus cases, louse-infested persons, and suspect louse-infested persons, shall be disinfested.

(2) Persons.

(a) Persons ill from, or suspected to be ill from, typhus shall be isolated and kept under medical observation until known to be non-infectious.

(b) Contacts whose anti-typhus vaccinations are not up to date shall be vaccinated and held under surveillance or observation for 12 days from the date of their last contacts.

(c) Contacts whose anti-typhus vaccinations are up-to date may be released under 12-days' surveillance.

(d) All persons infested, or suspect-infested, with vermin shall be disinfested.

— 10 —

(e) Vermin-free persons who had no contact with typhus cases or vermin-infested persons, may be released under 12 days' surveillance, provided that their anti-typhus vaccinations are up to date, or that vaccination is given upon arrival.

e. Yellow Fever.

(1) Vessels Only.

(a) The infected vessel shall be moored not less than 400 meters from shore until disinsectization has been completed.

(b) An infected vessel shall be disinsected prior to discharge of cargo.

(2) Aircraft Only.

Aircraft arriving with yellow fever aboard shall be immediately disinsected under the supervision of the quarantine officer, regardless of any prior disinsectizations performed.

(3) Persons Only.

(a) Persons ill with yellow fever, and persons suspected of being infected with yellow fever, shall be isolated until known to be non-infectious.

(b) Persons from an infected vessel or aircraft who cannot present satisfactory evidence of immunity to yellow fever shall be vaccinated and placed under observation or surveillance for 6 days from the date of last possible exposure. Persons who present satisfactory evidence of immunity shall be released.

20. *Free Pratique.* A certificate of free pratique shall signify that the vessel may enter, discharge cargo, and land passengers.

21. *Provisional Pratique.* A certificate of provisional pratique shall signify that the vessel may enter, but that additional measures, as specified in such certificate, must be taken in connection with the discharge of cargo, or the landing of passengers, or the sanitary condition of the vessel. A certificate of free pratique shall be issued after such additional measures have been accomplished.

23. *Importation of Certain Things.*

    a. Animals and Animal Products.

        (1) The importation of animals, of animal products, including meat, meat products, hides, skins, furs, wool, hair, bristles, bones and animal fertilizers, and other items which by their nature or former use may act as mechanical carriers of animal diseases, shall be governed by applicable veterinary quarantine regulations of the Government of Korea.

        (2) Animals and other items as listed in the preceding sub-paragraph brought into Korea contrary to the provisions of the applicable regulations shall be held in quarantine on board the ship or aircraft of arrival, or in quarters approved by the quarantine officer, pending disposition.

    b. Rats.

        (1) Vessels Only.

            (a) Captains shall observe the measures against rats described in paragraph 7 *c* hereof.

            (b) Rat trapping shall be undertaken on board rat infested vessels. Trapping shall be supervised by the captain, or by a ship's officer designated by the captain.

            (c) Vessels shall be fumigated to destroy rats, when directed by the quarantine officer.

            (d) A "deratization exemption certificate" shall be issued to the captain when a thorough inspection of a ship reveals no rats to be aboard. A "deratization certificate" shall be issued to the captain subsequent to a properly performed fumigation undertaken to kill rats (as prescribed by Article 28 of the International Sanitary Convention, 1926, as amended by International Sanitary Convention, 1944).

    c. Disease Agents and Vectors.

        (1) The etiological agent or insect (as well as eggs, larvae, pupae, nymphs), animal or plant vector of human disease, or any

exotic living insect, animal or plant capable of being a vector of human diseases, shall not be imported except as specifically authorized by the Director of the Department of Public Health and Welfare.

(2) Application for permission to import any item mentioned in paragraph 22 *c (1)* above shall include a full description of the item, including common and scientific names, quantity, origin, destination, intended use, and other necessary matters.

(3) Items listed in paragraph 22 *c (1)* above which arrive in Korea without authorization as specified therein shall be held in quarantine aboard the aircraft or vessel of arrival, or ashore, pending disposition.

(4) Items denied entry shall be destroyed under the supervision of the quarantine officer or removed from the country by the person or agent responsible for importation.

d. Dead Bodies.

(1) Bodies may be embarked for Korea, regardless of cause of death, provided they have been properly embalmed and are contained within hermetically-sealed metal containers.

(2) No restriction shall be imposed upon the entry of ashes remaining from cremation.

(3) Bodies brought into Korea shall be dealt with in such manner as to ensure protection of public health.

23. *Mail.* The quarantine officer shall refrain from inspection of mail, except for quarantine reasons. When such inspection is necessary, inspection and treatment shall be performed only in the presence of a representative of the postal authorities.

24. *Disinfection of Imports.* When the freight manifest of a vessel or aircraft lists articles which may require disinfection, the quarantine officer may order such articles separated from the other freight pending disinfection or other appropriate disposition.

— 13 —

SECTION III. OUTGOING TRAFFIC

25. *Persons.*

a. Individuals destined for foreign ports shall present certificates to the quarantine officer showing that all vaccinations required for entry to the countries of transit and destination have been received.

b. Medical examinations shall be made when warranted by the presence or suspected presence of dangerous communicable disease or the vectors of such disease.

c. Persons known or believed to be infected with a quarantinable disease, or infested with the vector of such disease, shall not be embarked.

d. No person in recent contact with a quarantinable disease shall be embarked unless known to be adequately protected by vaccination or by a previous attack of such disease.

e. Persons embarked with a dangerous communicable disease shall be properly isolated and attended by a physician or trained nurse or orderly.

f. Dead bodies may be embarked if properly embalmed and contained within hermetically-sealed metal containers, regardless of cause of death.

26. *Animals.*

a. The requirements for entry to the countries of transit and destination shall be complied with, in so far as possible, prior to embarkation of any animal being shipped out of Korea.

b. Animals destined to be shipped out of Korea shall be examined within the week prior to departure by a veterinary officer who shall issue a certificate of health. This certificate shall be made available to port authorities.

c. Animals known or suspected to be infected with a serious communicable disease of man or animal shall not be embarked.

27. *Vessels and Aircraft.*

a. Vessels and aircraft destined for foreign ports shall be subject to health and sanitation inspection by the quarantine officer.

b. Only pure water and wholesome food shall be embarked.

c. Good sanitation shall be maintained.

d. The presence or suspected presence on board of the agents or vectors of quarantinable disease shall be sufficient reason for the quarantine officer to order such disinfection as he may deem necessary for the protection of public health.

e. Aircraft shall be disinsected immediately prior to departure from Korea.

SECTION IV. SPECIAL PROVISIONS.

28. *International Sanitary Convention.* Foreign quarantine problems not specifically provided for by this regulation shall be resolved in accordance with the provisions of the International Sanitary Convention, 1926, as amended by International Sanitary Convention, 1944.

29. *Port Quarantine Facilities.*

Existing city or provincial hospitals and laboratories shall be used, when required for isolation, observation, diagnosis and treatment of persons with quarantinable diseases who are removed from vessels or aircraft and are not entitled to treatment in United States Army hospitals.

30. *Reports.*

a. Quarantine officers shall submit weekly reports of quarantine activities which will include the following information:

(1) Number of ships entering the port from foreign ports.

(2) Number of persons debarked from such vessels.

(3) Number of ships on which quarantine restrictions have been placed during the week, and the diseases for which such restrictions were placed.

(4) Laboratory work accomplished in connection with port quarantine procedures during the week.

(5) Remarks.

b. One copy of such report shall be forwarded to the Department of

— 15 —

Public Health and Welfare. Headquarters, USAMGIK, Attention: Preventive Medicine Section. One copy shall be submitted to the authority in charge of the port, one copy to the provincial Bureau of Public Health and Welfare for information, and one copy shall be retained for file.

31. The master or any officer of any vessel who knowingly violates the provisions of paragraphs 2 or 10 hereof shall, upon conviction by a Military Occupation Court, suffer such punishment as the court shall determine.

32. This regulation shall be effective on the tenth day after the date appearing hereon.

BY DIRECTION OF THE MILITARY GOVERNOR:

LEE YONG SUL

Director

Department of Public Health and Welfare

軍政廳　官報　保健厚生部令　第二號　一九四七年八月二十五日

HEADQUARTERS
UNITED STATES ARMY MILITARY
GOVERNMENT IN KOREA
Department of Public Health and Welfare
Seoul, Korea

DEPARTMENT ORDER
NUMBER 3

21 June 1947

NARCOTICS REGULATION

1. *Purpose.* The purpose of this regulation is to implement the control of narcotics pursuant to Ordinance No 119 dated 11 November 1946 (*Narcotics Control*) and all other ordinances relating to narqotics.

2. *Unlicensed Transactions Prohibited.* Except as otherwise provided in Section VI *a-d* (inclusive) of Ordinance No 119 and paragraph 18 *d* of this regulation, no person shall without a license issued pursuant to this regulation and conspicuously displayed at his place of business, possess, produce, manufacture, compound, purchase or in any manner obtain, sell, transfer, send, ship, carry, transport or deliver, convey any interest in, or give away any narcotic drugs, or attempt, offer to do, cause or facilitate any of such acts; and no person shall permit any of said acts to be done without a license, in or upon any place owned, occupied, used, maintained or controlled by him. Nor shall any person sell or deliver narcotic drugs to another person not licensed under this regulation, except as otherwise provided in paragraph 18 hereof.

3. *Issuance of Licenses.* *a.* The Director of the Department of Public Health and Welfare may issue:

    (1) Manufacturers' licenses.
    (2) Repackagers' licenses.
    (3) Wholesalers' licenses.
    (4) Pharmacists' licenses.
    (5) Practitioners' licenses.
    (6) Research licenses.
    (7) Exempt narcotic preparations licenses.

b. Any person, except as otherwise specifically provided in this regulation, may receive more than one license.

c. A manufacturer's license may be issued to any person who, under the supervision of a duly-licensed pharmacist, produces, mixes or in any way processes narcotic drugs for sale to another manufacturer, to a repackager, or, to a wholesaler.

d. A repackager's license may be issued to any person who, under the supervision of a duly licensed pharmacist, buys narcotic drugs in bulk and repackages them into smaller packages without mixing or further processing for sale to manufacturers, repackagers or wholesalers.

e. A wholesaler's license may be issued to any person who, under the supervision of a duly licensed pharmacist, deals in medicines generally, buys narcotic drugs and sells them without repackaging to pharmacists, physicians, dentists, veterinary surgeons or research workers and, in the case of exempt narcotic preparations, to licensed drug merchants as defined in sub-paragraph *i* of this paragraph.

f. A pharmacist's license may be issued to a pharmacist otherwise duly licensed to practice pharmacy who dispenses narcotic drugs upon prescription.

g. A practitioner's license may be issued to any physician, dentist or veterinary surgeon duly licensed to practice, who administers or prescribes narcotic drugs in the course of professional treatment;- *provided however,* that no person shall prescribe or administer narcotic drugs for a person chronically poisoned by or addicted to the use of narcotic drugs for the purpose of relieving or curing such poisoning or addiction.

Every practitioner shall within ten days of the diagnosis of any person chronically poisoned by or addicted to the use of narcotic drugs report to the provincial Pharmaceutical Affairs Section for the province in which he has his facility, the name, address and diagnosis of such person.

h. A research license may be issued to any person who desires to use narcotic drugs in the course of scientific research projects.

— 2 —

i. A license to sell exempt narcotic preparations at retail may be issued to any drug merchant who sells medicines and exempt narcotic preparations at retail and who is not otherwise licensed to sell narcotic drugs. A license to sell narcotic drugs also licenses the sale of exempt narcotic preparations.

j. A license shall be issued to or renewed only for persons who, in the opinion of the Director of the Department of Public Health and Welfare:

(1) Possess good moral character; *and*

(2) Possess: (*a*) such experience in importing, manufacturing, administering, distributing, marketing or handling narcotic or other medical drugs as wholesale or retail dealers, practitioners, pharmacists, or laboratory workers duly licensed and lawfully entitled to engage in such activities, *and* (*b*) such means and facilities for manufacturing, handling, and safeguarding narcotic drugs, as to render reasonably probable the orderly and lawful distribution of narcotic drugs of suitable quality to supply medical and scientific needs, without diversion to illicit channels.

4. *Application for License.*

a. Any person may apply for a license pursuant to this regulation by filing an application with the appropriate provincial Pharmaceutical Affairs Section. Each application shall be accompanied by written statements made by the head of the city, county or island in which the applicant and his supervising pharmacist have their place of business, setting forth:

(1) Whether such persons are chronically poisoned by or addicted to the use of narcotics;

(2) Whether such persons have been convicted of any crime or offense in connection with narcotics, *and*

(3) Any reasons why in his opinion the license should be refused.

b. The appropriate provincial Pharmaceutical Affairs Section shall approve or disapprove applications within 30 days of receipt thereof. It shall forward all applications which it approves to the Director of the Department

— 3 —

of Public Health and Welfare. If an application is disapproved, the appropriate provincial Pharmaceutical Affairs Section shall forthwith notify the applicant in writing, stating the reasons therefor. The applicant may within 60 days of such disapproval appeal to the Director of the Department of Public Health and Welfare.

5. *Information in Application.*

a. The application for a license shall contain the following information, except where the context indicates that the information is inapplicable :

(1) Name and address of applicant.

(2) Name and address of applicant's supervising pharmacist.

(3) Complete details regarding the license of the supervising pharmacist, physician, dentist, veterinarian or drug merchant, including the place and date of its issuance and the issuing authority.

(4) Location and description of applicant's place of business, plants, warehouses, equipment, and other manufacturing, distribution and research facilities.

(5) Items which the applicant proposes to manufacture and the estimated amount of such items to be manufactured during the period for which the license is requested.

(6) Sources of material for manufacturing.

(7) Purpose of the applicant's research.

(8) Detailed personal history of the applicant and his supervising pharmacist, including the names and addresses of their present and former employers, a description of their present and all former employment, and reasons for changing employment; all periods of unemployment shall be explained and five references furnished. (*Not required in the application for a pharmacist's license, practitioner's license or exempt narcotic preparations license.*)

— 4 —

(9) Where the applicant is a juridical person, the applicant shall submit the articles of incorporation and a list of the names and addresses of the officers, directors, auditors, stockholders, bondholders, mortgagees or other security holders. (*Not required in the application for a pharmacist's license, practitioner's license or exempt narcotic preparations license.*)

(10) A complete inventory of all narcotic drugs and exempt narcotic preparations on hand as of the date of the application.

(11) Such other information as the Director of the Department of Public Health and Welfare requires.

b. No change shall be made by the licensee or his supervising pharmacist in any of the items listed in items (2)–(7) inclusive, or in the officers, directors, auditors, stockholders, bondholders, mortgagees or other security holders required to be listed by item (9) hereof, except with the prior written approval of the Director of the Department of Public Health and Welfare. An application for such change shall be filed with the appropriate provincial Pharmaceutical Affairs Section, accompanied by a fee of 5 won. Changes in items (1) and (2), or changes in item (9) not requiring prior approval, shall within ten days of the change be reported to the appropriate provincial Pharmaceutical Affairs Section which shall forward the application or report to the Director of the Department of Public Health and Welfare within ten days of its receipt.

6. *Fees.* The following fees shall be paid to the Director of the Department of Public Health and Welfare upon the issuance or renewal of a license:

a. Manufacturer's license	500	won
b. Repackager's license	300	won
c. Wholesaler's license	200	won
d. Pharmacist's or practitioner's license	50	won
e. Exempt narcotic preparations license	50	won
f. Research license	20	won

Such fees shall be paid to the appropriate provincial Pharmaceutical Affairs Section or other agency to which the license is transmitted for delivery to the licensee. The agency receiving such fees shall forward them forthwith to the Director of the Department of Public Health and Welfare.

7. *Duration of License:* All licenses shall terminate on the 31st day of December of the year for which they are issued.

8. *Renewal of License.* All applications for renewal shall be filed during the month of November preceding the expiration date, in accordance with the provisions of paragraphs 5 and 6 hereof.

9. *Lapse of License.* Any person whose license has expired or lapsed or whose license has been suspended or revoked shall, within ten days of such expiration, lapse, suspension or revocation, surrender to the appropriate provincial Pharmaceutical Affairs Section such license, together with a detailed inventory of all narcotic drugs on hand as of the date of such surrender. Such person shall not thereafter sell, transfer or otherwise dispose of such inventory or any part thereof, except by direction of the Director of the Department of Public Health and Welfare.

10. *License not Transferable.* No license shall be transferable. If a licensee transfers his business establishment, dies, is missing or becomes legally incapacitated, his license shall thereby lapse and the licensee or his legal representative, as the case may be, shall surrender the license in accordance with paragraph 9 hereof.

11. *Lost or Damaged License Certificates.* If a licensee has lost his license certificate or if his license certificate has been damaged, he may apply for a new license certificate to the appropriate provincial Pharmaceutical Affairs Section, which shall forward the application to the Director of the Department of Public Health and Welfare. The applicant shall surrender the damaged certificate with his application, and pay a filing fee of 5 won. Where the licensee finds his lost certificate subsequent to the issuance of a new certificate, he shall forthwith surrender the new certificate to the appropriate provincial Pharmaceutical Affairs Section.

12. *Suspension and Revocation.* a. The appropriate provincial Pharmaceutical Affairs Section or the Director of the Department of Public Health and

Welfare may, after due notice and hearing, suspend a license if the licensee has violated the provisions of this regulation or of Ordinance No 119, or is guilty of a substantial violation of law. All such suspensions made by the appropriate provincial Pharmaceutical Affairs Section shall be reviewed by the Director of the Department of Public Health and Welfare, who shall provide all interested parties an opportunity to be heard. Upon the expiration of the suspension period, the appropriate Pharmaceutical Affairs Section or the Director of the Department of Public Health and Welfare shall note on the certificate the period of suspension and the reasons therefor, and shall return the certificate to the licensee. No suspension under this paragraph shall exceed 60 days.

b. The Director of the Department of Public Health and Welfare may revoke a license if, after due notice and hearing, he finds such action to be in the public interest, or that the licensee has failed to comply with regulations or requirements of law relating to narcotics or is guilty of a substantial violation of law material to the licensee's qualification to sell narcotics.

13. *Packaging of Narcotics.* *a.* No manufacturer, repackager or whole saler shall sell or deliver narcotic drugs (other than exempt narcotic preparations) unless such narcotic drugs are packed in receptacles sealed with stamps approved by the Director of the Department of Public Health and Welfare. This subsection shall not apply to sales or deliveries of bulk narcotic drugs to manufacturers or repackagers.

b. The following information shall appear on all such receptacles and the wrappings for such receptacles:

(1) Name and address of principal place of business of the manufacturer or repackager.

(2) Date of packaging.

(3) The percentage by weight of each narcotic drug in the contents of the package; except that, where the contents are in tablet form, the number of tablets and weight and type of each narcotic drug per tablet shall appear.

(4) The character *ina.*

— 7 —

14. *Manufacturers' or Repackagers' Quota Permit.*

a. Each licensed manufacturer or repackager shall apply to the Director of the Department of Public Health and Welfare within the 30-day period preceding 1 January, 1 April, 1 July and 1 October, for the establishment of the amount and items of narcotic drugs he will be permitted to manufacture or repackage during the ensuing quarter year. The application shall be filed with the appropriate provincial Pharmaceutical Affairs Section, which shall forward such application to the Director of the Department of Public Health and Welfare within ten days of receipt thereof. Such application shall set forth the following:

(1) Name and present place of business of applicant.

(2) Items and quantity of narcotic drugs proposed to be manufactured or repackaged during the ensuing quarter.

(3) The types and number of each type of receptacle to be used.

b. The Director of the Department of Public Health and Welfare shall issue permits for the manufacturing and repackaging of such amounts and items as in his judgment are consistent with the public interest. He may issue such permits for the manufacture or repackaging of specific amounts at any time when the public interest so requires.

15. *Stamps.* Each manufacturer or repackager shall obtain stamps from the appropriate provincial Pharmaceutical Affairs Section or the Director of the Department of Public Health and Welfare submitting:

a. A copy of his permit to manufacture or repackage received pursuant to paragraph 14 hereof, and

b. The number of stamps on hand.

The number of stamps issued and the date of issuance shall be noted on the permit.

16. *Order Forms.* No licensee shall transfer or sell narcotic drugs to another licensee unless he receives from the buyer or other transferee an order in the form prescribed by the Director of the Department of Public Health and Welfare. Both the buyer and seller shall retain copies of such order form as part of their records.

— 8 —

17. *Inventory Reports.* *a.* Every manufacturer, repackager and wholesaler shall file monthly reports, setting forth:

(1) Quantity and items of narcotic drugs in his possession at the beginning of the calendar month for which the report is made.

(2) Names and addresses of all persons from whom he acquired narcotic drugs and the quantity and items acquired from each such person during the calendar month.

(3) Names and addresses of all persons to whom he sold or otherwise transferred narcotic drugs and the items and quantity sold or otherwise transferred to each such person.

(4) Quantity and items of narcotic drugs in his possession at the end of the calendar month.

(5) Explanation of any discrepancy between the sum of the quantities specified in Item (1) plus Item (2) and the sum of the quantities specified in Item (3) plus Item (4).

Such reports shall be filed with the appropriate provincial Pharmaceutical Affairs Section on or before the 20th day after the month for which the report is made. The provincial Pharmaceutical Affairs Section shall forward such reports to the Director of the Department of Public Health and Welfare.

b. Every pharmacist, practitioner, drug merchant licensed to deal in exempt narcotic preparations in accordance with paragraph 3 i hereof, and narcotic research worker shall file an inventory report for the balance of the calendar year in which the license is issued and thereafter for each calendar year, setting forth:

(1) Quantity and items of narcotic drugs in his possession at the beginning of the period for which the report is made.

(2) Quantity and items of narcotic drugs acquired during that period.

(3) Quantity and items he sold or otherwise transferred or used during that period.

— 9 —

(4) Quantity and items of narcotic drugs in his possession at the end of that period.

Such reports shall be filed with the appropriate provincial Pharmaceutical Affairs Section on or before the 20th day after the period for which the report is made. The provincial Pharmaceutical Affairs Section shall forward such reports to the Director of the Department of Public Health and Welfare.

18. *Prescriptions.* a. Every practitioner prescribing narcotic drugs other than exempt narcotic preparations shall write and sign a prescription in triplicate setting forth:

> (1) Name, sex, address and age of the patient, where the prescription is made by a physician or dentist. (A prescription made by a veterinary surgeon shall set forth the species of animal and the name and address of its owner)
>
> (2) Type and quantity of narcotic drugs and directions for use.
>
> (3) Date of prescription, and name, address and license number of the practitioner

The practitioner shall retain one copy as part of his records and give two copies to the patient.

b. Every practitioner administering narcotic drugs other than exempt narcotic preparations shall write and sign a prescription in duplicate setting forth items (1), (2) and (3) of preceding subparagraph. The practitioner shall retain one copy as part of his records, and within ten days after the end of each calendar month send the second copy of each such prescription written during that month to the appropriate provincial Pharmaceutical Affairs Section.

c. No pharmacist shall sell or otherwise dispense narcotic drugs other than exempt narcotic preparations unless he receives from the buyer two copies of a prescription in the form herein specified for such narcotic drugs. The pharmacist shall sign both copies of the prescription and indicate the date it was filled. He shall retain one copy of each prescription and he shall keep such copy separately from his other records.

d. Within ten days after the end of each calendar month, the pharmacist shall send the second copy of each such prescription filled during that

軍政廳　官報　保健厚生部令　第三號　一九四七年六月二四日

month to the appropriate provincial Pharmaceutical Affairs Section.

e. A pharmacist or drug merchant duly licensed hereunder to sell exempt narcotic preparations may sell such exempt narcotic preparations only to persons who present a prescription or a signed request, containing the name and address of the buyer, the date of the request, the type and amount of exempt narcotic preparations, and a statement that it is being bought for use by the buyer or other named person (whose address shall also be given). Each seller of exempt narcotic preparations shall retain such requests and prescriptions separately from his other records.

19. *Period of Retention of Records.* Any person obligated to keep records (including copies of prescriptions and requests for exempt narcotic preparations) pursuant to this regulation shall retain such records for at least five years from the date they are recorded or filed.

20. *Storage of Narcotics.* Each licensee shall keep his stock of narcotic drugs (other than exempt narcotic preparations) under lock and key apart from his other merchandise or materials.

21. *Inspection.* a. Each licensee shall keep his records, stocks of narcotic drugs and all facilities available for inspection by agents of the Department of Public Health and Welfare or of the appropriate provincial Pharmaceutical Affairs Section.

b. Any licensee shall upon request of such agent deliver to him a sample of any narcotic drug if such agent leaves a receipt for such narcotic drugs. The licensee shall within ten days report in writing the taking of such sample, to the appropriate provincial Pharmaceutical Affairs Section.

c. An inspector authorized by the Director of the Department of Public Health and Welfare may require the surrender of a license or of the licensee's stock of narcotic drugs, if he finds that the licensee has violated this regulation or Ordinance No 119. The inspector or any other person authorized to seize narcotic drugs pursuant to Section XI of Ordinance No 119, shall ofrthwith in writing report to the Director of the Department of Public Health and Welfare any such seizure and the reasons therefor. The Director of the Department of Public Health and Welfare shall within ten days after receipt of such report notify the licensee in writing of the charges against him and give him an opportunity to be heard.

22. *Other Instructions.* The Director of the Department of Public Health and Welfare may, by written order, whenever the public interest requires, prohibit or restrict the manufacture, repackaging, sale or transfer of narcotic drugs by any licensee.

23. *Former Regulations.* Any person who was entitled to deal in narcotic drugs prior to the effective date of this regulation may continue to deal in narcotic drugs for 30 days after the effective date of this regulation if he has filed an application for a license pursuant to this regulation and such application has not been denied. Any person who has not filed an application on or before the effective date of this regulation shall report his inventory of narcotic drugs within ten days of such effective date to the appropriate provincial Pharmaceutical Affairs Section and hold such narcotic drugs subject to the order of the Director of the Department of Public Health and Welfare.

24. *Definitions.* a. *Narcotic drugs* includes opium, morphine, heroin, codeine, cocaine, marihuana and any other component, derivative or preparation of any thereof. It includes exempt narcotic preparations unless the context otherwise requires.

b. *Exempt narcotic preparation* means any preparation or remedy which contains by weight not more than 0.4 per cent of opium, or not more than 0.05 per cent of morphine, or not more than 0.2 per cent of codeine, hydrocodeine, or any salt, or derivative of any of them, *and* which preparation contains active medicinal drugs other than narcotics which confer upon the preparation valuable medicinal qualities other than those provided by the narcotic drug alone.

c. *Appropriate Provincial Pharmaceutical Affairs Section* with reference to any person means the Pharmaceutical Affairs Section of the provincial

Bureau of Public Health and Welfare for the province in which such person has his place of business.

    d. Person includes natural and juridical persons.

    25. *Penalties.* Any person violating the provisions of this regulation shall, upon conviction by a Military Occupation Court, suffer such punishment as the Court shall determine.

    26. *Effective Date.* This regulation shall be effective on the twentieth day after the date appearing hereon.

BY DIRECTION OF THE MILITARY GOVERNOR:

LEE YONG SUL
Director
Department of Public
Health and Welfare

SOUTH KOREAN INTERIM GOVERNMENT
Department of Labor
Seoul, Korea

DEPARTMENT ORDER
NUMBER 1   15 September 1947

IMPLEMENTATION OF CHILD LABOR LAW

1. The purpose of this order is to implement Public Act No. 4 dated 16 May 1947 (*Child Labor Law*).

2. Pursuant to the prohibition against night work provided in Section IV of Public Act No. 4, children under eighteen (18) full years of age are prohibited from working before 7 A.M. and after 7 P.M.

3. *a.* The hours of work provided in Section V of Public Act No. 4 shall include at least one full hour for rest and/or meals.

b. Children under sixteen (16) full years of age shall be given such hour for rest and/or meals in the period after the end of the first hour and before the last hour at the work place *provided, however*, that at least one half of such hour shall be given for meals at the end of the fourth hour at the work place.

c. Children between sixteen (16) and eighteen (18) full years of age shall be given such hour for rest and/or meals in the period after the end of the second hour and before the last two (2) hours at the work place *provided, however*, that at least one half of such hour shall be given for meals at the end of the fourth hour at the work place.

4. The application for a permit to enter into an employment contract as provided in Section VII of Public Act No. 4 shall be submitted to the supervising authority on Form No. 1 attached, issued by the Department of labor. If the applicant cannot present this application in person, the reason for his absence shall be attached to the application.

5. The supervising authority shall issue the employment permit within thirty (30) days from the date the application is received on Form No. 2 attached, issued by the Department of Labor. If the permit is not issued or refused within such period, the applicant may assume that his application is

— 1 —

軍政廳　官報　勞動部令　第一號　　一九四七年九月十五日

granted and the copy of the application may be used in lieu of the permit until said permit or a notice of rejection of the application is received.

6. *a.* The twenty (20) day period provided in Section IX *c* of Public Act No. 4 is computed from the date of receipt of the application to dissolve the contract.

b. The supervising authority shall render a decision pursuant to Section IX *e* after the filing of Form No. 3 attached, requesting mediation and after full hearing.

c. The employer shall within sixty (60) days prior to the expiration of an employment contract notify the employee.

7. Education and recreation required by Section XIII of Public Act No. 4 shall not be required during working hours. Regulations regarding the education and recreation for children in authorized factories defined in Section X *c* of Public Act No. 4 shall be published in a separate order.

8. For the purpose of this order and Section XV of Public Act No. 4 the supervising authority means the head of the provincial Bureau of Labor and the head of the Bureau of Labor for the City of Seoul.

9. The following forms issued by the Department of Labor are attached:

a. Form No. 4—Registration of the names of employed children (Section X *a*).

b. Form No. 5—Schedule of working hours (Section X *b*).

c. Form No. 6—Health and physical examination certificate (Section XI *a*).

d. Form No. 7—Reports of violations of Section XVII.

e. Form No. 8—Reports of violations of Section XVIII.

10. This order shall be effective on the date appearing hereon.

BY DIRECTION OF THE MILITARY GOVERNOR:

LEE DAI WI

Director, Department of Labor

軍政廳　官報　　勞動部令　　第一號　　一九四七年九月十五日

Form No. 1

APPLICATION FOR PERMIT TO ENTER INTO A
CONTRACT INVOLVING THE EMPLOYMENT OF CHILD LABOR

1. Employee
   Permanent domicile
   Present address:
   Name: (Seal) Male
   Date of birth: Female

2. Person in parental authority
   Permanent domicile:
   Present address:
   Name: (Seal) Occupation relationship
            to the employee

3. Employer
   Present address:
   Name: (Seal) Occupation or profession
   Name of Organization and trade mark.

4. Contract period FROM: Year Month Day
          TO : Year Month Day

   Working place:

   Kind of job Character of Amount of work, Amount of
        the work   wage payment and wage
            method

   Renewal of contract. Period of   From_______
   Yes_______No__   renewal contract To_______

Thus I sign and seal this application to enter into an employment contract pursuant to the above agreement and in accordance with Public A.
No. 4 (Child Labor Law).

   Date:

TO: The supervising authority

Remarks: The applicant should attach the following documents:

  1. A letter of consent of the person in parental authority.
  2. A certificate of age. (An abstract of a census register or a certificate of the local government).
  3. A scholastic record, school certificate or a personal history.
  4. A health certificate.
  5. A contract of employment, or offer or acceptance.

— 3 —

軍政廳 官報 勞動部令 第一號 一九四七年九月十五日

Form No. 2

CHILD LABOR EMPLOYMENT CERTIFICATE

1. Employee
   Present address:
   Name:

   Male
   Female

   Date of birth

   External appearance;
   special marks

2. Person in parental authority
   Present address:
   Name:

   Occupation and relationship
   with employee

3. Employer
   Present address:
   Name:

   Organization and trade mark
   Occupation or profession

4. Period of contract **FROM:**   Month   Day   Year
   TO   :   Month   Day   Year

   Working place:

   Kind of job   Character of the job   Amount of work;
   wage payment method

   Amount of wage:

   A renewal   Yes__________   Period of From__________
   contract   No __________   Renewal: To__________
   Received: Month______Day______Year______

Thus the employment is approved pursuant to Public Act No. 4 (Child
Labor Law).

   Issuance date: Month______Day______Year______
   (The certified date)
   Permit No.__________
   Effective date __________

   Title
   Supervising Authority
   NAME: (Seal)

— 4 —

878

Form No. 3

APPLICATION TO MEDIATE AN EMPLOYMENT TERMINATION

Employer : Address :

    Name of Company

    Name of Supervisor

Employee : Name   Address

    Date of birth : Month_______ Day_________ Year __________

    Kind of Job and date of contract

    Contract : From_______   To_______

Reasons for Termination :

    Contract is not practical

    Employee is unable to work because of ill health or other physical
      trouble ·

    It is clear that there is no hope that employee will change his bad
      conduct

    Employee has been absent more than 20 days

    Employer requests too heavy work, treats employee too cruelly, or
      violates the law

    Thus I apply for mediation because both sides cannot continue the
contract for employment for the stated reasons.

Date :   ____________________(Seal)

                               Applicant

    To : The Directing Organization.

Form No. 4

REGISTRATION OF CHILD EMPLOYEES

Name of the Working place :   Date_________

Name of employee :   Sex :   Age :

Type of Work :

Details of job :

Date of employment contract :   Address :

Remarks :

— 5 —

軍政廳　官報　　勞動部令　　第一號　　一九四七年九月十五日

Form No. 5

REGULATION OF WORKING HOURS OF MINORS

Name of working place :   Date :

Age :   Under 16 years : .   Under 18 Years :

Weekly working hours: Starting Time :   Closing time :

Meal time :   Recess period :

Weekly hours for education, health and amusement :

Period this form is to be in effect :

Miscellaneous :

Form No. 6

PHYSICAL EXAMINATION

Name of working place

Date

Name of employee

Date of birth   Sex

Previous diseases

Body : Weight   Height

Head : Eyes   Ears

       Nose   Mouth

       Teeth   Throat

Chest : Percussion   Auscultation

Heart :

Abdomen : Palpitation   Tumor masses

Female : Menses

Pulse   Respiration   Temperature

Urine   Stools

Recommendation

Remarks

     I certify that I have examined the person named above in accordance with Section II, Ordinance No. 4, and have listed the results hereon.

                           Name of Physician   (Seal)

To : Directing Organization

軍政廳　官報　　勞動部令　　第一號　　一九四七年九月十五日

Form No. 7

REPORT OF VIOLATIONS BY CHILD EMPLOYEES

This is report of violations of Public Act No. 4.

Date_______________ Directing Organization and Occupation

Name___________(Seal)

To: The Head of the Labor Department

Date the violation occurred	Name and address of employee and factory or working place	Number of previous violations of law	Types of violations (Sections 3, 4, 5, 6, and miscellaneous)

Disposition of violations

Reprimand	Certificate from the person in parental authority or principal of school	Request for a written oath	Rejection or cancellation of the employment certificate

Form No. 8

REPORT OF VIOLATIONS BY EMPLOYERS

This is a report of violations of Public Act No. 4 by employers.

Date ________________________________

Name _________________________ (Seal)

Official Rank :

Directing Organization :

To : The Head of the Labor Department.

Date of violation	Name of the employer or company	Address of employer	Number of previous violations of this law

Disposition of the crime

Types of Violations			
Violations of sections 3 and 4 by employing children in unauthorized occupations	Violations of Sections 5 and 6	False statement or continual wilful violation	Rejection of order for investigation

— 8 —

SOUTH KOREAN INTERIM GOVERNMENT
Department of Labor
Seoul, Korea

DEPARTMENT ORDER
NUMBER 2   15 September 1947

EDUCATION AND RECREATION OF CHILD EMPLOYEES

1. The purpose of this order is to prescribe the details regulating education and recreation pursuant to Section XIII *d* of Public Act No. 4 dated 16 May 1947 (*Child Labor Law*). This order was framed in cooperation with the Director of the Department of Education and the Director of the Department of Public Health and Welfare.

2. The provisions of this order shall apply to children under the age of eighteen (18) full years.

3. Each employer shall supply for children in his employ:

a. Necessary equipment for the eduction and welfare of child labor in the factories and other working places as approved by the Director of the Department of Labor;

b. Textbooks published by the Department of Education and used in the national primary schools and middle schools;

c. Teachers holding permits or those with at least equivalent education. (The selection and dismissal of teachers shall be reported to the Director of the Department of Labor.)

d. A program of education consisting of three classes as follows:

(1) The first class will be for those who have not completed the third year in a national primary school.

(2) The second class will be for those who have completed the third year in a national primary school.

(3) The third class will be for those who are graduates of a national primary school.

軍政廳　　官報　　勞動部令　第二號　　一九四七年九月十五日

4. The curriculum and schedule of school hours attached hereto shall be used.

5. Education, sports and recreation for employees shall be supervised by the Director of the Department of Labor and the employer shall make reports once every four months to the Director regarding the progress of such education, sports and recreation.

6. "Employer" as used in this order means an employer who employs ten (10) or more children. (See Section X *c* of Public Act No. 4).

7. This order shall be effective on the date appearing hereon.

BY DIRECTION OF THE MILITARY GOVERNOR:

LEE, DAI WI
Director, Department of Labor

SEAL

軍政廳　　官報　　勞動部令　　第二號　　一九四七年九月十五日

EDUCATION AND WELFARE
Curriculum and Schedule of School Hours

	Hours	*Courses for* *Recreation*	Hours	*Courses for* *Education*
FIRST CLASS :	At least two (2) hours	Athletics, Music	At least four (4) hours	Korean Citizenship National History Arithmetic, Housekeeping, Sewing, Reading and Writing
SECOND CLASS :	At least two (2) hours	Athletics, Music	At least four (4) hours	Korean Citizenship National History. Arithmetic, Housekeeping, Sewing, Reading and Writing
THIRD CLASS :	At least two (2) hours	Athletics, Music	At least four (4) hours	Advanced Arithmetic. Elementary Science. Reading and Writing

REMARKS :

1. At least four (4) hours per week will be used for education at suitable hours and days.

2. At least two (2) hours per week will be used for welfare at suitable hours and days.

3. Housekeeping and sewing are for females only. One of the other courses may be omitted for such females.

軍政廳　　官報　　勞動部令　第二號　　一九四七年九月十五日

SOUTH KOREAN INTERIM GOVERNMENT
Department of Labor
Seoul, Korea

DEPARTMENT ORDER
NUMBER 3

11 March 1948

MAXIMUM WORKING HOURS FOR EMPLOYEES OF THE BUREAU OF RAIL TRANSPORTATION, DEPARTMENT OF TRANSPORTATION

SECTION I. *Purpose.* The purpose of this order is to regulate and limit maximum working hours for employees employed in the Bureau of Rail Transportation, Department of Transportation, in accordance with paragraph *d*, Section III, Ordinance Number 121, 7 November 1946.

SECTION II. *Maximum Hours.* Employees of the Bureau of Rail Transportation, Department of Transportation, shall not work more than sixty (60) hours in any work week (the work week means any period of seven (7) consecutive days) and provided that all employment in excess of forty (40) hours per work week, shall be compensated at the rate of one and one half times the regular rate of payment.

SECTION III. *Exemptions.* Any employee may be employed in excess of sixty (60) hours per work week, where such work is necessary in cases of urgency, provided that the Director of the Department of Transportation shall, before requiring any employment in excess of sixty (60) hours, in each case, first secure the consent of the Director of the Department of Labor or his authorized representative, also submitting the following information:

a. The place of work.

b. The period of work.

c. Positions of employees.

d. The number of employees to be engaged in the work.

e. Reasons for excess hours.

SECTION IV. *Effective Date.* This order shall be effective on the date appearing hereon.

BY DIRECTION OF THE MILITARY GOVERNOR:

LEE DAI WI
Director, Department of Labor

— 1 —

SOUTH KOREAN INTERIM GOVERNMENT
Department of Labor
Seoul, Korea

DEPARTMENT ORDER
NUMBER 4   **13 April 1948**

EXEMPTIONS FOR SEASONAL INDUSTRIES

1. The purpose of this order is to make provisions and exemptions for industries which are seasonal in nature, in accordance with Section III, *a*, of Ordinance Number 121, dated 7 November 1946.

2. The following industries are hereby declared seasonal and shall be exempt from the provisions of Section II, Ordinance Number 121; provided that in all cases where employment is in excess of forty-eight (48) hours in any work week, the employee shall be compensated for all hours in excess of forty-eight (48) at the rate of not less than one and one-half ($1\frac{1}{2}$) times his regular rate of compensation.

   a. The salt manufacturing industry shall be exempt during the months of 1 April to 31 October, inclusive.

   b. The silk worm industry, including making nest eggs of silk worms, the work of killing the pupas in cocoons, and drying the cocoons shall be so exempt during the months of 1 June to 31 October, inclusive.

   c. The picking and manufacture of ginseng shall be so exempt during the months of 1 September to 30 November, inclusive.

   d. The vegetable, fruit and marine processing and packing shall be so exempt, as follows:

     (*1*) Vegetable processing and packing during the months of 1 August to 30 November, inclusive.

     (*2*) Fruit processing and packing during the months of 1 June to 31 December, inclusive.

     (*3*) Processing and packing of marine products during the months of 1 May to 30 November, inclusive.

3. This order shall be effective on the date appearing hereon.

BY DIRECTION OF THE MILITARY GOVERNOR:

(S E A L)

LEE DAI WI
Director
Department of Labor

軍政廳　官報　勞動部令　第四號　一九四八年四月一三日

HEADQUARTERS
UNITED STATES ARMY MILITARY
GOVERNMENT IN KOREA
Office of the Military Governor
Bureau of Finance
Seoul, Korea

INSTRUCTIONS TO BANKS

NUMBER 5

25 February 1946

TO: All Banks, Trust Companies and Financial Institutions

1. No financial institution will exchange Bank of Japan notes for Bank of Chosen notes nor will Bank of Japan notes be accepted for deposit, subject to withdrawal at a later date, unless specifically authorized by this Bureau.

2. Bank of Japan notes will be accepted for safekeeping, upon request by a holder, and custody receipt issued for same. After acceptance for safekeeping, withdrawal will not be permitted and neither loans nor advances will be made against these funds.

BY DIRECTION OF THE MILITARY GOVERNOR:

CHARLES GORDON
Lt Col, FD
Director

HEADQUARTERS
UNITED STATES ARMY MILITARY
GOVERNMENT IN KOREA
Bureau of Finance
Seoul, Korea

INSTRUCTIONS TO BANKS
NUMBER 6
              27 February 1946

TO: All Banks, Trust Companies and Financial Institutions, including the Postal Savings Section of the Bureau of Communications.

1. Title to accounts of Japanese nationals was vested in the Military Government of Korea as of 25 September 1945 by Section II of Ordinance Number 33, dated 6 December 1945. You are hereby directed that withdrawals from such formerly owned Japanese accounts will be permitted as follows:

a. Any Japanese national having proper official permission in writing for immediate return to Japan may withdraw one thousand yen (¥1,000) in Bank of Japan notes for each person of his household who has been granted official permission to return to Japan; and

b. Any Japanese national remaining in Korea who represents the head of his household may withdraw five hundred yen (¥500) a week in Bank of Chosen notes for each adult dependent for living expenses.

BY DIRECTION OF THE MILITARY GOVERNOR:

CHARLES GORDON
Lt Col, FD
Director

HEADQUARTERS
UNITED STATES ARMY MILITARY
GOVERNMENT IN KOREA
Department of Finance
Seoul, Korea

INSTRUCTIONS TO BANKS
NUMBER 12

2 July 1946

TO: Bank of Chosen
     Choheung Bank
     Chosen Commercial Bank
     Chosen Industrial Bank
     Chosen Savings Bank
     Chosen Trust Company
     Federation of Financial Associations

1. In accordance with provisions of Ordinance 95, dated 1 July 1946, you are directed to accept for deposit supplemental military yen marked "A" in all denominations and to pay out at face value Bank of Chosen yen deposited. The supplemental military yen marked "A" will be forwarded to the Head Office of the Bank of Chosen which will, in turn, forward such currency to this Department.

2. Inasmuch as Bank of Chosen currency will be paid out, no deposit receipts will be issued. However, banks accepting such currencies will use the same forms issued under Ordinance Number 57 and will record the serial numbers of the currency deposited as well as other information called for thereon. Only part 2 and 3 of the receipts will be used. Part 3 of the custody receipt will be forwarded with the supplemental military yen currency marked "A" to the Head Office of the Bank of Chosen.

3. You will accept such currency starting 1 July 1946. No supplemental military yen marked "A" will be received after 10 July 1946.

4. The Head Office of the Bank of Chosen will transfer credit at face value for all supplemental military yen marked "A" turned in by other financial institutions and will render a claim to this Department at the time the notes are forwarded.

CHARLES GORDON
Lt Col, FD
Director

CUSTODY ORDER
NUMBER 1

14 November 1945

CONSTRUCTION OF ORDINANCE NUMBER 2 : EXTENSION OF PERIOD : REGISTRATION OF TRANSFERS : GENERAL PROHIBITION.

1. Construction of 60-Day Period Referred to in Section 3*d* of Ordinance Number 2: Extension of Period.— a.- Buyers and sellers reporting proposed transactions under Section 3*c* of Ordinance Number 2 are failing to file reports in writing with the Property Custodian of the Military Government of Korea as required. Others are filing incomplete or defective reports. Still others are filing reports indicating a lapse of a considerable period after the date which appears on the alleged contract of sale. b.- The words "sixty (60) days of the date of the report" in Section 3*d* of Ordinance Number 2 are construed to mean sixty (60) days after receipt by the Property Custodian of the Military Government of Korea, Seoul, Korea, of a final, complete and accurate report, as required by the Property Custodian. c.- The sixty-day period, as defined in Section 1*b* hereof, may at any time be extended by the Property Custodian by noting such fact on his records. There is no requirement that the parties to the transaction be affirmatively advised by the Property Custodian of such extension. The status of the transaction may be ascertained by the parties at any time from the Property Custodian. d.- The sixty-day period mentioned in Section 3*d* of Ordinance Number 2 is hereby extended for a period of sixty (60) days from the date of the expiration thereof.

2. Registration of Transfers: General Prohibition.— a.- No transfer of property coming within the purview of Ordinance Number 2 may be registered at any Land Register, District Court, or elsewhere, unless and until Office of Property Custodian Form 104 is received from the Property Custodian

by the appropriate office of registry. All registrations of such property
effected at any time, heretofore and hereafter, without the receipt of such Form
104, are deemed to be invalid registrations. b.- To the extent necessary to
accomplish the intent of this provision, this Section shall be construed as a
general prohibition under the terms of Section 3d of Ordinance Number 2.

3. Effective Date.— This order is effective at midnight 14 November 1945.

BY DIRECTION OF THE MILITARY GOVERNOR :

JOHN B. LAPSLEY
Lt Col Inf
Property Custodian

HEADQUARTERS
UNITED STATES ARMY FORCES IN KOREA
Office of the Military Governor
Property Custody Section
Seoul, Korea

CUSTODY ORDER
NUMBER 2   14 December 1945

REPORT OF JAPANESE PROPERTY
TITLE TO WHICH IS IN THE MILITARY GOVERNMENT OF KOREA
AND OPERATION, OCCUPANCY OR USE OF SUCH PROPERTY

SECTION I REPORT OF PROPERTY

The title to property of every type and description, and the proceeds thereof, owned or controlled, directly or indirectly, in whole or part on or since 9 August 1945 by the Government of Japan, or any agency thereof, or by any of its nationals, corporations, societies, associations, or any other organization of such government or incorporated or regulated by it has been vested in the Military Government of Korea as of 25 September 1945 under Ordinance Number 33, dated 6 December 1945. Every person who has possession or control of any such property will file a written report regarding such property. The report will be prepared in triplicate. All three copies will be certified as correct by the person filing the report and will be filed on or before 31 December 1945 with the Property Custody Officer, care of the Governor of the province in which such person is residing. A report will be deemed to be filed when it is received by appropriate Property Custody Officer or when it is properly addressed and mailed and bears a postmark date prior to midnight of the date upon which the report is due. The report will contain the following information:

    a. Name, address and nationality of person making the report.

    b. Relation of person making the report to the property reported.

        (1) If former Japanese owner, state when ownership of the property was acquired.

        (2) If not former Japanese owner, state under what authority and the date possession or control of the property was taken, attaching a copy of any written authorization.

    c. Name and address of former owner of property.

d. If property reported is real property, state:

(1) Location of each parcel.

(2) Size of each parcel.

(3) Description of buildings and appurtenances on each parcel.

(4) Use being made of each parcel.

(5) Name of tenants, if any, of each parcel.

(6) Value of each parcel on 9 August 1945,

(7) Location of Land Registry where each parcel is registered, the date of registration and the name of the person in which registered.

e. If property reported is tangible personal property, furnish:

(1) A list of all such property by types.

(2) Identity, location and value of each type of such property and basis of valuation.

f. If property reported is intangible personal property and evidences thereof, furnish:

(1) Detailed description of each item.

(2) Location of each item.

(3) Value of each item.

g. Receipts and expenditures in connection with such property on and after 9 August 1945.

SECTION II OPERATION, OCCUPANCY AND USE OF PROPERTY

All industrial, financial, commercial, agricultural, residential, and other properties or enterprises will be operated, occupied or used under existing or future agreements or arrangements authorized by the Property Custodian of the Military Government of Korea acting:

a. Through Property Custody Officers; or

b. Through officers of other Bureaus, Sections or agencies of the Military Government of Korea.

SECTION III EFFECTIVE DATE OF ORDER

This order is effective upon publication in the Official Gazette.

BY DIRECTION OF THE MILITARY GOVERNOR:

JOHN B LAPSLEY
Lt Col Inf
Property Custodian

Date of Publication: 26 December 1945

HEADQUARTERS
UNITED STATES ARMY FORCES IN KOREA
Office of the Military Governor
Seoul, Korea

CUSTODY ORDER
NUMBER 3   19 December 1945

    1. Agricultural lands which have been requisitioned by tactical units of USAFIK or by agencies of the Military Government of Korea will remain under the control and management of the requisitioning unit or agency, which will be responsible to the Property Custodian for the proper accounting for, and conservation and utilization of, such property.

    2. All agricultural lands formerly owned by Japanese nationals (natural or juridical persons) or by the Japanese Government or any of its agencies, not so requisitioned by any tactical unit or any agency of the Military Government, but owned as of 25 September 1945 by the Military Government of Korea as a result of the vesting of title thereto by Ordinance Number 33 dated 6 December 1945, will be managed by the New Korea Company, which is designated the responsible agent of the Property Custodian for the conservation, utilization, and accounting for such property.

    3. This order is effective upon publication in the Official Gazette.

BY DIRECTION OF THE MILITARY GOVERNOR

JOHN B. LAPSLEY
Lt Col Infantry
Property Custodian

Date of Publication: 14 January 1946

895

CUSTODY ORDER
NUMBER 4   8 March 1946

1. Forest lands which have been requisitioned by tactical units of USAFIK or by agencies of the Military Government of Korea will remain under the control and management of the requisitioning unit or agency, which will be responsible to the Property Custodian for the proper accounting for, and conservation and utilization of, such property.

2. All forest lands formerly owned—by Japanese nationals (natural or juridical persons) or by the Japanese Government or any of its agencies, not so requisitioned by any tactical unit or any agency of the Military Government, but owned as of 25 September 1945 by the Military Government of Korea as a result of the vesting of title thereto by Ordinance Number 33, dated 6 December 1945, will be managed by the Bureau of Agriculture (Forestry Department), which is designated the responsible agent of the Property Custodian for the conservation, utilization and accounting for such property.

3. This order shall be effective on the tenth day after the date appearing hereon.

BY DIRECTION OF THE MILITARY GOVERNOR:

JOHN B. LAPSLEY
Lt. Col. INF
Property Custodian

CUSTODY ORDER
NUMBER 5   **REPRINTED COPY**   21 July 1946

REPORTS BY VESTED COMPANIES

1. *Purpose.* The purpose of this order is to require the submission of financial reports with emphasis on total receipts and total expenditures of corporate, company, business, organizational, etc.,* funds to the end that adequate control may be exercised to preserve property and assets vested in USAMGIK. Reports shall be in English.

2. *Submission of Reports.*

    a. Quarterly Reports.

(1) Corporations, companies, businesses and other organizations, etc.,* operated by or under the control of Military Government, shall submit quarterly balance sheets and profit and loss statements to their respective Department or Office Director who shall, in turn, forward a copy of each report to the Office of Property Custody before the 20th day of the second month succeeding the quarterly reporting date. Quarterly reporting dates are hereby fixed as follows: The 31st day of March, 30th day of June, 30th day of September and 31st day of December, respectively.

(2) Corporations, companies, businesses and other organizations, etc.,* operated by or under the control of Military Government, not responsible to any Department or Office but acting, in part, as an agency of the Property Custodian, shall submit quarterly balance sheets and profit and loss statements to the Office of Property Custody before the 30th day of the second month succeeding the quarterly reporting date. Quarterly reporting dates are hereby fixed as follows: The 31st day of March, 30th day of June, 30th day of September and 31st day of December, respectively.

(3) Corporations, companies, businesses and other organizations, etc., * with offices in more than one province, operated by Koreans for Military Government, shall submit a quarterly balance sheet and profit and loss

---

*Including *all* types of vested business organizations (industrial, mercantile, banking and financial, insurance, transportation, shipping, etc.).

statement, before the 30th day of the second month succeeding the quarterly reporting date, to the Provincial Property Custodian in the Province where the head office is located. Quarterly reporting dates are hereby fixed as follows: The 31st day of March, 30th day of June, 30th day of September and 31st day of December, respectively.

(4) Corporations, companies, businesses and other organizations, etc.,* located wholly within one province, shall submit monthly balance sheets on the date set by the Provincial Property Custodian, or by the Department or Office to which the corporation, company, business or organization, etc.,* is responsible, but in no event later than the 20th day of the month succeeding the monthly reporting date. The monthly reporting date shall be considered as of the close of business of the last day of the month.

(5) In cases where a corporation, company, business, or other organization, etc.,* has reporting dates other than these stated in this order, that corporation, company, business or other organization, etc.,* shall report such fact to the Provincial Property Custodian, or the Department or Office to which that corporation, company, business or organization is responsible. The Provincial Property Custodian, or the Department or Office to which the corporation, business or organization is responsible, may authorize the use of such other reporting dates, and shall report such action to the Office of Property Custody.

b. Monthly Profit and Loss Statements.

(1) Monthly profit and loss statements shall be submitted by all corporations, companies, businesses and other organizations, etc.,* to the Provincial Property Custodian, or to the Department, Office or independent agency to which the corporation, company, business or organization, etc.,* is responsible. Statements shall be submitted on the date set by the Provincial Property Custodian, the Department, Office or independent agency responsible for the operation of the corporation, company, business or organization, etc.,* but in no event later than the 20th day of the month succeeding the monthly reporting date.

(2) Copies of all monthly profit and loss statements shall be forwarded to the Office of Property Custody on the 20th day of the month succeeding such monthly reporting date.

3. *Definitions.*

a. *Balance sheet.* A *balance sheet* means a record showing all assets and all liabilities of a corporation, company, business, or other organization etc.,* at the close of business of a specified day.

---

*Including *all* types of vested business organizations (industrial, mercantile, banking and financial, insurance, transportation, shipping, etc.).

(1) *Assets* means the entire property of a corporation, company, business or other organization, etc.,* normally applicable or subject to the payment of its debts. Assets on the balance sheet show the book or cost value of resources such as real property, cash, inventories, supplies, accounts receivable, etc.

(2) *Liabilities* means debits. Balance sheet debits include total indebtedness, capital investment, and surplus or deficit.

(3) Balance sheet will show inventories at *Cost*. Approximate current market prices of inventories will be shown as footnotes to the balance sheet. Unused overdrafts and/or credit lines will be shown as footnotes on the balance sheet.

b. Profit and Loss Statement A *profit and loss statement* means a report covering total receipts and total expenditures of a corporation, company, business, organization, etc.,* during a specified period. e.g., one month, one quarter, etc.

(1) *Total receipts* means all assets or items affecting asset value, received by a corporation, company, business, organization, etc.,* during a specified period, e.g., one month, one quarter, etc.,

(2) *Total expenditures* means all disbursements whether in money or property, by a corporation, company, business, organization etc.,* during a specified period, e.g., one month, one quarter, etc.

4. *Enterprises Operating at a Loss or on a Financially Unsound Basis.*

a. Provincial Property Custodians, Departments, Offices or Independent Agencies shall submit lists in duplicate of corporations, companies, businesses or other organizations, etc.,* operating at a loss for two successive months, or on the borderline, to the Office of Property Custody with case histories recommendation as to whether corporation, company, business or other organization, etc.,* (including in each case specific reasons for opinions) should be continued in operation.

b. The Office of Property Custody shall forward lists, case histories and recommendations to the Military Governor through the National Economic Board for determination of future policy as to the continuation or deactivation of the individual corporations, companies, business or other organization, etc.*

---

*Including *all* types of vested business organizations (industrial, mercantile, banking and financial, insurance, transportation, shipping, etc.)

Custody Order No. 5

5. *Initial Date For Submission of Reports.*

c. The first quarterly balance sheet shall be for the quarter ending 30 June 1946.

The first monthly balance sheet shall be for the month ending 30 June 1946.

b. The first quarterly profit and loss statement shall be for the quarter ending 30 June 1946.

The first monthly profit and loss statement shall be for the month ending 30 June 1946.

C H MENGER
Lt Col TC
Property Custodian

---

*Including *all* types of vested business organizations (industrial, mercantile, banking and financial, insurance, transportation, shipping, etc.)

CUSTODY ORDER
NUMBER 6   21 July 1946

<u>REPRINTED COPY</u>

FISHING AND OTHER VESSELS

1. *Purpose.* The purpose of this order is to clarify the responsibilities of various elements and agencies of the Government of Korea in connection with utilization, conservation, and accountability for vessels vested in the Military Government by Ordinance Number 33, dated 6 December 1945.

2. *Reports of Vessels.* In order to effectuate the purpose of this order it is first necessary to ascertain the existing facts concerning the possession and control of vessels vested in the Military Government. To ascertain such facts reports are required, and they shall be made within fifteen days after the effective date of this order, as hereinafter provided.

   a. *Persons, etc. Required to Report:* Every person, natural or juridical, and every political subdivision, agency and instrumentality of the Government of Korea having possession or control of any vessel or vessels vested in the Military Government of Korea by Ordinance Number 33, dated 6 December 1945, shall file a dated and signed report in triplicate.

   b. *Contents of Reports.* Each such report shall contain the following information:

   (1) Name and address of person making the report and relation to each vessel reported, including a statement of authority for or circumstances under which possession or control was acquired and the date thereof.

   (2) Description of each vessel reported, including its type, name, number and signal letters, length, loading capacity if over 20 meters in length, port of registry, date of registration, present location and purposes for which it is being used.

   (3) Estimate of the value of each vessel reported and a statement

of all expenditures and receipts in connection therewith on and after 9 August 1945, showing the purpose of each such expenditure and the disposition of each such receipt.

(4) Statement as to whether or not person or agency reporting desires to continue in possession or control of each vessel reported.

c. Submission of Reports. Reports on fishing vessels* shall be delivered personally or mailed by registered mail addressed to the Department of Agriculture of the Government of Korea (Attention: Fisheries Bureau). Reports on other vessels shall be delivered personally or mailed by registered mail addressed to the Department of Transportation of the Government of Korea (Attention: Marine Bureau). The Marine Bureau and the Fisheries Bureau shall, in turn, submit a report broken down by provinces, in duplicate, to the Office of Property Custody.

3. *Disposition of Vessels.* After the reports have been received decisions shall be made, and specific orders issued with respect to the disposition of each vessel reported, by the Property Custodian (or by provincial Property Custodians under his direction) in accordance with directives of the Military Governor and other existing legal requirements. Disposition orders shall provide either for continuance of possession or control by the person reporting the vessel, or for a turn-over of such vessel to a designated person or agency at a specified time and place. The following general principles shall be observed in connection with such orders:

a. All fishing vessels* shall either be surrendered and delivered up to the Korean Fisheries Corporation, or operated in accordance with its orders concerning utilization, conservation and accounting.

b. All other vessels shall either be surrendered and delivered up to the Marine Bureau of the Department of Transportation, or operated in accordance with its orders concerning utilization, conservation and accounting.

---

*For purposes of this order, fishing vessels are defined in accordance with *Article 3* of Order Number 4 of the Communications Department of the Government of Japan, dated February 1934 (applied in Korea by Government General Order Number 20 dated February 1935), as revised, which provides that fishing vessels shall include any of the following:

1. Vessels exclusively engaged in catching fish.
2. Vessels engaged in catching fish and equipped for preserving fish and manufacturing fish products.
3. Vessels exclusively used for transportation of fish or fish products.
4. Vessels exclusively engaged in experimentation, examination, training or controls connected with fishing and equipped for fishing.

c. A copy of each disposition order shall be furnished by the Property Custodian (or provincial Property Custodian) to the Korean Fisheries Corporation or to the Marine Bureau of the Department of Transportation.

4. *Penalties.* Persons who fail to report as required herein, or who fail to surrender and deliver up possession of vessels as required by the Property Custodian (or provincial Property Custodian), or who continue in unauthorized possession of such vessels without proper legal instruments evidencing right to possession, or who otherwise violate the provisions or attempt to evade the intent of this order, shall be charged with violating Ordinance Number 33 and other applicable laws and regulations.

5. *Limitations on Scope of Order.* This order has no effect on existing and future requirements of law relating to registration, inspection, licensing, marking, traffic and clearance, customs, navigation and similar matters pertaining to vessels and personnel employed in connection therewith.

6. *Effective Date.* This order shall be effective on the tenth day after the date appearing hereon.

CHARLES H MENGER
Lt Col   TC
Property Custodian

6. *MGPC Form 19: Cash Receipt Voucher to Accompany Deposit Slip.*

This form will be submitted by both the depositor and the bank receiving the deposit to the Property Custodian, together with the deposit slip (MGPC Form No 14) of each deposit made to the account of the Military Government, Property Custodian Deposit Account.

7. *MGPC Form 20: Profit and Loss Statement and Balance Sheet.*

All corporations, companies, partnerships or other vested business organizations shall file with the Property Custodian, through the appropriate controlling department, office, agency or instrumentality (*in addition to* the reports required to be filed by Custody Order Number 5), semi-annual and annual profit and loss statements and balance sheets on MGPC Form 20 not later than 60 days after the semi-annual or annual reporting date under Custody Order Number 5.

8. *MGPC Form 21: Short Form Profit and Loss Statement and Balance Sheet.*

Effective for all reporting periods subsequent to 31 December 1946, all profit and loss statements and balance sheets submitted to the Property Custodian pursuant to Custody Order Number 5 will be submitted on MGPC Form 21 (*Short Form Profit and Loss Statement and Balance Sheet*).

BY DIRECTION OF THE MILITARY GOVERNOR:

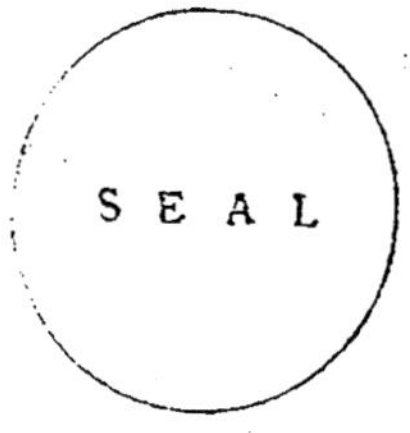

HARRY D BISHOP
Lt Col FA
Property Custodian

HEADQUARTERS
UNITED STATES ARMY MILITARY
GOVERNMENT IN KOREA
Office of Property Custody
Seoul, Korea

CUSTODY ORDER

NUMBER 8   31 December 1946

OPERATION OF ALL TYPES OF VESTED BUSINESS ORGANIZATIONS

1. *Purpose and Scope*

  a. The purpose of this order is to define the responsibilities of various officials, elements and agencies of the Military Government in connection with utilization and conservation of and accountability for the operation of vested business organizations.

  b. The Property Custodian will transfer the control of management, including the appointment of directors and/or managers, and the operation of all vested corporations, companies, partnerships and other vested business organizations (industrial, mining, mercantile. contracting, fishing, banking, agriculture, forestry, financial, private schools, insurance, transportation, shipping, *etc*) now under the jurisdiction of the Property Custodian, to appropriate advisory officials of departments, offices or agencies of the Military Government in accordance with the procedures outlined by the Office of Property Custody in procedural memoranda.

  c. Control of management of any vested business organization will be exercised by such appropriate advisory official appointed by the Property Custodian to operate the specific vested business organization.

  d. The appropriate advisory official appointed to control the management of a specific vested business organization will be accountable and responsible to the Property Custodian for the appointment of proper management, procurement of raw materials, disposal of finished products (subject to applicable

— 1 —

軍政廳　官報　管財令　　第八號　　一九四六年十二月三十一日

directives, laws and regulations, and maintenance and preservation of the assets of such business organization.

e. Control of management of vested business organizations not hitherto disclosed and/or considered as hidden assets will be similarly transferred to appropriate advisory officials as such organizations or assets are discovered.

2. *Responsibility and Accountability; Amendment of Prior Orders, etc.*

a. All Custody Orders, procedural memoranda and other orders, directives and requirements of the Property Custodian pertaining to the custody of vested business organizations or vested property, which require any action, including the filing of any reports, by the department, office or agency of the Military Government having control over such vested business organizations or vested property, are hereby amended so as to require such action, including the filing of such reports, by the adviser to the director or other head of such department, office or agency. Such advisers are in all cases considered the responsible and accountable officials for the maintenance, preservation, care, security and proper administration and custody of such vested business organizations or property. All provisions of this Custody Order shall apply to the management of all vested business organizations and property heretofore under the control of any department, office or agency of the Military Government.

3. *Records and Accounts.*

a. For purposes of this order, the Property Custodian of the City of Seoul is considered as a provincial Property Custodian.

b. In accordance with *Custody Order Number 5,* all vested business organizations operating under the control of the accountable and responsible advisory official of any department, office or agency of Military Government and physically located entirely within the boundaries of one province will maintain proper records and accounts and will forward five copies of required reports to such controlling advisory official. The advisory official will forward three copies to the provincial Property Custodian and one copy to the adviser of the corresponding national department, office or agency. The provincial Property Custodian will in every case forward two copies of required reports to the Office of Property Custody.

c. All vested business organizations operating under the control of the accountable and responsible advisory official of any department, office or agency of Military Government and having offices in two or more provinces or otherwise being operated nationally will maintain proper records and accounts and will forward in duplicate required reports for the business organization as a whole to such controlling advisory official. The controlling advisory official will forward one copy in every case to the Office of Property Custody.

d. Balance sheets, income statements, requests for approval of capital expenditures, reports of change of status of capital assets and other reports will be submitted on forms provided in Custody Order Number 7 and other Custody Orders now or hereafter issued.

e. The Property Custodian reserves the right to make such inspections and audits, and to require such further reports as he deems necessary.

4. *Capital Expenditures.*

a. No capital expenditures will be made without the prior written approval of the Property Custodian. All requests therefor will be submitted to the Property Custodian in triplicate by the accountable and responsible advisory official of the department, office or agency of Military Government concerned.

b. Requests for capital expenditures by vested business organizations located entirely within one province will be routed in the following manner: The advisory official accountable and responsible for the vested business organization will forward five copies of the request for capital expenditures to the provincial Property Custodian. The provincial Property Custodian will submit four copies of the request with his recommendations to the advisory official of the national department, office or agency of Military Government corresponding to the provincial bureau or agency concerned. The advisory official of the national department, office or agency of Military Government will submit the request to the Property Custodian in triplicate, with his recommendations.

c. Capital expenditures made prior to the date of this order will be reported by the appropriate advisory official as required in paragraphs *a* and *b* of paragraph 4 hereof. One copy will be forwarded in every case by such advisory official to the Property Custodian. Where applicable, a negative report will be made concerning capital expenditures from 9 August 1945 to the date appearing on this Custody Order.

— 3 —

軍政廳　官報　管財令　第八號　一九四六年十二月三十一日

d. "Capital expenditures" as used herein is limited to cash expenditures for the acquisition or extraordinary repair or maintenance of fixed, permanent or capital assets.

e. Approvals for capital expenditures will be limited by the Property Custodian so as to be within the scope of his legal authority.

5. *Discontinuance of Operations.*

a. The appropriate advisory official will be responsible and accountable for all operations of vested enterprises and properties as outlined above.

b. If the appropriate advisory official of any national department, office or agency of Military Government considers that a portion or all of a vested business organization should cease operations, he shall forward five copies of his recommendations to the Office of Property Custody. The Property Custodian will forward these recommendations with an indorsement thereon to the National Economic Board for review and recommendations. The Property Custodian will thereafter take such action as is appropriate, within the scope of his legal authority.

c. Channels employed in routing requests for capital expenditures of vested business organizations located entirely within one province as set forth in paragraph 4 hereof will also apply in requesting the discontinuance of operations, partial or complete, of such vested business organizations.

d. The accountable and responsible advisory official will at all times exercise utmost care and caution in the operation of vested enterprises so as to prevent disposition or diminution in value of assets, unauthorized liquidation, or improper payments or transfers of any kind.

e. When discontinuance of operations has been approved by the Property Custodian, the Chairman of the Board of Directors and/or manager of the vested business organization concerned will report personally to the provincial Property Custodian or, in the case of nationally operated vested business organizations, to the Property Custodian, for written instructions on discontinuing operations of the whole or part of the business of such vested business organization. The Chairman of the Board of Directors and the manager and the appropriate advisory official will be accountable and responsible for orderly discontinuance of operations. Discontinuance of operations will be limited by

— 4 —

the Property Custodian so as to be within the scope of his legal authority.

f. Reports of the discontinuance of operations of vested business organizations or parts thereof prior to the date appearing on this Custody Order will be submitted as outlined in paragraph 3 hereof, to the provincial Property Custodian, or, in the case of nationally operated vested business organizations, to the Office of Property Custody, by the appropriate advisory official. One copy will be forwarded in every case to the Office of Property Custody. Where applicable, a negative report will be submitted covering discontinuances authorized or unauthorized, from 9 August 1945 to the date appearing on this Custody Order.

6. *Profits.*

a. Profits of all corporations, companies, partnerships or other vested business organizations will be preserved in the organization treasury until further order of the Property Custodian.

b. No dividends, stock issues in lieu of dividends, bonuses, emoluments, gifts or similar payments or transactions will be made, paid or declared without the prior written approval of the Property Custodian. Approval will be limited by the Property Custodian so as to be within the scope of his legal authority.

c. A report of dividends declared or paid, of stock issued in lieu of dividends and of any bonuses, emoluments, gifts or similar payments made prior to the publication of this Custody Order will be submitted as outlined in paragraph 3 by the appropriate advisory official to the provincial Property Custodian or, in the case of nationally operated vested business organizations, to the Office of Property Custody. Where applicable, a negative report will be submitted for each vested business organization for the period 9 August 1945 to the date appearing on this Custody Order.

d. No manager, director or employee of any vested business organization shall, by himself or through another, enter into or derive any profit or advantage from any transaction involving such enterprise except with the prior written permission of the Property Custodian, based upon a complete disclosure of the circumstances and nature of the proposed transaction.

— 5 —

7. *Cannibalization.*

a. If the accountable and responsible advisory officials concerned determine that, in the process of operations, certain machinery or equipment belonging to a vested business organization may be used to the best interest of Korea in another vested business organization, the following procedure will be followed:

(*1*) The Chairman of the Board of Directors and/or manager of the vested business organization owning the machinery or equipment will cause an appraisal to be made to determine the value of such machinery or equipment. The Chairman of the Board of Directors and/or manager of the vested business organization proposing to receive the machinery or equipment will likewise have an appraisal made of the value of such machinery or equipment. The accountable and responsible advisory officials concerned will be responsible for an independent appraisal. All such appraisals shall be forwarded to the Property Custodian. The Property Custodian will thereupon in each case approve or disapprove the proposed transfer of use and fix the rental and other terms and conditions of such proposed use within the scope of his legal authority.

(*2*) Upon receipt of approval from the Property Custodian, the accountable and responsible advisory official into whose jurisdiction the machinery or other equipment is removed will immediately report in writing all such removals, date of removals, and former and present location thereof to the provincial Property Custodian or, in the case of nationally operated vested business organizations, to the Office of Property Custody. One copy will be forwarded in every case to the Office of Property Custody.

(*3*) Requests for approval of such proposed changes in use will be forwarded to the Office of Property Custody in triplicate. The routing of such requests will be as outlined in paragraph 4 hereof for requests for capital expenditures.

(*4*) Vested business organizations receiving such machinery or equipment will be prepared to pay an acceptable rental for the use of such machinery or equipment.

(*5*) All proposed changes in use of such machinery or equipment as outlined above must be approved in writing by the Property Custodian before they are initiated. Forms provided in *Custody Order Number 7* will be used. All proposed changes in use will be limited L the Property Custodian so as to be within the scope of his legal authority. Care will be exercised to prevent interested persons from acting in a dual capacity, such as managers or chairmen of boards or directors representing two different companies involved in any transaction.

b. Vested business organizations which have purchased, requisitioned, sold or had machinery and/or equipment removed or taken away prior to the issuance of this Custody Order, whether authorized or unauthorized, will report the facts fully as outlined in paragraph 3 hereof. One copy will be forwarded in every case to the Property Custodian. The appropriate advisory officials are responsible for the forwarding of such reports. Where applicable, a negative report will be submitted for the period 9 August 1945 to the date appearing on this Custody Order.

8. *Bonding of Chairman of Board of Directors and/or Manager.* The advisory official of any department, office or agency of Military Government who is appointed by the Property Custodian to operate a specific vested business organization will require a bond of the Chairman of the Board of Directors and/or manager appointed by him, as outlined in procedural memoranda of the Office of Property Custody.

9. *Date for Reports.* All reports required hereunder shall be submitted in writing within 30 days from the date appearing hereon; and, thereafter, in writing, from time to time, within 10 days after facts arise which require further reports. The appropriate advisory official shall in each case be responsible for the forwarding of such reports and the accuracy of the information therein contained.

10. *Effective Date.* This order shall be effective on the date appearing hereon.

BY DIRECTION OF THE MILITARY GOVERNOR:

HARRY D BISHOP
Lt Col   FA
Property Custodian

— 7 —

軍政廳　官報　管財令　　第八號　　一九四六年十二月三十一日

HEADQUARTERS
UNITED STATES ARMY MILITARY
GOVERNMENT IN KOREA
Office of Property Custody

Seoul, Korea

CUSTODY ORDER
NUMBER 9

31 March 1947

REVISING CUSTODY ORDER NUMBER 8

Custody Order Number 8 is hereby revised to read as follows:

OPERATION OF ALL TYPES OF VESTED BUSINESS ORGANIZATIONS

1. *Purpose and Scope.*

   a. The purpose of this order is to define the responsibilities of various officials, elements and agencies of the Military Government in connection with utilization and conservation of and accountability for the operation of vested business organizations.

   b. The Property Custodian will entrust the management, including the appointment of directors and/or managers, and the operation of all vested corporations, companies, partnerships and other vested business organizations (industrial, mining, mercantile, contracting, fishing, banking, agriculture, forestry, financial, private schools, insurance, transportation, shipping, *etc*) now under the jurisdiction of the Property Custodian, to appropriate heads of departments, offices or agencies of the Military Government, in accordance with the procedure outlined by the Office of Property Custody in procedural memoranda.

   c. Management of any vested business organization will be exercised by such appropriate heads of departments, offices or agencies, appointed by the Property Custodian to operate the specific vested business organizations. In all cases relating to management or control of vested property, orders, directives and appointments will require the written concurrence of the appropriate advisers.

— 1 —

d. The appropriate officials appointed to control the management of specific vested business organizations will be accountable and responsible to the Property Custodian for the appointment of proper management, procurement of raw materials, disposal of finished products (subject to applicable directives, laws and regulations), and maintenance and preservation of the assets of such business organizations.

e. Control of management of vested business organizations not hitherto disclosed and/or considered as hidden assets will be similarly transferred to appropriate officials as such organizations or assets are discovered.

2. *Responsibility and Accountability; Amendment of Prior Orders, etc.*

All Custody Orders, procedural memoranda and other orders, directives and requirements of the Property Custodian pertaining to the custody of vested business organizations or vested property, which require any action by the department, office or agency of the Military Government having control over such vested business organizations or vested property, are hereby amended so as to require such action by the head of such department, office or agency with the written concurrence of the appropriate adviser. Such officials are in all cases considered the responsible and accountable officials for the maintenance, preservation, care, security and proper administration and custody of such vested business organizations or property. All provisions of this Custody Order shall apply to the management of all vested business organizations and property heretofore under the control of any department, office or agency of the Military Government.

3. *Records and Accounts.*

a. For purposes of this order, the Property Custodian of the City of Seoul is considered as a provincial Property Custodian.

b. In accordance with Custody Order Number 5, all vested business organizations operating under the control of the accountable and responsible officials of any department, office or agency of Military Government and physically located entirely within the boundaries of one province will maintain proper records and accounts and will forward four copies of required reports to

軍政廳 官報 管財令 第九號　　　一九四七年三月三十一日

such controlling officials. The officials will forward two copies to the provincial Property Custodian and one copy to the officials of the corresponding national department, office or agency. The provincial Property Custodian will in every case forward one copy of required reports to the Office of Property Custody.

c. All vested business organizations operating under the control of the accountable and responsible officials of any department, office or agency of Military Government and having offices in two or more provinces or otherwise being operated nationally will maintain proper records and accounts and will forward in duplicate required reports for the business organization as a whole to such controlling officials. The controlling officials will forward one copy in every case to the Office of Property Custody.

d. Balance sheets, income statements, requests for approval of capital expenditures, reports of change of status of capital assets and other reports will be submitted on forms provided in Custody Order Number 7 and other Custody Orders now or hereafter issued.

e. The Property Custodian reserves the right to make such inspections and audits, and to require such further reports as he deems necessary.

4. *Capital Expenditures.*

a. No capital expenditures will be made without the prior written approval of the Property Custodian. All requests therefor will be submitted to the Property Custodian in triplicate by the accountable and responsible officials of the department, office or agency of Military Government concerned.

b. Requests for capital expenditures by vested business organizations located entirely within one province will be routed in the following manner: The officials accountable and responsible for the vested business organization will forward five copies of the request for capital expenditures to the provincial Property Custodian. The provincial Property Custodian will submit four copies of the request with his recommendations to the officials of the national department, office or agency of Military Government corresponding to the provincial bureau or agency concerned. The officials of the national department, office or agency of Military Government will submit the request to the Property Custodian in triplicate, with their recommendations.

c. Capital expenditures made prior to the date of this order will be reported by the appropriate officials as required in paragraphs *a* and *b* of section 4 hereof. One copy will be forwarded in every case by such officials to the Property Custodian. Where applicable, a negative report will be made concerning capital expenditures from 9 August 1945 to the date appearing on this Custody Order.

d. "Capital expenditures" as used herein is limited to cash expenditures for the acquisition or extraordinary repair or maintenance of fixed, permanent or capital assets.

e. Approvals for capital expenditures will be limited by the Property Custodian so as to be within the scope of his legal authority.

5. *Discontinuance of Operations.*

a. The appropriate officials will be responsible and accountable for all operations of vested enterprises and properties as outlined above.

b. If the appropriate officials of any national department, office or agency of Military Government consider that a portion or all of a vested business organization should cease operations they shall forward five copies of their recommendations to the Office of Property Custody. The Property Custodian will forward these recommendations with an indorsement thereon to the National Economic Board for review and recommendations. The Property Custodian will thereafter take such action as is appropriate, within the scope of his legal authority.

c. Channels employed in routing requests for capital expenditures of vested business organizations located entirely within one province as set forth in section 4 hereof will also apply in requesting the discontinuance of operations, partial or complete, of such vested business organizations.

d. The accountable and responsible officials will at all times exercise utmost care and caution in the operation of vested enterprises so as to prevent disposition or diminution in value of assets, unauthorized liquidation, or improper payments or transfers of any kind.

— 4 —

軍政廳　官報　管財令　第九號　　　　一九四七年三月三十一日

e. When discontinuance of operations has been approved by the Property Custodian, the Chairman of the Board of Directors and/or manager of the vested business organization concerned will report personally to the provincial Property Custodian or, in the case of nationally operated vested business organizations, to the Property Custodian, for written instructions on discontinuing operations of the whole or part of the business of such vested business organization. The Chairman of the Board of Directors and the manager and the appropriate officials will be accountable and responsible for orderly discontinuance of operations. Discontinuance of operations will be limited by the Property Custodian so as to be within the scope of his legal authority.

f. Reports of the discontinuance of operations of vested business organizations or parts thereof prior to the date appearing on this Custody Order will be submitted as outlined in section 3 hereof, to the provincial Property Custodian, or, in the case of nationally operated vested business organizations, to the Office of Property Custody, by the appropriate officials. One copy will be forwarded in every case to the Office of Property Custody. Where applicable, a negative report will be submitted covering discontinuances authorized or unauthorized, from 9 August 1945 to the date appearing on this Custody Order.

6. *Profits.*

a. Profits of all corporations, companies, partnerships or other vested business organizations will be preserved in the organization treasury until further order of the Property Custodian.

b. No dividends, stock issues in lieu of dividends, bonuses, emoluments, gifts or similar payments or transactions will be made, paid or declared without the prior written approval of the Property Custodian. Approval will be limited by the Property Custodian so as to be within the scope of his legal authority.

c. A report of dividends declared or paid, of stock issued in lieu of dividends and of any bonuses, emoluments, gifts or similar payments made prior to the publication of this Custody Order will be submitted as outlined

— 5 —

916

in section 3 by the appropriate officials to the provincial Property Custodian or, in the case of nationally-operated vested business organizations, to the Office of Property Custody. Where applicable, a negative report will be submitted for each vested business organization for the period 9 August 1945 to to the date appearing on this Custody Order.

d. No manager, director or employee of any vested business organization shall, by himself or through another, enter into or derive any profit or advantage from any transaction involving such enterprise except with the prior written permission of the Property Custodian, based upon a complete disclosure of the circumstances and nature of the proposed transaction.

7. *Cannibalization.*

a. If the accountable and responsible officials concerned determine that, in the process of operations, certain machinery or equipment belonging to a vested business organization may be used to the best interest of Korea in another vested business organization, the following procedure will be followed :

(1) The Chairman of the Board of Directors and/or manager of the vested business organization owning the machinery or equipment will cause an appraisal to be made to determine the value of such machinery or equipment. The Chairman of the Board of Directors and/or manager of the vested business organization proposing to receive the machinery or equipment will likewise have an appraisal made of the value of such machinery or equipment. The accountable and responsible officials concerned will be responsible for an independent appraisal. All such appraisals shall be forwarded to the Property Custodian. The Property Custodian will thereupon in each case approve or disapprove the proposed transfer of use and fix the rental and other terms and conditions of such proposed use within the scope of his legal authority.

(2) Upon receipt of approval from the Property Custodian, the accountable and responsible officials into whose jurisdiction the machinery or other equipment is removed will immediately report in writing all such removals, date of removals, and former and present location thereof to the

軍政廳　官報　管財令　第九號　　　一九四七年三月三十一日

provincial Property Custodian or, in the case of nationally operated vested business organizations, to the Office of Property Custody. One copy will be forwarded in every case to the Office of Property Custody.

(3) Requests for approval of such proposed changes in use will be forwarded to the Office of Property Custody in triplicate. The routing of such requests will be as outlined in section 4 hereof for requests for capital expenditures.

(4) Vested business organizations receiving such machinery or equipment will be prepared to pay an acceptable rental for the use of such machinery or equipment.

(5) All proposed changes in use of such machinery or equipment as outlined above must be approved in writing by the Property Custodian before they are initiated. Forms provided in Custody Order Number 7 will be used. All proposed changes in use will be limited by the Property Custodian so as to be within the scope of his legal authority. Care will be exercised to prevent interested persons from acting in a dual capacity, such as managers or chairmen of boards or directors representing two different companies involved in any transaction.

b. Vested business organizations which have purchased, requisitioned, sold or had machinery and/or equipment removed or taken away prior to the issuance of this Custody Order, whether authorized or unauthorized, will report the facts fully as outlined in section 3 hereof. One copy will be forwarded in every case to the Property Custodian. The appropriate officials are responsible for the forwarding of such reports. Where applicable, a negative report will be submitted for the period 9 August 1945 to the date appearing on this Custody Order.

8. *Bonding of Chairman of Board of Directors and/or Manager.* The officials of any department, office or agency of Military Government who are appointed by the Property Custodian to operate specific vested business organizations will require a bond of the Chairman of the Board of Directors and/or manager as outlined in procedural memoranda of the Office of Property Custody.

— 7 —

軍政廳　官報　管財令　第九號　　　　一九四七年三月三十一日

9. *Date for Reports.* All reports required hereunder shall be submitted in writing within 30 days from the date appearing hereon; and, thereafter, in writing, from time to time, within 10 days after facts arise which require further reports. The appropriate officials shall in each case be responsible for the forwarding of such reports and the accuracy of the information therein contained.

10. *Effective Date.* This order shall be effective on the date appearing hereon.

BY DIRECTION OF THE MILITARY GOVERNOR:

HARRY D BISHOP
Lt Col FA
Property Custodian

HEADQUARTERS
UNITED STATES ARMY MILITARY
GOVERNMENT IN KOREA
Office of Property Custody
Seoul, Korea

CUSTODY ORDER
NUMBER 10
                                                                6 December 1947

OPERATION OF ALL TYPES OF JURIDICAL PERSONS ORGANIZED IN KOREA IN WHICH THE PROPERTY CUSTODIAN OWNS A STOCK OR OTHER INTEREST

1. *Purpose and Scope.* *a.* The purpose of this order is to implement the directive of the Military Governor dated 17 September 1947, subject, "Control of Juridical Persons Organized in Korea" and to accomplish the objectives stated therein.

b. This order shall be applicable to all juridical persons organized in Korea in which a stock or other interest is vested in USAMGIK under Ordinance No. 33, dated 6 December 1945, except such as may be specifically excepted by directive of the Property Custodian or higher authority.

c. The Property Custodian will entrust the management of said juridical persons to Boards of Directors duly elected and qualified in accordance with the charter, articles of incorporation or organization and by-laws of such juridical persons and in accordance with procedural memoranda of the Office of Property Custody.

2. *Responsibility and Accountability.* *a.* The appropriate heads of departments, offices or agencies of SKIG to whom management of such juridical persons is presently entrusted under Custody Order No. 9 dated 31 March 1947, as agents of the Property Custodian, are hereby entrusted with and are responsible for the exercise of all stockholders' rights in connection with such vested stock or other interest therein, as agents of the Property Custodian, including but not limited to, the voting of such stock or other interest at stockholders' meetings. In the exercise of such powers and the discharge of the duties

— 1 —

incident thereto, such officials shall be directly accountable and responsible to the Property Custodian.

b. In all cases in which vested stock or other interests are voted by such officials, written concurrence of the appropriate advisor will be required.

3. *Transfer of Control. a.* Transfer of management of juridical persons to Boards of Directors will be accomplished in an orderly manner. Before management of any juridical person is entrusted to a Board of Directors action will be taken as follows:

> (*1*) A preliminary study of each juridical person will be made in accordance with procedures established by the Property Custodian.
>
> (*2*) A general meeting of stockholders will be held on call of the Property Custodian.
>
> (*3*) The elected directors shall be organized and be qualified as such.
>
> (*4*) Management of the juridical person will be formally transferred to the Board of Directors by the Property Custodian in writing.

4. *Records and Reports. a.* Boards of Directors to whom management of juridical persons is entrusted pursuant to this order shall be responsible for establishing and maintaining records and accounts and for making reports as follows:

> (*1*) All records, accounts and reports required by the law of Korea and/or by the charter, articles of incorporation, organization or by-laws of such juridical person.
>
> (*2*) All records, accounts and reports required or to be required by the Property Custodian or other appropriate authority.

b. Such Boards of Directors shall submit to the Property Custodian all reports required under orders or directives of the Property Custodian, but the Property Custodian may, upon the recommendation of the official entrusted with the exercise of stockholders' rights in such juridical persons, concurred in by the appropriate advisor, reduce the number of reports required from a particular juridical person, when such juridical person has demonstrated the capacity for the establishment and maintenance of sound business practice.

c. The Property Custodian reserves the right to make such inspections and audits of such juridical persons as he deems necessary. This right may be exercised by the Property Custodian or by the official entrusted with the exercise of stockholders' rights pursuant to paragraph 2a hereof, or by the appropriate advisor.

5. *Additional Controls. a.* Boards of Directors entrusted with the management of juridical persons pursuant hereto shall have no power to perform the following acts without prior written approval of the Property Custodian :

(*1*) Incur liability for or make capital expenditures.

(*2*) Discontinue or change the type of operations.

(*3*) Authorize or make any division or payment of profits in the form of dividends, bonuses or otherwise.

(*4*) Authorize or make any gifts or contributions of any kind or character, or loan the funds of the juridical person except where in the business of loaning money.

(*5*) Authorize or make any expenditure for entertainment except as provided for in existing directives of the Military Governor or other appropriate authority and in accordance with the conditions therein set forth.

(*6*) Acquire or dispose of capital assets.

(*7*) Effect merger, consolidaton or dissolution, or change the structure of the juridical person, or its by-laws.

(*8*) Establish or pay salaries or fees of directors, auditors or managers.

b. Application for approval of any of the acts above set forth shall be made in accordance with procedures established by the Property Custodian for the management of vested business organizations and business organizations in which the Property Custodian has an interest.

6. *Interim Management.* Until such time as management of such juridical persons is formally entrusted to Boards of Directors as provided for herein, such juridical persons will continue to be managed by heads of departments, offices and agencies of SKIG as agent of the Property Custodian with the concurrence of the appropriate advisors in accordance with the provisions of Custody Order No. 9.

7. *Effective Date.* This order shall be effective on the date appearing hereon.

BY DIRECTION OF THE MILITARY GOVERNOR :

HARRY D BISHOP

Lt Col FA

Property Custodian

HEADQUARTERS
UNITED STATES ARMY MILITARY GOVERNMENT
IN KOREA
Office of Property Custody
APO 235 Unit 2

CUSTODY ORDER

NUMBER 11   4 May 1948

REVISING CUSTODY ORDER NUMBER 4

Custody Order Number 4 is hereby revised to read as follows:

1. Forest lands which have been requisitioned by tactical units of USAFIK or by agencies of the Military Government of Korea, will remain under the control and management of the requisitioning unit or agency, which will be responsible to the Property Custodian for the proper accounting for and conservation and utilization of such property.

2. All forest lands formerly owned by Japaneses nationals (natural or juridical persons) or by the Japanese Government or any of its agents, or so requisitioned by any tactical unit or any agency of the Military Government, but owned as of 25 September 1945 by the Military Government of Korea as a result of the vesting of title thereto by Ordinance No. 33, dated 6 December 1945, will be managed by the Director of Agriculture who is designated the responsible agent of the Property Custodian for the conservation, utilization and accounting for such property and who will act with the advice and concurrence of the appropriate advisor.

3. The Director of Agriculture with concurrence of his Advisor is authorized to expend out of the annual income derived from the sale of timber and other products from vested forest lands and rents from vested lands such sums as are necessary to protect, conserve, and promote good forestry practices on these properties all major expenditures to be subject to prior approval by the Property Custodian.

4. The Director of Agriculture will submit to the Property Custodian forty-five days after expiration of each calender quarter, an accounting of income and expenditures by province for the three previous months. Also, there will be prepared within sixty days after each three-month period, a statement covering

— 1 —

significant forest activities on vested forest lands, including progress reports on reforestation, timber sales, and effectiveness of measures for protection of the tree cover against fire, insects, and theft.

5. No part of the income from vested forest lands shall be used for any purpose not covered in Paragraph 3 above.

6. This order supersedes Custody Order Number 4, 8 March 1946, and shall become effective on the date appearing hereon.

BY DIRECTION OF THE MILITARY GOVERNOR:

HARRY D BISHOP
Lt Col   FA
Property Custodian

NATIONAL FOOD REGULATION
NUMBER 1

29 May 1946

COLLECTION OF SUMMER GRAINS

1. *Purpose.* The purpose of this regulation is to implement the national plan for the collection of summer grains in order to provide for the food needs of Koreans during the four-month period between the summer and fall grain harvests of the year 1946.

2. *Scope.* The objective is to collect a total of approximately 2.094.830 suk of summer grains (primarily by government purchase from farmers at fair and equitable prices), and the rationing of such collected grains to Koreans through government controlled channels. The National Food Administration is charged with the responsibility for overall coordination and direction of the execution of the collection program. Provincial Governors shall be responsible for the successful execution of the program in their respective provinces. The Korean Agricultural Association, the Controller of Commodities, the Provincial Food Service, and all other political subdivisions, agencies and instrumentalities of the Government of Korea shall assist and co-operate in the execution of the program.

3. *General Duties and Functions.* The following general duties and functions are assigned to the agencies indicated:

    a. National Food Administration. The National Food Administration shall:

    (1) Establish summer grain collection quotas for each province based upon production statistics supplied by the Department of Agriculture; and

    (2) Direct all movement of summer grains and the release thereof for retail distribution.

    b. Provincial Governors. Each Provincial Governor shall:

    (1) Break down his provincial summer grain collection quota into county, city, town and village quotas based upon reports from city and county heads and Agriculture Association and Provincial Food Service officials;

— 1 —

(2) Supervise the collection and delivery of the summer grain
quota to warehouses designated by the city and county heads;
and

(3) Supervise inspection of summer grains collected, as to weight
and quality.

c. Controller of Commodities; Provincial Food Services. The Controller
of Commodities, acting through the Provincial Food Services, shall:

(1) Make prompt payment to farmers upon presentation of their
collection receipts;

(2) Transport summer grains from town and village collection
points to designated warehouses;

(3) Process, store, move and release collected summer grains for
retail distribution, at the direction of the National Food Administrator;

(4) Receive funds from sale of summer grains to retailers at approved government prices; and

(5) Report and account as required.

4. *First Phase: Preparation for Collection Program.* The first phase of
the plan, involving preparation for the collection program in each province,
shall be completed at least ten days prior to the commencement of the harvest.

a. Explanation of Operational Details. Explanation of operational
details within each province, to county, city, town, village and block heads
and to local representatives of the Korea Agricultural Association and Provincial Food Service, shall be the responsibility of each Provincial Governor.

b. Summer Grain Crop Reports and Quotas. Not later than 31 May
1946, town, village and block heads shall make summer grain crop estimate
reports to the county or city heads, who shall forward them immediately to
the appropriate Provincial Governor. Upon receipt of such reports the Provincial Governor shall pro-rate the provincial quota among the respective
counties and cities. County and city heads shall assign a quota to each town,
village and block. Town, village and block heads shall in turn assign a
quota to each farm in their respective areas of jurisdiction. Collection
quotas shall be fixed so that each farmer is permitted to retain sufficient
summer grains for planting the next ensuing crop and a combined aggregate
or a total amount of summer grains not exceeding 300 hop per person for
each person regularly and actually residing with him as a member of his
household. County and city heads shall designate warehouses as summer
grain collection points. Such warehouses shall be suitable storage places,

adequately protected against fire and theft, and located as centrally as possible in each production area. Prices to be paid to farmers for summer grains shall be published in the press and posted at each collection point. Non-negotiable triplicate receipts shall be made available at collection points for presentation to farmers. Design, printing and distribution of such receipts to the collection points, shall be the joint responsibility of the Controller of Commodities and the Korea Agricultural Association. Expenses of printing and distribution thereof shall be borne by the Korea Agricultural Association, subject to reimbursement by the Government of Korea. All preparation shall be completed so that collections provided for herein may commence not later than 15 June 1946.

5. *Second Phase: Collection of Summer Grains.* The second phase of the plan, involving the actual collection of summer grains, shall commence as soon as such grains are harvested.

a. *Delivery by Individual Owner—Farmers.* Each individual owner—farmer shall deliver his quota of summer grains (and any amount in excess thereof that he may desire) to the designated collection point in his area and shall receive a non-negotiable receipt therefor. No sale of summer grains shall be made be made except to the Government or an authorized agency thereof. In case of necessity, town, village, or block authorities may collect grains at individual farms, in which case transportation and labor charges incident thereto shall be deducted from the amount paid to such farmers for such grain, and the amount of such charges shall be entered on receipts issued.

b. *Delivery by Tenant Farmers.* Each tenant farmer shall either deliver his quota of summer grains (and any amount in excess thereof that he may desire) directly to the collection point, where any amount of such crops due and owing to his landlord shall be noted on the collection receipt issued therefor, or shall first pay such proper crop rentals to his landlord.

In those cases where an individual landlord receives summer grains as rental, he shall be permitted to retain for his own use (regardless of the number of farms from which he derives such income) a combined aggregate or total amount of such grains not exceeding 300 hop per person for each person regularly and actually residing with him as a member of his household. All amounts in excess thereof shall be turned in at the collection point, and payment therefor made to the landlord at established prices.

In those cases where the landlord is not an individual person, the entire portion of such crops (except the necessary amount of grain needed as seed for planting the next ensuing crop) received by such landlord as rental shall be

delivered to the appropriate collection point, and payment to such landlord
made at established prices. No sale of summer grains shall be made except
to the Government or an authorized agency thereof.

    c. Inspection of Summer Grains; Receipts; Payment. Summer grains
received at collection points shall be inspected for weight and grade by regular government employed inspectors, and graded in accordance with current
grain inspection regulations. Non-negotiable receipts shall be prepared in
triplicate by the town, village, or block official at each collection point. The
original receipt shall be delivered to the farmer, who shall present it to the
Provincial Food Service for payment. One copy shall be forwarded immediately to the Provincial Food Service, and the remaining copy retained by the
town, village or block office. The Provincial Food Service shall compare
the original receipt presented by the farmer for payment with the copy received by it from the town, village, or block officials, and, if the two
correspond, shall pay the farmer the amount due and mark the receipt paid.
The farmer shall retain the original receipt after it has been marked paid.
The Provincial Governor may utilize the services of other agencies, including
civilian agencies, within the province in order to expedite payment to the
farmer for such grains.

    6. *Third Phase: Reports and Accounting.* The third phase of the plan,
involving reports and accounting, shall commence with the collection of summer grains, and shall be completed not later then 16 September 1946.

    a. Weekly Collection Reports. Each Saturday during the collection
program the town, village, or block heads shall forward to the appropriate
county or city heads a consolidated report showing the amount and kinds of
grains collected during the past seven days, in the following categories:

        (1) From owner—farmers;
        (2) From tenant farmers;
        (3) From individual landlords;
        (4) From government agencies;
        (5) From corporate farms; and
        (6) From others (specifying);

and the amounts and kinds of such grains delivered to the Provincial Food
Services during the week. Each county and city head shall consolidate these
reports and submit them to the Provincial Governor for further consolidation
and immediate forwarding to the Military Governor in Korea, attention National Food Administrator.

    b. Final Consolidated Collection Reports. Each county and city head
shall prepare a final consolidated report of the information submitted to him

by town, village and block heads and shall submit such report to the appropriate Provincial Governor not later than 10 September 1946. Each Provincial Governor shall consolidate his county and city reports and forward a final consolidated report so as to reach the Military Governor in Korea, attention National Food Administrator, not later than 16 September 1946.

c. Additional Reports and Accounting. The Controller of Commodities, the Provincial Food Services, and the Korean Agricultural Association shall keep records of their respective operations, and shall furnish reports on their activities when and as required. They shall be reimbursed by the Government of Korea for losses sustained by transactions authorized hereunder, and shall account for all moneys disbursed by them.

7. *Definition of "Summer Grains".* "Summer grains", as used herein, means wheat, rye, barley and naked barley.

8. *Specific Offenses Enumerated.* Without limiting the provisions of this regulation or of any ordinance, the following are specifically declared to be violations hereof:

a. Hoarding, illegally storing or concealing, or wrongfully failing to deliver up summer grains in accordance with collection quotas assigned by town, village or block heads;

b. Obstructing or contravening the announced plan for the collection of summer grains, or any announced program, policy, orders or requirements of the Military Government or authorized agencies thereof in connection therewith;

c. Selling, offering for sale, buying or offering to buy summer grains, except by permission or authority of the Military Government, or at other than officially established prices, or otherwise illegally or improperly dealing in such grains;

d. Failing or refusing to receive, collect, inspect, grade, store, safeguard, transport, issue proper receipts, keep accurate records and accounts, or make payments for or reports concerning summer grains delivered under the collection program, or other official misconduct or failure or wilful neglect of duty in connection therewith;

e. Any transaction or act, or attempted transaction or act, for the purpose or which has the effect of evading or avoiding the provisions of this regulation or the program described herein.

9. *Penalties.* Any person violating the provisions of this regulation shall, upon conviction by a Military Occupation Court, suffer such punishment as the court shall determine.

10. *Effective Date.* This regulation shall be effective on the date appearing hereon.

BY DIRECTION OF THE MILITARY GOVERNOR:

CARROLL V HILL
Lt Col · CE
National Food Administrator

Food Regulation No 1   — 5 —

HEADQUARTERS
UNITED STATES ARMY MILITARY
GOVERNMENT IN KOREA
Office of the National Food Administration
Seoul, Korea

NATIONAL FOOD REGULATION
NUMBER 2

12 August 1946

COLLECTION OF RICE

1. *Purpose.* The purpose of this regulation is to implement the national plan for the collection of rice in order to provide for the food needs of Koreans during the eight-month period between 1 December 1946 and 1 August 1947.

2. *Scope.* The objective is to collect the maximum possible amount of rice (by government purchase from farmers at fair and equitable prices), and the rationing of such collected rice to Koreans through government-controlled channels. The National Food Administration is charged with the responsibility for overall coordination and direction of the execution of the collection program. Provincial governors shall be responsible for the successful execution of the program in their respective provinces. The Korean Agricultural Association, Federation of Financial Associations, Controller of Commodities, Provincial Food Service, and all other political subdivisions, agencies and instrumentalities of the Government of Korea shall assist and cooperate in the execution of the program.

3. *General Duties and Functions.* The following general duties and functions are assigned to the agencies indicated:

a. *National Food Administration.* The National Food Administration shall:

(1) Establish rice collection quotas for each province based upon production statistics supplied by the Department of Agriculture; and

(2) Direct all movement of rice and the release thereof for retail distribution.

b. *Provincial Governors.* Each provincial governor shall:

(1) Break down his provincial rice collection quota into county, city, town and village quotas based upon reports from city and

county heads and Agriculture Association and Provincial Food Service officials;

(2) Supervise the collection and delivery of the rice quota to warehouses designated by the city and county heads; and

(3) Supervise inspection of rice collected, as to weight and quality.

c. *Controller of Commodities; Provincial Food Services.* The Controller of Commodities, acting through the Provincial Food Services shall:

(1) Make prompt payment to farmers upon presentation of their collection receipts;

(2) Transport rice from town and village collection points to designated warehouses;

(3) Process, store, move and release collected rice for retail distribution, at the direction of the National Food Administrator;

(4) Receive funds from sale of rice to retailers at approved government prices; and

(5) Report and account as required.

4. *First Phase: Preparation for Collection Program.* The first phase of the plan, involving preparation for the collection program in each province, shall be completed at least twenty days prior to the commencement of the harvest.

a. *Explanation of Operational Details.* Explanation of operational details within each province, to county, city, town, village and block heads and to local representatives of the Korea Agricultural Association, Federation of Financial Associations and Provincial Food Service, shall be the responsibility of each provincial governor.

b. *Rice Crop Reports and Quotas.* Not later than 30 August 1946, town, village and block heads shall make rice estimate reports to the county or city heads, who shall forward them immediately to the appropriate provincial governor. Upon receipt of such reports the provincial governor shall pro-rate the provincial quota among the respective counties and cities. County and city heads shall assign a quota to each town, village and block. Town, village and block heads shall in turn assign a quota to each farm in their respective areas of jurisdiction. Collection quotas shall be fixed so that each farmer is permitted to retain sufficient rice for planting the next ensuing crop plus at least 60% *hop* per person for each person regularly and actually residing with him as a member of his household. Individual landlords shall report to the collection points the amount of the rental rice to be supplied by their respective tenants.

c Collection Points. County and city heads shall designate warehouses as rice collection points. Such warehouses shall be suitable storage places, adequately protected against fire and theft, and located as centrally as possible in each production area.

d. Receipts. Non-negotiable triplicate receipts shall be made available at collection points for presentation to farmers. Design, printing and distribution of such receipts to the collection points, shall be the responsibility of the Controller of Commodities. Expenses of printing and distribution thereof shall be borne by the office of the Controller of Commodities, subject to reimbursement by the Government of Korea. All preparations shall be completed so that collections provided for herein may commence not later than 15 September 1946.

e. Prices. The prices to be paid to farmers for rice shall be published in the press and posted at each collection point.

5. *Second Phase: Collection of Rice.* The second phase of the plan, involving the actual collection of rice, shall commence as soon as such rice is harvested.

a. Delivery by Individual Owner-Farmers. Each individual owner-farmer shall deliver his quota of rice (and any amount in excess therof that he may desire) to the designated collection point in his area and shall receive a non-negotiable receipt thereof. No sale of rice shall be made except to the government or authorized agency thereof. In case of necessity, town village, or block authorities may collect rice at individual farms, in which case transportation and labor charges incident thereto shall be deducted from the amount paid to such farmers for such rice, and the amount of such charges shall be entered on receipts issued.

b. Delivery by Tenant Farmers. In those cases where the landlord is an individual, the tenant farmer shall deliver the quota for his farm directly to the collection point where he will be issued a collection receipt therefor. The individual landlord shall collect his rent in cash from the tenant on the basis of the established government price paid for rice, any provision in his rental contract requiring payment to the landlord in kind to the contrary notwithstanding. The individual landlord shall receive the rice for his household in the same manner as any other non-supplier through established rationing and distribution channels in an amount not to exceed allowances permitted to each farmer as set forth in paragraph 4*b* hereof.

In those cases where the landlord is not an individual person, the entire portion of such rice (except the necessary amount of rice needed as seed for planting the next ensuing crop) received by such landlord as rental shall be delivered to an appropriate collection point, and payment to such landlord made at established prices. No sale of rice shall be made except to the government or an authorized agency thereof.

c. *Inspection of Rice; Receipts; Payment.* Rice received at collection points shall be inspected for weight and grade by regular government-employed inspectors, and graded in accordance with current grain inspection regulations. Non-negotiable receipts shall be prepared in triplicate by the town, village, or block official at each collection point. The original receipt shall be delivered to the farmer, who shall present it to the Provincial Food Service for payment. One copy shall be forwarded immediately to the Provincial Food Service, and the remaining copy retained by the town, village, or block office. The Provincial Food Service shall compare the original receipt presented by the farmer for payment with the copy received by it from the town, village, or block officials, and, if the two correspond, shall pay the farmer the amount due and mark the receipt paid. The farmer shall retain the original receipt after it has been marked paid. The provincial governor may utilize the services of other agencies, including civilian agencies, within his province in order to expedite payment to the farmer for such rice.

6. *Third Phase: Reports and Accounting.* The third phase of the plan, involving reports and accounting, shall commence with the collection of rice, and shall be completed not later than 31 January 1947.

a. *Weekly Collection Reports.* Each Saturday during the collection program the town, village, or block heads shall forward to the appropriate county or city heads a consolidated report showing the amount of rice collected during the past seven days, in the following categories:

   (1) From owner-farmers;

   (2) From tenant farmers;

   (3) From government agencies;

   (4) From corporate farms; and

   (5) From others (specifying);

and the amounts delivered to the Provincial Food Service during the week. Each county and city head shall consolidate these reports and submit them to the provincial governor for further consolidation and immediate forwarding to the Military Governor in Korea, attention National Food Administrator.

b. Final Consolidated Collection Reports. Each county and city head shall prepare a final consolidated report of the information submitted to him by town, village and block heads and shall submit such report to the appropriate provincial governor not later than 10 February 1947. Each provincial governor shall consolidate his county and city reports and forward a final consolidated report so as to reach the Military Governor in Korea, attention National Food Administrator, not later than 20 February 1947.

c. Additional Reports and Accounting. The Controller of Commodities, Provincial Food Service, Korea Agricultural Association and Federation of Financial Associations shall keep records of their respective operations, and shall furnish reports on their activities when and as required. They shall be reimbursed by the Government of Korea for losses sustained by transactions authorized hereunder, and shall account for all moneys disbursed by them.

7. *Transportation.* All transportation of rice will be controlled by the government. Except for the delivery of rice to collection points as herein provided, no rice shall be transported by other then an authorized governmental agency or an individual person who possesses a valid government permit in writing for such purpose.

8. *Specific Offenses Enumerated.* Without limiting the provisions of this regulation or of any ordinance, the following are specifically declared to be violations hereof:

a. Hoarding, illegally storing or concealing, or wrongfully failing to deliver up rice in accordance with collection quotas assigned by town, village or block heads;

b. Obstructing or contravening the announced plan for the collection of rice, or any announced program, policy, orders or requirements of the Military Government or authorized agencies thereof in connection therewith;

c. Selling, offering for sale, buying or offering to buy rice, except by permission or authority of the Military Government, or at other than officially established prices, or otherwise illegally or improperly dealing in rice;

d. Failing or refusing to receive, collect, inspect, grade, store, safeguard, transport, issue proper receipts, keep accurate records and accounts, or make payments for or reports concerning rice delivered under the collection program, or other official misconduct or failure or wilful neglect of duty in connection therewith;

e. Any transaction or act, or attempted transaction or act, for the purpose of which has the effect of evading or avoiding the provisions of this regulation or the program described herein.

9. *Penalties.* Any person violating the provisions of this regulation shall, upon conviction by a Military Occupation Court, suffer such punishment as the court shall determine.

10. *Effective Date.* This regulation shall be effective on the date appearing hereon.

BY DIRECTION OF THE MILITARY GOVERNOR:

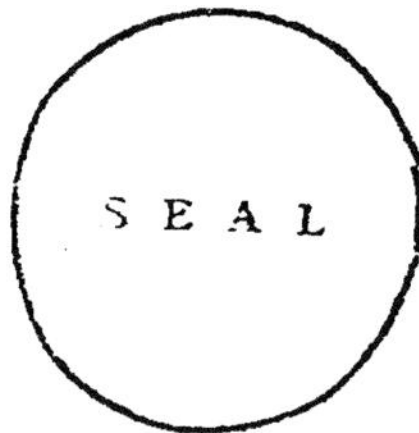

CARROLL V HILL
Lt Col   CE
National Food Administrator

HEADQUARTERS
UNITED STATES ARMY MILITARY
GOVERNMENT IN KOREA
Office of the National Food Administration
Seoul, Korea

NATIONAL FOOD REGULATION
NUMBER 3   23 September 1946

AMENDING NATIONAL FOOD REGULATION
NUMBER 2: COLLECTION OF RICE AND OTHER FOODSTUFFS:
RENTAL PAYMENT PROCEDURES

1. *Purpose and Scope.* The purpose of this regulation is to amend National Food Regulation Number 2, dated 12 August 1946, and to broaden the scope thereof so as to provide rental payment procedures for rice and other foodstuffs, necessary in the interest of the people of Korea.

2. *Paragraph 4b Amended.* Paragraph 4b of National Food Regulation Number 2, dated 12 August 1946, is hereby amended to read as follows:

"4. *Rice Crop Reports and Quotas.* Not later than 30 August 1946, town, village and block heads shall make rice estimate reports to the county or city heads, who shall forward them immediately to the appropriate provincial governor. Upon receipt of such reports the provincial governor shall pro-rate the provincial quota among the respective counties and cities. County and city heads shall assign a quota to each town, village and block. Town, village and block heads shall in turn assign a quota to each farm in their respective areas of jurisdiction. All landlords shall report to the collection points the amount of the rental rice to be supplied at each collection point by their respective tenants, and shall furnish a consolidated report, listing all their tenants wherever located, to the local Financial Association nearest the legal residence of the landlord."

3. *Paragraph 5b Amended.* Paragraph 5b of National Food Regulation Number 2, dated 12 August 1946, is hereby amended to read as follows:

"5. *Delivery by Tenant Farmers.* Each tenant-farmer shall deliver the quota of rice for his farm (and any amount in excess thereof that he may desire) directly to the designated collection point in his area; *provided, however,* that this provision shall not affect rice rental collection from tenants of the New Korea Company, Limited, pursuant to Department of Agriculture Order Number 2, dated 15 June 1946. No sale of rice shall be made except to the government or an authorized agency thereof."

4. *Paragraph 5c Amended.* Paragraph 5c of National Food Regulation Number 2, dated 12 August 1946, is hereby amended to read as follows:

"c. *Inspection of Rice; Receipts; Payment.*

(1) Rice received at collection points shall be inspected for weight and grade by regular government-employed inspectors, and graded in accordance with current grain inspection regulations. Inspection certificates shall be issued by the grain inspector at each collection point. Non-negotiable receipts shall be prepared in triplicate by the town, village, or block official at each collection point. Except in those cases where a tenant-farmer delivers rental rice directly to the collection point, the original receipt for all rice received shall be delivered to the farmer, who shall present it to the Provincial Food Service for payment. One copy shall be forwarded immediately to the Provincial Food Service, and the remaining copy retained by the town, village, or block office. The Provincial Food Service shall compare the original receipt presented by the farmer for payment with the copy received by it from the town, village, or block officials, and, if the two correspond, shall pay the farmer the amount due and mark the receipt paid. The farmer shall retain the original receipt after it has been marked paid. The provincial governor may utilize the services of other agencies, including civilian agencies, within his province, in order to expedite payment to the farmer for such rice. In those cases where rental rice is delivered to the collection point by the tenant-farmer, the procedure shall be as follows:

(a) Collection receipts shall be executed in the name of the landlord;

(b) One copy shall be forwarded immediately to the Provincial Food Service and one copy retained by the town, village, or block office;

(c) The original receipts shall be held by the collection official at the collection point;

(d) When all of the rental rice due to the landlord of such tenant has been turned in at the collection point, in accordance with the report of amounts to be supplied made by such landlord pursuant to paragraph 4i hereof*, all the original collection receipts shall be given to the landlord.

The inspection certificate issued to the tenant for his rental rice shall serve as evidence of his delivery of such rice to the collection point, and such delivery shall discharge him from any and all obligation to make such rental payment in kind to his landlord, any provision in his rental contract or other agreement to the contrary notwithstanding.

(2) If the landlord is a juridical person owned or controlled by the government, special arrangements for payment shall be made between such landlord and the National Food Administration. All other landlords shall obtain payment of the total rice rentals due them by presenting all collection receipts for rice delivered by all their tenants to the local Financial Association nearest the legal residence of the landlord.

(*a*) The local Financial Association shall:

1. Check such receipts against the consolidated reports previously furnished to it by such landlord;

2. Present such receipts from all tenants of such landlord, located within any one province, to the provincial Bureau of Banking and Taxation of such province for verification and approval.

(*b*) The provincial Bureau of Banking and Taxation shall:

1. Verify the validity of such receipts;

2. Compare such receipts with tax and land records; and verify that such receipts constitute receipts for rice from all tenants of such landlord, located within that province;

3. Report to the Treasury Bureau of the Department of Finance of the Government of Korea, the yen value (at the price officially established by the government) of the total amount of rice represented by receipts presented by such landlord;

4. Return such receipts by messenger or registered mail to the local Financial Association, marked "Approved for payment". In no case shall the "Approved for payment" receipts be delivered to the landlord.

(*c*) When all of the receipts for the rental rice of any landlord have been received from the provincial Bureaus of Banking and Taxation,

---

*See paragraph *2, above.*

marked "Approved for payment", the local Financial Association shall pay the landlord for such rice at the price officially established by the government, as follows:

1. Amounts under ¥ 20,000, in legal tender;

2. Amounts of ¥ 20,000, or over, partly in legal tender and partly by credit to a blocked account in the name of the landlord in the local Financial Association, in accordance with the following schedule:

Total Payment Due Landlord	Amount To Be Credited To A Blocked Account
¥ 20,000 — 25,000	20% of amount over ¥ 10,000
¥ 25,001 — 30,000	¥ 1,000 plus 25% of amount over ¥ 25,000
¥ 30,001 — 40,000	¥ 2,250 plus 30% of amount over ¥ 30,000
¥ 40,001 — 50,000	¥ 5,250 plus 40% of amount over ¥ 40,000
¥ 50,001 — 75,000	¥ 9,250 plus 50% of amount over ¥ 50,000
¥ 75,001 — 100,000	¥ 21,750 plus 60% of amount over ¥ 75,000
¥ 100,001 — 150,000	¥ 36,750 plus 75% of amount over ¥ 100,000
¥ 150,001 — 200,000	¥ 74,250 plus 85% of amount over ¥ 150,000
¥ 200,001 — 300,000	¥ 116,750 plus 95% of amount over ¥ 200,000
¥ 300,001 — 400,000	¥ 211,750 plus 97% of amount over ¥ 300,000
¥ 400,001 — 500,000	¥ 308,750 plus 98% of amount over ¥ 400,000
¥ 500,001 — 1,000,000	¥ 406,750 plus 99% of amount over ¥ 500,000
¥ Over ¥ 1,000,000	Amount credited to blocked account shall be 99% of total payment due the landlord, plus ¥ 1,750.

(d) Withdrawals from such blocked accounts shall be limited to the following:

1. For payment of taxes, by check drawn in favor of a tax-collecting authority;

2. For any purpose specifically approved in advance in writing by the Department of Finance of the Government of Korea; and

3. In the case of blocked accounts of an original amount of ¥ 50,000 or over, in addition to *1* and *2* above, monthly withdrawals of ¥ 5,000 or 5% of the amount in excess of ¥ 50,000, whichever is greater; *provided, however*, that the aggregate amount of such withdrawals* shall not exceed 20%

---

* *i. e.*, under paragraph *(d)* 3.

nominated as a candidate, by submitting to the Election Review Board (*Seoul, Capitol Building*) not later than 9 July 1948 a written application stating:

 a. The name of the electoral district concerned

 b. His own name and address and the name of the elected candidate

 c. A brief summary of the alleged violations of the Election Law and /or the Election Regulations.

The application may contain the name and address of witnesses and other evidence tendered.

W. F. DEAN
Major General, United States Army
Military Governor in Korea·

軍政廳　官報　行政命令　第二三號　・一九四八年六月二二日

SOUTH KOREAN INTERIM GOVERNMENT
Seoul, Korea

EXECUTIVE ORDER
NUMBER 24

26 July 1948

EMERGENCY WATER CONSERVATION BOARD

The Commanding General, United States Army Forces in Korea, recognizing that an acute shortage of potable water presently exists in South Korea, and will continue to exist, and further recognizing the necessity for a better and more equitable distribution of the available water for the essential needs of the people of South Korea, has directed the establishment of an Emergency Water Conservation Board with authority to regulate the use and distribution of potable water until such time that a more adequate supply of such potable water becomes available.

In compliance with such directive, there is hereby created and established a Board to be known as the "Emergency Water Conservation Board," generally referred to hereinafter as the "Board."

1. The Board shall consist of eleven (11) members. Seven (7) of these members shall represent the National interest, and four (4) shall represent local interests. Two (2) local members shall represent the Seoul-Inchon Water District and two (2) shall represent the Kyongsang Namdo-Pusan Water District.

The National members shall be three (3) persons named by the Commanding General, United States Army Forces in Korea, and four (4) persons to be named by the Military Governor. The local members shall be named by the Military Governor.

2. Meetings of the Board will be on the call of the Chairman. A quorum shall consist of four (4) of the National members, together with one (1) member each from the Local water districts specified when dealing with general matters affecting the whole of South Korea; and of four (4) of the National members and one (1) member from a specified water district when dealing with matters solely affecting such specified water district; and of four (4) National members when dealing with matters affecting water districts other than those specified. A member of the Board, in case of unavoidable absence, may be represented by an alternate named by him.

3. The Board shall inform the Commanding General, United States Army

— 1 —

Forces in Korea, of each directive issued by it, and shall request issuance of appropriate directives to the military establishment when and where applicable.

4. The Board is empowered to, and may:

a. Issue orders, instructions, priorities, and restrictions, both general and specific, dealing with the distribution and use of potable water, directed to all persons in South Korea, individual, corporate and governmental. Any such action by the Board shall have the effect of law.

b. Call upon and receive the services of any department or agency of the South Korean Interim Government, for such assistance as may be deemed necessary in carrying out its duties.

c. Ration the use of potable water for all domestic purposes, for limited but necessary business and industrial purposes, and with the consent of the Commanding General of United States Army Forces in Korea, for the essential needs of the military forces.

d. To the extent deemed necessary, prohibit the use of potable water for stated purposes, other than for use against fires.

e. Inspect or order the inspection of, any water supply and distribution system, including all taps and outlets wherever found, and peremptorily to order whatever remedial measures may be necessary or desirable to stop the voluntary or involuntary wastage of water, or to order repair of leaky water mains, water lines, water taps, or outlets.

f. Inspect all Korean warehouses and storehouses, whether governmental, private or otherwise, for the purpose of locating equipment, materials, supplies and whatever other maintenance items may be necessary to repair pumping plants, treatment plants, water mains and distribution system for highest operating efficiency and to requisition such equipment, materials, supplies, etc., for immediate delivery to the several water systems of South Korea, unless held for military use of equal or greater importance or priority.

g. Hold hearings, administer oaths, take testimony, admit evidence, and make whatever investigations as may be necessary or desirable in all matters affecting the use or wastage of potable water, and to prepare and promulgate rules and regulations based on the findings of such hearings and investigations.

h. Call upon the water companies or agencies to deliver promptly to it any operating and financial reports.

i. Define violations of any order or directive issued by it

— 2 —

043

j. Issue its orders in writing. verbally, by telephone, telegraph. radio. or by any other form of communication.

k. Make studies and recommendations for enlargement of existing water plants and facilities or for new and additional water facilities. if and where needed.

5. The Board shall, as soon after completing its organization as possible. determine the availability of potable water in each water supply and distribution area, and proceed to establish ration quotas for the civil population; the military establishment as stipulated by the Commanding General, United States Army Forces in Korea; the welfare institutions. such as hospitals, encampments. prisons, jails, etc.; business and industrial entities; and other public purposes; and to establish such prohibitions as may be warranted on all unnecessary use of water, or on the use of potable water for irrigating purposes; but the use of such water for fire protection and extinguishment shall not be impaired. The Board may vary such quotas, from time to time, as it may determine to be necessary or desirable.

6. Any person who may be convicted in any duly constituted court of violating this order, or of any order or directive the Board may issue hereunder shall suffer such punishment as the court shall determine. The Board may issue schedules of suggested punishments for violations of this order or of any order or directive issued by it.

7 All decisions. directives, or orders of the Board, shall be subject to the review of the Military Governor, and to the further review of the Commanding General, United States Army Forces in Korea, but shall have full force and effect until such time as changed, modified or voided on such review.

8. All laws, ordinances, orders, regulations, directives, and instructions, or parts thereof, which are inconsistent herewith, or in conflict with the provisions hereof, are, to the extent to which they may be so inconsistent or in conflict, hereby suspended during the period of this emergency

9. The provisions of this order, and the powers of the Board hereunder, shall continue in effect until such time as a permanent water supply and conservation board shall be brought into being, or a public proclamation declaring

— 3 —

HEADQUARTERS
UNITED STATES ARMY MILITARY
GOVERNMENT IN KOREA
Office of the National Food Administration
Seoul, Korea

NATIONAL FOOD REGULATION
NUMBER 5   8 May 1947

COLLECTION OF SUMMER GRAINS

1. *Purpose.* The purpose of this regulation is to provide for the collection of summer grains in 1947 in accordance with Public Act No 2, duly enacted by the Korean Interim Legislative Assembly on 19 April 1947, issued simultaneously herewith, which provides as follows:

"SECTION I. The summer grains collection program, submitted by the National Food Administration, is approved, subject to the modifications set out in Section II. These modifications shall be made part of the program which shall be published simultaneously herewith.

SECTION II. *Modifications.*

   a. Since summer grains are customarily used for food by the farmers, the collection shall be limited to one-fifth of the estimated total production. Individual quotas must be just and fair and based on a careful estimate of production.

   b. In order to insure just quotas, each eup and myun head shall organize an investigation committee in each dong or ri in order to check farmers' production estimates.

   c. Compulsory methods shall be prohibited in making the collection. Only civilian administrative officials shall be used; they shall encourage the farmers to turn in their quotas as a matter of conscience and duty to their country; *provided, however,* that penalties set out in National Food Regulation No 5, "Collection of Summer Grains", shall be enforced by the duly constituted authorities in accordance with law.

— 1 —

軍政廳　官報　中央食糧規則　第五號　一九四七年五月八日

d. Prices for summer grains shall be as fair as possible and shall take into consideration costs of production and the latest price index.

e. Farmers who have turned in their quota shall be accorded preferential rights in the rationing o consumers' goods.

f. Summer grains collected shall be adequately protected and provision shall be made for proper transportation, warehousing and administration".

2. *Scope.* The objective is to collect summer grains (by government purchase from farmers) for rationing to Koreans through government-controlled channels. The National Food Administration is charged with the responsibility for overall coordination and direction of the execution of the collection program. Provincial governors shall be responsible for the successful execution of the program in their respective provinces. The Korea Agricultural Association, the New Korea Company Ltd, Federation of Financial Associations, Controller of Commodities, Provincial Food Services and all political sub-divisions, agencies and instrumentalities of the government shall assist and cooperate in the execution of the program.

3. *Definition of Summer Grains.* *Summer Grains,* as used herein, means wheat, rye, barley and naked barley.

4. *General Duties and Functions.* The following general duties and functions are assigned to the agencies indicated:

a. *National Food Administration.* The National Food Administration shall:

> (1) In collaboration with the Department of Agriculture, and after consideration of the recommendations of the provincial governors, establish a summer grain collection quota for each province; *provided, however,* that the total of the provincial quotas shall not exceed 20% of the estimated total production of summer grains in 1947; and

> (2) Establish summer grain shipping quotas, and direct all movement of summer grains and the release thereof for retail distribution.

b. Each provincial governor shall:

— 2 —

945

(1) Break down his assigned provincial summer grain collection quota into county, city, town, village and block quotas based upon reports from county and city heads and Agricultural Association and Provincial Food Service officials. All such quotas shall be given wide publicity;

(2) Supervise the collection and delivery of the summer grain quota to warehouses designated by the county and city heads;

(3) Supervise the inspection of summer grains collected as to weight and quality;

(4) Be responsible for meeting the collection and shipping quotas established for his province; and securing the full cooperation of the appropriate administrative officials;

(5) Initiate prosecution of violators. So far as practicable civilian administrative officals shall be utilized in collecting summer grains quotas.

c. *Controller of Commodities; Provincial Food Services.* The Controller of Commodities, acting through the Provincial Food Services, shall:

(1) Make prompt payment to farmers upon presentation of their collection receipts;

(2) Transport summer grains from town and village collection points to designated warehouses;

(3) Process, store, move and release collected summer grains for controlled retail distribution, at the direction of the National Food Administrator;

(4) Receive funds from authorized sales of summer grains;

(5) Report and account as required.

d. *New Korea Company Ltd.* The New Korea Company Ltd shall:

(1) Jointly with local officials, make summer grain crop estimates pertaining to its tenants for the purpose of assigning collection quotas. Final decision as to the amount of the quota for each

— 3 —

individual farm shall rest with the government officials;

(2) Act as the official collection agency for the entire quota of grain for its tenants, and turn over such grain promptly to the Provincial Food Service; and

(3) Make recommendations to the county and city heads as to which of its warehouses should be used as government collection points.

5. *First Phase. Preparation for Collection Program.* The first phase of the plan, involving preparation for the collection program in each province, shall be completed at least ten days prior to the commencement of the harvest.

a. *Explanation of Operational Details.* It shall be the responsibility of the provincial governor to explain the operational details to his county, city, town, village and block heads, to police officials and to local representatives of the Korea Agricultural Association, New Korea Company Ltd, Federation of Financial Associations and Provincial Food Service.

b. *Summer Grain Collection Quotas.*

(1) Upon receipt of the summer grain collection quota from the National Food Administration each provincial governor shall promptly prorate his provincial quota among his respective counties and cities. County and city heads shall assign quotas to each town, village and block without delay. Not less than ten days prior to the harvesting of summer grains in each province, the town, village or block heads shall assign in writing and deliver a quota to each farm in their respective areas of jurisdiction. They shall publicly and conspicuously post on a list the name of each farmer with the amount of his quota. The date the farmer delivers his quota to the collection point shall be entered on the list at the time of delivery.

(2) The assignment of a collection quota to a farm, and the amount of such quota, shall be made only after due consideration has been given to the estimated summer grain production

of the farm, the amount of seed needed to plant the ensuing crop, and the amount of summer grain required for each person regularly and actually residing with the farmer as a member of his household.

(*3*) Each town, village or block head shall shall establish review boards (to serve without compensation in his area) to hear complaints of farmers who believe they have been assigned an unfair collection quota and make recommendations thereon to such town, village or block head.

c. *Collection Points.*

(*1*) County and city heads shall designate warehouses as summer grain collection points. Such warehouses shall be suitable storage places, adequately protected against fire and theft and located centrally and as conveniently as possible to milling facilities in each production area.

(*2*) Wherever the local government official uses New Korea Company Ltd or Korea Agricultural Association warehouses and designates them as collection points, at least one government official shall be present at such collection point and shall take part in the collection.

(*3*) All grains collected or stored at warehouses designated pursuant to subparagraphs (*1*) and (*2*) of this paragraph shall be under the jurisdiction and control of the Provincial Food Service which shall make distribution therefrom in accordance with plans and policies of the National Food Administration.

d. *Inspection of Summer Grains.*   Summer grains received at collection points shall be inspected for weight and grade by regular government employed inspectors and graded in accordance with current grain inspection regulations. The grain inspector shall give the farmer an inspection certificate and retain a duplicate thereof.

e. *Receipts and Payment.*   Non-transferable triplicate receipts shall be made available at collection points. Design, printing and distribution to

— 5 —

the collection points of such receipts shall be the responsibility of the Controller of Commodities. Expenses of printing and distribution thereof shall be borne by the Office of the Controller of Commodities, subject to reimbursement by the government. All preparations shall be completed so that collections provided for herein may commence not later than 15 June 1947. Receipts shall be prepared in triplicate by the town, village or block official at each collection point. The original shall be given to the farmer, the duplicate delivered to the representative of the Provincial Food Service and the third copy retained at the town, village or block office. A representative of the Provincial Food Service shall be at the collection point and shall then and there pay the farmer. He shall mark the original receipt paid and return it to the farmer.

f. Prices. Prices to be paid to farmers for summer grains shall be conspicuously posted at each collection point. Such prices shall be established with due regard to the general price level and costs of production generally. So far as practicable incentive goods will be distributed to farmers cooperating in the summer grains collection program.

6. *Second Phase* The second phase of the plan, involving actual collection of summer grains, shall commence as soon as such grains are harvested.

a. Delivery of Quota Grains; Seizure of Undelivered Grains. All quota grains shall be delivered to the collection points prior to 1 September 1947. All quota grains not so delivered prior to 1 September 1947 shall be subject to seizure by the government. The government, in its discretion, may pay in whole or in part for such seized grains.

b. Delivery by Individual Owner-Farmers. Each individual owner-farmer shall deliver his quota of summer grains (and any amount in excess thereof that he may desire) to the designated collection point in his area and shall receive a non-transferable collection receipt therefor. No sale or transfer of summer grains shall be made except to the government or an authorized agency thereof. Where necessary, town, village or block authorities may collect summer grains at individual farms, in which case transportation and labor charges incident thereto shall be deducted from the amount paid to such farmers for such grain, and the amount of such charges shall be entered on receipts issued.

— 6 —

a. Hoarding, illegally storing or concealing or wrongfully failing to deliver up summer grains in accordance with   ᵗection quotas assigned by town, village or block heads. (All summer grains of any crop in the possession of persons who received them from unauthorized sources shall be subject to confiscation);

b. Selling, offering for sale, transferring or offering to transfer, buying or offering to buy summer grains, except by permission or authority of the government, or at other than offically established prices, or otherwise illegally or improperly dealing in such grains;

c. Failing or refusing to receive, collect, inspect, properly assign a quota, store, grade, safeguard, transport, issue proper receipts, keep accurate records and accounts, or make payments for or reports concerning summer grains delivered under the collection program, or other official misconduct or failure or willful neglect of duty in connection therewith;

d. Any transaction or act, or attempted transaction or act, for the purpose or which has the effect of evading or avoiding the provisions of this regulation.

10. *Penalties.* Any person violating the provisions of this regulation shall, upon conviction by a Military Occupation Court, suffer such punishment as the court shall determine.

11. *Regulation No 1 Revoked.* National Food Regulation No. 1, *Collection of Summer Grains*, dated 29 May 1946, is hereby revoked.

12. *Effective Date.* This regulation shall be effective on the date appearing hereon.

BY DIRECTION OF THE MILITARY GOVERNOR:

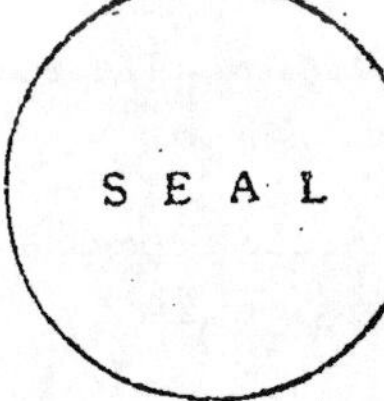

CARROLL V HILL
Adviser to National
Food Administrator

CHEE YOUNG EUN
National Food Administrator

SOUTH KOREAN INTERIM GOVERNMENT
Office of the National Food Administration
Seoul, Korea

NATIONAL FOOD REGULATION
NUMBER 6

18 August 1947

COLLECTION OF RICE

Pursuant to Section V, Ordinance Number 90, dated 2r May 1946 the following shall govern:

1. *Purpose.* The purpose of this regulation is to provide for the collection of rice during the fall and winter months, 1947-48.

2. *Scope.* The objective is to collect rice (by government purchase from farmers) for rationing, through government-controlled channels, to Koreans living south of 38° north latitude. The National Food Administration is charged with the responsibility for overall coordination and direction of the execution of the collection program. Provincial governors shall be responsible for the successful execution of the program in their respective provinces. The Korea Agricultural Association, the New Korea Company Ltd. Federation of Financial Associations, Controller of Commodities, Provincial Food Services and all political subdivisions, agencies and instrumentalities of the government shall assist and cooperate in the execution of the program.

3. *General Duties and Functions.* The following general duties and functions are assigned to the agencies indicated:

a. *National Food Administration.* The National Food Administration shall:

(1) In collaboration with the Department of Agriculture, establish rice collection quota for each province; and

(2) Establish rice shipping quotas and direct all movement of rice and the release thereof for retail distribution.

— 1 —

軍政廳　官報　中央食糧規則　第六號　一九四七年八月十八日

b. Each provincial governor shall:

(1) Break down his assigned provincial rice collection quota into equitable gun and pu quotas;

(2) Be responsible for meeting the collection and shipping quotas established for his province; and securing the full cooperation of the appropriate administrative officials;

(3) Initiate prosecution of violators;

(4) Supervise the collection and delivery of rice quotas to warehouses designated by the gun and pu heads; and

(5) Supervise the inspection of rice collected as to weight and quality.

c. Controller of Commodities; Provincial Food Services. The Controller of Commodities, acting through the Provincial Food Services, shall:

(1) Make prompt payment to farmers upon presentation of their collection receipts;

(2) Transport rice from eup, myun and New Korea Company collection points to designated warehouses, mills, or shipping points;

(3) Process, store, move and release collected rice for controlled retail distribution, at the direction of the National Food Administrator;

(4) Receive funds from authorized sales of rice; and
(5) Report and account as required.

4. New Korea Company Ltd. The New Korea Company Ltd shall:

(1) Jointly with local officials make rice crop estimates pertaining to its tenants for purpose of assigning collection quotas. Final decisions as to the amount of the quota for each individual farm shall rest with government officials. Each New Korea Company sub-branch office shall furnish each eup, myun or dong head with a list containing the name and address of each of its tenants in that area. As soon as quotas for New Korea Company tenants have been determined, the local official shall enter the amount of the quota of each tenant on the list and return it to the New Korea Company sub-branch office;

(2) Act as official collecting agency for the entire quotas of rice from its tenants, including both the quota from land which the tenants farm

— 2 —

for New Korea Company and the quota for all other land the tenants farm, and turn over such collections to the Provincial Food Service, which agency shall be responsible for hauling the rice promptly from New Korea Company collection points. The Provincial Bureau of Agriculture shall furnish grain inspectors for New Korea Company collection points on a full-time basis. The Provincial Food Service shall furnish a full time official at each New Korea Company collection point to represent the government, write official receipts and pay for the grain.

(3) Select and operate New Korea Company collection points for New Korea Company tenants. Provincial officials shall designate as official collection points those points operated by the New Korea Company;

(4) Be authorized to station one or more New Korea Company representatives at any government collection point, either full time or part time to facilitate collection of rice from New Korea Company tenants who are not located near a New Korea Company collection point. These New Korea Company representatives shall assist New Korea Company tenants to turn in their rice, and will keep a record of the amount of quota rice thus turned in by New Korea Company tenants;

(5) Collect New Korea Company rentals from tenants in cash only. The rice collected from New Korea Company tenants shall include all rice which they are required to turn in, and they shall be paid by the Provincial Food Service for the entire amount. The tenants' cash payments to New Korea Company for rent shall be entirely separate from the rice collection;

(6) Be authorized to hold for delayed exchange to tenants, not to exceed 50,000 (rough) suk of high quality seed rice, which has been produced on certified New Korea Company seed rice fields or selected from rice which has been received at New Korea Company collection points. The amount of seed rice collected shall be reported to the Provincial Food Service in each gun, so that the gun official can include this amount as a delayed collection; and

(7) Keep records of expenses incurred as a result of rice collection operations so that suitable arrangements for reimbursement may be made.

4. *First Phase. Preparation for Collection Program.* The first phase of

— 3 —

the plan, involving preparation for the collection program in each province shall be completed at least ten days prior to the commencement of the harvest.

    a. Explanation of Operational Details. It shall be the responsibility of the provincial governor to see that the operational details are thoroughly explained to the gun, pu, eup, myun, dong and lower level heads, to police officials, and to local representatives of the Korea Agricultural Association, New Korea Company Ltd, Federation of Financial Associations and Provincial Food Services.

    b. Rice Collection Quotas.

    (1) Upon receipt of the notice of the rice collection quota from the National Food Administration each provincial governor shall promptly prorate his provincial quota among his respective gun and pu. Gun and pu heads shall assign quotas to each eup, myun and dong without delay. Not less than ten days prior to the harvesting of rice in each province, the eup, myun and dong heads shall assign in writing and deliver a quota notice to each farm in their respective areas of jurisdiction. They shall publicly and conspicuously post on a list the name of each farmer with the amount of his quota. The date the farmer delivers his quota to the collection point shall be entered on the list at the time of delivery.

    (2) The assignment of a collection quota to a farm, and the amount of such quota shall be based on recommendations made by impartial committees (to serve without compensation) established in each dong or ri by the pu, eup or myun head. In determining the quota for a farm, due consideration shall be given to the estimated rice production of the farm, the amount of seed needed to plant the ensuing crop, and the amount of rice required for each person regularly and actually residing with the farmer as a member of his household.

    (3) Each eup, myun or dong head shall establish a review board (to serve without compensation in his area) to hear complaints of farmers who believe they have been assigned an unfair collection quota. The review board shall determine the validity of the complaint and shall have authority to change any farm quota when the facts warrant. The board shall be composed of officials and prominent citizens and shall have as chairman the eup, myun or dong head.

— 4 —

c. Collection Points.

(1) Gun and pu heads shall designate warehouses as rice collection points. Such warehouses shall be suitable storage places, adequately protected against fire. and theft and located centrally and as conveniently as possible to milling facilities in each production area.

(2) All rice collected or stored at designated warehouses shall be under the jurisdiction and control of the Provincial Food Service which shall make distribution therefrom in accordance with plans and policies of the National Food Administration.

d. *Inspection of Rice.* Rice received at collection points shall be inspected for weight and grade by regular government-employed inspectors and graded in accordance with current grain inspection regulations. The grain inspector shall give the farmer an inspection certificate and retain a duplicate thereof.

e. *Receipts and Payment.* Non-transferable triplicate receipts shall be made available at collection points. Design, printing and distribution to the collection points of such receipts shall be the responsibility of the Controller of Commodities. Expenses of printing and distribution thereof shall be borne by the Office of the Controller of Commodities, subject to reimbursement by the government. All preparations shall be completed so that collections provided for herein may commence not later than 1 October 1947. Receipts shall be prepared in triplicate by eup, myun or dong officials at each collection point. The original shall be given to the farmer, the duplicate delivered to the representative of the Provincial Food Service and the third copy retained at the eup, myun or dong office. A representative of the Provincial Food Service shall be at the collection point and shall then and there pay the farmer. He shall mark the original receipt paid and return it to the farmer.

f. *Prices.* Prices to be paid to farmers for rice shall be conspicuously posted at each collection point. Such prices shall be established with due regard to the general price level and costs of production generally.

5. *Second Phase.* The second phase of the plan, involving the actual collection of rice, shall commence as soon as such rice is harvested.

軍政廳　官報　中央食糧規則　第六號　一九四七年八月十八日

   a. Delivery of Quota Rice; Seizure of Undelivered Rice. All quota rice shall be delivered to the collection points prior to 15 January 1948. All quota rice not so delivered prior to 15 January 1948 shall be seized by the government and the owner shall be paid 50 percent of the established price.

   b. Delivery by Individual Owner-Farmers. Each individual owner-farmer shall deliver his quota of rice (and any amount in excess thereof that he may desire) to the designated collection point in his area and shall receive a non-transferable collection receipt therefor. No sale or transfer of rice shall be made except to the government or an authorized agency thereof.

   c. Delivery by Tenant-Farmers. The tenant-farmer shall deliver the quota for his farm directly to the collection point where he will be issued a non-transferable collection receipt therefor. In those cases where rent is due, the landlord, natural or juridical, shall collect his rent in cash from the tenant on the basis of the established government price paid for such rice, any provision in the rental contract requiring payment to the landlord in kind notwithstanding. An individual landlord shall receive the rice for his house hold in the same manner as any other non-supplier (through established rationing and distribution channels.).

   6. *Transportation.* All transportation of rice shall be controlled by the government. Except for the transportation of rice to collection points and to and from retail ration stores, no rice shall be transported by other than an authorized governmental agency. Rice in unauthorized transit or on sale in other than legal distribution channels shall be seized without payment and shall be delivered to the appropriate Provincial Food Service for sale through the regular legal distribution channels.

   7. *Third Phase. Reports and Accounting.* The third phase of the plan involving reports and accounting, shall commence with the collection of rice and shall be completed not later than 10 March 1948.

   a. Collection Reports by the Office of the Controller of Commodities Every five days during the collection period the Office of the Controller of Commodities shall submit to the National Food Administrator a report showing the total amount of rice, stated in polished suk purchased by the respective Provincial Food Services. The report on the tenth, twentieth and thirtieth of

軍政廳　官報　中央食糧規則　第六號　一九四七年八月十八日

of each month shall be broken down to show the amount of rice purchased in each gun and pu of each province. On 10 March 1948 a final report shall be submitted by the Office of the Controller of Commodities.

b. Report of Undelivered Quota Rice. Each eup, myun and dong head shall prepare and forward so as to reach the appropriate gun or pu head not later than 20 January 1948, a list containing the following information for each person who has failed to deliver to the government his rice collection quota :

    (1) Name,

    (2) Address,

    (3) Amount of rice quota, and

    (4) Amount of rice delivered to the government.

Local officials shall take appropriate action against violators who fail to deliver their quota rice to the collection point prior to 15 January 1948.

c. Additional Reports and Accounting. The Controller of Commodities, Provincial Food Services, Korea Agricultural Association and Federation of Financial Associations shall keep records of their respective operations and shall furnish reports on their activities when and as required. They shall be reimbursed by the government for losses sustained by transactions authorized hereunder, and shall account for all moneys disbursed by them.

d. Records and Reports by Government and Corporate Landlords. The New Korea Company Ltd, all other corporate landlords, and all governmental subdivisions, agencies and instrumentalities owning or operating agricultural property upon which rice is produced, shall maintain accurate records of the rice delivered by their tenants to the collection points under this regulation. Such records shall be open to inspection at any time by competent authority.

8. *Specific Offenses Enumerated.* Without limiting the provisions of this regulation or of any ordinance, the following are specifically declared to be violations hereof :

a. Hoarding, illegally storing or concealing, or wrongfully failing to deliver up rice in accordance with collection quotas assigned by eup, myun or dong heads. (All rice in the possession of persons who received it from unauthorized sources shall be subject to confiscation.) :

b. Selling, offering for sale. transferring or offering to transfer, buying or offering to buy rice, except by permission or authority of the government, or at other than officially established prices, or otherwise illegally or improperly dealing in such rice;

c. Failing or refusing to receive, collect, inspect, properly assign a quota, store, grade, safeguard, transport, issue proper receipts, keep accurate records and accounts, or make payments for. or reports concerning rice delivered under the collection program, or other official misconduct or failure or wilfull neglect of duty in connection therewith;

d. Any transaction or act, or attempted transaction or act, for the purpose or which has the effect of evading or avoiding the provisions of this regulation.

9. *Penalties.* Any person violating the provisions of this regulation shall, upon conviction by a duly constituted court, suffer such punishment as the court shall determine.

10. *Prior Regulations Revoked.* National Food Regulation No. 2: *Collection of Rice*, dated 12 August 1946; National Food Regulation No. 3, amending National Food Regulation No. 2: *Collection of Rice and Other Foodstuffs. Rental Payment Certificates*, dated 23 September 1946; and National Food Regulation No. 4, amending National Food Regulation No. 3: *Rental Payment Procedures*, dated 26 March 1947, are hereby revoked but the revocation of such regulations is not intended and shall not be construed to extinguish offenses under such regulations committed prior to the effective date hereof nor to abate or bar existing or future prosecutions based on such offenses.

11. *Effective Date.* This regulation shall be effective on the date appearing hereon.

BY DIRECTION OF THE MILITARY GOVERNOR:

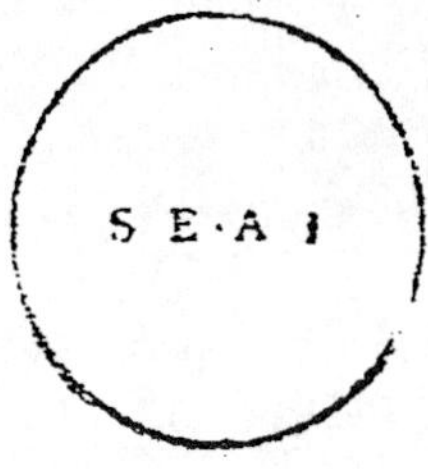

CHEE YONG EUN
National Food Administrator

— 5 —

SOUTH KOREAN INTERIM GOVERNMENT
Office of the National Food Administration
Seoul, Korea

NATIONAL FOOD REGULATION
NUMBER 7   4 May 1948

COLLECTION OF SUMMER GRAIN

1. Purpose. Pursuant to Section V, Ordinance Number 90, dated 23 May 1946, this regulation is published to provide for the collection of summer grain in 1948. "Summer grain," as used herein, means *wheat, rye, barley,* and *naked barley.*

2. Scope. The objective is to collect summer grain (*by government purchase from farmers*) for rationing, through government-controlled channels, to Koreans living south of thirty-eight degrees north latitude. The National Food Administration is charged with the responsibility for overall coordination and execution of the collection program. Provincial governors shall be responsible for the successful execution of the program in their respective provinces. The Korea Agricultural Association, National Land Administration, Controller of Commodities, Provincial Food Services, and all political subdivisions, agencies, and instrumentalities of the government shall assist and cooperate in the execution of the program.

3. General duties and functions. The following general duties and functions are assigned to the agencies indicated:

a. *National Food Administration.* The National Food Administration shall:

(1) In collaboration with the Department of Agriculture, establish a summer grain collection quota for each province; and

(2) Direct and control the distribution, storage, movement, and release of summer grain.

b. *Provincial governors.* Each provincial governor shall:

(1) Break down his assigned summer grain collection quota into equitable *gun* and *pu* quotas;

(2) Be responsible for meeting the collection quota established for his province and securing the full cooperation of the appropriate administrative officials;

— 1 —

軍政廳 官報 中央食糧行政處 食糧收集令 一九四八年五月四日

(3) Supervise the collection and delivery of the summer grain quota to collection points. So far as practicable, civilian administrative officials shall be utilized in collecting such quotas;

(4) Supervise the inspection of collected summer grain as to weight and quality;

(5) Limit the release of summer grain for rationing purposes in accordance with the rationing plan prescribed by the National Food Administration;

(6) Be responsible for the prosecution of violators;

(7) Insure that the farmer is not assigned an unduly heavy or unfair collection quota;

(8) Be authorized to make use of Korea Agricultural Association Warehouses and other warehouses necessary for the collection and storage of summer grain; and

(9) Make use of all available government trucks within his province as may be necessary for the collection of summer grain.

c. *Controller of Commodities: Provincial Food Services.* The Controller of Commodities, acting through the Provincial Food Services, shall:

(1) Make prompt payment to farmers upon presentation of their collection receipts;

(2) Transport summer grain from *myun, eup* or *dong (in a pu)* collection points to designated warehouses, mills, or shipping points;

(3) Process, store, move, and release collected summer grain for controlled retail distribution, at the direction of the National Food Administration;

(4) Receive funds from authorized sales of summer grain; and

(5) Report and account as required.

d. *National Land Administration.* The National Land Administration (*hereinafter referred to as "NLA"*) shall:

(1) Jointly with local officials make grain crop estimates of the farms of its tenants and purchasers of land for the purpose of assigning collection quotas. Final decision as to the amount of the quota for each individual farm shall rest with the responsible local officials;

(2) The purchasers of land from NLA under the provisions of Ordinance Number 173, shall deliver their assigned quotas (*and any amount in excess thereof that they may desire*), directly to government collection points. The assigned quota shall not be less than that provided for in the mortgage as the

— 2 —

required annual minimum payment. In case the annual installment on the mortgage is less than the assigned collection quota, the purchaser of the land may apply the difference (*or any part thereof*) between the collection quota and the annual installment due the NLA as an extra payment on the mortgage, so as to complete his payments more quickly;

(3) The tenants of NLA shall deliver their assigned quotas (*and any amount in excess thereof that they may desire*), directly to government collection points. The assigned quota shall not be less than that provided for in the rental agreement;

(4) A NLA official will be assigned full time at collection points where NLA land is involved. When a NLA purchaser (subparagraph (2) above) or a tenant (subparagraph (3) above) has turned in his quota, he shall take his official receipt, upon which is recorded in a space provided thereon the "amount due for the account of NLA," to the NLA official and have appropriate entries made and a NLA receipt issued to him. At the end of the day, the total amount of the receipts issued by the NLA shall be reconciled with the amount of the deductions for the account of the NLA on the National grain collection receipts.

(5) The purchasers of land (subparagraph (2) above) and the tenants (subparagraph (3) above) of the NLA shall be treated in the same manner as any other land owner and/or tenant. Their individual quotas shall be comparable to the quotas assigned to other farmers who till similar farms adjacent to them, except that the minimum quota to be assigned to NLA purchasers or tenants will be the amount specified in their purchase or rental contracts;

(6) NLA officials shall report to each *myun, eup,* or *pu* head in their respective areas the amount of grain required as rental or as payment required on the mortgage;

(7) The warehouses of NLA may be used as grain collection and storage points by the National Food Administration.

e. *Korea Agriculture Association.* The Korea Agricultural Association shall make available all its warehouses for the collection and storage of summer grain as directed by the provincial governor.

f. Summer Grain Collection Committees. Summer Grain Collection Committees (*to serve without compensation*) selected from landlords, owner-farmers, tenant-farmers, farming experts and other persons well acquainted with the

local situation, shall be appointed as hereinafter provided at least one month
prior to the summer grain harvest in each province. One Summer Grain Collection Committee in each *myun, eup,* or *pu* with the *myun, eup* or *pu* head as
chairman, and about fourteen other members, shall be appointed by the *gun* or
pu head on recommendations submitted by the respective *myun, eup* or *dong* (in
a *pu*) heads. This Committee shall assist the *myun, eup* or *pu* head by helping
him determine the *ku* or *dong* (*in a eup or pu*) quotas, and functioning as a
review board to act on objections to the individual quotas raised by individual
farmers within the respective *myun, eup* or *dong* (*in a pu*). One Summer Grain
Collection Committee in each *ku* or *dong* (*in a eup or pu*) with the *ku* or *dong*
head as chairman, and about nine other members shall be appointed by the *myun,
eup* or *pu* head on recommendations submitted by the respective *ku* or *dong* (*in
a eup or pu*) heads. This Committee shall assist the *ku* or *dong* (*in a eup or pu*)
head by investigating the crop estimates of the farms, investigating the food
needs of the farm families, estimating individual farm quotas, and performing
other important matters pertaining to the summer grain collection program. It
also shall assist the *ku* or *dong* (*in a eup or pu*) head by advising the approval
or modification of the individual farm quotas tentatively set up at the *ban* head
meetings.

5. *General rules for determining summer grain collection quota to be assigned
a farmer.* The assignment of summer grain collection quota to a farm and the
amount of such quota shall be made by the respective *myun, eup* or *pu* head,
assisted by the Summer Grain Collection Committees (*see paragraph 4 hereof*).
All officials and Committees who have any part in determining the quota, shall
take every precaution to insure that the quota assigned to the individual farmer
is determined according to the following procedure in a fair and equitable
manner:

a. Any farmer who produces the equivalent of one polished *suk* of
summer grain or less, shall be assigned a quota which shall not exceed ten per
cent of his production of summer grain. However, if he produces substantial
quantities of other summer food crops, his summer grain quota may be raised
above the stipulated ten per cent in consideration of the additional food which
is available to him. A NLA purchaser or tenant who is required by the terms
of his contract with NLA to deliver summer grain in excess of ten per cent
of his production shall be excluded from the provisions of this ten per cent

— 4 —

963

quota restriction.

b. In determining the quota of a farmer whose production of summer grain exceeds the equivalent of one polished *suk* of summer grain, due consideration shall be given to the estimated summer grain and other summer food crop of his farm, the amount of summer grain seed needed to plant the ensuing crop, and the amount of summer grain and other summer food required for each person regularly and actually residing with the farmer as a member of his household.

6. *First phase, preparation for collection program.* . The first phase of the plan involving preparation for the collection program in each province shall be completed at least 10 days prior to the commencement of the harvest.

a. Explanation of operational details. It shall be the responsibility of the provincial governor to insure that the detailed operational plans for the collection program are thoroughly explained to the *gun. pu. myun 'eup dong* and lower level. heads, to Summer Grain Collection Committees, to police officials, and to local representatives of the Korea Agricultural Association, National Land Administration, and Provincial Food Services.

b. Detailed instructions regarding collection quotas.

(1) *Assignment of quotas. in general.* Upon receipt of the notice of the summer grain collection quota from the National Food Administration, each provincial governor shall prorate promptly his provincial quota among his respective *gun* and *pu*. *Gun* and *pu* heads shall assign quotas to each *myun, eup* and *dong (in a pu)* without delay. The *myun* and *eup* heads shall assign a quota to each *ku (in a myun)* and *dong (in a eup)*. Each *ku* and dong *(in a eup or pu)* head shall assign a tentative quota to each *ban (small number of households)*. Each *ban* head shall make a tentative estimate of a quota for each individual farmer under his control. The *ban* head may determine such quotas at a meeting of the farmers of his *ban*;

(2) *Assignment of quotas to individual farmers.* The *ku (in a myun)* or *dong (in a eup or pu)* head on receipt of his summer grain collection quota from the *myun, eup* or *pu* head shall call a meeting of his *ban* heads and as a result of this meeting, tentative collection quotas shall be assigned to the individual farmers. The *ku* or *dong (in a eup or pu)* head with the advice of the *ku* or *dong* Summer Grain Collection Committee shall approve or modify the recommendations tentatively decided upon at the meetings of the *ban* heads

— 5 —

and send his signed recommendations, as to the quotas to be assigned the individual farmers, to the *myun, eup* or *pu* head. At least ten days prior to the harvesting of summer grain in each province, the *myun,- eup* or *pu* head with the advice of his *myun, eup* or *pu* Summer Grain Collection Committee shall, on the basis of the *ku* or *dong* head's signed recommendations, issue through the *ku* or *dong* (in a *eup* or *pu*) heads and through the *ban* heads, a written quota notice to each farmer located within his area;

(3) *Posting of quotas.* Each *myun, eup* or *pu* head shall publicly and conspicuously post on a bulletin board at his office a list showing the amount of the quotas of his respective *ku* or *dong* (in a *eup* or *pu*). Each *ku* (in a *myun*) or *dong* (in a *eup* or *pu*) head shall publicly and conspicuously post a list on a bulletin board in his office or public assembly hall showing the name of each farmer with the amount of his quota. The date the farmer delivers his quota of summer grain to the collection point shall be entered on the list at the time of delivery. The lists posted on the bulletin boards at each level of government will be posted throughout the entire period of the collection program.

c. Complaints by farmers. Within five days of the receipt of the quota notice by the farmer, he may appeal in writing to his *myun eup* or *pu* head for a reconsideration of his quota assignment. The appeal will state in detail the reasons for his objections. Within seven days, the *myun, eup* or *pu* head, after investigation and consultation with his *myun, eup* or *pu* committee, shall in writing communicate his decision, which shall be final, to the *ku* or *dong* (in a *eup* or *pu*) head concerned. The *ku* or *dong* head shall immediately notify the farmer of the result of his appeal. If the decision is favorable to the farmer it will be necessary for the *ku* or *dong* head to adjust some of the quotas already established.

d. Collection points.

(1) *Gun* and *pu* heads shall designate warehouses as summer grain collection points. Such warehouses shall be suitable storage places, adequately protected against fire and theft and located centrally and as conveniently as possible to milling facilities in each production area;

(2) All summer grain collected or stored at designated warehouses shall be under the jurisdiction and control of the Provincial Food Service which shall make distribution therefrom in accordance with plans and policies of the

-- 6

National Food Administration.

    e. Inspection of summer grain Summer grain received at collection points shall be inspected for weight and grade by regular government-employed inspectors and graded in accordance with current grain inspection regulations. The grain inspector shall give the farmer an inspection certificate and retain a duplicate thereof.

    f. Receipts and payment. Non-transferable triplicate receipts shall be made available at collection points. Design, printing and distribution to the collection points of such receipts shall be the responsibility of the Controller of Commodities. Expenses of printing and distribution thereof shall be borne by the Office of the Controller of Commodities. All preparations shall be completed so that collections provided for herein may commence not later than 15 June 1948. Receipts shall be prepared in triplicate by *myun, eup* or *pu* officials at each collection point. The original shall be given to the farmer, the duplicate delivered to the representative of the Provincial Food Service, and the third copy retained at the *myun, eup or pu* office. A representative of the Provincial Food Service shall be at the collection point and shall then and there pay the farmer. He shall mark the original receipt paid and return it to the farmer.

    g. Prices. Prices to be paid to farmers for summer grain shall be conspicuously posted at each collection point. Such prices shall be established by the government.

    7. *Second phase.* The second phase of the plan involving the actual collection of summer grain, shall commence as soon as such summer grain is harvested.

    a. Delivery of quota summer grain: seizure of undelivered grain. All quota summer grain shall be delivered to the collection points prior to 1 September 1948. All quota summer grain not so delivered prior to 1 September 1948, shall be seized by the government and the owner shall be paid 50 per cent of the established price.

    b. Delivery by individual owner-farmers. Each individual owner-farmer shall deliver his quota of summer grain (*and any amount in excess thereof that he may desire*) to the designated collection point in his area and shall receive a non-transferable collection receipt therefor. No sale or transfer of summer grain shall be made by a farmer except to the government or an authorized agency thereof until after 30 September 1948.

— 7 —

c. *Delivery by tenant-farmers.* The tenant-farmer shall deliver his quota, including rental grain, (*and any amount in excess thereof that he may desire*) for his farm directly to the collection point, where he will be issued a non-transferable collection receipt therefor. In those cases where rent is due, except rent due NLA, as explained in paragraph 3 *d* hereof, the landlord, natural or juridical, shall collect his rent in cash from the tenant on the basis of the established government price paid for such summer grain, any provision in the rental contract requiring payment to the landlord in kind notwithstanding. An individual landlord shall receive summer grain for his household in the same manner as any other non-supplier (*through established rationing and distribution channels*).

d. *Transportation.* During the period 15 June 1948 to 30 September 1948, inclusive, except for transportation to collection points, no summer grain shall be transported by other than an authorized governmental agency. Any summer grain in unauthorized transit during this period shall be seized without payment and shall be delivered to the Provincial Food Service of the province in which the seizure was made, for sale through regular legal distribution channels.

e. *Illegal sale.* Sale or offering for sale of summer grain during the period of 15 June 1948 to 30 September 1948, inclusive, except by government rationing agencies, is prohibited, and all summer grain sold or offered for sale in violation hereof shall be seized without payment and shall be delivered to the Provincial Food Service of the province in which the seizure was made, for sale through regular legal distribution channels.

8. *Third phase; Reports and accounting.* The third phase of the plan involving reports and accounting shall commence with the collection of summer grain and shall be completed not later than 30 September 1948.

a. *Collection reports.*

(1) *By the Office of the Controller of Commodities.* Every five days during the summer grain collection program, the Office of the Controller of Commodities shall submit to the National Food Administration a report showing the total amount and kinds of summer grain stated in polished *suk*, purchased by the respective Provincial Food Services. The report on the tenth, twentieth, and thirtieth of each month, shall be broken down to show the amounts purchased in each *gun* and *pu* of each province. The final summer grain report

— 8 —

shall be submitted by the Office of the Controller of Commodities on 30 September 1948.

(2) *By the provincial governors.* Every ten days during the summer grain collection program, each provincial governor shall forward to the National Food Administration a report showing, in polished *suk* the amount and kinds of summer grain collected up to date.

b. Report of undelivered quota summer grain. Each *myun, eup* and *dong* (*in a pu*) head shall prepare and forward so as to reach the appropriate *gun* or *pu* head, not later than 10 September 1948, a list containing the following information for each person who has failed to deliver to the government his summer grain collection quota:

(*1*) Name

(*2*) Address

(*3*) Amount of summer grain quota

(*4*) Amount of summer grain quota delivered to the government

Gun and *pu* officials shall take appropriate action against persons who fail to deliver their quota prior to 1 September 1948.

c. Additional reports and accounting. The Controller of Commodities, NLA. Provincial Food Services and Korea Agricultural Association, shall keep records of their respective operations and shall furnish reports on their activities when and as required.

d. Records and reports by government and corporate landlords. NLA, all other corporate landlords, and all governmental subdivisions, agencies and instrumentalities owning or operating agricultural property upon which summer grain is produced, shall maintain accurate records of the summer grain delivered by their tenants to the collection points under this regulation. Such records shall be open to inspection at any time by competent authority.

9. *Specific offenses enumerated.* Without limiting the provisions of this regulation, the following are specifically declared to be violations hereof:

a. Concealing or wrongfully failing to deliver summer grain in accordance with collection quotas assigned by *myun, eup* or *pu* heads.

b. Selling, offering for sale, transferring or offering to transfer, buying or offering to buy, summer grain, except by permission or authority of the government, or at other than officially established prices, or otherwise illegally dealing in such summer grain during the period 15 June 1948 to 30 September

— 9 —

1948, inclusive.

c. Failing or refusing to receive, collect, inspect, properly assign quota, store, grade, safeguard, transport, process, issue proper receipts, keep accurate records and accounts, or make payments for or reports concerning summer grain delivered under the collection program, or other official misconduct or failure or willful neglect of duty in connection therewith.

d. Any transaction or act, or attempted transaction or act, for the purpose or which has the effect of evading or avoiding the provisions of this regulation.

e. Hampering or preventing the Summer Grain Collection Committees or officials from investigating the amount of the Summer Grain Production in order to obstruct the summer grain collection program, or knowingly spreading false propaganda, or threatening the farmers in order to lessen the farmers willingness to deliver summer grain to the government collection points.

10. Penalties Any person violating the provisions of this regulation shall, upon conviction by a duly constituted court, suffer such punishment as may be particularized for the offense in ordinance or other law; and in the absence of such particularization, such punishment as the court may determine.

11. Prior regulation revoked. National Food Regulation Number 5. (*Collection of Summer Grains*) dated 8 May 1947, is hereby revoked, but the revocation of such regulation is not intended and shall not be construed to extinguish offenses under such regulation committed prior to the effective date hereof nor to abate or bar existing or future prosecutions based on such offenses

12. Effective date. This regulation shall be effective on the date appearing hereon.

BY DIRECTION OF THE MILITARY GOVERNOR:

CHEE YONG EUN
National Food Administrator

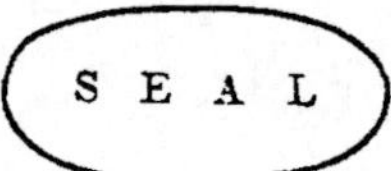

HEADQUARTERS
UNITED STATES ARMY MILITARY
GOVERNMENT IN KOREA
National Price Administration
Seoul, Korea

NATIONAL PRICE REGULATION
NUMBER 1

26 June 1946

1. *Establishment of Maximum Prices.* Pursuant to Section IV of Ordinance Number 90, dated 28 May 1946, the following maximum prices are hereby established:

ARTICLE	UNIT	MAXIMUM PRICE (Yen)
a. *Cotton Cloth*		
Unbleached sheeting	Yard	Y 33.00
Unbleached shirting	Yard	30.00
Twill, light weight	Yard	38.00
Twill, heavy weight	Yard	40.00
b. *Mixed Cloth*		
Hemp and cotton	Yard	23.00
c. *Cotton Socks*		
White	Pair	9.00
Colored	Pair	10.00
d. *Rubber Shoes*		
Men's	Pair	45.00
women's	Pair	35.00
Children's	Pair	30.00
e. *Workmen's or Zikatabi Shoes*		
White	Pair	55.00
Colored	Pair	60.00
f. *Duck or Canvas Shoes*		
Men's	Pair	45.00
Women's	Pair	35.00
Children's	Pair	30.00
g. *Soap*		
Laundry, 100 momme (Other weights, ¥ 0.12 per momme)	Bar	12.00
Toilet, 30-momme (Other weights, ¥ 0.30 per momme)		9.00

ARTICLE	UNIT	MAXIMUM PRICE (Yen)
h. *Matches*		
Small box	Box	Y 2.00
Large box	Box	15.00
i. *Electric Light Bulbs*		
5 to 30-watt	Bulb	30.00
31 to 65-watt	Bulb	40.00
66 to 100-watt	Bulb	100.00
Over 100-watt	Bulb	1.00 *per watt*
j. *Bicycle Tires*		
26-inch	Tire	160.00
28-inch	Tire	175.00
k. *Bicycle Tubes*		
26-inch	Tube	50.00
28-inch	Tube	60.00

2. *Posting of Price Lists; Furnishing Bills.* Persons offering for sale any of the items enumerated in Section 1 hereof shall have posted price lists thereof in a conspicuous place easily seen by prospective purchasers, and shall furnish to any purchaser, upon demand, a written bill stating the exact amount received for any of such items.

3. *Certain Offenses Enumerated.* It shall be unlawful for any person (natural or juridical) to sell or offer to sell to or buy or offer to buy from, any other person (natural or juridical), or display for sale or otherwise deal or offer to deal in any of the articles enumerated in Section 1 hereof, at prices higher than those established by Section 1, or to barter any of such articles for any commodity, service or economic benefit whatsoever, except in cases where specific authorization is given by the National Price Administrator in writing.

4. *Penalties.* Any person (natural or juridical) violating any of the provisions of this regulation shall, upon conviction by a Military Occupation Court or Special Judicial Officer, suffer such punishment as the court or Special Judicial Officer shall determine.

5. *Effective Date.* This regulation shall be effective at 2400 (midnight). 14 July 1946.

By DIRECTION OF THE MILITARY GOVERNOR:

T E ROBERTS
Major   FD
National Price Administrator

SOUTH KOREAN INTERIM GOVERNMENT
National Price Administration
Seoul, Korea

NATIONAL PRICE REGULATION
NUMBER 2   30 June 1947

NATIONAL PRICE REGULATION NUMBER 1 REVOKED

1. *Purpose.* The purpose of this regulation is to create a free-market in those items produced in private facilities with raw materials procured from non-governmental sources and hitherto controlled by National Price Regulation No 1, dated 26 June 1946. Prices for sales of items distributed through governmental channels will be established in accordance with instructions to appropriate agencies issued by the Natonal Price Administration.

2. *Revocation of National Price Regulation No 1.* National Price Regulation No 1, dated 26 June 1946, is hereby revoked, but this revocation is not intended and shall not be construed to extinguish offenses under National Price Regulation No 1 committed prior to the effective date hereof nor to abate or bar existing or future prosecutions based on such offenses.

3. *Agreed Resale Prices.* No person, natural or juridical, who has agreed with the government on the price at which or the formula by which he will sell any item shall sell or deliver such item at a price higher than the agreed price or otherwise in violation of his agreement.

4. *Definition.* Government, as used in this regulation, includes any department, agency, instrumentality or element of the government and any business, corporation or other juridical person controlled directly or indirectly by the government.

5. *Evasion.* The provisions of this regulation shall not be evaded either by direct or indirect methods, by way of any commission, barter, or service, transportation, container or other charge or by tying agreement or otherwise.

6. *Penalties.* Any person violating any of the provisions of this regulation shall be subject to the penalties provided in Ordinance No 90, dated 28 May 1946, and Ordinance No 120, dated 24 October 1946, and in appropriate

cases, to penalties provided in Ordinance No 19, dated 30 October 1945, for profiteering.

7. *Effective Date.* This regulation shall be effective on the date appearing hereon.

BY DIRECTION OF THE MILITARY GOVERNOR:

CHOI TAI WOOK

Director

National Price Administration

HEADQUARTERS
UNITED STATES ARMY MILITARY
GOVERNMENT IN·KOREA
National Economic Board
Seoul, Korea

REGULATION
NUMBER 1   14 September 1946

NATIONAL DISTRIBUTION OF CONTROLLED COMMODITIES

1. *Application.* This regulation supersedes the "Plan for National Distribution of Controlled Commodities" published in letter, this Headquarters, dated 5 March 1946.

2. *Purpose.* The purpose of the plan for the national distribution of controlled commodities is:

   a. To provide fair allocations to the provinces and fair distribution within the provinces of all essential commodities on hand or being held or produced under the control of Military Government (hereinafter referred to as *controlled commodities*), through Military Government controlled channels at prices established by Military Government. Agencies holding or producing commodities under the control of Military Government include but are not limited to vested companies, New Korea Company, governmental organizations and companies, Materials Control Corporation, and other similar organizations holding or producing commodities for sale or disposition.

   b. To provide for a centralized policy of distribution coordinated at the national level under policies established by the National Economic Board in accordance with Ordinance Number 90, dated 28 May 1946.

   3. *Scope.*

   a. The Department of Commerce of the Government of Korea shall designate as *controlled commodities* the type, kind, and classification of goods falling within the scope of this plan. This list may be revised from time to time.

   b. The various provinces shall obtain and report, as hereinafter described, to the Department of Commerce of the Government of Korea, information concerning production, production costs. stocks on hand, and estimated requirements in the respective provinces.

c. On the basis of information so furnished and based upon the overall requirements of the economy of Korea, the Department of Commerce shall make equitable allocations among the provinces of the available stocks and production of *controlled commodities*. The Department of Commerce may use national commercial agencies (a list of which shall be published periodically), the services of existing national government agencies, or for particular provinces, such agencies as may be designated by the respective provincial governors, to perform the actual buying, selling, transportation, contracting, etc., in the various provinces.

4. *Details of Plan.*

a. *Reports.*

Each provincial governor shall prepare monthly reports as of the first of each month covering the following information for the past month:

(1) All production of *controlled commodities* produced by factories under Military Government control, plus any available information on the same commodities produced in Korean factories not under Military Government, within his province. Such report shall include production costs, reported in sufficient detail to enable price determinations to be arrived at, and the name and location of each factory, listing separately those factories controlled from the national level, with the production figures of each.

(2) All stocks of *controlled commodities* on hand and not previously allocated by the Department of Commerce, including Materials Control Corporation stocks of *controlled commodities*. Negative reports shall be submitted when there are no stocks of such items. *Stocks on hand* shall not include commodities covered in the *production report*, nor commodities previously allocated by the Department of Commerce.

(3) The minimum requirements for all *controlled commodities* and for other items urgently required in his province, such report to indicate priorities of requirements and destinations desired, if and when shipment is made (See attached forms *A*, *B*, and *C*.)

(4) All reports shall be forwarded so as to reach the Bureau of Domestic Commerce, Economic Section, Seoul, Korea, not later than the sixth day of each month.

b. *Allocations.*

(1) Allocations of *controlled commodities* to the various provinces shall be made by the Department of Commerce. Such allocations, based on the monthly reports forwarded from the various provinces and on the overall

economy of Korea, shall be considered final. The allocations shall indicate the type and quantity of goods allocated; the province from which they are to be shipped: the province to which they are to be shipped; and the agency that will handle the distribution to the provinces. One copy of the allocation shall be forwarded to the shipping province; one copy to the receiving province; and one copy to the agency or agencies designated to effect distribution, with further instructions and authority to arrange for the picking up and shipping of the commodities to the location or locations in the province or provinces designated, and to sell such commodities to the provincial governor or his agents, subject to price regulations of the Military Government. Variations in this procedure may be authorized by the Department of Commerce, but in such cases instructions will be contained in the allocation form (See form D) at the time the allocation is made.

(2) Allocations of *controlled commodities* produced or held under Military Government control will not be made by factories, provincial governors, or other agencies, unless and until such commodities are allocated by the national Department of Commerce to the agency concerned.

(3) Allocations shall be made monthly on the basis of the *estimated production during the next calendar month*, corrected to account for the amount actual production exceeds or falls short of the estimated amount for the preceding month. This procedure will enable commodities to be moved as they are produced.

(4) Unfilled commitments for commodities by factories, made prior to placing such commodities under control, may be reported so that individual consideration can be given to such commitments by the Department of Commerce. Failing such report and specific permission by the Department of Commerce authorizing the fulfillment of such commitment, such commodities may be subjected to control without notice.

c. Distribution.

(1) National commercial agencies or other designated agents shall be responsible for picking up the allocated commodities either at the factories which produce them or at other designated locations within the "producing" province. After picking up the allocated commodities, the national commercial agencies of other designated agents shall transport them, under guard whenever necessary, to the provinces indicated in the allocation. They will cooperate with the provincial governor concerned in delivering allocated commodities to the receiving province as may be agreed upon between them and the provincial governor or his agent.

(2) The Department of Commerce may authorize the specific agent to be used within a province whenever it is necessary in order to carry out the centralized control policy for distribution. The provincial governor may use a *Provincial Commercial Agent* (or agents) to distribute, transport, sell, etc., *controlled commodities* within his province, when authorized by the Department of Commerce.

(3) The provincial governor or his agent shall be responsible for safeguarding and distributing the allocated commodities from the time he or his agent accepts delivery thereof until the commodities reach the hands of the ultimate consumers.

(4) The Chief of the Commerce Bureau of each province shall act as the agent for the governor of his province for coordinating the movement of *controlled commodities* to, from, and within his province. Direct communication between the national Bureau of Domestic Commerce and the chiefs of the provincial Commerce Bureaus is authorized to expedite and facilitate effective operation of his plan.

b. Prices.

(1) It is the objective that controlled commodities reach the hands of ultimate consumers at the lowest prices possible.

(2) Prices of controlled commodities are subject to the control of the Military Governor of Korea, and are determined by the National Price Administrator. Prices will be fixed by the Price Administrator at the time of allocation or published separately. The prices may be either the prices for such commodities at the factory or warehouse, or the prices to the ultimate consumer, or both.

5. *Effective Date.* This regulation shall be effective on the date appearing hereon.

BY DIRECTION OF THE MILITARY GOVERNOR:

JAMES H. SHOEMAKER
Chairman, National Economic Board

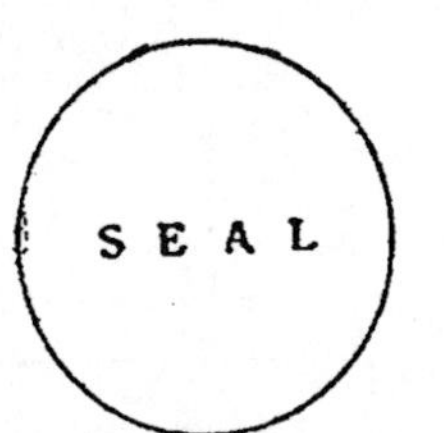

4 Incls: Forms *A*, *B*, *C*, and *D*.

NATIONAL DISTRIBUTION OF CONTROLLED COMMODITIES
FORM A (REV.): REPORT OF PROVINCIAL PRODUCTION

No.___________________________   Date___________194____

TO: Economic Section
    Bureau of Domestic Commerce
    Department of Commerce
    Hq, USAMGIK
    Seoul, Korea

1. Production within___________________________Province, for the
current month of_____________________, is reported as follows:

Commodity	Factory location	Unit	Actual production during past calendar month	Estimated production during current calendar month	Estimated production during next calendar month	Mfg Actual for past month	Cost Estimated for current month

REMARKS: BY:_______________________

                                  ____

Page____ ____of________Pages

NATIONAL DISTRIBUTION OF CONTROLLED COMMODITIES
FORM B (REV.): REPORT OF PROVINCIAL STOCKS ON HAND

NO______________________   Date__________ 194____

TO: Economic Section
Bureau of Domestic Commerce
Department of Commerce
Hq. USAMGIK
Seoul, Korea

1. Stocks on hand within____________________________Province as of
________________194____are reported as follows:

Commodity	Unit	Quantity	Location and Remarks

REMARKS: BY:________________________

____

Page__________of________Pages

NATIONAL DISTRIBUTION OF CONTROLLED COMMODITIES
FORM C (REV.): REPORT OF PROVINCIAL REQUIREMENTS

No._______________________   Date__________194___

TO: Economic Section
  Bureau of Domestic Commerce
  Department of Commerce
  Hq, USAMGIK
  Seoul, Korea
  1. The minimum requirements of________________________Province
as of________________________194____are as indicated below:

Commodity	Unit	Quantity	Destination	Remarks *

NOTE: Indicate by numerals the
   priority of items required. By:________________________

* State what period of time, by months, is covered by the requirement.

Page__________of________Pages

NATIONAL DISTRIBUTION OF CONTROLLED COMMODITIES
FORM D (REV.): ALLOCATION OF CONTROLLED COMMODITIES

Allocation Number________

HEADQUARTERS UNITED STATES ARMY MILITARY GOVERNMENT IN KOREA

Department of Commerce
Bureau of Domestic Commerce
Seoul, Korea

SUBJECT: Allocation of Controlled Commodities   ________________ 1946

TO:

   1. Reference:

   2. The following allocations for (production) & (stocks on hand) & (MCC stocks) of___________
are authorized to: (check one)

Item No. (1)	Commodity (2)	Quantity Allocated (3)	Quantity Received (4)	Unit (5)	Price (6)	Source of Stocks Allocated (7)

   3. Transportation   (see other side)

4. Remarks:

5. Request acknowledgment by indorsement hereon of goods received in accordance with column (4) of paragraph 2 to be filled in by receiving agency. In the event that subject commodities are not received within_______ weeks after receipt of this allocation, notify Economic Section, Bureau of Domestic Commerce.

FOR THE DIRECTOR:

FRED W. LOUIS
Major,   QMC
Chief, Bureau of
Domestic Commerce

1st Indorsement

To: Economic Section, Bureau of Domestic Commerce, Dept of Commerce, Hq, USAMGIK, Seoul, Korea
1. Subject commodities received.
2. Remarks:

HEADQUARTERS
UNITED STATES ARMY MILITARY
GOVERNMENT IN KOREA
National Economic Board
Seoul, Korea

REGULATION
NUMBER 2   14 September 1946

CONTROL OF TEXTILE MATERIALS AND EQUIPMENT

1. *Purpose.* The purpose of this regulation is to facilitate the revival of textile production in Korea.

2. *Scope.* This regulation applies to all materials and equipment named therein which are in use or storage in, or which are produced in or imported into Korea south of 38° north latitude.

3. *Controlling Agency.* The Department of Commerce of the Government of Korea, acting through the Textile Section of the Bureau of Industry, shall be responsible for accomplishing the purpose of this regulation.

4. *Materials Controlled.* All sales of the materials described below shall be made to or under the control of the Textile Section of the Bureau of Industry of the Department of Commerce:

a. Ginned cotton produced by gins owned by or under the control of USAMGIK, or sold by any corporation or enterprise owned or controlled by USAMGIK;

b. Cotton yarn of any description, except yarn actually required in the further production of textiles by the producing factory, *provided* that the Textile Section of the Bureau of Industry shall in all cases regulate or determine the amount of such yarn so excepted; and cotton yarn of any description, sold by any corporation or enterprise owned or controlled by USAMGIK;

c. Raw hemp offered for sale by any producer, association, broker, storage company, or corporation or enterprise owned by USAMGIK;

d. Skein silk produced by any filature or other enterprise owned or controlled by USAMGIK.

5. *Equipment Controlled.* All sales of the equipment described below shall be made to or under the control of the Textile Section of the Bureau of Industry of the Department of Commerce:

a. Machinery or other equipment, other than human-powered, design-ed or primarily serviceable for use in spinning or weaving cotton, wool, hemp, or cotton mixed yarns, sewing thread, cloth or tissue;

b. Machinery or other equipment, other than human-powered, design-ed or primarily serviceable for weaving silk cloth;

c. Machinery or other equipment, other than human-powered, design-ed or primarily serviceable for ginning of cotton;

d. Machinery or other equipment, other than human-powered, design-ed or primarily serviceable for any type of garment manufacture;

e. Machinery or other equipment, other than human-powered, designed or primarily serviceable for any type of knitting manufacture;

f. Machinery or other equipment, other than human-powered, design-ed or primarily serviceable for silk thread manufacture;

g. Replacement, repair or maintenance parts for the machinery and equipment described above.

6. *Price Control.*

a. Prices for materials, machinery and equipment described above shall be established by the National Price Administrator, from time to time, as required;

b. At the request of the National Price Administrator, the Department of Agriculture and the Department of Commerce shall submit price recommendations and supporting data to assist the National Price Adminis-trator in fixing prices;

c. At the request of the National Price Administrator, corporations or other enterprises owned or controlled by USAMGIK shall furnish sales records and other data deemed by the National Price Administrator to be necessary in establishing prices of equipment.

7. *Effective Date.* This regulation shall be effective on the date ap-pearing hereon.

BY DIRECTION OF THE MILITARY GOVERNOR:

JAMES H SHOEMAKER
Chairman, National Economic Board

CHAPTER 6 KOREAN INTERIM LEGISLATIVE ASSEMBLY

SOUTH KOREAN INTERIM LEGISLATIVE ASSEMBLY
*Rules of the Assembly**

TABLE OF CONTENTS

CHAPTER I. OPENING, CLOSING, SESSION, ADJOURNMENT.

Sec. 1. The opening and closing of sessions shall be determined by the Assembly itself. The Speaker shall publish the date fixed for opening fifteen days in advance. The Assembly shall be opened and closed with due ceremony.

Sec. 2. The Assembly shall be in continuous session until a democratic Provisional Government is established after the North and South are unified, or until a new Legislative Assembly is set up by National popular election.

Sec. 3. Adjournment of the Assembly for any period up to thirty

---

*English translation, Office of Adviser to Director, Department of Justice, Headquarters, United States Army Military Government in Korea.

軍　政　廳　　　官　報　　　　朝鮮過渡立法議院院法

days shall be within the power of the Assembly; the Speaker shall have the power to adjourn the Assembly for any period up to seven days. But in any case the Assembly shall be reconvened within three days at the request of the Executive Head of the government, or of three or more committees, or of one fourth of the members of the Assembly duly qualified and seated.

CHAPTER II. MEMBERS.

Sec. 4. Debates in the Assembly, on the floor thereof and in committees shall be free, and no member shall be questioned anywhere for his utterances or voting, nor be answerable for them. But if a member publicizes his utterance in speech, printed form or otherwise, outside of the Assembly, he shall be responsible in accordance with existing general law.

Sec. 5. While the Assembly is in session a member shall not be arrested without its consent, unless he has committed treason or is apprehended in the act of committing a crime.

Sec. 6. Each member shall report to the Assembly on the date it convenes. Each member shall register three days before the opening day by presenting his credentials and personal history. A member shall lose his membership if he has failed to register for seven days after the opening date without proper reason. The Assembly shall determine whether or not a member shall be disqualified if he is absent more than fifteen days after it has convened, though he has proper reason. Until a new election law is enacted, the vacancy of any member shall, upon notification by the Speaker, be filled in accordance with the method by which the member was originally chosen.

Sec. 7. Qualifications of the members shall be reported by the Qualifications Review Committee at the beginning of the meeting.

Sec. 8. If there is an objection to the qualifications of any member, five members or more may file with the Speaker an application for review. The Speaker then shall refer the matter to the Qualifications Review Committee for review and report.

Sec. 9. Such member shall not be deprived of the right to speak and vote until the final decision is made, but he shall not vote on his own case.

CHAPTER III. SPEAKER AND ASSISTANT SPEAKERS.

Sec. 10. The Assembly shall elect a Speaker and two Assistant Speakers.

Sec. 11. The Speaker and Assistant Speakers shall be elected by secret ballot. The members shall vote for one in the case of the Speaker, and may vote for two in the case of the Assistant Speakers. The election shall be determined by a majority vote of the members present, and for this purpose two-thirds of the members duly qualified and seated shall constitute a quorum. If a majority vote is not received on the second ballot, a plurality will decide on the third ballot. When two or more persons have the same number of votes on the third ballot, the choice shall be made by lot.

Sec. 12. The Speaker shall keep order within the Assembly and shall regulate the proceedings and represent the Assembly outside the Assembly; he shall supervise and control administration within the Assembly and personnel attached to it.

Sec. 13. The Assistant Speakers shall assist the Speaker and shall substitute for him when the Speaker is absent.

Sec. 14. When both the Speaker and Assistant Speakers are absent, a temporary Speaker shall be elected and he shall act in place of the Speaker.

Sec. 15. If the Speaker or Assistant Speaker violates the Rules of Procedure of the Assembly, the matter may be referred to the Petition and Discipline Committee on motion of one-fifth or more of the members duly qualified and seated, and he may be removed by a majority vote of the members present. For this purpose two-thirds of the members duly qualified and seated shall constitute a quorum.

CHAPTER IV. COMMITTEES.

Sec. 16. The Assembly shall have two types of committees, as follows:
a. Standing Committees.
b. Special Committees.

Sec. 17. The Assembly shall elect a Nominating Committee of fifteen members which shall nominate members of the various committees. The members of the Nominating Committee shall be elected by a plurality of votes on secret ballot. Each member may vote for fifteen. When two or more persons have the same number of votes, the choice shall be made by lot.

Sec. 18. A committee shall take up only matters which have been referred to it by the Assembly.

Sec. 19. A committee may, with the consent of the Speaker, have its own secretaries, clerks and/or specialists.

Sec. 20. Each committee shall elect a chairman. The chairman of the committee shall keep order and regulate the proceedings of the committee.

Sec. 21. A committee shall not be convened unless a majority of its members is present.

Sec. 22. All the actions of a committee shall be determined by majority vote of the members present, and in case of a tie vote, the chairman of the committee shall cast a deciding vote.

Sec. 23. The chairman of a committee shall make written report to the Assembly concerning matters acted upon.

Sec. 24. When the chairman of a committee is absent, the members shall elect a temporary chairman.

Sec. 25. A chairman of a committee may call meetings of the committee, but he shall not schedule them while the Assembly is meeting, without permission of the Assembly.

Sec. 26. The committee shall keep a minute book and record the names of members present, number of votes, a summary of the matters acted upon, and other important matters. The minute-book shall be signed by the chairman of the committee and shall be kept in the Assembly.

Sec. 27. Two or more committees by agreement may hold a joint meeting. The chairman for the joint meeting shall be the chairman of the committee listed first in Section 31 or Section 32.

Sec. 28. A sub-committee shall elect one of its members as *Kansa* in order to regulate its proceedings.

Sec. 29. The Speaker of the Assembly shall have the right to speak in committee meetings, and a member shall have the same right when requested to appear or when so permitted by the committee.

— ÷ —

Sec. 30. The Assembly may require a committee to report its findings by a designated date, and may elect new committee members if it fails to report within such period without proper reason.

Sec. 31. Standing committees shall be as follows:

   a. Legislative Policy and Judiciary Committee (15 members).

     (1) First sub-committee (10 members).
       Legislative Policy.
     (2) Second sub-committee (5 members).
       Judiciary.

   b. Home Affairs and Police Committee (12 members).

     (1) First sub-committee (6 members).
       General internal affairs.
       Administrative organization.
       Provincial and local government.
       Civil service examination.
       Inspection of government officials.
       Census.
     (2) Second sub-committee (6 members).
       Police matters.

   c. Finance and Economic Committee (21 members).

     (1) First sub-committee (9 members).
       Budget.
       Accounts.
     (2) Second sub-committee (6 members).
       Economy in general.
       Currency.
       Banking.
     (3) Third sub-committee (6 members).
       Taxation.
       Monopolies.

d. Industry, Labor and Agriculture Committee (15 members).

(1) First sub-committee (7 members).
Industrial engineering.
Mining.
Commerce
Foreign trade.
(2) Second sub-committee (8 members).
Farm land.
Agriculture and forestry.
Fisheries.
Labor organization.

e. Foreign Affairs and National Defense Committee (12 members).

(1) First sub-committee (6 members).
Foreign affairs.
(2) Second sub-committee (6 members).
National defense.

f. Education, Public Health and Welfare Committee (12 members).

(1) First sub-committee (6 members).
Educational institutions.
Religion.
General culture and education.
(2) Second sub-committee (6 members).
Public Health.
Welfare.

g. Transportation and Communications Committee (10 members).

(1) First sub-committee (5 members).
Transportation.
(2) Second sub-committee (5 members)
Communications.

h. Petition and Discipline Committee (10 members).

(1) First sub-committee (5 members).
Petitions.
(2) Second sub-committee (5 members).
Discipline.

Sec. 32. Special committees shall be as follows, provided that other committees may be added, as necessary, by resolution of the Assembly:

    a. Qualifications Review Committee (18 members).

        (1) First sub-committee (6 members).

        Qualifications of members of the Assembly.

        (2) Second sub-committee (12 members).

        Qualifications of government officials above Grade 4.

    b. Provisional Constitution and Provisional Election Law Drafting Committee (9 members).

    c. Government Organization Law Drafting Committee (9 members).

    d. Food and Price Committee (9 members).

    e. Enemy Property Committee (9 members).

    f. Committee for Drafting a Special Law Concerning Pro-Japanese Collaborators, Traitors, War Criminals and Profiteers (9 members).

Sec. 33. The Assembly may increase or decrease the numbers of members of the committees or sub-committees prescribed in the two preceding sections.

CHAPTER V. SESSIONS OF THE ASSEMBLY.

Sec. 34. The meetings of the Assembly shall be open to the public, but may be held in secret by resolution of the Assembly.

Sec. 35. The Speaker may declare the opening, adjournment, extension and suspension of any session.

Sec. 36. A printed calendar, including agenda, shall be distributed to the members at least two days before the meeting to which it is applicable.

Sec. 37. Printed copies of the minutes shall be distributed to the members within three days after each meeting, and corrections shall be made by the next meeting.

Sec. 38. A bill must be signed by ten or more members.

Sec. 39. A motion will be taken up as the subject of discussion if seconded by two or more members, except as otherwise provided in the rules of procedure of the Assembly.

— 7 —

Sec. 40. A bill shall not be voted on until it has gone through three readings. Attendance of two-thirds of the whole membership duly qualified and seated, and the vote of a majority of the members present shall be necessary to pass such a bill. A two days interval is necessary between readings, but the interval may be shortened or one of the readings may be omitted by resolution of the Assembly.

Sec. 41. A bill shall be read and explained generally by the sponsor at the first reading and shall be referred to the Legislative Policy and Judiciary Committee, which shall make a report on the bill. Thereupon it shall be determined whether the bill should be continued to the second reading or not.

Sec. 42. At the second reading the bill will be read and passed article by article.

Sec. 43. Any member may make a motion for amendment of the bill or file with the Speaker an amended draft before the reading. The concurrence of five or more members is required for a motion to amend.

Sec. 44. An amendment proposed by the committee may be taken up as a subject of discussion as presented.

Sec. 45. The Speaker may for purposes of discussion change the order of articles, combine two articles or break down an article.

Sec. 46. At the third reading the bill as a whole will be voted upon. No motion for amendment shall be made except for correction of wording; provided, that a motion for amendment may be made if it is found that the bill contains inconsistent provisions, or is contrary to or in conflict with other laws, and an amendment is required thereby.

Sec. 47. A member who wishes to speak on a subject entered in the calendar shall notify the Speaker before the meeting whether he is for or against the bill.

Sec. 48. As the Speaker is notified, he shall schedule speeches in order of notification, beginning with an opponent of the bill and alternating between opponents and proponents.

Sec. 49. A member who has not notified the Speaker shall not have an opportunity to speak until all the listed members have spoken; provided, that when there are no more speakers for one side but speakers for the other side remain, unlisted members may ask to speak for the former side.

Sec. 50. When a member wishes to speak, he shall stand and ask permission. When two or more request permission the Speaker will recognize the person who stood first and permit him to speak.

Sec. 51. When the Speaker wishes to take part in discussion, he shall leave the chair and an Assistant Speaker shall substitute for him. He shall not resume the chair until the subject under discussion has been put to a vote.

Sec. 52 No one shall speak more than twice on the same subject except to ask or answer questions.

Sec. 53. The Speaker shall declare the closing of discussion.

Sec. 54. When a motion for closing discussion is made and concurred in by four or more members, the discussion will be suspended while the motion is voted on.

Sec. 55. No member who is not present shall participate in the voting.

Sec. 56. The Speaker shall summarize the main points of the subject matter before voting and no one shall speak after the summary is made.

Sec. 57. The Speaker may ask members to vote by standing or by a show of hands, and shall declare the result. If it is difficult to determine the count, or if there is an objection seconded by four or more members, a secret ballot shall be taken.

Sec. 58. No member shall change his vote.

Sec. 59. A motion to reconsider may be made by the sponsor with the concurrence of two-thirds of the members present.

CHAPTER VI. BUDGET AND ACCOUNTS.

Sec. 60. The Finance and Economic Committee shall divide the budget or accounts into several parts. After the deliberation on any part has been completed, the committee may hold a meeting. After the determination on each part of the budget has been completed, the committee shall vote on the budget as a whole.

Sec. 61. If the Assembly in considering the budget discovers some item

requiring further study, the Finance and Economic Committee may be requested to re-examine the specified item.

CHAPTER VII. RELATIONS AMONG THE ASSEMBLY, ADMINISTRA-TIVE OFFICES AND THE PEOPLE.

Sec. 62. In approving appointments, two thirds of members present
shall be required.

Sec. 63. The Assembly or any committee may request through the
executive head the presence, answer, or explanation of the government officials
concerned or responsible, when deemed necessary.

Sec. 64. The Assembly may request from an administrative office
through the executive head, in any case under investigation, reports and
documents. An administrative office shall not refuse such request by the
Assembly.

Sec. 65. A proposed request or resolution calling for administrative
action may not be filed unless it is signed by five or more members. It shall
be sent to the Administrative office within five days after it is approved.

Sec. 67. The names of all officials holding civil service status above
Grade 4 shall be referred to, and their qualifications reviewed by the Quali-
fictions Review Committee. The review shall be based on special laws concerning pro-Japanese collaborators, traitors, war criminals, and profiteers, and
on laws and regulations concerning government officials.

Sec. 68. The Assembly shall have no authority to issue either proclamations or public notices directly to the people, or to summon and question the
people.

CHAPTER VIII. PETITIONS.

Sec. 69. A petition must be in written form, signed and sealed by the
petitioner with three or more members of the Assembly sponsoring it. A
petition by a juridical person must be signed by its representative and sealed
with the seal of the juridical person.

Sec. 70. A petition shall be referred to the Petition and Discipline
Committee for review.

— 10 —

Sec. 71. When the Petition and Discipline Committee has decided to present to the Assembly a petition involving the enactment of an ordinance, the committee may do so with a draft of a proposed ordinance. In such a case, the chairman of the Petition and Discipline Committee shall be the sponsor of the bill.

Sec. 72. If the Petition and Discipline Committee has reported its decision not to present a bill to the Assembly, the decision of the committee shall be final unless request by a member or members of the Assembly be made within five days, to present the bill to the Assembly.

Sec. 73. Petitions of the following nature shall not be accepted:

    a. Petitions presented by a representative in the name of a group other than a juridical person.

    b. Petitions contrary to the provisions of laws and regulations.

    c. Petitions to interfere in court proceedings.

    d. Petitions which are contemptuous of the Assembly or administrative branch of the government.

CHAPTER IX. ABSENCE AND RESIGNATION.

Sec. 74. When a member desires to take leave of absence, he shall file with the Speaker an application for leave stating the reason and the length of time. Approval for leave up to five days shall be given by the Speaker, and up to ten days by the Assembly. No leaves of absence shall be approved for four consecutive periods.

Sec. 75. If a member attends the Assembly before his leave expires, the permission for leave will be canceled from the date of his return.

Sec. 76. A member may file with the Assembly a letter of resignation stating the reason. Such resignation shall be voted on without discussion.

CHAPTER X. ADMINISTRATION OF THE ASSEMBLY; EXPENDITURES.

Sec. 77. In order to administer affairs of the Assembly, the office of the secretary and office of clerk are created. The office of secretary shall have one chief secretary, one assistant chief secretary, several secretaries, clerks and stenographers who will take care of all documents, records, editing, stenography, official seals, correspondence and other such matters. . The office of clerk shall have one chief clerk, one assistant chief clerk, several section chiefs, clerks, guards and employees who will take care of finance, accounts, printing, statistics, guards and other such matters.

Sec. 78. A description of each section and personnel attached to it is shown on a separate list. The Chief of each office and personnel attached to it shall be appointed and removed by the Speaker according to civil service principles. There shall be secretaries in the office of the Speaker.

Sec. 79. Expenditures of the Assembly shall be disbursed from the national treasury.

Sec. 80. The Assembly shall determine its budget and accounts.

CHAPTER XI. GUARDS, ORDER, SPECTATORS.

Sec. 81. The Speaker shall control the guarding of the Assembly while it is in session.

Sec. 82. The Chief Guard and other guards shall perform their duties under the supervision of the Speaker.

Sec. 83. The Speaker may interrupt a speech or require a member to retract his statement or expel a member from the Assembly Hall when a member violates the rules of or disturbs the order in the Assembly while it is in session.

Sec. 84. The Speaker may adjourn or suspend the meeting when it is in disorder. The Speaker will handle all other matters concerning maintenance of order within the Assembly.

Sec. 85. No one shall be allowed in the Assembly Hall who has firearms or weapons, or who is drunk or insane or a person of suspicious action.

Sec. 86. Spectators shall be admitted by ticket. Tickets shall be distributed by the Assembly to foreign officials, government offices and the press; and to the general public through the members of the Assembly.

Sec. 87 Spectators shall observe the rules and instructions of the Speaker concerning the maintenance of order within the Assembly.

CHAPTER XII. DISCIPLINE.

Sec. 88. The Assembly shall have disciplinary power over its members.

Sec. 89. Members who violate rules or disobey the Speaker's order twice or more shall be subject to discipline.

Sec. 90. Such cases shall be referred to the Petition and Discipline Committee. After hearing the report of the committee, the Assembly will decide such cases and will announce its judgment.

Sec. 91. The methods of discipline shall be as follows:

    a. Apology in an open session of the Assembly.
    b. Suspension of the right to speak for a designated period of time.
    c. Suspension of the right to attend meetings for a designated period of time.
    d. Disqualification as a member.

Sec. 92. When a member is absent from the Assembly or from a Committee without proper reason, the Speaker shall issue a notice urging his attendance. If he does not attend within ten days after such notification, he may be disqualified by resolution of the Assembly.

Sec. 93. Meetings concerning disciplinary matters shall be held in secret.

Sec. 94. A disciplinary motion must be made within five days of the occurrence of the act on which the motion is based. A disciplinary motion must be made by the Speaker or by three or more members of the Assembly.

Sec. 95. The party concerned shall not attend the disciplinary meeting, except that, with the permission of the Speaker or committee chairman, he may attend the meeting and explain his case, or may request another member to do so. A resolution for disqualification must be concurred in by two thirds of the members present.

CHAPTER XIII. SUPPLEMENTARY PROVISIONS.

Sec. 96. These rules may be amended upon motion of the Legislative Policy and Judicial Committee or of the Speaker or of ten or more members;

a majority vote of the members present shall be required. For this purpose a quorum shall be two thirds of the members duly qualified and seated.

Sec. 97. These rules shall be effective on the day they are adopted.

KOREAN INTERIM LEGISLATIVE ASSEMBLY
Seoul, Korea

LEGISLATIVE RESOLUTION

At the 59th Session of the Korean Interim Legislative Assembly, held 25 April 1947, with reference to the protection of property owned by Koreans in Japan, it was RESOLVED:

"1. That the Korean Interim Legislative Assembly request General MacArthur, Supreme Commander Allied Powers, through General Hodge, that General MacArthur have the following put into practice by the Japanese Government in treating Koreans in Japan:

(*a*) Grant Koreans equal rights with nationals of other countries in Japan holding and managing general property necessary in the operation of industrial, commercial, agricultural and fishery enterprises;

(*b*) Cancel the policy of charging individual Koreans who own property with robbery, profiteering, and disturbing the economy and the policy of banishing them;

(*c*) Provide Koreans with every facility in removing their industrial and commercial equipment from Japan to Korea;

(*d*) Exempt Koreans from the Japanese property tax since they were not responsible for the war.

2. That, at the same time it be requested that the investigating team which the Assembly intends to dispatch be given the full cooperation of General MacArthur's headquarters."

KIM KIU SIC
Chairman, Korean Interim
Legislative Assembly

HEADQUARTERS
UNITED STATES ARMY MILITARY
GOVERNMENT IN KOREA
Office of the Military Governor
Seoul, Korea

DECLARATION
NUMBER 2

21 February 1946

<u>Declaration of National Holiday</u>

Upon the recommendation of the Korean Representative Democratic Council of South Korea, March 1, 1946 is declared a <u>National Holiday</u>. It is the 27th anniversary of the first declaration of Korean <u>Independence</u> from Japanese oppressive rule now happily ended. This day is dedicated to the memory of the Korean patriots who gave their lives to the cause of Korean Independence. On this day let all Koreans honor such patriots as the fathers of the new Korea in which liberty and human rights will blossom into the flower of Nationhood.

ARCHER L. LERCH
Major General United States Army
Military Governor in Korea

CHAPTER **7** USAMGIK DECLARATION

HEADQUARTERS
UNITED STATES ARMY MILITARY
GOVERNMENT IN KOREA
Office of the Military Governor
Seoul, Korea

DECLARATION
NUMBER 3   15 November 1946

PUBLIC UTILITIES DECLARED ESSENTIAL

WHEREAS, continuous and uninterrupted operation of transportation facilities and electric power output are vital to the welfare of the people of Korea, and

WHEREAS, the possibility of disputes over terms and conditions of employment may result in suspension or diminution of transportation facilities and electric power output,

NOW, THEREFORE, I HEREBY DECLARE the following to be industries essential to the people's livelihood, within the meaning of Section II of Ordinance Number 19, dated 30 October 1945:

1. All plants, factories, establishments, installations and facilities engaged in or used for the production or transmission of electricity or electric power for public or general use, and all manufacture, repair or maintenance facilities thereof;

2. All railroads (including interurban and street railways) and railroad operation, manufacture, repair and maintenance facilities;

3. All public shipping, and public shipping and marine operation, manufacture, repair and maintenance facilities, including docks, wharves and and piers;

4. All public trucking, bus and highway transportation facilities, including operation, manufacture, repair and maintenance facilities thereof;

5. All air transport and airports, and operation, manufacture, repair and maintenance facilities thereof;

— 1 —

1001

6. All reservoirs, waterworks, water supply and transmission or distribution systems, installations or facilities, and operation, repair and maintenance facilities thereof;

7. All postal, telephone, telegraph, radio and other public or general communication or communication transmission systems, installations or facilities, and operation, repair and maintenance facilities thereof;

8. All other public utilities and operation, repair and maintenance facilities thereof.

ARCHER L LERCH
Major General United States Army
Military Governor in Korea

軍政廳 官報 布告 第三號   一九四六年十一月十五日

HEADQUARTERS
UNITED STATES ARMY MILITARY
GOVERNMENT IN KOREA
Office of the Military Governor
Seoul, Korea

DECLARATION
NUMBER 4

30 November 1946

DESIGNATION OF FIRE PREVENTION WEEK

WHEREAS fires continue to exact a great toll of lives and material resources in Korea, *and*

WHEREAS destructive fires endanger safety and lay waste houses, food and other necessities essential to Korean economy, *and*

WHEREAS the great majority of fires can be prevented by the exercise of greater care and caution by all people,

NOW, THEREFORE, I HEREBY designate the week commencing 8 December 1946 as FIRE PREVENTION WEEK.

I urge every person to help in protecting lives and property from fires by learning to discover and by eliminating all fire hazards in and around the home, dormitory, barracks and place of business.

I also urge places of worship, schools, the press, radio, business, labor and other organizations to cooperate to the **fullest** extent possible in the observance of FIRE PREVENTION WEEK.

The various agencies of Military Government, including provincial and local governments, will assist to the maximum in arousing the public to the dread consequences of carelessness in causing and failing to prevent unnecessary fires.

C G HELMICK
Brigadier General United States Army
Acting Military Governor in Korea

軍政廳　官報　布告　第四號　　一九四六年十一月三十日

HEADQUARTERS
UNITED STATES ARMY MILITARY
GOVERNMENT IN KOREA
Office of the Military Governor
Seoul, Korea

DECLARATION
NUMBER 5

6 December 1946

DECLARATION OF NATIONAL HOLIDAY

To mark and commemorate the opening session of the Korean Interim Legislative Assembly, the 12th day of December, 1946, is hereby declared a *National Holiday*.

On this day let all patriotic Koreans pledge themselves to cooperate with and assist their representatives in the weighty tasks which confront them in their coming deliberations, so that this Assembly may be truly responsive to the needs and aspirations of the Korean people.

S E A L
USAMGIK

C G HELMICK
Brigadier General United States Army
Acting Military Governor

軍政廳　官報　布告　第五號　　　　一九四六年十二月六日

SOUTH KOREAN INTERIM GOVERNMENT
Seoul, Korea

DECLARATION
NUMBER 6   15 March 1948

ILLEGAL STRIKE OF EMPLOYEES OF THE SEOUL ELECTRIC COMPANY

The continuous and uninterrupted operation of transportation facilities and electric power output are vital to the welfare of the people of Korea. Declaration Number 3, United States Army Military Government in Korea dated 15 November 1946, specifies, among other things, that all plants, factories, establishments, installations and facilities engaged in or used for the production or transmission of electricity or electric power for public or general use, and all railroads, including interurban and street railways, and railroad operation, and all manufacture, repair and maintenance facilities thereof are essential to the people's livelihood, within the meaning of Section II of Ordinance Number 19, United States Army Military Government in Korea, dated 31 October 1945. The present dispute between the management and the employees of the Seoul Electric Company is an attempt on the part of certain persons to interfere in the management of the company by dictating to the company which management personnel it shall employ. The company has already agreed to pay to its employees accrued overtime pay to the extent of fifteen million won. Although the employees have not exhausted all peaceful means of settlement of any existing grievance, some of them have left their employment in violation of Ordinance Number 19, United States Army Military Government in Korea, thereby interrupting the operations of an essential industry.

NOW, THEREFORE, I HEREBY DECLARE, by virtue of powers vested in me as Military Governor in Korea, that:

1. The normal operations of the Seoul Electric Company are vital to the welfare of the people and essential to their livelihood.

2. The strike of employees of the Seoul Electric Company beginning on or about 13 March 1948 is unlawful.

3. All employees of said company are directed to return to their regular

— 1 —

employment with said company at the regular hour of work on Wednesday, 17 March 1948.

4. Any person who incites the employees of said company to leave their employment, interferes with their peaceful return thereto, or interferes with the operation of said company will be prosecuted for violation of Ordinance Number 19, United States Army Military Government in Korea, in accordance with the provisions of law.

WILLIAM F DEAN
Major General, United States Army
Military Governor in Korea

SOUTH KOREAN INTERIM GOVERNMENT
Seoul, Korea

DECLARATION

NUMBER 7　　　　　　　　　　　　　　　　28 May 1948

DECLARATION OF NATIONAL HOLIDAY

At the request of elected members of the National Assembly, the 31st day of May 1948 is hereby declared a National Holiday, to mark and commemorate the convening of the duly elected Representatives of the Korean people.

On this day, let all patriotic Koreans pledge themselves to cooperate with, to support, and to assist their Representatives in the weighty tasks which confront them in their coming deliberations, so that this Assembly may fulfill its mission of establishing a National Government in Korea.

W. F. DEAN
Major General, United States Army
Military Governor in Korea

軍政廳　官報　宣布令　第七號　　　　一九四八年五月二八日

CITY CHARTER

AUTHORIZED BY THE
MILITARY GOVERNOR IN KOREA
10 AUGUST 1946

軍 政 廳　官 報　서울市憲章　一九四六年八月十日

CHAPTER 8　OTHERS

INDEX

— a —

軍　政　廳　官　報　서울市憲章　一九四六年八月十日

Powers and Duties of Elective Officers

Departments

— b —

Commissions

CHARTER
OF THE CITY OF SEOUL

THE CITY AND ITS POWERS

Name and Boundaries of the City

SECTION 1.   The City of Seoul is hereby constituted a municipal corporation to be known as SEOUL.   The boundaries of the municipal corporation are the present limits of the City of Seoul consisting of the following eight districts: Choong Koo, Chong No Koo, Sur Tai Moon Koo, Tong Dai Moon Koo, Sung Tong Koo, Ma Po Koo, Yong San Koo, and Yung Doung Po Koo, and as such may be extended as provided by law.

Powers of the City

SECTION 2.   The City of Seoul shall have perpetual succession; may appear, sue and defend in all courts and places in all matters and proceedings; may have and use a common seal and alter the same at pleasure; subject to the restrictions contained in this charter, may purchase, receive bequests, gifts and donations of all kinds of property in fee simple, or in trust for charitable and other purposes; and do all acts necessary to carry out the purpose of such gifts, bequests and donations, with power to manage, sell, lease or otherwise dispose of the same in accordance with the terms of the gift, bequest or trust.

— 1 —

All rights and titles to property, all rights and obligations under contracts or trusts, and all causes of action of any kind in any court or tribunal vested in the City of Seoul or in any officer or employee thereof in his official capacity, at the time this charter becomes effective, as well as all liabilities in contract or tort and causes of action involving the same insofar as they affect the city or any officer or employee thereof in his official capacity, which shall be outstanding at the time this charter becomes effective, shall continue without abatement or modification by reason of any provision hereof.

All public improvements or other proceedings legally authorized under the authority superseded by this charter shall be carried to completion under previously existing laws or under this charter. The powers or duties vested in city officers, boards or commissions by law superseded by this charter shall be exercised, continued and carried out by their successors or by other city officers, boards or commissions, consistent with the provisions of this charter.

All functions of the city, and the powers and duties of officers and employees charged with the performance thereof, as these shall have been apportioned among departments and offices, and institutions, utilities, bureaus or other subdivisions thereof, as existing at the time this charter shall go into effect, shall continue to be the functions of such departments and offices and the powers and duties of officers and employees assigned thereto, except as in or under authority of, this charter otherwise specifically provided.

— 2 —

The legally authorized officers and employees of each of said departments and offices or subdivisions thereof shall continue as the officers and employees of said departments and offices or subdivisions thereof, subject to the conditions governing their respective appointments to such positions, and except as in this charter otherwise provided; and where part of the functions and duties of any department or office are, by this charter, transferred or placed in any other department or office, the persons performing such functions and duties shall be transferred therewith. The compensations legally authorized for the several officers and employees shall be continued subject to the other provisions of this charter.

The city may make and enforce all laws, ordinances and regulations necessary, convenient or incidental to the exercise of all rights and powers in respect to its affairs, officers and employees, and shall have all rights and powers appropriate to a province and a city, subject only to the restrictions and limitations provided in this charter, including the power to acquire and construct plants, works, utilities, areas, highways and institutions outside the boundaries of the city, and maintenance and operation of the same and the exercise of functions or maintenance of services outside the boundaries of the city, including the expenditure of funds therefor through any agency. The specification or enumeration in this charter of particular powers shall not be exclusive. The exercise of all rights and powers of the city granted by this charter shall be as provided by ordinance or

軍 政 廳　官 報　서울市憲章　一九四六年八月十日

resolution of the board of supervisors.

General Law Procedure

SECTION 3. Where a procedure for the exercising of any rights or powers belonging to a city, or a province, is provided by statute of Korea, said procedure shall control.

ELECTIVE AND APPOINTIVE OFFICERS

City Officers

SECTION 4. The officers of the city shall be the officers elected by vote of the people, members of the board of education, members of boards and commissions appointed by the mayor, the superintendent of schools, the executive appointed by each board or commission as the chief executive officer under such board or commission, the chief administrative officer, the head of each department under the chief administrative officer and the coroner, assessor, treasurer, tax and license collector, and such other officers as may hereafter be provided by law or so designated by ordinance.

Elective Officers and Terms

SECTION 5. The mayor, the members of the board of supervisors, an auditor, a city attorney, and a director of finance shall be elected by the voters of the city. At the general municipal election in 1946, and every four years thereafter, there shall be elected a mayor, eight supervisors, a city attorney, a director of

軍　政　廳　官　報　서울市憲章　一九四六年八月十日

finance and an auditor; and at the general municipal election in 1947, and every fourth year thereafter, there shall be elected seven supervisors. All terms of office of elective officials shall begin at twelve o'clock noon on the 30th day following the date of their election.

Any appointive officer or employee of the city who shall become a candidate for election by the people to any public office shall automatically forfeit such appointive city office or position.

No person elected as mayor or supervisor shall be eligible, for a period of one year after his last day of said service as mayor or supervisor, for appointment or election to any full-time position carrying compensation in the city service.

Absence from Country, and Vacancies

SECTION 6. No officer of the city shall absent himself from the country, except by permission of the mayor and board of supervisors. Violation of this section shall be sufficient cause for removal of any officer violating the same.

An office becomes vacant when the incumbent thereof dies, resigns, is adjudged insane, convicted of a crime involving moral turpitude or felony, or of an offense involving a violation of his official duties, or is removed from office, or ceases to be a resident of the city, or neglects to qualify within the time prescribed by law, or within twenty days after his election or appointment, or shall have been absent from Korea without leave for more than sixty consecutive days.

— 5 —

Qualifications of Officers and Employees

SECTION 7. No person shall be a candidate for any elective office, nor shall be appointed as a member of any board or commission or as an officer of the city unless he shall have been a resident of the city for a period of at least five years. Unless otherwise specifically provided in this charter, all employees of the city shall be citizens and shall have been residents thereof for at least one year prior to appointment, unless otherwise provided in this charter, and members of the fire department hereafter appointed shall be citizens and shall have been residents of the city for at least five years next preceding appointment; provided, that appointees whose duties are performed outside the city shall not be subject to residence requirements of this section; and provided also that positions requiring expert or technical training may, on the recommendation of the department head and the mayor, and with the approval of the civil service commission and the board of supervisors, be exempted from the requirements of this section.

Bonds of Officers and Employees

SECTION 8. Unless otherwise provided in this charter, all officers and such employees as may be specified by ordinance, on the recommendation of the chief administrative officer or the board or commission concerned, shall give bond in such amounts as may be required by ordinance. The surety on such bonds shall always be such as is specified by and approved in the manner provided by ordinance. The supervisors may, by ordinance, provide for group bonding of officers and employees. The premiums on all official bonds shall be paid by the city.

Powers Vested in Board of Supervisors

SECTION 9. The powers of the city, except the powers reserved to the people or delegated to other officials, boards or commissions by this charter, shall be vested in the board of supervisors and shall be exercised as provided in this charter. It shall be the duty of the board of supervisors to canvass the votes cast at each election in the city and certify the offici count of such balloting. Subject to Korean Civil Service regulations the supervisors shall determine the maximum number of each class of employment in each of the various departments and offices of the city and shall fix rates and schedules of compensation therefor in the manner provided in this charter. On the recommendation of the mayor and the chief administrative officer, the board of supervisors may create or abolish departments which are now or may hereafter be placed under the chief administrative officer or under commissions appointed by the mayor.

The board of supervisors may, by ordinance, confer on any officer, board or commission such other and additional powers as the board may deem advisable, provided only that such action is not inconsistent with the other provisions of this charter.

軍 政 廳　官 報　서울市憲章　一九四六年八月十日

Number, Compensation and Meetings of Supervisors

SECTION 10. The board of supervisors shall consist of fifteen members elected at large, provided, however, that until the general municipal election of 1947 the board shall consist of eight members. Each member of the Board shall be paid a salary of nine thousand three hundred won (₩9,300) per year and each shall execute a bond to the City in the sum of five thousand won (₩5,000).

At twelve o'clock noon on the 30th day next following their election the newly elected and continuing members of the board of supervisors shall meet at the legislative chamber in the City Hall, and thereafter regular meetings shall be held as fixed by resolution. The supervisors constituting the new board shall, on the 30th day next following their election, and annually thereafter, elect one of their number as president of the board for a term of one year. The president shall preside at all meetings, shall appoint all standing and special committees of the board and shall have such other powers and duties as the supervisors may provide.

The meetings of the board shall be held in the City Hall provided that, in case of emergency, the board, by resolution, may designate some other appropriate place as its temporary meeting place. The board shall cause a calendar of the business scheduled for each meeting to be published and shall keep and publish a journal of its proceedings. Notice of any special meeting shall be published at least twenty-four hours in advance of such special meeting.

— 8 —

Suspension and Removal

SECTION 11. Any elective municipal officer, including any member of the civil service commission or public utilities commission or school board, may be suspended by the mayor and removed by the board of supervisors for official misconduct, and the mayor shall appoint a qualified person to discharge the duties of the office during the period of suspension. On such suspension the mayor shall immediately notify the supervisors thereof in writing and of the cause therefor, and shall present written charges against such suspended officer to the board of supervisors at or prior to its next regular meeting following such suspension, and shall immediately furnish copy of same to such officer who shall have the right to appear with counsel before the board in his defense. Hearing by the supervisors shall be held not less than five days after the filing of written charges. If the charges are sustained by not less than a three-fourths vote of all members of the board, the suspended officer shall be removed from office; if not so sustained, or if not acted on by the board of supervisors within thirty days after the filing of written charges, the suspended officer shall thereby be reinstated. The mayor must immediately suspend from office any elective official guilty of official misconduct or convicted of a felony or a crime involving moral turpitude, and failure of the mayor so to act shall constitute official misconduct on his part.

Any appointee of the mayor, exclusive of civil service and

public utilities commissioners and members of the school board, may be removed by the mayor. Any nominee or appointee of the mayor whose appointment is subject to confirmation by the board of supervisors, except the chief administrative officer, may be removed by a majority of such board and with the concurrence of the mayor. In each case, written notice shall be given or transmitted to such appointee, of such removal, the date of effectiveness thereof, and the reasons therefor, a copy of which notice shall be printed at length in the journal of proceedings of the board of supervisors, together with such reply in writing as such official may make. Any appointee of the mayor or of the supervisors, guilty of official misconduct or convicted of a felony or a crime involving moral turpitude, must be removed by the mayor or the board of supervisors, as the case may be, and failure of the mayor or any supervisor to take such action shall constitute official misconduct on his or their part.

LEGISLATIVE PROCEDURE

Clerk of the Board of Supervisors

SECTION 12. The board of supervisors shall appoint a clerk, who shall be designated as clerk of the board of supervisors. The clerk shall have charge of the office and records of the board and its committees, and the personnel employed to handle the business, affairs and operations of the board, its committees, and members, when engaged in official duty. The clerk shall keep a journal of proceedings of the board and files of all ordinances, resolutions and other matters acted on by the board for which publication is specified. He shall have such other duties and responsibilities as the board shall prescribe.

Action by Resolution or Ordinance

SECTION 13. Action of the board of supervisors shall be by ordinance or resolution introduced in writing and passed or adopted by at least a majority of all the members present and voting at each reading. Every legislative act shall be by ordinance. The enacting clause of all ordinances shall be, "Be it ordained by the people of the City of Seoul". Every ordinance and resolution, except ordinances making appropriations, shall be confined to one subject which shall be clearly expressed in the title, and ordinances making annual or supplemental appropriations shall be confined to one subject which shall be clearly expressed in the title, and ordinances making annual supplemental appropriations shall be confined to the subject of appropriations. If any subject is embraced in an ordinance and is not expressed in the title thereof, the ordinance shall be void only as to so much thereof as is not expressed in the title. No ordinance shall be amended in whole or in part unless each section of such ordinance be printed in full, together with the proposed amendment or amendments thereto at the time such amendment or amendments are introduced and voted upon.

An ordinance shall be passed by the board of supervisors only after references to and report thereon from committee, unless it be an ordinance prepared and reported out by committee, and after two readings and votes at separate meetings of the board, which meetings shall be at least 10 days apart, provided, however, that as to an emergency measure, reference to committee or the readings and votes at separate meetings may be waived by a three-fourths vote of all members of the board. The existing or impending emergency as defined in such ordinance shall be declared by specific vote of the board. No other resolution shall be adopted by the board of supervisors on the date of its introduction and without reference to committee, except by the unanimous consent of the supervisors present. Annual budget and appropriation ordinances shall be passed only after two readings, not less than five days apart, and the second or final passage shall be not less than fifteen days after the introduction of each such ordinance.

Except as otherwise provided in this charter, proposed ordinances which are introduced and referred to committee shall be published in full within three days after their presentation to the board and, if modified, also upon passage on first reading, or as an emergency measure as herein defined. The term "published" as used in this charter shall mean publication in the official newspaper of general circulation, published in the city and which has a bona fide daily circulation of at least 8000 copies.

The vote on all ordinances and resolutions upon each reading shall be by ayes and noes. The vote by ayes and noes on all measures shall be recorded in the journal of the proceedings of

— 12 —

the board.

Approval by Mayor—Reconsideration and Veto

SECTION 14.   Each proposed resolution or ordinance, voted on by the supervisors and failing of passage, and each ordinance or resolution adopted by the supervisors, shall, within twenty-four hours of such action, be transmitted to the mayor by the clerk of the board, with appropriate notation of the action of the board thereon.   Any resolution acted upon by the board of supervisors by unanimous consent of those present on the date of the introduction of such resolution and any ordinance adopted by the board as an emergency measure shall be acted upon by the mayor within three days after receipt thereof by him from the clerk of the board.   All other ordinances or resolutions shall be acted upon by the mayor within ten days of such receipt.

The mayor shall either approve each resolution or ordinance adopted by the supervisors, by signing and returning same to the clerk of the board within the time limit, or he shall disapprove and veto any resolution or ordinance, or veto or reduce any separate appropriation item therein and shall return each such resolution or ordinance to the clerk of the board with his written objections within the time limit.   His failure to make such return shall constitute approval and such ordinance or resolution shall take effect without the mayor's signed approval.   The clerk of the board shall note such fact on the official copy of such resolution or ordinance.

The board of supervisors **may** reconsider any resolution or ordinance vetoed or disapproved, or any separate appropriation item vetoed or reduced by the mayor, and if, after such reconsideration, two-thirds of all the members of the board shall vote in favor of passage thereof, it shall become effective notwithstanding the mayor's veto. If a larger vote is required for the adoption of a measure by the provisions of this charter, such larger vote shall be required to overcome the veto of the mayor. The vote of reconsideration of each such vetoed resolution, ordinance or separate appropriation item therein shall be taken at the convenience of the board. If the ordinance, resolution or separate appropriation item is not passed over the mayor's veto within thirty days, the measure or item shall be lost.

In the event of any absence of the mayor for which he or the board of supervisors has failed to designate an acting mayor, no resolution or ordinance adopted by the board of supervisors shall take effect by reason of the failure of the mayor to approve, or disapprove and return, such resolution or ordinance within the time limits applicable thereto, and in such case, the time periods or limitations as fixed by this section shall not start until an acting mayor is appointed by the mayor or elected by the supervisors, as in this charter provided, or until the return of the mayor. Any proposed resolution or ordinance voted on by the board of supervisors and failing of passage shall be reconsidered by the board on the written request of the mayor, stating his reasons therefor, and filed with the clerk of the board by the mayor

— 14 —

within ten days of the board's action on such resolution or
ordinance. The board shall reconsider such measure at its
convenience, but not later than thirty days after the filing of
the mayor's request therefor.

Record Publication and Effect of Ordinances and Resolutions

SECTION 15. All ordinances, after final passage or upon their
becoming effective, shall be certified by the clerk of the board
and recorded in a book kept for that purpose, and resolutions
adopted shall be certified and recorded in like manner. Except
in case of an emergency measure passed and not previously
published, and except as otherwise specified in this charter,
publication of ordinances and resolutions in full shall not be
required after final passage. Notice that an ordinance or resolution has passed or become final shall be published once within
five days of such final passage. To amend an ordinance which
has proceeded to second reading shall require publication of the
ordinance as amended and proceeding de novo.

Emergency Measures

SECTION 16. No ordinance shall become effective until ten
days after final passage, unless adopted by a three-fourths vote
of all members of the board as an emergency measure as defined
in this section.

No ordinance affecting franchises, grants, bond issues or the
sale, lease or purchase of land shall ever be passed as an emer-

gency measure. Immediate necessary preservation of public peace, property, health or safety, provision for the uninterrupted operation of any city department or office, or action required to comply with time limitations as established by law, shall be emergencies within the meaning hereof; provided, however, that such emergency shall actually exist and shall be specifically voted on as provided in Section 13 of this charter.

GENERAL POWERS AND DUTIES OF BOARDS, COMMISSIONS. DEPARTMENT HEADS AND OFFICERS

Powers and Duties of Boards and Commissions

SECTION 17. The board of supervisors and each board and commission appointed by the mayor, or otherwise provided by this charter, shall have powers and duties as follows:

(a) To prescribe reasonable rules and regulations not inconsistent with this charter for the conduct of its affairs, for the distribution and performance of its business, for the conduct and government of its officers and employees, and for the administration, custody and protection of property under its control and books, records and papers appertaining to its affairs. The board of supervisors, by ordinance, may provide that rules and regulations of any board or commission, or general orders of any department head, issued by authority of any board or commission, that are of general public concern shall be published or posted.

軍 政 廳　官 報　서 울 市 憲 章　一九四六年八月十日

(b) To appoint one of its members as president to hold office for such term as each such board or commission, by its rules or regulations not inconsistent with this charter, may prescribe.

(c) To establish such standing or special committees as it shall deem necessary.

(d) To receive on behalf of the city, gifts, devises and bequests for any purpose connected with or incidental to the department or affairs placed in its charge, and to administer, execute and perform the terms and conditions of trusts, or any gift, devise or bequest which may be accepted by vote of the people or by the board of supervisors for the benefit of such department or purpose, and to act as trustees, under any such trust when so authorized to do by the board of supervisors. The title to all real and personal property now owned or hereafter acquired by gift, devise, bequest or otherwise, by and for the purposes of any board or commission, shall vest in the city.

(e) To require such periodic or special reports of departmental operations, costs and expenditures under its control as may be necessary and, exclusive of the board of supervisors, to submit an annual report to the mayor.

(f) To hold meetings at regular fixed dates and at regular meeting places, which dates or places shall not be changed except as in the manner provided by Section 10 for the meeting times and places of the board of supervisors. All such meetings shall be open to the public.

(g) To hold special meetings for the purposes and in the

manner provided by the board of supervisors by ordinance, provided that no matter may be considered at any special meeting unless specifically designated in the notice calling such special meeting.

(h) To appoint a secretary, a superintendent, or other executive to be the administrative head of the affairs under its control, who, unless otherwise specifically provided, shall not be subject to the civil service provisions of this charter, and shall hold office at its pleasure.

(i) To require a bond or other security from each such executive officer and from any employee in such form as the board of supervisors may authorize and in such amount as the mayor may approve, the premiums on such bonds to be paid by the city.

A quorum for the transaction of official business shall consist of a majority of all the members of each board or commission, but a smaller number may adjourn from time to time and compel the attendance of absent members, in the manner and subject to penalties to be provided by ordinance. A majority, two-thirds, three-fourths, or other vote specified by this charter for any board or commission, shall mean a majority, two-thirds, three-fourths, or other vote of all the members of such board or commission. Each board or commission shall keep a record of the proceedings at each meeting and a copy thereof shall be forwarded promptly to the mayor.

軍　政　廳　　官　報　　서울市憲章　　一九四六年八月十日

Powers and Duties of Department Heads

SECTION 18. Each elective officer in charge of an administrative office, the chief executive appointed by each board or commission, the chief administrative officer, and each department head appointed by the chief administrative officer shall have the powers and duties of a department head, except as otherwise specifically provided in this charter. Each appointive department head shall be immediately responsible to the chief administrative officer or the board or commission, as the case may be, for the administration of his department, and shall file an annual report and make such other reports, estimates and recommendations at the time and in the manner required by law, or as required by the chief administrative officer, board or commission.

He shall issue or authorize all requisitions for the purchase of materials, supplies and equipment required by such department, provided that, on the written approval of the chief administrative officer or the board or commission in charge of any department, the head of any utility, institution, bureau or other subdivision of a department may likewise be vested with such power. Each department head or the head of a utility, institution, bureau or other subdivision of each department shall be responsible for the proper checking of all materials, supplies and equipment ordered for its purposes, and for the approval or disapproval of bills for claims rendered for such materials, supplies or equipment.

The head of any department, through the chief administrative officer or the board or commission in charge thereof, shall recommend to the board of supervisors such ordinances as may be required to carry out the powers vested and the duties imposed, and to establish or readjust fees or charges for permits issued to or work performed for persons, firms or corporations when these are subject to his or its jurisdiction.

Each department head may suggest the creation of positions subject to the provisions of this charter and may reduce the forces under his jurisdiction to conform to the needs of the work for which he is responsible, any other provisions of this charter to the contrary notwithstanding.

The mayor, the chief administrative officer, or the board or commission concerned, on the recommendation of any department head or on his or its own motion, may combine or may transfer and redistribute among departments or offices under his or its authority, respectively, any function or duty assigned to or continued by this charter in any department.

Power of Hearing, Inquiry and Subpoena

SECTION 19. The mayor, the board of supervisors, the chief administrative officer, or any board or commission appointed by the mayor, relative solely to the affairs under its control, may require periodic or special reports of departmental costs, operation and expenditures, examine the books, papers, records and accounts of, and inquire into matters affecting the conduct of any department or office of the city and for that purpose may

軍　政　廳　官　報　　서울市憲章　　一九四六年八月十日

hold hearings, subpoena witnesses, administer oaths and compel
the production of books, papers, testimony and other evidence.
Any person refusing to obey such subpoena and the other
requirements hereof, or to produce such books, shall be deemed
in contempt and subject to proceedings and penalties as provided
by general law in such instances.

Non-interference in Administrative Affairs

SECTION 20. Except for the purpose of inquiry, the mayor
and the board of supervisors shall deal with the administrative
service for which the chief administrative officer is responsible,
solely through such officer, and for administrative or other
functions for which elective officials or boards or commissions
are responsible, solely through the elective official, the board
or commission or the chief executive officer of such board or
commission concerned. Except for the purpose of inquiry, each
board or commission, in its conduct of administrative affairs
under its control, shall deal with such matters solely through its
chief executive officer. The board of supervisors, and each board
or commission relative to the affairs of its own department, shall
deal with administrative matters only in the manner provided by
this charter, and any dictation, suggestion or interference herein
prohibited on the part of any supervisor or member of a board or
commission shall constitute official misconduct; provided, however,
that nothing herein contained shall restrict the power of hearing
and inquiry as provided in this charter.

Administrative Code

SECTION 21. The powers and duties of the departments and offices which by this charter are established or continued as departments or offices under elective officers, boards or commissions, or the chief administrative officer, as such powers and duties exist at the time this charter shall go into effect, shall be continued as powers and duties of each such department or office, except as otherwise provided in this charter.

The board of supervisors may enact and provide for the publication in printed form of an administrative code, which shall specify or detail the powers, duties, methods and procedure in the several departments and offices.

Permits and Inspections

SECTION 22. The board of supervisors shall regulate by ordinance the issuance and revocation of licenses and permits for the use of, obstruction of or encroachment on public streets and places, and for the operation of businesses or privileges which affect the health, fire-prevention, fire fighting, crime, policing, welfare or zoning conditions of or in the city and for such other matters as the board of supervisors may deem advisable. Such ordinance shall fix the fees or licenses to be charged, which shall be not less than the cost to the city of regulation and inspection; and shall specify which department shall make the necessary investigations and inspections and issue or deny or revoke the

軍　政　廳　　官　報　　서울市憲章　　一九四六年八月十日

permits and licenses therefor. The chief of police in the perform ance of police duties shall have power to examine at any time the books and the premises of pawnbrokers, peddlers, junk and second–hand dealers, auctioneers and other businesses designated by the board of supervisors, and the tax collector shall have power to examine the books of any business for which a license is issued and a fee charged on the basis of the receipts of such business, and for these purposes such officials shall have the power of inquiry, investigation and subpoena, as provided by this charter.

Permits and licenses shall be issued by the departments as designated by ordinance, only after formal application for such permit or license. If any application for a permit or license is denied by the department authorized to issue same, the applicant may appeal to the board of permit appeals.

POWERS AND DUTIES OF ELECTIVE OFFICERS

The Mayor

SECTION 23. The mayor shall be the chief executive officer of the city upon whom process issued by authority of law shall be served. He shall be an elective officer and shall be paid a salary of twenty thousand won (₩20,000) per year. He shall furnish an official bond in the sum of twenty-five thousand won (₩25,000). He shall appoint, and at his pleasure may remove, an executive secretary, one confidential secretary, and one stenographer.

The board of supervisors may annually appropriate additional sums to be expended by the mayor for purposes and duties incidental to the administration of the office of mayor, which shall be subject to the provisions of this charter relative to appropriations and the payment of claims. He shall, at the first meeting of the board of supervisors each year, communicate by message to the supervisors a general statement of the condition of the affairs of the city and recommend the adoption of such measures as he may deem expedient and proper.

The mayor shall be responsible for the enforcement of all laws relating to the municipality and for the review and submission of the annual executive budget; he shall supervise the administration of all departments under boards aud commissions appointed by him; he shall receive and examine, without delay, all complaints relating to the administration of the affairs of the city and immediately inform the complainant of findings and actions thereon; and he shall coordinate and enforce cooperation between all departments of the city.

The mayor shall appoint such members of boards or commissions and other officers as provided by this charter. He shall also make an ad interim appointment of a qualified person to fill any vacancy occurring by reason of the expiration of a term. He shall appoint for the unexpired term of the office vacated, a qualified person to fill any vacancy occurring in any elective municipal office.

The mayor shall have a seat but no vote in the board of supervisors and in any board or commission appointed by him, with the right to report on or discuss any matter before such board or commission concerning the departments or affairs in his charge. He shall have power to designate a member of the board of supervisors to act as mayor in his absence. Should he fail, neglect or refuse so to do, the supervisors shall elect one of their members to act as mayor during his absence. When a vacancy occurs in the office of mayor, it shall be filled, for the unexpired portion of the term, by the supervisors.

In case of a public emergency involving or threatening the lives, property or welfare of the citizens, or the property of the city, the mayor shall have the power, and it shall be his duty, to summon, organize and direct the forces of any department in the city in any needed service, to summon, marshal, deputize or otherwise employ other persons, or to do whatever else he may deem necessary for the purpose of meeting the emergency. The mayor may make such studies and surveys as he may deem advisable in anticipation of any such emergency.

City Attorney

SECTION 24. The city attorney shall be an elective officer and shall receive an annual salary of eleven thousand one hundred won (₩11,100). Subject to the rules of Korean Civil Service he shall appoint and at his pleasure may remove, all assistants and employees in his office. He must, at the time of his election, be qualified to practice in all the courts of this country, and he

軍 政 廳 官 報 　 서 울 市 憲 章 　 一九四六年八月十日

must have been so qualified for at least five years next preceding his election.

The city attorney must represent the city in all actions and proceedings in which it may be legally interested, or, for or against the city, or any officer of the city, in any action or proceeding, when directed so to do by the supervisors, except where a cause of action exists in favor of the city against said officer. Whenever any cause of action exists in favor of the city, the city attorney shall commence the same when within his knowledge or when directed so to do by the supervisors. He shall give his advice or opinion in writing to any officer, board or commission of the city when requested. He shall not settle or dismiss any litigation for or against the city unless, upon his written recommendation, he is ordered so to do by ordinance.

The city attorney shall prepare, or approve as to form, all ordinances before they are enacted by the supervisors. He shall approve, by endorsement in writing, the form of all official or other bonds required by this charter or by ordinance, before the same are submitted to the proper commission, board or office for final approval, and no such bonds shall be finally approved without such approval, as to form, by the city attorney. Except as otherwise in this charter provided, he shall prepare in writing the draft or form of all contracts before the same are executed on behalf of the city. He shall examine and approve the title of all real property to be acquired by the city.

He shall keep on file in his office copies of all written

communications and opinions, also all papers, briefs and transcripts used in matters wherein he appears; and books of record and registers of all actions or proceedings in his charge in which the city or any officer or board thereof is a party or is interested.

Director of Finance

SECTION 25. The director of finance shall be an elective officer and shall receive an annual salary of twelve thousand nine hundred won (₩12,900). He shall furnish official bond in the sum of one hundred thousand won (₩100,000). Subject to Korean Civil Service regulations he shall appoint, and at his pleasure may remove, a tax collector, a treasurer, an assessor, an attorney, and one chief assistant.

The director of finance shall be the chief accounting officer of the city, and shall exercise general supervision over the accounts of all officers, commissions, boards and employees of the city charged in any manner with the receipt, collection or disbursement of city funds or of other funds, in their capacity as city officials or employees. He shall have the power and duty of prescribing the method of installing, keeping, and rendering accounts of, and the financial reports to be rendered by, the several officers, boards and employees of the city.

The director of finance shall keep accounts showing the financial transactions of all departments, offices and other subdivisions of the city. Such accounts and the accounting procedure shall be adequate to record:

(a) all budgeted revenues and appropriations, together with additions or transfers thereto, and to show at all times the amount of encumbrances, expenditures or transfers therefrom, and the balances therein;

(b) all revenues accrued and liabilities incurred;

(c) all cash receipts and disbursements; and

(d) in general, all transactions affecting the acquisition, custody or disposition of values.

Subject to the provisions of this section, the public utilities commission shall maintain separate accounts for each utility·in such manner as to exhibit exact and complete financial results of ownership, management and operation; the actual cost of each utility; all cost of maintenance, extension and improvement; all operating expenses of every description; the general expenses of the commission and bureaus thereof apportioned to each such utility; the amount paid or set aside for depreciation, insurance, interest and sinking fund; and estimates of the amount of taxes that would be chargeable against such property and the revenue thereof if privately owned and operated.

It shall be the duty of the director of finance to determine, where practicable, the unit cost of work done by the city for the purpose of determining whether similar work could be done under public contract at a lower cost. The director of finance shall devise adequate systems of internal check of all departments and offices of the city relative to the custody, collection or disbursement of moneys.

— 28 —

The director shall make a monthly report, not later than the 20th day of each month, showing a summary statement of revenues and expenditures for the preceding calendar month and for that portion of the fiscal year ending on the last day of such preceding month. Such statments shall be detailed as to assets, liabilities, income, expenditures, appropriations and funds in such manner as to show the financial condition of the city and of each department, office, bureau or division thereof, for that portion of the fiscal year to and including the preceding calendar month, and with comparative figures for the similar period in the preceding fisca year. The director of finance shall at the same time prepare statements showing the cash position of the ci y and the unencumbered balance in each fund. A copy of each such report and annual reports compiled therefrom, and special fiscal reports as requested, shall be transmitted to the mayor, the board of supervisors, the administrative officer and all department heads concerned, and shall be kept on file in the office of the director of finance.

Auditor

SECTION 26. The auditor shall be an elective officer and shall receive a salary of twelve thousand nine hundred won (₩12,900) per year. He shall appoint, and at his pleasure may remove, one chief assistant.

He shall audit the accounts of all boards, officers and employees of the city charged in any manner with the custody, collection, or

軍 政 廳 官 報　서울市憲章　一九四六年八月十日

disbursement of funds.. When requested by the mayor, the board of supervisors, the chief administrative officer, or any board or commission for its own department, he shall audit the accounts of any officer or department and, on the death, resignation, removal, expiration of term or retirement of the head of any department, or any officer or employee charged with the receipt, collection or disbursement of money, shall make an audit of the accounts of such department, officer or employee.

He shall make, an annual audit of the director of finance's ounts, records and transactions. The report of such audit shall he made to the mayor, the board of supervisors, the chief administrative officer, the director of finance, and to such citizens as may apply therefor.

The auditor shall be the custodian of all official bonds and of the sufficiency thereof and the solvency of the sureties thereon, and forthwith report in writing to the board of supervisors any insufficiency or irregularity. The board of supervisors shall take such action as shall be necessary to protect the city.

Assistants and Employees in Executive Offices

SECTION 27. Subject to the rules of Korean Civil Service the elective officers of the city may appoint such assistants and employees as are authorized by the supervisors upon the recommendation of the mayor, in the annual budget and annual or supplemental appropriation ordinances, and may discipline and remove the same, except as otherwise specifically exempted by the provisions of this charter.   The salaries, wages and compensations of every kind and nature, for assistants and employees in such elective offices, shall be fixed by ordinance.   Pending passage of such ordinance, the compensation schedules in effect at the time this charter takes effect shall prevail

Chief Administrative Officer

SECTION 28.   The mayor shall appoint as administrative officer a qualified person who shall have been a resident of Korea for at least five years immediately preceding his appointment.   The requisite qualifications of such appointees shall be administrative and executive ability and experience for the position to be filled. He shall be paid an annual salary of fourteen thousand seven hundred won (W14,700).

He shall be subject to suspension and removal in the same manner as elective officers.   He shall also be subject to removal by a vote of not less than two-thirds of the board of supervisors, on the basis of written charges, and if he so request, only after a public hearing on such charges before the board of supervisors not less than five days nor more than fifteen days after the filing thereof, and prior to the date on which the supervisors shall vote on the question of his removal, but on the filing of written charges, and pending and during such hearing, the supervisors, by majority vote, may suspend him from office.   The written charges and any reply thereto by the chief administrative officer shall be entered at length in the journal of the board of supervisors.

The action of the board of supervisors in removing the chief administrative officer shall be final.

'Powers and Duties of Chief Administratrative Officer

SECTION 29. The chief administrative officer shall be responsible to the mayor and to the board of supervisors for the administration of all affairs of the city that are placed in his charge by the provisions of this charter and by ordinance, and to that end he shall have power and it shall be his duty to exercise supervision and control over all administrative departments which are under his jurisdiction; to prescribe general rules and regulations for the administrative service under his control; to have a voice but no vote in the board of supervisors, with the right to report on or to discuss any matter before the said board concerning the affairs of the department in his charge; to make such recommendations and propose such measures to the mayor, the board of supervisors, or committees thereof, concerning the affairs of the city in his charge as he may deem necessary; to coordinate the functioning of the several departments of the city; and to provide for the budgeting and control of publicity and advertising expenditures of the city.

DEPARTMENTS

Public Works Department.

SECTION 30. The Department of Public Works shall be in

charge of a director of public works, who shall be paid an annual salary of eleven thousand one hundred won (₩11,100). He shall be appointed by and hold office at the pleasure of the mayor. The director of public works shall appoint a city engineer, who shall hold office at the pleasure of said director. He shall possess the same power in the city in making surveys, plats and certificates as is or may from time to time be given by law to city engineers and to surveyors, and his official acts and all plats, surveys and certificates made by him shall have the same validity and be of the same force and effect as are or may be given by law to those of city engineers and surveyors.

All examinations, plans and estimates required by the supervisors in connection with any public improvements, exclusive of those to be made by the public utilities commission, shall be made by the director of public works, and he shall, when requested to do so, furnish information and data for the use of the supervisors.

General Law and Ordinance Procedure for Public Works

SECTION 31. Where a procedure for the exercising of any rights and powers belonging to a city, or a province, relative to the establishment or change of grades and the lay-out, extension, opening, widening, changing, closing, vacating, paving, repaving or otherwise improving streets and highways and public places, and constructing sewers, drains, conduits and culverts, subways, tunnels, viaducts and bridges, or other public improvements incidental

or appurtenant thereto, to planting trees, constructing parks
and removing weeds or the executing of any other public work
or improvement hereby or hereafter placed under the jurisdiction
ofthe department of public works, and the payment of damages,
or levying of special assessment to defray the whole or part of
the cost of such works or improvements is provided by statute,
such procedure shall control.

Sewer, Water and Other Connections

SECTION 32.   The director of public works shall have authority
in the manner provided by ordinance by the board of supervisors,
(1) to order the laying of sewer, water, gas and other mains,
conduits or connections, whenever, in view of contemplated
street improvements or as a sanitary regulation, such construction
is recommended by the city engineer, and (2) to order that
excavations, fences, embankments or grades on private property in a
condition deemed by him as endangering the persons or property
of those using the abutting streets, shall be put in such condition
as to insure the safety of the public.

Spur Tracks

SECTION 33.   The board of supervisors shall refer all requests
for spur track permits to the director of public works who shall
grant such permits in all cases where the spur track is to be
located within a heavy industrial zone, as classified by the city
planning commission, provided that such spur track shall be so

constructed and operated as not to establish an unreasonable interference with the public use of the street affected.

Repair and Maintenance Bureau

SECTION 34. The director of public works will establish and maintain a repair and maintenance bureau for the repair of departmental supplies and equipment, and the operation of a central garage and repair shop.

Real Estate Bureau

SECTION 35. The director of public works will establish and maintain a real estate bureau for the acquisition, management and control and leasing of city-owned real property. In the acquisition of property required for street opening, widening or other public improvements, the bureau shall make preliminary appraisals of the value of the property sought to be condemned or otherwise acquired, and report thereon to the responsible officer.

The bureau shall maintain complete records and maps of all real property owned by the city, which shall show the purchase price, if known, the department in charge of each parcel and the estimated value of each parcel and improvement. The bureau shall be responsible for the collection of all rentals on real property belonging to the city.

Police Department

.be in

SECTION 36. The police department shall have all the pow

軍 政 廳　官 報　서 울 市 憲 章　一九四六年八月十日

and duties that are now or that may hereafter be conferred on the police department at provincial level.

Medical Examiner

SECTION 37. The city attorney shall appoint a duly qualified physician to act as medical examiner. The medical examiner will investigate and determine the cause of death in each case where there is reasonable suspicion that a crime may have been committed. The medical examiner shall have the usual power of inquest. He shall be paid an annual salary of nine thousand three hundred won (₩9,300) per year, and shall hold office at the will of the city attorney.

Fire Department

SECTION 38. The fire department shall be in charge of a fire chief, who shall be paid an annual salary of eleven thousand one hundred won (₩11,100). He shall be appointed by and hold office at the pleasure of the mayor.

The fire department is charged with the principal duty of fire prevention, fire protection and fire-spread control, and the protection of persons and property from fire.

The chief engineer, or in his absence any assistant chief engineer or battalion chief in charge, may during a conflagration, cause to be cut down or otherwise removed any buildings or structures for the purpose of checking the progress of such conflagration.

The fire chief may call upon police officers to assist in the protection or salvaging of property and shall have such other powers and duties as by ordinance may be prescribed relative to the protection of property at fires and the storage of property salvaged therefrom. He shall have such duties appertaining to the enforcement of laws relative to the storage, sale and use of oils, combustible materials and explosives as the fire chief by rule, or the supervisors by ordinance, may prescribe.

The fire chief shall establish and maintain a bureau of fire prevention and public safety. The bureau shall inspect all structures and premises to determine whether or not compliance is being had with statutes and ordinances relative to fire protection and fire prevention.

Duties of Bureau of Fire Prevention

SECTION 39. The bureau shall examine the application, plans and specifications for the erection, and for alterations or repairs estimated to exceed ten thousand won (₩10,000) in cost, of any structure or premises subject to the statutes and ordinances referred to in this charter. The bureau shall by written report, filed with the superintendent of building inspection, approve such plans and specifications, or report to said superintendent the particulars wherein noncompliance exists, and upon modification of the application, plans and specifications to comply therewith, the bureau shall inform the superintendent of its approval. No permit for alteration or repair exceeding ten thousand

— 37 —

won (₩10,000) in cost, or for erection, shall be issued unless said approval is given.

Any structure or premises wherein there exists any violation of statutes or ordinances referred to herein, or which is maintained or used in such manner as to endanger persons or property by hazard of fire explosion or panic and any structure or premises hereafter constructed, altered or repaired in violation of said statutes or ordinances is hereby declared to be a public nuisance, and it shall be the duty of the bureau to prosecute abatement proceedings.

An appeal and advisory board is hereby created, consisting of the chief engineer of the fire department, the fire chief, and the heads of the bureau of building inspection and the assistant city engineer, and a lay member to be appointed by the mayor.

An appeal may be taken from any act, determination or order of the bureau, by filing a written appeal with the appeal and advisory board. Pending action on such appeal any construction, alteration or repair embraced therein may proceed if a building permit therefor has been issued, but no such permit may be issued while the appeal is pending. The advisory and appeal board may affirm, reverse or modify the determination of the bureau. If the appeal is determined adversely to the appellant, said structure or premises shall be made to comply with such decision.

Department of Education

SECTION 40. All of the public schools of the city, excluding

軍 政 廳 官 報　　서 울 市 憲 章　　一九四年六八月十日

universities and normal colleges, shall be under control and management of a board of education, composed of seven commissioners, who shall be nominated by the mayor and be subject to confirmation or rejection by vote of the board of supervisors, and who shall be subject to recall and to suspensions and removal in the same manner as elective officers, as provided in this charter. The term of each member shall be seven years, except that the initial seven appointed shall be for 1, 2, 3, 4, 5, 6, and 7 year terms respectively, and each member shall receive a salary of ten won per year.

Powers and Duties of the Board of Education

SECTION 41. The board of education shall have power to establish and maintain such schools as are authorized by law and as the board may determine, and to change, modify, consolidate or discontinue the same as the public welfare may require.

Subject to the rules of Korean Civil Service the board shall also have power to employ such teachers and other persons as may be necessary to carry into effect its powers and duties; to fix, alter and approve their salaries and compensations, except as in this charter otherwise provided; to withold for good and sufficient cause any part of the compensation of any person employed as aforesaid; and to promote, transfer and dismiss teachers for insubordination, immoral or unprofessional conduct, or evident unfitness for teaching. All promotions of teachers shall be based solely on merit. Charges against teachers must be made in

軍政廳 官報　서울市憲章　一九四六年八月十日

writing by the superintendent after investigation and shall be finally passed upon by the board after giving the accused teacher a fair and impartial hearing before the board.

The board of education shall have power to grant and to renew and, for cause, as specified herein, to revoke teachers' certificates.

Superintendent of Schools

SECTION 42.   The superintendent of schools shall be the executive officer of the board of education.   He shall be appointed by said board to serve at its pleasure, and he shall receive such salary as may be fixed by the board.   He shall have the powers, and duties specified by this charter for department heads, in addition to such powers and duties as are fixed by general law. He shall observe and enforce the regulations of the board of education, inform the board of the condition of schools, school houses and of other matters connected therewith, and to recommend such measures as he may deem necessary for the advancement of education in the city and for the care and improvement of the property of the school department.

Library Bureau

SECTION 43.   Subject to the rules of Korean Civil Service the board of education shall appoint a chief librarian and such assistants as they shall deem necessary, to serve at the pleasure of the board at such salary as shall be fixed by the board.

The library bureau is charged with observing and enforcing the regulations of the board of education relative to the operation, management and control of city-owned libraries.

Public Utilities Department

SECTION 44. It is the declared purpose and intention of the people of the city, when public interest and necessity demand, that public utilities shall be gradually acquired and ultimately owned by the city.

General Power and Duties

SECTION 45. The public utilities department shall be in charge of a manager of utilities, who shall be paid an annual salary of nine thousand three hundred won (₩9,300). He shall be appointed by and hold office at the pleasure of the mayor.

The public utilities department shall have charge of the construction, management, supervision, maintenance, extension, operation and control of all public utilities and other properties used, owned, acquired, leased or constructed by the city, for the purpose of supplying any public utility service to the city and its inhabitants, to territory outside the limits of the city and to the inhabitants thereof. The commission shall also have power to enter into contract for the furnishing of heat, light and power for municipal purposes, and to supervise the performance and check the monthly bills under such contract.

The public utilities department shall have power to fix, change

..nd adjust rates, charges or fares for the furnishing of service by any utility under its jurisdiction, maintaining wherever possible rates uniform and consistent with those in other areas in Korea and to collect by appropriate means all amounts due for said service, and to discontinue service to delinquent consumers and to settle and adjust claims arising out of the operation of any of said utilities. The board of supervisors shall have power to establish by ordinance the requisite conditions and procedure to effect any such change of rates or fares.

In the absence of a franchise to the contrary, the public utilities department shall have power to regulate street railroads, cars and track, and to regulate rates of speed and propose such ordinances as are necessary to protect the public from danger or inconvenience in the operation of such street railroads.

Public Health and Sanitation Department

SECTION 46. The public health and sanitation department shall be in charge of a director of health and sanitation, who shall be paid an annual salary of eleven thousand one hundred won (₩11,100). He shall be appointed by and hold office at the pleasure of the mayor. He shall be a regularly licensed physician or surgeon, with not less than ten years practice in his profession immediately preceding his appointment thereto.

The department shall perform all duties and functions of the department of public health and sanitation as existing at the time this charter shall go into effect, including the maintenance and operation of public health institutions, clinics, dispensaries, hospitals and health centers.

The department shall observe and enforce all ordinances and regulations relative to safeguarding the public health of the people and the sanitation of their environment. The department shall maintain an adequate staff of inspectors to inform the department of the health and sanitation conditions and to take such measures as shall be prescribed for the correction or improvement thereof.

Public Welfare Department

SECTION 47. The public welfare department shall be in charge of a director of public welfare, who shall be paid an annual salary of eleven thousand one hundred won (₩11,100). He shall be appointed by and hold office at the pleasure of the mayor.

The department shall perform all duties and functions of the department of public welfare existing at the time this charter shall go into effect, including the supervision of welfare, refugee and charitable organizations, societies and institutions. The department shall supervise and recommend standards to be observed by all persons engaged in relief, welfare, refugee, orphanage or similar activities purporting to assist the indigent and needy.

The department shall supervise and administer the distribution of all relief and social welfare supplies and contributions on behalf of the city.

軍　政　廳　官　報　서울市憲章　一九四六年八月十日

Commerce and Industry Department

SECTION 48. The commerce and industry department shall be in charge of a director of commerce and industry, who shall be paid an annual salary of eleven thousand one hundred won (₩11,100). He shall be appointed by and hold office at the pleasure of the mayor.

The department shall perform all duties and functions of the department of commerce and industry existing at the time this charter shall go into effect. This shall include studies and recommendations for the promotion and encouragement of new industrial enterprises and commerce and trade, surveys on labor capacity and requirements, incentive practices, cooperative and community enterprises.

The department shall supervise and administer all rationing and price control measures as may be provided by ordinance or law, insofar as concerns the city and the control and distribution of goods and commodities by the city, other than medical, school, relief or other supplies falling within the scope of other departments as herein or hereinafter specified.

The department shall establish and maintain a permit bureau for the issuance of permits and licenses according to regulations pertaining thereto.

Central markets and other retail goods and commodities outlets operated, managed or conducted by the city shall be under the supervision of the commerce and industry department.

The department shall establish and maintain a purchasing bureau. The bureau shall purchase all materials, supplies and equipment of every kind and nature required by the several departments and offices of the city, except as otherwise provided herein. Purchases of books, magazines and periodicals for the library bureau and other things of unusual character and peculiar to a particular department may, on the recommendation of the purchasing bureau, be purchased directly by the department head. Purchases for construction operations may be made by the department head, but in accordance with regulations established by the purchasing bureau. The bureau shall have authority to exchange, trade, or sell used or unfit materials, supplies and equipment belonging to the city, on the recommendation of a department head that it is to the advantage of the city to so do.

All purchases shall be by written purchase order or written contract. The bureau shall establish specifications and tests to cover all recurring purchases of material, supplies and equipment. Purchases of equipment shall be made in accordance with specifications furnished by the department requiring such equipment in case the use of such equipment is peculiar to such department. The bureau shall require departments to make adequate inspection at all purchases, and shall make such other inspection as it deems necessary. The bureau shall maintain an inventory of all material, supplies and equipment purchased for and in use in all departments of the city and, shall be responsible for a periodic check of such property and in case of loss or damage deemed by the bureau to be due to negligence, it shall report thereon to the mayor and the chief administrative officer. The bureau shall have authority to require the transfer of surplus property in any department or office to stores or to other departments or offices.

Board of Permit Appeals

SECTION 49. The mayor shall appoint five qualified electors, other than city officials or employees, for terms of four years, to constitute a board of permit appeals. The compensation for each member shall be one hundred won (₩100) per meeting of the board actually attended by such members, provided that the total amount paid all members of the board shall not exceed six thousand won (₩6000) per year. Any applicant for a permit or license who is denied such permit or license by the department authorized to issue same; or whose license or permit is ordered revoked by any department; or any person who deems that his interests or property or that the general public interest will be adversely affected as the result of operations authorized by, or under any permit or license granted or issued by any department; may appeal to the board of permit appeals. Such board shall hear the applicant, the permit-holder, or other interested party, as well as the head or representative of the department issuing or refusing to issue such license or permit, or ordering the revocation. After such hearing and such further investigation as the board may deem necessary, it may concur in the action of the department authorized

— 46 —

to issue such license or permit or, by the vote of four members, may overrule the action of such department and order that the permit or license be granted, restored or refused.

COMMISSIONS

Park Commission

SECTION 50. The park commission shall consist of five members who shall be appointed by the mayor for a term of four years, and shall serve without compensation. The commission shall appoint a secretary and a superintendent of parks, who shall hold office at its pleasure and shall receive the salary fixed by the board of supervisors. The commission shall have the complete and exclusive control, management, and direction of the parks, squares, avenues, grounds and recreation centers and playgrounds, now or hereafter placed under charge of the commission, including exclusive right to erect and to superintend the erection of buildings and structures thereon, except as in this charter otherwise provided. The commission shall have the power to lease any stadium or recreation field under its jurisdiction for athletic contests and exhibitions and may permit the lessee to charge an admission fee.

Art Commission

SECTION 51. An art commission for the city is hereby created consisting of ten members appointed by the mayor.

In appointing the members, the mayor shall solicit nominations from ' architectural, art, musical, literary and other cultural organizations of the city. The members of the commission shall serve without compensation.

No work of art shall be contracted for or placed or erected on property of the city or become the property of the city, except for any museum or art gallery, unless such work of art, or a design or model of the same as required by the commission, together with the proposed location of such work or art, shall first have been submitted to and approved by the commission. The commission shall supervise and control the expenditure of all appropriations made by the board of supervisors for music and the advancement of art or music.

City Planning Commission

SECTION 52. The city planning commission shall consist of five members who shall be appointed by the mayor for the term of four years, at the compensation to be fixed by the board of supervisors. The commission may appoint a city planning engineer who shall hold office at its pleasure and who shall be a person of expert and technical training.

It shall be the duty of the commission to make, maintain and adopt a master plan of the physical development of the city, which plan, including maps, plats, charts and descriptive matter, shall make recommendations for the development of all areas within the city and approaches thereto, and including, among other

軍 政 廳 官 報　서 울 市 憲 章　一九四六年八月十日

things, the general location, character and extent of streets, viaducts, subways, bridges, boulevards, parkways, playgrounds, parks, squares, aviation fields, and other public ways, grants and open spaces, the general location of public buildings and other public property and the removal, relocation, widening, narrowing, vacating, abandonment or extension of any of the foregoing ways, grants, open spaces or buildings. In the preparation of such plans, the commission shall consult and cooperate with all other departments of the city which, by this charter, are vested with responsibility for or control over any of the matters hereinbefore enumerated and shall make such additional studies as it may deem necessary.

The city planning commission, from time to time, shall consider and hold hearings on proposed changes in the classification of the use to which property in the city may be put, and the establishment or changing of building set-back lines. The board of supervisors, by ordinance, shall establish procedure for action on such matters.

No ordinance shall be considered by the supervisors, the purpose or intent of which is the classification, regulation or control of the height, area, bulk, location or use of any building or premise; or the classification of any property into any district or zone for said purposes, or the establishing of a set-back line along any street, without being first submitted to the city planning commission for report and recommendations.

Civil Service Office

SECTION 53. Pursuant to rules and regulations of Korean Civil Service pertaining to establishment of civil service offices in each provincial government of Korea south of 38° North Latitude, a civil service office of like degree and dignity is hereby established in the City of Seoul. The City of Seoul Civil Service Officer shall be appointed by the Director of Office of Korean Civil Service in accordance with rules for Korean Civil Service. The responsibilities and duties of the Civil Service Officer for the City of Seoul shall be the same for the city as are prescribed for the Provincial Civil Service Officers for the provinces.

Budget and Fiscal Procedure

SECTION 54. The fiscal year for the city shall begin on the first day of April of each year.

The budget estimate for every department and office of the city and every board and commission shall be filed by the executive of such department with, and shall be acted upon by, the chief administrative officer not later than the 1st day of December. All budget estimates shall be compiled in such detail as shall be required on uniform blanks furnished by the director of finance.

The chief administrative officer shall obtain, in ample time to pass thereon, budget estimates from the heads of departments,

軍 政 廳 官 報　서 울 市 憲 章　一九四六年八月十日

offices, boards or commissions, and after adjusting and revising the same, not later than the 15th day of December, he shall transmit such budget estimates to the director of finance. By not later than the 30th day of December, the director of finance shall check such estimates, consolidate same and transmit such budget estimates to the mayor.

The mayor shall hold such public hearings on these budget estimates as he may deem necessary and may increase, decrease or reject any item contained in the estimates, excepting that he shall not increase any amount nor add any new item for personal services, materials, supplies or contractual service.

Subject to the rules of Korean Civil Service all increases in salaries or wages of officers and employees shall be determined at the time of the preparation of the annual budget estimates and the adoption of the annual budget and appropriation ordinances, and no such increase shall be effective prior to the fiscal year for which the budget is adopted.

Not later than the 1st day of February of each year, the mayor shall transmit to the board of supervisors a consolidated budget estimate for all departments and officers of the city for the ensuing fiscal year, including a detailed statement of the estimated revenues and receipts of each department and an estimate of the amount required to meet the expenses of conducting each department, bureau, division, office, board or commission for the ensuing fiscal year, together with a separate schedule of the proposed work schedule.

He shall, by message accompanying such proposed budget, comment upon the financial program incorporated in such budget, and the important changes as compared with the previous budget, recommended by him.

The mayor shall submit to the board of supervisors, at the time that he submits the annual budget estimate, a draft of the annual appropriation ordinance for the ensuing fiscal year, which shall be prepared by the director of finance. This shall be based on the proposed budget and shall be drafted in such detail as to furnish an adequate basis for accounting control by the director of finance of each appropriation item for the ensuing fiscal year.

The board of supervisors shall fix the date for consideration of and public hearings on the budget estimates and proposed appropriation ordinance.

The board of supervisors may decrease or reject any item contained in the budget estimates, but shall not increase any amount or add any new item for personal services or materials, supplies, or contractual services for any department, unless requested in writing so to do by the mayor, on the recommendation of the chief administrative officer, board, commission or elective officer, in charge of such department.

The board of supervisors may increase or insert appropriations for capital expenditures and public improvements.

After public hearing, and not earlier than the 15th day of February, nor later than the first day of March, the board shall adopt the budget estimates as submitted or as amended and shall pass the necessary appropriation ordinance.

The several amounts of estimated revenue and proposed expenditure contained in the annual appropriation ordinance as adopted by the board of supervisors shall be and become appropriated for the ensuing fiscal year to and for the several departments, bureaus, offices, utilities, boards or commissions, and for the purposes specified. The appropriation ordinance shall constitute authority for the director of finance to set up the required revenue and expenditure accounts. Each department for which an appropriation has been made shall be authorized to use the money so appropriated for the purposes specified in the appropriation ordinance, and within the limit of the appropriation. Appropriation items for bond interest, bond redemption, fixed charges and other purposes not allocated to a specific department, shall be subject to the administration of and expenditure by the chief administrative officer for the respective purposes for which such appropriations are made.

Annual Salary Ordinance

SECTION 55. Subject to the rules of Korean Civil Service, the number and rates of compensation for all positions continued or created by the supervisors in adopting each annual budget, and each annual or supplemental appropriation ordinance, shall be established and enumerated in an ordinance continuing and creating positions in the city departments and offices, and providing the rates of compensation therefor, which ordinance shall be passed at the same time as the annual or supplemental appropriation ordinance is passed.

Transfers

SECTION 56. Upon written recommendation of the chief administrative officer, or a board or commission for the use of which funds have been appropriated by the city, and the approval of the mayor, the board of supervisors may transfer an unencumbered balance, or part thereof, of an appropriation made for the use of the one department, to another. No such transfer shall be made on utility, bond, school, pension or trust funds, except by way of loans as in this charter provided.

Limitation on Incurrence of Liabilities

SECTION 57. No ordinance or resolution for the expenditure of money, except the annual appropriation ordinance, shall be passed by the board of supervisors unless the director of finance first certify to such boards that there is a sufficient unencumbered balance in a fund that may legally be used for such proposed expenditure, and that, in the judgment of the director of finance, revenues as anticipated in the appropriation ordinance for such fiscal year and properly applicable to meet such proposed expenditure will be available in the treasury in sufficient amount to meet the same as it becomes due.

No obligation involving the expenditure of money shall be incurred or authorized by any officer, employee, board or commission of the city unless the director of finance first certify that there is a valid appropriation from which the expenditure may be made, and that sufficient unencumbered funds are available in the treasury to the credit of such appropriation to pay the amount of such expenditure when it becomes due and payable.

Every officer who shall approve, allow or pay any demand on the treasury not authorized by law, ordinance or this charter, shall be liable to the city individually and on his official bond for the amount of the demand so illegally approved, allowed or paid.

Each such certification shall be immediately recorded by the director of finance. Each sum so recorded shall be an encumbrance for the purpose certified until such obligation is fulfilled, cancelled or discharged, or until the ordinance or resolution is repealed by the board of supervisors.

All obligations incurred, all ordinances passed, and resolutions and orders adopted, contrary to the provisions of this section, shall be void and any claim or demand against the city based thereon shall be invalid.

Limitation on Claims for Damages

SECTION 58. All claims for damages against the city must be presented to the director of finance within six months after the occurrence from which it is claimed the damages have arisen; otherwise there shall be no recovery thereon.

軍　政　廳　官　報　서울市憲章　一九四六年八月十日

Effective Date

SECTION 59. This charter shall take effect at twelve o'clock noon on the tenth day of August, 1946.

ARCHER L. LERCH
Major General, United States Army
Military Governor in Korea

REGULATIONS FOR IMPLEMENTING THE LAW FOR THE ELECTION OF REPRESENTATIVES OF THE KOREAN PEOPLE

CHAPTER I–GENERAL RULES

Section 1.   These regulations have been formulated in accordance with Section *56* of the Law for the Election of Representatives of the Korean people (*which shall be called "Election Law" hereinafter*).

Section 2.   A *citizen* in the meaning of Section *1,* Election Law, is any person who meets any of the following requirements:

(*1*)   Any person who is registered in a Korean HOJUK (family register);

(*2*)   Any person born of Korean parents;

(*3*)   Any person born of a Korean father and not possessing a foreign citizenship.

Section 3.   The status of persons to whom any of the categories set forth in Section 2, Election Law, is applicable shall be determined as of the date of registration as provided in Section *15,* Election Law.

Section 4.   "Expenses incurred in the election of assembly members", as defined in S   n 7, Election Law, mean the following:

Office expenses of election committees and other expenses necessary in the conduct of elections;

(*2*) *Expenses covering postage as provided in Section 44.*

CHAPTER II–ELECTION DISTRICTS

Section 5.   After the official announcement of the date of election, any change in administrative districts shall have no effect on election districts.

Section 6.   The election committees of each voting district shall establish a registration place (*polling place*) in each voting district and publicly announce them by posters and newspapers prior to the date of registration. Wherever possible, such registration places (*polling places*) shall be established in a school, a building of the Korean Agricultural Association, etc.

Section 7.   The election committees of the electoral districts shall publicly announce the place at which the ballot boxes shall be opened not later than five days prior to the date of election.

— 1 —

CHAPTER III—POLL REGISTER

Section 8. Registration papers prescribed in Section 15, Election Law, shall be prepared in accordance with the form attached in Appendix 1. Except for signature (or thumbprint if substituted for signature) the registration papers may be filled by people other than the voter.

The National Election Committee shall provide the election committees on all levels with copies of the Election Law and the Regulations. At the special request of a voter, each voting district election committee shall give an opportunity to such voter to check during office and voting hours, a copy of the Election Law and Election Regulations.

Section 9. When signing the registration paper, the voter shall exclusively use either Chinese or Korean characters or a mixture of both. If registration is accomplished by a thumbprint, this fact must be recorded on the registration paper in the column "Remark" and two witnesses have to sign this entry.

Section 10. A voter "Who has been in residence in a district," as defined in Section 16, Election Law, means a person who has his HOJUK in such district or who registered in KIRU-PU of such district and made his home there.

Section 11. Poll registers shall be prepared in accordance with the form attached in Appendix II.

Section 12. At the time at which the poll register is established the qualification of each registered voter shall be examined and decided on the basis of clear and definite evidence, wherever possible on the basis of official documents. Unless clear and definite evidence has been received that a registrant is ineligible to vote, he shall be recorded in the poll register.

Section 13 If anyone is denied the right to vote on the basis of the examination provided in the preceding section, he shall not be recorded in the poll register and shall be informed immediately of the reason in writing.

If the person affected objects to the decision, he may present an objection in accordance with Section 17, Election Law.

Section 14. An objection or request for review made in accordance with Section 17, Election Law, must state in writing the reasons for it, and wherever possible documentary evidence should be attached.

Request for review shall be presented through the election committee which decided on the objection. The latter shall forward without delay all

— 2 —

pertinent documents to the election committee of the electoral district.

Section 15. When the objection or request for review is decided in favor of the objector, the committee shall without delay revise the poll register or shall direct its revision and shall announce such decision publicly. When the objection or request for review is decided against the objector, he must be notified without delay in writing.

The notification must be served either directly to the objector or to a person living with him.

Section 16. Poll registers shall be open for public inspection from 9 A. M. to 4 P. M. daily, including public holidays during the period of such inspection.

Section 17. A person who moved from the place where he has been registered shall return to vote to the polling place of the voting district, where he is registered.

Section 18. If any documentary evidence concerning election is required of a public office, it must be prepared and presented without delay.

CHAPTER IV-ELECTION COMMITTEES

Section 19. The names of the members of election committees of all levels shall be publicly announced.

Section 20 Without prejudice to Section 6, the headquarters of election committees of all levels may be located at the office of the corresponding administrative agency unless otherwise determined by the election committee concerned.

If a special location is determined, the superior committee must be notified without delay and the change must be duly publicized.

Section 21. Election committees of all levels shall be summoned by the respective chairmen. When more than one-third of the members requests a conference, the committee must be convened.

Section 22. The chairman of the election committees of all levels shall represent their respective committees and perform necessary administrative functions.

Section 23. The chairman or a member of an election committee of all levels shall not participate in the discussion of election matters which affect himself, his grandparents, parents, spouse, children and grandchildren, brothers

— 3 —

and sisters, nephews, nieces, paternal uncles and aunts and their respective spouses. He may, however, have a voice in such matters if permission is given by the committee.

Section 24. Membership of election committees of all levels is honorary with appropriate allowances and travel expenses only.

The provisions concerning allowances and travel expenses shall be issued by the National Election Committee with the approval of the Chief Executive.

Section 25. The provisions concerning number, pay and travel expenses of secretaries and clerks of election committees of all levels shall be determined by the National Election Committee with the approval of the Chief Executive.

Section 26. Secretaries and clerks of election committees of all levels shall be appointed and removed by the respective chairmen in accordance with the resolution of the election committee concerned. However, officials and employees of the corresponding administrative agency may be appointed to such positions while in their regular offices.

Section 27. The duties of the National Election Committee are as follows:

(1) To appoint the election officials (*chairman, alternate chairman, members, and alternate members*) of the provincial election committees and to recall and replace them for violation of their duties;

(2) To instruct and supervise the subordinate election committees in all matters relating to the election;

(3) To inspect in general the election;

(4) To prepare and distribute registration papers and poll registers and to supervise the printing of ballots and envelopes;

(5) To issue and enforce rules and regulations necessary to put the Election Law and this regulation into effect;

(6) To prepare and carry out a budget for the election;

(7) To take such additional steps as are necessary to carry out the Election Law and this regulation.

Section 28. The duties of DO (*province*) and Seoul City election committees are as follows:

(1) To review the appointment of the election officials (chairman, alternate chairman, members and alternate members) of electoral districts concerned and to recall and replace them for violation of their duties;

(2) To instruct and supervise the subordinate election committees

軍政廳　官報　選舉細則　　　　　　一九四八年三月二二日

in all matters relating to the elections and to inspect the election;

(3) To prepare and present various reports concerning the election to the National Election Committee;

(4) To perform such additional functions as are necessary to conduct the elections within their jurisdiction.

Section 29. The duties of the election committees of the electoral districts are as follows:

(1) To review the appointment of the election officials (*chairman, alternate chairman, members and alternate members*) of the voting district election committees concerned and the election committees on the MYUN, EUP and DONG level, wherever established, and to recall and replace them for violation of their duties;

(2) To divide the electoral districts into voting districts;

(3) To decide on the request for review concerning registration in the poll register;

(4) To register candidates for representatives of the National Assembly;

(5) To open ballot boxes and to determine those elected and to prepare records of the election;

(6) To instruct and supervise the subordinate election committees in all matters relating to the election and to inspect the election;

(7) To prepare and present various reports concerning the election to the higher election committees;

(8) To issue public announcement and notices concerning the election;

(9) To perform such additional functions as are necessary to conduct the election within their jurisdiction;

Section 30. The duties of the election committees of voting districts are as follows:

(1) To administer all matters relating to the registration of voters and voting;

(2) To prepare the poll register and to decide on the objections to the poll register;

(3) To prepare ballot boxes and polling places and publicize their location,

(4) To prepare voting records and to safeguard ballot boxes and deliver them to the election committees of the electoral districts;

(5) To prepare and present various reports concerning the election to the higher election committees;

(6) To perform such additional functions as are necessary to conduct the election within their jurisdiction.

Section 31. The duties of the election committees on the MYUN, EUP and DONG level, when established by the National Election Committee. are as follows:

(1) To assist, instruct and supervise the subordinate voting district election committees;

(2) To prepare and present various reports concerning election to the higher election committees;

(3) To perform such additional functions as are assigned to them by the National Election Committee.

Section 32. When no newspaper is published in an electoral district, the electoral district election committee shall publish the items prescribed in Sections *28, 30, 35,* Election Law, in the newspaper or newspapers which, in its opinion, are most widely circulated in the district.

Section 33. Election committees of all levels may request all governmental agencies for necessary assistance in matters relating to the election.

CHAPTER V-CANDIDATES AND ELECTION CAMPAIGNING

Section 34. A person who registers as a candidate need not be a resident of that particular district nor have his BONJUK in that particular district.

Section 35. The application for registration of a candidate for the National Assembly shall be in writing and include the following: the name and signature of the candidate and those who recommend him (*in case of a recommended registration, the consent of the candidate*), their occupation, address, date of birth and party or organization affiliation, if any.

Those who recommend a candidate must have registered on the poll register in the electoral district concerned and must attach a certificate of registration of the election committee concerned.

Section 36. A candidate who desires to withdraw his candidacy shall submit in person to the election committee of the electoral district where he is registered a written statement of his withdrawal and the reasons therefor.

Section 37. When a candidate registers or withdraws, the election committee of the electoral district concerned shall immediately notify the head of the City of Seoul, PU, EUP or MYUN where such candidate resides.

When a candidate dies, the head of the city of Seoul, PU, EUP or MYUN where such candidate resides shall immediately inform the election committee of the electoral district where such person is registered as a candidate.

Section 38. When the election committee of the electoral district concerned learns of the withdrawal or death of a candidate, it shall immediately cancel his registration.

Section 39. After registration, each candidate shall without delay report in writing to the election committee of the electoral district concerned the location of his election office, and name, address, occupation, party or organization affiliation of a person who will represent him in all matters relating to the election.

Section 40. The term *"public official"* as used in Section 29, Election Law, shall not include members of the Korean Interim Legislative Assembly.

CHAPTER VI—USE OF PUBLIC BUILDINGS

Section 41: Unless special circumstances make it impossible, public buildings such as public halls and similar buildings dedicated to public use shall be at the disposal of any candidate upon his application at the regular rental rate.

Such application shall be presented to the manager of the building through the election committee of the voting district where the building is located.

Section 42. Unless schooling or other regular functions would be specially interfered with, the principal of every non-private school must permit the use of the school building as a place of public-election addresses of candidates.

The application for the use of a school building shall be presented to the principal of the school through the election committee of the voting district where the school is located.

Section 43. Permission for the use of buildings prescribed in the preceding two sections shall be granted in accordance with the following standards:

(1) When in several applications the use of the building is desired for the same time of the same day, the application first received shall have priority; when such applications are received at the same time, use shall be awarded to the candidate who has used public buildings less frequently, provided

— 7 —

that the joint use of a building may be permitted:

(2) Not more than one permission for use of a public building shall be granted at each time.

CHAPTER VII – MAILS FREE OF CHARGE

Section 44. For the purpose of his election campaign, each candidate may mail free of charge one of the following types of mail to the registered voters of his electoral district, but not more than once:

(1) Letters of the same content in open or half-open envelopes, weighing not more than 10 grams;

(2) private postcards.

Section 45. A candidate who wants to mail one of the types of mail mentioned in the preceding section shall apply to a postoffice in the electoral district concerned with a certificate of the election committee of the electoral district giving the following information:

(1) The postoffice the candidate desires to use;

(2) The date on which the mail will be delivered to the postoffice;

(3) Number and type of mail;

(4) Name and address of the candidate.

Section 46. Franked election mail must be marked *"election"* on the upper left hand corner.

CHAPTER VIII – ELECTION PROCEDURE AND SUCCESSFUL CANDIDATES

Section 47. Ballots shall be prepared in accordance with the form attached in Appendix III.

when the electoral district election committee determines the symbols before the name of each candidate in accordance with Section 33, Election Law, it shall assign *one stroke* to the candidate whose name has been placed first on the ballot, *two strokes* to the candidate whose name has been placed second on the ballot, etc.

Section 48. The election committee of each voting district shall keep the ballots under its seal until the voting begins.

Section 49. The ballot box shall be made in accordance with instructions set forth in Appendix IV. Each ballot box must have double lids and be under lock and key.

Section 50. Each voting district shall establish one or several sign boards and allocate to each candidate running in the electoral district adequate and equal space on each sign board. The National Election Committee shall publish detailed regulations concerning the size and material of the sign boards.

Each electoral district election committee and each voting district election committee shall print the word "*sample*" on handbills, posters. and other publications which reproduce the ballot for purposes of the information of the voter in accordance with Section 35, Election Law.

Section 51. On the day of the election no posters and sign boards for the purpose of campaigning will be permitted within 100 meters distance from the entrance of each polling place. nor shall any pamphlets or handbills be distributed within that distance.

Section 52. The chairman of each voting district election committee shall during the whole time of voting post at least one member of the election committee at the entrance of the separate room (*booth*) in which the ballots are marked. Each voting district election committee is responsible that the place where the voter marks the ballot is so arranged that no election official and no voter or other person may observe the marking of the ballot. The place where the ballots are marked will be provided only with tables, chairs, and writing materials.

When a candidate fails two days prior to the election, to furnish the voting district election committee with photographs as prescribed in section 35, Election Law, the voting district election committee shall display only the name and the symbol assigned to such candidate at the entrance of the polling place and in the balloting room. The photograph to be furnished by each candidate for display at the entrance of each polling place shall not exceed *12 × 16.5 cm.*, the photograph for display in the balloting room shall not exceed *4 × 6.5 cm.*

The mark to be used by a voter to indicate the candidate of his choice shall be a cross, a circle. a stroke or similar symbol. Ballots filled by a voter with pictures or words are invalid.

Section 53 Time necessary for registration of voters and voting shall not be deemed to be absence from the usual duties of public officials and persons employed by others,

Section 54. Immediately before balloting commences. in the presence of more than half of the members of the committee and of voters who are present in the

— 9 —

polling place, the chairman of the election committee of the voting district shall open the ballot box and show it to be empty and then lock the inner lid.

The chairman of each voting district election committee shall designate a member of the committee to observe during the whole time of voting the ballot box, to see that each voter properly casts the envelope into the ballot box.

Section 55. The "*notification of decision,*" provided for in Section *84*, Election Law, means the notification made in accordance with Section *17*, Election Law.

Section 56. More than half of the members of the election committee of the voting district shall be present at the polling place during the period of voting.

Section 57. Decisions concerning questions dealt with in Section *34*, paragraph 2, Election Law, shall be noted in the voting record.

Section 58. When a voter who has received a ballot is ordered to leave the polling place in accordance with Section *39*, paragraph *1*, Election Law, he shall return the ballot to the chairmen of the voting district election committee and such fact shall be noted in the voting record.

Section 59. When the balloting is over, the chairman of the election committee of the voting district shall without delay lock the lids of the ballot box and place his seals on both in the presence of voters.

Section 60. After the balloting is over, the unused ballots shall be counted and forwarded in closed envelopes to the election committee of the electoral district.

Section 61. The election committee of the voting district shall prepare a voting record describing the procedure of the voting with signatures and seals of the chairman and the members of the committee who have been present at the polling place.

Section 62. The following shall be recorded in the voting record:

(*1*) Outline of the procedure of voting;

(*2*) Total number of voters recorded in the poll register;

(*3*) Number of the ballots distributed;

(*4*) Number of unused ballots.

Section 63. When the ballots are being counted, the chairman of the election committee of the electoral district shall examine each ballot and have two persons engaged in the work of opening the ballots, separately compute the ballots each candidate has received. Voters may observe the opening of the ballot boxes at a designated place.

— 10 —

Section 64. After the computation prescribed in the preceding section has been completed, the chairman of the election committee of the electoral district shall announce the votes each candidate has received in each voting district and thereafter he shall announce the total number of votes received by each candidate.

Section 65. When the validity of a ballot is in question, the decision shall be made by the chairman of the election committee of the electoral district with the concurrence of the majority of the members of such committee present.

Section 66. The invalidation of each vote shall be announced at such time as it is decided.

Section 67. The drawing of a lot as prescribed in Section 43, paragraph 1, Election Law, shall be recorded in detail in the election record.

Section 68. After the counting of the ballots has been completed, the election committee of the electoral district shall without delay prepare in three copies a record of the election, with the signatures and seals of the chairman and the members present of the election committee.

Section 69. The records of election shall contain the following:

(1) Outline of the procedure of the ballot box opening;

(2) Total number of voters recorded in the poll register;

(3) Total number of ballots distributed;

(4) Total number of ballots cast;

(5) Number of invalid ballots;

(6) Number of votes each candidate received;

(7) Determination of the candidate elected.

Section 70. In case actual voting is not necessary in accordance with Section 43. Election Law, the election committee of the electoral district concerned shall immediately make announcement to this effect.

Section 71. The candidate elected without actual voting in accordance with Section 43, Election Law, shall be determined by the election committee of the electoral district concerned at 9A. M. of the day of the election and such fact shall be announced publicly. When a second election takes place, the poll register prepared at the general election shall be used for such election.

CHAPTER IX-PROCEDURE IN CASE OF VIOLATION OF ELECTION PROVISIONS

Section 72. The police authorities and public prosecutors shall immediately report to the National Election Committee when they have started investigations concerning violations of Section 53 Election Law.

Section 73. Except in case of a clear and present **danger to the public** peace, no person shall be arrested, confined or otherwise **detained prior to com**mencement of prosecution by the police or public prosecutor **for violation of** Section 53, Election Law, without a writ of a court; but even **if a person is** arrested in case of clear and present danger, a writ of **a court must im**mediately be secured.

Section 74. Criminal cases against candidates and their **accomplices for** violation of Section 53, Election Law, shall be tried in **the first instance before** a collegiate court of the district court.

Section 75. When a judgment is rendered in accordance **with Section 53,** Election Law, the chief judge of the court concerned shall **without delay for**ward a copy of such judgment to the Chief Executive, **the National Election** Committee, and the election committee of the electoral **district concerned.**

SUPPLEMENTARY RULES

Section 76. These regulations shall be effective as of **22 March 1948.**

APPENDIX I

Form of Voter's Registration Paper for the Election of Members of the National Assembly

.......nt Domicile:

Present Address:

Name of Householder and Relation
between Householder and Registrant: *Householder   Relation*

Occupation, Sex, Name: *Occupation* ____Male ____Female *Name*

Date of Birth:

Age: Full years

Period of Residence in the Electoral District

I register as indicated above

Date:

Name of Registrant:

Seal:

Qualification for Voting:

(Entries in this column shall be signed and sealed by the chairman of
the election committee.)

(over)

— I —

APPENDIX I (*Cont*)

REMARKS:

(*1*) Each registrant shall be prepared to certify by his family register or temporary residence register as to his residence, present address, name, date of birth, and period of residence.

(*2*) Age and period of residence shall be counted as of the date of election.

(*3*). The name of the registrant shall be signed or thumbprinted. In case a registrant is illiterate a clerk may sign the name of the registrant in the presence of a witness, the registrant shall thumbprint the registration paper and both the clerk and the witness shall sign and seal it.

(*4*) When there are reasons for the disqualification of a voter the chairman of the voting district shall state briefly the reasons for the disqualification and check them in the column "qualification for voting." The chairman shall seal entry in the column "qualification for voting."

(*5*) Each registrant on paper shall indicate by a number the DONG in which the registrant resides.

— 2 —

APPENDIX II

Form of Poll Register for the Election of Members of the National Assembly

No .1	Permanent Domicile	Present Address	Relation between Householder and voter	Date of Birth	Sex	Name	Remarks	Date of Registration

(Back)

No .1	Permanent Domicile	Present Address	Relation between Householder and Voter	Date of Birth	Sex	Name	Remarks	Date of Registration

投　票　函　（大）
BALLOT BOX

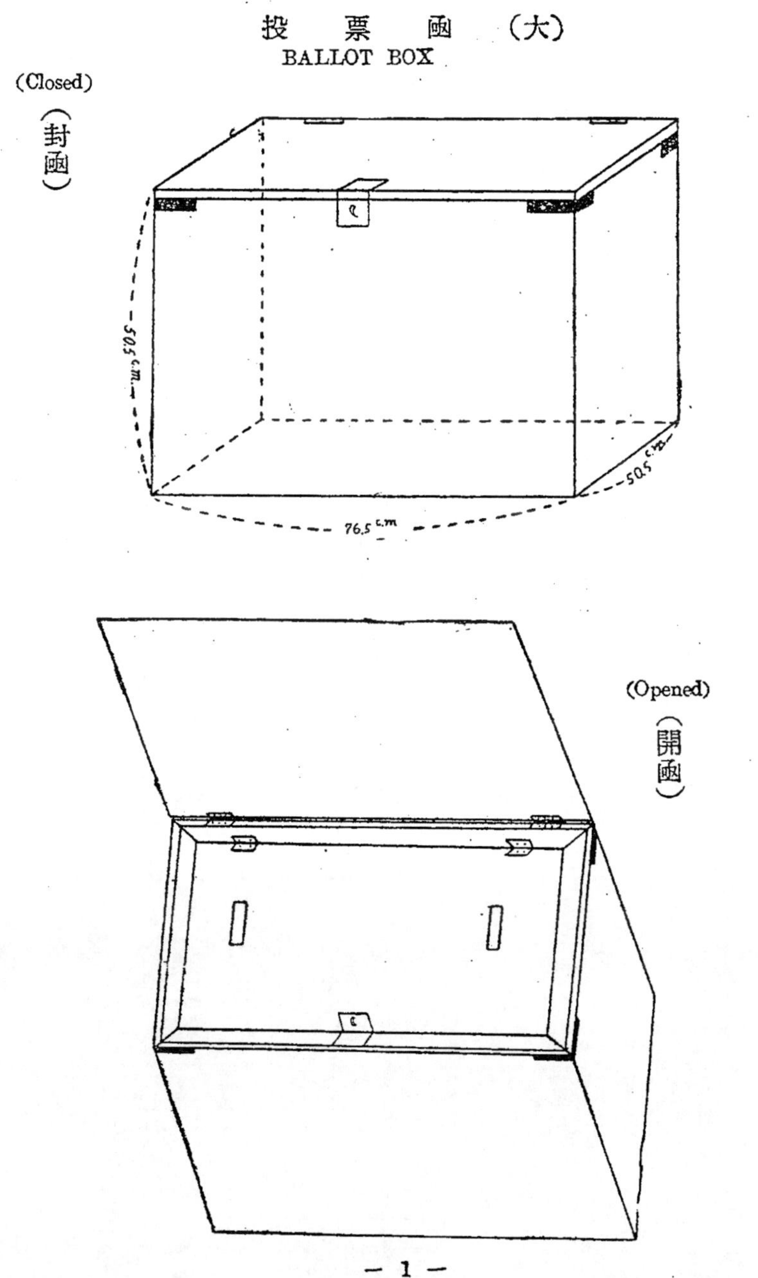

— 1 —

投　票　函　（小）
BALLOT BOX

(Closed)　　（封函）

(Opened)　　（開函）

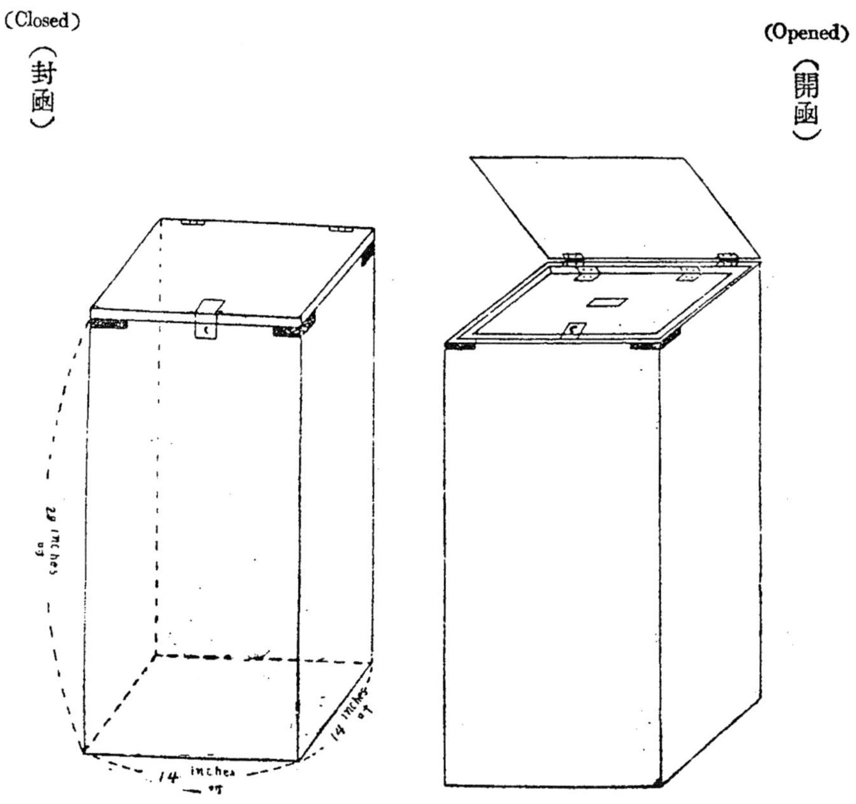

軍政廳　　官報　　選舉細則附錄第四號　　一九四八年三月二二日

APPENDIX III

FORM OF BALLOT

1	2	3	4	5
Name of Candidate	Name of Candidate	Name of Candidate	Name of Candidate	Name of Candidate

SOUTH KOREAN INTERIM GOVERNMENT
Seoul, Korea

MGJUS

16 July 1948

SUBJECT: Jurisdiction of Korean Courts Over Litigation Involving Corporations in Which the Property Custodian
Has An Interest

TO: Property Custodian

Director, Department of Justice

Chief Justice of the Supreme Court

1. *Purpose.* The purpose of this directive is to state conditions under which lawsuits may be instituted against corporations whose stock has vested wholly or in part in Military Government.

2. *Allowable suits.* A corporation whose stock has vested wholly or in part in Military Government, pursuant to Ordinance Number 33, dated 6 December 1945 (Vesting Title to Japanese Property Within Korea), may hereafter be sued in the Korean Courts on claims arising out of a transaction entered into with the corporation on or after 25 September 1945. The requirements and conditions hereinafter set forth shall apply to such lawsuits.

3. *Notice to be given by plaintiff.* Where any party to an action making claims against the corporation, knows that all or any part of such corporation's stock has vested in Military Government, such party shall within five (5) days after filing suit in which such claim is made, give written notice thereof, personally or by registered mail, to the provincial property custodian of the province in which the action is pending. Such written notice shall state:

a. The name and address of the plaintiff.

b. The name and address of the defendant corporation.

c. The court where the action is pending and date upon which claim was made.

d. The amount of the claim and a brief statement of the claim, including the date upon which it arose.

e. If plaintiff is represented by a lawyer, the name and address of the lawyer.

f. Whether title to property is involved in the suit.

4. *Notice to be given by defendant.* Where claim is made in any action against a corporation whose stock has vested wholly or in part in Military

— 1 —

軍政廳　官報　軍政長官의書翰　　　一九四八年七月一六日

Government, the highest responsible official of the corporation, whatsoever called, even if such an official is not the director called for in the Civil or Commercial Code, shall, within five (5) days from the date when a complaint is served, give written notice of such claim, personally or by registered mail, to the provincial property custodian of the province in which the action is pending. Such written notice shall state:

a. The name and address of the plaintiff.

b. The name and address of the defendant corporation

c. The court where the action is pending and the date upon which claim was made.

d. The amount of the claim and a brief statement of the claim, including the date upon which it arose.

e. If defendant is represented by a lawyer, the name and address of the lawyer.

f. Whether title to any property is involved in the action.

g. Whether the defendant corporation admits or denies the claim; and if admitted, the reasons why admitted.

h. The total number of issued shares of stock of the defendant corporation and the number of shares of stock which has vested in Military Government.

5. *Notice to be filed in court.* In addition to giving the written notice set forth in Paragraph 4 above, the highest responsible official of the defendant corporation at the time it filed answer to the suit, or if it does not file answer or appear within the time given by law for appearing or filing answer, shall file with the court a written statement, setting forth the percentage of the issued shares of stock which has vested in Military Government. Where such statement has been filed, no judgment shall be entered in the case until at least forty (40) days have elapsed from the date the complaint was served, so that the provincial property custodian shall have had sufficient time to investigate the facts.

6. *Provisional Execution.* No provisional execution shall issue in any case in which a corporation whose stock has vested wholly or in part in Military Government pursuant to Ordinance Number 33, dated 6 December 1945 (Vesting Title of Japanese Property Within Korea).

7. *Execution of judgment.* Where a corporation has filed notice with the court as required by Paragraph 5 of this directive, of the fact that its stock

— 2 —

is vested wholly or in part, and judgment is thereafter entered against the corporation, no execution of judgment shall be taken against the land, buildings, or other tangible property of the corporation, until such time as the judgment-creditor has given written notice of such judgment personally or by registered mail, to the provincial property custodian of the province in which judgment is entered, and until thirty (30) days have elapsed since the giving of such notice. The written notice shall state:

    a. The name and address of the judgment-creditor.

    b. The name and address of the defendant corporation.

    c. The amount of the judgment of the court in which entered.

    d. If the judgment-creditor is represented by a lawyer, his name and address.

    e. Whether title to any property was involved in the suit.

    f. If known, the percentage of the outstanding shares of stock of the judgment-debtor corporation which is vested in Military Government.

8. *Statute of limitation.* Any statute limiting the time for bringing action upon any claim, shall be deemed not to have run as to persons or corporations having claims against corporations whose stock is vested wholly or in part in Military Government during any period in which suit was prohibited against any such corporation,

9. *Effective date.* This Directive shall be effective on the date appearing hereon.

W. F. DEAN
Major General, United States Army
Military Governor in Korea

SOUTH KOREAN INTERIM GOVERNMENT
Seoul, Korea

MGJUS   28 July 1948

SUBJECT: Limited Jurisdiction of Korean Courts in Cases Involving Movable and Immovable Property Affected or Claimed to be Affected by Ordinance Number 2 and Ordinance Number 33.

TO: Property Claims Commission
Office of Property Custody
Chief Justice, Supreme Court

1. *Purpose.* The purpose of this directive is to expedite the determination of certain claims filed with the Property Claims Commission against United States Army Military Government in Korea under Ordinance Number 103, dated 31 August 1946 (Establishment of Property Claims Commission), for the return of property vested in Military Government under Ordinance Number 2, dated 25 September 1945 (Concerning Property Transfers), and Ordinance Number 33, dated 6 December 1945 (Vesting Title to Japanese Property Within Korea).

2. *Claims which may be referred to Korean courts.* Where it has been determined by the Property Claims Commission that any claim filed with the Commission under Ordinance Number 103 cannot be decided administratively under Letter Directive of the Military Governor, dated 17 April 1948, MGJUS 040.2, Subject: "Simplified Procedure in Disposing of Claims for Vested Property," or where a claim has been returned to the Property Claims Commission by the Office of Property Custody for the reason that it cannot be decided administratively, the Property Claims Commission may immediately refer such claim to the appropriate Korean court having jurisdiction thereof, for judicial determination.

3. *Effect of reference of claim to Korean court.* The transfer of such a claim to the Korean court shall have the same effect as the filing of a suit in the Korean court by the claimant, as plaintiff, against the Office of Property Custody, as defendant. No filing fee shall be required to be paid in such cases. Upon transfer of a case to the court, copies of the petition need not be served on the parties, but the court shall notify the parties that suit has been filed. The notice upon the Office of Property Custody shall be sent to the National Office of Property Custody, Seoul, and to the Provincial Office of Property

Custody in the Province where the property is located. After the service of such notices, the court shall handle the suit pursuant to the ordinary provisions of law, as modified by the provisions of this directive.

4. *Date of hearing; defense by Office of Property Custody.* In the cases referred to Korean courts under Paragraphs 2 and 3, above, the court shall fix a date for hearing not less than thirty (30) days after the service of notification of filing of suit referred to in Paragraph 3, above, upon the National Office of Property Custody and the Provincial Office of Property Custody. The Office of Property Custody, within such time or at the hearing, shall answer the petition stating one of the following: (1) that the Office of Property Custody does not desire to defend the suit and consents to the entry of a judgment divesting Military Government of any title which it had in the claimed property by virtue of Ordinance Number 2 and Ordinance Number 33; (2) that the Office of Property Custody is ready to defend the suit, in which case its defense shall be stated. If the Office of Property Custody needs further time to investigate in order to answer the petition, it shall ask for a postponement of trial and give reason therefor, and the court shall grant the Office of Property Custody reasonable time in which to complete its investigation and file its answer.

5. *Action to be taken by the court where Office of Property Custody enters no defense.* In cases where the Office of Property Custody does not appear and answer as required by paragraph 4, above, or where the Office of Property Custody does appear and states that it does not desire to defend the suit under Paragraph 4, above, it shall be the duty of the court to examine the plaintiff and his proofs, and the court shall not award judgment to the plaintiff unless it appears that, on the facts and the law, the plaintiff is entitled to judgment.

6. *Limitation on court costs and disbursements of plaintiff.* Expenses which the plaintiff has previously incurred in preparing documents for and in filing his petition with the Property Claims Commission shall not be allowed as court costs.

7. *Limitation of time within which claims or action against vested property may be filed.* No claim or action against United States Army Military Government in Korea to determine the ownership of property affected or claimed to be affected by the provisions of Ordinance Number 2, 25 September 1945, and Ordinance Number 33, 6 December 1945, shall be filed with the Property Claims Commission under Ordinance Number 103, 31 August 1946, or in any

— 2 —

1090

Korean court after 31 August 1948, and all claims, or actions, if not filed on or before that date, shall be forever barred.

8. *Effective date.* This directive shall be effective on the date appearing hereon.

W. F. DEAN
Major General, United States Army
Military Governor in Korea

軍政廳　　官報　　　軍政長官에書翰　　　　一九四八年七月二八日

UNITED STATES ARMY MILITARY
GOVERNMENT IN KOREA
Office of the Military Governor
Seoul, Korea

19 January 1946

SUBJECT : Publication of Ordinances and Other Regulatory Measures; Publicity Thereon.

TO : All Military Government Personnel.

1. Effective immediately, the practice of preparing and distributing mimeographed copies of ordinances and other regulatory measures prior to publication in the Official Gazette will cease. No further mimeographed or unprinted copies will be prepared or distributed except by the prior written approval of the Military Governor.

2. Effective immediately, all ordinances and other regulatory measures printed in the Official Gazette will by their terms be operative ten (10) days after the date of publication. Exceptions to this practice will be authorized, when necessary, only by the prior written approval of the Military Governor.

3. Effective immediately, publicity in any form or through any medium, with relation to any ordinance or other regulatory measure printed or to be printed in the Official Gazette :

a. Is forbidden prior to the effective date thereof..

b. Is forbidden at any time without prior written approval of the General Affairs Section of the Secretariat as to the legal accuracy thereof, except that such approval is not required in the case of issuance of exact copies or the making of direct quotations.

ARCHER L. LERCH
Major General, United States Army
Military Governor in Korea

TFYMG 386.2 (OCA)   27 April 1946

SUBJECT: Payment of Obligations by Juridical Persons Affected by Vesting
          of Japanese Property under Ordinance No. 33.

TO   : Directors of all Departments, Offices, and Government Controlled
          Corporations.
          All Provincial Military Governors.

1. All juridical persons registered as such in a District Court in Korea south of 38° North Latitude, the ownership of all or a majority of the stock or other ownership interest of which was vested in the Military Government of Korea, are referred to in this directive as "Vested Companies." Juridical persons registered in Korea south of 38° North Latitude in which the Military Government of Korea owns, by reason of vesting, a minority of the stock or other ownership interest but over which the Military Government of Korea is exercising management control are referred to in this directive as "Requisitioned Companies." The property in Korea south of 38° North Latitude of juridical persons registered in Japan which was operated as a business enterprise and which property was vested in the Military Government of Korea is referred to in this directive as "Vested Corporate Property of Japanese Companies." Vested Companies, Requisitioned Companies, and Vested Corporate Property of Japanese Companies owed obligations and held unpaid receivables at the time of becoming vested.

2. Except as otherwise provided by this directive, all Vested Companies, Requisitioned Companies and Vested Corporate Property of Japanese companies will collect all receivables in Korea and pay obligations due to persons in Korea south of 38° North Latitude in accordance with their actual ability to pay.

3. When the officer designated by the Military Governor of Korea or his authorized representative as responsible for the management control of any such organization determines that the payment of obligations in accordance with present actual, ability to pay fo any Vested Company, Requisitioned Company, or Vested Corporate Property of Japanese Companies would

apparently result in the inability of such organization to pay its creditors in Korea on an equitable basis, such officer shall file with the Property Custodian a certificate, supported by all pertinent data, that in his opinion such organization should operate under the provisions of paragraph 4 hereof. The Property Custodian shall approve or disapprove such basis of operation.

4. When approved by the Property Custodian, any organization to referred in paragraph 3 hereof may operate on the basis of paying only obligations incurred by it or which represent goods or services received by it after 25 September 1945.

5. Regarding the effect of vesting Ordinance No. 33 upon juridical persons it is the opinion of General Counsel that the stock or other ownership interest therein owned by Japan or a Japanese national (natural or juridical) was vested in the Military Government of Korea, but that title to the assets of such juridical person remained in such juridical person. Such opinion is hereby approved as the basis of this directive.

6. This directive does not affect the treatment of claims by persons in Korea against anyone other than Vested Companies, Requisitioned Companies, Vested Corporate Property of Japanese Companies.

7. This directive does not affect or supersede the policies previously announced by this headquarters concerning the payment of taxes to the government and interest to financial institutions.

J. R. SHEETZ
Brigadier General, United States Army
Deputy Military Governor

HEADQUARTERS
UNITED STATES ARMY FORCES IN KOREA
Office of the Deputy Military Governor
Seoul, Korea

17 May 1946

SUBJECT: Control and Distribution of Petroleum Products.

TO   : Provincial Military Governors.

1. <u>Mission</u>: The Military Government of Korea has the responsibility of supplying to the Korean people the minimum quantities of petroleum products essential to prevent a break-down of civilian economy which would hamper the attainment of Military Government objectives.

2. <u>Organization</u>:

a. General Notice No. 5, dated 15 December 1945, established the Petroleum Distributing Agency and granted operating control of that agency to a board of directors to be appointed by the Military Governor.

b. The Petroleum Distributing Agency Board of Directors will be subject to the staff supervision of the Director of the Department of Commerce.

c. Each provincial Military Governor will appoint an officer from his staff as a provincial Petroleum Control Officer.

3. <u>Procedure</u>:

a. The Petroleum Distributing Agency will carry out the functions enumerated in General Notice No. 5.

b. The Provincial Petroleum Control Officer will perform the following functions:

    (1) Become thoroughly familiar with civilian needs for petroleum products with particular emphasis on municipal and provincial agencies, essential transport, essential industry and civilian illuminating requirements.

    (2) Approve in writing all sales of petroleum products stored in Petroleum Distributing Agency warehouses in the Province.

    (3) Forward to the Petroleum Distributing Agency at the end of the month:

1095

(a) Records of quantities of petroleum products :   :or
sale.

(b) Information necessary for determining the total quantities of
petroleum products to be allocated to the province.

(4) Make certain that minimum stock levels are maintained in each
warehouse within the Province to meet emergency requirments.

4. Function stated in 3 b (2) will be eliminated as soon as practical, upon orders from this headquarters. Upon the elimination of this function, the Provincial Petroleum Control Officer will be responsible for making periodic checks to determine that the Provincial Agent of the Petroleum Distributing Agency allocated petroleum products to dealers in accordance with allocation lists developed in detail by the Provincial petroleum Control Officer.

5. Provincial Military Governors will submit the name, rank, and serial number of the appointed Provincial Petroleum Control Officers to the Department of Commerce, this headquarters.

6. Provincial petroleum Control Officers and the Petroleum Distributing Agency are authorized direct communication on technical matters pertaining to this directive, through the Department of Commerce, this headquarters.

7. Upon approval of the Military Governor, regulations issued by the Petroleum Distirbuting Agency will be published in the Official Gazette.

BY DIRECTION OF THE MILITARY GOVERNOR:

J. R. SHEETZ
Brigadier General, United States Army
Deputy Military Governor of Korea

2 April 1946

REPRINTED COPY

SUBJECT: Political Parties
TO   : Director of Department of Public Information
            All Provincial Governors

Reference: Ordinance No 55, Registration of Political Parties, dated 19 February 1946.

1. Provincial Governors will:

a. Receive, translate, analyze, consolidate, file and prepare to report when and as directed to the Department of Public Information concerning provincial political party registrations.

One copy of all information required to be registered by each party will be furnished the Department of Public Information in either English or Korean within a week after receipt of that party's registration. in accordance with letter, HQ USAMGIK, Bureau of Public Information, Subject: "Information of political parties Registered at Provincial Level" dated 21 February 1946.

b. Receive and file quarterly accounting returns and furnish the Department of Public Information one copy of each within one week following receipt thereof.

c. Immediately recommend to the Department of Public Information one or more persons to be authorized by the Department of Public Information to conduct examinations of provincial political party account books and records, as directed.

d. Cause examinations of provincial political party account books and records to be made by the examiners authorized by the Department of Public information wherever there is cause to suspect irregularities.

e. Report results of accounting examinations and of any prosecutions resulting therefrom to the Department of Public Information.

— 1 —

HEADQUARTERS
UNITED STATES ARMY MILITARY
GOVERNMENT IN KOREA
Office of Civil Administrator
Seoul, Korea.

TFYMG 231. 4 (Property)   27 May 1946
    X 095 (Materials Control Corporation)
SUBJECT: Concerning Relationship between Property Custodian and Materials Control Corporation.

TO: All Provincial Governors (Attn:. Property Custody Officers) and Materials Control Corporation.

1. With respect to former Japanese-owned personal property now vested in the Military Government, Materials Control Corporation is both an independent agent and an agent of the Property Custodian. It is an independent agent in dealing with surrendered Japanese Army property and former Japanese abandoned property (by which is meant former Japanese property, the ownership of which is unknown or concerning which an intention to abandon may reasonably be inferred). It is an agent of the Property Custodian in dealing with such vested property of known former Japanese ownership where such property is not abandoned.

2. Concerning such surrendered or abandoned property, Materials Control Corporation may request information of local Property Custodians, as to whether any records of ownership of a given piece of property are available. If no record exists, local Property Custodians will so inform the representatives of Materials Control Corporation. Appropriate action thereafter will be taken by Materials Control Corporation.

3. Materials Control Corporation is hereby authorized to act as agent of the Property Custodian, with reference to sales of vested property within the scope of Opinion of General Counsel No. 5 (last sentence of first paragraph, dated 5 February 1946. Sales of such property will be accomplished as follows:

a. Where such property is in the possession of, or first brought to the attention of, representatives of Materials Control Corporation, such corporation will perpare and submit inventories, together with all available information as to the former ownership of the property, to the local Property Custodian. The local Property Custodian will thereupon

1200

renewal of registration of dissolved parties. Such orders and regulations will be prepared by the Department for publication in the Official Gazette upon approval of the Military Governor.

b. Order examination of political party accounts books and records at any time. Authorize examiners in addition to those recommended by provincial governors and withdraw authorization of any examiner at any time.

c. Order appropriate investigations of any political party or person suspected of violating any of the provisions of Ordinance No 55 in addition to any investigations ordered by Provincial Governors.

d. Report any suspected violations of Ordinance No 55 to the proper authorities for police investigation and public prosecution.

e. Prepare additional regulations concerning political parties for publication in the Official Gazette upon approval of the Military Governor in Korea.

BY DIRECTON OF THE MILITARY GOVERNOR.

WILLIAM A GLASS JR
Lt Col   Sig C
Acting Civil Administrator

TFYMG 231. 4 (Property)   27 May 1946
  X 095 (Materials Control Corporation)

SUBJECT: Concerning Relationship between Property Custodian and Materials Control Corporation.

TO: All Provincial Governors (Attn:, Property Custody Officers) and Materials Control Corporation.

1. With respect to former Japanese-owned personal property now vested in the Military Government, Materials Control Corporation is both an independent agent and an agent of the Property Custodian. It is an independent agent in dealing with surrendered Japanese Army property and former Japanese abandoned property (by which is meant former Japanese property, the ownership of which is unknown or concerning which an intention to abandon may reasonably be inferred). It is an agent of the Property Custodian in dealing with such vested property of known former Japanese ownership where such property is not abandoned.

2. Concerning such surrendered or abandoned property, Materials Control Corporation may request information of local Property Custodians, as to whether any records of ownership of a given piece of property are available. If no record exists, local Property Custodians will so inform the representatives of Materials Control Corporation. Appropriate action thereafter will be taken by Materials Control Corporation.

3. Materials Control Corporation is hereby authorized to act as agent of the Property Custodian, with reference to sales of vested property within the scope of Opinion of General Counsel No. 5 (last sentence of first paragraph) dated 5 February 1946. Sales of such property will be accomplished as follows:

a. Where such property is in the possession of, or first brought to the attention of, representatives of Materials Control Corporation, such corporation will perpare and submit inventories. together with all available information as to the former ownership of the property, to the local Property Custodian. The local Property Custodian will thereupon

investigate the question of ownership, consulting the local Legal Officer where
necessary, and, if satisfied that title to the property vested in the Military
Government under Ordinance 33, will approve the sale. Materials Control
Corporation and not the Property Custodian will determine to whom the
sale will be made.. A report of the sale will be made by Materials Control Corporation to the local Property Custodian immediately upon completion of the sale.

b. Where such property is in the possession of, or first brought to
the attention of, the local Property Custodian, he will prepare the necessary inventories. Other action shall be as described in paragraph 3 a,
above.

4. Letter OCA dated 19 January 1946, Subject: "Authority to Act
as Agent", is hereby rescinded.

ARTHUR S. CHAMPENY
Colonel Infantry
Acting Military Governor

29 May 1946

SUBJECT: Determination and Execution of Policy.

TO   : Directors of Departments and Offices, Provincial Governors and Officers in Charge of Agencies under the Control of the Military Government.

1. Staff officers referred to in this directive will make recommendations regarding policy determinations affecting operations for which they are responsible as part of completed staff work. Such recommendations will be made through the Civil Administrator in the normal course of staff work. When irreconcilable conflicts of opinion or devergences in viewpoints exist between two or more Offices, Departments or Agencies, the Civil Administrator will refer the matter to the Deputy Military Governor for conciliation prior to submission, when necessary, to the Military Governor.

2. When determinations of policy are made by the Deputy Military Governor or by the Military Governor in oral form or incident to conferences, the staff officers whose operations are affected by such determinations will prepare in written form a report of such policy determinations and file such written report in duplicate with the Civil Administrator before acting upon the execution of such policy. This will enable the Civil Administrator to perform his function of coordination and expediting of the execution by staff officers of policies determined by the Military Governor of Korea in the operations for which they are responsible.

3. Each staff officer at the national level of the government will submit through the Civil Administrator for purposes of approval before issuance all orders, instructions, and regulations proposed to be issued by him having governmental and legal force and effect or having application to operations beyond his Department or Office or which are of more general application than in a particular instance, situation or a single regulated or operated agency. Each such document having governmental and legal force and effect or having such general application will also be subject to approval by the Department of Justice as to language, form and legality, including

legal authority for its issuance and legal effectiveness in the accomplishment of the policy as determined by the Military Governor of Korea or directives of higher authority. Each such document will become effective only after being placed under seal, assigned an official document number and published in the Official Gazette of Korea.

4. The sealing, documenting and publication in the Official Gazette of Korea of all ordinances, orders, instructions and regulations having governmental and legal force and effect and having general application whether issued by the Military Governor of Korea, the Deputy Military Governor of Korea, the Civil Administrator or any staff officer at the national level of government, when so authorized, will insure the existence of a controlled body of public law and regulations governing the civil population of Korea south of 38° North Latitude.

5. All documents of the character described in paragraphs 3 and 4 above, heretofore issued, which have not been handled in the manner prescribed in such two paragraphs and have not been published in the Official Gazette of Korea will be progressively so processed, and when in proper form will be reissued, sealed, documented and published in the Official Gazette of Korea. The Department of Justice is responsible for reviewing all such documents to determine which documents are characterized by this directive.

6. Each Provincial Governor when so authorized by an ordinance issued by the Military Governor of Korea may in his name issue local ordinances, orders, instructions and regulations having governmental and legal force and effect within his Province. The language, form and legal effect of such documents will be approved by the Provincial Legal Officer, sealed, documented and published in an Official Gazette of the Province. Documents heretofore issued by Provincial Governors will be progressivly so processed, and when in proper form will be reissued, sealed, documented and published in the Provincial Official Gazette, which will be entitled "The Official Gazette of (name of Province)." Each Provincial Governor will furnish to the Department of Justice the number of copies of each document requested from time to time.

7. After the effective date of this directive, copies of each document published in the National and Provincial Official Gazettes will be furnished free of charge to the representative of each registered newspaper upon his request. At the national level such distribution will be the responsibility of the Department of Public Information. In each Province such distribution will be the responsibility of the Provincial Public Information Section.

8. All documents, except legal notices of court orders and other court matters, published in the Official Gazette of Korea, or in a Provincial Official Gazette, will be printed in both the English and Korean languages. The English version shall be the official version.

9. This directive is effective on the date appearing hereon.

ARCHER L LERCH
Major General United States Army.
Military Governor in Korea

United States Army Forces in Korea
Seoul, Korea

PROCLAMATION

OF ELECTION OF REPRESENTATIVES OF THE KOREAN PEOPLE

TO THE PEOPLE OF KOREA:

The General Assembly of the United Nations, having established a United Nations Temporary Commission on Korea, recommended that elections be held to choose representatives with whom the Commission may consult regarding prompt attainment of the freedom and independence of the Korean people and which representatives, constituting a National Assembly, may establish a National Government of Korea: and

The United Nations Temporary Commission on Korea having consulted the Interim Committee of the United Nations, which expressed the view that it is incumbent upon the United Nations Temporary Commission on Korea to implement the program as outlined in the Resolution of the General Assembly in that part of Korea which is accessible to the Commission; and

The United Nations Temporary Commission on Korea having concluded to observe such elections in those parts of Korea accessible to it and the territory occupied by the Armed Forces of the United States of America being accessible to the Commission:

NOW, THEREFORE, by virtue of the power vested in me as Commanding General of the United States Army Forces in Korea, I do hereby proclaim as follows:

1. That election of the representatives of the Korean people, under the observance of the United Nations Temporary Commission on Korea, shall be held within the territory of this command on 9 MAY 1948.

2. That such election is being held under the terms and provisions of Public Act Number 5, dated 3 September 1947, *Law for the Election of Members of the Korean Interim Legislative Assembly,* with such changes, additions, and emendations as, after consultation with the United Nations Temporary Commission on Korea, may be deemed necessary.

Given under my hand at Seoul, Korea, on 1 March 1948.

JOHN R. HODGE
Lieutenant General, United States Army
Commanding

HEADQUARTERS
UNITED STATES ARMY FORCES IN KOREA
Seoul, Korea

PROCLAMATION ON NATIONAL ASSEMBLY

1. In accordance with recommendations of the Resolution of the General Assembly of the United Nations, 14 November 1947, as amplified by Resolutions of the Interim Committee of the United Nations on 27 February 1948, and under the Law for the Election of Representatives of the Korean People, issued 17 March 1948, an election was held on 10 May 1948, under the observation of the United Nations Temporary Commission on Korea, to choose representatives with whom the Commission may consult regarding the prompt attainment of the freedom and independence of the Korean people, and which representatives constituting a National Assembly, may establish a Government.

2. The National Election Committee having informed me that the successful candidates have been notified and public notice of the results of the election having been given, I, John R, Hodge, Lieutenant General, United States Army, Commanding General, United States Army Forces in Korea, hereby authorize the chairman of the National Election Committee to determine and publicly to announce the date on which the first meeting of the elected representatives of the Korean people shall be held in the capital city of Seoul, to call the first meeting to order, and to designate the oldest member as temporary chairman who shall preside over the meeting until such time as the National Assembly has elected a chairman and determined its own organization.

3. The Law for the Election of Representatives of the Korean People, issued 17 March 1948, requires by Section 43 that the Election Committee of the Electoral District shall inform the successful candidate of his election. It is hereby provided that such notification of elections shall constitute credentials of the successful candidates, and shall entitle them and each of them to assume their representative seats in the National Assembly.

4. Given in the capital city of Seoul on this 25th day of May 1948.

JOHN R. HODGE
Lieutenant General, United States Army
Commanding

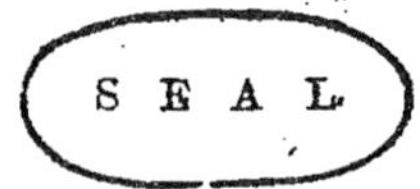

— 1 —

1106

HEADQUARTERS
UNITED STATES ARMY FORCES IN KOREA
Seoul, Korea

PROCLAMATION

PROVIDING FOR DAYLIGHT-SAVING TIME

TO THE PEOPLE OF KOREA:

By virtue of the power vested in me as Commanding General of the United States Army Forces in Korea, I do hereby proclaim as follows:

Daylight-saving time shall be effective from midnight, 31 May 1948, to midnight, 22 September 1948, and shall be known as "Standard Daylight-Saving Time."

At midnight, 31 May 1948, all clocks shall be set up one (*1*) hour; and at midnight, 22 September 1948, all clocks shall be set back one (*1*) hour.

Given under my hand at Seoul, Korea on the *20th* day of May 1948.

JOHN R. HODGE
Lieutenant General, United States Army
Commanding

(S E A L)

재조선미국육군사령부 군정청 법령집

(전2권)

인쇄일: 2025년 6월 01일
발행일: 2025년 6월 16일
지은이: 재조선 미육군 사령부
발행인: 윤영수
발행처: 한국학자료원
서울시 구로구 개봉본동 170-30
전화: 02-3159-8050 팩스: 02-3159-8051
문의: 010-4799-9729
등록번호: 제312-1999-074호

잘못된 책은 교환해 드립니다.

정가   350,000원